FUNDAMENTALS OF Environmental Engineering

Danny D. Reible

CRC Press is an imprint of the
Taylor & Francis Group, an informa business

Contact Editor: Bob Hauserman
Project Editor: Maggie Mogck
Marketing Manager: Arline Massey
Cover design: Dawn Boyd

CRC Press
Taylor & Francis Group
6000 Broken Sound Parkway NW, Suite 300
Boca Raton, FL 33487-2742

First issued in paperback 2019

ISBN-13: 978-1-56670-047-4 (hbk)
ISBN-13: 978-0-367-40027-9 (pbk)

Library of Congress Card Number 98-22372

Library of Congress Cataloging-in-Publication Data

Fundamentals of environmental engineering / Danny D. Reible.
p. cm.
Includes bibliographical references and index.
ISBN 1-56670-047-7 (alk. paper)
1. Environmental engineering. I. Title.
TD146.R45 1998
628—dc21

98-22372
CIP

Visit the Taylor & Francis Web site at
http://www.taylorandfrancis.com

and the CRC Press Web site at
http://www.crcpress.com

Preface

ENVIRONMENTAL ENGINEERING — AN EMERGING NEW FIELD

Environmental engineering as a profession and as a formal course of academic training has long suffered from lack of a well-defined identity. The field is broad and the science on which it is founded includes, in nearly equal measures, all of the traditional scientific disciplines including chemistry, geology, physics, and biology. Similarly, the technology which an environmental engineer might be expected to understand and apply is equally broad. The problems that an environmental engineer might face may require him or her to simultaneously exhibit the knowledge and experience expected of a chemical engineer, a civil or sanitary engineer, and a mechanical engineer. The design of an undergraduate curriculum to adequately balance the breadth and depth demanded of an environmental engineer is a daunting task. For some institutions of higher learning, the best balance has been identified as a graduate level curriculum that builds upon undergraduate training, preferably in a core engineering discipline. Increasingly, however, increased demand for undergraduate training in environmental engineering from both potential students and potential employers has led to growth in the number of undergraduate programs.

For many institutions, this represents a second attempt at developing an undergraduate environmental engineering program. During the early 1970s a number of environmental programs were developed. Many of these curricula never developed into viable programs and were ultimately eliminated. In my view, many of the problems of these curricula were the result of a lack of clearly defined achievable objectives and a focus on the breadth of the field without adequate depth. The field has its roots in sanitary engineering and developed by expanding on the approach and technology of this discipline. Environmental engineering has grown beyond its roots in sanitary engineering, however, and must emerge as an entity in its own right to survive.

TECHNOLOGY VS. FUNDAMENTALS

I feel that a significant factor hindering the development of environmental engineering programs both then and now is the focus of many of the programs on the technology of environmental engineering. Because of the breadth of the technology required of an environmental engineer, I feel that this necessarily limits the time available for the study of fundamentals and problem-solving techniques. I believe that a better approach for an undergraduate level environmental engineering program is one that makes no attempt to provide a comprehensive exposure to all facets of environmental engineering technology but instead focuses on fundamentals. That is, one focused on providing students with a basic understanding of environmental processes and traditional engineering problem-solving techniques and relegates the technology to illustrative examples.

For me, the best example that illustrates the technology orientation of many environmental engineering programs is an examination of the introductory textbooks in the field. With few exceptions, the introductory textbooks are surveys of environmental pollutants, their effects, and qualitative discussions of control or treatment devices. This is very different from the introductory textbooks in most engineering fields which tend to be quantitative and focused on problem solving and problem-solving techniques at the same time as they use examples that illustrate the application of these techniques and methods to that engineering discipline. The education of an engineer has always been defined by a focus on exercise and example problems, i.e., "learning through doing." Admittedly, environmental engineering is so broad that it is difficult to define the unifying quantitative techniques and procedures that define the discipline. If an attempt is not made, however, environmental engineering can never grow into a mainstream engineering discipline.

A NEW APPROACH

It is toward this objective that the current textbook is directed. My approach to the introduction of environmental engineering is one borrowed from chemical engineering and reflects my bias of being trained in that discipline. I believe that the technology associated with the unit operations for waste treatment, control, and cleanup are naturally taught from a background similar to that of a chemical engineer. In addition, the fate and transport behavior of pollutants and the pharmacokinetics which describe their action on receptors are clearly dependent on chemical and dynamic principles that are central to the education of a chemical engineer. This suggests to me that the natural starting point for an environmental engineer is a firm grounding in dimensional analysis, physical chemistry, mass, energy, and component balances, and application of these topics to environmental engineering problems.

This textbook is oriented around these topics. It is designed for a one semester course aimed at second- or third-year college students in an environmental engineering curriculum. It also could be used by students in traditional engineering disciplines to provide an introduction to environmental applications of their discipline. The text also could be used at the graduate level for students without an undergraduate engineering background. Since many environmental engineering students come to the discipline in this manner, this may be the most common use of the book. It assumes that students will have a working knowledge of introductory calculus and some exposure to first order ordinary differential equations. In addition, introductory chemistry and physics is assumed.

As with any text, the topics covered are the results of the bias of the author. The same can be said for the errors of commission and omission that undoubtedly fill the book. I would appreciate the comments and criticisms of the reader and hope that if I am lucky or unlucky enough to produce a second edition that I can correct some of these problems.

Danny D. Reible
Louisiana State University

Author

Danny D. Reible, Ph.D., is the Chevron-Endowed Professor of Chemical Engineering at Louisiana State University and Director of the Hazardous Substance Research Center/South and Southwest, a consortium of LSU, Rice University, and Georgia Tech. Dr. Reible joined the university after receiving his Ph.D. in Chemical Engineering from the California Institute of Technology for research on atmospheric flow and transport in 1982. He has published more than 60 journal articles and book chapters on environmental dynamics and transport phenomena.

In addition to his work at LSU, Dr. Reible served as an AAAS Environmental Science and Engineering Fellow assigned to the U.S. EPA Office of Environmental Processes and Effects in 1987. During 1991 he was a Senior Visitor to the Department of Applied Mathematics and Theoretical Physics at Cambridge University. Between 1993 and 1995, he was the Shell Professor of Environmental Engineering at the University of Sydney in Australia.

Dr. Reible has received the New Engineering Educator Excellence Award from the American Society of Engineering Education, and in 1993 was named a Fellow of the Institution of Engineers/Australia. He also is an external examiner for the National University of Singapore.

Nomenclature

Units notation and typical units

L Length — meters (m)
M Mass — kilograms (kg)
mol Moles — g/MW (moles)
P Pressure — atmospheres (atm)
E Energy — ($M\ L^2/t^2$)
T Temperature — Kelvin (K)
t Time — seconds (s)

Variables notation [Units]

A Area [L^2]
B Buoyancy flux [L^4/t^3]
C Cover factor for erosion
C_i Concentration of component of interest in phase i [M/V]
C_D Drag coefficient
$C_{S_{1/2}}$ Half substrate concentration microbial reaction rate
C_T Total concentration [M/V]
d_c Diameter of collector [L]
d_p Diameter of particle [L]
D_b Effective biotubation diffusion coefficient [L^2/t]
D_i Molecular diffusivity in phase i [L^2/t]
D_{sv} Molecular diffusivity in soil vapor [L^2/t]
D_{sw} Molecular diffusivity in soil or sediment water [L^2/t]
D_t Turbulent or eddy diffusivity [L^2/t]
E_i Internal Energy [E]
f_i Fraction of i
f_X Fractional rate of removal of biomass [1/t]
f_A Fugacity of component A [P]
f_{BV} Brunt-Vaisala frequency [1/t]
F_B Buoyancy force [$M\ L/t^2$]
F_D Drag force [$M\ L/t^2$]
F_e Faradays (96,500 amp/s or Co)
F_p Pressure force [$M\ L/t^2$]
F_v Viscous force [$M\ L/t^2$]
F/M Food-to-microorganism ratio
g Gravitation constant [L/t^2]
G Gibb's free energy [E]
ΔG_f Gibb's free energy of formation [E]
h Height of free surface in river or hydraulic head [L]
h_a Air entry pressure head [L]
h_c Capillary rise [L]
h_e Elevation head [L]
h_p Pressure head [L]
H_m Mixing height [L]

H_c	Height of collector [L]
H_r	Height of rainfall [L]
ΔH_r	Enthalpy change of reaction [E]
ΔH_v	Enthalpy change in evaporation [E]
ΔH_f	Enthalpy of formation or fusion [E]
H_s	Effective stack height [L]
H_0	Physical stack height [L]
H	Enthalpy [E]
k_d	Microbial death rate [1/t]
k_i	Overall mass transfer coefficient in phase i [L/t]
k_{if}	Film mass transfer coefficient in phase i [L/t]
k_{ij}	Overall mass transfer coefficient between phase i and j [L/t]
k_m	Maximum microbial growth rate [1/t]
K_D	Dispersion coefficient [L^2/t]
K_{ij}	Equilibrium partition coefficient between phase i and j (concentration units)
k_p	Permeability [L^2]
K_p	Hydraulic conductivity [L/t]
k_r	Reaction rate constant [t^{-1}]
K_r	Reaction equilibrium constant
k_T	Thermal conductivity [$E \cdot M^{-1} \cdot T^{-1} \cdot t^{-1}$]
L	Length [L]
L_c	Length of a capillary
L_m	Mixing length [L]
L_{MO}	Monin–Obukhov length scale [L]
m_i	Mass of component of interest in phase i [M]
$\tilde{M}_A$.	Molarity of component A (mol/L)
n_i	Moles of component of interest in phase i [mol]
N	Normality (equivalent weights/L)
Mw	Molecular weight of component of interest [M/mol]
Mw_A	Molecular weight of component A [M/mol]
p_i	Partial pressure of component of interest in phase i [P]
p_A	Partial pressure of component A in phase of interest [P]
P	Pressure [P]
π_o	Osmotic pressure [P]
Q_i	Volumetric flowrate of fluid i [V/t]
Q_m	Mass flowrate of component/fluid of interest [M/t]
q	Volumetric flowrate of fluid per unit area (superficial velocity) [V/t]
q_h	Energy flux (E/L^2/t]
q_m	Mass or molar flux of material [M/L^2/t]
q_A	Mass or molar flux of component A [M/L^2/t]
r_H	Hydraulic radius [L]
R	Reaction rate [M/t]
RH	Relative humidity
s	Slope of free surface
S	Entropy [E]
S_i	Solubility in phase i [M/V]
S_r	Storage capacity for rainfall [L]
t	Time [s]
t_L	Lagrangian time scale — time following motion [s]
T	Temperature [K or, if absolute temperature not required, °C]
u	Velocity

u_i	Interstitial velocity
U	Average velocity of fluid [L/t]
U_s	Specific utilization rate of substrate [1/t]
u_*	Friction velocity, $(\tau_0/\rho)^{1/2}$ [L/t]
V	Volume [L^3]
v	Velocity perpendicular to mean velocity [L/t]
w,W	width [L]
w_*	Vertical velocity scale [L/t]
W_i	Mass ratio of component of interest in phase i [M/M]
x	Mole fraction of component of interest in a liquid or primary phase [mol/mol]
y	Mole fraction of component of interest in a vapor, gaseous, or secondary phase [mol/mol]
x_A	Mole fraction of component A in liquid or primary phase [mol/mol]
X	Biomass concentration [M/L^3]
y_A	Mole fraction of component A in vapor, gaseous, or secondary phase [mol/mol]
Y	Microbial yield coefficient [M/M — mass of biomass formed per mass of substrate]
x,y,z	Position in a Cartesian coordinate system [L]

Greek Characters [Units]

α_C	Cunningham correction factor
α_D	Dispersivity [L]
α_f	Porous bed friction factor
α_M	Manning roughness factor [$t/L^{1/3}$]
α_o	Osmotic coefficient
α_p	Particle shape factor
α_{Sh}	Shield's parameter
α_T	Thermal diffusivity [L^2/t]
α_t	Turbulent or eddy thermal diffusivity [L^2/t]
α,β,γ	Constant of integration or other arbitrary constant
β_T	Coefficient of thermal expansion [1/T]
δ	Small length scale typically a film or boundary layer thickness [L]
δ_T	Thickness of the thermocline in a lake [L]
Δ_i	Difference or deficit in property i
ε	Ratio of void volume to porosity [V/V]
ε_i	Volume of phase i to total volume [V/V]
γ	Activity coefficient of a compound in a liquid phase
κ	von Karman constant
κ_0	Permittivity of free space
κ_r	Relative permittivity (dielectric constant)
λ	Mean free path of molecules [L]
ϕ_i	Saturation — volume of phase i to total void volume = $\varepsilon_i/\varepsilon$
Φ	Fugacity coefficient of a compound in a vapor phase; also association factor for estimation of liquid diffusivites
η	Efficiency
η_0	Single collector collection efficiency
θ	Potential temperature; also used as a measure of angle and as dimensionless concentration
ρ	Density [m/V]
τ	Tortuosity, L_e/L
τ_i	Characteristic time scale of process i [t]
τ_{ij}	Shear stress or flux of i momentum in the j direction
τ_0	Interfacial shear stress
μ_i	Chemical potential of component i

ω_i	Mass ratio of component of interest in phase i [m/m]

Superscripts

*	Equilibrium value
ref	Reference value
^	Caret indicates per unit mass or moles
~	Tilde indicates molar units

Subscripts

ads	Adsorption	
adv	Advection	
diff	Diffusion	
exch	Ion exchange	
i	Indicates phase i (lower case)	
	a	Air
	b	Biota
	c	Collector
	doc	Dissolved organic carbon
	e	Epilimnion or effluent
	g	Gas
	h	Hypolimnion
	l	Liquid
	L	Lipid
	n,o	Non-aqueous or oily/organic phase
	oc	Organic carbon
	p	Particle
	s	Soil, sediment, or sludge
	soc	Suspended organic carbon
	sv	Soil vapor
	sw	Soil water
	v	Vapor
	w	Water
	S	Stratosphere
	T	Troposphere
A	Indicates particular component A (upper case)	
f	Film property or friction	
r	Reaction	
S	Substrate	
X	Biomass	
0,in	Indicates inlet conditions	
1,out	Indicates outlet conditions	

Contents

Chapter 5 Rate Processes

Chapter 6 Air Pollution and Its Control

Chapter 7 Water Pollution and Its Control

Chapter 8 Soil Pollution and Its Control

Dedication

to
my daughters and my wife, and to my students, colleagues, and friends

1 Introduction

1.1 WHAT IS ENVIRONMENTAL ENGINEERING?

Engineering involves the application of fundamental scientific principles to the development and implementation of technologies needed to satisfy human needs. For environmental engineering the body of knowledge whose application defines the discipline is environmental science and the goal of the discipline is satisfying present and future human needs through protection of the environment. Such a broad definition, however, does little to define the actual function of an environmental engineer. Even the core science, environmental science, includes aspects of each of the physical, natural, and life sciences. This has made it difficult to characterize environmental engineering and has led to widely varying views as to its focus and responsibilities.

The environmental engineer is often presumed to focus on technologies for the elimination of environmental pollution. Of growing interest and concern, however, are the broader issues of sustainable development, environmental equity, habitat loss, and biodiversity. The protection of the environment thus involves social, political, economic, and legal issues far beyond the domain of any single scientific or engineering discipline. These issues are broader in scope than environmental engineering, but they are issues about which an environmental engineer should be knowledgeable and are issues which are likely to be a component of the work of the environmental engineer.

This text, however, remains largely directed toward the more narrow issue of the technical basis for assessing and eliminating the effects of pollutants in the environment. The engineering science which underpins this effort represents the fundamental knowledge base required of environmental engineers. Environmental pollution is the contamination of the environment with substances that are potentially injurious to human, plant, and animal life or the quality of that life. The polluting substances may arise naturally, for example, in the eruption of volcanoes, or artificially, through the actions of humankind. Generally, we will reserve the term pollution for substances introduced as a result of human activities, but it is important to recognize that natural sources of pollution exist and, on a global scale, account for the majority of many important pollutants. The term contaminant is sometimes interchangeably used with pollution, but generally we will reserve the term pollutant for a substance that has a demonstrated adverse effect on human or ecological health. A contaminant is a substance that is not a component of the natural or "clean" system, but it may or may not pose a hazard. One task of an environmental engineer is to help differentiate between harmless contaminants and pollutants.

Environmental pollution might be in the form of hazardous chemicals released into the environment from a chemical processing plant, by transportation spills, or during the application of pesticides on an agricultural property. Environmental pollution might also be the result of erosion and sediment-laden runoff to a water body as a result of agricultural, residential, or commercial development. It also might be odor or noise problems associated with an industrial or commercial activity. We will often employ pollutants from industrial facilities as examples of pollution, but environmental pollution and environmental problems are much broader. Regardless of the form of the pollution, Seinfeld (1975) pointed out that it involves three important components.

$$\text{Pollutant Release} \Rightarrow \text{Environmental Processes} \Rightarrow \text{Receptors/Effects}$$

The most important of these components is the receptors, i.e., plants, animals, and/or people. Pollution only occurs when adverse receptor effects exist, whether it be toxicity, habitat loss, or elimination of an important resource. Pollutant emissions, at the opposite end, are the source of the problem and generally the most convenient element to control. This is especially true of point sources, where the pollutants are released from a single location or facility. This is also true, however, if the pollutant source is distributed in space; for example, nonpoint source emissions such as agricultural runoff. Linking the source to be controlled and the receptors are the mixing and transformation processes that occur in the environment.

The identification, evaluation, and resolution of a particular pollution problem typically evolves in the reverse of the direction implied by the above, that is

1. Identification of an adverse receptor effect.
2. Determination of the substance of substances causing that effect and estimation of the threshold concentration below which the effect is no longer important.
3. Identification of the source of the polluting substance or its precursors.
4. Estimation of the mixing and transformation processes between source and receptor so that the threshold receptor level (the no-effects or acceptable effects level) can be translated into a safe emission level.
5. Control of the source to achieve the safe emission level.

An environmental engineer is the type of engineer that must work with each of these problems and assist in their evaluation and resolution. Although an environmental engineer should be able to contribute to each of these areas, some of these may be better addressed by other professionals. If the adverse receptor effect is a human health problem, for example, physicians and the affected public can generally more effectively identify the problem. Similarly, life scientists are better equipped to identify the causes and effects of environmental pollution on the flora and fauna of a particular ecosystem. An environmental engineer might be asked to help assess the cause of the problem, however, both as to the chemical and its source.

Source control is often thought of as the defining activity of an environmental engineer, but it is also sometimes better addressed by someone other than an environmental engineer. It is often the engineer trained in a core engineering discipline that has the knowledge and experience to identify, design, and implement a control strategy or technology within an industry served by that discipline. If the pollutant source is a chemical production facility requiring, for example, a gas scrubber to reduce emissions of a particular pollutant, it is often a chemical or mechanical engineer that may design and construct the treatment system. It is also a chemical engineer that is likely to have sufficient knowledge about a chemical facility to identify process changes that will lead to minimization or reuse of the waste streams. Similarly, it is probably inappropriate to expect all environmental engineers to develop the detailed design of a drinking water piping system, a rainwater collection system, or a municipal wastewater treatment system, traditional activities of civil or sanitary engineers.

An environmental engineer would, however, be expected to have a greater understanding of the environmental impact of engineering activities than traditionally trained engineers. In addition, the environmental engineer should exhibit a greater understanding of the availability and feasibility of control and waste minimization technologies than an environmental scientist. Thus an environmental engineer serves in an integrating role, meshing traditional engineering activity with environmental concerns. This is depicted in Figure 1.1 where the environmental engineer is seen to hold a central position between the environmental scientist with a traditional focus on the ecosystem and the impacts of development and the industry engineer with a traditional focus within the fenceline of such a development. The greater breadth of the ideal environmental engineer encourages them to see on both sides of the fence. It is from this perspective that the environmental engineer may be best able to resolve environmental issues while balancing all external constraints, whether

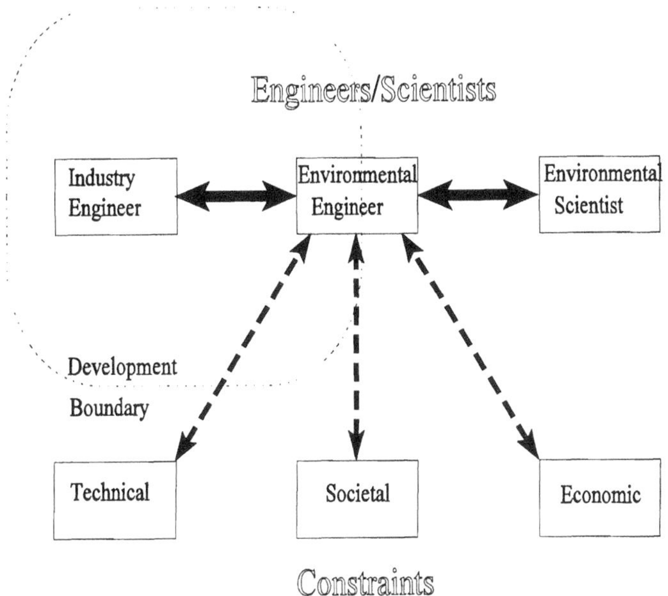

FIGURE 1.1 Relationship between an environmental engineer and other disciplines and constraints.

they be technical, economic, or societal constraints such as moral, social, political, or legal constraints.

The defining activity of an environmental engineer is thus the application of engineering science to the analysis of environmental processes and effects and the design of control systems designed to minimize adverse effects on those processes. Among the technical functions that fit this definition of an environmental engineer are

- Identifying sources of a pollution problem on the basis of a knowledge of viable migration pathways in the environment
- Evaluation of the fate and transport processes of a pollutant between source and receptor to assess exposure or identify a rate of emission that would achieve desired exposure goals
- Evaluation, design, and implementation of systems designed to remediate contaminated soil, sediments, or water
- Interacting with science disciplines in the identification of a pollution problem and the human or ecological response
- Interacting with other engineering disciplines in the evaluation, design, and implementation of systems designed to control pollutants in industrial facilities

In addition to these technical issues are the legal, societal, political, and economic issues alluded to above. Although these issues are not completely within the control of the environmental engineer, the nature of the issues confronting an environmental engineer are such that a significant fraction of his activity is likely to be associated with addressing these issues. Although not the primary focus of this text, these issues will be included where appropriate.

It is important to recognize that complete elimination of pollution is an unachievable goal. L.J. Thibodeaux (1992) stated the problem succinctly when he stated, "I am, therefore I pollute." Control can only approach but not achieve 100% efficiency. It thus becomes the task of the engineer to balance the cost of environmental degradation with the cost required to control that degradation. Because these costs are often paid by different segments of the community, they have rarely been considered together. It is also generally much more difficult to identify and assess the costs

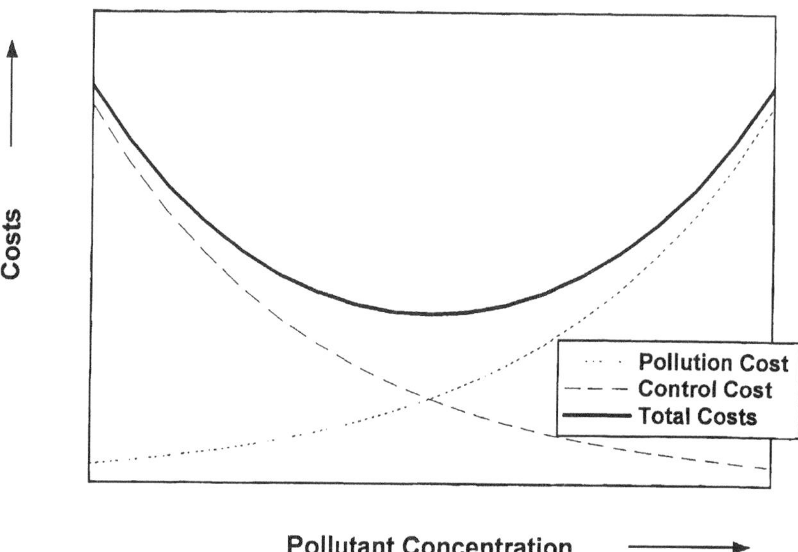

FIGURE 1.2 Cost of pollution vs. cost of control. (From Seinfeld, J.H. (1975) *Air Pollution, Physical and Chemical Fundamentals*, McGraw-Hill, New York. With permission.)

associated with uncontrolled pollution or environmental degradation. Costs of pollution might include increased medical costs for sensitive people or loss of a quantifiable resource or material. It might also include the less easily quantifiable factors of premature death and reduction in the quality of life or habitat for sensitive species. Especially difficult is establishing these "costs" to future generations. The long-term environmental effects of any activity are difficult to estimate. This information is needed, however, if the goal of producing maximum environmental benefit at least cost is to be realized.

Typically, most of the pollution from a particular source may be controlled relatively inexpensively. The increasing cost of further control must be balanced against the comparatively small environmental benefit to be realized. There exists, at least in principle, a minimum total cost for any activity as depicted in Figure 1.2. A desirable goal for the environmental engineer is to define that minimum cost. If these total costs were defined in relation to a particular industrial or commercial development it would then be possible, again in principle, to make an informed decision whether to go ahead with the development. For the reasons outlined above and discussed in more detail in Chapter 2, however, this ideal is rarely, if ever, achieved.

1.2 THE ENVIRONMENTAL ENGINEERING PROCESS — MODELING

As indicated above, environmental engineering is largely about evaluating environmental processes and effects and designing or constructing systems to minimize any potential adverse effect of more traditional engineering activities. A key component of this process is modeling. Modeling is used to demonstrate understanding of past system behavior and to project that understanding for the prediction of future behavior or to design appropriate control measures. A model can be conceptual and qualitative, but generally it is not possible to demonstrate understanding of a process and make appropriate decisions influencing that process if there is no quantitative measure of success or failure. Thus, the environmental engineer's preference is for models that are as quantitative as possible given the uncertain nature of many of the processes with which he or she must deal.

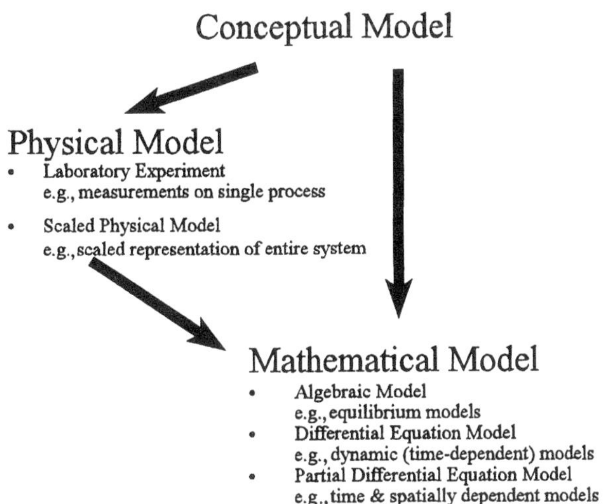

FIGURE 1.3 Types of models for environmental systems.

Models can be used to assess the impact of a particular action on the environment or to evaluate the effectiveness of an intervention. Much of the material presented in this text can be viewed as models developed to describe the behavior of processes in the environment or for waste treatment.

A model is nothing more than a representation of a system for the purposes of analyzing behavior and decision making. A system is simply the entire domain affected by the environmental problem in question while a model is a description of the processes of importance (to the particular problem) within the system. The system may be the world for pollutants of global import. It may also be an individual waste treatment process. A variety of models exist depending on the level of understanding of the system and the objectives of the specific modeling effort. Models of waste treatment processes are especially useful in that they provide a basis for design of engineered systems. These models can always be tested against the performance of the constructed systems. Models of processes and effects in the natural environment, however, are subject to large uncertainty due both to the inability to exactly model particular processes and the difficulty of identifying what processes are applicable to a particular situation. The type of model to be developed and used must be matched with the information available. The basic types of models are shown in Figure 1.3 and include:

Conceptual models — A conceptual model is literally a mental picture of the processes of the system. This model is all that is possible when sufficient information about a system to enable quantitative descriptions is unavailable. A conceptual model is also the first step in any successful modeling effort. The qualitative understanding required to form a conceptual model is required to produce a more sophisticated quantitative one.

Physical models — Often one of the first improvements over a conceptual model is a laboratory simulation of the processes and systems under investigation. These physical models can be used to explore the behavior of the system or to test mathematical models. They are especially useful for very complex systems in that the physical model can be used to explore specific important processes that cannot be isolated in the full system.

Mathematical models — Mathematical models are the most common modeling tool. They range in sophistication from very simple algebraic models to lumped parameter models composed of ordinary differential equations to dynamic spatially distributed models described by partial differential equations. The form of the model employed is dependent upon the level of knowledge available about the system being modeled and the objectives desired of the model.

One of the reasons that models, and especially mathematical models, are widely used by environmental engineers is that they provide a tool for systematic analyses of system behavior. To illustrate the various levels of mathematical modeling tools of interest to an environmental engineer it is useful to consider the need to predict environmental concentrations of a particular environmental contaminant. The various forms of mathematical models and their abilities are illustrated below.

Algebraic models — They may be used to predict properties such as concentrations under steady or time-independent conditions assuming that the contaminant is uniformly distributed or well mixed over some region. Such models might be useful to determine the phases (air, water, or soil) where a particular pollutant might reside or to define a "global" balance on the sources and sinks of a particular pollutant.

Ordinary differential equation models — These may be used to extend the algebraic models to conditions that are dependent on time or position. A contaminant might still be assumed to be uniformly mixed over some region and the objective of the model would be the prediction of an average or constant contaminant concentration. Such models are termed "lumped parameter models" in that constant or average parameters can be used to describe such systems. A parameter is simply a physical constant such as a rate constant that characterizes an important process within the system. This is the minimum level of model capable of describing the basic features of many systems of environmental interest. It might be useful in describing or predicting the dynamics of the recovery of a particular environmental system after a pollutant source is removed, and when designing the steady operation of pollution control devices.

Partial differential equation models — These can be used to describe either spatially variable, or time and spatially variable properties such as the concentration of the contaminant within a region. The model includes terms characterizing the change in concentration with distances in more than one direction or with both distance and time. Because the objective is no longer an average or assumed constant concentration, the model is termed a spatially distributed or a distributed parameter model. These models are generally capable of describing more realistic phenomena in both the environment and in waste treatment processes, but they are also the most difficult to develop and require the most information to use.

It is important to recognize that large uncertainties may exist in attempting to quantify or model some environmental processes or effects. In such cases, the objective of modeling may be limited to identifying weaknesses in understanding or in attempting to assess the relative significance of various processes or effects. This information may be used to guide further investigation or to choose an action from among various alternatives. Quantification or modeling does not necessarily mean assigning an absolute value or rank. Still, without a measure of significance as provided by modeling, the appropriate intervention to an environmental problem or the need for such intervention cannot generally be determined.

The process of modeling incorporates a number of steps. These include:

- Definition of the system to be modeled
- Development of a conceptual model of the system
- Testing of the conceptual model and revision if necessary
- If appropriate, translation of the conceptual model to a physical or mathematical model of the relevant processes
- Testing and validation of the model perhaps partly through use of a physical model to support a mathematical model
- Use of the model to assess performance of the system

A key ingredient for successful application of a model is appropriate selection of the system and the level of sophistication of the model to be employed. There is no clear guidance for these selections, but the level of sophistication of the model should be consistent with (1) the nature and quality of the predictions desired of the model and (2) the quality of the data available to test the model.

Especially important is the selection of the region to be modeled and the processes to be included. Clearly if the objective is to determine trends in behavior with time, a steady-state model cannot be used. To illustrate the selection of a system for modeling, consider the dynamics of ozone in the stratosphere. The loss of stratospheric ozone, due in large part to the movement of halogenated compounds from the earth surface, is a global concern and is difficult to understand and analyze without consideration of the global system. As will be seen in Chapter 6, however, it is possible to model the global behavior of certain stratospheric pollutants by assuming that their concentration is constant over large regions of the atmosphere. Thus, lumped-parameter ordinary differential equation models can be used to describe this macroscale behavior. In contrast, certain pollutants are found in significant quantities only near individual emission sources. Large transients with both time and space may occur requiring a distributed parameter (or partial differential equation) model to adequately describe system behavior.

The environmental engineer should be cautioned, though, against defining an overly small domain of influence or system. Engineers, as well as other professionals, have been faulted for failing to consider the wider implications of their work. This results from defining overly small "system" boundaries that do not adequately incorporate all processes that influence the phenomena or problem of interest. Engineers tend to solve problems via decomposition, by breaking problems into their component parts. While this approach has served the profession well, it can result in a focus on minor components of a system rather than key "externalities" that can control system behavior.

This has implications for the environmental engineer beyond system modeling. The environmental engineer must always strive to see the broad implications of his and others work on the environment. Focus on the components of a problem rather than its broader implications has led to many of the environmental problems we encounter today. Traditionally engineers at an industrial facility, for example, have tended to limit their view to activities within the fenceline and have not considered environmental or other processes beyond that fenceline, except as required by law. This single-minded focus on the primary activity has often led to severe environmental problems such as those observed at many weapons production facilities in the U.S. where defense and security issues were paramount over environmental concerns and where extensive remediation of the damage is now required. A better recognition of the broader implications of an engineering activity is one of the foremost objectives of an environmental engineer.

1.3 EXAMPLE ACTIVITIES OF ENVIRONMENTAL ENGINEERS

It is perhaps easiest to illustrate the activities of an environmental engineer through examples. A few such examples are given below. They are presented to provide both illustrations of tasks that an environmental engineer might be asked to perform and to provoke thought about common environmental problems and their solutions. The reader is encouraged to consider the steps that an environmental engineer might take in addressing these problems and the types of problems that might arise.

Example 1.1: Remediation of soil contaminated by a gasoline leak

Soil may become contaminated with gasoline as a result of a transportation accident. Often the first response is removal of some of the contaminated soil. Even after soil removal, however, it is possible that some portion of the spilled gasoline will remain. Gasoline can move rapidly through soils and any soil that contacts the gasoline will likely retain a residual equal to 10 to 20% of the soil volume. The major health concern is often the aromatic compounds benzene, toluene, ethyl benzene, and xylene (the BTEX fraction), which may compose 10 to 20% of the gasoline. These compounds are relatively mobile in soils and even very low concentrations can render drinking water unusable.

The technology used to remediate the soil may be selected, designed, and operated by an environmental engineer. Because these compounds are relatively volatile, a popular means of remediating or cleaning near-surface soils not saturated with water is by applying a vacuum and forcing air through the soil. A vapor extraction system, any required above-ground treatment of the withdrawn air, and the in-ground (or *in situ*) transport processes might be modeled by the environmental engineer to define the design and estimate its effectiveness.

Example 1.2: Permit application based on atmospheric dispersion estimates

Industrial facilities are normally required to receive permission from a regulatory body before releasing contaminants into the air or water. The permission is given in the form of a license or permit that defines the amount of pollutant that can be released. An environmental engineer might be asked to support an application for such a permit by estimating or measuring the amount that needs to be released, assessing the ability to reduce that amount through some on-site treatment, and finally by modeling the impact of the released material on the surrounding environment. Only by making a convincing case for the need for the pollutant releases and demonstrating that the impact on the environment is minimal will the facility be likely to receive permission for the emissions.

Example 1.3: Improving environmental performance after an audit

Often industrial firms identify potential environmental problems and solutions by conducting an environmental audit. This is similar to an accounting audit in that individuals, usually external to the company or facility, examine and evaluate the company's performance, in this case their environmental performance. External review and evaluation is an excellent way to identify problems or solutions that might otherwise be missed. An environmental engineer might be expected to conduct such an audit, especially in cooperation with an engineer or other professional that has intimate knowledge of the individual industry or process. With the assistance of such an expert, it should be possible to identify potential problems as well as their solution by revising on-site activities.

Example 1.4: Site assessment after plant decommissioning

An industrial facility removed from service is often a potential problem because of spills or leaks of environmental pollutants during the plant operation. This is especially true for older facilities where stewardship of the environment improved over time. Practices considered quite appropriate even 5 to 10 years ago may now be considered environmentally unsound. As a result, an industrial facility is likely to require site assessment including on-site sampling after decommissioning. The range of subsequent uses available to the facility will depend on the degree of contamination and the ease of returning the site to more pristine conditions. An environmental engineer may be involved in both the assessment of the site and the design and operation of any subsequent remediation process.

Example 1.5: Estimation of volatile emissions from surface impoundment

Trace quantities of volatile contaminants may be found in industrial process waters. Often, these are treated by using the contaminated water stream as feed for bacteria in a biological treatment system. Volatile contaminants, however, may not be degraded and instead released by evaporation. It is often important to estimate the quantity of any such losses, especially in areas where such compounds (typically volatile organic compounds) may contribute to other air pollution problems such as photochemical smog or ozone pollution. An environmental engineer may be asked to estimate the fate of any of the variety of compounds that might pass through such a system and, if necessary, design an alternative control process that reduces the evaporation rate of the volatile compounds. Often such an effort would involve a combination of direct measurements and modeling to help interpret those measurements.

Example 1.6: Emergency response

Some of the worst environmental problems have resulted from fires or other accidents in chemical processing or storage facilities. Often a variety of chemicals might be stored in such a facility and questions arise as to the appropriate means of fighting the fire or responding to the accident. Some chemicals react violently with water while others might be transported off-site and into bodies of water or to sensitive habitats through the use of water. An environmental engineer might be asked to provide advice to a fire department to ensure that the response to the fire was appropriate and that the safety of the firefighters was adequately safeguarded. An environmental engineer might also be asked to estimate the impact of any combusted chemicals including predicting the form in which they would be transported from the site, the concentrations likely to be observed at a receptor location, and the potential adverse effects at that receptor.

Example 1.7: Mediator between industry and citizen groups

Local citizens are often concerned about environmental problems in their community. Often as a result of questionable past activities, there is very limited trust of the polluting individual or firm. The issues are often complex and not fully understood by the community and in some cases by the polluter. An environmental engineer may be asked to provide an assessment of the conditions and help the community understand the real risks associated with the environmental problems.

Example 1.8: Preparation of an environmental impact assessment

Much of the focus of the previous examples has been on industrial facilities which are largely assumed to be the major causes of pollution. Any planned development whether commercial or industrial, is increasingly being asked to assess the environmental impact of the facility during construction and upon completion. If we consider a golf course development, as an environmental engineer you might be asked to evaluate such things as:

- Habitat loss for native animals, especially endangered or threatened species
- Loss of native grasses or other flora
- Downstream or groundwater table impact of irrigation of the golf course
- Changes in the sediment loading of adjacent streams
- Traffic and noise problems associated with visitors to the course

Some of these issues are not easily addressed by an environmental engineer without appropriate experience or without the assistance of specialists in these areas. The size of the development, however, may be such that independent expertise in each of the important areas may not be economically justified. Increasingly it is the more broadly trained environmental engineer that might be judged most capable of adequately addressing these concerns.

These examples have illustrated the breadth of problems that an environmental engineer is likely to face over the course of a career. While many of these examples illustrate a need for knowledge of the technology of environmental control, they should indicate that any professional preparation based entirely on current technologies is unlikely to assist in addressing the great variety of problems that an environmental engineer must address. Instead, the examples indicate that the environmental engineer requires knowledge of biological and chemical behavior, basic thermodynamics, transport phenomena, and their application to environmental processes and conditions. That knowledge, *combined* with knowledge of pollution control technologies, is what should define the training of an environmental engineer. It is in that direction that this text is aimed. The text will focus on providing the beginning environmental engineer a firm grounding in the basic tools of the analysis of environmental processes. It is hoped that subsequent courses will build on that foundation and develop an environmental engineer appropriately trained in environmental science and its technology.

It is useful to provide a historical context for environmental engineering by providing a short history of environmental pollution. The discussion will point out that anthropogenic pollution has existed for at least several centuries, but that serious attempts at understanding and resolving the problems are relatively recent. The relatively recent development of the field provides both great challenges and great opportunities to the environmental engineer. The text will then turn to identifying current environmental problems and their regulation. In subsequent chapters, the basic thermodynamics and transport phenomena that governs the behavior of these pollutants in the environment will be introduced. The final chapters of the text will use this knowledge of the chemical and physical properties of the pollutants and basic thermodynamics and transport phenomena to evaluate environmental problems and their solutions in each of the environmental media, i.e., air, water, and soil.

1.4 HISTORY OF ENVIRONMENTAL POLLUTION AND ITS CONTROL

1.4.1 Early History

The environmental "problem" has only recently been recognized as such. Humankind's early efforts were directed toward simple survival. The development and exploitation of the available resources were both necessary and useful. In the sparsely populated and primarily agrarian past, environmental problems were problems of nature that resisted human efforts at survival. As the population of humankind grew and with the development of urban centers, however, the reverse became true. People began to have a deleterious effect on their environment; for example, contamination of water supplies through their activities and those of domesticated animals. With the development of heavy industry and the nearly universal dependence on fossil fuels for energy, such problems have been exacerbated. Although these problems have existed for several centuries in densely populated zones and to a lesser extent in rural areas, it was not until very recently that control of these environmental problems became an important issue for much of the world's population. Until the mid 20th century the pressure to develop and exploit natural resources far exceeded the concern for the effects of such actions. This remains true even today in many of the developing nations. In the more developed nations, especially in North America, Western Europe, and Japan, the last three decades have seen growing recognition of the problem of environmental degradation and a better understanding of means to combat it. In recent years, there also has been an increasing recognition that some types of pollution represent a global problem requiring global responses.

While widespread recognition of the problem of environmental degradation has come relatively recently, its existence throughout history is undeniable. The early pollution problems were undoubtedly associated with agriculture. Clearing of the land for agricultural purposes contributed to erosion while irrigation ditches and canals improved crop yields at the expense of other water users. Domestication of sheep and cattle resulted in overgrazing followed by erosion and contamination of drinking water supplies. Several biblical references to pollution exist, indicating the age of the problem. As cities were built, more severe forms of environmental damage began to occur. The problem of disposal of municipal waste in Rome led to the construction of the Cloaca Maxima (literally, great sewer) for the collection and discharge of wastes into the Tiber River. Even so, Otto of Freising described Rome in the following manner in 1167. "The ponds, caverns, and ruinous places around the city were exhaling poisonous vapors and the air in the entire vicinity had become densely laden with pestilence and death." The German Rhine River, which means clear river, was considered polluted in the middle ages. St. Hildegarde wrote in the thirteenth century that the waters of the Rhine, if drunk unboiled, "would produce noxious blue fluids in the body." In 1800, Samuel Taylor Coleridge "counted two and seventy stenches" in the city of Cologne and wondered what could "wash the river Rhine" of the city's contamination. The Thames River in 18th and 19th century London was another example of water pollution due to urbanization.

Although a number of problems were associated with urbanization and the release of raw sewage into bodies of water, industrialization brought the most easily recognizable effects of pollution. An early example is the mining and smelting of lead by the Romans in Britain. This resulted in soil and water contamination by lead, the signs of which are still evident today.

Because England was the center of the Industrial Revolution, it provides the best historical examples of industrial pollution. Many rivers lined by early factories were contaminated by their activities. The earliest pollution control efforts were undoubtedly directed at sources of water contamination due to their effect on drinking water supplies. Because people have long recognized the importance and limitations of drinking water supplies, the initial steps toward water pollution control were relatively straightforward. The atmosphere, like the ocean, was generally considered an infinite repository of wastes. Air pollution in England was the result of industrialization and urbanization combined with the use of coal as fuel. Coal use in England was recorded as early as 852 AD and over the subsequent centuries coal gradually replaced wood as a fuel. Occasional complaints about coal smoke were recorded during these times. In 1306, a citizen was reportedly executed for defying a royal ban on the burning of coal during sessions of Parliament. Despite some recognition of the hazards of coal smoke, however, its use grew until about 3 million tons were consumed in 1700. The primary use of coal prior to the early 17th century was domestic heating, but within a very few years, industrial pollution became important in cities such as London. With the identification of significant individual sources of pollution, it became possible for people to focus the blame for the air quality degradation on small segments of the community. John Evelyn wrote a pamphlet in 1661 identifying the industrial sources of coal smoke and some methods for alleviating the problem. Evelyn felt that the smoky industries should be forced to locate away from the city of London. In addition to smoke and soot, the combustion of coal results in the oxidation of sulfur constituents to sulfur dioxide (SO_2). The term acid rain was first used in the 19th century in recognition that sulfur dioxide can be absorbed by water forming sulfurous and sulfuric acids.

Despite the slowly increasing public awareness, it was not until 1814 that legislation concerning air pollution appeared in England, and it was not until the mid 1800s with the Alkali Acts that any real control methods were attempted and there were no truly effective efforts to reduce pollutant levels until this century. It was not until a clear link between human mortality and normal releases of air pollution could be identified that widespread recognition of the problem was possible and effective control measures began to be implemented.

1.4.2 Air Pollution Episodes

The proof of a link between human mortality and air pollution came in a series of well-documented severe air pollution episodes in the middle of this century. Table 1.1 summarizes these episodes where people at risk (the old or sick) died due to sharp increases of pollutant levels during adverse weather conditions. Note that there have also been a number of industrial accidents that have led to deaths from the resulting pollution, most notable the more than 2000 people who died as a result of the release of methyl isocyanate in Bhopal in 1985. These are not included in the table in that they represent deaths clearly attributable to an abnormal level of emission. Such episodes lead to control over individual sources and better planning to avoid or cope with industrial accidents but do not generally lead to efforts to reduce the general level of normal emissions. The episodes in Table 1.1, however, serve to illustrate that normal emissions can be deadly under certain meteorological conditions over which people have no control. The episodes tended to focus the attention

TABLE 1.1
Air Pollution Episodes

Location	Year	Pollutants	Excess Deaths
Meuse Valley, Belgium	1930	SO_2	63
Donora, PA	1948	SO_2, particles	20
Poza Rica, Mexico	1950	H_2S	22
London	1952	SO_2, particles	4000
	1962	SO_2, particles	340
New York	1953	SO_2, particles	200
	1966	SO_2, particles	168

of people in the industrialized world upon the effects of pollution (both air and water), and ultimately started strong interest by the public in environmental problems and the development of effective control measures.

The first of these episodes, in Belgium, was little understood at the time. As indicated in the table, all of the episodes except that in Poza Rica, Mexico involved sulfur dioxide and particulate pollution. These pollutants are associated with heavy industry and the combustion of coal. The episode in Mexico involved hydrogen sulfide, a poisonous gas which has a characteristic rotten egg smell. Although the early episodes were not well understood, they did raise concern about the potential impact of similar conditions in a large city and impacting a large area.

Just such an episode occurred in London during December of 1952. Between the fifth and the ninth of December of 1952, very little atmospheric mixing resulted in a rapid buildup of pollutants. The stagnant conditions were the result of a strong temperature inversion over the city. A temperature inversion is a condition in which air temperature increases with height rather than decreases. The warmer and less dense air aloft dampens vertical motions from below and limits the vertical dilution of pollutants from the surface. Such a limited ventilation condition occurs often in Los Angeles, CA accounting for much of their air quality problems. When this occurred in London during 1952, the visibility within the city was very poor and unusually smoky and sooty conditions were apparent to everyone. The smoke (particulate) and sulfur dioxide (SO_2) concentrations observed during this period are shown in Figure 1.4. It should be noted that although this figure shows a maximum SO_2 concentration of about 0.7 parts per million (ppm), some measurements indicated levels as high as 1.34 ppm. People at risk, such as emphysema victims, asthmatics, bronchitis sufferers, and the aged, readily fell victim to the heavy "fog."

The total number of deaths in London during each day in December is also included in Figure 1.4. Note the very rapid increase in the death rate as the pollutant concentrations began to increase. The death rate in London in December of 1952 increased from approximately 250 per day to close to 1000 per day during the peak of the episode. The difference between the actual and expected number of deaths indicated that about 4000 people died as a direct result of the pollution episode. Other supporting information included the death of an unusual number of animals on display at the Smithfield Cattle Show.

This death rate would not be expected on the basis of the SO2 concentrations alone, suggesting synergistic effects with the high particulate concentration. Larsen (1970) found a very good correlation between deaths and the product of the SO2 and particulate concentrations observed during a number of pollution episodes in New York and London, as shown in Figure 1.5. It is believed that the presence of the particles accelerated a complex series of reactions that resulted in oxidation of the SO2 to sulfuric acid.

While the atmospheric conditions which led to the excess deaths were unusual, there was no reason to expect that they would not occur again. At the time of the "killer fog," several thousand tons per day of both particulate and SO_2 were emitted into the atmosphere over London. The

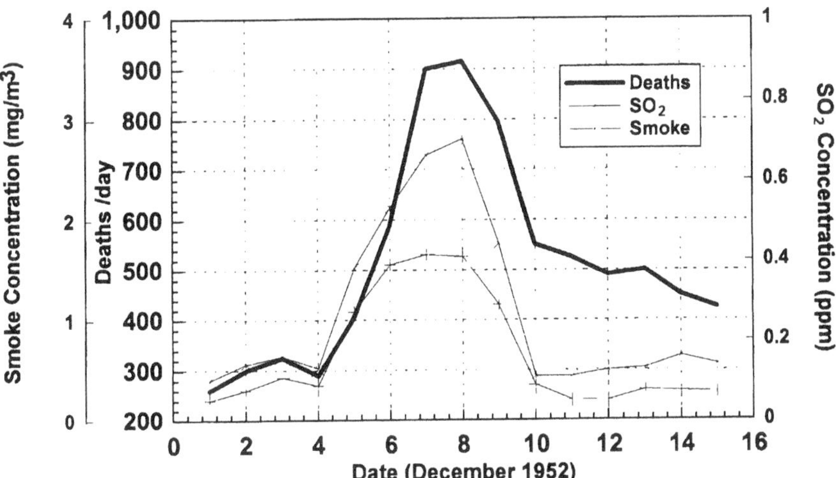

FIGURE 1.4 Smoke and SO_2 concentrations compared to deaths per day in London during December 1952.

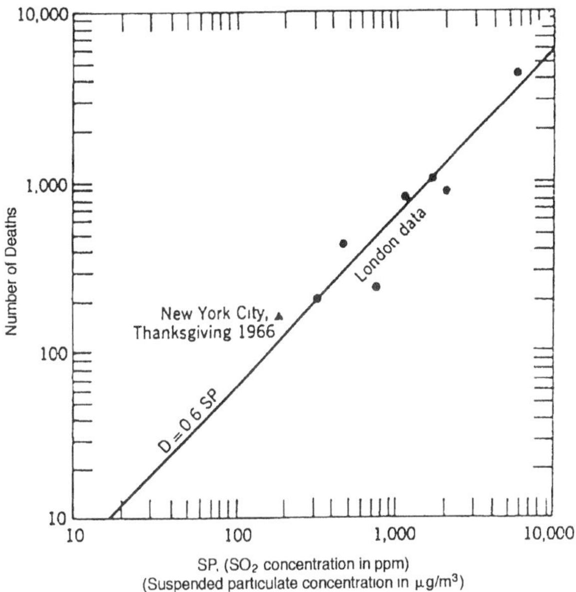

FIGURE 1.5 Number of deaths in London or New York air pollution episodes as a function of the product of SO_2 and suspended particulate concentrations. (From Larsen, R.I. (1970) Relating air pollutant effects to concentration and control, *J. Air Pollut. Control Assoc.*, 20, 4, 214–225. With permission.)

realization of the potential human cost of air pollution led to action by the government of England, culminating with the passage of the Clean Air Act of 1956. This laid the groundwork for pollution control in England and ensured that a serious episode such as that observed in 1952 would not occur again. The effectiveness of these control measures is demonstrated by the rapid decline in smoke (or particulates) in London as depicted in Figure 1.6.

1.4.3 Water Pollution Regulation in the U.S.

An examination of the history of pollution regulation in the U.S. is also instructive. Let us first focus on water pollution regulation. As indicated previously, attempts to control water pollution to

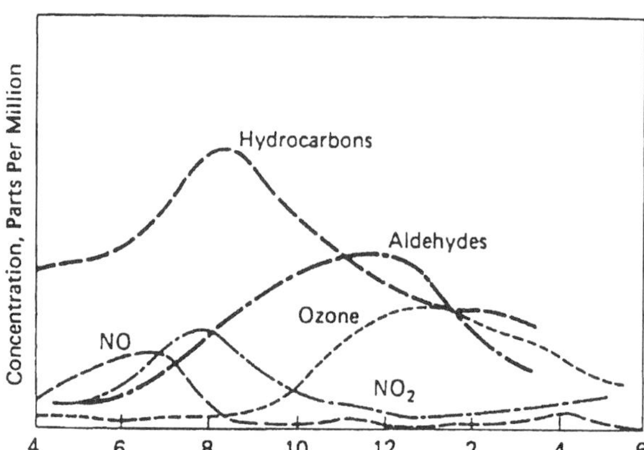

FIGURE 1.7 Average concentration during smog in downtown Los Angeles: hydrocarbons, aldehydes, and ozone for 1953 to 1954; NO and NO_2 for 1958. (From Seinfeld, J.H. (1975) *Air Pollution, Physical and Chemical Fundamentals*, McGraw-Hill, New York. With permission.)

$$HCHO + h\nu \rightarrow 2HO_2\cdot + CO$$
$$HO_2\cdot + NO \rightarrow NO_2 + OH\cdot$$

The reactions and their products are more numerous with more complicated organic molecules. For example, alkylperoxy radicals may be formed from more complicated aldehydes. These can also oxidize NO and in the process form additional aldehydes and other peroxy radicals, contributing further to ozone production.

The hydroxyl radical, OH·, produced by these atmospheric reactions also can react with organic molecules to produce more hydroperoxyl radicals or can directly oxidize NO_2 to form nitrates. For formaldehyde, these reactions can take the form

$$HCHO + OH\cdot \rightarrow HO_2\cdot + CO + H_2O$$
$$OH\cdot + NO_2 \rightarrow HNO_3$$

Some nitrates, such as peroxyacetylnitrate (PAN), are important contributors to the hazards of photochemical smog despite their presence in low concentrations.

The complexity of the processes that produce ozone and photochemical smog have made it very difficult to achieve a solution to the problem. In addition to automobiles, oxides of nitrogen are generated by essentially all combustion processes and complex hydrocarbons are released by a variety of natural as well as anthropogenic sources. Most of the worldwide emissions of methane are from natural sources, although methane is of limited reactivity and therefore does not contribute significantly to urban photochemical smog. More importantly, trees release complex organic compounds called terpenes that are a significant factor in the formation of ozone in the eastern U.S. Even improvements in automobile emission control have a limited impact on photochemical smog since the number of automobiles in urban areas continue to grow and several years are required before new automobiles constitute a significant portion of the automobile fleet.

The fundamental question in resolving pollution problems is how to allocate finite resources to achieve the maximum improvement in environmental quality at the least cost. For photochemical smog, the chemistry is sufficiently complex to make the answer to that question ambiguous. Most urban atmospheres exhibit a behavior similar to that shown in Figure 1.8. The highest ozone and smog levels occur when the ratio of reactive hydrocarbons to oxides of nitrogen is neither very large nor very small. At high ratios of hydrocarbons to oxides of nitrogen, the maximum ozone levels obtained are essentially independent of hydrocarbon emissions. That is, increased control

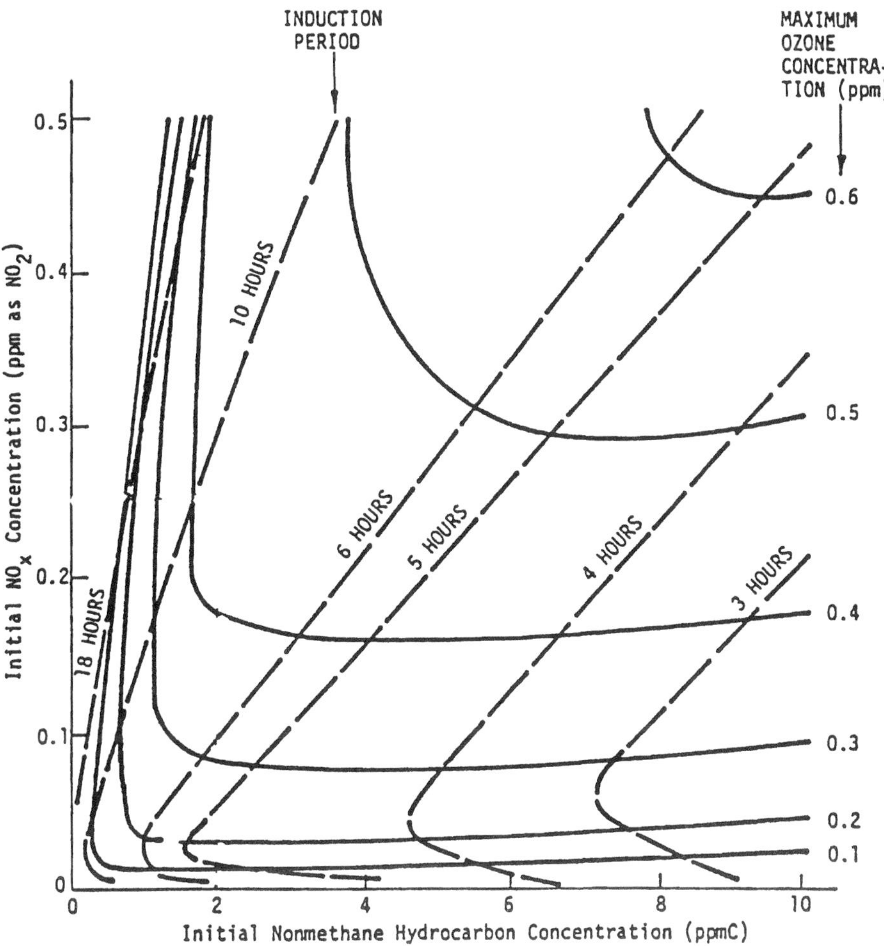

FIGURE 1.8 Isopleths of maximum 1-h average ozone concentrations and induction periods as a function of initial concentrations of oxides of nitrogen and nonmethane hydrocarbons. (From Seinfeld, J.H. (1980).)

of hydrocarbon emissions will have very little affect on maximum ozone levels. Similarly, at low ratios, control of oxides of nitrogen has very little affect on maximum ozone levels. Thus, the optimum or even a minimally effective control strategy depends significantly upon the understanding of the chemistry of the local atmosphere. Even after extensive research in Los Angeles and other urban areas, the mix of emissions and compounds present in any particular location make it difficult to generalize about appropriate control strategies. In most urban areas, it has been generally assumed that the ratio of hydrocarbons to oxides of nitrogen is relatively low and as a result emission control efforts have been focused on hydrocarbons as the most effective means of ozone reduction. In heavily wooded areas, however, where significant amounts of hydrocarbons are emitted naturally, this assumption may be inappropriate and more effective ozone control may be affected by focusing control efforts on oxides of nitrogen. This may be true for much of the eastern U.S., for example.

Thus, the work of Dr. Haagen-Smit and the subsequent research into photochemical smog marked a turning point in the history of the fight against air pollution in that

1. It identified a major pollutant which could not be controlled by regulation of strictly industrial point sources.

2. It indicated the potential complexities of pollutant behavior in the atmosphere, complexities that meant that no clear and universally applicable solution to the pollution problem existed.
3. It demonstrated that research and process understanding was a necessary prerequisite to the development of effective pollution abatement strategies.

The work of Dr. Haagen-Smit set the stage for the modern era of controlling pollution. Its lesson of no easy answers is still true today and is recognized by regulator and polluter alike. Since Dr. Haagen-Smit's beginning, the first step in control of any complex environmental pollution problem has been proper investigation through research. This research has built upon the existing body of knowledge and provides an ever clearer picture of the effects and means of control for at least some pollution problems. Unfortunately this may often lead to delays in responding to pollution problems. Currently there is significant controversy over whether our understanding of global warming and ozone depletion in the upper atmosphere is sufficient to justify the expenditure of vast sums. This controversy has significantly delayed the development of a solid response to these problems.

The examples given in this chapter illustrate environmental problems, their historical significance, and the approach to their resolution. In Chapter 2, a more comprehensive summary of the environmental problems that we currently face is provided.

PROBLEMS

1. For one of the example activities of an environmental engineer, consider the following questions. What is the system of interest and its boundaries? What steps might you suggest to assess and resolve the problems with the system? Where might modeling be employed to assist in the assessment or design of a response? What form might this modeling take, physical or mathematical? What would be the objective of the model? What are the key externalities or outside influences on the system that should be considered? What should be the goal of the environmental engineer? What are other constituencies and "stakeholders" and would they express a different goal?

2. Is pollution more or less of a problem than it was 40 years ago? What pollution problems are worse? What environmental issues may be less of a problem?

3. How are the lessons learned from the London killer fog episode different from those learned from the Bhopal incident in 1985?

4. Can you postulate a mechanism for the observation that SO_2 and particulates together can be more hazardous than separately at similar concentrations?

5. Figure 1.6 shows a direct relationship between reductions of "smoke" emissions and observed concentrations. Would you expect a similar relationship between maximum ozone levels and oxides of nitrogen or reactive hydrocarbons?

6. If you were Dr. Haagen-Smit, for what clues might you search to demonstrate that particulates and SO_2 were not the primary cause of Los Angeles smog?

7. Based on Figure 1.8, is it possible to encounter conditions in which increases in nitrogen oxides will decrease ozone?

8. Name and discuss some current environmental issues for which the research is not yet complete and the scientific basis for an environmental management decision is lacking.

9. The fundamental components of the Los Angeles ozone problem were recognized by 1970 and it was felt at that time that the problem could be resolved by the mid 1980s. What factors may have caused this projection to be overly optimistic?

REFERENCES

Haagen-Smit, A.J. (1952) Chemistry and physiology of Los Angeles smog, *Ind. Eng. Chem., 44*, 1342–1346.

Haagen-Smit, A.J. and M. Fox (1956) Ozone formation in photochemical oxidation of organic substances, *Ind. Eng. Chem.,* 48, 1484–1487.

Hov, O., I.S.A. Isaksen, and E. Hesstvedt (1976) Diurnal variations of ozone and other pollutants in an urban area, *Atm. Environ.*, 10, 1095.

Larsen, R.I. (1970) Relating air pollutant effects to concentration and control, *J. Air Pollut. Control Assoc.*, 20, 4, 214–225.

Leighton, P.A. (1961) *Photochemistry of Air Pollution*, Academic Press, New York.

Seinfeld, J.H. (1975) *Air Pollution, Physical and Chemical Fundamentals*, McGraw-Hill, New York.

Seinfeld, J.H. (1986) *Air Pollution*, John Wiley & Sons, New York.

Seinfeld, J.H. (1980) *Lectures in Atmospheric Chemistry*, AIChE Monograph Series 12, American Institute of Chemical Engineers, New York.

Thibodeaux, L.J. (1992) *The Four Laws of Hazardous Wastes,* LSU Hazardous Substances Research Center, Baton Rouge, LA.

2 Environmental Hazards and Their Management

2.1 INTRODUCTION

In this chapter, a number of environmental hazards will be discussed. These include the historical hazards identified in Chapter 1 as well as other problems that have been identified more recently. The environmental hazards discussed include a variety of conditions that are not easily linked to individual pollutants or sources. These conditions are driven by a variety of factors that stress the environment and lead to imbalance and potential environmental degradation. They are sometimes global in scope which increases their importance but makes it even more difficult to define and address the causes. Some of these hazards include:

- Population growth
- Loss of biodiversity
- Global warming
- Ozone depletion in the stratosphere

In contrast to these global problems, localized environmental hazards associated with specific environmental pollutants and sources are more easily understood and controlled. As illustrated by photochemical smog in Chapter 1, they may still represent formidable challenges. Many scientists and engineers differentiate between environmental contaminants, which may or may not have an adverse environmental impact, and an environmental pollutant, which has an observable negative impact on human health or the ecosystem. The adverse effects of pollutants depend on the availability of suitable transport pathways between source and receptor. Thus, the emphasis is on contaminants that are found or may partition to a mobile phase such as air or water. We will separate our discussion of these pollutants by the media impacted. Much of the work of the environmental engineer is aimed at identification, assessment, and control of these pollutants.

In the subsequent sections of this chapter, the causes and affects of global, atmospheric, and waterborne environmental hazards will be discussed and the means of assessing and managing these hazards will be examined. The reader is encouraged to seek additional information on these hazards through the worldwide web site of the U.S. Environmental Protection Agency at *http://www.epa.gov.*

2.2 GLOBAL HAZARDS

2.2.1 Population Growth and Standard of Living

The rapid growth in human population in the past century has placed strains on the environment. Each member of the population has needs for food and shelter that can only be met at some expense to the environment. The environmental impact of human civilization depends on the needs and desires of the population (their standard of living) and the efficiency with which these needs can be met. Let us define the per capita standard of living in terms of the environmental impact of achieving that standard of living. The *ideal per capita resource usage* is the minimum amount of resources and environmental degradation required to achieve that standard of living. The ideal per capita resource usage is nonzero regardless of the degree of environmental control or the desired standard of living. We could define the minimum ideal per capita resource usage as the actual per capita usage multiplied by an *environmental efficiency*. The environmental efficiency measures the effectiveness of translating environmental degradation and resource usage into the maximum possible standard of living. It simply measures the effectiveness of environmental control efforts and the environmental impact of the approaches and technologies employed to achieve the desired standard of living. These are not easy concepts to quantify but they suggest

$$\text{Environmental Impact } \alpha \ \frac{(\text{Population}) \times (\text{Ideal Per Capita Resource Usage})}{(\text{Environmental Efficiency})} \tag{2.1}$$

Thus, reduction in the environmental impact of the Earth's population can be achieved by reducing population, decreasing their average standard of living (as measured by resource usage), or minimizing the impact of a given standard of living on the environment through environmental controls and planning. The simplicity of this relationship belies the considerable complexity of its components. Equation 2.1 simply expresses that the meeting of human needs and desires always results in some negative environmental impact, the magnitude of which depends on the desired standard of living, the number of people aspiring to that standard of living, and the effectiveness of efforts to mitigate the resulting environmental impact.

Let us consider the problem of providing shelter for a population. If a woodframe house for every family represented the desired standard of living this would entail a minimum environmental damage associated with the use of wood and other materials used to construct the dwelling. The environmental efficiency, and thus the environmental impact, of that choice for a standard of living, however, would be different depending on the degree of environmental control and planning in the harvesting and preparation of the wood and other materials.

Note that the three factors of population, ideal resource usage, and environmental efficiency are not independent. In a developing country the lack of quality medical care (a component of standard of living) can impact population, while the lack of access to appropriate environmental control technologies in the same population can decrease the environmental efficiency. Despite these limitations, the relationship does provide a means of conceptualizing the three most important factors in degradation of the environment as the result of human activities.

Minimization of the impact of human activities on the environment can be achieved with the least controversy by improving the environmental efficiency; namely, minimizing the environmental impact per unit resource usage or standard of living. This can occur primarily through the use of practices and technologies that minimize the environmental impact or through technological advances in waste treatment. This is clearly a desirable goal and is the primary effort of the environmental engineer. Relatively simple and inexpensive changes can sometimes achieve dramatic improvements in environmental efficiency, especially in the less developed portions of the world. The application of basic wastewater treatment can, for example, dramatically lessen the environmental impact of an industry or small city on an adjacent river. In the more developed countries, more sophisticated solutions may be required to further improve the environmental efficiency. The other factors of population and standard of living have proven more difficult to change.

2.2.1.1 Population

Developed western nations have tended to view global environmental degradation as a problem strongly associated with population growth. Population growth in less developed countries is about 2.1% per year, 2.4% if the growth rate in China is not included in this total (Population Reference Bureau, 1989). In the more developed countries, the average population growth rate is only about 0.6% per year (Population Reference Bureau, 1989). In addition, the less developed countries, including China, contain about 80% of the world's population. With an average growth rate of

2.1% per year, the more than 4 billion people in less developed countries would double to more than 8 billion in about 33 years. Growth that is proportional to the current population is referred to as exponential growth. Mathematically it can be written

$$\frac{dP}{dt} = kP \tag{2.2}$$

where P is the population at any time, *dP/dt* is the rate of population growth with time, t, and k is a growth rate constant in fraction per unit time. The solution to this differential equation is

$$P(t) = P(0)\, e^{kt} \tag{2.3}$$

The time required for the population to double is simply given by

$$\frac{\ln 2}{k} \approx \frac{70}{k(\%)}$$

where k(%) is the growth rate constant in % per year. Thus, a 2.1% per year rate of growth results in a doubling in 33 years, as indicated above. At the lower growth rate of the developed countries, more than 116 years would be required before their population would double. The more rapid growth rate as well as related statistics such as fertility and infant mortality rates are included in Table 2.1 for both the more developed and less developed countries.

TABLE 2.1
Population Statistics

Statistic	World	Developed Countries	Less Developed Countries[a]	China
Population (millions)	5234	1206	2924	1104
%	100	23	56	21
Birth rate (#/1000 population)	28	15	35	21
Death rate (#/1000 population)	10	9	11	7
Fertility rate (births/female)	3.6	1.9	4.7	2.4
Infant mortality rate (#/1000)	75	15	93	44
% Population under 15	33	22	40	29
Growth rate (%/year)	1.8	0.6	2.4	1.4
Est. population in year 2020 (millions)	8330	1339	5468	1523

[a] Excluding China

Source: Population Reference Bureau (1989) *1989 World Population Data Sheet.* Washington, D.C.

There is a finite capacity of the Earth to accommodate the needs of this growing population. Given a level of technological development and environmental efficiency there exists a corresponding human carrying capacity of the world. If this carrying capacity is exceeded, either a reduction in numbers or a decrease in resource usage or standard of living must result. Unfortunately, it is not possible to accurately estimate the world's human carrying capacity. A system that is limited by a maximum carrying capacity is described by logistic growth dynamics. Mathematically, Equation 2.2 must be modified to indicate the slowing of growth near the carrying capacity.

$$\frac{dP}{dt} = kP\left(1 - \frac{P}{C}\right) \tag{2.4}$$

Here, C is the carrying capacity, the maximum allowable value of P. The factor P/C is resistance to continued population growth as the population nears the carrying capacity. For small values of the population the resistance factor is near 0 and the exponential growth of Equation 2.3 is followed. As the population nears the carrying capacity, however, the rate of growth slows as environmental factors reduce the standard of living and force a steady-state population. The solution to Equation 2.4, which describes the population as a function of time is given by,

$$P(t) = \frac{C\,P(0)e^{kt}}{P(0)(e^{kt} - 1) + C} \tag{2.5}$$

Equation 2.5 clearly has the appropriate limits of exponential growth at short times and at long times when e^{kt} is large, P approaches C, the carrying capacity. A plot of Equation 2.5 is shown in Figure 2.1. This plot assumes that the population when the period of logistic growth began was 5% of the carrying capacity.

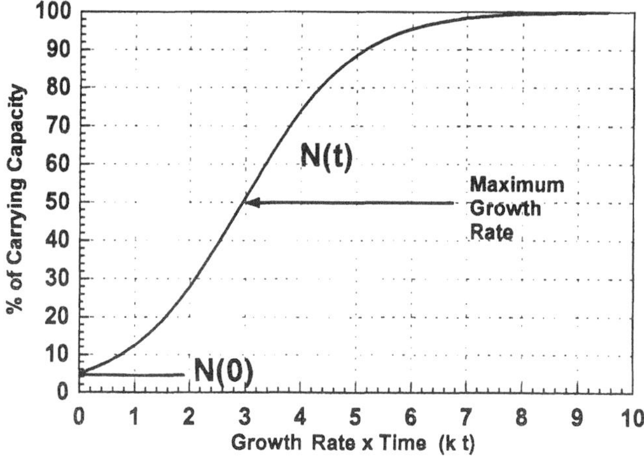

FIGURE 2.1 Illustration of logistic growth kinetics assuming that the period of logistic growth begins when the population is 5% of the carrying capacity.

Given a growth rate of 2.1% per year, as observed in the less developed countries, 100 years would be required to move from 50% capacity to slightly more than 90% capacity. It is also important to recognize that population growth continues long after corrective action is required since the average lifetime exceeds 70 years in the developed nations, and a rapid increase in average life expectancy is being observed in developing nations as medical care improves. Note that according to the logistic growth model, the maximum growth rate occurs at half of the carrying capacity. The maximum growth rate is an indication of conditions which are most favorable to growth and thus might represent a potential for the highest standard of living. Its value as half of the carrying capacity is dependent on the applicability of the logistic model to human population statistics, but clearly the optimum population is less than the carrying capacity of the world. For populations of animals subject to harvesting (e.g., fish), the maximum growth rate corresponds to what is termed the *maximum sustainable yield* in that removal or harvesting at this growth rate can be continued indefinitely without depletion of population stocks.

It is clear that population is a potentially significant factor in environmental degradation. Population must be stabilized prior to achieving the carrying capacity of the Earth or else natural forces will force stabilization of the population. Unfortunately no one knows what the carrying capacity of the Earth might be. It is instructive to recognize that the entire world's population could be housed in an area the size of New Zealand if it had the population density of a modern city. Viewed in that manner, it would seem possible that a clean, modern city could be built for all of the world's inhabitants that would have a minimum impact upon the Earth's environment. Under current economic, social, and political systems, however, this is unlikely to occur or even be considered desirable, and there remain many regions of the Earth subject to drought and famine because the local carrying capacity of the Earth has been exceeded.

2.2.1.2 Standard of Living

Population is not the only means of controlling the environmental impact of humankind. As indicated previously, the standard of living is also a significant factor, at least as measured by the per capita resource usage or degradation of the environment caused by a given standard of living. This is often the view of the developing world which sees the high standard of living of the more developed nations as a significant factor in environmental degradation. The population of the U.S., for example, is approximately 250 million people (1990 U.S. Census). This is less than 5% of the world's population, but in 1987 the U.S. accounted for almost 22% of the carbon released from fossil fuels. As will be seen, the combustion of fossil fuels is linked to most of the observed local and regional air quality problems and also is seen as a major contributor of greenhouse gases to the atmosphere that are responsible for global warming. Thus, carbon emissions are well correlated with at least the potential for significant air quality degradation of the environment. Because the combustion of fossil fuels provides power and heat for homes and factories and fuel for automobiles, carbon emissions are also closely linked with the productivity and standard of living in the U.S. and other more developed nations.

Table 2.2 shows the carbon emissions in a number of countries. The total carbon emissions vary dramatically between the more developed countries and the less developed countries, as does the carbon usage per capita. The carbon emissions per dollar gross national product (GNP), a measure of the total economic output of a country, however, vary much less. This again illustrates the correlation between economic productivity, at least as measured by the GNP, and fossil-fuel carbon emissions. To their credit, the more developed nations have significantly improved their efficiency of using fossil fuels with the carbon emissions per dollar of GNP dropping about 25% between 1960 and 1987. The total carbon releases, and presumably the resulting stresses on the global environment, have increased dramatically due to the expansion of the economies of all nations. Recent efforts to reduce carbon usage and carbon dioxide emissions by developed countries may reverse this trend.

For the foreseeable future, however, improvements in the economies and standard of living of developing countries are likely to result in increased carbon emissions from fossil fuels. It is unlikely that increases in the efficiency of carbon usage (for example, releases per dollar of GNP) will be able to offset the entire increase without significant political will to change the fuel mix or reduce fuel usage in the developed countries. This leaves, therefore, either additional environmental degradation in the form of increased worldwide carbon releases, reduction in the standard of living and production in the more developed countries, or limited economic and standard of living growth in developing countries. This will be a difficult and contentious issue between developing and more developed countries for some years to come.

As the standard of living and population in the currently undeveloped world continues to grow, is the carrying capacity of the Earth sufficient to meet this increased need for resources given current levels of technology? Can technological improvements fill the gap or is degradation in

TABLE 2.2
Carbon Emissions from Fossil Fuels

	Carbon (10^6 tons)		per $ GNP (g)		per capita (tons)	
Country	**1960**	**1987**	**1960**	**1987**	**1960**	**1987**
U.S.	791	1224	420	276	4.38	5.03
Soviet Union	396	1035	416	436	1.85	3.68
China	215	594	NA	2024	0.33	0.56
Japan	64	251	219	156	0.69	2.12
West Germany	149	182	410	223	2.68	2.98
India	33	151	388	655	0.08	0.19
Poland	55	128	470	492	1.86	3.38
Nigeria	1	9	78	359	0.04	0.03
World	2547	5599	411	327	0.82	1.08

Adapted from: Flavin, C. (1990) *State of the World 1990*, Worldwatch Institute, Norton, NY.

the standard of living in the developed world a necessary consequence? Some would suggest that the development of a western-style civilization in the currently undeveloped world will have devastating consequences for the environment and ultimately the standard of living throughout the world. Issues of equity, however, suggest that it is unlikely that a higher standard of living in the developed world can be maintained indefinitely. It is clear that the social and political pressures of these issues should be minimized (but they are unlikely to be eliminated) by the use of environmentally friendly technologies bringing about improvements in environmental efficiency in the human population.

2.2.2 Loss of Biological Diversity

Human development and population growth has resulted in rapid extinction of a number of plant and animal species. This has led to concern about the ability of the Earth to sustain the biological diversity upon which all life depends. *Biodiversity* refers to the wide variations seen in plant and animal life on the planet. At least three types of diversity exist.

- Genetic diversity — Variation between individuals of the same species
- Species diversity — Variation within an ecosystem by the presence of different species
- Ecosystem diversity — Variations within and among species in different ecological environments

Many reasons exist to protect biological diversity. Moral, ethical, and aesthetic reasons are commonly cited to protect and preserve the beauty of the natural environment for present and future generations. There are a number of practical reasons to work to maintain biodiversity. These include:

- Genepool preservation — A broad genepool provides a source of plant and animal traits that may be introduced into valuable agricultural products
- Genepool diversity — Biodiversity preserves traits that may be needed to adapt to a changing environment or conditions

- Important products — Many important medicines are extracted from natural plants and, in addition, many plants have never been evaluated for commercial or medical benefits; retention of biodiversity ensures that these products will be available when found
- Ecosystem stability — Ecosystems depend on a variety of interdependent organisms to survive and thrive, and elimination of any one organism could threaten the survival of the entire ecosystem

The causes of loss of biodiversity are many. Included among the causes are stresses induced in plants and animals from environmental pollutants. For some species, overharvesting has eliminated species from the wild or caused complete extinction of certain species. In sensitive ecosystems, the introduction of non-native species has resulted in dislocation of natural species reducing biodiversity. This is easily seen in Australia and New Zealand which were largely isolated until the last century when human immigration led to the introduction of a number of non-native and feral species into these countries. The large rabbit and possum populations in Australia and New Zealand, respectively, are just two examples of the result of introducing species into an environment with no natural predators. The rapid growth of these species has stressed the environment and limited the availability of food for native species.

The most important cause of loss of biodiversity, though, is the physical alteration of the environment. These alterations can be placed into one of three categories.

1. Conversion of the natural environment to other uses (e.g., residential development).
2. Fragmentation of the ecosystem resulting in smaller ecosystems that are not themselves sustainable.
3. Simplification of an ecosystem by selective harvesting or creating conditions leading to dominance by a single species.

The goal of maintaining biodiversity is central to minimizing the impact of human activities and the concept of sustainable development. Achieving the goal requires knowledge of the processes operative in an ecosystem and recognition of the role that individual plant and animal species play in those processes. The importance of biodiversity to an environmental engineer is the need to recognize the interdependence of organisms within an ecosystem and the attempt to understand the broad, long-term impact of an engineering development. An environmental engineer may not lead or control the effort to maintain biodiversity, but he needs to ensure that he does not become overly focused on the goals of the particular project and not the broader environmental implications.

2.2.3 Global Warming

The ability of humankind to exceed the carrying capacity of the Earth on a local or regional basis has long been recognized. Even hunter-gatherer societies recognized that overhunting in a particular area required the tribe to move to a new area when game stocks were reduced. This problem was aggravated by early agriculturally based societies and aggravated further by early industrial societies. In the past few decades, however, it has become recognized that the activity of humankind can have a **global** impact on the environment. With this recognition comes the chilling realization that humankind can seriously endanger their ability to survive on the planet. Population growth and development have always been recognized as environmental challenges. Identification of global environmental changes as a result of human activities provides evidence of the potential severity of those challenges.

The first large-scale environmental change that seemed to be linked to human activities is the rapid increase in atmospheric carbon dioxide levels that has been observed since the beginning of

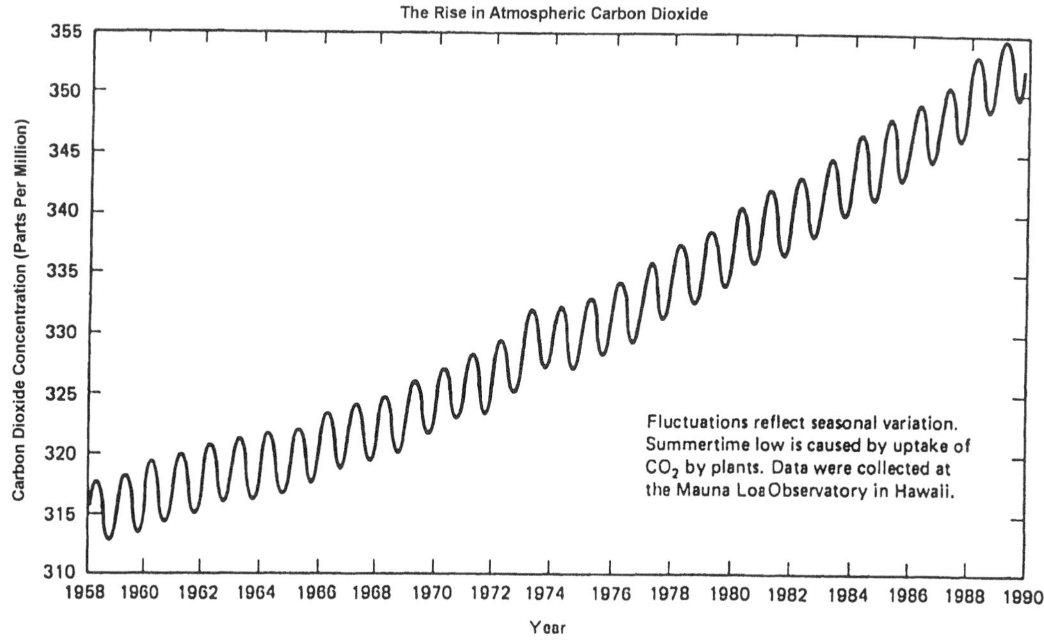

FIGURE 2.2 Concentration of atmospheric carbon dioxide at Mauna Loa Observatory. (From Keeling, C.D. et al. (1989) *Aspects of Climate Variability in the Pacific and Western Americas*, Geophysical Monograph, American Geophysical Union, Vol. 55. With permission.)

the industrial revolution. Examination of air bubbles trapped in ice in Antarctica has suggested that carbon dioxide levels fluctuated between 180 and 300 ppm over the past 150,000 years (Barnola et al., 1987). While carbon dioxide levels were 280 ppm as recently as the year 1750, they had increased to more than 330 ppm in the southern hemisphere and to more than 350 ppm in the northern hemisphere by 1990. This is illustrated in Figure 2.2. Because CO_2 is the natural end product of fossil fuel combustion, the anthropogenic sources of this pollutant have grown drastically in this century. In 1950, about 6.4 billion metric tons of CO_2 were emitted and the ambient CO_2 level was 306 ppm. By 1975, the emissions had tripled to about 18 billion metric tons and the ambient CO_2 level was almost 330 ppm. It has been estimated that about 50% of the CO_2 emitted from anthropogenic sources each year remains in the atmosphere. The remainder is presumably absorbed in the oceans or incorporated into the biosphere.

Further evidence that this recent increase is the result of human activities, and specifically the increasing use of fossil fuels since the advent of the industrial revolution, can be found by comparing the historical record of carbon dioxide concentration with temperature. Both are recorded in ice cores — carbon dioxide by the composition of trapped gas bubbles and temperature by the ratio of element isotopes in the ice. The general pattern of air movement is toward the pole during which the air cools and higher boiling point fractions are condensed. Lighter isotopes have a lower boiling point and tend to be found in somewhat higher concentrations in the air or water vapor from which ice is formed near the poles. Thus, isotopic ratios in ice cores at the poles can be related to mean atmospheric temperatures. Figure 2.3 indicates the high degree of correlation between carbon dioxide levels in the trapped gases and the ambient temperatures estimated in this way. The only significant deviation has occurred in recent times, presumably due to the large increase in carbon dioxide emissions into the atmosphere during the Industrial Revolution.

The importance of the increased emissions of carbon dioxide is the result of the *greenhouse effect*. The greenhouse effect is the ability of atmospheric gases to absorb the energy radiating from the Earth, reducing energy losses to space and ultimately increasing the Earth's temperature. It is

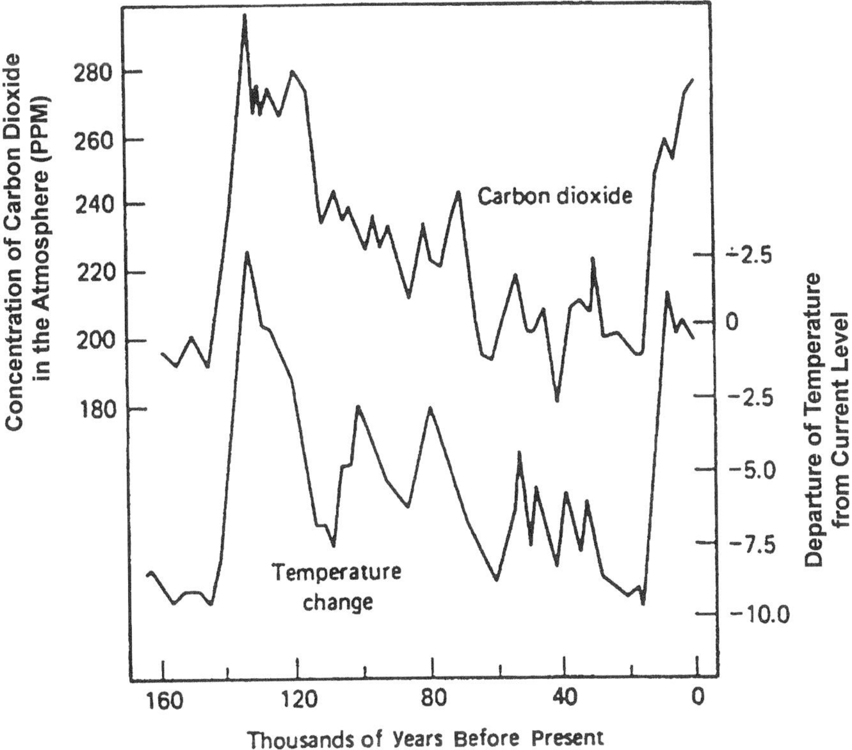

FIGURE 2.3 CO_2 concentrations (ppm) and Antarctic temperatures (°C) plotted against age in the Vostok record. Temperatures are referenced to current Vostok surface temperature. (From Barnola, J.M., et al. (1987) *Nature*, 329, 1 October. With permission.)

so named because of the similar energy containment process associated with greenhouses. The Sun's energy incident upon the Earth is heavily focused in high energy short wavelength light with a maximum intensity at a wavelength of about 0.5 μm. The energy radiating from the Earth, however, is at much lower energies and longer wavelengths with a maximum intensity around 10 μm. It is these low energy, longer wavelengths that are effectively absorbed by carbon dioxide and other trace gases in the atmosphere. Carbon dioxide, for example, strongly absorbs light with a wavelength of 13 to 18 μm. Oxygen (O_2), ozone (O_3), and water vapor (H_2O) tend to absorb higher energy light with a wavelength of less than 10 μm. Thus, the absorption of energy radiating from the Earth's surface tends to be largely associated with the presence of specific gases present at much lower concentrations than oxygen or water vapor. Figure 2.4 depicts the resulting greenhouse effect.

The greenhouse effect can be illustrated quantitatively by assuming that the Earth's temperature is approximately constant. Then the net energy absorbed by the Earth and its atmosphere must equal the net incident radiation. In particular, the energy absorbed by the atmosphere must equal the difference between the energy radiated from the surface and the incident radiation that is not directly reflected by the Earth.

$$\begin{matrix}\text{Energy Absorbed by}\\ \text{Atomosphere and}\\ \text{Greenhouse Gases}\end{matrix} \approx \begin{matrix}\text{Energy}\\ \text{Radiated}\\ \text{from Surface}\end{matrix} - \left[\begin{matrix}\text{Incident}\\ \text{Solar}\\ \text{Radiation}\end{matrix} - \text{Energy Reflected}\right] \tag{2.6}$$

The incident solar radiation averages about 343 W/m^2 of the Earth's surface. The *albedo* of the Earth, or the fraction of the incident radiation that is reflected, is about 30%, or about 103 W/m^2.

The average temperature of the Earth's surface is about 15°C or 288 K. The energy radiated from a body at this temperature is expected to be about 390 W/m^2 from the formula

$$\text{Energy Radiated from Surface} = \sigma T^4 = \left(5.67\times 10^{-8}\,\frac{W}{m^2K^4}\right)(288\,K)^4 = 390\,\frac{W}{m^2} \tag{2.7}$$

σ is termed the Stefan-Boltzmann constant. Note that radiation from a surface is a strong function of temperature (fourth power dependence). Using these estimates in Equation 2.6, the energy that must be absorbed by greenhouse gases in the atmosphere must be about 150 W/m^2, or about 38.5% of the energy radiating from the surface.

Thus, an impressive amount of the Earth's radiated energy is absorbed in the atmosphere by the trace greenhouse gases. Table 2.3 summarizes the most important greenhouse gases.

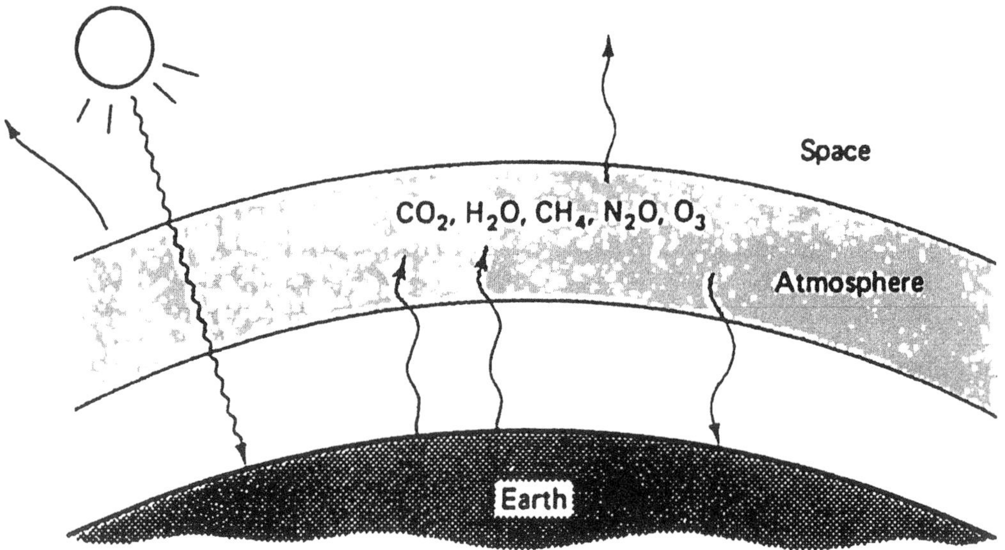

FIGURE 2.4 Greenhouse gases trap long-wavelength energy from the Earth's surface, heating the atmosphere, which, in turn, heats the Earth. (From Masters, G.M. (1991) *Introduction to Environmental Science and Engineering*, Prentice-Hall, Englewood Cliffs, New Jersey. With permission.)

TABLE 2.3
Major Greenhouse Gases

Gas	Air Conc. (ppm)	Annual Increase (%)	Relative Efficiency	Total Absorption (%)
Carbon dioxide	351	0.4	1	57
Chloroflurocarbons	0.00225	5	15,000	25
Methane	1.675	1	25	12
Nitrous oxide (N_2O)	0.31	0.2	230	6

Source: Flavin (1990)

Although carbon dioxide is not an inherently strong greenhouse absorber, its presence in such relatively high concentrations means that it accounts for more than half of the total greenhouse effect. Looking at this table and the proportion of energy that is absorbed by the atmosphere, it is perhaps surprising that the rapid recent increase in carbon dioxide levels has not already led to significantly higher global average temperatures.

The actual degree of warming expected to result from the increase in carbon dioxide levels is still controversial. Complicating factors include environmental sinks for carbon dioxide, such as the ocean and atmospheric changes in cloud cover and global weather patterns. Between 1880 and 1940, the mean temperature of the Earth's surface increased about 0.5 K. Surprisingly, however, the Earth's average temperature decreased about 0.2 K between 1940 and 1975, the period of the greatest increases in CO_2 emissions. Although the global climatic effects of this pollutant may be overshadowed by other processes, there is concern about its potential effect if the emissions remain unchecked. The climatic effects of CO_2 may be the limiting factor on the use of fossil fuels in the next century. It is generally expected that a doubling of carbon dioxide levels will lead to a temperature increase of 1.5 to 4.5°C (Masters, 1991). Although these temperature changes may seem small, a 4.5 °C temperature would result in temperatures not observed on Earth for the past 100 million years. Temperature changes of that magnitude would likely lead to significant dislocations of current productive agricultural areas and extensive melting of the polar ice caps with an accompanying rise in sea level of a few meters.

In order to avoid these changes, international efforts are now underway to reduce fossil fuel use and global deforestation. Both of these activities lead to the release of carbon into the environment, ultimately leading to the formation of more carbon dioxide. As indicated previously by Table 2.2, there is generally a good correlation between carbon usage and the size of the economy. Because the size of the economy as measured by GNP is often used as a crude indicator of standard of living, the political difficulties of controlling carbon usage become clear. It is doubtful that meaningful reduction in the volume of carbon released to the atmosphere can be achieved without some economic dislocations. Clearly there are differences in efficiency between countries, as also shown in Table 2.2. In addition, it should be remembered that the GNP is a measure of the size of the economy and is, at best, a misleading indicator of standard of living. The costs of environmental degradation and advantages of energy efficiency, among other factors, are not appropriately included in the total.

The reduction of carbon released into the environment does not necessarily mean reduction in energy usage and reduction in GNP. Natural gas contains less than 60% of the carbon per unit energy produced during combustion than coal. Thus, use of natural gas can significantly decrease greenhouse gas emissions. Other sources of energy such as solar, wind, or nuclear energy produce no greenhouse gases during the production of electricity. Unfortunately, natural gas is a nonrenewable resource with limited supplies and problems remain in providing cost-effective and safe electricity from the other sources mentioned. In addition, most of the world's supply of fossil fuels is in the form of coal which has the highest carbon dioxide emissions per unit of energy produced and exhibits a variety of other severe environmental challenges as well.

2.2.4 Stratospheric Ozone Depletion

Although ozone near the Earth's surface is the cause of photochemical smog, such as that found in Los Angeles, ozone at an altitude of 30 to 40 km acts as a beneficial filter of harmful ultraviolet (UV) radiation. Figure 2.5 displays the number density and mixing ratio of O_3 (ozone) molecules as a function of altitude. At the base of the stratosphere between 30 and 40 km above the Earth's surface, the fraction of ozone reaches a maximum. Because of the low air densities at this altitude, the concentration in molecules per unit volume is less than 10^{12} per cm^3, but this is still sufficient to cause significant sorption of short wavelength UV radiation from the Sun. This is accomplished by the chemical reaction

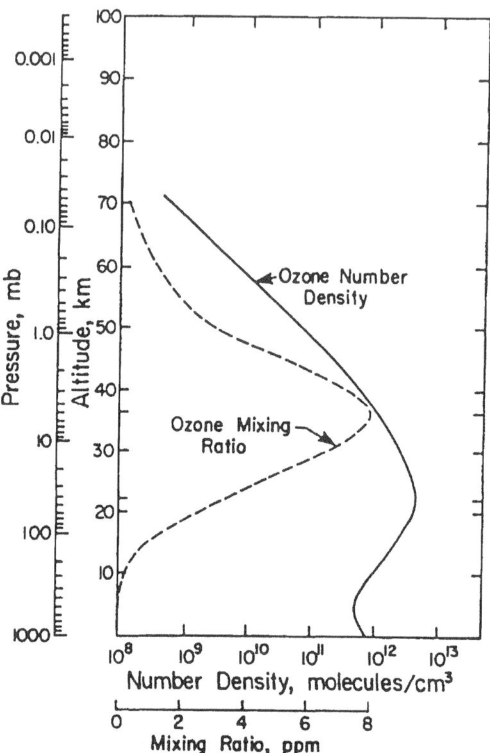

FIGURE 2.5 Atmospheric ozone concentration profiles. The ozone concentration is shown in terms of both number density and mixing ratio. (From Seinfeld, J.H. (1986) *Atmospheric Physics and Chemistry of Air Pollution*, Wiley-Interscience, New York. With permission.)

$$O_3 + h\nu \text{ (radiation)} \rightarrow O_2 + O \quad (2.8)$$

which describes the reaction of ozone (O_3) with electromagnetic radiation or light (h being Planck's constant and ν representing the wavelength of that light) to molecular oxygen and atomic oxygen. Atomic oxygen is extremely reactive and can react with additional oxygen to once again form ozone. The absorption of sunlight reduces the energy reaching the Earth, easing the pressure of greenhouse gases on global warming and, more importantly, eliminates much of the most harmful, higher energy radiation from the sun. Ozone absorbs especially effectively in the UV range of the energy spectrum between 200- and 320-nm wavelengths, so-called UV-B radiation. UV-A radiation, with a wavelength between 320 and 400 nm, passes through essentially without absorption. The effectiveness of the ozone layer at removing UV-B radiation is depicted in Figure 2.6.

Chlorine, bromine, nitrogen, and hydrogen, however, can interfere in this process by removing ozone from the process represented by Reaction 2.8. For example, consider the following sequence of reactions

$$\begin{aligned} Cl + O_3 &\rightarrow ClO + O_2 \\ OH + O_3 &\rightarrow HO_2 + O_2 \\ ClO + HO_2 &\rightarrow HOCl + O_2 \\ HOCl + h\nu &\rightarrow Cl + OH \end{aligned} \quad (2.9)$$

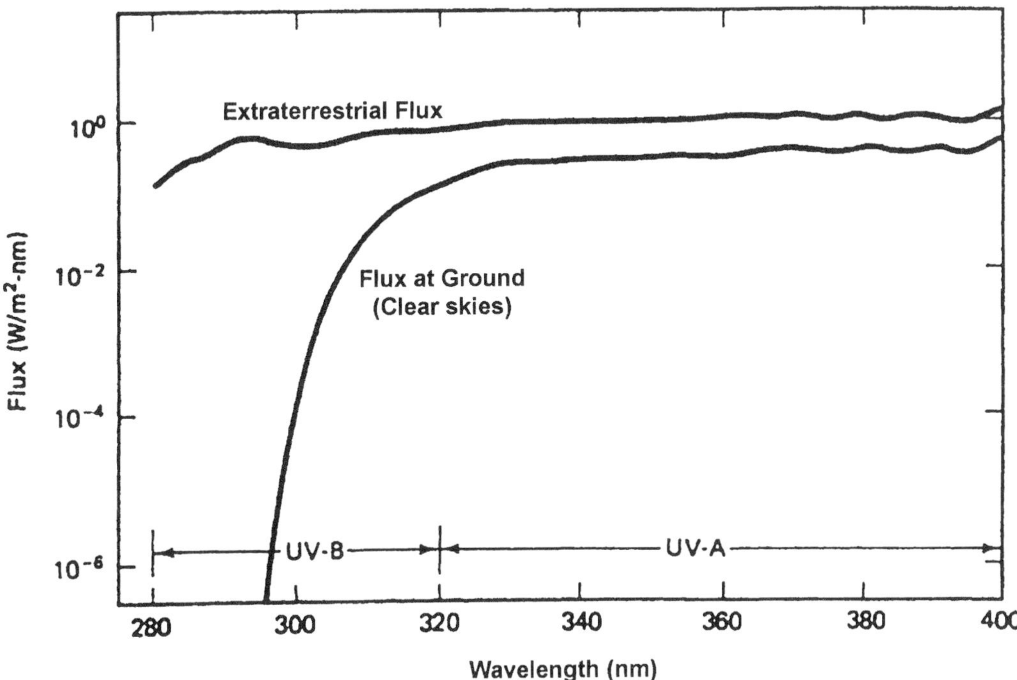

FIGURE 2.6 Extraterrestrial solar UV flux and expected flux at the Earth's surface on a clear day with the Sun 60° from the zenith. (From Frederick, J.E. (1986) *Effects of Changes in Stratospheric Ozone and Global Climate*, U.S.EPA and UNEP, Washington, D.C.)

The net effect of this series of reactions is that two ozone molecules are transformed into three oxygen molecules. Note also that chlorine and the hydrogen oxide or hydroxyl radical are not destroyed in this reaction but serve as *catalysts* that remain available to destroy more ozone. The chlorine monoxide, ClO, formed by Reactions 2.9 ultimately react with nitrogen dioxide to form chlorine nitrate ($ClONO_2$), a relatively inert molecule that serves to protect the ozone from further damage.

Most compounds containing chlorine or bromine are nonvolatile or are destroyed in the lower atmosphere before they can impact the stratospheric ozone. Chlorofluorocarbons (CFCs) and some brominated organic compounds have sufficiently long lifetimes in the atmosphere that they can reach the stratosphere before the chlorine or bromine is released. The stability, lack of toxicity, and excellent physical properties of CFCs have led to their wide use in a number of products and processes. The compounds have seen wide use as refrigerants, solvents, blowing agents in foamed plastics, and until recently, as aerosol propellants. The worldwide use of the three most common CFCs was approximately 1 million tons in 1985. The two most common chlorofluorocarbons are CFC-11 (with chemical formula $CFCl_3$) and CFC-12 (CF_2Cl_2). CFC-11 has been used as an aerosol propellant, a blowing agent in foams, and as a refrigerant in large building air conditioning systems. CFC-12 also has been used as an aerosol propellant and is the most common refrigerant in refrigerators and automobiles.

Some CFCs have a lifetime approaching 100 years. As we will see in Chapter 6, compounds released at the surface can mix throughout the lower stratosphere in less than 5 years. It is for this reason that long-lived CFCs pose a significant danger to stratospheric ozone and the absorption of UV radiation. The link between CFCs and stratospheric ozone destruction is sufficiently strong that use of CFCs as an aerosol propellant was banned in the U.S. in 1979 and a worldwide agreement to reduce and ultimately eliminate the use of CFCs was reached with the Montreal Protocol on Substances that Deplete the Ozone Layer adopted in 1987. The Montreal Protocol called for rapid elimination of the use of ozone-depleting compounds. All compounds with an ozone-depleting potential greater than 0.2 were identified for elimination. The ozone-depleting potential is a measure of the rate of ozone-depletion for a particular chemical compared to the rate of ozone depletion of CFC-12. It is estimated by a combination of compound lifetime in the atmosphere and its effectiveness in depleting ozone.

Other greenhouse gases such as carbon dioxide, methane, and nitrous oxide also affect the ozone layer but, in the case of these compounds, ozone production rather than destruction occurs. These effects must be considered in ozone budget models, but they do not completely offset the negative impact of CFCs.

The depletion of stratospheric ozone has received worldwide attention in recent years as a result of the "ozone hole" first observed over the southern pole during the Antarctic spring and summer. By 1993, the minimum ozone volume present south of 45° S latitude had decreased by about 50% over its value in 1980. By 1993, the maximum size of the ozone hole had grown to more than 20 $(10)^6$ km^2, approximately the size of North America.

The cause of the ozone hole has been linked to the polar vortex which traps extremely cold air above the Antarctic and forms ice despite the low average moisture content. It is believed that under these conditions the chlorine nitrate which serves to eliminate free chlorine from the ozone destruction cycle can react with the ice water forming hypochlorous acid (HOCl) and nitric acid (HNO_3). The HOCl acid then decomposes upon exposure to sunlight during the Antarctic spring and summer resulting in freeing Cl and the hydroxyl radical (OH). The free Cl is then once again available to attack ozone. The Cl cycle in the Antarctic ozone hole can thus be summarized by

$$\begin{aligned} Cl + O_3 &\rightarrow ClO + O_2 \\ ClO + NO_2 &\rightarrow ClONO_2 \\ H_2O + ClONO_2 &\xrightarrow{ice} HOCl + HNO_2 \\ HOCl + h\nu &\rightarrow Cl + OH \end{aligned} \tag{2.10}$$

The significance of the Antarctic ozone hole is that the southern pole was affected by CFCs, whose source is predominantly the Northern Hemisphere. Ozone depletion over the southern pole was the first clear evidence of a global impact of human pollution and served to accelerate efforts to eliminate production of CFCs and other ozone-depleting gases. It has generated worldwide concern and action in a way that the more theoretical and controversial prediction of global warming has not been able to. Largely as a result of the evidence of significant ozone depletion presented by the southern ozone hole, the phase-out of CFCs called for in the Montreal Protocol was accelerated by an agreement reached in 1990 calling for the elimination of many of these compounds by the year 2000.

The removal of many of the ozone-depleting fluorocarbons has been even more rapid than these accelerated schedules would suggest. This has been hailed by many as an example of good cooperation between government, environmentalists, and industry and there have been calls for similar efforts in other areas. While the reduction in the use of ozone-depleting substances has been impressive it is important to recognize that there are special circumstances that suggest that other environmental issues will not be as easily resolved. In particular, the production of fluorocarbons for refrigerants and other uses was a competitive, mature industry that produced what had become a high volume, low profit margin chemical. Non-ozone depleting chemicals that are being used to replace the fluorocarbons are new products with limited capacity and enormous profit potential. If there is a lesson to be learned from the rapid elimination of CFCs, it is that rapid action is possible when the science, popular will, and economics all push toward a common goal.

TABLE 2.4
Common Air Pollutants

Pollutants	Primary	Secondary	Major Source
Carbon monoxide	CO		Combustion
Oxides of S	SO_2,SO_3	SO_3, MSO_4	Combustion
Oxides of N	NO	NO_2, MNO_3, O_3	Combustion
Organic compounds (VOC)	C1–C5 Compounds	O_3, organics	Combustion
Oxidants	None	O_3, PAN	—
Particulates (TSP)	Same	Same	Combustion
Lead	Same		Transportation (w/leaded fuel)

2.3 AIR POLLUTANTS

2.3.1 Introduction

In addition to global warming and stratospheric ozone depletion, air pollutants also pose local and regional hazards. In this section, each of the most important air pollutants will be introduced. Remember that pollutants are defined as substances emitted through the actions of humans that result in adverse effects upon the environment. The sources, sinks, and potential adverse effects for each pollutant will be discussed. Among the air pollutants to be examined are oxides of carbon, sulfur, and nitrogen; organic compounds; photochemical oxidants; particulates; and lead. Oxides of sulfur and particulates are associated with heavy industry and resulted in the first widely recognized industrial pollution problems as noted in Chapter 1. Oxides of carbon are associated with the combustion of fossil fuels and can be toxic at high concentrations (carbon monoxide) and may significantly affect the global climate at low concentrations (CO_2). Photochemical oxidants, organic compounds, and oxides of nitrogen play an important part in the "smog" found in modern cities, as also discussed in Chapter 1. Certain organic and other compounds also have become important in recent years as potential carcinogens or air toxins. Table 2.4 contains a summary of the major pollutants to be examined. Examples of both primary and secondary pollutants are noted in each class. A primary pollutant is one that is emitted directly to the atmosphere while a secondary pollutant is one that is formed from the primary pollutants in the atmosphere.

As indicated in the table, the major source of essentially all of the primary pollutants is combustion of fossil fuels, whether in the automobile, industrial boilers, or in conventional power plants. To further illustrate this point, Table 2.5 indicates the amounts of the major primary pollutants emitted during 1989 in the U.S. from four types of anthropogenic sources. Fuel combustion primarily refers to conventional power plant emissions while transportation refers to all mobile sources but is usually dominated by automobile emissions. SO_x and NO_x are oxides of sulfur and oxides of nitrogen, respectively. VOC refers to reactive volatile organic compounds (essentially all organic vapors except methane) and TSP refers to total suspended particulates. Table 2.5 also includes the amount and year of the largest estimated emissions. Note that the greatest emissions occurred during the 1970s for all pollutants except particulates. Early and effective control of particulates occurred because the largest airborne particles are the easiest to remove and contribute the bulk of the emitted mass. The rapid reduction of particulates was earlier demonstrated by reductions in smoke in London after the killer fog of 1952 (see Chapter 1, Figure 1.6). Attention is now focused on the finer particulates less than 10 μm in diameter that are not as easily controlled and pose the greatest health problems.

TABLE 2.5
Pollutant Emissions in the U.S. — 1989
Millions of Metric Tons Per Year

Source Type	CO	SO_x	NO_x	VOC	TSP	Lead
Fuel combustion	7.8	16.8	11.1	0.9	1.8	0.5
Transportation	40.0	1.0	7.9	6.4	1.5	2.2
Industrial processing	4.6	3.3	0.6	8.1	2.7	2.0
Solid waste	1.7	0.0	0.1	0.6	0.3	2.3
Miscellaneous	6.7	0.0	0.2	2.5	1.0	0.0
Total	60.9	21.1	19.9	18.5	7.2	7.2
Maximum emissions	101.42	28.3	21.72	27.52	24.92	203.82
Year		1970				
% Reduction	40	25	8.3	33	71	96

Source: Environmental Quality (1990) 21st Annual Report of the Council of Environmental Quality, Washington, D.C.

Table 2.5 also shows that lead has been effectively controlled in recent years. Lead in the form of tetraethyl lead was added to gasoline to improve octane number, that is to improve the early ignition characteristics of the fuel. During combustion, the tetraethyl lead was converted to inorganic lead and emitted into the air. With the advent of catalytic converters to control hydrocarbons and oxides of nitrogen, however, tetraethyl lead was rapidly phased out of gasolines because it poisons the catalyst, destroying its effectiveness. As a result, lead emissions to the environment were reduced 96% between 1970 and 1989.

Other pollutants, such as oxides of nitrogen, have experienced much more limited emission reductions. This is partially due to the focus on hydrocarbon controls for control of photochemical smog in urban areas but also has been the result of the offsetting effects of population increases (22% between 1970 and 1989) and increases in production (72% increase in gross domestic product between 1970 and 1989). The emission reductions reported in Table 2.5 would be much larger if 1989 emissions were compared to the projected emissions if no pollution control efforts had been implemented. Significant improvements in pollution control efforts or environmental efficiency are needed simply to restrain the rate of growth of emissions due to expansion of the population and the economy.

Because some pollutants are always present in the atmosphere due to natural emissions, it will be useful to describe the structure and composition of the "clean" atmosphere. Table 2.6 compares "clean" air with polluted air. The concentration units employed in Table 2.6 are the part per billion, ppb, and the gram per cubic meter, g/m^3. Parts per billion is the ratio of moles of pollutant (with one mole being 6.023×10^{23} molecules) per billion (10^9) moles air. Another common unit that is defined similarly is the parts per million (1 part in 10^6). For an ideal gas such as air, the mole ratio is identical to the volume ratio, that is

$$\frac{\text{(mole pollutant)}}{\text{(mole air)}} \approx \frac{\text{(volume pollutant)}}{\text{(volume air)}} \quad \text{for an ideal gas}$$

Note, however, that a mass ratio, which is the usual definition for water pollutants, is not equal to the mole or volume ratio. These terms are defined more fully in Chapter 3.

TABLE 2.6
A Comparison of Clean and Polluted Air

Pollutant	Clean Troposphere[a]	Polluted Air[a]
SO_2	1–10	20–200
CO	120	1000–10,000
NO_x	0.1–0.5	50–250
O_3	20–80	100–500
NH_3	1	10–25
Organics		
HCHO — formaldehyde	0.4	20–50
CH_4	1650	-
Non-methane organics	-	500–1200
Particulates	20 μg/m^3	200–2000 μg/m^3

[a] Concentrations in ppb except as noted.

Source: Seinfeld, J.H. (1975) *Air Pollution. Physical and Chemical Fundamentals*, McGraw-Hill, New York. With permission.

The other common measure of air pollutant concentration, g/m^3 or μg/m^3, is also somewhat ambiguous in that it is a function of temperature. Due to the limited range of temperatures experienced in the atmosphere, this rarely poses a problem. Conversion between concentration in μg/m^3 and ppb can be accomplished via the following formula, valid near the temperature of 298 K (25°C).

$$C[ppb] = C\left[\frac{\mu g}{m^3}\right]\frac{24.5}{MW} \tag{2.11}$$

Here, C represents the pollutant concentration in the indicated units and MW represents the molecular weight of the pollutant compound which is needed to convert between moles and mass.

Excluding trace constituents such as these air pollutants, the composition of dry air at sea level is given in Table 2.7. In addition, up to 3% of the air is water vapor. As noted, the above distribution is found at sea level in the troposphere. As can be seen from Table 2.7, air is essentially a mixture of 78% nitrogen, 21% oxygen, and slightly less than 1% argon. While these compounds define most of the important properties of air, the trace constituents are, of course, the primary focus of the work of the environmental engineer.

2.3.2 Oxides of Carbon

The oxides of carbon are an ideal starting point for our analysis of air pollutants in that they exhibit not only directly toxic effects when present at high concentrations close to the source (carbon monoxide, CO), but also potentially important adverse effects when present over large areas at low concentrations (carbon dioxide, CO_2). Both CO and CO_2 are emitted during the combustion of fossil fuels. CO_2 is the natural end product of hydrocarbon combustion while carbon monoxide is an intermediate combustion product. Thus, the environmental problems associated with oxides of carbon have grown as the usage of fossil fuels has increased in this century. As indicated in Table 2.5, more oxides of carbon are emitted to the atmosphere than any other pollutant. Since the global warming effects of carbon dioxide have already been discussed, let us consider the effects of CO.

Natural sources of CO account for at least 10 times the global anthropogenic emissions (Seinfeld, 1975). The major natural source of CO is the oxidation of ambient methane by OH

TABLE 2.7
Composition of Dry Air at Sea Level

Gas	Concentration, ppm
Nitrogen (N)	780,840
Oxygen (O)	209,460
Argon (Ar)	9340
Carbon dioxide (CO_2)	315
Neon (Ne)	18
Helium (He)	5.2
Methane (CH_4)	1.5
Krypton (Kr)	1.1
Nitrous oxide (N_2O)	0.5
Hydrogen (H_2)	0.5
Xenon (Xe)	0.08

Note: ppm = parts per million, 10^3 ppb.

radicals in the atmosphere. Offsetting the natural production of CO, however, is a variety of natural depletion mechanisms including atmospheric oxidation of CO with OH radicals and biological oxidation of CO in soils. The balance between the natural CO sources and sinks results in a worldwide CO concentration of 0.01 to 0.2 ppm. An examination of CO concentration data in cities or industrialized areas reveals local CO concentrations which may far exceed this level. Thus, the concentrated nature of anthropogenic CO emissions causes local imbalances between the natural sources and sinks.

The adverse health effects of CO are due to its affinity for bonding with hemoglobin, which normally carries oxygen throughout the body. Hemoglobin that has combined with CO, forming carboxyhemoglobin (COHb), is unavailable for oxygen transport. Because the affinity of CO for attachment to the hemoglobin molecule is more than 200 times greater that of oxygen, an ambient CO concentration of only 30 ppm can lead to 5% of the body's hemoglobin forming COHb. A 5% blood hemoglobin level is common in cigarette smokers in clean air. Most people begin to feel the adverse effects of the reduced oxygen supply when the COHb concentration exceeds 10% (an ambient CO concentration of about 60 ppm), although some effects have been noted at COHb levels as low as 2.5%. The initial effects are headaches and reduced mental acuity. These effects rapidly become life-threatening when the ambient CO concentrations exceed 300 ppm which results in COHb levels of 40% or more. These effects as a function of concentration are shown graphically in Figure 2.7. Note that 6 to 10 h of exposure to elevated CO levels are required to achieve equilibrium COHb levels in the bloodstream. If the exposed person is active (running or heavy work), the equilibration time may be only 3 to 4 h. Thus, the adverse effects of the pollutant are a function of both concentration and time. As we shall see, this behavior also is found in many other pollutants. A useful indicator of pollutant levels which combines both concentration and time is pollutant dosage, the time integral of the pollutant concentration. Where appropriate we will employ this quantity to describe pollutant levels.

2.3.3 Oxides of Sulfur

Oxides of sulfur, which are commonly denoted by SO_x, are some of the most important air pollutants. Despite significant anthropogenic sources, natural sources of oxides of sulfur are at least three times larger on a global scale. The natural sources include volcanoes, the ocean, and natural biological decay. Much of the natural emissions of sulfur compounds are in the form of hydrogen

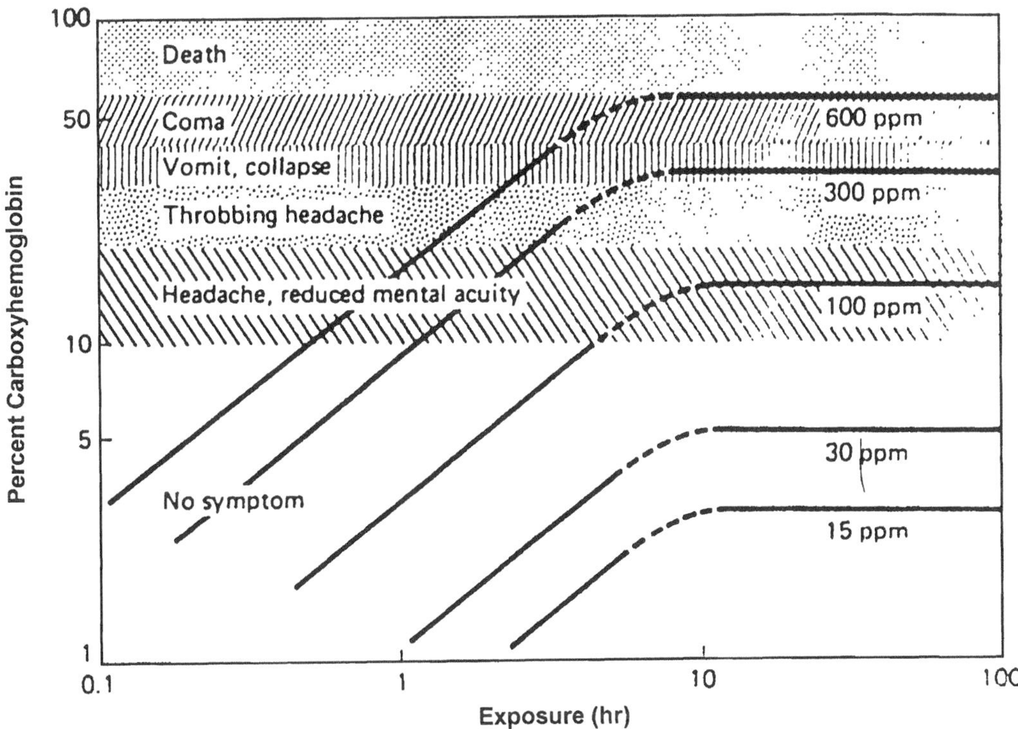

FIGURE 2.7 Effects of exposure to CO on man. (From Seinfeld, J.H. (1975) *Air Pollution, Physical and Chemical Fundamentals,* McGraw-Hill, New York. With permission.)

sulfide (H_2S), which quickly oxidizes to SO_2 in the atmosphere. The anthropogenic emissions of SO_x are also large and trail only the emissions of CO and CO_2 in amount. As with most pollutants, it is the anthropogenic emissions that are of more concern in industrial or urban areas. As noted in Chapter 1, SO_x represented a problem early in history due to the presence of sulfur in coal used in home heating and to power the early stages of the Industrial Revolution. Even today, the vast majority of the anthropogenic emissions of SO_x are due to the combustion of sulfur-containing fossil fuels. The majority of the sulfur emitted in a combustion process is as sulfur dioxide (SO_2). In the atmosphere, however, the SO_2 can be further oxidized to sulfur trioxide (SO_3) and sulfate compounds (SO_4^-). As in oxides of carbon, SO_x compounds result in directly toxic effects at high concentrations and potentially adverse modification of the environment when spread at low concentrations over a large area. The latter problems refer to the acidity of the sulfite (SO_3^-) and sulfate (SO_4^-) ions which can alter the pH balance of natural ecosystems.

Because of the importance of the combustion fuel sulfur content on SO_x emissions, let us examine the typical makeup of coal, oil, and natural gas fuels. Natural gas contains essentially no significant quantities of sulfur compounds while fuel oils can contain as much 2% sulfur (by weight) and coal can contain as much as 6% sulfur. The average sulfur content of all coals used in the U.S. prior to 1970 was about 2.5%. Limitations on SO_x emissions in the early 1970s caused a reduction in this number, but due to shortages of cleaner fuels during the 1970s, high sulfur fuels were again used. Because much of the coal reserves contain in excess of 3% sulfur, it is unlikely that only low sulfur coals will be used in combustion sources.

The adverse health effects of SO_2 have been alluded to previously with respect to the London "killer fogs." Constriction of the bronchial passages occurs in humans and other animals in response to SO_2. Detectable responses in man usually require an SO_2 level of at least 5 ppm, although

TABLE 2.8
Effects of SO_2 at Various Concentrations

Concentration	Effect
0.03 ppm, annual average	1974 air quality standard, chronic plant injury
0.037-0.092 ppm, annual mean	Accompanied by smoke at a concentration of 185 $\mu g/m^3$, increased frequency of respiratory symptoms and lung disease may occur
0.11-0.19 ppm, 24-h mean	With low particulate level, increased hospital admission of older persons for respiratory diseases may occur; increased metal corrosion rate
0.19 ppm, 24-h mean	With low particulate level, increased mortality may occur
0.25 ppm, 24-h mean	Accompanied by smoke at a concentration of 750 $\mu g/m^3$, increased daily death rate may occur (British data); a sharp rise in illness rates
0.3 ppm, 8 h	Some trees show injury
0.52 ppm, 24-h average	Accompanied by particulate, increased mortality may occur

sensitive individuals may show symptoms of SO_2 exposure at 1 to 2 ppm. At concentrations below about 20 ppm, the adverse effects of SO_2 apparently exist only as long as exposure to the compound is continued. No chronic or cumulative effects have been found. The presence of particulates in the air, however, can result in adverse effects at much lower SO_2 concentrations, suggesting that a synergism exists. Table 2.8 indicates the effect of SO_2 when particulates are also present.

The severalfold increase in the effects of SO_2 in the presence of particulates can probably be attributed to the ability of the smaller particles to penetrate deeply into the lung. In the absence of particulates, SO_2 would normally be absorbed in the mucous membranes of the upper respiratory tract.

As indicated previously, SO_2 oxidizes in the atmosphere to SO_3^- and SO_4^-. Absorption of oxidized SO_3^- and SO_4^- species into water droplets results in the formation of sulfurous and sulfuric acids. Acidic rainfall due to the presence of these compounds can then result in acidification of surface water bodies and adversely affect exposed plantlife. It should be noted that acid deposition also can occur by dry mechanisms. Acid rain and acid deposition has affected parts of Scandinavia and the northwestern U.S. and Canada, but increasing evidence exists that the problem is essentially global or at least hemispheric in scope. Acidic rainfall, as well as acidic dew and fog, has been observed in much of the U.S. Figure 2.8 indicates that in 1985, the annual average pH of rainfall was less than 5.0 over the entire eastern U.S. Slightly acidic rainfall is normal due to the presence of CO_2 in the atmosphere. Absorption of the CO_2 to form carbonic acid should result in an equilibrium rain water pH of about 5.6. The presence of natural and anthropogenic sources of SO_x, as well as NO_x sources to a lesser extent, have resulted in much lower rain water pH levels, sometimes less than 2. The average pH of rainfall in The Netherlands during the 1970s was below 4.0. Remember that the pH scale is logarithmic indicating that a reduction in pH by 1 unit is a tenfold increase in acidity as measured by the hydrogen ion concentration.

The oxidation of SO_2 can occur either homogeneously entirely within the gas phase or heterogeneously within atmospheric aerosols or particulates. Some of these processes are depicted in Figure 2.9. The homogeneous oxidation path is the result of molecular excitation by sunlight. Seinfeld (1975) pointed out that potentially the most important oxidation step for SO_2 in clean air is the reaction of an SO_2 molecule in the triplet, or first excited state, with molecular oxygen, as symbolized by

$$SO_2 + O_2 \rightarrow SO_3 + O \quad (2.12)$$

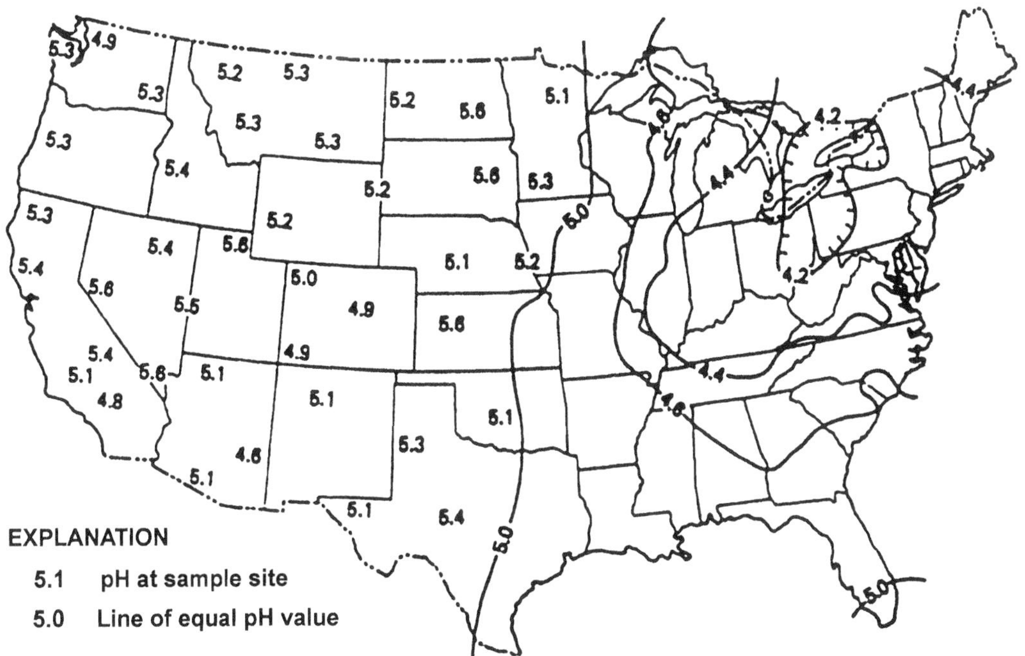

FIGURE 2.8 pH of wet deposition in 1985. Numbers represent a precipitation weighted average based on NADP/NTN data. (From the National Acid Precipitation Assessment Program, Annual Report, 1985.)

Typical SO_2 oxidation rates in clean air are of the order of 0.1%/h. Atmospheric oxidation rates are typically much faster due to the presence of other pollutants such as NO_x and hydrocarbon compounds. These pollutants also can generate free radicals, thus providing a more concentrated pool of reaction initiators and resulting in a more rapid SO_2 oxidation. The oxidation rate of SO_2 is further enhanced by catalytic action along heterogeneous reaction paths. The overall reaction with the hydroxyl radical can be written

$$SO_2 + 2OH \rightarrow H_2SO_4 \tag{2.13}$$

The oxidation is catalyzed by metal salts which are normally found in small concentrations in the atmosphere. Water or pollutant aerosol droplets provide a medium for contacting dissolved SO_2 with the metal salt catalyst. Thus, polluted atmospheres which exhibit high aerosol concentrations or abnormally high metal salt concentrations can more rapidly oxidize SO_2. Observed SO_2 oxidation rates range from less than 1%/h in relatively "clean" locales to 1 to 10%/h or more in heavily polluted areas or in pollutant source plumes (e.g., a power plant plume).

The oxidation of SO_2 to sulfurous and sulfuric acids and subsequent deposition on the Earth's surface has been blamed for many reductions in fish and plant life populations in affected areas. The addition of strong acids to lakes eventually results in loss of the natural buffering action of the lake and a rapid increase in lake acidity. Reproduction in many fish species is halted at a pH level of about 5.5. Fish kills can occur at lower pHs. The death of forest tree populations also has been blamed on acid rain. Some varieties of trees are more susceptible to acid damage than others, probably resulting in a shift of the forest ecosystem rather than its complete destruction. In order to relieve the acute problems of lake acidification, bases such as lime or buffering agents have been added to lakes at risk until a more permanent solution could be found. A permanent solution

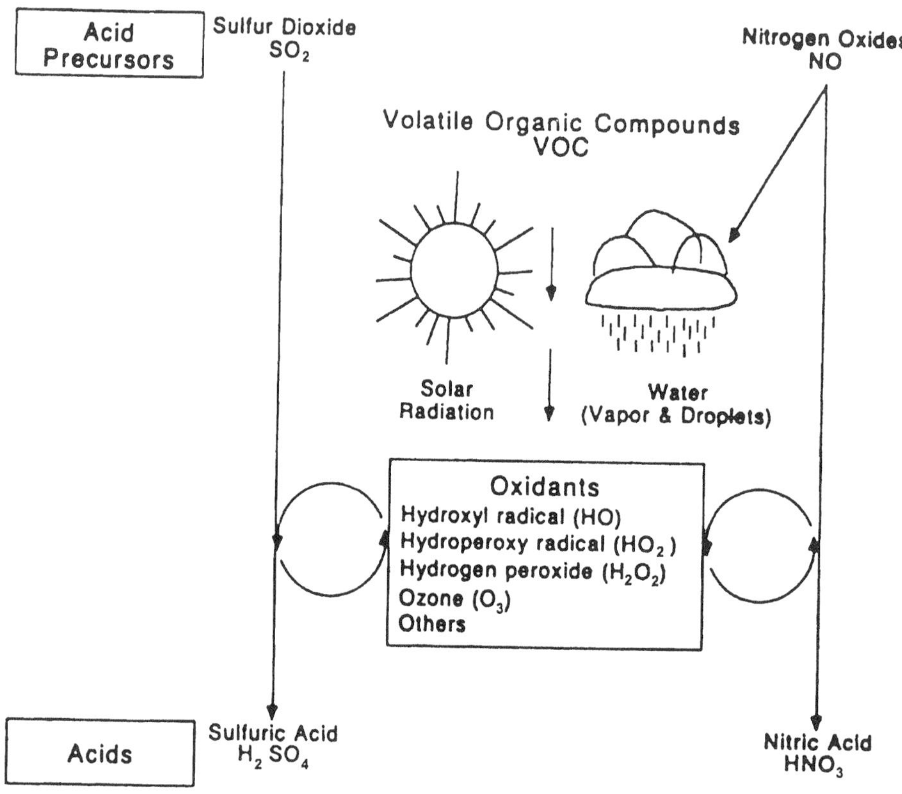

FIGURE 2.9 Transformation of acid precursor gases into acids: a coupled chemical system. (From National Acid Precipitation Program, Annual Report, 1984.)

probably lies in the strict control of SO_x pollution sources and significant strides have been made toward that end.

2.3.4 Oxides of Nitrogen

Oxides of nitrogen, which are commonly denoted by NO_x, include nitric oxide (NO) and nitrogen dioxide (NO_2). N_2O (laughing gas) also is commonly found in the atmosphere but is not generally considered a pollutant. NO and NO_2 are formed at the high temperatures of combustion when oxygen and nitrogen molecules can break down to allow recombination with each other. As was indicated in Table 2.5, essentially all of the oxides of nitrogen emitted from anthropogenic sources are from combustion sources, either mobile sources such as the automobile or stationary sources such as power plants. As with the other pollutants discussed previously, natural sources of oxides of nitrogen, which include photolytic dissociation of N_2O and atmospheric electrical discharges, far exceed anthropogenic sources on a global scale by a factor of 10 or more. The vast majority of the anthropogenic sources are in the Northern Hemisphere, and the majority of those are concentrated in heavily populated, industrialized regions. Thus, it appears that once again the density of the relatively small quantity of anthropogenic sources results in the potential for significant local degradation in air quality. In "clean air," typical concentrations of NO are 0.2 to 2 ppb, and NO_2, 0.5 to 4 ppb. Yearly average NO_2 concentrations during the mid 1960s in Chicago, Los Angeles, and Washington, however, were about 0.05 ppm (50 ppb). Hourly average values as high as 1 ppm were observed. The emissions and concentrations of oxides of nitrogen in cities have remained approximately constant with increased control efforts largely offset by growth in population and emission sources. The inability to control oxides of nitrogen has been aggravated as well by the historic focus on hydrocarbon control to reduce ozone levels in cities.

The health effects of NO_x compounds are not as well understood as those for the previously described pollutants. In general, direct adverse health effects require concentrations of NO and NO_2 that far exceed those found in polluted air. Exposure to concentrations of 0.7 to 5 ppm NO_2 over 10 to 15 min has resulted in increased airway resistance, while exposure to 15 ppm NO_2 has caused eye and nose irritation. In the atmosphere, NO_x compounds are generally associated with other pollutants making the identification of adverse health impacts difficult. An epidemiological study of the population of Chattanooga, TN found that NO_2 levels exceeding 0.06 ppm resulted in increased illness rates. The conclusions of this study have been faulted, however, because of the existence of significant quantities of other pollutants.

While the direct adverse health effects of NO and NO_2 appear to be small in urban atmospheres, some of the secondary pollutants formed from these compounds are very important. Nitrous and nitric acids formed from nitrogen-containing compounds add to the problem of acid rain, accelerating the acidification of ecosystems. In Los Angeles, for example, much of the acidity of rainfall can be attributed to nitrogen-based acids rather than the sulfur-based acids which affect the northeastern U.S. and Scandinavia. As indicated earlier, NO and NO_2 also play an important part in the formation of ozone and other photochemical oxidants in the atmosphere. The role of NO_x in the ozone problem of Los Angeles was not fully understood when automobile emission controls were first implemented. Unfortunately, the methods employed to reduce the CO and hydrocarbon emissions in these cars resulted in increased NO_x emissions. By the early 1970s, this problem had been corrected and NO_x also was controlled in automobile emissions. The strategies to control ozone pollution in major cities such as Los Angeles remain focused on hydrocarbons; NO_x emissions are not as strictly controlled.

2.3.5 Organic Pollutants

Hydrocarbons and other organic contaminants are emitted from a variety of anthropogenic sources. The majority of the near 20 million metric tons of organic compounds emitted yearly in the U.S. is through evaporative processes of various types. The remainder is the result of incomplete combustion of fossil fuels. The largest single source is incomplete combustion of gasoline in the automobile. Even the process of filling an automobile gas tank results in significant gasoline emissions. All of the anthropogenic sources of organic compounds pale in comparison to natural sources, however. Robinson and Robbins (1968) estimated that natural emissions of methane exceed the anthropogenic emissions by a factor of 20. Even the natural emission of high molecular weight organic compounds from vegetation exceeds the total anthropogenic output by a factor of 2. This has led to some suggestion that ozone can never be controlled "while a single tree stands." As with the previously described pollutants, anthropogenic sources result in local organic concentrations that often far exceed levels attributable to natural sources. In addition, natural sources predominantly emit methane, which does not appreciably influence ozone levels, while anthropogenic sources emit a variety of more reactive organic compounds.

Some organic compounds borne by the air have been linked to increases in cancer rates, even when found at low concentrations. Vinyl chloride and benzene are perhaps the most widely recognized potential hazards, but other chemicals also may be important airborne carcinogens. Air toxics have received increasing attention in recent years culminating in a number of regulations in the Clean Air Act Amendments of 1990. The effects of the hydrocarbon air toxics will be discussed in more detail later in this chapter.

Historically, the most important adverse effect attributable to organic contaminants in the atmosphere is their ability to encourage the formation of ozone. Olefins, ketones, aldehydes, and higher molecular weight hydrocarbons are very important in the formation of ozone. About one third of the total anthropogenic sources of organic compounds are related to transportation (i.e., the automobile) indicating the link between the automobile and atmospheric ozone levels as observed in Los Angeles.

2.3.6 Photochemical Oxidants

The term photochemical oxidants refers to the strong oxidizing agents formed in the atmosphere due to the photochemically stimulated reactions that occur in the presence of air, reactive hydrocarbons, and oxides of nitrogen. As indicated previously, the major adverse effects associated with oxides of nitrogen and organic compounds in the atmosphere are the result of their ability to form ozone and other oxidants in the presence of sunlight. There are essentially no significant natural or anthropogenic sources of ozone in the lower troposphere and yet it represents the single most important pollutant in many areas of the U.S. In many areas, control of the pollutants generating photochemical smogs has been limited by population growth, natural sources, and the diffuse nature of the sources. Thus, essentially all urban areas in the U.S. and its entire eastern region are troubled by high ozone levels. It should be emphasized that there is no link between tropospheric ozone pollution and stratospheric ozone, which aids removal of harmful UV radiation. Due to its short atmospheric lifetime, tropospheric ozone cannot replace stratospheric ozone and the protection from UV radiation provided by the thin layer of tropospheric ozone is negligible.

The most commonly observed effect of tropospheric ozone pollution is eye and throat irritation. Eye irritation results from exposure to 0.1 ppm levels, and at 0.3 ppm, nasal and throat irritation and constriction result. These effects are compounded in sensitive people, such as those that suffer from respiratory problems like bronchitis and emphysema. The level that causes significant adverse health effects is a subject of some controversy. The legal maximum allowable ozone level has been 0.12 ppm in the U.S. based upon a single hourly average, but there is evidence for impaired lung function and other problems at much lower levels. A move to 0.08 ppm averaged over 8 h is being implemented.

Increasing evidence has surfaced that crop and vegetation damage can result from continued exposure to levels as low as 0.03 ppm. The effect on vegetation varies, but reductions in the yield of certain crops at maximum ozone concentrations of 0.1 ppm may be as much as 30%. The effects of ozone and other photochemical oxidants are summarized in Table 2.9.

TABLE 2.9
Health Effects of Ozone and Photochemical Oxidants

Concentration		Ozone	
(ppm)	(μm/m³)	Exposure	Effects
0.02	40	1 h	Cracked, stretched rubber
0.03	60	8 h	Vegetation damage
0.10	200	1 h	Increased airway resistance
0.30	590	Continuous working hours	Nose and throat irritation, chest constriction
2.00	3900	2 h	Severe cough
		Photochemical Oxidants	
0.05	100	4 h	Vegetation damage
0.10	200		Eye irritation
0.13	250	Maximum daily	Aggravation of respiratory diseases
0.03	240	1 h	Impaired performance of athletes
0.12		1 h maximum	Air quality standard

Source: National Air Pollution Control Administration (1970) *Air Quality Criteria for Photochemical Oxidants*, AP-63, HEW, Washington, D.C. 1970.

2.3.7 Particulates

Particulates are non-gaseous pollutants. We include in this category any substance that exists in the form of microscopic liquid or solid particles in the atmosphere. Dusts, smokes, mists, and aerosols are all different names for particulates. There is not a commonly accepted definition for some of the terms referring to particulates, but dust is usually used to refer to solid particles formed by mechanical disintegration, while smokes result from the condensation of vapor. Aerosols generally refer to a cloud of particulates. As indicated by Figure 2.10, particulates occur in a large range of particle sizes. The effects and control processes appropriate for particulates are a strong function of particle size.

Because particulates tend to be emitted by industry and represent larger stationary point-sources of pollution, progress toward their elimination has been rapid. This is demonstrated in Table 2.10. Notice that more than half of the total emissions of particulates are from industrial processes and the combustion of coal. This is still a significant improvement over the early years of the table when these sources accounted for some 80% of the total particulate emission. The overall reduction in emissions of about 75% has been almost totally from these sources.

The adverse effects of particulate pollution are a strong function of the size of the particles. Large, heavy particulates (e.g., greater than 100 μm in diameter) tend to settle to the ground quickly. Particles close to molecular dimensions in size tend to move randomly due to Brownian motion and not settle. Brownian motion is caused by collisions with gaseous molecules. Only particles less than 0.1 to 1 μm in size are influenced by molecular collisions and exhibit significant Brownian motion. It is the intermediate sized (0.1 to 10 μm) particles that are the biggest air quality concern. In the human respiratory system, for example, larger particles tend to be removed efficiently by the cilia of the nasal passages. Very small particles exhibit sufficient Brownian motion to limit their collection by any part of the respiratory system. Particles of the order of 0.1 to 10 μm in diameter are too small to be collected efficiently in the upper respiratory system, and thus they are largely collected in the alveoli of the lung. This behavior is depicted graphically in Figure 2.11. SO_2 and other pollutants absorbed in the particulates thus have an effective entry mechanism to the body. The ability of the appropriately sized particles to act as a carrier of other pollutants is the cause of the apparent synergistic effect of particulates and other pollutants. SO_2 is readily absorbed by water and thus the moist passages of the upper respiratory system. Without the carrier action of particulates, SO_2 could not penetrate deeply within the lung. This relationship was also indicated in Chapter 1, during the discussion of the London "killer fog" episode. Other important pollutants that may be associated with particulates are heavy metals, such as lead, and atmospheric acids. The health effects of particulates are summarized in Table 2.11.

Particulates also result in the degradation of visibility or visual range in the atmosphere. Loss of visual range occurs due to the combination of three mechanisms: (1) light scattering by atmospheric gases; (2) light scattering by particulates; and (3) light absorption by particulates. The attenuation of light rays by particle scattering is usually the most important mechanism in polluted atmospheres. Particles whose characteristic size is roughly that of the wavelength of visible light (~0.5 μm) are most effective at scattering light. In clean air, scattering of light by gases is important and this effect limits the visual range in clean air to about 200 km. Visibility reduction due to light scattering is usually correlated in terms of a scattering coefficient, b_{sp},

$$\text{Visual Range} \approx \frac{3.9}{b_{sp}} \tag{2.14}$$

b_{sp} is also called an extinction coefficient. An approximate relationship between b_{sp} and the mass concentration of particulates is given by Charlson et al. (1968),

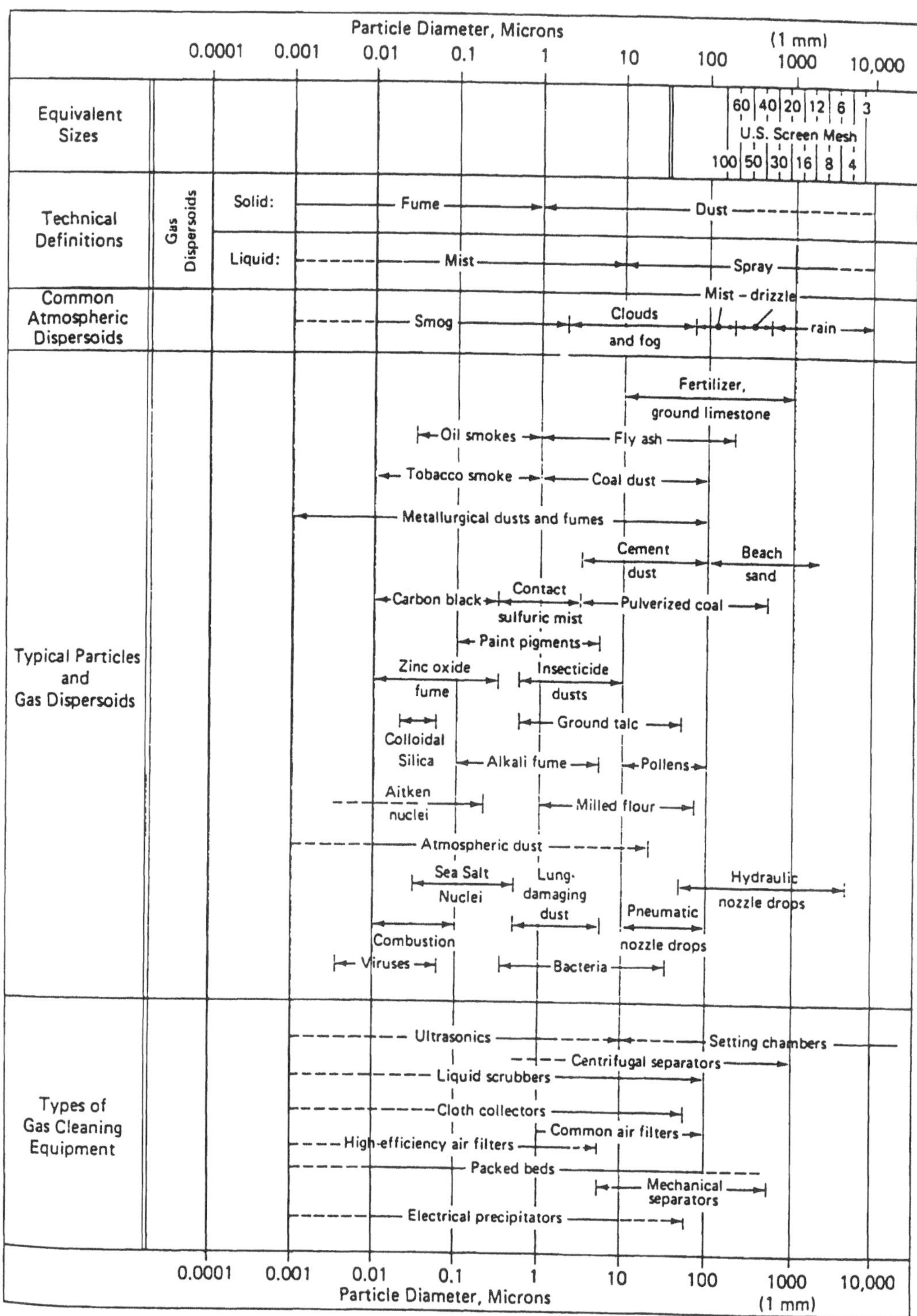

FIGURE 2.10 Characteristics of particles and particle dispersoids. (From Wark, K. and C.F Warner (1981) *Air Pollution: Its Origin and Control*, Harper Collins, New York. With permission.)

TABLE 2.10
Total Suspended Particulate Emissions

Year	Transport	Fuel Combustion	Industrial Processes	Solid Waste	Misc.	Total
1940	2.7	7.5	8.7	0.5	3.7	23.1
1950	2.1	7	12.7	0.6	2.5	24.9
1960	0.7	5.7	12.5	0.9	1.8	21.6
1970	1.2	4.6	10.5	1.1	1.1	18.5
1980	1.3	2.4	3.3	0.4	1.1	8.5
1989	1.5	1.8	2.7	0.3	1	7.2

Note: In millions of metric tons.

Source: Environmental Quality (1990) 21st Annual Report of the Council of Environmental Quality, Washington, D.C.

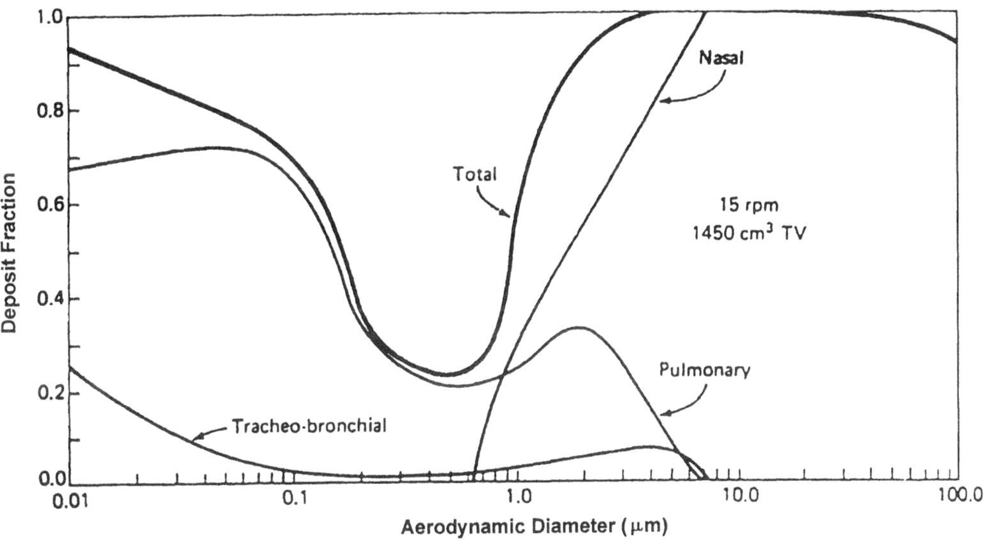

FIGURE 2.11 Deposition of monodisperse aerosols of various diameters in the respiratory tract of man (assuming a respiratory rate of 15 respirations per minute and a tidal volume of 1450 cm^3). (From Task Group on Lung Dynamics, 1966. *Health Phys.*, 12:173. As referenced in Seinfeld, 1975.)

$$b_{sp} = \frac{(\text{mass concentration})}{0.31\ \text{g/m}^2} \tag{2.15}$$

While not adversely affecting health directly, the loss of visual range does limit our ability to preserve the beauty of the wilderness and national forest areas that the American public prize so deeply. In recognition of the concern over the degradation of visibility in such areas, Congress passed legislation in 1977 requiring action to prevent significant deterioration (PSD program) in air quality and visibility in certain areas of the country. The legislation divided the country into three types of areas depending upon their current air quality and the perceived need to retain the air in a pristine state. Wilderness areas and national parks are class I areas and subject to the most severe restrictions on pollution sources.

TABLE 2.11
Observed Effects of Particulate

Concentration	Measurement Time	Effects
60–180 μg/m^3	Annual geometric mean with SO_2 and moisture	Acceleration of corrosion of steel and zinc panels
150 μg/m^3	Relative humidity less than 70%	Visibility reduced to 5 miles
80–100 μg/m^3	With SO_2 > 120 μg/m^3	Increased death rate of persons over 50 may occur
100–130 μg/m^3	24-h average and SO_2 > 250 μg/m^3	Children likely to experience increased incidence of respiratory disease
200 μg/m^3	24-h maximum and SO_2 > 630 μg/m^3	Illness of industrial workers may cause an increase in absences from work
300 μg/m^3	24-h average and SO_2 > 715 μg/m^3	Chronic bronchitis patients will be likely to suffer acute worsening of symptoms
750 μg/m^3		Excessive number of deaths and considerable increase in illness may occur

Source: Abridged from data presented in National Air Pollution Control Administration (1969), *Air Quality Criteria for Particulate Matter*. AP-49, HEW, Washington, DC. 1969.

TABLE 2.12
Example Water Quality Criteria for Streams

Class	Use	Quality Criteria
A	Potable water supply	Microbial counts, color, turbidity, pH, dissolved O_2, toxicity, taste, odor, temperature
B	Bathing, other recreation	Same as above, but less stringent
C	Industrial, agricultural fishing	Dissolved O_2, pH, suspended solids, temperature
D	Cooling water, navigation	Floating material, pH, suspended solid

2.4 WATER POLLUTANTS AND THEIR EFFECTS

2.4.1 Introduction

This section will identify the major water pollutants and their effects. Unlike the atmosphere, bodies of water can be segregated by their use. That is, water used for navigation can be much more loosely controlled than drinking water. In recognition of this fact, bodies of water have often been divided into different use categories, each of which has different criteria for the evaluation of water quality. An example categorization is included in Table 2.12. Table 2.12 also lists measures of quality that might indicate suitability for each of the uses. In the following sections, we will examine the problems measured by these criteria and identify the specific pollutants responsible. The pollutants which will be discussed include organic matter, toxic inorganic contaminants, suspended and floating solids, and bacteria.

Before examining the individual water pollutants, it is instructive to look at the uses of water. Water usage in the U.S. is in excess of 400 BGD (billion gallons per day). This amounts to an average of more than 1600 gallons per day per person, 150 times the water usage rate of a typical developing country. However, only 50 to 75 gallons per person per day is used by individuals in the home. The balance is used in manufacturing and commerce, power generation, and agriculture.

TABLE 2.13
Water Usage and Consumption in the U.S.

Water User	Usage — BGD (%)	Consumption — BGD (%)
Industry and power generation	225 (53%)	17 (14%)
Agriculture	168 (40%)	95 (79%)
Public supplies	24 (6%)	6 (5%)
Rural supplies	4 (1%)	3 (2%)
Total	421 BGD	121 BGD

Source: Adapted from Corbitt, R.A. (1989) *Standard Handbook of Environmental Engineering*, McGraw-Hill, New York.

In addition, only about 120 BGD is consumed, i.e., ingested, made into products, or lost via evaporation to the atmosphere. Table 2.13 identifies the users of this water. Note that the high consumption figure for agricultural irrigation is the result of a large amount of evaporation. More than 50% of the 166 BGD used for agricultural irrigation is lost by evaporation.

Of the water used, 19% is withdrawn from ground water supplies, while 59% is fresh surface water, and 22% saline surface water. The proportion of saline water withdrawn has grown from about 14% in 1970. This is a reflection of the fact that 90% of the world's water supplies are in the salty oceans, 2% lie in the polar ice caps, and only 8% can be found in the ground and surface fresh water supplies on which we depend for the vast majority of our water requirements. Continuously refilling the fresh water resources is rain runoff, which is estimated at 1200 BGD in the U.S. Unfortunately, only 42%, or 515 BGD, is retained in a usable form, the remainder is unavailable, for example, due to drainage to the sea. Thus, the U.S. uses approximately two-thirds of its available freshwater supply on a nationwide basis. In parts of the arid southwest, however, demand outstrips supply. In southern Colorado, demand is more than seven times supply. These estimates clearly indicate the importance of proper care for our water supplies.

2.4.2 Organic Materials

Of the various measures of water quality, one is especially common and important. Organic matter can be oxidized by aerobic microbial processes that remove dissolved oxygen from the water. Dissolved oxygen content is perhaps the single most important indicator of water quality. Many fish, for example, cannot live in an environment containing less than 3 to 5 mg O_2/L. Thus, the oxygen demand of a waste stream, i.e., its oxygen requirement in mg O_2 per liter of water, is a direct indicator of its potential for degrading the ecological health of the receiving body of water. The effect of oxygen demanding wastes on streams depends upon the assimilation capacity of the stream. Included in this is the initial concentration of dissolved oxygen, the concentration and type of microorganisms present, and the temperature of the water. The microorganisms play a crucial part in the depletion of oxygen since the depletion occurs within microorganisms via the following reaction pathway:

$$\text{Organic matter} + O_2\ (aq) \Rightarrow CO_2 + H_2O + NH_3 + \text{Biomass} \tag{2.16}$$

This is basically a growth reaction for the bacteria in that the product *biomass* is new bacteria. Temperature plays an important role primarily due to the strong dependence of the oxygen gas-

TABLE 2.14
Oxygen Solubility in Water

Water Temperature (°C)	Oxygen Solubility[a] (mg/L)
0	14.6
5	12.8
10	11.3
15	10.1
20	9.1
25	8.2
30	7.5
35	6.9

[a] Assuming contact with dry air (20.9% oxygen) at one atmosphere total pressure (101.3 kPa).

liquid equilibrium on temperature and the temperature dependence of the biochemical growth rates. As shown in Table 2.14, the solubility of oxygen in water is only about 9 mg/L at 20°C. When it is remembered that some fish species cannot survive at dissolved oxygen levels below about 5 mg O_2/L, it is clear that very little margin exists for the assimilation of oxygen-demanding wastes. Note that by describing the organic concentration in terms of the oxygen demand, it is quite easy to estimate the potential effect of a waste stream on a water body. If 1000 L of 1000 mg/L oxygen-demanding waste is dumped into a 1×10^6 L lake, the potential reduction in dissolved oxygen in the lake would be 1 mg O_2/L. The time required to achieve this reduction or the time required for recovery of the lake depends upon the microbial growth rate (i.e., the oxygen utilization rate), and the time required for oxygen restoration by *reaeration*. These factors will be evaluated in subsequent chapters. The qualitative behavior of stream oxygen levels after the introduction of organic material is depicted in Figure 2.12. Initially the oxygen levels decrease due to the bacterial action, but eventually reaeration begins to dominate and leads to recovery of the stream. Until the improvements in domestic and industrial wastewater treatment in the 1960s and 1970s, dilution and reaeration were depended upon to limit the impact of the discharge of oxygen-demanding wastes upon bodies of water. The improvements in wastewater treatment techniques resulted in rapid and obvious improvements in water quality.

Oxygen demand is measured in one of three basic ways: biochemical oxygen demand, chemical oxygen demand, or total oxygen demand. The biochemical oxygen demand (BOD) measures oxygen removal after inoculating a waste sample with representative bacteria and nutrients. Because the oxidation of compounds in lakes and streams is generally the result of biological action, the BOD gives perhaps the best indication of the actual oxygen demand of the waste. Measurement of the dissolved oxygen content of the sample before inoculation and again at the conclusion of the experiment allows estimation of the oxygen demand. A 5-day incubation period is normally employed giving rise to the measurement, BOD_5. The 5-day period is arbitrary as indicated in Figure 2.13. It gives a good indication, however, of carbonaceous demand, or the oxygen demand associated with organic compounds in the wastewater. After 5 days, however, depletion of the oxygen continues, partially as a result of ammonia nitrification reactions

$$\begin{gathered} 2NH_3 + 3O_2 \rightarrow 2NO_2^- + 2H + 2H_2O \\ 2NO_2^- + O_2 \rightarrow 2NO_3^- \end{gathered} \qquad (2.17)$$

The oxygen consumed by this process is termed the nitrogenous biochemical oxygen demand.

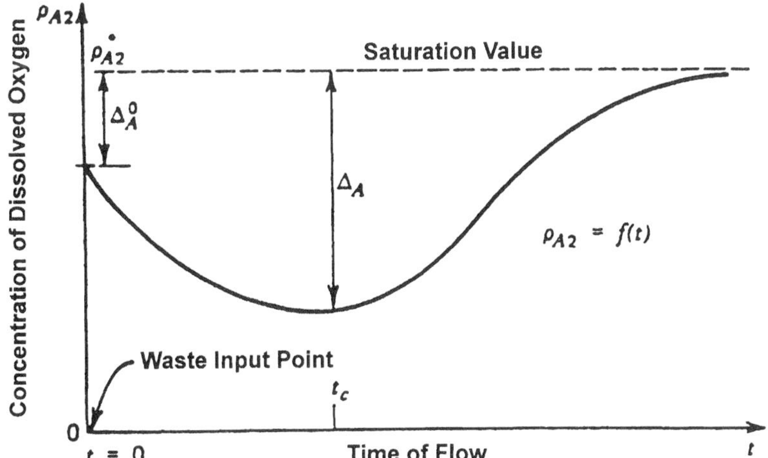

FIGURE 2.12 Dissolved oxygen sag curve. P_{A2} represents the concentration of dissolved oxygen and Δ_A the deficit from saturation conditions. (From Thibodeaux, L. J. (1996) *Air, Water, and Soil,* John Wiley & Sons, New York. With permission.)

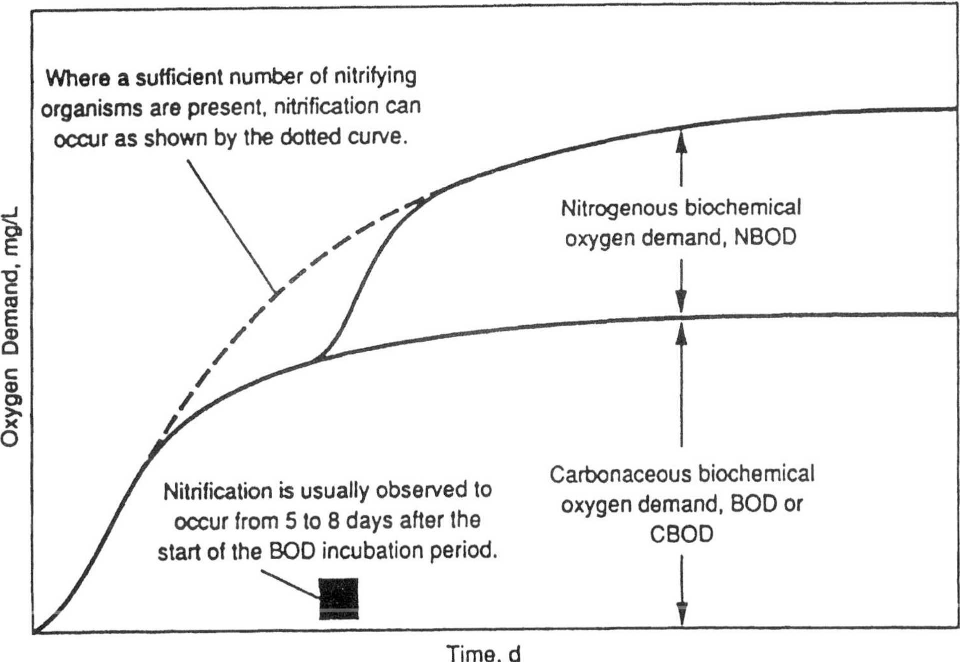

FIGURE 2.13 Definition sketch for the exertion of the carbonaceous and nitrogenous biochemical oxygen demand in a waste sample. (From Metcalf and Eddy Staff (1991) *Wastewater Engineering,* 3rd ed., McGraw-Hill, New York. With permission.)

The sources of organic materials include both industrial and municipal point sources as indicated by the BOD levels in Table 2.15. About 90% of the industrial sources of BOD are the metals, chemicals, paper, petroleum, and food products industries. In addition, to the point sources included in Table 2.15, urban and industrial runoff and other nonpoint sources account for a significant fraction of the total BOD generated in this country. The fraction associated with nonpoint sources is also growing due to increasingly stringent controls on the point sources.

TABLE 2.15
Typical BOD Levels from Industrial and Municipal Sources

Source	BOD mg/L
Domestic sewage	350
Brewery	550
Petroleum refinery	125–850
Tannery waste	8400–16,800
Textile mill waste	140–13 MM
Paper mill waste	25,000

Note: Population equivalent = persons per unit of daily production.

Another means of measuring oxygen demand, the chemical oxygen demand (COD) is measured by refluxing the waste sample with hot potassium dichromate for 2 to 3 h. The primary advantage of this measure of oxygen demand is its rapidity, although the process results in the oxidation of many compounds not normally biodegradable. An even quicker measurement procedure is the combustion of the sample by reaction with oxygen over a catalyst at 900°C. If the oxygen requirements are measured in this manner, the total oxygen demand (TOD) can be estimated. If the rate of evolution of CO_2 is measured, the total organic carbon (TOC) can be estimated. Because organic materials provide the largest source of oxygen-demanding compounds, TOC is also a useful measure of oxygen demand. Each of the oxygen demand measures, BOD, COD, and TOD, are generally expressed as milligrams of oxygen required per liter of the waste stream. Historically, oxygen demand by one or more of these measures has been the most important indicator of industrial effluent quality.

2.4.3 Suspended and Floating Solids

Along with oxygen demand, the suspended solids content of a lake or stream is commonly used as an indicator of overall quality. Here we speak of suspended innocuous solids. Solids which are toxic or otherwise directly hazardous are discussed elsewhere. Suspended solids affect the turbidity, color, and taste of water. At high suspended solid levels, bathing and other recreational uses also may not be acceptable. Suspended solids also may adversely affect the lake or stream ecosystem. Increased turbidity and coloration of a stream can reduce photosynthetic activity and reduce the plant population. Sufficiently high suspended solid levels can damage the gills of fish as well. Suspended solids result in corrosion and clogging problems in industrial equipment. Settling of solids in pipelines, etc. also causes material handling problems. Many of the same sources of oxygen demanding wastes are also sources of suspended solids. Table 2.16 shows that paper mill wastes, for example, can be as much as 6100 times as concentrated as municipal wastes. Perhaps the largest sources of suspended solids, however, are nonpoint sources. This includes natural sources and urban and rural rainfall runoff. The Council of Environmental Quality (1978) indicated that sediments from nonpoint sources exceed the sediments from municipal and industrial point sources by a factor of 360, a ratio that has likely grown since that time. Agriculture was estimated to provide at least 50% of all sediment loadings. On a per acre basis, construction results in the most sediment discharges, averaging 1100 tons per acre per year. The relationship between sediment erosion and land usage in uncontrolled situations is indicated in Table 2.17. Current techniques for controlling erosion and other water pollution problems associated with construction and other land uses can reduce these releases but not eliminate them entirely.

TABLE 2.16
Suspended Solids — Domestic vs. Industrial Wastes

Sources	Suspended Solids Population Equivalent
Domestic sewage	1
Paper mill waste	6100
Tannery waste	40–80
Textile mill waste	130–580
Cannery waste	3–440

TABLE 2.17
Estimated Soil Erosion Rates in Connecticut

Land Use	Tons/Acre/Year
Natural woodland	0.11
Adequately managed cropland	2.73
Inadequately managed cropland	10.77
Urban area	0.86
Construction site	185.20

2.4.4 Priority Pollutants

In contrast to the nonspecific water quality measures of oxygen demand and suspended and floating solids, a number of pollutants, both organic and inorganic, have been designated by the EPA as priority pollutants. Control of these pollutants must be accomplished in addition to the control of the nonspecific measures of water quality such as BOD. The priority pollutants can be divided into 10 basic classes: pesticides, polychlorinated biphenyls, metals and inorganics, halogenated aliphatics, ethers, phthalate esters, monocyclic aromatics, phenols, polycyclic aromatics, and nitrosamines. Some examples of pollutants in these classes are included in Table 2.18 with their physical properties. In some cases, the pollutants may not pose as great a hazard when intake occurs via a particular pathway, but often these compounds are presumed to pose a hazard dependent only upon the amount of intake regardless of the route of exposure. Thus, the volatile pollutants of those below also are assumed to be toxic air pollutants.

2.4.4.1 Pesticides

The pesticide priority pollutants are generally chlorinated hydrocarbons. They include such compounds as aldrin, dieldrin, DDT, DDD, endosulfan, endrin, heptachlor, lindane, and chlordane. They are readily assimilated by aquatic animals and *bioaccumulate* in body fats and are subject to *biomagnification*. Biomagnification is the concentration of pollutants by natural processes such as the food chain. Greater amounts of the pollutant are accumulated higher in the food chain. They are also persistent contaminants. DDT (dichlorodiphenyldichoroethane), for example, remains an environmental contaminant many years after being banned in the U.S.

DDT is an excellent example of the potential problems associated with these compounds. It is not considered extremely toxic but its usage was decreasing during the 1960s and its use was

TABLE 2.18
Toxic Pollutants and Selected Physical Properties

Compound Class	Examples	Vapor Pressure (atm)	Solubility (mg/L)
Pesticide	Aldrin	$3\ (10)^{-8}$	0.017
	DDT	$2\ (10)^{-10}$	0.005
	Heptachlor	$5.3(10)^{-7}$	0.12
Polychlorinated biphenyls and related compounds	Aroclor 1242	$5.3(10)^{-7}$	0.24
	Aroclor 1254	$1.0(10)^{-7}$	0.057
	2,3,7,8-TCDD	$1.8(10)^{-12}$	0.0193
Metals	Lead	—	—
	Cadmium	—	—
	Chromium	—	—
Halogenated aliphatics	1,2-dichloroethane	0.24	5500
Ethers			
Phthalate esters			
Monocyclic aromatics	Benzene	0.125	1780
	Toluene	0.0374	515
	Xylene	0.0087	175
Phenols	Phenol	0.00046	80000
	Chlorophenol	0.0019	20000
	Pentachlorophenol	0.0020	25
Polycyclic aromatics	Naphthalene	0.00011	34
	Phenanthrene	$4.5(10)^{-6}$	1.2
	Pyrene	$1.6(10)^{-7}$	0.15
Nitrosamines			

effectively banned in 1972 as a result of its persistence and potential for biomagnification. Hickey et al. (1966), for example, reported biomagnification of DDT by more than 170,000 times between Lake Michigan sediments and fish-eating birds. Bottom-feeding crustacea exhibited concentrations about 30 times sediment concentrations while the fish that fed off the crustacea exhibited concentrations about 10 times still higher. Finally fish-eating birds exhibited concentrations of DDT 500 times higher than observed in the fish. The magnification at each level is dependent on the feeding habits and animal metabolism and, because the organochlorines tend to build up in the lipid or fat fraction of the body, the proportion of body fat.

As a result of the persistence and potential for biomagnification of these pesticides, organochlorines have been largely replaced by organophosphates, such as parathion, malathion, diazinon, and carbamates, such as aldicarb. Although these pesticides are not persistent they are generally more toxic than the organochlorines and greater care is required in their use.

2.4.4.2 Polychlorinated Biphenyls

Polychlorinated biphenyls, or PCBs, are complex mixtures of organochlorines that are extremely stable and therefore found wide industrial use. Their stability made them especially useful as electrical capacitor and transformer oils. Unfortunately, their stability also means that they are persistent in the environment. In addition, as with the organochlorine pesticides, they are readily assimilated by aquatic animals, soluble in body fats, and will biomagnify in the food chain. Although the toxicity of many of the individual PCBs is relatively low, specific isomers plus trace contamination with other chlorinated species gave rise to significant health concerns. As a result, PCB production was banned in the U.S. in 1979. It should be emphasized that PCBs are a complex mixture of compounds and, in fact, are generally named only by the total percentage of chlorine in the mixture. Specific PCB mixtures are referred to as Aroclors and Aroclor 1254 is 54% chlorine and Aroclor 1260 is 60% chlorine.

Again, as a result of the material's persistence, significant quantities remain in the environment. Industrialized harbor areas in the Great Lakes and northeastern U.S. are often contaminated with PCBs. Now that the production has long been halted, the sediments in the Great Lakes and northeastern harbors represent the most significant repository of PCBs. Fish advisories exist in many of the Great Lakes as a result of health concerns from eating PCB-contaminated fish. Because of the potential for PCBs to sorb onto organic materials in sediments and in fish lipids, such advisories are aimed primarily at fatty, bottom-feeding fish where PCB concentration is the highest.

2.4.4.3 Metals and Inorganics

The toxic elements include antimony, arsenic, beryllium, cadmium, copper, lead, mercury, nickel, silver, thallium, and zinc. These pollutants are important in that they are non-biodegradable, toxic in solution, and subject to biomagnification. The metals processing industries as well as urban and rural runoff provide the most significant sources of these elements. Lead was widely distributed in the environment as a result of the use of tetraethyl lead in gasoline to control premature ignition (knocking). As indicated previously, a reduction in lead content in gasoline has led to significant reductions in environmental exposure to lead.

Especially important sources of such compounds are leeching from abandoned mining sites and urban runoff. The problems of heavy metal and acid mine drainage pollution are not easily controlled. Also, unlike oxygen-demanding wastes, these pollutants are not easily neutralized by natural processes. Because of their importance let us examine some illustrative case histories in more detail.

Mercury from a vinyl chloride manufacturing plant in Minamata, Japan contaminated fish which were the main source of food to local residents. Biomagnification of the mercury levels in the residents eventually led to irreversible brain damage in 116 people and the death of 43 others. In 1972, 500 deaths and 7000 injuries resulted from the consumption of mercury-treated seed in Iraq. Elevated levels of mercury also have been found in swordfish and tuna in the U.S. The toxicity is a strong function of the chemical form of the metal. Methyl mercury is a highly toxic organic form of mercury which was responsible for the effects in Minamata. Since many industrial sources of mercury lead to emission in the less toxic inorganic form, there had historically been little concern over such emissions. It has been found, however, that various bacteria can convert the inorganic mercury to methyl mercury.

Cadmium is another heavy metal that has resulted in well-documented adverse health effects. Between 1946 and 1965, itai-itai (literally, ouch-ouch) resulted in the death of approximately 100 people in the Toyama prefecture on the west coast of Japan. This disease is the direct result of cadmium poisoning. The source of the cadmium was drainage from a lead-zinc-cadmium mine. The drainage contaminated a river which was used to flood the local rice fields. Ingestion of the rice led to biomagnification of the cadmium and the onset of the disease.

2.4.4.4 Halogenated Aliphatics

Halogenated aliphatics include a large number of industrial solvents, refrigerants, fire extinguishers, and propellants. This is the largest single class of priority toxics and includes such compounds as chlorinated methanes, ethanes, ethenes, propanes, and propenes. Because of their low molecular weight relative to the organochlorine pesticides and PCBs, these compounds are more soluble in water and more volatile. They are therefore less persistent and less likely to biomagnify. They are rarely found in surface soils or waters. They are common groundwater contaminants, however, where their isolation from the surface and low rates of groundwater movement provide them a degree of persistence not observed at the surface. Some of these compounds are believed to cause cancer while others damage the central nervous system and liver.

Vinyl chloride (chloroethylene) is a known human carcinogen, or cancer causing agent, that often dominates the suspected health risks of exposure to contaminated groundwater. It is a byproduct of the production of polyvinyl chloride products, but it is also, unfortunately, a natural degradation product of other chlorinated ethenes in the environment. Under low oxygen or anaerobic conditions, these compounds can undergo consecutive dechlorination as a result of microbial processes. Thus, tetrachloroethylene can form trichloroethylene, then dichloroethylene and finally, vinyl chloride. Each of these other compounds are also priority pollutants and are common groundwater contaminants. Tetrachloroethylene and trichloroethylene are two of the most common organochlorine found in groundwater, as a result of their past widespread use as industrial solvents.

The chlorinated methane and ethanes are also important groundwater contaminants. Some of the chlorinated methanes are formed during the chlorination of drinking water, raising significant public health concerns. One of the most important chlorinated ethanes is 1,2-dichloroethane, a degreasing solvent that is also a widely used chemical intermediate used in the manufacture of other organochlorine compounds. It can cause damage to the central nervous system, the liver, and the kidneys. Like other low molecular weight organochlorine compounds, 1,2-dichloroethane is much more soluble in water than non-chlorinated organics. Organochlorine compounds are typically soluble in the range of 0.1 to 1% by weight in water while the solubility of most non-chlorinated, nonpolar organic compounds is less than 0.001%. Their greater solubility makes the halogenated aliphatics less persistent and bioaccumulative than PCBs and organochlorine pesticides. Offsetting this, they tend to be more mobile in groundwater (as a result of their increased solubility) and more toxic.

2.4.4.5 Ethers

Ethers of concern are also chlorinated organic compounds that are used primarily as solvents for polymers and plastics. Examples include bis (chloromethyl) ether and bis (2-chloroethoxy methane). Some of these compounds are considered potent carcinogens. They are released to the environment during production and use of the materials for which they serve as solvents. They are not as widely distributed at levels of concern as the previous compounds and less is known about their aquatic toxicity and fate.

2.4.4.6 Phthalate Esters

Phthalate esters, such as bis (2-ethylhexyl) phthalate are used chiefly in the production of polyvinyl chloride and as a plasticizer in thermoplastics; that is, plastics that are heated to allow extrusion or injection molding and that set when cool. Their wide-spread use as plasticizers makes them a common analytical contaminant, introduced during laboratory handling. This makes detection and analysis of these compounds difficult. They are moderately toxic but they do exhibit mutagenic properties which suggest that they have a potential for causing cancer. They are relatively persistent and can be biomagnified. Aquatic invertebrates appear to be especially sensitive to their toxic effects.

2.4.4.7 Phenols

Phenols are large volume industrial compounds that are extensively used as chemical intermediates in the production of pesticides and herbicides, polymers, and dyes. They have objectionable odors and affect the taste of drinking water at low concentrations. Many are very soluble in water. Chlorinated phenols are more toxic than non-chlorinated phenols. Phenol itself (C_6H_5OH) has antiseptic properties and its odor is largely associated with antiseptics. Phenols are naturally occurring in fossil fuels and are a common contaminant in wastewater from sites where these fuels are processed or handled.

2.4.4.8 Monocyclic Aromatics

This is a common class of industrial chemicals and are widely used as industrial solvents and as chemical intermediates. In addition, fossil fuels such as gasoline contain significant amounts of monocyclic aromatics. As lead levels were reduced in gasoline, the aromatic content was often increased to offset the reduction in knocking performance. Because these compounds are volatile and exhibit relatively high water solubilities, additional air and water pollution problems were the result. In the atmosphere, these compounds assist in the formation of photochemical smog. In drinking water, many of these compounds cause health risks and, for example, a strict drinking water quality standard exists for benzene. As with low molecular weight organochlorines, persistence or contamination of surface waters with these compounds is not normally a problem due to their mobility and volatility. In groundwater, however, the health risk of spilled or leaked gasoline is generally governed entirely by the so-called BTEX fraction. BTEX refers to benzene, toluene, ethylbenzene, and xylene, all monocyclic aromatics.

2.4.4.9 Polycyclic Aromatics

The polycyclic aromatic hydrocarbon compounds (PAHs) are also used as chemical intermediates and are present in fossil fuels. Although the monocyclic aromatics tend to be present in the lighter oil stocks, the polycyclic aromatics tend to be present in coal liquids and the heavier oil stocks as a result of their lesser volatility. Many of the compounds have been found to be carcinogenic in animals and are assumed to be carcinogenic in humans. They tend to be intermediate in persistence and bioaccumulation potential between the monocyclic aromatics/halogenated aliphatics and the PCBs. Their presence in industrial fuels and oils (diesel, coal liquids, heavy fuel oils) has resulted in their presence at old industrial sites where contamination levels might be especially high. Examples of PAHs include naphthalene, fluoranthene, pyrene, and chrysene. The compounds composed of two aromatic rings (naphthalene) tend to be the most volatile, soluble, and mobile, while solubility and volatility tend to decrease as the number of rings increases.

2.4.4.10 Nitrosamines

Nitrosamines, such as *N*-nitrosodimethylamine, are also used as chemical intermediates and are sometimes produced during the cooking of food. These compounds are considered to be potent carcinogens.

Although these pollutants have been identified as requiring special consideration, the relationship between concentration or dosage of these pollutants and adverse health effects is often unknown. It is not known, for example, whether a threshold concentration exists, below which no adverse health effects occur. If no threshold concentration exists, then very low concentrations result in some damage and over a period of years may result in significant health problems. Radiation is assumed to work in this manner, in that its effects are cumulative and long exposure to low levels can be as hazardous as short exposure to high levels. Further complicating the uncertainty with respect to the determination of the response vs. pollutant concentration is the fact that many of these compounds are biomagnified and have complex and subtle effects on host populations. For example, DDT has been linked to reductions in the population of the peregrine falcon. The buildup of DDT, and a daughter product DDE, in the adult falcons resulted in thin shelled eggs which were easily broken, thus reducing the numbers of surviving offspring.

Of increasing interest over the past few years is the effects of endocrine disruptors, compounds that influence the endocrine system regulating hormonal levels in the body. Birth defects and aberrant sexual behavior are often indications of the effects of endocrine disruptors and there is evidence that these effects may occur after long exposures to very low levels of some of these pollutants. This problem clearly indicates the complex interrelationships among plant and animal life and the subtle, but nevertheless, damaging impact of even low levels of some of the priority pollutants.

2.4.5 Inorganic Pollutants

Some of the priority pollutants are inorganic contaminants. In addition to the inorganic elemental species identified above, inorganic pollutants include inorganic acids, bases and salts, nutrients such as nitrogen and phosphorus, and sulfide. Municipal, industrial, and nonpoint sources are all responsible for one or more of these water pollutants. Nitrogen and phosphorus are not normally considered hazardous, but they are nutrients for plant life and result in rapid growth of unwanted algae and other aquatic plants. The addition of such nutrients to a stable ecosystem can rapidly change its structure. Several sulfides, such as hydrogen sulfide, H_2S, are toxic to fish at levels of 1 to 6 ppm (mg/L) in water. The inorganics acids and bases are generally introduced into water bodies from industrial facilities where such compounds are employed. Waters in which aquatic life should be maintained require a pH in the range of 6.0 to 9.0, with 7.0 to 9.0 being the optimum for most water ecosystems. The resulting pH of a lake or stream due to the addition of acids or bases is a strong function of the buffering capability of the water. The presence of acids, bases, or salts in water supplies can also cause a wide variety of industrial processing problems, such as scaling, corrosion, and degradation of product quality in the food processing industries. Salts in water are often characterized by the nonspecific measure of total dissolved solids.

2.4.6 Pathogens

Pathogens such as bacteria are perhaps the original water quality problem. The decomposition of plant and animal tissue leads to the growth of microorganisms in the form of bacteria. These bacteria are of two basic forms: (1) bacteria which assist in the degradation of organic material and (2) pathogenic bacteria which may lead to disease in humans and other animals. Bacteria live and grow by absorbing organic matter through their cell walls, thus they are potentially a problem anytime organic pollutants are present. The pathogenic bacteria include, for example, Rickettsia bacteria, which cause typhoid and spotted fever. As the name suggests, this bacteria is the result of the decomposition of fecal matter, and is thus the result of rural wildlife or municipal discharges. Industrial sources of such bacteria include tanneries and slaughterhouses. Tanneries, for example, may spread the anthrax bacillus by processing contaminated hides.

2.5 MANAGEMENT OF ENVIRONMENTAL HAZARDS

2.5.1 Approaches

Any engineering activity or development action involves the potential for exposure to one or more of the environmental hazards discussed above. A need exists, therefore, to assess the significance of the potential environmental degradation associated with such activity. Ultimately, evaluation of the need for and effectiveness of various environmental mitigation measures is required. These analyses constitute the *environmental decision making process*. It might be preferred that the process were an organized, structured process. De Garmo et al. (1990) see this decision making process as one including

- Problem recognition and formulation
- Search for feasible alternatives
- Selection of criteria

- Analysis
- Selection/decision
- Monitoring and feedback

Historically, environmental decision making has tended to focus on use of a single criteria and problem formulation. The single criteria might have been a guideline or standard on allowable emissions from a particular facility or an environmental quality standard. Relatively narrow, single criteria assessments remain the rule for many environmental decisions. It remains easy to make environmental decisions on the basis of a single criteria. Broader issues, such as trading pollution for jobs, however, are not easily addressed using individual criteria. Questions involving such issues remain largely determined by political considerations.

The environmental impact assessment process that began with the passage of the National Environmental Policy Act in the U.S. in 1969 is just such an attempt to make a comprehensive assessment of the positive and negative impacts of engineering or development activity. In many countries, environmental impact statements are required for almost any type of development. Although not used as extensively in the U.S., at least informal evaluations of environmental impacts are typically required for new or significant changes to industrial, agricultural, and commercial developments. Since the inception of environmental impact statements, the analyses have improved and the criteria used to judge the results have become more comprehensive. In addition, alternative tools and criteria have become available to judge environmental suitability and to select and design developments to minimize wastes or prevent pollution.

2.5.2 Environmental Assessment Criteria

Regardless of the basic approach employed to make environmental decisions, a variety of criteria or standards are used to evaluate individual actions. Standards define appropriate levels of environmental performance and provide a basis for comparing various activities or alternatives that impact the environment. Standards serve to measure the significance of the environmental impact of an activity and to indicate if it is appropriate and acceptable.

Environmental standards can be categorized as being in one of the following categories

- Source control standards
- Environmental quality standards
- Risk-based standards
- Cost-benefit based standards
- Community-based standards

Standards defining the degree of source control required or simple measures of the resulting quality of the environment surrounding the source tend to be set by statute. Examples include simply limiting a certain type of facility to so many tons per year of a particular pollutant or limit the concentrations in the area surrounding the source to so many mg/m^3. Such standards are unambiguous and widely used. By not linking the cause (source) with effect (environmental quality), however, either of these approaches are inherently incomplete. Risk-based and cost-benefit based approaches tend to link the source to the resulting environmental quality and therefore provide, at least in principle, a better guide to the actions required to achieve a certain environmental outcome. Applied generically, these approaches can be used to define appropriate source control or environmental quality standards. Community-based standards are a rather vague grouping of approaches based on meeting the needs and desires of the community and are generally not defined by statute. Instead, civil and common law are often used to define these standards or they may be loosely defined by the expressed needs and desires of the community. Each of these approaches will be discussed in more detail below.

2.5.2.1 Source Performance Standards

Source control or source performance standards are perhaps the easiest to implement and regulate. They are nothing more than a limit to the rate of release of contaminants from an activity. The source might be an easily characterized emission point in a process, such as a stream outfall or exhaust stack, or it might be poorly characterized or controlled such as runoff from an agricultural field or a development site. Performance against such a standard is easily gauged. The resulting improvement in the environment is not. Often source performance standards are based on what is achievable rather than what is needed to achieve a given environmental outcome.

Generally source control standards are written on the basis of the concentration of a particular pollutant in the effluent stream. This can result in the situation where a large stream is treated to the same standard as a smaller volume effluent despite its much larger impact on environmental quality.

Stricter source performance standards are often placed on new or significantly modified sources of pollution. This recognizes that new equipment can be designed to be more effective with less cost than modifications of older, existing equipment. This is illustrated in Table 2.19, in which selected new source performance standards for a coal-burning power generation facility are summarized.

TABLE 2.19
New Source Performance Standards for Coal-Burning Power Generating Plants

Pollutant	New Source Performance Standard
Sulfur dioxide (SO_2)	70% reduction (if less than 0.6 lb $SO_2/10^6$ BTU heat input)
	90% reduction (if less than 1.2 lb $SO_2/10^6$ BTU heat input)
Particulate matter	0.03 lb particulates/10^6 BTU heat input
	20% opacity of exhaust gases

Sometimes a source-oriented standard may require that a certain technology be implemented for control. This may be done either directly, by specification of a technology, or indirectly by specification of a certain clean-up level or effluent concentration that can be achieved by only selected technologies. Technology-oriented standards work best when there is a large base of similar processes that can all make use of the same technology. It has long been used to control wastewater treatment in the U.S. Recent amendments to the Clean Air Act in the U.S. have also applied this approach to atmospheric pollution sources in areas that suffer the worst air quality. Despite the long experience with technology-oriented standards, they have proven difficult to set because of the difficulty of defining appropriate processes. Selection of unproven but promising technologies may result in an expensive, poorly performing source treatment facility. On the other hand, using only demonstrated technologies stifles innovation and limits the ultimate effectiveness of the treatment system.

2.5.2.2 Environmental Quality Standards

Environmental quality standards attempt to define the levels of pollutants that are acceptable in the air, water, and soil of the environment. These standards are generally health-based, that is they are designed to be protective of human and/or ecological health. Generally such standards are set on the basis of just human health concerns, but the response to individual pollutants varies dramatically even within the human population. Some individuals respond to pollutants at levels far below that which are considered safe for the vast majority of the population. Attempting to set reasonable standards protective of a wide variety of plant and animal species is even more difficult.

Environmental quality standards also are difficult to use effectively to evaluate the environmental performance of an engineering activity. They are not directly linked to individual sources of the pollutant and therefore provide little guidance on the strategy necessary to achieve adherence to the standards.

Despite these problems, environmental quality standards are widely used because of their simplicity and, once set, they become unambiguous indicators of unacceptable performance. Because they do not identify the source of environmental degradation nor a means of solution, however, they are best used conservatively to indicate satisfactory environmental performance. For example, they are often used to indicate soil contamination levels that require further assessment. Soils that contain higher contaminant levels may not pose a significant environmental risk because of the lack of mobility of the contaminants, but further assessment is normally required before such a determination can be made.

Environmental quality standards also have been used to define air quality standards. Table 2.20 lists the ambient air quality standards in the U.S. Note that several of the air quality standards included in Table 2.20 have different standards for different periods of exposure. This recognizes that effects of some pollutants are related to the total intake or time integrated exposure.

TABLE 2.20
Ambient Air Quality Standards in the U.S.

	Primary Standard[a]	
Pollutant	**ppm**	**μg/m³**
Particulates <2.5 μm		
Annual mean		15
24-h Average		65
(proposed 7/97)		
Particulates <10 μm		
Annual mean		50
24-h Average		150
Sulfur oxides		
Annual mean	0.03	80
24-h Average	0.14	365
3-h Average		
Carbon monoxide		
8-h Average	9	10,000
1-h Average	35	40,000
Oxidants		
1-h Average (<7/97)	0.12	260
8-h Average (>7/97)	0.08	173
Hydrocarbons		
3-h Average	0.24	160
Nitrogen oxides		
Annual mean	0.053	100
Lead		
3-month Average		1.5

Note: Primary standards are protective of human health.

Environmental quality standards are often tiered. In The Netherlands, soil contamination has been evaluated on the basis of three levels for many common pollutants. The "A" level represents a soil concentration level below which the soil is assumed to be essentially clean. The "B" level represents slight contamination and suggests that the soil should be evaluated further for contamination, the degree of sophistication of evaluation being dependent on whether the observed concentration level is above or below the "B" level. The soil may or may not pose an environmental hazard depending on access to the soil, the mobility of the contaminants, and their *bioavailability*, or the availability of the chemicals to plant and animal life. Concentrations in excess of a "C" level represent contaminant levels that are likely to pose a significant risk to any exposed individuals and requires certain remedial actions to be initiated. Table 2.21 illustrates the three levels of environmental quality standards for selected pollutants in both soil and in the associated groundwater. These levels were set on the basis of evaluation of human health and ecological risks. An approach to analyzing these risks is discussed in the next section.

Drinking water standards in the U.S. are also tiered for some contaminants. For these contaminants, the standards define a *maximum contaminant level goal (MCLG)* and a *maximum contaminant level* (MCL). The maximum contaminant level represents an environmental quality standard for drinking water while the maximum contaminant level goal is a desirable concentration level which is currently not enforceable. Drinking water standards in the U.S. also include *secondary maximum contaminant levels* (SMCL), which are non-enforceable guidelines for aesthetic quality. The maximum contaminant levels represent health-based levels that are also required to take into account the technical limitations to drinking water treatment. Selected U.S. drinking water standards are included in Table 2.22.

TABLE 2.21
Dutch Soil Contaminant Criteria

	Soil (mg/kg dry weight)			Groundwater (μg/L)		
Substance	A	B	C	A	B	C
Arsenic	20	30	50	10	30	100
Chromium	100	250	800	20	50	200
Mercury	0.5	2	10	0.2	0.5	2
Lead	50	150	600	20	50	200
Benzene	0.01	0.5	5	0.2	1	5

Note: <A — Uncontaminated soil; >A,<B — preliminary investigation required; >B — detailed investigation required; and >C — soil removal or remediation required.

Unlike ambient air, exposure to contaminated soil and water depends upon the use to which the medium is put. As a result, use-based environmental quality standards have been developed that recognize, for example, that water used for industrial cooling and process water need not be of as high a quality as community drinking water. Use-based environmental quality standards allow finite resources to be expended more efficiently on meeting environmental goals by focusing attention on the medium that requires the greatest degree of control. The approach retains, however, the remaining problems of conventional environmental quality standards and introduces new ones of classifying environmental media by current or expected use. The potential uses of a property or water body are also not static but, in fact, are likely to change with time.

In summary, environmental quality standards are useful to indicate potential problems if the standards are set conservatively, that is such that they are protective of human or ecological health. A conservative standard may not be useful in the practical allocation of resources to the solution

TABLE 2.22
Selected Primary and Secondary U.S. Drinking Water Standards

Pollutant	MCL/[SMCL] (mg/L)	MCLG (mg/L)
Metals		
Arsenic	0.05	—
Chromium	0.1	0.1
Iron	[0.3]	—
Lead	0.015	0
Manganese	[0.05]	—
Mercury	0.002	0.002
Selenium	0.05	0.05
Zinc	[5.0]	
Inorganic		
Chlorides	[250]	—
Nitrate	10	10
Nitrite	1	1
Solids (total dissolved solids)	[500]	—
Biological		
Fecal coliform	No positive repeat sample	0
Volatile organics		
Trihalomethanes	0.10	0
Benzene	0.005	0
Carbon tetrachloride	0.005	0
1,2-Dichloroethane (EDC)	0.005	0
Dichloromethane	0.005	0
Tetrachloroethylene	0.005	1
Toluene	1	0
Trichloroethylene	0.005	0
Vinyl chloride	0.002	10
Xylenes	10	
Herbicides and pesticides		
Chlordane	0.002	0
2,4-D	0.07	0.07
Ethylenedibromide	0.00005	0
Heptachlor	0.0004	0
Lindane	0.0002	0.0002
Pentachlorophenol	0.001	0
Toxaphene	0.003	0
Semi-volatile organics		
Hexachlorobenzene	0.001	0
Benzo(a)pyrene	0.0002	0
PCBs (as decachlorobiphenyl)	0.0005	0
2,3,7,8-TCDD (dioxin)	0.00000003 (3.0×10^{-8})	0

Note: MCL — maximum contaminant level, SMCL — secondary MCL for aesthetics, and MCLG — maximum contaminant level goal.

of environmental problems. In addition, large potential problems may be identified without any guidance as to their solution.

Air quality in Los Angeles, for example, has exceeded the ambient air quality standard for photochemical oxidants since their inception in the U.S. in 1970 and there is little likelihood that the standard will be attained soon. The recent change in ambient air quality standard for ozone means that more sites will not attain the standard and that this noncompliance is likely to continue indefinitely.

2.5.2.3 Risk-Based Standards

Use-based environmental quality standards are related to risk-based standards in that certain uses imply less risk of exposure. In risk-based standards, that risk is explicitly estimated and compared to "acceptable" risks. Any engineering activity leads to some risk of an adverse environmental impact or effect. In principle, evaluation of that risk can provide a direct indication of the environmental impact of a particular engineering activity and thus guidance on its appropriateness or the need for environmental controls. The primary difficulty of the approach is the need for accurate and extensive data to guide the analysis of the risk evaluation. This approach is time-consuming and sensitive to the necessarily incomplete database that is generally available.

The deficiencies of the approach are most apparent when attempting to estimate an absolute risk associated with a particular activity. Using the predicted absolute risk also requires definition of an acceptable level of risk. This has proven exceedingly difficult since the risk that a community is willing to accept tends to vary depending on whether (1) they or someone else receives the potential benefits associated with the activity causing the risk, and whether (2) they control exposure to that risk.

Location of a high-risk industrial facility in an area that does not benefit from the facility through jobs, for example, is likely to meet with significant opposition. People have also been much more willing to accept the higher risks of smoking or driving an automobile than the much lower risks (statistically) of being a passenger in an airplane or living next to a nuclear power plant or chemical plant.

A risk-based approach can be much more meaningful when common methodologies are used to evaluate the *relative* risks of components of potential activities. It may be possible to determine an alternative of minimum risk, for example, without being overly dependent on the absolute value of risk and the many assumptions that go into its estimation. A still-simpler approach to the definition of relative risk is to consider the contaminant losses, or the rate of release of contaminant to the environment, rather than the risk that those releases pose. The environmental concentrations, dose, and ultimately the health effects, depend on the rate of a contaminant's release to the environment. Alternatives that result in a minimum release rate, therefore, will tend to minimize the risk. Thus, a relative measure of risk can be determined on the basis of estimation of contaminant losses without worrying about the uncertainties associated with the health risk assessment. The advantage of this approach is that the physical-chemical processes that govern the rate of release are generally known with greater precision and accuracy than the dose-response function which governs the health risk assessment.

2.5.2.4 Cost-Benefit Analysis

None of the above methods for setting environmental standards recognize the potential cost of achieving appropriate environmental performance. Because cost is always an important consideration and poses real limitations on any project, it may be appropriate to include it directly in setting an environmental standard. Without consideration of costs, unrealistic environmental goals may be set that are impossible to attain or obtainable only at a cost that individuals, companies, or society is unwilling or unable to pay. Some feel that the revised standards for ozone exhibit this problem.

It is also this problem that has led to the slow progress of the risk-based Superfund approach to cleanup of contaminated sites.

Cost-benefit approaches attempt to overcome this by detailing the real costs and benefits of environmental control. The benefits to the environment can be averted costs, like medical costs, but they do not need to be expressed in dollar terms. The benefit must be quantifiable such as the number of cancer deaths avoided.

Cost-benefit analysis is, in principle, easily incorporated into conventional engineering analysis which is also largely an evaluation of cost vs. benefit tradeoffs. The approach is somewhat in conflict with the professional codes of conduct of organizations such as the Institute for Engineers — Australia, Institution for Chemical Engineers, or the American Society of Civil Engineers, however, which often require that the environment and safety of the public be held paramount in all engineering activities. Taken at face value, these codes of conduct require that no engineering activities will be undertaken that may adversely impact the health and safety of members of the public. This is an impossible standard for a practicing engineer. All engineering activities attempt to achieve maximum "benefit" at minimum "cost." There is neither infinite benefit nor zero cost of any activity. Engineers have often focused their attention solely on quantifiable benefits and costs and defined them solely from the perspective of their client or employer rather than the society as a whole. It is this wider perspective that the codes of conduct attempt to encourage.

It remains difficult, though, to define all costs or benefits of any activity in an absolute sense. It may again be more appropriate to set environmental priorities by assessing the *relative* cost and benefits of an activity relative to others. Even then, however, it remains difficult to fully assess the costs and benefits in that many are difficult to quantify and for others no means of assessing value have been developed. Often, the environmental damage associated with an engineering activity is generally borne by an entity different from the one that will economically benefit. These features make cost-benefit analysis a very difficult tool to apply in the practical evaluation of environmental performance.

It is largely for this reason that traditional measures of economic performance have not included environmental factors. The GNP of a nation, perhaps the most used indicator of wealth and economic activity, for example, does not include the cost of environmental damage. Instead, the cost of correcting that damage represents a positive component of the GNP, in effect suggesting that environmental damage increases the wealth of a nation. Most traditional accounting systems for businesses will at least include environmental control as a cost when realized but uncharged environmental damage is not considered. Thus, there are significant economic incentives to devalue and degrade the environment using traditional economic systems.

There have been attempts to incorporate environmental costs and benefits more properly within the framework of economic evaluation. One of the most difficult problems is valuation of a clean environment. Among the methods that have been proposed to set the value of such an environment are

- Contingent valuation — In this approach, the amount an individual might pay to preserve or improve the environment, or the amount they might accept in compensation for loss of the environment, is assessed. This approach depends on surveys of individuals to make this assessment. Responses to such a survey may not represent the true value of the environmental resource to the individual and, in addition, any such valuation is based on incomplete information that might change the individuals valuation significantly.
- Opportunity costing — Opportunity costing sets the value of an environment by its potential value if it were developed for alternative land use. This approach does not answer the question of the value of the pristine environment, only the potential unrealized benefit if it remained pristine.
- Proxie valuation — In this approach, the value of a pristine environment is judged on the basis of differences in value of related assets between the pristine and other areas.

For example, a commonly used approach is to estimate the effect of the pristine environment on the value people place on homes and land. The problem with this approach is the difficulty of isolating other factors that influence the values of assets such as property.

A somewhat different valuation problem is appropriate costing of a product to reflect its total costs. In the *life cycle analysis* approach the total material, energy, and waste requirements of a particular activity or product is included in the analysis. The total cost of a particular product, for example, includes the traditional costs of raw materials and production but, in addition, the costs of disposal of the product and repairing the environmental damage associated with the product would be included. The price of the product would then be set on the basis of this total cost. An alternative is the use of taxes on the producer to force the product to be valued at its true cost to the community. This approach is fundamentally a very appropriate way to regulate environmental activity and judge standards of environmental performance, but it can be quite difficult to implement. This approach also does not take into consideration excess capacity, resources, or energy that could be used or "recycled" into a product that would make its production more efficient and environmentally friendly. Although this approach is difficult to implement, the ability to do life cycle analyses is expected to increase as experience is gained with the approach.

Related to this approach is the use of taxes to encourage appropriate environmental performance. A tax would be levied on business, industry, and potentially individuals related to the waste they generate. It is not a system of judging environmental performance, but it reduces the environmental decision making process to one of simply optimizing the net income from an activity. While this approach is easily implemented by businesses, it simply shifts the problem of setting an appropriate value on adverse environmental impacts to the tax system. All of the difficulties of employing the cost-benefit analysis to set environmental standards remain.

A form of a cost-benefit analysis has been used in the U.S. to evaluate the cost-effectiveness of health and safety regulations. In this approach, the costs of regulations were compared to the premature deaths that were averted by those regulations without attempting to place a dollar value on those premature deaths. More than 50 different health and safety regulations were rank-ordered by cost per premature death averted. Some of the "unit costs " of the regulations are summarized in Table 2.23.

Accepting these estimates as accurate, it can be seen that easily implemented safety regulations that impact a large fraction of the population are the most cost effective. Expensive environmental regulations that influence small numbers of people are the least cost effective. The cost-benefit analysis, if applied rigorously, might suggest that many such environmental regulations should not be implemented in preference to more cost-effective measures. Despite this, many of the least "cost-effective" measures have received wide support from all segments of the community. The table illustrates the problems of an overly narrow approach to cost-benefit analysis in that it uses only a single measure of benefit, the number of premature deaths averted, and does not recognize other quality of life benefits that may be valued by certain segments of society. In summary, cost-benefit analyses can assist (but not replace) environmental decision making. Although this potential is hindered by the difficulty of valuing the costs and benefits, defining these competing factors is the key to the decision making process.

2.5.2.5 Community-Based Standards

Included in this category are standards based on common law and standards that are based on the interests of the community defined in its broadest form; for example future generations or one that includes nonhuman members.

TABLE 2.23
Cost-Effective of Selected Regulations

Regulation	Cost/Death Averted (million 1990 dollars)
Unvented space heater ban	0.1
Trihalomethane water standards	0.2
Children's sleepware flammability ban	0.8
Air traffic alert and avoidance system	1.5
Rear lap/shoulder belts for autos	3.2
Benzene occupational exposure limit	8.9
Refinery sludges listed as hazardous	27.6
Asbestos ban	111
Hazardous waste land disposal ban	4190
Atrazine/alachlor water standard	92,100
Wood preservative wastes listing	5,700,000

Source: Council on Environmental Quality, 1990. Report Numbers rounded to ≤3 significant digits.

Common law standards are those not based on statutory requirements found in laws or in rules promulgated by responsible government bodies. These standards are instead based on civil court actions and precedence. Individuals and companies are liable for damages caused by their actions through either negligence, or willful action. In addition, under certain conditions, especially if an activity is considered to be abnormally hazardous, an individual may be *strictly liable*, and liability depends on neither demonstration of negligence nor intent. Standards set by civil common law are difficult standards to follow in that there are no clear guidelines and the standards depend largely on the ability to prove damages and responsibility in a court of law. In addition, the standard of proof required in civil actions is normally "more likely than not." Both the criteria for defining damages and criteria for being found liable are likely to change with time as the standards of the community change. The dependence of common law on precedence, however, suggests that what constitutes damage and the standards for determination of liability change only slowly.

Subject to more rapid change is the perception of the community as to what constitutes appropriate behavior toward the environment. Although the entire community may change only slowly, a sufficiently large segment can be motivated on personal, moral, or ethical grounds to push for different environmental standards of performance. Although these standards do not carry the weight of the law, they represent powerful forces for changing the behavior of individuals and businesses. The recent efforts by many companies to reduce packaging of their products and encourage recycling is largely the result of these pressures rather than statutory requirements.

In general, it is quite difficult to consider the changing community standards of environmental performance. Sometimes these community-based standards are based on unrealistic or incorrect perceptions of environmental processes. It often seems unfair to the quantitatively minded engineer that personal opinions and beliefs can interfere with otherwise safe and environmentally sound planning. An environmental engineer, in particular, must be aware of this dynamic and ensure that community standards are incorporated into all engineering activities in an appropriate manner. It is necessary for the environmental engineer to recognize that the standards of environmental performance may not be set by minimization of risk or cost or the maximization of benefits. Instead, the community may hold the engineer or firm to a higher standard based either on the grounds of the broader community interest or based on moral or ethical grounds.

Perhaps this is best illustrated by the increasingly important concept of *sustainable development*. According to the World Commission on Environment and Development (1990), sustainable development *is development that meets the needs of the present without compromising the ability of future generations to meet their own needs.*

Sustainable development is playing an increasingly important role in the debate over global development. It is expected that sustainable development will soon be playing a more important role in decisions on individual industrial, commercial, and residential developments. If this occurs, engineers will be asked to evaluate the sustainability of their designs and proactively select and adapt development plans such that they are sustainable. This is much broader than the traditional view of engineering design and one that is inherently multifaceted and involves multicriteria analysis. Unfortunately, what constitutes sustainable development is not clearly defined.

One of the difficulties of sustainable development is the definition of the needs of a future generation. If they have needs identical or greater than our own, sustainable development would be quite difficult. Mining, or the use of non-renewable resources such as fossil fuels, could not meet such a standard regardless of how well designed to minimize the environmental impact of the activity. If the technological improvements developed by this generation lessen the need of future generations for certain resources, however, it may be possible to have sustainable development that employs non-renewable resources. The sustainable use of non-renewable resources, therefore, involves balancing the loss of this natural capital with investments in human capital. This is clearly far beyond the role of the traditional engineering decision making process that tends to be focused on balancing profits with quantifiable material and construction costs.

A further difficulty with the concept of sustainable development is defining what constitutes a need for our own generation. The standard of living in portions of the world are far higher than in developing countries. It is likely that the attainment of the current western standard of living for all peoples of the world would result in an environmental catastrophe.

Related to this problem is the question of the geographical limitations inherent in sustainable development. It is most appropriate to define sustainability on the basis of a single ecosystem. It is, of course, difficult to define an ecosystem in that there are no sharply defined boundaries. In addition, any appropriately defined ecosystem boundaries are unlikely to coincide with political boundaries. Finally, certain portions of civilization cannot be made sustainable in a strict sense of the word. One definition of a city, for example, is as a population center that is inherently unsustainable without importation of resources. A city requires importation of food and other goods from outside the city. It exports its wastes and degrades the environment in and near the city. It reduces the habitat available for native flora and fauna, generally to the point where these are no longer sustainable near the city. From a different perspective, however, cities may be the only *economically* sustainable form of community in the world today despite their lack of environmental sustainability.

2.5.3 Assessment of Environmental Effects

All of the above measures of environmental performance are ultimately linked to the ability to evaluate the environmental impact or effects of a current or proposed engineering activity. A sound evaluation of environmental risks is perhaps the best answer to perceived problems within the community and the best means to evaluate environmental damage for which an individual or company may be liable.

Assessment of environmental risks is often limited to the evaluation of human health risks but pollutants generally affect nonhuman members of the ecosystem as well. The number of species at risk and the sensitivity and specific response of those species are essentially infinite and, as a result, our ability to conduct ecological risk assessments is quite limited. Human health risks are characterized by a relatively small number of exposure pathways, for example, ingestion with food

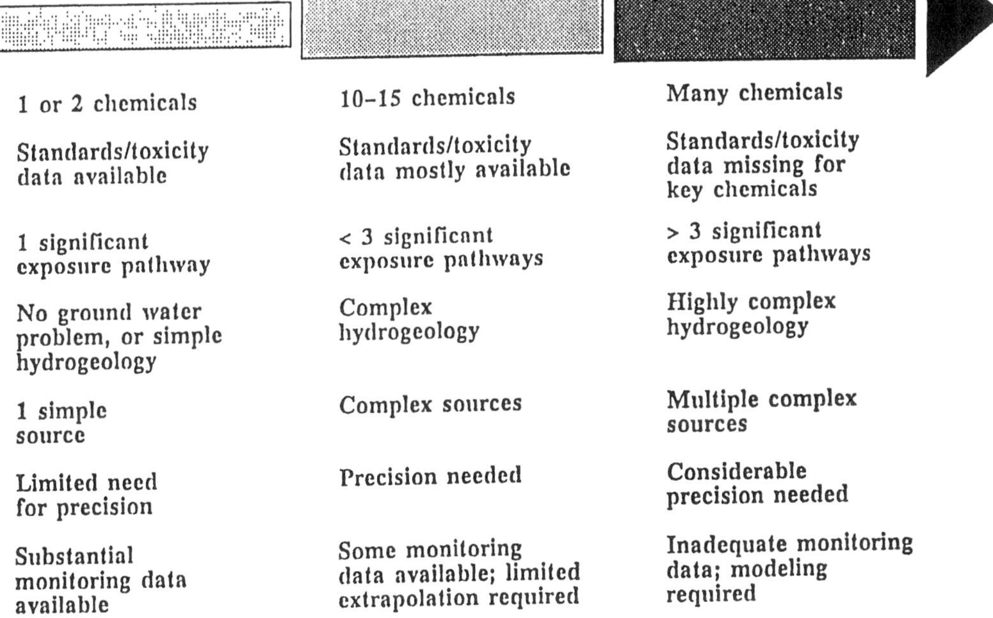

FIGURE 2.14 Continuum of analytical complexity for Superfund public health evaluations. (From Environmental Protection Agency (1986) Superfund Public Health Evaluation Manual, Office of Emergency and Remedial Response, Washington, D.C.)

or drink, inhalation from the air, and adsorption through the skin and, of course, are limited to the effects on a single species.

Although human health risk assessment is far more limited and better understood than a complete ecological risk assessment, it remains subject to large uncertainty. Both of the primary components of a human health risk assessment (assessment of exposure to human populations and, assessment of the health risk associated with that exposure) are subject to significant errors and an analysis of that uncertainty is an integral part of any exposure assessment. Some of the sources of complexity and uncertainty are illustrated in Figure 2.14. Despite this problem, human health risk assessment is one of the most used tools to guide environmental decision making.

Significant improvements in the methodology of risk assessment have resulted from the extensive use of the approach during the selection of remedial technologies at Superfund sites in the U.S. *Superfund* is the common name given to the Comprehensive Emergency Response and Community Liability Act (CERCLA) originally enacted in 1980 to ensure that abandoned contaminated waste sites were cleaned up. The Superfund name is a reference to the large amounts of money collected as taxes, primarily from the petroleum and chemical industries, to fund the bulk of the cleanup activities. The approach was enacted primarily as a response to Love Canal in New York State and many of the features of the bill were a direct response to the problems associated with the cleanup of that site. In particular, there existed no readily available mechanism for the government agencies cleaning up the site to seek reimbursement of the associated costs. The Superfund bill thus includes provisions for "strict, joint, and several liability" for the producers of the waste placed in the abandoned site. This meant that anyone with partial or total responsibility for the contaminated site (potentially responsible parties or PRPs) could be found liable to pay the cleanup costs. This has resulted in high stakes legal challenges which have often increased the costs and significantly delayed site cleanups under the Superfund program. As a result of difficulties and delays in the implementation of the initial cleanups, the act was amended to specify a very strict process for the evaluation of cleanup technologies which included the following steps:

1. Site Remedial Investigation (RI)
2. Remedial Feasibility Study (FS)
3. Record of Decision (ROD)
4. Remedial Design (RD)
5. Remedial Action (RA)

The importance of this to the present discussion is that an assessment of risks was required to compare alternative remedial technologies such that risk minimization became the primary goal of the approach. Because more than 30 billion (10^9) dollars have been spent by both government and PRPs on Superfund cleanups since 1980, there was a large financial incentive for the development of risk assessment methodologies to assist in the appropriate selection and design of remedial technologies.

Although the Superfund public health evaluation process was designed for the evaluation of contaminated sites under the Superfund program, in the absence of acceptable alternatives it has become largely the standard procedure for the assessment of risks associated with any environmental activity. The two basic steps to any risk assessment is an exposure assessment followed by a health risk assessment. The discussion will be organized around these two basic steps.

2.5.3.1 Exposure Assessment

Exposure assessment is the process whereby the contaminant levels influencing individuals within a human population (or other at-risk group) are defined. Definition of the exposure is conducted in the following steps:

1. Identification of the contaminants and contaminated media potentially available for transport to the receptor.
2. Estimation of the rate or quantity of contaminants released.
3. Identification of the exposed populations and the exposure pathways that link the source and receptor.
4. Evaluation of the transport and fate processes influencing the contaminants along each of the exposure pathways.
5. Analysis of the uncertainty associated with the integrated exposure assessment.
6. Comparison of the estimated exposure to applicable environmental quality standards and criteria.

2.5.3.1.1 Identification of the Contaminents

The first step, identification of the contaminants, is rarely as simple as it may seem. Contaminants present at large concentrations may pose small risks compared to trace contaminants. This led, for example, to misleading assertions that PCBs were not a health hazard because many of the pure PCB congeners appear to pose little risk. Unfortunately, commercially available PCBs were mixtures and not pure congeners and some trace components of these mixtures may have posed the greatest health risks. Similarly, much of the human health risk associated with Agent Orange, a herbicide used during the Vietnam War, has been linked to trace contamination by compounds such as dioxins. Dioxins are also present as products of incomplete combustion in some industrial and municipal incinerators and may represent a significant fraction of the total human health risk associated with these facilities. Another example is that small amounts of vinyl chloride, considered

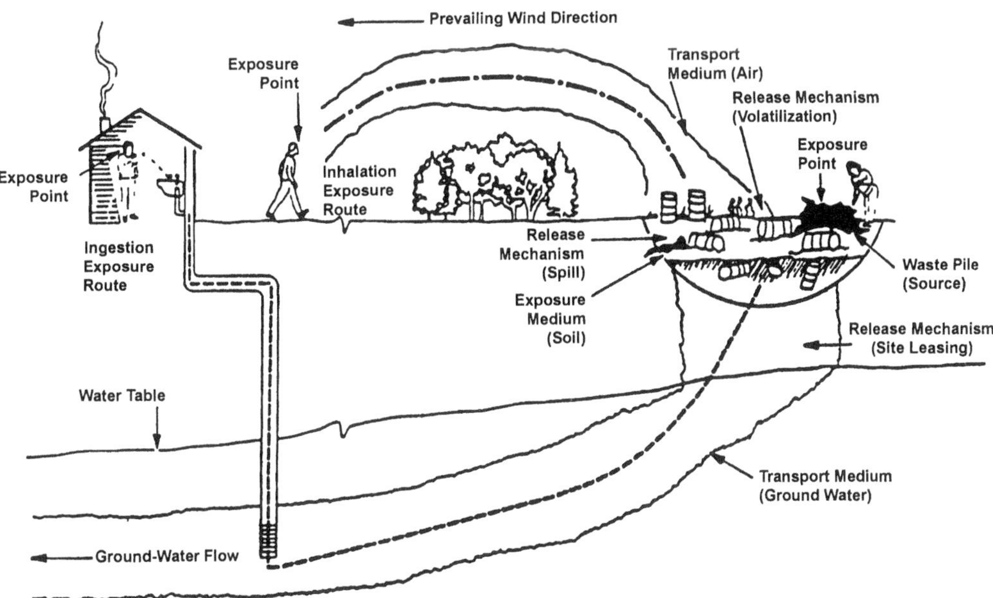

FIGURE 2.15 Pathways of exposure to contaminants released from a contaminated site. (From Environmental Protection Agency (1986) Superfund Public Health Evaluation Manual, Office of Emergency and Remedial Response, Washington, D.C.)

to be a potent human carcinogen, can dominate the health risks associated with drinking waters despite the presence of other hazardous compounds at much higher concentrations. The selection of the most important contaminant(s) from a health risk perspective is most difficult for a contaminated site containing a variety of hazardous compounds. It should be emphasized, however, that the primary contaminant of concern also may not be obvious in a seemingly well-characterized industrial exhaust or outfall; the presence of dioxins in incinerator stack gases being a case in point.

The selection of contaminants that may be released to the environment and therefore considered in a human health risk assessment should thus be done carefully. Compounds should not be excluded at this stage of the analysis on the basis of immobility or because of the lack of a direct exposure pathway. Such decisions should be deferred until a later stage when the mobility and environmental partitioning into other phases will be quantified.

2.5.3.1.2 Estimation of the Quantity Released

Even if the contaminant or contaminants of most concern have been identified, they must be released to the environment in a form capable of impacting human health. In the simplest situation, the source of the contaminant is a well-defined industrial effluent. The contaminant loading and the distribution among air, water, and particulate phases in the effluent can, in principle, be measured. This information can then be used to identify exposure pathways, dilution, and fate processes in the environment and ultimately, an estimate of concentrations in each medium influencing the exposed population.

It is much more difficult, however, to define the rate of contaminant release in an open system such as an industrial impoundment or at a contaminated site. The variety of processes and phases that must be considered are illustrated in Figure 2.15 which outlines pathways for contaminant release analysis from soil. It should be emphasized that if there is no release or exposure of the contaminants, there is no health risk to the surrounding community. Contaminants present in storage drums do not normally pose a health hazard until the drums leak into the environment through either accident or normal deterioration. Similarly, contaminants strongly sorbed to soils may not be an important source of risk unless the soil is mobile (through runoff or wind erosion), or unless the contaminant can partition to a mobile fluid phase such as the air or water. Many metals are less mobile and therefore often of less concern from a health risk perspective, under low-oxygen, anaerobic conditions. Thus, heavy metal contaminated sediments buried just a few centimeters below the surface may not pose a health risk, while identical concentrations exposed at the surface of the sediments may be of significant concern.

It is important to recognize, however, that it is rarely possible to completely isolate contaminants. Care is required to avoid elimination of a particular release mechanism in the risk assessment process on the basis of inaccurate or incomplete information or overly confident assumptions about lack of mobility. Every environmental phase should be considered a potential source or sink of contaminants including air, water, soil, suspended particulate matter in air or water, and, if present, a separate nonaqueous liquid phase. Contaminants deeply buried in sediments, for example, might be considered isolated, but these sediments may be brought to the surface by erosional processes or by the action of sediment-dwelling or benthic organisms that mix sediments by burrowing or feeding activities. While most benthic organisms are limited to the upper few centimeters of sediment, some organisms may penetrate to a depth of 30 to 40 cm or more. Another example of incomplete isolation of contaminants is the surprisingly high volatility of some low vapor pressure contaminants.

Compounds such as PAHs and PCBs are often thought to be immobile in stable soil environments because of their low solubility in water and their low vapor pressure. Such compounds are primarily characterized, however, by their hydrophobicity or their preference to partition to phases other than water, namely air or soil. In particular, the partitioning of these low vapor pressure compounds to air can be as significant as many of the very volatile solvents such as benzene. The partitioning to air from water is governed by the Henry's Law Constant , which is basically the ratio of the pure component vapor pressure to water solubility. For low solubility compounds this constant can be large despite a very low pure component vapor pressure.

In summary, the rate of release of a contaminant introduced directly to the environment by an industrial effluent can be measured. The rate of release of a contaminant contained within an open system such as a surface impoundment or at a contaminated soil site is not easily measured nor estimated.

2.5.3.1.3 Identification of Exposure Pathways

Determination of the release rate and environmental phases impacted by the contaminants allows definition of the pathways of potential human exposure. The direct pathways of exposure for well-defined industrial effluent streams are easily defined. Air emissions will lead directly to airborne exposure to the contaminants. The direct airborne exposure can be to gaseous pollutants or to sorbed contaminants on suspended particulate matter. In addition indirect means of exposure may exist that involve other environmental phases. Deposition of airborne contaminants onto surface waters may lead to exposure through dermal adsorption during bathing, ingestion by drinking, or ingestion through the consumption of contaminated fish. Deposition of airborne contaminants onto surface soils may lead to exposure through inhalation after subsequent resuspension by winds or via ingestion through the consumption of contaminated soil. Young children, for example, are assumed to consume as much as 100 mg of soil per day. For strongly sorbing contaminants, this can prove to be a major mechanism of exposure. Processes analogous to these exist for releases directly to surface waters.

Each of these possible migration pathways should be considered when identifying sources of exposure to a receptor population. Note that risk to the human population occurs only when the pathways are complete, that is when a pathway is not broken by mitigating factors. For example, contamination of groundwater does not lead to a complete pathway until people come in direct contact with the groundwater. Generally this would be via ingestion of the groundwater as drinking water, but it could also be as a result of seepage of the groundwater into basements or to the surface.

In a human health risk-based analysis, contamination of the environment is unimportant unless a complete pathway for exposure to a receptor population can be identified. Thus, emissions to the air generally require control because atmospheric contaminants cannot be isolated from a receptor population. Contaminant releases to soil and water may, under some conditions, be acceptable on the basis of insignificant risk to a human population.

2.5.3.1.4 Evaluation of Exposure

Any complete pathway that can result in exposure of a human population requires further analysis to define the level of exposure. This generally requires technical evaluation and mathematical modeling of the transport and fate processes along each pathway between the point of release and the receptor location. The degree of sophistication of the model used to describe these processes is normally limited by the number of processes and contaminants, the understanding of the applicable transport processes, and the data defining the environmental conditions on which the model simulations are based. Normally, the estimation of chronic exposures is the primary focus and therefore the models must be extrapolated far into the future. The uncertainties associated with the process, combined with the need for rapid calculation under a number of conditions and contaminants over long time periods, suggests that relatively simple models are most appropriate.

2.5.3.1.5 Analysis of Uncertainty

A crucial part of any exposure assessment is estimation or evaluation of the uncertainty in the predicted exposure. Sources of uncertainty include:

- Input variable uncertainty
- Model simplification uncertainty
- Modeling operational uncertainty
- Scenario uncertainty

Model simplification uncertainty is associated with the errors introduced by the simplifications of the actual environmental processes in the model. Modeling operational uncertainties are associated with the errors arising from the use of the model such as the discretization errors in using finite time and space difference steps in a distributed parameter model. Scenario uncertainty is uncertainty associated with the particular conditions selected for the simulation which may not adequately represent the true contaminant release and transport conditions. Scenario uncertainty may be a source of bias in the results, rather than a random uncertainty. Bias could lead to significant understatement or overstatement of risk and should be avoided with great care. Where possible, field data should always be used to validate models or to fit parameters in the model. The resulting model will provide improved estimates of contaminant exposures. It should be emphasized that the objective of the risk assessment is to make the *best* estimate of risk and conservatism is only used to the minimum extent necessary to offset the uncertainty in the prediction.

It is often not possible to quantitatively estimate the uncertainty in a risk assessment. Where a quantitative assessment is impossible, however, the sources of error and a qualitative measure of the magnitude of those errors should be identified in order to assess the accuracy of the resulting risk estimates. The influence of uncertainty in particular model parameters can be estimated through an analysis of the sensitivity of the model to those parameters. This also can be done quantitatively or qualitatively. For complicated mathematical models with many parameters, it is sometimes convenient to employ *Monte Carlo* techniques to estimate the uncertainty in the final estimated exposure. Monte Carlo techniques involve random selections of model input parameters from an assumed distribution of possible conditions. The element of chance in the selection of parameters is the reason for the name Monte Carlo. By conducting the model simulations numerous times it is possible to estimate the statistics of the predicted model outputs or exposures. If the input distributions are realistic representations of the likely values and their uncertainty, the output distributions are equally good representations of the likely exposure and its uncertainty.

In order to minimize the effects of uncertainty on the conclusions of the risk assessment, both an average and a reasonable maximum exposure are often estimated. Depending on the degree of uncertainty in the calculations, the reasonable maximum exposure may be significantly larger or very similar to the average or best estimate exposure.

2.5.3.1.6 Comparison to Standards

Before using the calculated exposures to predict inhaled, ingested, or adsorbed doses in the at-risk population, the concentrations should be compared to applicable environmental quality criteria or standards. In the *baseline risk assessment*, the risk associated with current activities is estimated in order to indicate whether further source control or remediation is necessary. If the predicted environmental concentrations in air or water (or soil), however, already exceed applicable standards, this question is answered affirmatively even without assessment of risks. Alternatively, if a proposed remedial option is expected to result in environmental concentrations exceeding quality standards, the option would be rejected on the basis of unacceptable risks.

Applicable standards might be ambient air quality standards for atmospheric pathways or surface or ground water quality standards for waterborne pathways of exposure. As with any use of environmental quality standards, however, it sometimes becomes necessary to recognize the potential uses of the environmental medium. Potential drinking water, for example, might be required to meet drinking water standards whereas navigable, nonpotable surface waters may be allowed to meet a much less restrictive water quality criteria.

If applicable environmental quality standards are not exceeded or do not exist, the risk assessment should continue into the health risk assessment phase. Instead of applicable environmental quality standards, the standard of comparison will become a target or acceptable level of risk.

2.5.3.2 Human Health Risk Assessment

The human health risk assessment translates the contaminant exposure estimate into an estimate of health hazards or risk. The steps in the human health risk assessment include:

1. Estimation of contaminant intake or dose as a result of the exposure at the estimated concentrations of each contaminant.
2. Evaluation of the potential for non-carcinogenic health effects by comparison of the intake dose to a reference dose.
3. Evaluation of the lifetime carcinogenic risk (for carcinogenic compounds) by combining the intake dose with a cancer potency factor.
4. Integration of the estimated risks or hazards for each contaminant and characterization of the overall risk.

In order to conduct a human health risk assessment, the dose-response relationship for the contaminant and the at-risk population must be known. This dose-response relationship defines the adverse environmental impact of a contaminant; the response, as a function of the exposure. Before examining the process of risk assessment, let us first discuss what is meant by a dose-response relationship.

2.5.3.2.1 Dose-response relationships

The effects of environmental pollutants can often be separated into two different categories, acute effects and chronic effects. Acute effects are associated with high concentrations and the effect may be observed even after short periods of exposure. It is normally assumed that there is a concentration threshold below which no acute effects are discernible. The threshold concentration may be a function of duration of exposure with higher concentrations being tolerated during shorter periods of exposure. Chronic effects are associated with low concentrations and may require long periods of exposure. Chronic environmental effects are normally considered to be a function of the time-integrated concentration to which a receptor is exposed, or dose, which is formally defined by,

$$D = \int_0^t C(t)\,Q\,dt \tag{2.18}$$

Here D is the dose in milligrams of contaminant adsorbed, inhaled, or ingested; C(t) is the time dependent concentration in the medium to which the individual is exposed; and Q is the volumetric rate of intake of that medium by the individual. The inhaled dose, for example, is the integral of the air concentration times the rate of inhalation. If the air concentration were constant over time, then the inhaled dose would be simply the product of the concentration times the inhalation rate. It is generally assumed that the response to a particular dose scales with the body weight of the individual at risk. It is thus convenient to define the dose on a per unit mass basis. It has also proven convenient to relate the adverse human health effects to an average daily dose which is simply the total dose (per unit mass) divided by the number of days of exposure.

A hypothetical dose-response curve is shown in Figure 2.16. At sufficiently low doses, those within region A in the figure, there may be no discernible effect of a pollutant. In most cases, it is assumed that non-carcinogenic, or non-cancer related responses, exhibit a threshold dose. The ultimate goal of any environmental standard would be to ensure that the resulting exposure would be less than this threshold dose. It should be emphasized, however, that because of the varied sensitivities of individual members of any population, an adverse response may occur regardless of the threshold dose employed. An environmental quality standard, for example, would be set at a conservative estimate of this threshold dose to ensure that the vast majority of the at-risk population was protected from an adverse response. The threshold dose, so defined, is normally called the reference dose. Protection of 90% of the at-risk population is often used as the criteria for selection of the reference dose. Due to the normalization of the dose with body weight and use of the average daily dose, the units of reference dose are typically mg/kg/day.

For a carcinogenic response, it is normally assumed that the dose-response curve is linear with no threshold. This is expected to be valid only under chronic low dose exposures. Acute responses are not well represented by this model. A measure of the potency of the carcinogenic agent is the slope of the dose-response curve which is often called the cancer potency factor (PF). The potency factor would represent the probability of contracting cancer per unit dose of exposure. As a result of the normalization with body weight and reference to an average daily dose, it would typically have units of cancer probability per milligram of contaminant of exposure per kilogram body weight per day. Note that if no threshold exists the dose-response curve shown as region B in Figure 2.16 would intercept the origin.

In principle, a dose-response curve could be produced for any environmental contaminant. In reality, there is essentially no dose-response information for ecological or nonhuman responses and even among humans, the dose-response relationships that are "known" are generally crude extrapolations of limited data. Very few chemicals are demonstrated human carcinogens, for example, because of the need for large populations exposed for many years to identify a statistically significant number of cancers. The U.S. EPA considers less than 10 chemicals as proven human carcinogens on the basis of epidemiological studies, including asbestos, benzene, benzidine, and vinyl chloride, although a much larger number are suspected human carcinogens. It has been estimated that

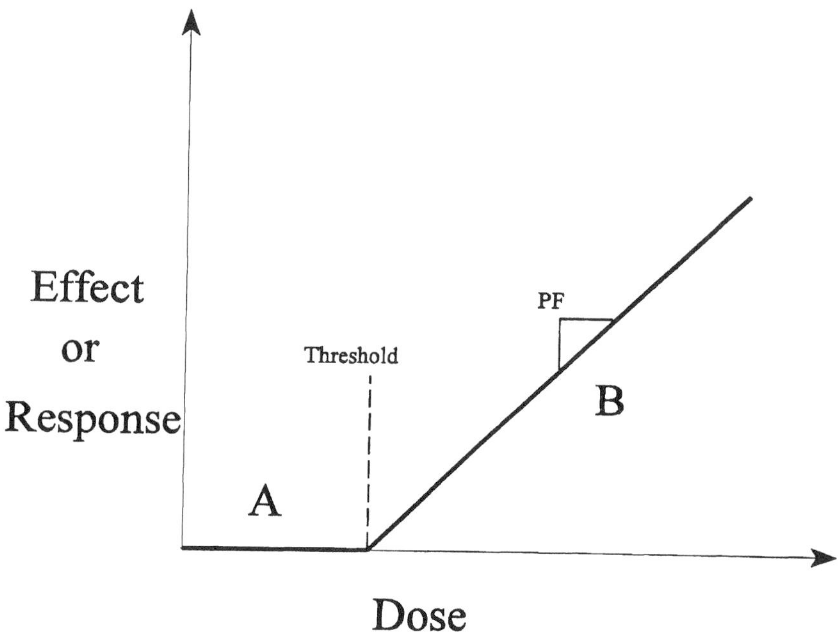

FIGURE 2.16 Typical assumption for dose-response relationship for carcinogenic pollutants.

TABLE 2.24
U.S. EPA Weight of Evidence Categories for Potential Carcinogens

New Category	Old Category	Description	Evidence
Known or likely carcinogen	Group A	Human carcinogen	Epidemiological studies support causal association between exposure and cancer
	Group B	Probable human carcinogen	B1 — Limited epidemiological support B2 — Carcinogenicity in animals
	Group C	Possible human carcinogen	Limited evidence of carcinogenicity in animals
Cannot be determined	Group D	Not classified	Inadequate evidence of carcinogenicity in animals
Not likely to be carcinogenic	Group E	No evidence of carcinogenicity	Two or more animal studies indicating no carcinogenic effects

approximately 100,000 people are expected to die prematurely from cancer as a result of their close proximity and exposure to the Chernobyl nuclear reactor explosion. These excess cancers still represent less than 1% of the total number of cancers expected in the region and their influence on the total number of cancer deaths will likely not be observed, although very high cancer rates will be observable in specific locations.

As a result of this fact, suspected human carcinogens are instead generally identified by animal tests in mice, rats, or other mammals assumed to have a similar metabolism to our own. Some carcinogens identified in this manner, however, have not been proven to cause cancers in other species suggesting that the extrapolation to humans is not necessarily valid. In recognition of this ambiguity, the U.S. EPA categorizes compounds by the weight of evidence of carcinogenicity. Both old categories and newer descriptors are included in Table 2.24. The newer descriptors would normally include a narrative that provides indications of the strength and type of evidence for the categorization.

Extrapolation of animal studies to humans is often criticized and the use of this approach to define cancer potency and this classification system are currently recommended for change. This will likely cause very little change in the short term since data to support a more sophisticated approach is lacking for many compounds.

Because of the difficulties in defining a dose-response curve, the "natural" background concentration is sometimes assumed as being equal or below the no-effects concentration level. Rarely would source reduction, waste stream control, or site cleanup be required beyond that which would maintain the surroundings at natural levels. The natural background is also difficult to define, however, and the best estimate is subject to change as improved environmental performance standards are implemented.

For the balance of this discussion, it will be assumed that either a threshold level is known below which no adverse ecological and/or human health effects occur, or that a potency factor for the adverse response is known. The human health risk assessment then proceeds by estimation of the intake dose of the contaminant, use of the dose-response information to define the risk or a potential hazard, and, finally, integration and characterization of this risk.

2.5.3.2.2 *Estimation of the Intake Dose*

The estimation of the intake dose involves translation of the exposure concentration into the rate of mass uptake by members of the at-risk population. The basic relationship required to estimate intake dose can be written

$$ID = \frac{C \cdot CR \cdot EF \cdot ED}{Wt \cdot \tau} \tag{2.19}$$

where,

ID = Intake dose ($mg \cdot kg^{-1} \cdot day^{-1}$)
C = Exposed concentration (e.g., mg/L)
CR = Contact rate with medium containing contaminant (e.g., L/day)
EF = Exposure frequency (days/year)
ED = Exposure duration (years)
Wt = Body weight (kg)
τ = Averaging time (days)

The contact rate is different depending on the medium of intake. Assumptions have to be made as to the contact rate for the particular pathway and medium under consideration and the units may vary. For example, ingestion of drinking water is characterized by an average water consumption rate in liters per day. Ingestion of contaminants via food (e.g., fish) is typically characterized by an average consumption rate in grams per day. A summary of selected contact rates suggested by the U.S. EPA can be found in Table 2.25. The averaging time for non-carcinogenic risks is normally taken as equal to the exposure duration (i.e., ED = τ), while the averaging time for lifetime carcinogenic risks is normally taken as 70 years. The body weight of an adult is normally taken as 70 kg, while the body weight of children is adjusted accordingly if special pathways exist that require specific calculations of the risk to children. The most common example of this is soil ingestion which is assumed to be as much as 100 mg/day in children aged 1 to 5 (average weight 13.2 kg), 50 mg/day in children aged 6 to 15 (average weight 37.2 kg), and 25 mg/day in adults. Soil ingestion can pose a significant source of risk to hydrophobic, organic, and metal contaminants which tend to be strongly associated with the soil phase.

TABLE 2.25
Selected Human Contact Rates with Environmental Media

Exposure Pathway	Exposure Mechanism	Contact Rate
Air	Inhalation	20 m^3/day
Surface Soil	Ingestion	25 mg/day
		100 mg/day (<5 yrs)
Food (contaminated fish)	Ingestion of fish	22.5 g/day
Water	Ingestion	2 L/day

Source: Exposure Factors Handbook (U.S. EPA, 1989).

2.5.3.2.3 Estimation of non-carcinogenic effects

For non-cancer effects, the dose calculated via the above means is compared to a reference dose which represents an expected threshold for adverse health effects. It is, of course, difficult to set a threshold for adverse health effects given the variety of potential responses and the variations in sensitivity among the population. There also is concern that exposure to extremely low concentrations or doses of some compounds triggers responses in the endocrine or hormonal system that have negative implications for reproduction. As a result of this uncertainty, setting a reference dose for non-carcinogenic effects is done conservatively.

In evaluating the response of humans or the ecosystem to a particular contaminant, two important threshold levels are defined:

1. *Lowest observable effects level* (LOEL) is the lowest concentration or dose to which a receptor population has been exposed resulting in observable adverse effects.
2. *No observable effects level* (NOEL) is the highest concentration or dose to which a receptor population has been exposed that has resulted in no observable adverse effects.

These dosage levels might be set on the basis of large-scale epidemiological studies or in small laboratory tests. Due to the inherently incomplete datasets that define these levels and the other sources of uncertainty described above, the reference dose is generally set far lower than the NOEL. That is, an uncertainty factor of 10 to 1000 is generally applied to the no observable effects level to produce an estimate of the reference dose that is expected to be protective of the health of essentially all members of the receptor population.

The comparison of the intake to reference dose is most simply represented by a ratio of the intake dose to the reference dose, which is often called a hazard quotient.

$$HQ = \frac{ID}{RfD} \qquad (2.20)$$

where,

HQ = Hazard quotient
ID = Intake dose ($mg \cdot kg^{-1} \cdot day^{-1}$)
RfD = Reference dose ($mg \cdot kg^{-1} \cdot day^{-1}$)

Defined this way, adverse non-carcinogenic health effects are expected to be caused if this quotient exceeds unity. A different hazard quotient is estimated for each contaminant and sometimes for each pathway if pathway specific reference doses are known. The hazard quotients are combined to form a hazard index which assumes that the effects of the different compounds and effects are additive.

$$HI = \sum_i (HQ)_i \qquad (2.21)$$

where,

HI = Hazard index

i = Contaminant and pathway index

HQ_i = Hazard quotient for contaminant or pathway i

As with the hazard quotient, a hazard index that exceeds unity implies that there is a significant risk to the exposed population of non-carcinogenic adverse health effects. The assumption of additivity of effects is normally assumed to be a conservative assumption in lieu of chemical-specific information on effect interactions. This may not be a conservative assumption, however, when synergistic effects exists between compounds.

2.5.3.2.4 *Carcinogenic risks*

Unlike non-carcinogenic risks in which only a binary (i.e., yes or no) indicator of health effects is used, the actual probability or risk of cancer during a lifetime is calculated for carcinogenic risks. The lifetime cancer risk is assumed to be modeled by the equation

$$\text{Risk} = 1 - e^{-(ID\,PF)} \qquad (2.22)$$

where,

Risk= Risk of contracting cancer over a lifetime

ID = Intake dose ($mg \cdot kg^{-1} \cdot day^{-1}$)

PF = Cancer potency factor from the slope of the dose-response relationship ($(mg \cdot kg^{-1} \cdot day^{-1})^{-1}$)

For low levels of risk (risk $<10^{-3}$), this relationship is approximated by

$$\text{Risk} \approx ID \cdot PF \qquad (2.23)$$

This calculation is repeated for all contaminants, and in principle, all pathways if different PFs exist for different exposure pathways. The cumulative effect of all contaminants and pathways is again summed, as it was for non-carcinogenic risks.

$$\text{Risk} = \sum_i (\text{Risk})_i \qquad (2.24)$$

This is generally assumed to be a conservative assumption, but again synergistic effects may result in an actual cancer risk greater than that predicted by a simple sum.

If the calculated total risk exceeds 10^{-4} – 10^{-6} excess lifetime cancer risk, then the risk is normally considered unacceptable and additional remediation or mitigation steps must be taken. This standard, however, only serves to ensure that the estimated cancer risk is small enough that it is insignificant compared to the total cancer risk experienced by the population. It is impossible to prove the source of any particular case of cancer and, if an individual is exposed to carcinogens, even at low levels and ultimately contracts cancer, they can be expected to assume that the exposure was the cause. Contrary to popular belief, a relatively small proportion of cancers are suspected to be caused by environmental pollutants. The most common estimates are of the order of 4% of the total number of cancers but some estimates range up to 20%. The most common cause of cancer remains smoking. As indicated previously, however, the opportunity for personal choice in smoking and automobile usage means that these risks are generally considered much more acceptable than risks imposed by external forces, such as sources of environmental pollution. It should be emphasized that the risk assessment process can attempt to identify the risks associated with an activity, but it cannot indicate an acceptable level of risk.

Risk assessment does provide a quantitative measure of at least the relative risk of an environmental pollutant or activity. It is a quantitative framework that allows decisions to be made such as degree of control required and acceptability of an activity. It is simply a framework, however, and knowledge of or ability to estimate the exposure and the health effects of particular pollutants is necessary for its application. The estimation of carcinogenic and non-carcinogenic risks for a particular compound is illustrated in Example 2.1.

Example 2.1: Illustration of the calculation of carcinogenic and non-carcinogenic risks

Consider the carcinogenic and non-carcinogenic risks of exposure to chloroform under 2 scenarios:

1. Exposure to drinking water containing 50 mg/m^3 of chloroform as a result of chlorination to remove pathogens (potency factor ~$6.1 \cdot 10^{-3}$ $\{(mg/kg)/day\}^{-1}$
2. Exposure to air containing 50 mg/m^3 of chloroform which is the threshold limit value allowable for short-term workplace exposures. (Potency factor ~0.081 $\{(mg/kg)/day\}^{-1}$ and a non-carcingenic reference dose of 0.01 $\{(mg/kg\}/day\}^{-1}$)

Drinking water exposure

Assuming consumption of 2 L/day of water containing the 50 mg/m^3 chloroform and exposure for a full 70-year lifetime of a 70 kg person, the daily intake is

$$ID_w = \frac{C_w \cdot CR \cdot EF \cdot ED}{W_\tau \cdot \tau}$$

$$= \frac{50\ \text{mg/m}^3 \cdot 2\text{L/day} \cdot 365\ \text{days/yr} \cdot 70\ \text{yrs}}{70\ \text{kg} \cdot 365\ \text{days/yr} \cdot 70\ \text{yrs}}$$

$$= 0.00143 \frac{\text{mg}}{\text{kg} \cdot \text{day}}$$

This results in a lifetime cancer risk of $6.1(10)^{-3} \cdot 0.00143 = 8.71\ (10)^{-6}$ or approximately 1 case of cancer would be expected per 100,000 persons with this exposure. This might be considered an insignificant risk, but other factors may cause this risk to be viewed as unacceptable.

Airborne exposure

A threshold limit value is designed to be protective of human health for short-term workplace exposures. Continuous exposure at these levels would be expected to pose some long-term health risk. Let us assume a 20-year, 8h per day exposure. Assuming inhalation of 20 m^3/day of air at 50 mg/m^3 over this period gives an intake dose of

$$ID_a = \frac{50\ \text{mg/m}^3 \cdot 16.7\ \text{m}^3\text{/day} \cdot 250\ \text{days/yr} \cdot 20\ \text{yrs}}{70\ \text{kg} \cdot 365\ \text{days/yr} \cdot 70\ \text{yrs}}$$

$$= 0.932 \frac{\text{mg}}{\text{kg} \cdot \text{day}}$$

This implies a lifetime cancer risk of 0.073 and a non-carcinogenic hazard quotient of 93. This would be clearly unacceptable for lifetime exposure but, as indicated above, the threshold limit value is designed to protect against short-term exposure and acute effects.

2.6 MANAGEMENT OF POLLUTANT RELEASES

Assuming that the environmental impact of a particular development or activity can be assessed, how can that impact be minimized? If environmental or source emission standards have been defined, how can these standards be met? These are generally assumed to be the primary goal of the environmental engineer. There are generally three types of interventions or waste management strategies designed to minimize the environmental impact of an engineering activity. These include:

1. Pollution prevention.
2. Recycling of waste materials.
3. Treatment, destruction, or storage of generated wastes.

In general, these interventions can be considered to be written in order of preference. That is, the most preferred option is elimination of the pollution at the source or prevention of its production. If wastes must be generated, then recycling of the waste stream is desired. If wastes are produced that cannot be recycled, treatment to eliminate the environmental consequences of the waste stream is needed. Finally, if none of the above options are feasible, disposal or storage, perhaps in a landfill, is required.

The purpose of the environmental impact assessment and environmental decision-making process is to define the minimum standard of management required. That is, the most severe environmental hazards require source reduction or pollution prevention. Materials, products, or activities that pose a smaller environmental hazard can be recycled, while destruction, or even storage, may be acceptable if minimal environmental hazards are posed. Much of this text is oriented around developing a fundamental description of environmental processes or waste treatment technologies designed to mitigate impacts on those processes. That is, the text is focused on the last of the management strategies. The first two strategies, generally referred to as waste minimization, tend to be very specific to individual products, materials, or manufacturing processes, and it is difficult to generalize or provide fundamental guidance. Often specialists in the specific technical discipline central to the engineering activity are required to identify and take full advantage of opportunities to eliminate or recycle produced wastes at the source. For example, a chemical engineer might be better trained to recognize opportunities for waste minimization in a chemical production facility. There are, however, general rules for the identification and evaluation of opportunities in waste minimization and these are presented here.

2.6.1 Pollution Prevention

Pollution prevention is the reduction or elimination of waste at the source. It aims to minimize the impact of development and engineering activities by avoiding the generation of waste. While complete elimination of waste is not possible, it is clearly better to avoid its generation, as far as possible, rather than be forced to implement treatment or disposal after generation. The avoidance of costs associated with waste treatment, disposal, and liability for environmental risk or damage to public or employee health can be powerful incentives for the implementation of effective pollution prevention programs. In addition, pollution prevention normally is the result of reducing wastage of raw materials or by increasing efficiency of production. Thus pollution prevention also contributes directly to reduction in facility operating costs. Better utilization of raw materials or energy benefits economic performance as well as the environment. Industries that strive to prevent the generation of wastes and pollution are also likely to improve the relationship of that industry with both employees and the public. This may improve employee performance and loyalty and help maintain community support for plant activities and proposed changes including expansion.

Pollution prevention involves changes in materials, processes, or procedures to avoid generation of wastes. Modifications in any of these follows a process of

1. Audit or assessment of current or planned environmental performance at a facility.
2. Identification of opportunities for pollution prevention.
3. Evaluation of the technical, environmental, and economic feasibility of these opportunities.
4. Implementation of a pollution prevention plan taking advantage of the opportunity.
5. Monitoring of performance to determine adherence and effectiveness of plan.

Changes in raw materials or the manufacturing process requires the most detailed knowledge of the technical aspects of the industry under evaluation. Sometimes these are the most difficult pollution prevention opportunities to identify since they require creativity and imagination and are not simple revisions of the existing system. For such problems the thought process that is needed is sometimes referred to as *reverse engineering*. That is, define first the objective that you desire whether it be a particular product or any other outcome, and then work backward to define the most technically feasible, environmentally appropriate and least cost means of achieving that outcome. In reality, this is really not reverse engineering. That is working backward from the desired outcome is the normal *modus operandi* of engineers and this, therefore, constitutes conventional or forward engineering. The key is to not overly constrain the possible opportunities for pollution prevention within the design decisions that were made for the existing process, facility, or development. The original design decisions were made subject to a different set of objective criteria that may not have even included environmental considerations or did so in a limited manner.

2.6.1.1 Lifecycle Analysis

One means of incorporating environmental considerations more explicitly into the design process is through life cycle analysis. Life cycle analysis examines a product from "fertilization to dust." This is an extension of the commonly used "cradle-to-grave" because life cycle analysis must extend beyond the life of the product. It includes evaluation of the raw materials and raw material processing required prior to production of the product, the production of the product itself, the use and reuse of the product, and the consequences of its ultimate disposal. Life cycle analysis is an effort to avoid optimization to minimize costs or environmental impact within the artificial constraints of short-term needs or current practice. A farmer wishing to find an alternative outlet for corn awaiting harvest may view the production of automobile fuels mixed with corn-derived alcohol as an obvious win-win situation. On the other hand, a more detached view which also examines the energy and environmental costs of producing that corn, including seed, fertilizer, irrigation, and government subsidy of the entire process, may come to a less supportive conclusion. A life cycle analysis is an attempt to start with a clean slate and define the best alternative. Any conclusion is obviously still constrained by a variety of factors, but it attempts to take a much broader view of the life cycle of the product than has heretofore traditionally been the case.

The keys to life cycle analysis are twofold:

1. Extension of product evaluation from raw material acquisition and processing through to ultimate product disposal.
2. Quantification of not only the explicit costs of raw materials and processing but the implicit costs of product disposal and environmental consequences through each step in the product's extended life.

By taking the much broader view to a product's value and negative environmental impacts required by a life cycle analysis, it is expected that selection among design alternatives can be placed on a more sound basis. While true in principle, the analysis involves quantification of costs that may not be well understood and the additional uncertainty may not improve the quality of the ultimate environmental decision making. Certain costs are so uncertain that they are generally not included in the analyses; for example, the cost of employing renewable vs. non-renewable resources in a product. This is commonly seen in comparisons of wood and paper products to plastic products. Wood is a renewable resource, although potentially severe environmental consequences result from its harvesting and use. Plastic employs a non-renewable resource, oil, as a raw material, but the environmental consequences of its use are often less, and more easily controlled, than for wood. Allen et al. (1992) presented abridged life cycle analyses on four important consumer product issues. These life cycle inventories attempted to compare the environmental impact of two alternatives on the basis of energy usage or emissions to the air or water. These issues and a short summary of the conclusions suggested by the analysis of each are summarized below.

The use of polyethylene vs. paper grocery bags — Energy and atmospheric emissions for the production of paper bags were greater than that for polyethylene bags even if twice as many polyethylene bags were required. High recycling rates decreased the differential between the two types of bags, but greater atmospheric emissions were expected from the production of paper bags even at high recycle rates. The advantage of paper bags, of course, is that they are made from a renewable resource. The use of a durable bag made of fabric may have advantages over both but because of the much larger energy requirements of the production of the durable bag, reuse for at least 10 to 20 times would be required before any environmental benefit would be realized.

The use of plastic vs. aluminum or glass beverage containers — A 64-oz bottle made from polyethylene terphthalate (PET) requires less energy to make, results in lower atmospheric emissions during production, and produces less solid waste than an equivalent volume of 12-oz aluminum cans or 16-oz glass bottles. If 100% recycling of all containers as achieved, however, the emissions, energy requirements, and solid waste generated for aluminum and PET containers would be approximately equal and less than that for glass.

The use of polystyrene vs. paper cups — Polystyrene cups were found to produce slightly less waste to air or water and less industrial and post-consumer waste than either low density polyethylene coated paperboard cups or wax coated paperboard cups. In addition, of course, polystyrene is 100% recyclable although there are no economically viable recycling systems in operation and this was not considered.

The use of disposable vs. cloth diapers — Table 2.26 compares the various environmental impacts of disposable and either commercially laundered or home laundered cloth diapers as found by Allen et al. (1992). Note that all impacts are given relative to home laundered cloth diapers and incorporate usage and emission assumptions as given in their analysis. No attempt was made to compare the cost of solid waste vs. energy requirement or air or water wastes. The calculations suggest that with the exception of the solid waste generated, disposable diapers compare favorably to their cloth counterpart in terms of environmental consequences.

The life cycle inventory of these problems generated a number of interesting results. They are incomplete life cycle analyses, but they indicate that full accounting for environmental consequences during raw material processing and final production can suggest a very different environmental impact than consideration of only the product. Often the publicly held view of the most "environmental-friendly" option does not hold up under a more sophisticated analysis.

2.6.1.2 Operation for Pollution Prevention

The opportunity to rethink and retool the process to maximize pollution prevention, however, is not always available. Often, improvement in the environmental performance of an existing facility

TABLE 2.26
Ratio of Impact to Home Laundered Impact

Impact	Disposable	Cloth Laundering Commercial	Cloth Laundering Home
Energy	0.5	0.55	1
Solid waste	4.1	1	1
Air emissions	0.48	0.47	1
Water emissions	0.14	0.95	1
Water required	0.27	1.3	1

Source: Adapted from Allen, D.T., N. Bakshani, and K.S. Rosselot (1992) *Pollution Prevention: Homework and Design Problems for Engineering Curricula*, Center for Waste Reduction Technologies, AIChE, New York.

or development is required. Again, the specific pollution prevention activities that might be undertaken are quite specific to that facility, but general rules can be used to guide the identification of procedural changes that might aid the effort. A successful pollution prevention program should consider the following essential attributes:

- Waste segregation to minimize waste volumes ensure proper handling
- Preventative maintenance programs to avoid wastes generated as a result of equipment failure or repair
- Training programs to ensure employee use of procedures designed to minimize waste and ensure proper handling of materials
- Effective supervision to emphasize the importance placed on pollution prevention and insure following of best practice procedures
- Employee participation to assist in the identification of pollution prevention opportunities
- Scheduling of activities to minimize waste production; for example, by minimizing equipment cleaning requirements
- Appropriate cost allocation and accounting procedures to make sure that waste treatment and disposal costs are properly attributed to their source

2.6.2 Recycling of Waste Materials

Even the best pollution prevention cannot eliminate all waste from any engineering activity. Some waste will be generated and discharged to the environment. The next consideration in the management strategy is then to consider recycling this material to avoid its treatment and disposal. Much of the same creativity applied to the pollution prevention effort must also be applied to the investigation of opportunities for waste minimization. Recycling paper and aluminum cans and other metals are well known and comparatively easy to implement. Similarly segregation of wastes can ensure that the recycled materials are of the highest quality and retain as much value as possible. More difficult is recognizing what is generally considered to be waste as a valuable product. Since recycling is process- and product-dependent, we will not focus on it in this text, but examples will be provided where appropriate.

2.6.3 Destruction, Treatment, or Storage of Wastes

Waste not eliminated by pollution prevention or waste recycling must be treated or disposed. This is the major focus of an environmental engineer and of this textbook. The foundations of the design and implementation of treatment and disposal options are considered in Chapters 6 through 8. Generally, the preferred choice is permanent destruction of the waste. If this is not possible or feasible, treatment to render the wastes permanently harmless is the preferred option. Finally, long-term storage may be necessary if no other feasible alternatives exist. Although this may not result in a permanent solution, many wastes are still landfilled or face other long-term storage simply because economic alternatives simply do not exist. The material in Chapters 6 through 8 will provide an indication of the technologies that do exist and their effectiveness for disposal or treatment of various wastes.

PROBLEMS

1. Prove that the approximate doubling time for a process governed by exponential kinetics is given by 70 divided by the rate of growth in percent.

2. Prove that the maximum rate of growth of a process following logistic kinetics occurs at half of the carrying capacity and that the rate of growth at that time is twice the growth rate constant in the logistic kinetics equation.

3. Approximately 100 years was required to move from 50 to 90% of carrying capacity assuming logistic growth and 2.1% growth rate per year. Show that this time is independent of the assumed initial concentration when logistic growth began.

4. Do you feel that the current population of the Earth is sustainable? Support your answer.

5. Do you feel that population growth in the undeveloped countries or high resource use in developed nations is the greater world environmental problem?

6. Consider the implications of the fact that 40% of the population of underdeveloped countries is under the age of 15. Consider demand for resources, the environmental consequences, and socioeconomic factors such as employment and care for the aged. Project the implications through the year 2060.

7. Develop the concept of the bulk of the world's population being housed in an area the size of New Zealand. Even if the political and social consequences could be overcome, is it feasible? Would there be a significant reduction in the infrastructure required to conduct resource mining (lumber, metals, and fossil fuels) and agricultural operations? Where could production facilities be located? What technological advances would be required to make the concept viable?

8. What are some of the limitations of using gross national product (GNP) as a measure of standard of living? In particular, what effect does environmental degradation have on GNP?

9. Estimate the carbon emissions if the world's per capita GNP were to rise to the level of the U.S. Clearly state your assumptions as to population growth and projected carbon emissions per dollar GNP.

10. What do you feel is a more realistic estimate of carbon emissions in the year 2050? What are the differences in assumptions that were used in the estimate made in the previous problem?

11. Examine a housing, commercial, or industrial development in your area. What are the potential effects of that development on reduction in biological diversity? How might these effects be mitigated?

12. Estimate the mass of carbon dioxide held in the world's oceans and compare it to the mass of carbon dioxide in the atmosphere. Could the movement of carbon dioxide between the atmosphere and the ocean be a significant source of uncertainty in the estimation of global warming?

13. Considering the proportion of the atmospheric absorption of energy that is due to carbon dioxide, estimate the change in the global mean temperature that would result from a doubling of atmospheric carbon dioxide concentrations. Assume that the albedo of the Earth remains unchanged. What might cause the albedo to change to reduce this estimate?

14. Why has the reduction of carbon emissions proven to be much more difficult than the reduction in chlorofluorcarbon usage?

15. What are the advantages and disadvantages of employing a carbon tax to reduce carbon emissions? Such a tax would be leveled on the use of fossil fuels providing an incentive for the use of alternative energy sources.

16. Methyl bromide is an agricultural fumigant that has been planned for elimination to avoid its small but significant contribution to ozone depletion. It has significant natural sources, and it is used by humans in relatively small amounts for high value crops in the field and to remove parasites and other biological contaminants from finished food crops. Alternative chemicals have significantly greater health problems including cancer causation. How should these competing factors be weighed in deciding whether the chemical's production should be banned?

17. Derive Equation 2.11.

18. Assume that the blood in the lungs absorbs 100% of the carbon monoxide which is inhaled. Is this consistent with the observed 6 to 10 h required to equilibrate the blood with a given atmospheric concentration of carbon monoxide? Assume that a person breathes 20 times per minute with a tidal volume of 500 ml per respiration.

19. If gasoline has a vapor pressure of half an atmosphere, 50% of the vapor space in an automobile gasoline tank is gasoline vapors. What is the mass of gasoline emitted to the atmosphere in the U.S. if refilling of these tanks is uncontrolled (i.e., vapors are emitted directly to the atmosphere) and gasoline usage is about 15 million gallons per day?

20. What is the visual range expected for a concentration of 100 $\mu g/m^3$ of particulate matter? What is the effect of reducing the particulate concentration to 50 $\mu g/m^3$?

21. How many particles per m^3 are associated with the 100 $\mu g/m^3$ concentration in Problem 20 if the average particle diameter is 1 μm and the density is 1 g/cm^3?

22. Assume that you are employed by a mining company. What arguments could you make that the activities of your company represent sustainable development?

23. Identify one of the approaches to setting environmental standards and summarize the advantages and disadvantages of this approach.

24. Which of the approaches to setting environmental standards do you feel would be the most effective? Defend your answer.

25. Compare the three costing approaches that have been proposed to set the value of the environment.

26. Identify the possible pathways of exposure associated with an industrial wastewater outfall.

27. Are there populations for which the human contact rates presented in Table 2.25 are clearly unrealistic?

28. Show that Equation 2.22 reduces to Equation 2.23 if the risk is less than about 1 in 1000.

29. Benzene has an oral slope factor (ingestion potency factor) of about 0.029 $\{(mg/kg)/day\}^{-1}$. Estimate what drinking water concentration would result in a lifetime cancer risk of 1 in 1,000,000.

30. Locate and access the IRIS (Integrated Risk Information System) on the Worldwide Web site, http://www.epa.gov. Use it to identify the reference dose for non-carcinogenic risks and the carcinogenic oral slope or potency factor for chloroform, 1,1-dichloroethylene, 1,1,1- trichloroethane, methylene chloride, and vinyl chloride.

REFERENCES

Allen, D.T, N. Bakshani, and K.S. Rosselot (1992) *Pollution Prevention: Homework and Design Problems for Engineering Curricula,* Center for Waste Reduction Technologies, AIChE, New York.

Barnola, J.M., D. Raynaud, Y.S. Korotkevich, and C. Lorius (1987) Vostok ice core provides 160,000 year record of atmospheric CO_2, *Nature, 329,* 1 October.

Charlson, R.J., N.C. Ahlquist, and H. Horvath (1968) On the generality of correlation of atmospheric aerosol mass concentration and light scatter, *Atmos. Environ.*, 2:455.

Corbitt, R.A. (1990) *Standard Handbook of Environmental Engineering,* McGraw-Hill, New York.

Council on Environmental Quality, 1990 Report.

De Garmo, E.P., W.G. Sullivan, and J.A. Bontadelli (1990) *Engineering Economy,* Macmillan, New York.

ECO JAPAN 1995 (1994) Keiazi Koho Center (Japan Institute for Social and Economic Affairs), Tokyo, Japan.

Exposure Factors Handbook, U.S. Environmental Protection Agency, Washington, D.C., 1989.

Environmental Protection Agency (1986) *Superfund Public Health Evaluation Manual,* Office of Emergency and Remedial Response, Washington, D.C.

Environmental Quality (1978), Annual Report of the Council of Environmental Quality, Washington, D.C.

Environmental Quality (1990), 21st Annual Report of the Council of Environmental Quality, Washington, D.C.

Flavin, C. (1990) Slowing global warming, *State of the World 1990,* Worldwatch Institute, Norton, NY.

Frederick, J.E. (1986) The ultraviolet radiation environment of the biosphere, *Effects of Changes in Stratospheric Ozone and Global Climate,* Vol. 1, USEPA and UNEP, Washington, D.C.

Hickey, J.J. et al. (1966) Concentration of DDT in Lake Michigan, *J. Appl. Ecol., 3,* p. 141.

Keeling, C.D. et al. (1989) A three-dimensional model of atmospheric CO_2 transport based on observed winds: observational data and preliminary analysis, *Aspects of Climate Variability in the Pacific and Western Americas,* Geophysical Monograph, American Geophysical Union, Vol. 55.

Lapple, C.E. (1961) *Stanford Res. Inst. J., 5:94.*

Masters, G.M. (1991) *Introduction to Environmental Science and Engineering,* Prentice-Hall, Upper Saddle River, NJ.

Metcalf and Eddy Staff (1991) *Wastewater Engineering, 3rd Ed.* , McGraw-Hill, New York, 1334.

National Acid Precipitation Program (1985) Annual Report, Washington, D.C.

National Air Pollution Control Adminstration (1969) *Air Quality Criteria for Particulate Matter,* AP-49, HEW, Washington, D.C.

National Air Pollution Control Adminstration (1970) *Air Quality Criteria for Photochemical Oxidants,* AP-63, HEW, Washington, D.C.

Neftel, A., H. Oeschger, J. Schwander, B. Stauffer, and R. Zumbrunn (1982) Evidence from polar ice cores for the increase in atmospheric CO_2 in the last two centuries, *Nature, 315,* 2 May.

Population Reference Bureau (1989) *1989 World Population Data Sheet,* Washington, D.C.

Robinson, E., and R.C. Robbins (1968) Sources, Abundance and Fate of Gaseous Atmospheric Pollutants, Final Report of Project PR-6755, Stanford Research Institute, Menlo Park, CA.

Seinfeld, J.H. (1975) *Air Pollution, Physical and Chemical Fundamentals,* McGraw-Hill, New York.

Seinfeld, J.H. (1986) *Atmospheric Physics and Chemistry of Air Pollution,* Wiley-Interscience, New York.

Siegenthaler, U., and H. Oeschger (1987) Biospheric CO_2 emissions during the past 200 years reconstructed by deconvolution of ice core data, Tellus, 39B, 140–154.

Task Group on Lung Dynamics (1966), *Health Phys.,* 12:173.

Thibodeaux, L. J. (1996) *Environmental Chemodynamics: Movement of Chemicals in Air, Water, and Soil,* John Wiley & Sons, New York.

3 Introduction to Environmental Engineering Calculations

The subject of this chapter is the review of basic physical quantities and the tools used to describe and interpret them. It is on this basis that all quantitative evaluation of environmental processes is built. In addition to this text, the reader is referred to introductory texts by Himmelblau (1974) and Felder and Rousseau (1986) for an introduction to basic chemical process calculations.

3.1 SYSTEMS OF UNITS

Central to all engineering calculations is a system of units. All physical quantities are not pure numbers but instead have dimensions expressible in a system of units. Dimensions are a measure of a physical quantity and define its magnitude. Dimensions, for example, indicate the size of an object, its mass, its temperature, or its speed. The particular numerical value of a dimension is determined by the system of units used for measurement. Throughout the world, the most common system of units is the *International System of Units,* or SI units. This is closely related to and often commonly referred to as the metric system or MKS system. MKS refers to the fundamental units of measure employed in the system, i.e., meters (length), kilograms (mass), and seconds (time). Most systems of units used today use length, mass, and time (and temperature) as the fundamental dimensions. All other quantities are derived from these fundamental dimensions. Thus, volume is $\{\text{length}\}^3$, acceleration is $\{\text{length}\}/\{\text{time}\}^2$, and force is, by Newton's law, equal to mass times acceleration, or $\{\text{mass}\}\{\text{length}\}\cdot\{\text{time}\}^{-2}$. Other systems of units that use mass, length, or time as the fundamental dimensions include cgs (centimeter, gram, second) and the English absolute system, Fps (foot, pound, second). The British engineering system of units assumes that the fundamental dimensions are length, time, and force while the American engineering system of units assumes that length, time, force, and mass are all fundamental dimensions. The units of the fundamental and derived dimensions in each of the common system of units are listed in Table 3.1.

American engineering units remain common in the U.S. despite the awkward requirement of a conversion factor in the definition of Newton's law because both the units of force and mass are defined. This is because 1 pound-mass (lb_m) in the American engineering system is defined to weigh 1 pound-force (lb_f) when subjected to the normal acceleration of gravity. Writing Newton's law, force equals mass times acceleration, with a constant, g_c, required to achieve this equality,

$$F = m\frac{a}{g_c}$$
$$\text{lb}_f = \text{lb}_m\frac{g}{g_c} \tag{3.1}$$

1 lb_m =1 lb_f when the magnitude of g_c equals the magnitude of g. The average acceleration of gravity at 45° latitude is 32.174 ft/s^2 (9.807 m/s^2) and therefore g_c is

$$g_c = 32.174\frac{(\text{ft})(\text{lb}_m)}{(\text{s}^2)(\text{lb}_f)} \tag{3.2}$$

TABLE 3.1
Fundamental and Common Derived Dimensions and Conversions

Dimension (symbol)	SI Unit (symbol)	cgs Unit (symbol)	Conversion[a]	AEU[b] (symbol)	Conversion[c]
		Fundamental Units			
Length (l)	meter (m)	centimeter (cm)	100	foot (ft)	3.281
Mass(m)	kilogram (kg)	gram (g)	0.001	pound (lb_m)	2.205
Time (t)	second (s)	second (s)	1	second (s)	1
Temperature (T)	Kelvin (K)	Kelvin (K)	1	Rankine (°R)	1.8
Force (F)	(Derived)	(Derived)	—	pound (lb_f)	—
		Derived Units			
Area (A = l_2)	m^2	cm^2	10^4	ft^2	10.76
Volume (V = l_3)	m^3	cm^3	10^6	ft^3	35.31
Volume (common)	liter (L)	liter (L)	1	gallon (gal)	0.2642
Density (ρ = m/V)	kg/m^3	g/cm^3	0.001	lb_m/ft^3	0.06245
Force (F = m·l/t²)	Newton (N) ($kg \cdot m \cdot s^{-2}$	dyne ($g \cdot cm \cdot s^{-2}$)	10^5	lb_f (fundamental)	0.2248
Pressure (p = F/A)	Pascal (P) ($N \cdot m^{-2}$)	dyn/cm^2	10	lb_f/ft^2	0.0209
Energy (E = F·l)	Joule (J) (N·m)	dyn·cm	1000	$lb_f \cdot ft$	0.7376
Power (P = E/t)	Watt (W) (J/s)	dyn·cm/s	1000	$lb_f \cdot ft$ /s	0.7376

[a] Value in cgs unit = value in SI unit * conversion
[b] American Engineering Units
[c] Value in AEU = value in SI unit * conversion

It is important to recognize that while the magnitude of g_c is equal to the magnitude of the gravitational constant, the units are quite different. The quantity g/g_c does not equal unity but in fact equals the ratio of lb_m to lb_f. This ratio of units of mass to units of force can be determined for any system of units. In a mass, length, and time system, however, $g_c = 1$ since the units of force are derived. Only in a system of units where both force and mass are fundamental units does g_c have units and a numerical value other than 1.

The reader is cautioned to avoid a system of units requiring the use of g_c. The minor convenience of 1 lb_m weighing 1 lb_f at the surface of the Earth in the American Engineering System (AES) is more than offset by the confusion caused by the need for g_c in the calculation of force (lb_f), energy (ft-lb_f), and power (ft-lb_f/s) using American Engineering Units (AEUs). This is illustrated in Example 3.1.

Example 3.1

In the AES, force is a fundamental unit and the interrelationship of force and mass must be redefined as

$$F = \frac{ma}{g_c}$$

and thus a mass of 1 lb_m subject to acceleration by gravity of 32.174 ft/s² exerts a force of 1 lb_f.

$$F = \frac{(1\ lb_m)(32.174\ ft/s^2)}{32.174 \frac{(lb_m)(ft)}{(lb_f)(s^2)}} = 1\ lb_f$$

Similarly, the pressure, or force per unit area, at the base of a column of liquid is given by the product of the density of the liquid (ρ), the acceleration of gravity (g), and depth of liquid (h), but in the AES this must be modified to

$$P = \frac{\rho g h}{g_c}$$

Thus, a 10-ft column of water (ρ = 62.4 lb_m/ft^3) exerts a pressure of

$$P = \frac{(62.4\ \text{lb}_m/\text{ft}^3)(32.174\ \text{ft/s}^2)(10\ \text{ft})}{32.174\dfrac{(\text{lb}_m)(\text{ft})}{(\text{lb}_f)(\text{s}^2)}} = 624\ \frac{\text{lb}_f}{\text{ft}^2}$$

Note that all units were carried through the calculations in Example 3.1 in the same manner as the numerical values. By doing so, it was clear that to achieve the units of energy or work in the AEUs ($ft \cdot lb_f$), division by g_c was required. This is required in any calculation involving force, including pressure, energy, or work and power. Recognition of the need for g_c depends on careful attention to the units of the calculation. Luckily, this system of units is less used for such calculations. Watts, for example, have become a standard unit of power regardless of the system of units used, although British Thermal Units per second (1 Btu/s = 1055 W) and horsepower (1 hp = 745.7 W) are still used. Both of the latter units are directly convertible to watts and use of lb_f can be avoided. Note that a system of units using SI units for mass, length, and time also exists, but it defines a force unit, kg_f in a manner analogous to lb_f. This system of units also requires the use of g_c.

The examples also illustrate that units can be carried through mathematical manipulations just like numerical values. The rules of unit mathematics can be summarized by the following.

1. The sum or difference of two quantities cannot be calculated unless they have identical units. As a result, two quantities cannot be equated unless they have identical units.
2. The product or *quotient* of two quantities is equal to the product or quotient of the magnitudes of the properties times the units similarly formed by the product or quotient. That is, 10 m traveled in 5 s implies a speed of 2 m/s. The latter rule implies that the algebra of units carried to powers also follows the rules of ordinary numbers. Thus, the product of 2 m/s (or 2 $m \cdot s^{-1}$) and 5 s is 10 m with seconds canceling.

In addition to the elimination of g_c, another reason for the preference of MKS or cgs units is that a naming convention exists that relates different units by powers of 10, making interconversion easy and less confusing. Millimeters, for example, refers to 0.001 meters and kilometers refers to 1000 meters. Table 3.2 indicates the prefixes and their meanings.

Note that the prefixes mega and milli both use the same letter of the alphabet (M and m, respectively). Improper capitalization can result in a small error of 10^9. To add to the confusion, M in AEUs, also is often used to indicate a factor of 1000 after the Roman numeral for 1000. For example, MBtu refers to 1000 Btu and MMBtu refers to 10^6 Btu.

In addition to the fundamental and derived units defined in Table 3.1 a variety of other units are in common usage. Appendix A summarizes the conversion factors between various units for a variety of physical properties. Larger conversion tables are available from a number of sources.

Note that all environmental properties measure a physical quantity and therefore have units. Certain quantities appear dimensionless and others appear to have simple units when their meaning might be more clear with more complete units. A student is encouraged to diligently trace the units through a calculation. It will assist greatly in the avoidance of common errors. The importance of this will become clear after the examination of a few examples during the subsequent discussion of environmental properties of interest.

TABLE 3.2
Common SI Unit Prefixes

Prefix	Symbol	Factor	Example
pico	p	10^{-12}	1 ps = 10^{-12} s
nano	n	10^{-9}	1 ns = 10^{-9} s
micro	μ	10^{-6}	1μm = 10^{-6} s
milli	m	10^{-3}	1 mm = 10^{-3} m
centi	c	10^{-2}	1 cp = 10^{-2} poise (viscosity)
deci	d	10^{-1}	1 dL = 0.1 L
kilo	k	10^{3}	1 kg = 1000 g
mega	M	10^{6}	1 MW = 10^{6} W
giga	G	10^{9}	1 Gton = 10^{9} tons

3.2 UNITS AND NOMENCLATURE FOR BASIC ENVIRONMENTAL VARIABLES

3.2.1 Density and Concentration

The preceding discussion defined the systems of units that allow us to characterize the state of a system. In this section we will discuss those variables that are most important in that characterization. The most important environmental measurement is generally the density or concentration of a pollutant. The impact of a particular pollutant is typically assumed or observed to have a direct relation to its prevalence in the system, or concentration.

Density or concentration is measured by the ratio of the quantity of a substance per unit volume or mass of the system in which the substance resides. For this purpose, the system is generally defined as a single phase, either air, water, soil, or oil or a nonaqueous phase. To minimize confusion, let us assign letters to each of these phases as shown in Table 3.3. A subscripted letter will be used to indicate that phase. Thus, the volume of an arbitrarily defined air phase system might be denoted as V_a, and that of a similarly defined soil system, V_s. It becomes important to differentiate between the concentration of a component in a phase vs. the density of the phase itself. Let C_a represent the concentration of a component of interest in the air, while ρ_a represents the density of the bulk air phase.

Density or concentration is a continuum quantity. A fluid is a collection of individual molecules. Density or concentration only has a meaning when a measuring volume is sufficiently large that it contains a large number of molecules as shown in Figure 3.1. The spacing between molecules is of the order of the mean free path between collisions, and measuring volumes of this size would be influenced dramatically by the random motion of molecules in and out of the region. Figure 3.1 also depicts the density that would be measured as a function of measuring volume if successively larger measuring volumes were collected in a stationary or time-independent system. The measuring volume must be very much larger than the mean spacing between molecules before the random motion of the molecules is statistically averaged such that consistent measures of the number of molecules per unit volume are found. Luckily this does not generally pose a problem as the number of molecules in 1 cm^3 of a gas at 25°C approaches 2.5×10^{19}. Densities in liquids are typically of the order of 1000 times greater, and densities in solids are typically slightly greater still.

A mole of a substance is the quantity of that substance that contains Avogadro's number of molecules, that is 6.023 (10^{23}) molecules. This measure of the quantity of a substance is very useful when considering processes or behavior that depend on the number of molecules of a substance rather than the mass of that substance. Chemical reactions, for example, generally occur at a reaction rate that is a function of the number of molecules per unit volume. The ratio of reactants to products

TABLE 3.3
Nomenclature for Density and Concentration

Concentration Symbols		Phase Indicator Subscripts	
φ	Volume fraction	a	Air
ω	Weight fraction	w	Water
x,y,z	Mole fraction	s	Soil
C	Moles/mass contaminant per volume of fluid	o,n	Oily/nonaqueous phase
W	Moles/mass contaminant per mass of solid/sorbent	b	Animal or human phase (biological phase)
ρ	Moles/mass solvent/sorbent per volume	v	Vapor phase (e.g., in soil)
		pw	Porewater phase (e.g., in soil)
*	Superscript indicates equilibrium	oc	Organic carbon

Some common examples:

C_{w*}	Concentration of a contaminant in water, mg/L
C_w	Concentration of a contaminant in water in equilibrium with an adjacent phase (e.g., air), mg/L
ρ_w	Density of water, mg/L
W_s	Mass fraction of contaminant on soil, mg/kg
C_v	Concentration of a contaminant in soil vapor, mg/L
ω_{oc}	Fraction organic carbon in soil, g/g
ρ_{oc}	Density of organic carbon (e.g., in water), mg/L

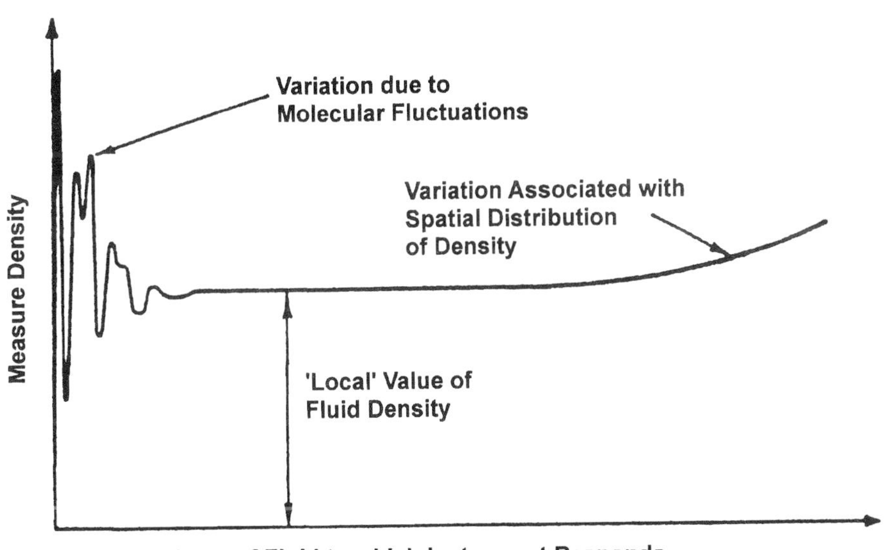

FIGURE 3.1 Fluid density as a function of averaging volume indicating that a minimum volume is required to define a continuum. (Batchelor, G.K. (1970) *An Introduction to Fluid Dynamics,* Cambridge University Press, Cambridge, UK.)

in a chemical reaction is also a function of molecular ratios and more easily measured in a quantity unit that measures number of molecules rather than mass. In addition, a molecule of a gas at low pressure and a given temperature will fill approximately the same volume, regardless of the mass of the molecule. In particular, at 0°C and 1 atm pressure, the volume of 1 mole of any gas is about 22.4 L. At 25°C and 1 atm pressure, the volume of 1 mole of any gas is about 24.5 L. Under each of these conditions, it is often more convenient to work in molecular units, that is moles.

Another way of defining a mole of a substance is that quantity of the substance whose weight in grams is equal to the molecular weight of the substance. Just 1 mole of atomic oxygen has a mass of about 16 g while 1 mole of atmospheric oxygen (O_2) has a mass of about 32 g. Therefore the number of moles of any substance, n, is its mass m, in grams divided by its molecular weight.

$$n_A = \frac{m_A}{\mathrm{MW}_A} \tag{3.3}$$

As indicated by Avogadro's number (N_{Av}), even small quantities of a substance contain a very large number of molecules. Example 3.2 illustrates the implications of this for natural recycling of the air you breathe and water you drink.

Example 3.2: "Recycling" of air and water molecules

The total mass of the atmosphere is equivalent to that in a layer 7 km deep with density equal to the density at the surface. Assuming that air is uniformly mixed (at least over a period of years), how many molecules of air that you have previously exhaled over your lifetime (of, say 20 years) will be inhaled in a single breath?

The Earth's surface area is $5.11(10^9)$ km^2 and thus the volume of a layer 7 km deep is $3.58\ (10^{19})$ m^3. The volume of air in one breath is 500 cm^3 and that breathed in 20 years is $1.05\ (10^5)$ m^3. The number of molecules in one breath, assuming 24.5 L/mol as air density (25°C and 1 atm pressure) is then

$$n = \frac{\left(6.02 * 10^{23}\ \dfrac{\text{molecules}}{\text{mole}}\right)}{\left(0.0245\ \dfrac{\text{m}^3}{\text{mole}}\right)} * \left(500\ (10^{-6})\ \text{m}^3\right) = 1.23\ (10)^{22}\ \text{molecules}$$

The ratio of exhaled volume over the 20 years to the total volume of air defines the proportion of the atmosphere associated with your exhalations. Thus, in a single breath the average number of air molecules that you are recycling is

$$n_{inhaled} = \frac{1.05\ (10^5)\ \text{m}^3}{3.58\ (10^{19})\ \text{m}^3}\left[1.23\ (10^{22})\frac{\text{molecules}}{\text{breath}}\right] = 3.6\ (10)^7\ \frac{\text{recycled molecules}}{\text{breath}}$$

Assuming that water is also mixed over its entire volume on the earth, $1.385\ (10^9)$ km^3, how many molecules of a 70-kg Viking (75% water) who died in battle 1000 years ago will you drink in a 250 cm^3 glass of water?

$$n = \frac{\left[\dfrac{70000\ \text{g}\ (0.75)}{18\ \text{g/mole}} N_{Av}\right]\left[(250\ \text{cm}^3)\left(\dfrac{1\ \text{g/cm}^3}{18\ \text{g/mole}}\right) N_{Av}\right]}{1.385 * 10^9\ \text{km}^3\ \dfrac{\left(10^{-15}\ \text{cm}^3/\text{km}^3\right)\left(1\ \text{g/cm}^3\right)}{18\ \text{g/mole}} N_{Av}}$$

$$= 317{,}000\ \frac{\text{molecules}}{\text{glass of water}}$$

Density or concentration of a substance is generally indicated by a volume, mass, or mole fraction. The volume fraction that was employed several times in the above example is the ratio of the volume of the contaminant to the total volume. Since the contaminants of interest are often present in very small quantities, the total system volume is effectively the volume of the solvent or surrounding phase. As long as the concentration (here measured by volume fraction) is less than about 5%, the contribution of the contaminant volume to the total volume is negligible. The volume fraction is a convenient unit only when the phase of the contaminant is the same as the phase of the bulk or surrounding phase. If the concentration of a contaminant in the atmosphere is expressed as one part per million (ppm), this refers to one part contaminant per million parts air by volume. For a single constituent, A, in a bulk fluid, B, the volume fraction is given by

$$\varphi_A = \frac{V_A}{V_A + V_B} \tag{3.4}$$

If V_A were less than about 5% of V_B, then $\varphi_A \sim V_A/V_B$.

The mass fraction is defined as mass of contaminant per unit mass of solvent or sorbent phase. This measure of concentration is commonly used for contaminants on solid sorbents. A typical unit might be milligram of contaminant per kilogram of dry soil or sediment. This might also be referred to as ppm, a very different ppm than described above for volume fraction in gases and vapors. Weight fraction is also used for describing contaminant concentration in a liquid phase, but it is rarely if ever used in describing contaminant concentration in a gas phase. The weight fraction of a component A in a bulk phase also containing a component B is defined as

$$\omega_A = \frac{m_A}{m_A + m_B} \tag{3.5}$$

Again, if the mass of A, m_A, is less than about 5% of the mass of B, m_B, then $\omega_A = m_A/m_B$.

The specific gravity is also a ratio of masses, the ratio of the mass of a substance relative to the mass of the same volume of a reference substance, typically water or air. The density of water at 4°C is 1.0000 g/cm^3 and this is often used as a reference for liquid specific gravity. The specific gravity of a substance relative to water at 4°C is thus just equal to its density in g/cm^3.

In addition to volume and mass fraction, the mole fraction is often used to indicate concentration. The mole fraction is defined as the moles of a contaminant per mole of bulk solvent phase. The number of *moles per liter of solution* is called the *molarity* of the solution. The molarity is a molar concentration of A and we shall use the symbol $\tilde{M}_A$. The tilde emphasizes that the only acceptable units are molar units, i.e., moles of A per liter. The number of *moles per kilogram of solvent (or sorbent)* is the *molality*, and since it is the number of moles of A per kilogram we will use the symbol $\hat{M}_A$. Note that the mass units implied by the caret refer to the solvent. The solute must be in molar units for both molarity and molality. For dilute solutions and ambient conditions, molarity and molality are essentially equivalent. If A represents the solute and B the solvent or sorbent, these quantities can be defined as

$$\begin{aligned} &\text{molarity} \quad \tilde{M}_A = \frac{n_A}{V_{AB}} \qquad V_{AB} = 1 \text{ L of } A + B \\ &\text{molality} \quad \hat{M}_A = \frac{n_A}{m_B} \qquad m_B = 1 \text{ kg of } B \end{aligned} \tag{3.6}$$

The tilde will be used to indicate molar quantities if we need to differentiate between mass and molar quantities or if we need to emphasize the molar quantity. The mole fraction is the ratio of the moles of contaminant to total moles in the system. For a component A in a bulk phase B, the mole fraction is defined as

$$y_A = \frac{\dfrac{m_A}{\mathrm{MW}_A}}{\dfrac{m_A}{\mathrm{MW}_A} + \dfrac{m_B}{\mathrm{MW}_B}} = \frac{n_A}{n_A + n_B} \tag{3.7}$$

where m again represents mass and MW represents the molecular weight of the substance. If the moles of component A, m_A/MW_A, represent less than about 5% of the total moles in the denominator of Equation 3.7, then

$$y_A \approx \frac{m_A}{m_B}\left(\frac{\mathrm{MW}_B}{\mathrm{MW}_A}\right) \quad \text{if} \quad \frac{m_A}{\mathrm{MW}_A} << \frac{m_B}{\mathrm{MW}_B} \tag{3.8}$$

and the molar ratio of A to B is essentially identical to the mole fraction. This is termed the solute free mole fraction because the solute (A) is not considered in volume, moles, or mass of the solution.

It should be emphasized that the molecular weight table represents a relative weight of the elements. All element weights are given relative to a single isotope of carbon, ^{12}C. The average molecular weight of natural carbon is normally taken to be 12.01. This simply reflects the small amounts of ^{14}C in natural carbon. Normally, the molecular weights in routine calculations will be rounded to the nearest integer.

Compounds are composed of more than one element, and the molecular weight of the compound is simply the sum of the molecular weights of the elements of which it is composed. Thus, the molecular weight of calcium carbonate, $CaCO_3$, is 41 + 12 + 3(16) = 101. The calculation of molecular weight and the determination of mass and mole fraction from a volume fraction is illustrated in Example 3.3 for beer, which is assumed to be essentially alcohol and water. Note that in the calculation of weight or mole fractions, the absolute quantity of material present is not important, so it is convenient to define a basis that makes the calculations easier. A basis of 100 mL of beer is assumed so that the given volume percentages are identical to the volume of each constituent.

Example 3.3: Volume, weight, and mole fraction

Estimate the mass and mole fractions of ethanol and water in beer, assuming that the beer is 5.7% ethanol and 94.3% water (volume basis).

Data:

Compound	Formula	Volume Fraction, φ	Density, ρ g/cm³	MW g/mol
Ethanol	C_2H_5OH	0.057	0.7	2*12+5*1+16+1 = 46
Water	H_2O	0.943	1.0	2*1+16 = 18

Basis: V = 100 cm³ of beer

Compound	Volumes, V_i (φ*V, cm³)	Mass, m_i (V_i *ρ g)	Mass Fraction ($m_i/\Sigma m_i$)	Moles, n_i (m_i/MW)	Mole Fraction (n_i /total)
Ethanol	5.7	3.99	0.0406	0.0867	0.0163
Water	94.3	94.3	0.9594	5.2389	0.9837
Sum (Σ)	100	98.29	1.0000	5.326	1.0000

Note that the mass fraction of ethanol on a solute free basis would be 3.99/94.3 or 4.2% rather than 4.06%. Similarly, the mole fraction on a solute free basis would be 0.0867/5.2389 = 1.66% rather than 1.63%. Both of these illustrate that the ethanol is present in a sufficiently low concentration that solute free basis estimates are essentially identical to actual mass and mole fractions.

Example 3.4, which indicates the composition of clean dry air, illustrates that because 1 mol of any gas at ambient conditions (e.g., 25°C and 1 atm pressure) occupies the same volume, the mole fraction and volume fraction of a compound in air are identical. The average molecular weight of air is also calculated in Example 3.4 to be about 29. Sometimes the relative densities of gases are compared to that of air to indicate relative buoyancy. Because 1 mole of any gas occupies the same volume, the specific gravity of gas, A is simply given by the ratio of molecular weight

$$s.g.|_{air} = \frac{MW_A}{MW_a} \tag{3.9}$$

Example 3.4: Mass and mole fractions of constituents and average molecular weight of air

Estimate the constituent mass and mole fraction and the average molecular weight of air. At 25°C and 1 atm the molar density of all gases, $\tilde{\rho}$, is about 0.04082 mol/L (24.5 L/mol).

Data:

Compound	Formula	Volume Fraction, φ	MW g/mol	Mass Density (ρ*MW, g/L)
Nitrogen	N_2	0.7808	14*2 = 28	1.143
Oxygen	O_2	0.2095	16*2 = 32	1.306
Argon	Ar	0.0093	40	1.633
Carbon dioxide	CO_2	315 ppm	12+16*2 = 44	1.796
Neon	Ne	18 ppm	20	0.816
Helium	He	5.1 ppm	2	0.0816
Methane	CH_4	1 ppm	12+1*4 = 16	0.653
Krypton	Kr	1 ppm	84	3.429

Basis: V = 100 L of air

Compound	Volumes$_i$ (φ*V, L)	Mass, m_i (V_i *ρ g)	Mass Fraction ($m_i/\Sigma m_i$)	Moles, n_i (m_i/MW)	Mole Fraction (n_i /total)
Nitrogen	78.08	89.23	0.7551	3.187	0.7808
Oxygen	20.95	27.36	0.2316	0.855	0.2095
Argon	0.93	1.518	0.0128	0.038	0.0093
Carbon dioxide	0.0315	0.0565	0.00048	0.001	315 ppm
Neon	0.0018	0.00147	$12(10^{-6})$	$73(10^{-6})$	18 ppm
Helium	0.00052	$42(10^{-6})$	$0.36(10^{-6})$	$21(10^{-6})$	5.1 ppm
Methane	0.0001	$65(10^{-6})$	$0.55(10^{-6})$	$4.1(10^{-6})$	1 ppm
Krypton	0.0001	$343(10^{-6})$	$2.9(10^{-6})$	$4.1(10^{-6})$	1 ppm
Sum (Σ)	100	118.17	1.0000	4.081	1.0000

Note: Average molecular weight = 118.17 g/4.081 moles = 28.95 g/mole. Note that mole fraction and volume fraction are identical for an ideal gas.

An example of a molecular weight calculation for mixtures of compounds including those with known elemental composition but unknown molecular formula is illustrated in Example 3.5. In Example 3.5 a basis of 100 g of mixture is selected because it provides mass fractions in percentage directly.

Example 3.5: Determination of empirical formula from known elemental mass fraction

Estimate the mole fractions of elements and the empirical formula of a compound containing (by weight) 52.1% carbon, 13.2% hydrogen, and 34.7% oxygen.

Basis: 100 g

Element	Weight (in 100 g)	MW (g/mol)	Moles (in 100 g)	Mole Fraction	Mole Ratio
Carbon	52.1	12	4.3	0.22	2
Hydrogen	13.2	1	13.2	0.67	6
Oxygen	34.7	16	2.17	0.11	1
Total	**100.0**		**19.7**	**1.00**	

The empirical formula is the ratio of the elements in the compound or C_2H_6O. This is the same as the molecular formula of ethanol, but because only the ratios of the elements are known, it could also be $C_4H_{12}O_2$, $C_6H_{18}O_3$, etc.

Occasionally, the number of moles of a substance is given in kg-moles or lb-moles. Moles without a prefix refer to gram-moles, that is the mass in grams divided by the molecular weight. A kg-mole is simply the mass in kilograms divided by molecular weight and therefore is 1000 gram-moles. Similarly, a lb-mole is 453.6 gram-moles. The use of anything other than gram-moles, or moles, however, should be avoided because of the potential for confusion.

3.2.2 Reaction Stoichiometry

A chemical reaction involves the combination or separation of individual molecules of a substance. For example, consider the chemical reaction,

$$\nu_A A + \nu_B B \leftrightarrow \nu_C C + \nu_D D \tag{3.10}$$

Stoichiometry refers to the quantities of the various substances taking part in the reaction. The stoichiometric coefficients, ν_i, are the molecules of component i required to complete the reaction per ν_A molecules of component A. Since there are always 6.02×10^{23} molecules per mole of a substance, the stoichiometric coefficient, ν_i, could also be recognized as the number of moles of component i required to complete the reaction per A moles of component ν_A.

Let us consider the following specific reaction which describes the process of combustion of methane to carbon dioxide and water,

$$CH_4 + 2O_2 \rightarrow CO_2 + 2H_2O \tag{3.11}$$

The reaction is said to be *balanced,* in that for every molecule of each element reacting there is a molecule of that element in the products. In the absence of nuclear reactions there must be 1 molecule or mole of carbon (C) in the products for every molecule or mole of C in the reactants.

The same is true for atomic oxygen (O) and hydrogen (H). Thus, 2 moles of molecular oxygen (O_2) are required and 1 mole of CO_2 and 2 moles of H_2O are formed per mole of methane. Because the molecular weight of the oxygen atom is 16, each mole of molecular oxygen has a mass of 32 g. Each mole of methane has a mass of 16 g. Thus, 64 g of oxygen are used to burn 1 mole or 16 g of methane. Determination of the amounts of reactants required to complete a reaction is demonstrated in Example 3.6.

For any given reaction the mass of reactants available could be less than or greater than that required for complete reaction. For the methane combustion reaction, Equation 3.11, 64 g of oxygen is required per mole of methane. If less than 64 g of oxygen are available per mole of methane, then oxygen would be referred to as the *limiting reactant* and only enough of the methane could be burned to use up that oxygen. If more than 64 g of oxygen is available per mole of methane, then oxygen would be referred to as the *excess reactant* and methane would be the limiting reactant. The methane reaction could occur only as long as there was methane available and some oxygen would remain unreacted. Most combustion reactions are conducted with excess oxygen to insure complete combustion of the fuel.

Note that the definitions of which compound is limiting or in excess depend on the reaction occurring exactly as it is written in Equation 3.11. Other reactions could occur that might result in the elimination of all of the reactant species even if one is present in insufficient quantities for the complete reaction. If oxygen were the limiting reactant, it is likely that some CO would be formed which only requires 1.5 moles of oxygen per mole of methane

$$CH_4 + \frac{2}{3}O_2 \rightarrow CO + 2H_2O \tag{3.12}$$

Example 3.6: Stoichiometric Requirements

Estimate the amount of oxygen required to burn a hydrocarbon with chemical formula C_xH_y. The approximate carbon:hydrogen ratio for common fuels is

Natural gas (methane)	1:4
Petroleum liquids	1:2
Coal	1:1

The combustion reaction using air (3.73 moles N_2 per mole of O_2) is given by

$$C_xH_y + \alpha(O_2 + 3.73N_2) \rightarrow \beta CO_2 + \gamma H_2O + \delta N_2$$

Although nitrogen does not take part in the reaction, its presence as a diluent gas is very important and the quantity present must be estimated.

Elemental balances to determine the stoichiometric coefficients:

Moles of reactant elements		= moles of product elements
Carbon	x	= β
Hydrogen	y	= $2*\gamma$
Oxygen	$2*\alpha$	= $2*\beta + \gamma$ or, $\alpha = \beta + \gamma/2$
Nitrogen	$2*\alpha*3.73$	= $2*\delta$

Required oxygen per mole of fuel:

Natural gas (CH_4)	$\beta = 1$ and $\gamma = 2$, therefore, $\alpha = 2$ and 2 moles O_2 required Products: 1 mole CO_2, 2 moles H_2O, and 7.5 moles N_2

Petroleum liquids (C_nH_{2n}) n = # of carbons	β = n and γ = n, therefore, α = O_2 required = 3/2 n Products: n moles CO_2 and H_2O, and 5.6 moles N_2
Coal (C_nH_n)	β = n and γ = n/2, therefore, α = 5/4 n Products: n moles CO_2, n/2 moles H_2O, and 4.7 moles N_2

Because less oxygen is required by this reaction, all of the methane could react to form some product (either CO or CO_2) even if oxygen is the limiting reactant with respect to the complete combustion described by Equation 3.11. If complete combustion is desired, however, stoichiometric calculations determine the minimum amount of oxygen required to allow the reaction to occur completely.

Often additional compounds that don't take part in the reaction can complicate the stoichiometric calculations. Again using the example of methane combustion, let us consider air as an oxygen source. For every mole of air there is about 3.73 moles of nitrogen (N_2) present in air. Thus, the balanced reaction has the form

$$CH_4 + 2(O_2 + 3.73\ N_2) \rightarrow CO_2 + 2H_2O + (2)(3.73)\ N_2 \tag{3.13}$$

The nitrogen does not take part in the reaction as written, so 7.46 moles (2 × 3.73) of nitrogen must be included in the combustor feed (and in the products) per mole of methane. This has important consequences for the volume and temperature of the product gases that will be explored later. This is even more important if the oxygen is added in excess to help ensure complete reaction of methane to carbon dioxide and water. For example, if the combustion is conducted with 15% *excess air*, the reaction is then

$$CH_4 + 2(1.15)\ (O_2 + 3.73\ N_2) \rightarrow CO_2 + 2H_2O + (2)(3.73)(1.15)\ N_2 \tag{3.14}$$

and 8.65 moles of nitrogen (N_2) is included in the feed and products of the combustor per mole of methane.

3.2.3 Temperature

Temperature is a measure of the energy level in the random motion of molecules in the system. High temperatures are associated with energetic random motions of molecules while low temperatures are associated with the opposite. At high temperatures, therefore, the energy of the molecular motions may be sufficient to essentially overcome all attractive forces between molecules and a fluid will boil and vaporize. At low temperatures, the energy level may be insufficient to overcome attractive forces and the molecules may become trapped in a crystalline or solid state. At that point, random molecular motions are limited to oscillations within a confined area. Ultimately, the temperature may be reduced to a sufficiently low level that all motion effectively ceases. The energy level when motion ceases (absolute zero), the energy level when there is a phase transition between solid and liquid (melting or freezing point), and the energy level when there is a transition between liquid and vapor (boiling or condensation point) represent the three reference points in all scales of measurement of temperature in use today. The Celsius or centigrade temperature scale is referenced to the freezing point of water (0°C) and the boiling point of water (100°C). The Fahrenheit scale is not so clearly linked to phase transitions, but it is now also referenced to the freezing point of water (32°F) and the boiling point of water (212°F). The absolute temperature scales are referenced to absolute 0. The Kelvin scale is the measure of absolute temperature with the same energy interval per degree as the Celsius scale. Because the zero point on the Kelvin scale is absolute zero, however, the freezing point of water is 273 K (actually 273.15 K) and the boiling point is 373

K. The Rankine scale is the measure of absolute temperature with the same energy interval per degree as the Fahrenheit scale. Because the zero point on the Rankine scale is also absolute zero, the freezing point of water is 492°R (actually 491.58°R) and the boiling point of water is 672°R. Table 3.4 lists these four units of temperature and the interrelationships between them.

TABLE 3.4
Temperature Scale Conversions

	°F	°R	°C	K
°F	1	°R – 459	1.8 °C + 32	1.8 K – 459
°R	°F + 459	1	1.8 °C + 491	1.8 K
°C	5/9 (°F – 32)	5/9 (°R – 491)	1	K – 273
K	5/9 °F + 255	5/9 °R	°C + 273	1

Note: Read down to desired temperature scale, horizontally to given scale.

In many problems of interest it is only the difference in temperature that is important. In these cases, either differences in absolute (e.g., K, or °R) or relative (°C or °F) temperature can be used. In estimating the density of a gas or other temperature-dependent thermodynamic quantity, the use of absolute temperature is essential.

3.2.4 Pressure

Pressure is a force per unit area that acts perpendicular or normal to the surface of the fluid. There are two common sources of pressure, collisions of molecules with a wall in a confined system and the weight of a fluid above a surface. The momentum of molecules moving randomly within a container tends to place a net force on the walls of the container. Normalized (or divided) by the area of the container, this force is a pressure tending to cause spreading of the fluid if the retaining walls were not present. The pressure acting at the base of a column of liquid, an example of the second source of pressure, is given by the force or weight of the liquid on the base divided by the area of the base. The weight of such a column is equal to its mass times the acceleration of gravity by Newton's law. Thus, a 10-m high water column exerts a pressure at its base given by

$$P = \rho_w\, g\, h = (977\ \text{kg/m}^3)(9.8\ \text{m/s}^2)(10\ \text{m}) = 97{,}700\,\frac{N}{\text{m}^2} \tag{3.15}$$

where a temperature of 25°C has been assumed. The column of air above the Earth's surface also has weight and therefore exerts a pressure on the Earth's surface which is very close to the pressure exerted by a column of water 10 m high. Atmospheric pressure varies over a narrow range. The pressure at ground level in the standard atmosphere is equal to, in various units of measure,

1.000 atmospheres (atm)
33.91 feet of water (ft H_20)
10.34 meters of water (m H_2O)
14.696 pounds per square inch (lb_f/in^2)
29.921 inches of mercury (in. Hg)
760 millimeters of mercury (mmHg)
1.013×10^5 N/m^2 (Pascals-Pa)

1.013×10^6 dyn/cm^2
1.013 bar

Because any pressure measurement system at the Earth's surface is also influenced by atmospheric pressure, a measured pressure (termed a gauge pressure) is generally the actual or absolute pressure minus the atmospheric pressure. When the gauge pressure is negative, the actual pressure is below atmospheric and termed a vacuum. The same units that measure pressure also can be used to measure vacuum. The equivalencies between these measures of pressure are given by the examples below

$$\begin{aligned} \text{0 absolute pressure} &= \text{1 atm vacuum} &&= -1\text{ atm gauge pressure} \\ \text{atmospheric pressure} &= \text{0 vacuum} &&= \text{0 gauge pressure} \\ 10.696\ \text{lb}_f/\text{in}^2 &= 4\ \text{lb}_f/\text{in}^2\text{ vacuum} &&= -4\ \text{lb}_f/\text{in}^2\text{ gauge pressure} \end{aligned} \tag{3.16}$$

The calculated pressure at the base of a 10-m column of water at 25°C, 97,700 Pa (N/m^2) is approximately 96% of the pressure at the ground in a standard atmosphere, consistent with the 10.3-m column required to equal 1 atm pressure. Because the column of water is also affected by the weight of the atmosphere above it, this pressure is a gauge pressure and the absolute pressure is 97,700 Pa + 101,300 Pa = 199,000 Pa = 199 kPa.

In many problems of interest it is only the difference in pressure that is important, for example, the flow resulting from a pressure difference between two ends of a tube. In these cases, the use of a gauge or absolute pressure is unimportant as long as all pressures are referenced to the same standard. As with temperature scales, however, it is essential that absolute pressures be employed for certain calculations such as the estimation of gas density.

The weight of a column of liquid provides a convenient means of measuring pressure. A U-tube manometer is simply a U-shaped tube with each end connected to a pressure. If there is a pressure difference between the two ends the fluid will rise to different heights in each leg of the tube. If one end is open to the atmosphere, the pressure at that end is 1 atm, or 0 atm gauge pressure. The difference in liquid height to the other end of the U-tube then provides a pressure difference between the two taps of

$$\Delta P = \rho\, g\, \Delta h \tag{3.17}$$

A number of examples of various applications of this principle can be found below.

Example 3.7: Pressure in a U-tube manometer

Estimation of differential pressure using a U-tube manometer filled with liquid mercury (ρ = 13.33 g/cm^3).

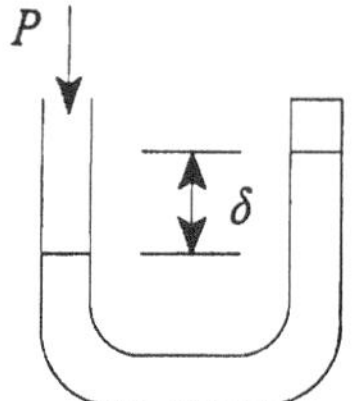

δ is the difference in height of liquid mercury between the right-hand stem (zero pressure) and the left-hand stem, exposed to the atmosphere.

If δ = 76 cm, the differential pressure, in this case the pressure in the atmosphere, is

$$\begin{aligned} \Delta P &= \rho g \delta \\ &= 13.33\ \text{g/cm}^3\ (1000\ \text{cm/s}^2)(76\ \text{cm}) \\ &= 101.3\ Pa \end{aligned}$$

Example 3.8: Multiple fluids in a U-tube manometer

Considering the oil (ρ_o= 0.7 g/cm^3) and water manometer in the adjacent figure, what is the height of water, δ_2, if the oil height is 15 cm lower than that of the water, i.e., $\delta_2 - \delta_1$ = 15 cm? P_1 is 0.05 atm or 0.506(10)^5 dyn/cm^2. P_2 is 0.031(10^5) N/m^2 or 0.31(10)^5 dyn/cm^2 due to the vapor pressure of water at the temperature of the U-tube.

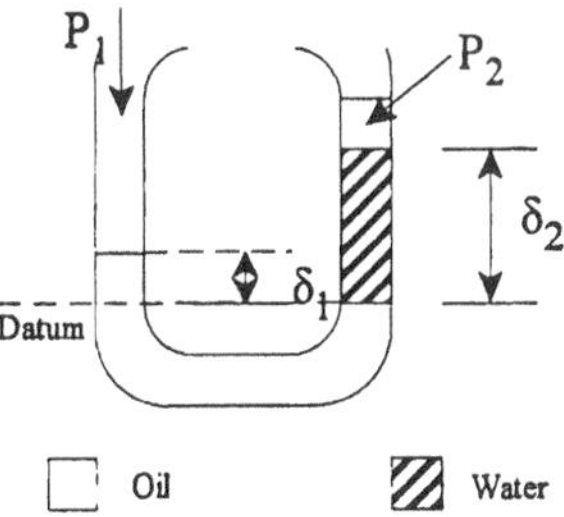

The pressure is identical in the two legs of the manometer at the location of the datum. At equilibrium, the pressure acting above this datum on the left-hand side is equal to the pressure acting on the right-hand side.

$$+\rho_w g\delta_2 = P_1 + \rho_o g\delta_1$$

$$+\rho_w g\delta_2 = P_1 + \rho_o g(\delta_2 - 15\text{ cm})$$

$$\delta_2 = \frac{P_1 - \rho_o g(15\text{ cm}) - P_2}{\rho_w g - \rho_o g}$$

$$\delta_2 = \frac{0.506(10)^5\text{ dyn/cm}^2 - (0.7\text{ g/cm}^3)(1000\text{ cm/s}^2)(15\text{ cm}) - 0.31(10^5)\text{ dyn/cm}}{(1000\text{ cm/s}^2)(1\text{ g/cm}^3\ -\ 0.7\text{ g/cm}^3)}$$

$$\delta_2 = 30.3\text{ cm}$$

To achieve the desired readings, 30.3 cm of water is required in the manometer. The oil in the opposite leg will rise about half this distance, 15 cm, above the bottom of the water.

The last example illustrates that most liquids exist in equilibrium with a gas phase that exerts a *vapor pressure*. In this case liquid water exerted a vapor pressure of 0.95 (10)^5 dyn/cm^2. Random motion of molecules results in some molecules leaving a liquid or solid phase and entering the vapor or air phase. Random motion of the same type of molecules in the vapor or air phase also leads to the return of some of these molecules to the liquid or solid surface where they can be trapped by attractive forces and incorporated within the liquid or solid phase. Thus, a dynamic equilibrium exists between the liquid or solid phase and the surrounding air as shown in Figure 3.2. In the last example, the net effect of water molecules leaving the water surface is a pressure of 31,700 dyn/cm^2 (0.0313 atm) in the sealed right-hand leg of the manometer.

Typically, a solid phase holds molecules very effectively and evaporation to the air is negligible. Liquids, however, exhibit a significant vapor pressure, i.e., the equilibrium pressure exerted by the evaporating molecules. At the normal boiling point of the liquid, the pressure exerted by these molecules is equal to atmospheric pressure. At this temperature, even molecules moving with the mean random velocities have sufficient energy to overcome the attractive forces and escape to the atmosphere. At temperatures below the normal boiling point, the vapor pressure is less than atmospheric. Figure 3.3 summarizes the vapor pressure of several compounds as a function of temperature. Methylene chloride (CH_2Cl_2) boils at about 40°C and benzene (C_6H_6) boils at 80°C, when their vapor pressure equals the total pressure of 1 atm (760 mmHg). The vapor pressure is a very strong function of temperature with a doubling of vapor pressure with temperature changes of less than 20°C for all three compounds. Vapor pressure data for a variety of compounds can be found in Appendix A.

As will be seen in Chapter 4, the concentration of a compound in the vapor state is in direct proportion to its vapor pressure under steady-state or equilibrium conditions. The vapor pressure curves in Figure 3.3 thus indicate directly the concentration of the evaporating compound that would be present in the vapor phase at equilibrium. At the boiling point, the vapor pressure is equal

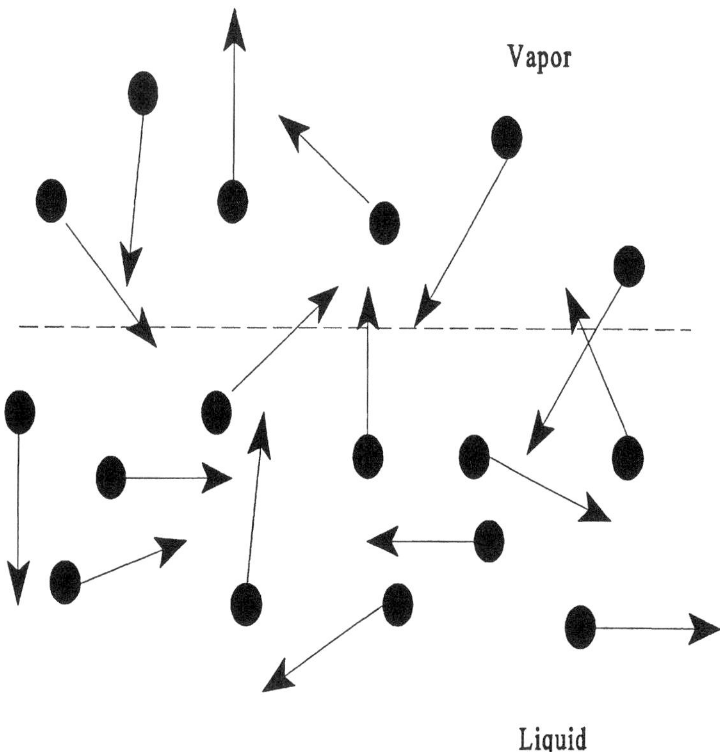

FIGURE 3.2 Depiction of dynamic equilibrium between a liquid and gas phase.

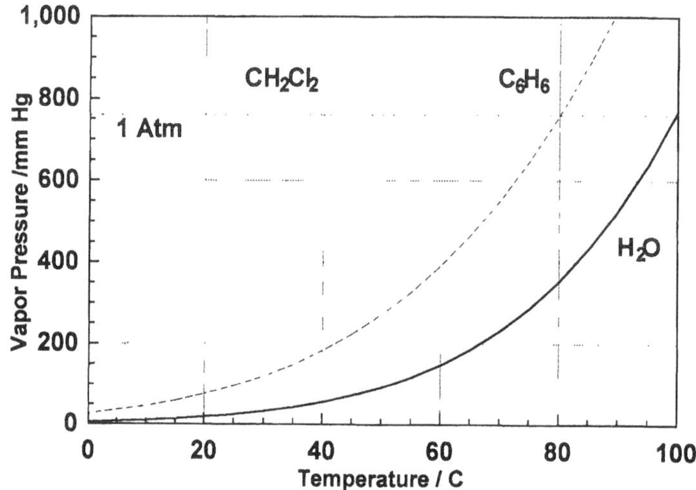

FIGURE 3.3 Vapor pressures of selected compounds.

to the total pressure and the vapor (at equilibrium) is entirely composed of the boiling liquid. Thus, at 100°C, the vapor state above boiling water would be 100% water vapor. Of course, dryer, cooler air will continually pass over an open container of boiling water diluting the water vapor and eventually causing all of the water to evaporate before this equilibrium can be reached. At 25°C the vapor pressure of liquid water is 0.0313 atm (31,700 dyn/cm^2). Air in contact with a body of water at 25°C will thus contain, at equilibrium, 3.13% water vapor.

3.3 DIMENSIONAL ANALYSIS

As indicated previously, it is not possible to compute the sum or difference of individual terms in an equation unless they have identical units. Nor can two quantities be equated unless they have identical units. This further implies that we can write any (correct) equation in dimensionless form by dividing both sides of an equation by a quantity that has the appropriate units. Let us consider as a simple example the position of an object with initial velocity, V_0, and position, x_0, after being subjected to a uniform acceleration, a, for a time, t.

$$x = \frac{1}{2}at^2 + V_0 t + x_0 \tag{3.18}$$

Each term in the equation has dimensions of length. Each term could be made dimensionless by dividing by some length, for example, x. It is thus possible to rewrite Equation 3.18 in the dimensionless form

$$\begin{aligned} 1 &= \frac{1}{2}x^{-1}a^1t^2V_0^0x_0^0 + x^{-1}a^0t^1V_0^1x_0^0 + x^{-1}a^0t^0V_0^0x_0^1 \\ &= \frac{1}{2}\frac{at^2}{x} + \frac{tV_0}{x} + \frac{x_0}{x} \end{aligned} \tag{3.19}$$

Writing the equation in this form illustrates that it is always possible to write an algebraic model of a system as the sum of dimensionless terms each involving all variables that arise in the system. For a general system governed by an algebraic model involving, for example, the four variables, a, b, c, and d, this implies that the model can be written in the form

$$\kappa_1 = f(a^\alpha b^\beta c^\gamma d^\delta,\ a^\varepsilon b^\zeta c^\eta d^\theta,\ ...) \tag{3.20}$$

where each of the quantities separated by commas within the parentheses are dimensionless. f represents some (possibly unknown) functional dependence on the dimensionless variables, while κ_1, α, β, etc., are simply coefficients or constants. The number of terms or dimensionless variables is equal to the number of different (and, as shown in Example 3.9, independent) dimensionless groups that can be formed from the physical variables. If there are n variables and m dimensions, the number of dimensionless groups or variables that can be formed is n-m. Thus, if four variables were important in the description of a physical problem and these variables included the complete set of units, mass, length, and time, only one dimensionless group would be required to represent the solution. That is

$$\begin{aligned} \kappa_1 &= f(a^\alpha b^\beta c^\gamma d^\delta) \\ a^\alpha b^\beta c^\gamma d^\delta &= f^{-1}(\kappa_1) = \kappa_2 \end{aligned} \tag{3.21}$$

Here f^{-1} represents the inverse of the function f. Note that when only one dimensionless group can be formed from the variables in the problem the function f can possess only a single value. That is, the combination of the variables is always equal to a specific constant that can be determined from a single experiment. The relationship between the variables is known, purely on the basis of dimensional consistency and without reference to a description of the physics of the system.

The implications of this result can be illustrated by considering the simple situation of the accelerated object initially at the origin ($x_0 = 0$) with zero initial velocity ($V_0 = 0$). On physical grounds, it should be expected that the position of the object at some later time would be a function of the acceleration. Thus, the three variables that appear in this problem are x, a, and t, which have dimensions of {length}, {length/time^2}, and {time}, respectively. Because there are three variables and only two dimensions (length and time), it should be possible to write the description of the object's motion in terms of a single dimensionless group. The general relationship can be written

$$\kappa_1 = f(x^\alpha a^\beta t^\gamma) = f\left(\text{length}^\alpha \left[\frac{\text{length}}{\text{time}^2}\right]^\beta \text{time}^\gamma\right) \tag{3.22}$$

The determination of α, β, and γ follows by recognition that the units in this equation must each cancel separately. That is, the sum of the powers on length and time must separately equal 0, giving the two independent equations

$$\begin{aligned} \alpha + \beta &= 0 \\ -2\beta + \gamma &= 0 \end{aligned} \tag{3.23}$$

This is two equations in the three unknowns α, β, and γ. Thus, any one of these coefficients can be arbitrarily specified. Choosing $\alpha = 1$, therefore, $\beta = -1$, and $\gamma = -2$) and the relationship between these variables must be

$$\frac{x}{a\,t^2} = \kappa \tag{3.24}$$

The constant κ could be determined from a single measurement of position at any time (assuming the acceleration is known). From our knowledge of the physics, we know that $\kappa = 1/2$. Note that if we would have chosen a different value for α, the result would have been the same dimensionless group raised to a different power. For example, $\alpha = 2$, would have resulted in the conclusion

$$\frac{x^2}{a^2\,t^4} = \kappa \tag{3.25}$$

The previous result is recovered by taking the square root of both sides. Thus, the basic conclusion, that $x = \kappa a t^2$ is unchanged.

If we now consider the situation where the object exhibits an initial velocity and position, the relevant variables are x, a, t, x_0, and V_0. Three dimensionless groups to describe the system (5 variables – 2 dimensions = 3) should exist. The easiest way to form these groups is to select two variables which contain all of the relevant dimensions (length and time) and force these to appear in all three dimensionless groups. Each group is then determined by combining these two selected variables with each of the remaining variables in turn. Choosing x and t as the common variables, the three groups to be formed, Π_1, Π_2, and Π_3 are

$$\begin{aligned} \Pi_1 &= x^\alpha t^\beta a^\gamma \\ \Pi_2 &= x^\delta t^\varepsilon x_0^\zeta \\ \Pi_3 &= x^\eta t^\theta V_0^\iota \end{aligned} \tag{3.26}$$

These three groups are related by the, as yet unknown, relationship f = f(π1,π2, π3) or, alternatively, π1 = f(π2, π3). From Π_1, setting the powers on length and time to 0 separately,

$$\begin{aligned} \alpha + \gamma &= 0 \\ \beta - 2\gamma &= 0 \end{aligned} \tag{3.27}$$

Choosing $\alpha = 1$ recovers the previous group. From Π_2,

$$\begin{aligned} \delta + \gamma &= 0 \\ \varepsilon &= 0 \end{aligned} \tag{3.28}$$

Choosing $\delta = 1$, $\zeta = -1$. Finally, from Π_3,

$$\begin{aligned} \eta + \iota &= 0 \\ 0 - \iota &= 0 \end{aligned} \tag{3.29}$$

Choosing $\eta = 1$, gives $\iota = -1$ and $\theta = -1$. Thus, the relationship between the variables must be

$$f = f\left(\frac{x}{at^2}, \frac{x}{x_0}, \frac{x}{V_0 t}\right) \tag{3.30}$$

Which is, as expected, of the form of Equation 3.19. In general, determination of the form of the function f requires knowledge of the physics or experimental measurement. Note that only three variables need be measured in the experiments, as opposed to all five. That is, only the quantities x/at^2, x/x_0, and x/V_0t, need be measured and not the independent variables separately.

This trivial example of dimensional analysis is illustrative of a very powerful technique for analysis. Additional examples of the application of the technique can be found in Examples 3.9 and 3.10.

Example 3.9: Dimensional analysis of capillary rise

Surface tension causes water and other liquids to be pulled up into small diameter capillaries. The rise is dependent upon the balance between this force and gravity. Determine the dimensionless groupings that relate capillary rise to the process variables.

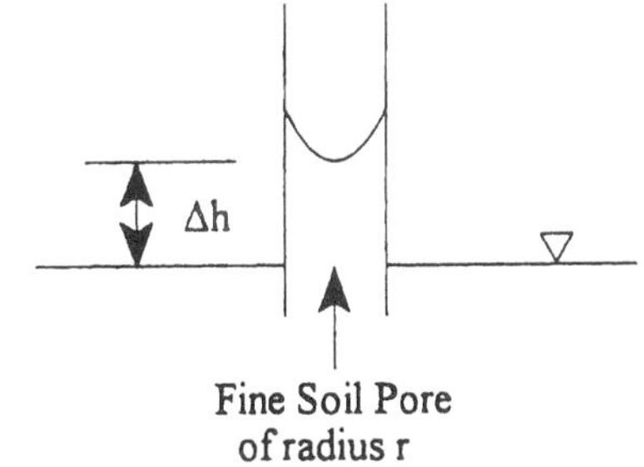

The downward force is dependent upon density, ρ, the gravitational constant, g, and the capillary rise height, Δh. The opposing capillary suction is dependent upon the liquid surface tension, σ, and the radius of the capillaries, r, assuming that the water fully wets the soil surface. Using the mass, length, time system of units there are five variables and three dimensions, suggesting two dimensionless groups are needed. Choosing ρ, g, and σ as the repeating variables (since they encompass all of the dimensions), the two groups can be formed by pairing the repeating variables with those remaining.

$$\Pi_1 = \rho^\alpha g^\beta r^\gamma \sigma \qquad \Pi_2 = \rho^\delta g^\varepsilon r^\varphi \Delta h$$

$$\left[\frac{M}{L^3}\right]^\alpha \left[\frac{L}{t^2}\right]^\beta [L]^\gamma \left[\frac{M}{t^2}\right] \qquad \left[\frac{M}{L^3}\right]^\delta \left[\frac{L}{t^2}\right]^\varepsilon [L]^\varphi [L]$$

$$\begin{aligned} M&: \quad \alpha + 1 = 0 \qquad && \delta = 0 \\ L&: \quad -3\alpha + \beta + \gamma = 0 \qquad && -3\delta + \varepsilon + \varphi + 1 = 0 \\ t&: \quad -2\beta - 2 = 0 \qquad && -2\varepsilon = 0 \end{aligned}$$

From these equations, $\alpha = -1$, $\beta = -1$, $\gamma = -2$, and, $\delta = 0$, $\varepsilon = 0$, $\varphi = -1$ and the expected form is

$$\Pi_1 = \frac{\sigma}{r^2 \rho g} \qquad \Pi_2 = \frac{\Delta h}{r}$$

and,

$$\frac{\Delta h}{r} = f\left(\frac{\sigma}{r^2 \rho g}\right)$$

Example 3.10: Dimensional Analysis of Bubble Rise in Water

A variety of water treatment systems are based on mass transfer that occurs as air bubbles rise through water. The time available for mass transfer depends on the rise velocity of gas bubbles, a balance between the buoyancy of the bubbles and the friction as it passes through the water. Determine the form of the relationship between the rise velocity, u, and the bubble diameter, d, the gravitational constant, g, the difference in density between the water and the air, Δρ, and the viscosity, or conductance to flow, of water, μ.

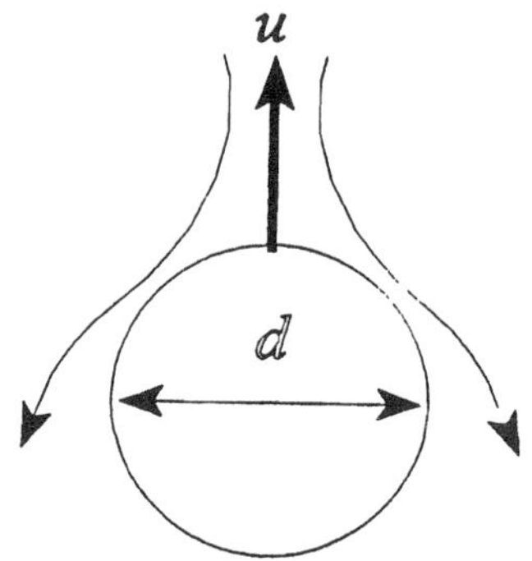

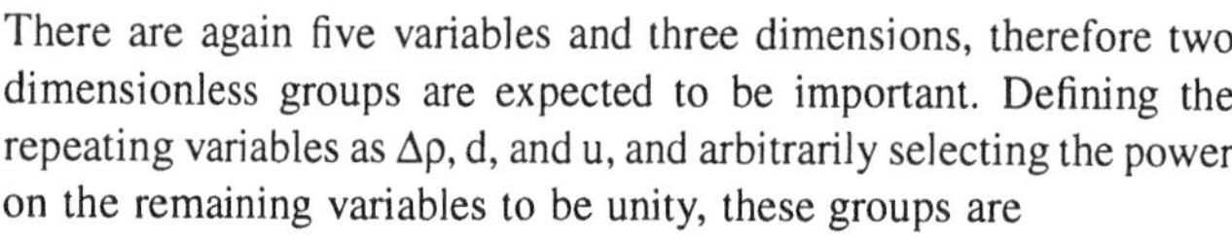

There are again five variables and three dimensions, therefore two dimensionless groups are expected to be important. Defining the repeating variables as Δρ, d, and u, and arbitrarily selecting the power on the remaining variables to be unity, these groups are

$$\Pi_1 = u^{\alpha} d^{\beta} \Delta\rho^{\gamma} g \qquad\qquad \Pi_2 = u^{\delta} d^{\varepsilon} \Delta\rho^{\varphi} \mu$$

$$\left[\frac{L}{t}\right]^{\alpha} [L]^{\beta} \left[\frac{M}{L^3}\right]^{\lambda} \left[\frac{L}{t^2}\right] \qquad \left[\frac{L}{t}\right]^{\delta} [L]^{\varepsilon} \left[\frac{M}{L^3}\right]^{\varphi} \left[\frac{M}{L*t}\right]$$

M:	$\gamma = 0$	$\sigma + 1 = 0$
L:	$\alpha + \beta + 3\gamma + 1 = 0$	$\delta + \varepsilon - 3\varphi - 1 = 0$
t:	$-\alpha + 2 = 0$	$-\delta - 1 = 0$

Solving these equations gives, $\alpha = -2$, $\gamma = 0$, $\beta = 1$, and $\delta = -1$, $\varphi = -1$, $\varepsilon = -1$. Thus the relationship must be of the form

$$\frac{gd}{u^2} = f\left(\frac{\mu}{du\Delta\rho}\right)$$

$u^2/(gd)$ and $(du\Delta\rho)/\mu$ are termed the Froude, N_{Fr}, and Reynold's, N_{Re}, numbers, respectively. Thus, this could be written as $N_{Fr} = f(N_{Re})$. Experiments have shown that this relationship takes on especially simple forms below $N_{Re} \sim 500$ and above $N_{Re} \sim 3000$. These forms are

$$N_{Fr} \approx 0.019 N_{Re} \qquad N_{Re} \leq 500$$

$$N_{Fr} \approx 0.5 \qquad N_{Re} \geq 3000$$

The lack of a dependence on Reynold's number above $N_{Re} = 3000$ is a reflection of the fact that liquid viscosity is largely unimportant at such high velocities.

In addition to these examples, the reader is referred to additional illustrations of dimensional analysis presented by Fox and McDonald (1978).

3.4 TREATMENT OF DATA

The application of dimensional analysis results in the need to determine constant values or universal functions from experimental data. To assist in that endeavor it is appropriate to discuss the analysis of such data. Collected in this section are a variety of common data analysis tools.

3.4.1 Mean and Standard Deviation

Determination of the value of a single dimensionless ratio of variables, for example, the determination that $f = 1/2$ in Equation 3.24, requires only a single measurement of the variables. Any single measurement, however, would not be expected to perfectly represent the value of any parameter. Any measurement is subject to *random errors,* which would be expected to provide estimates *both* above and below the actual value, and *systematic errors,* which might bias the measurements *either* above or below the actual value. It is often quite difficult to identify systematic errors and measurements often provide no indication of its existence. Only by a careful review and analysis of the experimental system and the parameter being measured can systematic errors be eliminated.

The measurements can, however, indicate the influence of random error on the estimate of an experimental parameter. To determine this influence, multiple measurements are required. Initially, each measurement would likely be different. After making a number of measurements, a given measurement, or a measurement within a small range of values, would likely be observed several times. A plot of the number of times a particular value or value within a particular range is observed vs. that value defines a probability distribution. A probability distribution is the function which describes the frequency of observation of a particular measurement. Examples of a symmetric and asymmetric probability distribution are shown in Figure 3.4. In a symmetric distribution, observations above and below the actual value are equally likely. In an asymmetric distribution, values above and below the actual value may be observed, but they are not equally likely. If we conducted the experiment an infinite number of times, the probability distribution generated would represent the *parent or population distribution.* The most likely value of the parameter, say μ, would be the value measured most often, or the value at the maximum of the probability distribution function. In a real experiment, however, the measurements represent a finite population and the resulting estimate of the probability distribution is the sample distribution. The challenge of the analysis of the experimental data is to get the best estimate of the most likely value of the parameter from the sample distribution.

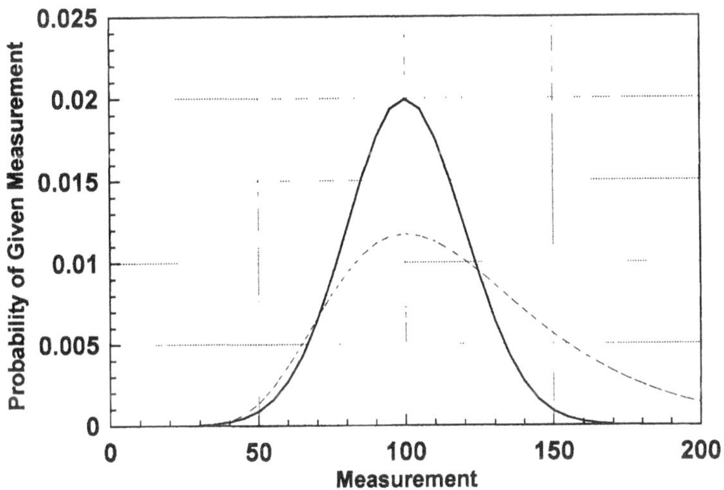

FIGURE 3.4 Symmetric and asymmetric probability distributions.

In many environmental measurements, it is often assumed that the probability distribution of those measurements is symmetric about the actual value. In such a situation, the most likely value is simply the mean of the population distribution, or, in any real and finite set of measurements, the mean of the sample probability distribution. With a large number of measurements, the calculated mean tends to the population mean,

$$\bar{X} = \lim_{N \to \infty} \frac{1}{N} \sum x_i \tag{3.31}$$

For finite N, the sample mean should be the best estimate of the population mean

$$\bar{X} \approx \bar{x} = \frac{1}{N} \sum x_i \tag{3.32}$$

It should be emphasized, however, that is x the most likely or most commonly observed value of the observation (i.e., $\bar{x} = \bar{X}$) only if the probability distribution is symmetric.

As indicated above, deviations about the mean ($d_i = x_i - \bar{x}$) are equally likely to be positive or negative for a symmetric distribution. Thus, the *average deviation* of the N measurements from the mean is zero.

$$\bar{d} = \frac{1}{N} \sum d_i = 0 \tag{3.33}$$

A measure of the magnitude of the deviations can be determined by squaring the individual deviations and then taking the square root of the result, the root mean square deviation or standard deviation. The mean square deviation is termed the variance,

$$\sigma^2 = \lim_{N \to \infty} \frac{1}{N} \sum d_i^2 \tag{3.34}$$

Note that taking only a single measurement requires us to use that estimate as the mean and calculation of even a single deviation is impossible. Thus the number of independent measurements of the deviation, d_i, is equal to the number of independent measurements of x_i minus one. Thus for any finite set of measurements, the best estimate of the parent or population variance is

$$\sigma^2 \approx s^2 = \frac{1}{N-1} \sum d_i^2 \tag{3.35}$$

Now the sum of the squared deviations can be expanded to the form

$$\begin{aligned} \sum d_i^2 &= \sum x_i^2 - 2\bar{x} \sum x_i + N\bar{x}^2 \\ &= \sum x_i^2 - 2N\bar{x}^2 + N\bar{x}^2 \\ &= \sum x_i^2 - N\bar{x}^2 \\ &= \sum x_i^2 - \frac{\left(\sum x_i\right)^2}{N} \end{aligned} \tag{3.36}$$

The last equation is generally the easiest way to calculate the variance from a given set of data. The error between an individual measurement and the "best" estimate of $\bar{X}$ can be written

$$e_i = x_i - \bar{X}_{est} \tag{3.37}$$

Writing a variance in terms of the mean squared error rather than the mean square deviations

$$\sum e_i^2 = \sum x_i^2 - 2\bar{X}_{est}\sum x_i + N\bar{X}_{est}^2 \tag{3.38}$$

Differentiating this equation with respect to the best estimate μ_{est} and setting the result equal to zero gives the minimum value of the error variance, or the best least square estimate. This recognizes the rule of calculus that the zeros of the differential of a function determine its minima, maxima, and inflection points. The sign of the second derivative indicates which, in this case, a minima. Differentiating, recognizing that the sums represented by the first two terms are simply constants,

$$0 = -2\sum x_i + 2N\bar{X}_{est} \tag{3.39}$$

This suggests that the best estimate (in the least square error sense) is given by the mean

$$\bar{X} \approx \bar{X}_{est} = \frac{\sum x_i}{N} = \bar{x} \tag{3.40}$$

Thus the variance or standard deviation between the observations and the estimate is a minimum when the estimate is taken as the mean of the observations, regardless of whether the population distribution from which the measurements are taken is symmetric or asymmetric. Although the mean may not be the most likely value if the probability distribution is asymmetric, it still represents the estimate that minimizes the variance with the observations.

Example 3.11: Mean and standard deviation in a set of measurements

Ten independent measurements of temperature and pressure in a chamber were made. Estimate the mean and standard deviation of the two sets of measurements. The final measurement of pressure and temperature seems anamolous. Recalculate the mean and standard deviation without that measurement.

Temperature measurements (T,°C) — 24.0, 23.8, 23.6, 24.1, 23.9, 24.0, 23.7, 23.9, 24.2, 24.5
Pressure measurements (P, mmHg) — 761, 759, 762, 763, 761, 761, 762, 761, 760, 757

With all data points (N = 10)

$$\bar{T} = \frac{\sum_{i=1}^{10} T_i}{N} = 23.97°C \qquad \sigma_T^2 = \frac{1}{N-1}\left[\sum_{i=1}^{10}(T_i)^2 - \frac{1}{N}\left(\sum_{i=1}^{10} T_i\right)^2\right] \qquad \sigma_T = 0.258°C$$

$$\bar{P} = \frac{\sum_{i=1}^{10} P_i}{N} = 760.7 \text{ mmHg} \qquad \sigma_P^2 = \frac{1}{N-1}\left[\sum_{i=1}^{10}(P_i)^2 - \frac{1}{N}\left(\sum_{i=1}^{10} P_i\right)^2\right] \qquad \sigma_P = 1.703 \text{ mmHg}$$

Note that the standard deviations are estimates of the population standard deviation. The standard deviation of the sample is divided by N rather than N – 1, giving $\sigma_T = 0.245°C$ and $\sigma_P = 1.616$ mmHg. Similarly, the mean and standard deviation with only the first nine measurements (N = 9) is given by

$$\bar{T} = \frac{\sum_{i=1}^{9} T_i}{N} = 23.97°C \qquad \sigma_T^2 = \frac{1}{N-1}\left[\sum_{i=1}^{9}(T_i)^2 - \frac{1}{N}\left(\sum_{i=1}^{9} T_i\right)^2\right] \qquad \sigma_T = 0.258°C$$

$$\bar{P} = \frac{\sum_{i=1}^{9} P_i}{N} = 761.1 \text{ mmHg} \qquad \sigma_P^2 = \frac{1}{N-1}\left[\sum_{i=1}^{9}(P_i)^2 - \frac{1}{N}\left(\sum_{i=1}^{9} P_i\right)^2\right] \qquad \sigma_P = 1.167 \text{ mmHg}$$

Note that although the best estimate of temperature and pressure (the mean) changed only slightly, the standard deviation, a measure of uncertainty in that estimate, was reduced by about 30% by removing the last measurement.

3.4.2 Probability Distributions

3.4.2.1 Gaussian Distribution

As indicated in the preceding section, the distribution of measurements may be distributed symmetrically or asymmetrically about the mean. One of the most important distributions in environmental work is the Gaussian or normal distribution. The Gaussian distribution is bell-shaped and symmetric as shown in Figure 3.4. Because the distribution is symmetric, the mean represents both the most likely value in such a distribution as well as the value that minimizes the variance with all possible measurements. The mathematical definition of the Gaussian distribution is

$$P(x;\bar{X},\sigma) = \frac{1}{\sigma\sqrt{2\pi}}\exp\left[-\frac{1}{2}\left(\frac{x-\bar{X}}{\sigma}\right)^2\right] \qquad (3.41)$$

where P(x; $\bar{X}$, σ) represents the probability of observation x given a Gaussian distribution with most likely value (and mean) $\bar{X}$, and standard deviation (= the square root of the variance) σ. Throughout the text, the semicolon in the definition of a function will be used to separate the independent variables from the parameters that appear in the function. Although the form of Equation 3.41 suggests that the probability is nonzero for any observation of x, the exponential causes the probability to decrease very rapidly away from the mean. Hence the bell-shaped form of the distribution. The probability of a single measurement 2.15 times the standard deviation away from the mean is only about 10% of the probability of a measurement equal to the mean. The distribution given by Equation 3.41 is normalized such that the cumulative probability or the sum of the probabilities of all possible measurements is unity.

A probability distribution defines the probability of observing a single measurement with a specified value. An alternative probability is the probability of that measurement being within a specified range of the mean or cumulative probability. This can be restated as a determination of the percentage of the total number of measurements that would be expected to fall within the specified range. It can be determined by integrating the Gaussian distribution function over the range of interest. That is, the area within the limits ± a beneath the Gaussian curve represents the probability that a single measurement would exhibit a value within those limits. Over 68% of the measurements would be expected to be within ±1 σ of the mean and more than 95% would be expected within ±2 σ. More than 96.8% of the measurements from a Gaussian distribution would be expected to fall within ±2.15 σ of the mean or within the zone where the probability of any given measurement is within 10% of that for the mean or most likely measurement.

The probability of a single measurement falling within a value x close to the population mean μ for a given standard deviation is given by

$$P\left(\leq \pm(x-\bar{X}),\sigma\right) = erf\left[\frac{x-\bar{X}}{\sqrt{2}\,\sigma}\right] \tag{3.42}$$

where erf (z) is the error function of argument z. The error function is simply the integral of the Gaussian distribution function in the symmetric interval about the mean defined by the value x. The error function arises in any problem involving a Gaussian distribution or process. As we shall see, models of this form often arise in problems involving transport by diffusion.

Many natural processes are expected or assumed to follow the normal distribution. As a result, you might expect, with 95% confidence, that any subsequent measurement would lie within ±2 standard deviations of the mean collected of a large set of previous measurements. The implications of this for the dataset in Example 3.11 are explored in the example below.

Example 3.12: Confidence limits in measurement

Consider the first nine measurements of temperature and pressure in Example 3.11. Assuming that the tenth measurement is an effort to determine the same temperature and pressure, the mean and standard deviation of the first nine measurements suggest that the tenth temperature measurement should be within the range defined by

$$\bar{T} - 2(\sigma_T) \leq T_{10} \leq \bar{T} + 2(\sigma_T)$$

$$23.53°C \leq T_{10} \leq 24.29°C$$

95% of the time. This tenth measurement is not within this range and you could state, with 95% confidence, that the tenth measurement is not the same measurement as the previous nine. Perhaps instrument drift or ambient temperature changes have modified conditions. Similarly, the tenth pressure measurement also exceeds the deviation from the mean that you would expect for 95% of the measurements. These data might be excluded from subsequent calculations as being subject to errors other than random measurement error.

This analysis cannot prove that the tenth data measurement is invalid, however,

- 5% of the data are expected to lie outside 95% confidence limits
- The calculated mean and standard deviations are only estimates of the most likely value and the standard error about that value
- The analysis assumes that the measurement errors are normally distributed about the mean, with only a limited amount of data to provide support

As indicated in the example, any finite number of measurements may not give the appearance of a Gaussian distribution even if the measurements are of a variable that is in fact distributed normally or in a Gaussian fashion. The shape of a Gaussian distribution would only be expected to be generated after an infinite number of measurements of the normally distributed parameter.

To determine if the measurements collected are described by a Gaussian distribution, plot the cumulative probability vs. measurement values on probability paper. The cumulative probability is the percentage of the measurements below a certain value and is related to the probability of a value within $x - \bar{X}$ of the mean by

$$\text{Cumulative } P\left(x, \bar{X}, \sigma\right) = \frac{1}{2} + \frac{1}{2} erf\left[\frac{x - \bar{X}}{\sqrt{2}\,\sigma}\right] \tag{3.43}$$

The example set of data is shown in Figure 3.5. A plot of Equation 3.43 on probability paper yields a straight line. Thus, if data plotted in this fashion yield a straight line, the data can be approximated with the Gaussian probability distribution (note that this does *not* necessarily prove that the data is Gaussian). The mean of the data set appears on the 50% probability line. Since 68% of the measurements from a Gaussian distribution fall within ±1 σ of the mean, the difference between the temperature corresponding to 50% cumulative probability (the mean) and the temperature corresponding to 84% cumulative probability (50% + 68%/2, actually 84.1%) represents one standard deviation. From the figure, the standard deviation is given by 24.15–23.9°C or 0.25°C, in excellent agreement with the calculated value in Example 3.11.

The Gaussian distribution has a very simple form and appears naturally in processes involving a large number of random, equally probable events. An important example of such a process is diffusion of molecules. Diffusion is the slow migration of substance resulting from the random motions between collisions of its component molecules. Because the number of molecules in a cm^3 of water is more than $3 \times (10)^{22}$, the molecular collisions represent an exceedingly large number of effectively random events. Thus, molecules moving by molecular diffusion would be expected to spread with a Gaussian or bell-shaped distribution away from the origin. Because movement in either direction would be equally likely, the mean location of the molecules would remain centered at the origin. The variance and standard deviation of the molecule's spread away from the origin, however, would increase with time. At any given time for which the standard deviation was equal to some value σ, more than 96.8% of the molecules would be within ±2.15 σ of the origin.

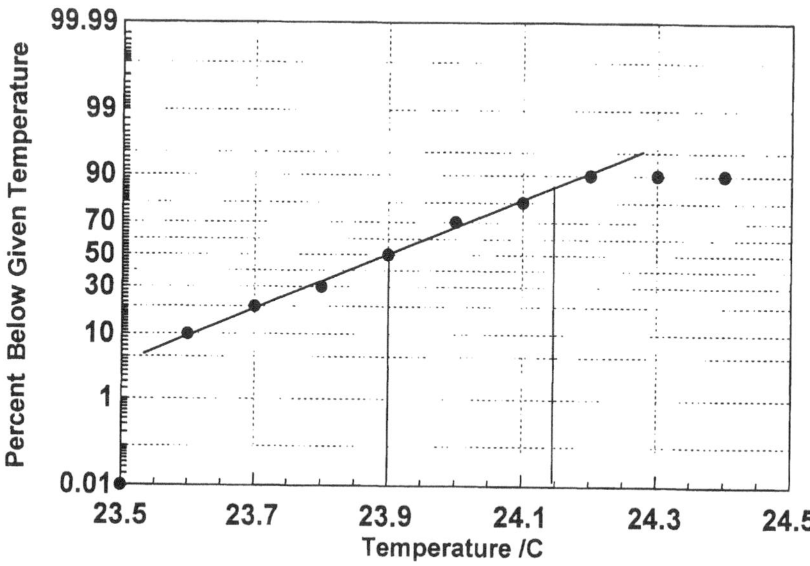

FIGURE 3.5 Linearization of a Gaussian distribution.

In a manner similar to that for diffusion, the spreading of a substance in the atmosphere and oceans is often governed by the seemingly random spreading of a large number of turbulent eddies in these fluids. The spreading of a substance through soils is also often approximated in this way because the soil pores represent a large number of random paths for the substance. For molecular diffusion, the expression of the resulting spread as a Gaussian or normal distribution is an excellent approximation. The approximation is much more limited and crude, however, in transport in soils, the atmosphere, or in bodies of water because of heterogeneity in those systems. All pores are not equally accessible in soils and near the ground or seabed, the intensity and scale of turbulence varies. These processes will be discussed in more detail in later sections of the text.

The Gaussian distribution can also be extended to multidimensional problems (as long as the measurements are Gaussian or normally distributed in both dimensions). The probability of measuring simultaneously a specific set of values x and y is given by

$$P(x,y;\bar{X},\bar{Y},\sigma_x,\sigma_y)=\frac{1}{2\pi\sigma_x\sigma_y}\exp\left(\frac{-(x-\bar{X})^2}{2\sigma_x^2}+\frac{-(x-\bar{Y})^2}{2\sigma_y^2}\right) \tag{3.44}$$

or simply the product of the probabilities of measuring x and y separately.

3.4.2.2 Log-Normal Distribution

Another very important distribution encountered in environmental engineering is the log-normal distribution. This is a modification of the Gaussian distribution in that the *logarithm of the observations* are distributed normally about the *logarithm of the mean.*

$$P_{\ln}(x;\bar{X},\sigma_{\ln})=\frac{1}{x\sigma_{\ln}\sqrt{2\pi}}\exp\left[-\frac{1}{2}\left(\frac{\ln x-\bar{X}_{\ln}}{\sigma_{\ln}}\right)^2\right] \tag{3.45}$$

Particle collectors are often used to remove suspended solids from an effluent air or water stream. The efficiency of these collectors is typically a strong function of the size of the particles. This distribution often describes the particle size distribution and is therefore useful in determining the overall efficiency of a particle collection device.

The most likely value $\bar{X}_{\ln}$ and standard deviation ($\sigma_{\ln}$) about that value are the mean and standard deviation, respectively, of the logarithm of the x measurements. The mean, for example, is

$$\begin{aligned}\bar{X}_{\ln}&=\frac{1}{N}\sum\ln x_i\\&=\frac{1}{N}[\ln x_1+\ln x_2+\ldots+\ln x_N]\\&=\ln(x_1x_2\ldots x_N)^{1/N}\\&=\ln\bar{X}_g\end{aligned} \tag{3.46}$$

$\bar{X}_g$ is the geometric mean which is defined by the Nth root of the product of N values.

$$\bar{X}_g=(x_1x_2x_3\ldots x_N)^{1/N} \tag{3.47}$$

The geometric mean is always smaller than the algebraic mean. If the particle sizes are log-normally distributed, however, more particles would be found in lower sizes then would be suggested by the arithmetic mean of the size range. Thus, the geometric mean is *weighted* to reflect the greater number of particles in the smaller size ranges.

A plot of the probability of measurements below a certain value (cumulative probability) vs. that value on log-probability paper (the log scale equivalent of probability paper) can be used to determine the parameters of the log-normal distribution. The geometric mean is that value corresponding to 50% probability and the geometric standard deviation is the ratio of the 84.1% value to the 50% value or the 50% value to the 15.9% value just as the mean and standard deviation could be estimated from the probability plot for the normal distribution. For a log normal distribution with a given geometric mean and standard deviation the error function can again be used to estimate values of the cumulative probability distribution, now with the form,

$$\text{Cumulative} \quad P_{\ln}\left(x;\bar{X}_g,\sigma_g\right)=\frac{1}{2}+\frac{1}{2}erf\left[\frac{\ln\frac{x}{\bar{X}_g}}{\sqrt{2}\,\ln\sigma_g}\right] \tag{3.48}$$

3.4.3 Propagation of Error

The estimation of any parameter by experimentation always gives rise to some error. This error can result in a greater or lessor error in any subsequently calculated quantity depending on the *sensitivity* of these calculations to the measured quantity. The mean and variance or standard deviation calculated above can be used to estimate the propagation of these errors through subsequent calculations.

If we consider a function f that is dependent upon experimental values of x,y,..., we can relate the value of the function from a single set of estimates x_i,y_i,... to the estimate based on the best or most likely value using $\bar{x}$, $\bar{y}$,... using a Taylor series expansion

$$f(x_i,y_i,...)\approx f(\bar{x},\bar{y},...)+\left(\frac{\partial f}{\partial x}\right)(x_i-\bar{x})+\left(\frac{\partial f}{\partial y}\right)(y_i-\bar{y})+... \tag{3.49}$$

The variance in f is given by

$$\sigma_f^2=\frac{1}{N}\sum\left[f(x_i,y_i,...)-f(\bar{x},\bar{y},...)\right]^2 \tag{3.50}$$

Substituting from Equation 3.49 and expanding the square

$$\sigma_f^2\approx\frac{1}{N}\sum\left[\left(\frac{\partial f}{\partial x}\right)^2(x_i-\bar{x})^2+\left(\frac{\partial f}{\partial y}\right)^2(y_i-\bar{y})^2+\left(\frac{\partial f}{\partial x}\right)\left(\frac{\partial f}{\partial y}\right)(x_i-\bar{x})(y_i-\bar{y})\right]+... \tag{3.51}$$

The partial derivatives reflect the sensitivity of the function f to changes in x or y and, as such, are called sensitivity coefficients. If these are assumed to be approximately constant they can be removed from within the summation sign and the first two terms can then be recognized as σ_x^2 and σ_y^2. The third term can be written in a similar fashion by defining the covariance, σ_{xy}^2,

$$\sigma_{xy}^2 = \frac{1}{N}\sum\left[(x_i - \bar{x})(y_i - \bar{y})\right] \tag{3.52}$$

then Equation 3.51 can be written

$$\sigma_f^2 \approx \sigma_x^2\left(\frac{\partial f}{\partial x}\right)^2 + \sigma_y^2\left(\frac{\partial f}{\partial y}\right)^2 + \sigma_{xy}^2\left(\frac{\partial f}{\partial x}\right)\left(\frac{\partial f}{\partial y}\right) + \ldots \tag{3.53}$$

This final equation provides a means of estimating the variance in the calculated quantity f as a result of the uncertainty, and thus nonzero variances, of the measured quantities, x and y. The ratio of the variances between the calculated quantity and measured quantities are given by the *sensitivity coefficients,* $\partial f/\partial x$ and $\partial f/\partial y$. If it is further assumed that the x and y measurements are completely independent then the product of the errors in x and y that appears in Equation 3.53 will tend to average to zero. The third term in the equation is often neglected.

Let us now examine the propagation of error in some common arithmetic operations. If f = axy, the sensitivity coefficients are

$$\left(\frac{\partial f}{\partial x}\right) = ay \qquad \left(\frac{\partial f}{\partial y}\right) = ax \tag{3.54}$$

and, after dividing by $a^2x^2y^2$, and assuming complete independence of the measurements of x and y, Equation 3.54 can be written

$$\frac{\sigma_f^2}{f^2} \approx \frac{\sigma_x^2}{x^2} + \frac{\sigma_y^2}{y^2} \tag{3.55}$$

Note that this suggests that if there is an estimated relative variance in both x and y of 10% (σ_x/x, $\sigma_y/y = 0.1$), that the expected relative variance in f is about 14%. If f were the product of three independent quantities each with a relative variance of 10%, the expected relative variance in f would be about 17%. The relative variance is usually referred to in this context as the relative uncertainty or relative error. The square root of the variance (without dividing by the magnitude of the variable) would normally be referred to as the absolute uncertainty or absolute error. A more precise definition of error can be obtained, if it is assumed that the distribution of the errors about the most likely value is known. For example, if the distribution of the measured values is assumed to be a Gaussian distribution about the most likely value, it is known that 95% of the measurements should lie with 1.96 standard deviations of the mean or most likely value. Thus, the *95% confidence limit* of the normally distributed variable with $\sigma_x/x = 0.1$, is $\sigma_x/x = 0.196$. That is, 95% of the time a single measurement will be within 19.6% of the mean value. Extending this to the estimation of the uncertainty of the product function f, the 95% confidence limit on f given this same uncertainty in x and y would be 27.7% of its mean value.

An identical result is found if f is formed by the division of two numbers. If f is formed by taking the exponential of x, $f = ae^{bx}$, it is left to an exercise to show that

$$\frac{\sigma_f}{f} \approx b\sigma_x \tag{3.56}$$

Similarly if f = a ln (bx),

$$\sigma_f \approx a\frac{\sigma_x}{x} \tag{3.57}$$

Note that taking the logarithm of a variable (or argument of a function) has the consequence of relating the *relative error* of the argument to the *absolute error* of the function. This has important consequences for the analysis of data manipulated by taking the logarithm (or exponential) that will be discussed in the analysis of curve fitting.

There exists another important consequence of the propagation of error analysis presented herein. If the mean of a set of (supposedly) identical measurements is desired, the error in the mean is given by

$$\begin{aligned}\sigma_f^2 &\approx \frac{1}{N^2}\left[\sigma_{x_1}^2 + \sigma_{x_2}^2 + \sigma_{x_3}^2 + \ldots\right] \\ &\approx \frac{1}{N^2}\left[N\sigma_x^2\right] \\ &\approx \frac{1}{N}\sigma_x^2\end{aligned} \tag{3.58}$$

Thus the variance in the estimate of the mean decreases linearly with the number of samples (or the standard deviation or uncertainty decreases with the square root of the number of samples).

Example 3.13: Propagation of error

The density of a gas at low pressure is proportional to pressure and inversely proportional to *absolute* temperature. Using the estimates of the variance in measurements of temperature and pressure from Example 3.11, estimate the 95% confidence limits in their ratio. Use the variance calculated for the first 9 measurements of temperature and pressure.

From Example 3.11, recognizing that the error in estimating absolute temperature is needed,

$$\frac{\sigma_T}{(\bar{T}+273)} = \frac{0.19}{296.9} = 0.00064 \qquad \frac{\sigma_P}{\bar{P}} = 0.00153$$

$$\begin{aligned}\frac{\sigma_{P/T}}{\bar{P}/\bar{T}} &= \sqrt{\left(\frac{\sigma_T}{\bar{T}}\right)^2 + \left(\frac{\sigma_P}{\bar{P}}\right)^2} \\ &= \sqrt{(0.00064^2 + 0.00153^2)} \\ &= 0.00166\end{aligned}$$

Thus the expected relative error, or normalized standard deviation, in the ratio is 0.166% or slightly higher than the relative error in the estimation of the pressure alone. The 95% confidence limits, assuming that the error is normally distributed about the mean, is twice this amount or ± 0.332%.

3.4.4 Fitting Data

Often it is not possible to arrange an experiment to allow direct measure of a particular parameter of interest. In such situations, the data must be interpreted through a model to allow determination of model parameters or to relate the experimental outcome (measured variables) to the control variables. This process can be conducted in two ways, functional approximation or data fitting. In functional approximation the existing data is assumed to be accurate and the development of a model that both describes this data and interpolates at points in between is desired. In data fitting, the model or functional form is assumed to be valid and the parameters that appear in the model are adjusted until the difference between the model and the data is minimized. Only the latter will be considered here.

The most common way of determining model parameters for this purpose is to minimize the mean square residual or error between model and experiment. Let us first consider fitting data to a straight line, e.g., $y = a_0 + a_1x$, given measurements of y and x. The sum of the square of the error between the measured and predicted value of y at a given value of x is given by

$$\sum e_i^2 = \sum \left[y_i - (a_0 + a_1 x_i)\right]^2 \tag{3.59}$$

The best estimates of a_0 and a_1 can be determined by minimizing the squared error or the variance between the observed and predicted values of y. Differentiating Equation 3.59 with respect to a_0 and setting the result equal to zero yields the so-called "normal" equation,

$$\frac{\partial}{\partial a_0}\sum e_i^2 = 0 = -2\sum \left[y_i - (a_0 + a_i x_i)\right] \tag{3.60}$$

while differentiating with respect to a_1,

$$\frac{\partial}{\partial a_1}\sum e_i^2 = 0 = -2\sum \left[x_i y_i - x_i(a_0 + a_i x_i)\right] \tag{3.61}$$

It is left to an exercise to show that the differentiation and setting equal to zero yields a minimum and not an inflection point or a maximum variance. Rearranging the two equations recognizing that $\Sigma a_0 = a_0N$, yields

$$\begin{aligned} a_0 N + a_1 \sum x_i &= \sum y_i \\ a_0 \sum x_i + a_1 \sum \left(x_i^2\right) &= \sum (y_i x_i) \end{aligned} \tag{3.62}$$

For a given set of data x and y, the sums are known and the value of the unknown a_0 and a_1 can be determined. The results can be written

$$\begin{aligned} a_1 &= \frac{N\sum (x_i y_i) - \sum x_i \sum y_i}{N\sum \left(x_i^2\right) - \left(\sum x_i\right)^2} \\ a_0 &= \bar{y} - a_1 \bar{x} \end{aligned} \tag{3.63}$$

The fitting of coefficients in a linear equation to data is illustrated in Example 3.14.

The approach outlined above can also be used for simple nonlinear functions that can be converted to linear form such as

$$\begin{aligned} f(z) = az^b \quad &\Rightarrow \quad \ln f(z) = \ln a + b \ln z \\ = a + \frac{b}{z} \quad &\Rightarrow \quad f(z) = a + b\left(\frac{1}{z}\right) \\ = a\,e^{bz} \quad &\Rightarrow \quad \ln f(z) = \ln a + bz \\ = \frac{a\,z}{1+bz} \quad &\Rightarrow \quad \frac{1}{f(z)} = a\left(\frac{1}{z}\right) + \frac{b}{a} \end{aligned} \tag{3.64}$$

In every case the equations were converted to a form where there is some function y set equal to $a_0 + a_1$ x for some function x. The definitions of the quantities in the linearized equation are shown in Table 3.5. The first form arises in fitting data to a power law, such as in sorption of a chemical onto a solid where f(z) represents the sorbed quantity and z represents the concentration in the adjacent fluid. The next form is also used to relate sorbed and fluid concentrations and is termed the Langmuir isotherm. Many common physical systems are described by exponential models of the third form, including time-dependent reaction or dilution. The final form might be used to correlate vapor pressure data where f(z) represents the logarithm of vapor pressure and z represents temperature.

TABLE 3.5
Forms of Linearized Functions

	Linearization of Equations			
f(z)	y	x	a_0	a_1
az^b	ln f(z)	ln z	ln a	b
az/(1+ bz)	1/f(z)	1/z	b/a	1/a
ae^{bz}	ln f(z)	z	ln a	b
a + b/z	f(z)	1/z	a	b

Example 3.14: Fitting measured velocity data to the "logarithmic" profile

The change in wind velocity (u) with height above the ground (z) can often be represented by a formula of the form

$$u = \frac{u_*}{\kappa}\left[\ln z - \ln z_0\right]$$

where u_* is termed the friction velocity and is related to the friction at the Earth's surface and z_0 is termed the roughness height and is a correction for the disruption caused by the uneven surface. κ is the von Karman constant and is approximately 0.4. This is a linear equation of the form $y = a_0 + a_1 x$ where y represents the wind velocity and x represents the *natural logarithm of height.* If a_0 and a_1 are fitted to a set of data, the model parameters u_* and z_0 can be determined from the relations

$$u_* = \kappa a_1$$

$$z_0 = \exp\left[\frac{\kappa}{u_*} a_0\right]$$

Given the following data and calculating the required sums in order to estimate these parameters, these parameters can be determined by substitution into the fitting equations.

z, cm	ln z = x	u = y m/s	xy	x^2
3	1.099	0.5	0.5493	1.207
7	1.946	1	1.946	3.787
20	2.996	1.5	4.494	8.974
55	4.007	2	8.015	16.059
148	4.997	2.5	12.493	24.972
Sums	15.045	7.5	27.50	55.00

$$a_1 = \frac{(5)(27.5\ \text{m/s}) - (15.05)(7.5\ \text{m/s})}{\text{N}(55.00) - 15.05^2} = 0.5067\ \text{m/s}$$

$$a_0 = \frac{1}{N}\left(7.5\ \text{m/s} - a_1 15.05\right) = -0.0245$$

Therefore, $u_* = (0.4)(0.5067\ \text{m/s}) = 0.203\ \text{m/s}$, and $z_0 = \exp[(0.4)(-0.0245)/(0.5067) = 0.953\text{cm}$. Note that the units are the same as those which were measured and which were used to generate the table. Figure 3.6 indicates the appropriateness of the form of the fitting equation.

3.4.4.1 Linearization and Errors

Let us illustrate the linearization process with the Langmuir model.

$$f(z) = \frac{a z}{1 + bz}$$

$$\frac{1}{f(z)} = \frac{1 + bz}{a z} \tag{3.65}$$

$$\frac{1}{f(z)} = \frac{b}{a} + \left(\frac{1}{a}\right)\left(\frac{1}{z}\right)$$

or

$$y = a_0 + a_1 x$$

A fit of the inverse of the data expected to fit a Langmuir model will yield a slope that defines the inverse of a and an intercept that defines the ratio of b/a.

Although algebraic manipulation may result in linearization of the function to be fitted to data, the linearization process influences the way errors in the data are reflected in the fitted parameters. This is most easily seen in the common practice of using the logarithm to linearize power law or exponential functions. In the linear regression process described here, all data points are valued, or weighted, equally. The x or independent measurements are implicitly assumed to be exact and the data for the dependent function (y) are assumed to contain all of the error. The residual error, $\Delta y = y_i - (a_0 + a_1\ x)$, between the observed and predicted (or fitted) values in the dependent variable (y) is assumed the same for all values of the independent variable, x. This is equivalent to assuming that the *absolute error* for each of these data is assumed approximately equal. If, however, the logarithm of both sides of the fitting formula is taken, the linear regression process implicitly assumes that $\Delta(\ln y)$ at each data point is treated equally. For small differences between the observed and predicted, the difference, Δ, is essentially identical to taking the differential and therefore

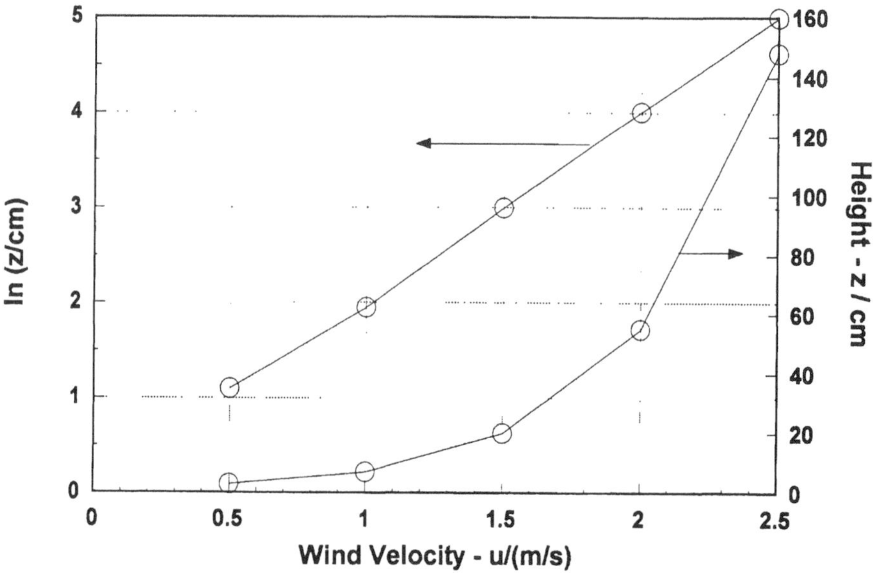

FIGURE 3.6 Plot of fitted velocity vs. height data from Example 3.14.

TABLE 3.6
Relative vs. Absolute Error

	Range of Measured Pressures (mb)	
True Pressure	**±5% relative error ($\Delta P/P = \pm 0.05$)**	**±50 mb absolute error ($\Delta P = \pm 50$ mb)**
1000 mb	950–1050	950–1050
500 mb	475-525	450–550
100 mb	95–105	50–150
50 mb	47.5-52.5	0–100

$$\Delta(\ln y) \approx \frac{\Delta y}{y} \tag{3.66}$$

Thus, while the fitting process treats Δy, or the absolute error, at each point equally, fitting the logarithm of the original function treats Δy/y, or the relative error, at each point equally. This effectively increases the weight associated with small values of y. If y represented pressure, for example, a constant relative error, Δy/y, implies that the absolute error of lower pressures is less than at higher pressures. This is illustrated in Table 3.6.

The effect of this behavior on fitting data is illustrated in Example 3.15. When the fitting procedure described above is employed directly on a function, the process serves to minimize the errors between the linearized function and the data and not the actual function. This is demonstrated in Figures 3.7 and 3.8 which plot the data of Example 3.15 in linearized and normal coordinates, respectively. The linear fit to the logarithm of the concentrations fits the entire range of data points as shown in Figure 3.7. Replotting in concentration vs. time coordinates as in Figure 3.8 shows that the linearized fit describes the low concentration data at the expense of the short time, high concentration data. The nonlinear fit in both figures is a direct approach that minimizes the error

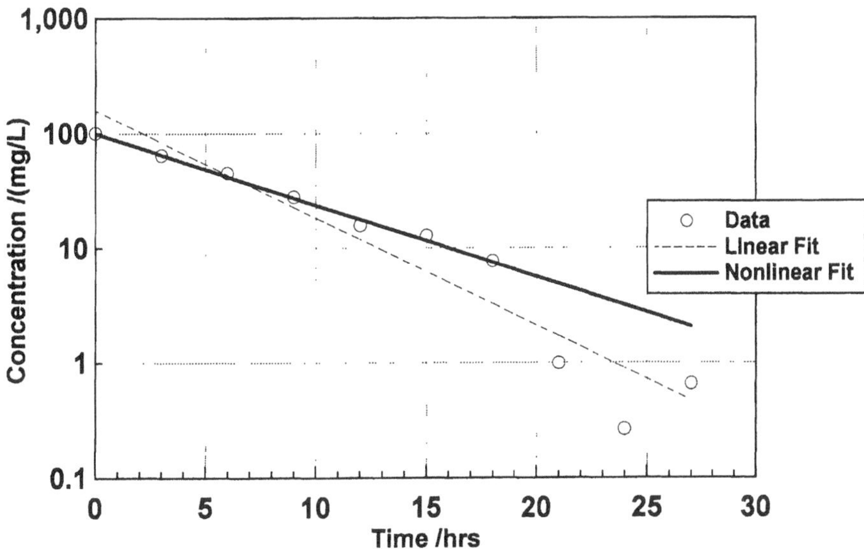

FIGURE 3.7 Plot of concentration vs. time data in Example 3.15 in linearized coordinates.

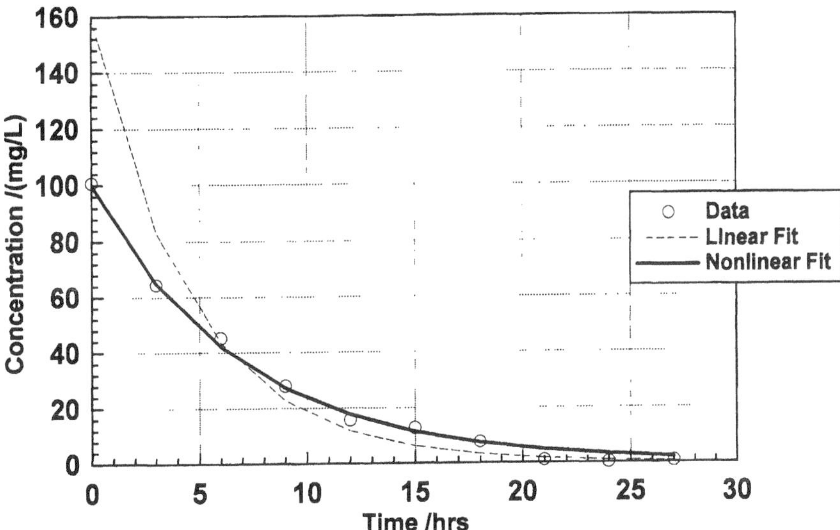

FIGURE 3.8 Plot of concentration vs. time data in Example 3.15 in normal coordinates.

in fitting the original exponential function. It describes the concentration vs. time data well, but it appears to be a poorer fit on the linearized plot (Figure 3.7).

Example 3.15: Fitting a linearized equation

The chemical degradation of a chemical was monitored in a laboratory cell using two different instruments simultaneously. A detailed analysis of the two instruments suggested that the first was capable of measuring the chemical concentration of the chemical with a standard deviation of 10% (relative error independent of concentration). The second instrument was capable of measuring the chemical concentration with a standard deviation of 2 mg/L (absolute error independent of concentration). The measured chemical concentrations with time, starting from an actual initial concentration of 100 mg/L, were

	Measured Concentration (mg/L)									
Instrument	0	3 h	6 h	9 h	12 h	15 h	18 h	21h	24 h	27 h
σ_c/C	114	60.8	46.6	27.3	15.3	13.8	7.78	4.95	2.87	2.09
σ_c	101	64.5	45.2	28.0	15.7	12.8	7.68	0.99	0.26	0.66

The chemical degradation is expected to be first order (and the data were, in fact generated from such a model with error introduced by random selection from a normal distribution with the standard deviations specified above). That is the model to be fitted is

$$C(t) = C_0 e^{-kt}$$

or

$$\ln C(t) = \ln C_0 - kt$$

The coefficients in the linearized equation can be determined as outlined previously. A nonlinear fit also was conducted that minimized the absolute errors between the data and the model directly. The results including error compared to the equation used to generate the data:

Instrument	$(C_0)_{est}$	k_{est}	% error (t = 0)	% error (t = 24 h)
Uniform relative error	105 mg/L	0.146 h^{-1}	5	–9.8
Uniform absolute error	159 mg/L	0.216 h^{-1}	59	–74.5
Nonlinear fit	99.95 mg/L	0.144 h^{-1}	–0.05	–9.2

The results of the fit are shown in Figures 3.7 in linearized form and Figure 3.8 in normal coordinates. Note that the linearized form fits the data better in the linearized coordinates, but in normal coordinates, the direct nonlinear fit is better.

3.4.4.2 General Linear Regression

The linear regression approach can be used to fit data to a much broader set of functions than those described above. Any function of the form

$$y = a_0\varphi_0(x) + a_1\varphi_1(x) + a_2\varphi_2(x) + \ldots \tag{3.67}$$

can be fit to data employing this approach. The key feature of this equation is that it is linear in the *coefficients*, a_1, a_2, etc. Thus, when partial derivatives are taken in the derivation of the normal equations during the application of the least square fitting approach, the coefficients appear only linearly. For example, differentiating with respect to some coefficient a_j yields

$$\frac{\partial}{\partial a_j}\sum e_i^2 = 0 = -2\sum \varphi_j[y_i - (a_0\varphi_0 + a_1\varphi_1 + \ldots)] \tag{3.68}$$

Each of the generated normal equations are of this form and if the equation has J terms with therefore J coefficients that need to be adjusted to fit the data, there will be J equations available for their determination. The general form of these equations can be written

$$\begin{bmatrix} \sum \varphi_0\varphi_0 \sum \varphi_0\varphi_1 \cdots \sum \varphi_0\varphi_J \\ \sum \varphi_1\varphi_0 \sum \varphi_1\varphi_1 \cdots \sum \varphi_1\varphi_J \\ \mathrm{M} \qquad\qquad \mathrm{M} \\ \sum \varphi_J\varphi_0 \sum \varphi_J\varphi_1 \cdots \sum \varphi_J\varphi_J \end{bmatrix} \begin{Bmatrix} a_o \\ a_1 \\ \mathrm{M} \\ a_J \end{Bmatrix} = \begin{Bmatrix} \sum y_i\varphi_0 \\ \sum y_i\varphi_1 \\ \mathrm{M} \\ \sum y_i\varphi_J \end{Bmatrix} \tag{3.69}$$

The solution of linear equations of this form is Gaussian elimination (no relation to the Gaussian distribution). The method converts the square matrix on the left-hand side to upper triangular form. At that point the bottom equation can be solved directly, and then via back substitution into the equations above, the entire system of equations can be solved. The procedure is detailed in any basic text on linear algebra.

Numerically this process can pose significant difficulties. There are no limitations on the form of the φ functions and they might include polynomials, trigonometric functions, or exponential or logarithmic functions. If the fitting equation is a polynomial of the form $y = a_0 + a_1x + a_2x^2 + ... + a_Jx^J$, the matrix is near singular. A singular matrix of coefficients is one whose determinant is zero, which implies that one or more of the equations that generated the matrix is not independent. A singular matrix cannot be solved and the solution with a near-singular coefficient matrix is very sensitive to small errors in the evaluation of the coefficient matrix. The net result is that even 6 or 7 digits of accuracy that are normally carried with single precision numbers on a modern digital computer are sometimes insufficient to achieve accurate results with J larger than 3 or 4. The fitted equation will adequately represent the data in such a situation, but small changes in the experimental measurements of the values of x will drastically change the estimated values of the fitted coefficients (the a's). This implies that this approach cannot be relied upon to make any physical interpretation on the values of the fitted coefficients.

This can be avoided by using σs which are orthogonal, that is when

$$\sum \varphi_i\varphi_j = \begin{cases} 0 & i \neq j \\ \neq 0 & i - j \end{cases} \tag{3.70}$$

In this situation, the equations for each of the a's are all independent of each other. Only the diagonal elements of the square matrix in Equation 3.69 are nonzero.

$$\begin{bmatrix} \sum \varphi_0\varphi_0 & 0 & \mathrm{K} & 0 \\ 0 & \sum \varphi_1\varphi_1 & \mathrm{K} & 0 \\ \mathrm{M} & & & \mathrm{M} \\ 0 & 0 & \mathrm{K} & \sum \varphi_J\varphi_J \end{bmatrix} \begin{Bmatrix} a_0 \\ a_1 \\ \mathrm{M} \\ a_J \end{Bmatrix} = \begin{Bmatrix} \sum y_i\varphi_0 \\ \sum y_i\varphi_1 \\ \mathrm{M} \\ \sum y_i\varphi_J \end{Bmatrix} \tag{3.71}$$

The solution for a_0, for example, can then be written directly as

$$a_0 = \frac{\sum y_i\varphi_0}{\sum \varphi_0^2} \tag{3.72}$$

The solution for the other coefficients is also immediate and not subject to the numerical errors introduced by insufficient precision. A set of orthogonal polynomials can be generated for any set of data, but it is beyond the scope of this text to discuss the process in detail. The reader is referred to the text by Conte and deBoor (1980) for further information on these procedures.

PROBLEMS

1. Convert the following quantities in common US units to SI units: Wind speed = 10 mi/h, Pressure = 14.8 lb_f/in^2, Temperature = –40°F, Weight = 150 lb_f, Mass = 150 lb_m, and Volume = 16 gal.

2. Find the following: (a) the acceleration when 5 lb_f acts on 5 lb_m, (b) the energy accumulated in the object in (a) after the force has acted for a distance of 5 ft, (c) the weight of an object with a mass of 5 lb_m located at a height where the acceleration of gravity is 30 ft/s², (d) the potential energy of the object in (c), (e) the kinetic energy of the object in (c) if the object is moving 1000 mi/h.

3. Find the kinetic energy of a ton of water moving at 60 mi/h expressed in ft·lb_f, joules, hp·s, L·atm, and ergs.

4. A rarely used system of units is a system that defines the kg-mass, meter, second, **and** kg-force as the fundamental units. What is the value of g_c in this system of units?

5. The frictional pressure drop for fluids flowing in a pipe is

$$\Delta p = \frac{2fL\rho v^2}{D}$$

where Δp is the pressure drop, f is the friction factor, L is the length of pipe, ρ is the density of the flowing fluid, v is the velocity of the fluid, and D is the diameter of the pipe. What are the units of the friction factor for dimensional consistency? Check dimensional consistency in both American Engineering units and SI units.

6. What is the likely definition of the following quantities: 20 MBtu, 20 MMBtu, 20 kJ, 20 MJ, and 20 MW?

7. The Reynolds number relates inertial to viscous forces in flowing fluids and is defined by

$$N_{Re} = \frac{\rho v D}{\mu}$$

where μ represents the fluid viscosity and the other terms are as defined in Question 4. Using the fact that the Reynolds number is dimensionless, answer the following problems. What are the units of viscosity in AEUs? What are the units of viscosity in SI units? What are the units of viscosity in cgs units?

8. A vessel has a volume of 22.4 L. At 0°C, how many moles of a gas can be contained in the vessel? At 25°C, how many moles of a gas can be contained? If the vessel is filled and sealed at 0°C and then heated to 25°C, how many moles of a gas are present?

9. If the vessel in problem 9 contains air at 0°C, how many moles of nitrogen does the vessel contain? How many moles of oxygen does the vessel contain? What is the mass of air in the vessel?

10. 26.5 g of sodium chloride and 88.1 g of water make a solution of 100 g of water. Determine the following quantities: mass concentration of sodium chloride (g/L), molarity of the solution, molality of the solution, average molecular weight of the solution, mole fraction water, and mole fraction sodium chloride.

11. Consider water that is 5 mol% sodium chloride (NaCl). Determine the mole ratio of NaCl to water. What is the error in estimating mole ratio as the mole fraction? What is the average molecular weight of the solution? Determine the weight ratio (weight NaCl to weight water) and concentration (weight NaCl to volume water) in the solution.

12. The concentration of fluoride in a drinking water system is to be held at 1.6 mg/L. How many grams of sodium fluoride (NaF) must be added to achieve this concentration in 1 million gallons of water?

13. Determine the amount (in grams) of oxygen required in the following combustion reactions: combustion of 1 mole of octane (C_8H_{18}), combustion of 100 g of octane, and combustion of 100 g of octane with 10% excess air.

14. Determine the volume of gaseous products (including nitrogen) formed in the reactions of Problem 13 at 1 atm pressure after cooling to 25°C.

15. Calculate the pressure at the base of a 33-ft column of water in atm, lb_f/in^2, and Pascals.

16. One leg of a U-tube manometer is connected to the combustion chamber of an incinerator while the other is open to the atmosphere. The water level in the leg attached to the combustion chamber is 25 cm higher than that exposed to the atmosphere. What is the vacuum in the combustion chamber in Pascals and atmospheres?

17. One leg of a U-tube manometer is sealed effectively with a total vacuum while the other is exposed to the atmosphere. The fluid in the manometer is mercury with a specific gravity of 13.7. What is the difference in height of mercury between the two legs?

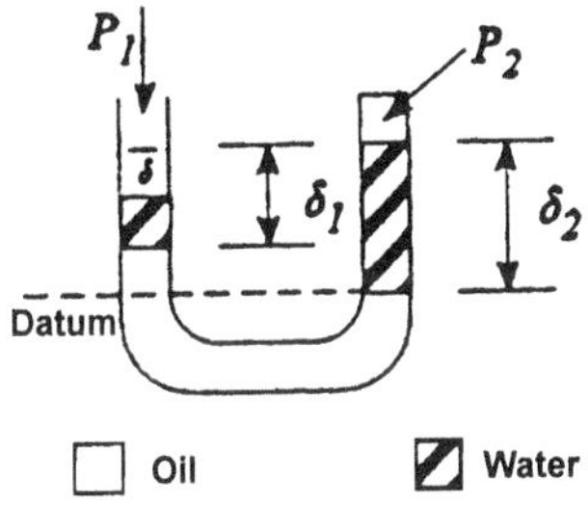

18. Consider the manometer in Problem 17. The oil has a density of 1.5 g/cm³. The height of the water columns in a leg exposed to the atmosphere is 5 cm while the height of the water column in the other leg is 15 cm. The pressure difference between the two legs, P_1-P_2, is 100 Pa. Find the difference in liquid height between the two legs, δ.

19. An airplane door is designed such that it must be opened inward. What force is required to open the door at altitude when the outside pressure might be 0.25 atm and the inside pressure 0.75 atm? The door is 1 m high by 50 cm wide.

20. What is the pressure and force acting on a 2 m × 2 m gate at the base of a dam? The water level behind the gate is 20 m while no water is present outside the gate.

21. Show that the drag force on a boat should be described by an equation of the form of

$$\frac{F}{\rho V^2 L^2} = f(N_{Re}, N_{Fr})$$

$$N_{Re} = \frac{\rho V L}{\mu}$$

$$N_{Fr} = \frac{V^2}{Lg}$$

where F is the drag force, ρ is the density of water, V is the velocity of the boat, L is the length of the boat, μ is the viscosity of water, and g is the acceleration of gravity. N_{Re} is the Reynolds number and N_{Fr} is the Froude number.

22. Experiments reveal that most of the resistance to surface ship motion is controlled by the Froude number. Thus any attempt to conduct scale model experiments of a ship in a towing tank should match the Froude number between the model and the ship. Is it possible to match both the Reynolds number and Froude number in an experiment with a model boat at 1:100 full scale?

23. The power input (P) to a surface aerator at a wastewater treatment pond is dependent upon the viscosity (μ) and density (ρ) of the water, the angular velocity (ω), diameter (d) of the aerator, and the gravitational acceleration at the air-water interface. Show that the relationship can be written,

$$\frac{P}{\rho\omega^3 d^5} = f(N_{\mathrm{Re}}, N_{\mathrm{Fr}})$$

24. Show that the mass transfer of oxygen across a surface aerator should follow

$$\frac{k_a d}{D_a} = f\left(\frac{P}{\rho\omega^3 d^5}, N_{\mathrm{Re}}, N_{\mathrm{Fr}}\right)$$

where, in addition to the terms defined in Problem 23, k_a is the mass transfer coefficient (length/time), and D_a is the diffusion coefficient of oxygen in air (length^2/time).

25. Calculate the mean and standard deviation of the following replicate measurements of pressure (0.982, 0.936, 0.928, 0.948, 1.056, 0.976, 1.009, 1.126, 0.929). All in atm.

26. Determine if the pressures in Problem 25 are normally distributed. What is the 95% confidence limits of the mean pressure?

27. Calculate the mean and standard deviation of the following measured PAH concentrations in a set of replicate sediment samples (41, 57, 104, 88, 76, 102, 115, 106, 100, 85). All in mg/kg.

28. Determine if the concentrations in Problem 27 are normally distributed. What is the 95% confidence limit on the concentration in the sediment?

29. Compare the normal and log-normal distributions as a description of the following PAH concentration data, also collected in a sediment (104, 111, 118, 120, 120, 105, 87, 110, 97, 84). All in mg/kg.

30. The mean and standard deviation of 10 measurements of the concentration of pyrene in a sediment were 10 mg/kg and 5 mg/kg, respectively. Assuming the applicability of a Gaussian distribution, could you state, with 95% confidence, that the sediment concentration is within what range?

31. The Reynolds number is given by $\rho Vd/\mu$. What is the estimated error in the Reynolds number under the following conditions: all parameters that make up the Reynolds number have an estimated error of ±10%, all parameters that make up the Reynolds number have an estimated error of ±10% except viscosity which has an estimated error of ±50%, and all parameters take on the following values, $\rho = 1 \pm 0.01$ g/cm^3; $V = 2 \pm 0.5$ m/s; $d = 10 \pm 0.01$ cm; and $\mu = 1 \pm 0.5$ cp?

32. The following concentration measurements were collected at the times indicated in a laboratory experiment. Are they data well represented by a linear relationship with time? What is the best-fit curve and correlation coefficient?

Time (min)	0	1	2	3	4	5
Concentration (mg/L)	0.884	4.32	7.91	11.6	13.6	16.8

33. The following concentration data were collected in a laboratory reaction with two different instruments. The first instrument exhibits an estimated relative error over the measured concentration range of ±25% while the second instrument exhibits an estimated absolute error of about ±20 mg/L. Fit the data to a first order reaction

$$C = C_0\, e^{-kt}$$

What are the best-fit values of k and C_0 estimated from the data collected from each instrument? Are the deviations from the best-fit curves consistent with the estimated errors of each instrument?

Time (h)	0	3	6	9	12	15	18	21	24	27
Instrument 1 (mg/L)	153	145	113	62	38	64	38	45	41	22
Instrument 2 (mg/L)	119	75	77	113	64	87	39	34	20	12

34. After a period of heavy rains, the river elevation data listed below was collected. You expect that the data represents a near linear rate of increase with a sinusoidal perturbation associated with the tides that have a 12-h period beginning at midnight. You thus expect the data to be fit by a function of the form

$$\text{Elevation} = a_0 + a_1 t + s_2 \sin\left(\frac{2\pi t}{12}\right) \qquad t \text{ in h}$$

Fit the data first to a linear curve and then to the full equation. Does your expectation accurately fit the data?

T,h (AM)	0	1	2	3	4	5	6	7	8	9	10	11
Elev (dm)	10.1	10.9	11.9	12.5	12.9	13.1	12.9	13	13.1	13.5	14.1	15
T,h (PM)	12	13	14	15	16	17	18	19	20	21	22	23
Elev (dm)	16.2	17	17.9	18.5	18.8	19	19.1	19	19.1	19.5	20.1	20.9

REFERENCES

Batchelor, G.K. (1970) *An Introduction to Fluid Dynamics,* Cambridge University Press, Cambridge, U.K., 615 pp.

Conte, S.D. and C. de Boor (1980) *Elementary Numerical Analysis: An Algorithmic Approach, 3rd ed.,* McGraw-Hill, New York, 432 pp.

Felder, R.M. and R.W. Rousseau (1986) *Elementary Principles of Chemical Processes, 2nd ed.,* John Wiley & Sons, New York, 571 pp.

Fox, R. W. and A.T. McDonald (1978) *Introduction to Fluid Mechanics, 3rd ed.,* John Wiley & Sons, New York, 684 pp.

Himmelblau, D.M. (1974) *Basic Principles and Calculations in Chemical Engineering, 3rd ed.,* Prentice-Hall, Inc., Upper Saddle River, NJ, 542 pp.

4 Physical and Chemical Equilibrium

A system that is at equilibrium is one whose behavior does not change with time. The forces that shape such a system are in balance, usually as a result of allowing a sufficient period of time such that deviations from equilibrium have dissipated. Of course, a system could be unstable such that there exists no steady or simple equilibrium condition. Even a system for which an equilibrium state exists, however, its attainment is generally impossible in that it requires an infinite time and maintenance of steady external conditions. Although true equilibrium may never be reached, a near-equilibrium state may be an excellent approximation to a real system. In addition, the deviation from equilibrium can be used to indicate the direction toward which the system is moving. Thus, the study of equilibrium is important and useful for both the understanding of environmental systems and environmental control and treatment systems. The quantitative description of equilibrium is the subject of this chapter.

4.1 PHASE BEHAVIOR

Central to the understanding of the equilibrium behavior of systems is assessment of their physical state and the behavior of that state. The environmental engineer has interest in contaminants in any of the three states of matter (solid, liquid, and gas) and the behavior of each of these phases will be discussed in turn.

4.1.1 PURE IDEAL GASES

It has long been recognized that gases under near ambient conditions exhibit simple relationships between volume and pressure and temperature. Jacques Charles indicated that a gas increases its volume in direct proportion to temperature in 1787. Boyle subsequently demonstrated that the volume of a gas is inversely proportional to pressure. Both of these relationships are exhibited by the ideal gas law

$$PV = nRT \tag{4.1}$$

where n represents the number of moles of the gas and P, V and T represent the pressure, volume and temperature, respectively. As stated in Chapter 3, a mole is the amount of gas that equals its molecular weight in grams. Volume is an *extensive* variable in that it depends on the quantity of gas present. It is often more convenient to work in *intensive* variables, that is, variables that do not depend on the quantity of the substance. Thus specific volume, or the volume per mole ($\hat{V} = v/n$), would be used instead of the total volume V. The caret will be used to indicate specific quantities per mole. R is the ideal gas constant which could in principle be determined from a single measurement of the *state* variables: pressure, specific volume, and temperature.

Since R is assumed to be a constant, the specific volume of gas is a constant at a given temperature and pressure. A suitable reference temperature and pressure (Standard Temperature and Pressure —STP) has been selected as 0°C and 1 atmosphere pressure. Under these conditions the specific volume of any gas that conforms to the ideal gas law is 22.4 L/mol. This is used to estimate the value of R in Example 4.1.

Example 4.1: Values of the ideal gas constant

Determine the ideal gas constant assuming that a gas at 0°C and 1 atm pressure fills a volume of 22.4 L.

Ideal gas law:

$$P\hat{V} = RT$$

or

$$R = \frac{P\hat{V}}{T}$$

Therefore,

$$R = \frac{(1\ \text{atm})(22.4\ \text{L/mol})}{273\ \text{K}}$$

$$= 0.08205 \frac{\text{atm} \cdot \text{L}}{\text{mol} \cdot \text{L}}$$

In other units,

$$R = \frac{760\ \text{mmHg}}{1\ \text{atm}} = 62.36 \frac{\text{mmHg} \cdot \text{L}}{\text{mol} \cdot \text{K}}$$

$$R \cdot \frac{1.013(10)^5\, Pa}{1\ \text{atm}} = 8.314 \frac{Pa \cdot \text{m}^3}{\text{mol} \cdot \text{K}}$$

$$= 8.314 \frac{\text{kg} \cdot \text{m}^2}{s^2 \cdot \text{K}}$$

$$= 8.314 \frac{\text{J}}{\text{molK}}$$

$$R \cdot \frac{2116\ \text{lb}_f/\text{ft}^2}{\text{atm}} \frac{454\ \text{mol}}{\text{lbmol}} \frac{K}{1.8°R} \left(\frac{3.28\ \text{ft}^3}{1\ \text{m}} \right) = 1547 \frac{\text{lb}_f/\text{ft}^2 \cdot \text{ft}^3}{\text{lbmol} \cdot °R}$$

Although R is given as a universal constant for all gases, it should be recognized that this is an approximation that is asymptotically valid as the pressure goes to zero and the temperature goes to infinity. Under these conditions, the average distance between molecules is sufficiently great that the intermolecular forces can be neglected and the interactions are essentially that of rare, nearly elastic collisions between hard spheres. Although technically only valid under these limiting conditions, the assumption of an ideal gas is an excellent assumption for gases under ambient conditions, that is at 1 atm pressure and temperatures in the range of −20 to +40°C.

The specific volume of the gas at any temperature (T_2) or pressure (P_2) can be determined from the ideal gas law or from the known specific volume at any other temperature (T_1) and pressure (P_1) (e.g., the known specific volume at STP) via the relationship

$$\hat{V}_2 = \hat{V}_1 \left(\frac{P_1}{P_2}\right)\left(\frac{T_2}{T_1}\right) \tag{4.2}$$

Note that the absolute (not gauge) pressure and the absolute temperature must be used in both Equations 4.1 and 4.2. Thus, the difference in density of air between 32 F and 77 F (0 and 25°C) changes 9.2% (not 240%) to about 24.5 L/mol.

Example 4.2: Estimation of molar volume at 25°C

Estimate the molar volume of air at 1 atm pressure and 25°C

Using the ideal gas law

$$\hat{V} = \frac{RT}{P}$$

$$= \frac{0.08205 \dfrac{\text{atm}\cdot\text{L}}{\text{mol}\cdot\text{K}} \cdot 298\ \text{K}}{1\ \text{atm}}$$

$$= 24.45\ \text{L/mol}$$

Using Equation 4.2

$$\hat{V} = (22.4\ \text{L/mol})\left(\frac{298\ \text{K}}{273\ \text{K}}\right)$$

$$= 24.45\ \text{L/mol}$$

Note that although the ideal gas law suggests that any gas will exhibit the same *molar* volume at a given temperature and pressure, the mass of that volume will be different because of the different molecular weights of gases. This is important in that concentrated gases with a molecular weight different from air (MW = 29) will exhibit a mass density different from that of air. A dense gas such as methyl bromide (MW = 95) will be heavier and tend to remain close to the ground, while a lighter than air gas such as methane (MW = 16) will tend to move upward. These comments assume that both gases are at the same temperature. If large amounts of methane were released to the atmosphere from liquefied natural gas (LNG), for example, the cooling during the evaporation of the LNG would tend to keep the methane cool and, by the ideal gas law, more dense than that of air. This could have unfortunate consequences in that the cool methane cloud would not mix vertically and could travel large distances to an ignition source before sufficient dispersal would occur to avoid an explosion.

A measure of the relative densities of gases is the specific gravity. Because the molar volume of two gases (ideal) would be identical, the relative densities or specific gravity would be just equal to the ratio of the molecular weights (Equation 3.9), assuming that both the gas and air (or other reference gas) both refer to the same temperature and pressure conditions.

4.1.2 Non-Ideal Gases

Although the ideal gas law is valid under typical environmental conditions, it is unlikely to be valid with gases under high pressure and for gases at low temperature such as the vapors above liquid nitrogen or oxygen. A basis for correction of the ideal gas law for low temperatures and high pressures can be found in the *law of corresponding states*. The law of corresponding states assumes that all substances exhibit similar state behavior at the critical point. The *critical point* is defined as the temperature and pressure at which the physical properties of the gaseous and liquid states are identical. At temperatures and pressures below the critical point, both a liquid and a gaseous phase can co-exist. Above the critical temperature and pressure, however, the phases are indistinguishable. The density of water and water vapor near the critical point is shown in Figure 4.1. Note that at 3200 psia (218 atm), the density of the gas and liquid phases of water approach the same value. This is the critical point. Example 4.3 examines the ideal gas law at the water critical point.

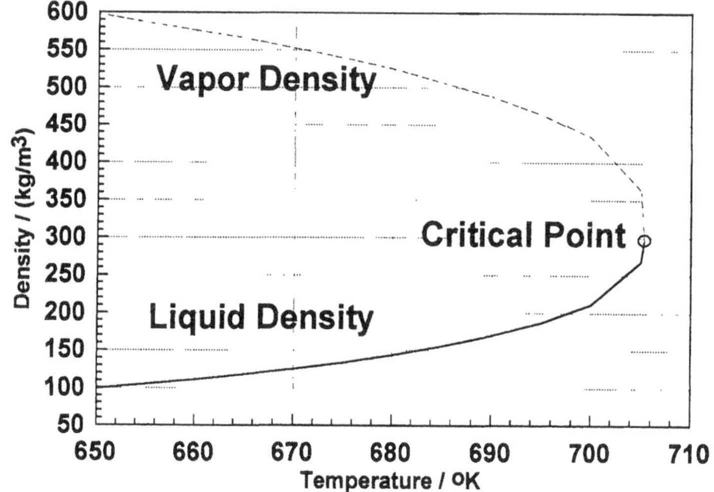

FIGURE 4.1 Water vapor and liquid density near the critical point.

Example 4.3: Ideal gases at the critical point

Evaluate the ideal gas law at the critical point of water ($T_c = 647$ K, $P_c = 218$ atm, and $V_c = 0.0608$ L/mol).

Evaluate the ratio of $P\hat{V}$ to RT (For the ideal gas law to hold this should equal unity)

$$\frac{P_c\hat{V}_c}{RT_c} = \frac{(218\ \text{atm})(0.0608\ \text{L/mol})}{\left(0.08205\,\frac{\text{atm}\cdot\text{L}}{\text{mol}\cdot\text{K}}\right)(647\ \text{K})}$$

$$= 0.25$$

Clearly, the ideal gas law does not hold near the critical point of water.

Although the ideal gas law is not satisfied at the critical point, an examination of a variety of gases shows that there is very little variation in the ratio of $P_c\ \hat{V}_c$ to RT_c. This is illustrated in Table 4.1. The average value of

$$\frac{P_c\hat{V}_c}{RT_c}$$

is 0.266 with a standard deviation of 0.02. This suggests that many real gases fill a volume only about 27% of that which would be expected of an ideal gas at their critical point. The approximate

TABLE 4.1
Critical Constants of Ratio of $P_c\hat{V}_c$ to RT_c

Substance	P_c (atm)	T_c (K)	V_c (L/mol)	$\frac{P_c\hat{V}_c}{(RT)_c}$
Water, H_2O	218	647	0.0608	0.23
Ammonia, NH_3	111	406	0.0725	0.243
Butane, C_4H_{10}	37.5	425	0.255	0.274
Carbon dioxide, CO_2	72.9	304	0.094	0.275
Hexane, C_6H_{14}	29.9	508	0.368	0.264
Nitrogen, N_2	33.5	126	0.090	0.291
Oxygen, O_2	49.7	154	0.074	0.290
Sulfur dioxide, SO_2	77.8	431	0.122	0.269
o-Xylene, C_8H_{10}	35.7	632	0.380	0.26
Range and average	33.5–218	126–647	0.06–0.255	0.266
				$\sigma = 0.02$

constancy of this value suggests that by relating conditions to the critical point an improved relationship between pressure, volume, and temperature for that gas can be achieved. This is the law of corresponding states.

If the temperature, pressure, and specific volume at the critical point are denoted by T_c, P_c, and $\hat{V}_c$, then a reduced temperature, pressure, and specific volume can be defined for any conditions as

$$T_r = \frac{T}{T_c}$$

$$P_r = \frac{P}{P_c} \tag{4.3}$$

$$\hat{V}_r = \frac{\hat{V}}{\hat{V}_c}$$

The concept of the law of corresponding states suggests that any gas will exhibit similar behavior at the same values of the reduced conditions. This has been incorporated into the ideal gas law by using a compressibility correction factor, z

$$P\hat{V} = zRT \tag{4.4}$$

The correction factor z (termed the compressibility factor) has been correlated with reduced temperature and pressure and the correlation has been found to describe essentially all gases equally well. As indicated by Table 4.1, the average value of z is about 0.27 at the critical point. At atmospheric pressure and ambient temperatures, when the ideal gas law is valid, its value is near 1. Figure 4.2 displays this correlation of compressibility factor over the range of reduced temperature between 1 and 3.5 and reduced pressure from 0 to 10. Note that as long as the reduced temperature is greater than 1.5 and the reduced pressure less than 1, the compressibility factor is less than 10% different from 1 and the gas behaves as an ideal gas. Thus, the criteria for the ideal gas law (within about 8%) is

$$\text{ideal gases: } \frac{T}{T_c} \geq 1.5 \text{ and } \frac{P}{P_c} \leq 1 \tag{4.5}$$

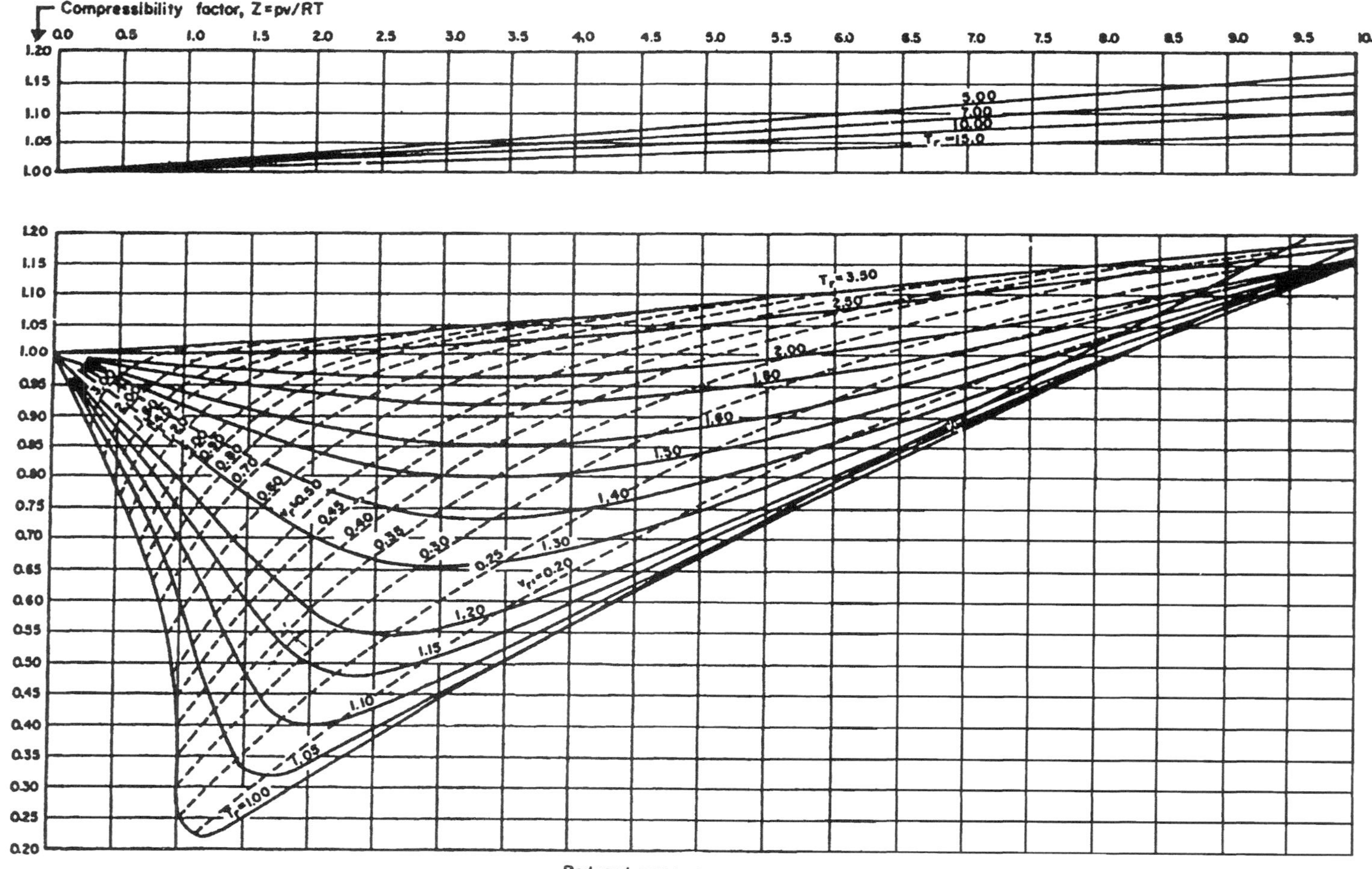

FIGURE 4.2 General compressibility chart, medium pressures. (From Himmelblau, D.M. (1974) *Basic Principles and Calculations in Chemical Engineering* 3rd ed., Prentice Hall, New York. With permission.)

It should be emphasized that this discussion is limited to a pure gas or, if a mixture, the properties of the bulk gas apply. For trace contaminants in air, the effect on the critical properties of the air are essentially negligible and the reduced temperature and pressure for the bulk air phase can be used to define the compressibility factor. The critical temperature for air is about 132.5 K and the critical pressure is 37.2 atm. Air behaves essentially as an ideal gas for pressures up to 37.2 atm and temperatures down to 199 K (–74°C). As indicated in Table 4.2, air is clearly well approximated as an ideal gas at standard temperature and pressure ($T/T_c = 2.06$, $P/P_c = 0.027$).

TABLE 4.2
Compressibility Factor of Air

T (K)	T_r	P (atm)	P_c	Z	% error (from ideal)
199	1.5	37.2	1	0.9	11
273	2.06	37.2	1	0.98	2
273	2.06	1	0.027	0.999	0.1

In the open environment, z can always be taken to be equal to unity and the atmosphere is ideal. Only in pressurized vessels or at low temperatures does the compressibility factor need to be considered. This is illustrated in Example 4.4.

Example 4.4: Use of compressibility factor to determine pressure in a tank

A 140-L compressed air tank is filled to a pressure of 170 atm at a temperature of 25°C. Determine the mass of air in the tank. If the tank's temperature rises to 75°C, will the rated pressure of 200 atm be exceeded?

The critical temperature of air is 132.5 K and the critical pressure 37.2 atm.

a. Evaluating the reduced temperature and pressure

$$T_r = \frac{T}{T_c} = \frac{298\ \text{K}}{132.5\ \text{K}} = 2.25 \qquad P_r = \frac{P}{P_c} = \frac{170\ \text{atm}}{37.2\ \text{atm}} = 4.573$$

For $T_r = 2.25$ and $P_r = 4.573$, z ~ 1.0. Therefore the molar volume of the air in the tank is

$$\hat{V} = z\frac{RT}{P} = (1.0)\frac{0.08205\frac{\text{atm}\cdot\text{L}}{\text{mol}\cdot\text{K}}\cdot(298\ \text{K})}{(170\ \text{atm})} = 0.144\ \text{L/mol}$$

Thus in 140 L, there is (140 L) · (0.144 mol/L) = 974 moles, which is (974) · (29) = 28.25 kg.

b. Since the mass of air in the tank and the volume of the tank remain the same, the molar volume is identical to that at the lower temperature, 0.144 L/mol. The reduced temperature, T_r, however, is now (273+75)/132.5 = 2.63. The reduced pressure is unknown but it can be estimated by

1. Assuming the reduced pressure remains as in part (a) to estimate z from T_r.
2. Calculating the pressure using the value of z from part 1.
3. Evaluating the new reduced pressure and z using the calculated pressure from 2.
4. Recalculation of the pressure using the value of z estimated from 3.
5. Repetition of 3 and 4 until the pressure used in 3 equals that calculated in 4.

$$\text{Iteration 1: } T_r = 2.63 \qquad P_r = 4.57, \qquad z = 1.06, \qquad P = z\frac{R \cdot T}{\hat{V}} = 210 \text{ atm}$$

$$\text{Iteration 2: } T_r = 2.63 \qquad P_r = 5.66, \qquad z = 1.07, \qquad P = z\frac{R \cdot T}{\hat{V}} = 212 \text{ atm}$$

The rating for the tank will be exceeded if the vessel were to warm to 75°C. Note that if the tank were assumed to contain an ideal gas (z = 1), the pressure of the hot tank would remain below 200 atm and the rating would not be exceeded.

4.1.3 Ideal Gas Mixtures

Ideal gas behavior allows gas mixtures to be specified simply. In particular the ideal gas law holds for each component of the mixture separately. Each component exerts its own pressure referred to as its *partial pressure.* The partial pressure is the pressure that would be exerted by the gas component if it alone were present in the system; that is, the pressure that would be exerted if the gas component filled the system volume at the same temperature as the system. Using p_A, p_B, etc., to represent the partial pressures of component A, B, etc., the ideal gas suggests that these pressures are given by.

$$p_A = \frac{n_A RT}{V} \qquad p_B = \frac{n_B RT}{V} \qquad \text{etc.} \tag{4.6}$$

Summing each of these partial pressures

$$\begin{aligned} p_A + p_B + \ldots &= \frac{n_A RT}{V} + \frac{n_B RT}{V} + \ldots \\ &= \frac{(n_A + n_B + \ldots)RT}{V} \\ &= \frac{nRT}{V} = P \end{aligned} \tag{4.7}$$

The last equation is a statement of *Dalton's law.* Dalton's law is the partial pressures for each component sum to the total pressure of the mixture.

As indicated in Chapter 3, all liquids exert a vapor pressure. The liquid boils or turns to vapor when this pressure exceeds the containing pressure in the system. The normal boiling point of the liquid is the temperature when this vapor pressure equals atmospheric pressure. The pressure exerted by a liquid at a given temperature, T, is termed the *saturation or vapor pressure* of the liquid, P_v. If a closed vessel contains only the single liquid, a vapor space would contain liquid vapors at this pressure. If the liquid surface was open to the air at 1 atm total pressure, the partial pressure of the liquid component at equilibrium, $p^* = P_v$. The molar concentration of the component, A, in the air is then given by

$$\frac{n_A}{V} = \tilde{C}_A = \frac{p_A^*}{RT} = \frac{P_v}{RT} \tag{4.8}$$

The tilde will be used to indicate molar quantities if we need to differentiate between molar and mass-based properties. Partial pressure and mole fraction are considered further in Example 4.5.

Example 4.5: Partial pressure of water vapor and relative humidity

What is the partial pressure, mole fraction, and volume fraction of water vapor in air in equilibrium with water at 25°C? At 25°C, water exerts a vapor pressure of 23.74 mmHg.

If water was the only component of the vapor phase, the pressure that would be exerted is the vapor pressure of the water. Thus, the partial pressure and molar concentration of water vapor is

$$p_W = 23.74 \text{ mmHG} = 0.0312 \text{ atm} = 3165\ Pa$$

$$\tilde{C}_a = \frac{p_W}{RT} = \frac{0.0312 \text{ atm}}{\left(0.08215 \frac{\text{atm} \cdot \text{L}}{\text{mol} \cdot \text{K}}\right)(298 \text{ K})} = 1.276 \text{ mol/m}^3$$

For an ideal gas, the mole fraction, y, is then given by

$$y = \frac{n_W}{n} = \frac{p_W}{p} = 0.0312$$

Also for an ideal gas, the mole fraction and the volume fraction are identical.

Note that air at 25°C containing 3.12% water vapor is said to be saturated with a relative humidity of 100%. Relative humidity is the ratio of the actual water vapor content to that which represents saturation. Thus air at 25°C containing 1.56% water vapor exhibits a relative humidity of 50%. If the air at 25 °C contained more that 3.12% water vapor, the air is supersaturated and water condenses (as "dew") until the partial pressure is again 0.0312 atm (or mole or volume fraction is again 3.12%).

Similarly, a partial or pure component volume can be defined that is the volume that the gas would fill if it were at the same temperature and pressure as the mixture. For the components A, B, etc., the partial volume is defined by

$$v_A = \frac{n_A RT}{P} \qquad v_B = \frac{n_B RT}{P} \qquad \text{etc.} \tag{4.9}$$

Adding the partial volumes as was done in Equation 4.7, we find the similar result that

$$v_A + v_B + v_c + \ldots = V \tag{4.10}$$

This is *Amagat's law* of additive volumes. Amagat's law is the partial volumes of each component sum to the total volume of the mixture. Note that the partial volume is defined as the volume that would occur if the component were at the mixture or total pressure. It must be recognized that the partial pressures and partial volumes have different defining conditions and cannot be combined.

$$p_A v_A \neq n_A RT \tag{4.11}$$

A useful result of Equations 4.7 and 4.10 is that the ratio of the partial pressure of a component to the total pressure, or the ratio of the partial volume of a component to the total volume, are both equal to the ratio of the moles of the component to the total number of moles in the system. This was illustrated in Example 4.5. The ratio of the moles of A to the total number of moles is the *mole fraction*, y_A.

$$\frac{p_A}{P} = \frac{v_A}{V} = \frac{n_A}{n} = y_A \tag{4.12}$$

Alternatively, the partial pressure or partial volume are defined by the mole fraction times the total volume or pressure

$$p_A = y_A P \qquad v_A = y_A V \tag{4.13}$$

Equation 4.12 also shows that the volume fraction (v_A/V) is equal to the mole fraction, as indicated previously. Thus air, which is about 21% oxygen, is both 21 mol% or 21 vol% oxygen. Air is not, however, 21% oxygen *by weight* nor is volume and mole percentage equal for any phase other than ideal gases.

The presence of gas mixtures poses problems with respect to the definition of the mixture properties since generally only pure component properties are available. Such properties might include the specific heat, transport properties such as viscosity and thermal conductivity, and critical temperature, pressure, and volume for compressibility factor determination. Typically, a mole fraction weighted combination of the pure component properties is sufficiently accurate for environmental engineering purposes. If $\psi_{mixture}$ represents the mixture property desired and ψ_A, ψ_B, etc., represent the pure component properties, the mixture property is given by

$$\Psi_{mixture} = y_A \Psi_A + y_B \Psi_B + \ldots \tag{4.14}$$

This approach is also often used with liquid mixtures, but the accuracy for liquid mixtures is far less certain. It is generally an excellent approximation for ideal and near-ideal gases, however.

4.1.4 General phase behavior

Many of the substances of interest to environmental engineers can stably exist in solid, liquid, and gaseous states. As a result of the limited temperature and pressure ranges normally observed in the environment, however, the bulk of the substance is normally present in only one of these phases and some fraction of the substance might partition into one of the remaining phases. For the moment we will not consider the possibility of reactions or multiple chemical states within a single phase.

Despite these limitations, the phase behavior or pressure-volume-temperature relationships for a pure substance can be quite complicated. A useful guide in the analysis of the phase behavior is the Gibbs' phase rule

$$\text{degrees of freedom} = \text{components} - \text{phases} + 2 \tag{4.15}$$

The degrees of freedom are the number of independent intensive (i.e., total mass independent) parameters that must be specified to define the physical state of the system. The system is simply the totality of the substances of interest and their physical setting. The parameters that describe the system are the properties such as state (liquid, gas, or solid), temperature, pressure, and specific volume. The number of the components in a nonreactive system is the number of different substances in the system, in our case of pure substances, one. Thus, the phase rule suggests that the number of phases plus the number of degrees of freedom in a system are always equal to 3. This suggests that a single phase system (e.g., a purely gaseous phase) has 2 degrees of freedom and

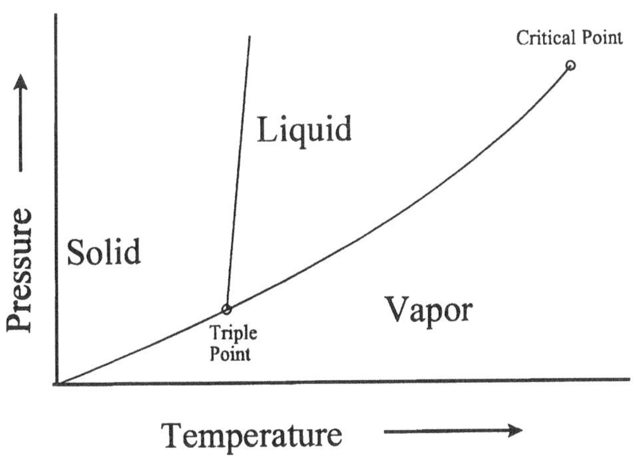

FIGURE 4.3 General pressure-temperature behavior of a substance showing phase boundaries.

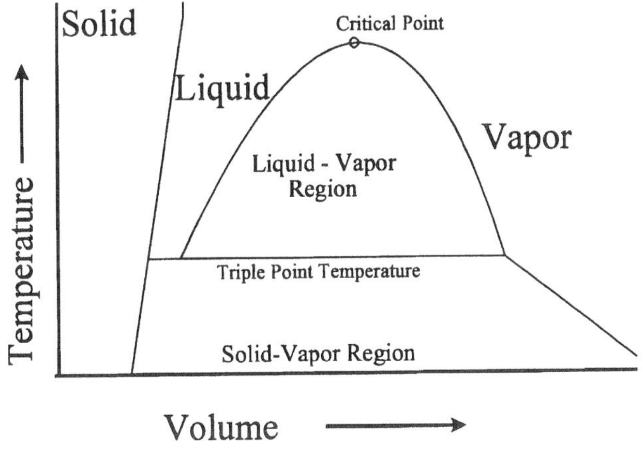

FIGURE 4.4 General temperature-volume behavior of a substance showing phase boundaries.

therefore requires two intensive parameters to completely specify the system. This has already been shown in that an ideal gas requires only the specification of any two of the three properties of pressure, temperature, and specific volume. The third can be determined from the ideal gas law. If a two-phase system occurs, for example, liquid and vapor, then only a single intensive property need be defined. If the partial pressure of the vapor phase is specified, for example, the vapor pressure vs. temperature requirements specifies the system temperature. If solid, liquid, and vapor of a pure substance are all simultaneously present in a system, the state is completely specified. That is, there must only be one temperature, pressure, and volume where all three phases can coexist. This particular point is called the triple point and it occurs at 0.01°C for water. If the temperature, pressure, or specific volume is changed in some fashion, one of the phases must be eliminated.

The structure of the pressure-volume-temperature relationships for a substance like water is shown in Figures 4.3 through 4.5. The full-phase diagram is three dimensional because of the presence of three state variables: pressure, specific volume, and temperature. In order to more clearly understand the phase behavior, two-dimensional views of each pair of state variables are shown.

Examining first the pressure-temperature behavior in Figure 4.3, there are zones below the critical point where liquid, solid, and gaseous water could exist. Above the critical point, of course, it is not possible to differentiate between the liquid and gaseous phases. Within each zone in the figure, only the pure phase material exists. At the boundaries between the phases shown by the lines, both phases can coexist. Higher temperatures and lower pressures tend to move the system toward vapor while lower temperatures and high pressure push the system

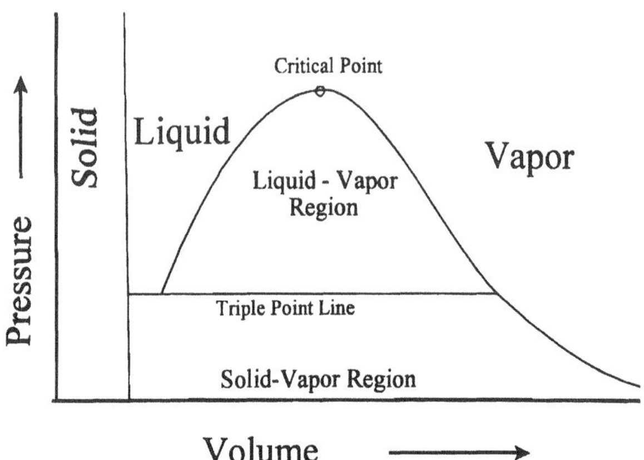

FIGURE 4.5 General pressure-volume behavior of a substance showing phase boundaries.

toward solids. Along the two-phase liquid-vapor line, temperature and pressure cannot be specified independently. This again indicates that only a single variable is needed to specify conditions in the mixed phase region. The three-phase region is shown as a single point indicating the triple point is already completely specified. Temperature or pressure cannot be changed without eliminating one of the phases.

Examining the temperature-volume construction next, the triple point now appears as a line. The line connects the specific volume of the vapor with the specific volume of the solid and liquid phases. Because the line is horizontal, it is clear that one temperature exists where liquid, solid, and vapor can coexist. The intersection of the line with the solid phase region defines the specific volume of the solid. Similarly, the intersection of the line with the fluid phases defines their specific volumes. As the temperature is increased, the liquid-vapor envelope separates from the solid phase region indicating that ice can no longer exist in equilibrium with the remaining phases. In the liquid-vapor region, lines of constant temperature connect the corresponding specific volumes of the saturated liquid (intersection with the liquid curve) and vapor (intersection with the vapor curve). As the temperature increases the specific volumes change. Again the pressure must also change if the system remains in the two-phase region because only a single variable defines the system state within this region.

The pressure-specific volume projection shows a structure similar to that of the temperature-specific volume relationship. The intersection of the constant pressure line with any pure phase region defines the specific volume of that phase.

4.2 THERMODYNAMICS OF PHASE CHANGES

The transitions between the gas, liquid, and solid phases are dependent on the energy changes in the system. Let us consider a transition taking place at constant pressure, specifically the transition

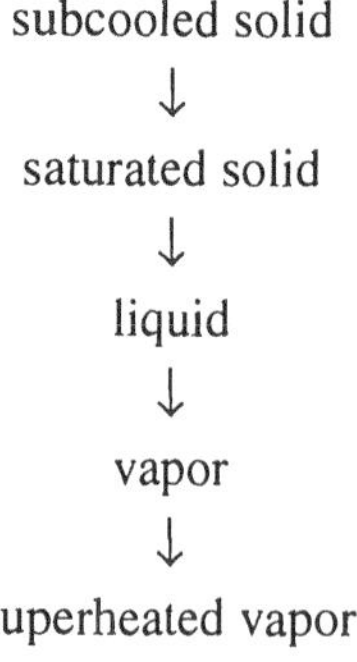

An example of this transition would be ice at –10°C, warming to the melting point, melting, finally evaporating, and the subsequent heating of the vapors. The phases in this transition include:

1. Subcooled solid — Solid cooled below the melting point.
2. Saturated solid — A saturated phase is one that is at the point of transition to another phase.
3. Liquid — Transition between melting point and saturated vapor.
4. Vapor — Transition to saturated vapor.
5. Superheated vapor — Vapor heated above the boiling point.

The transition between subcooled solid to saturated solid involves a change in the sensible heat or temperature of the solid. The transition between the saturated solid and the saturated liquid state occurs at constant temperature but involves latent heat, or the change in energy of the substance required to achieve the phase transition. With sensible heat energy changes involve just temperature changes. With latent heat energy changes involve phase transitions.

Energy transitions in the pure liquid and vapor states are sensible heat transitions while the transition between liquid and vapor occurs at constant temperature and involves a latent heat change.

Description of the energy changes associated with these transitions requires the definition of E_i several thermodynamic quantities. Internal energy is a measure of the intramolecular energy content (= E_i). Enthalpy is the sum of the internal energy and the available "pressure work" (H = E_i + PV). A force F acting through a distance x involves the application of the work, F·x. Similarly, the work or energy available from a constant pressure (P) transition from zero volume to volume V is PV. Pressure is the force per unit area while volume is area times distance. The work done by a system during an expansion of the system from an initial volume, V_1, to a final volume, V_2, is thus

$$\delta W = \int_{V_1}^{V_2} P dV \tag{4.16}$$

The positive sign for work in this text will be reserved for the work done by the *system on the surroundings*. The right-hand side of Equation 4.16 reduces to PV when the pressure is constant, the initial volume is 0, and the final volume is V. The pressure work or PV work represents the extra energy available for work on the surroundings. The sum of the two (that is, enthalpy) represents a "total energy content." Let us examine the change in the internal energy and enthalpy during both sensible heat transitions such as the warming of a subcooled solid to the melting temperature, and latent heat transitions such as the melting of a solid at constant temperature. The enthalpy change in water as a result of these transitions is shown in Figure 4.6.

4.2.1 Sensible Heat Transitions

If we consider a pure single phase material, the phase rule indicates that only two intensive properties completely specify the system. Selecting these two properties as temperature and volume, the internal energy change in any process has the form

$$\partial E_i = \left(\frac{\partial \hat{E}_i}{\partial T}\right)_{\hat{V}} dT + \left(\frac{\partial \hat{E}_i}{\partial \hat{V}}\right)_T d\hat{V} \tag{4.17}$$

That is, the net change in internal energy is the sum of the change due to a change in temperature at constant specific volume and the change due to a change in specific volume at constant temperature. Under essentially all conditions of interest to the environmental engineer, the influence of specific volume on internal energy is negligible in the absence of phase changes. The term is

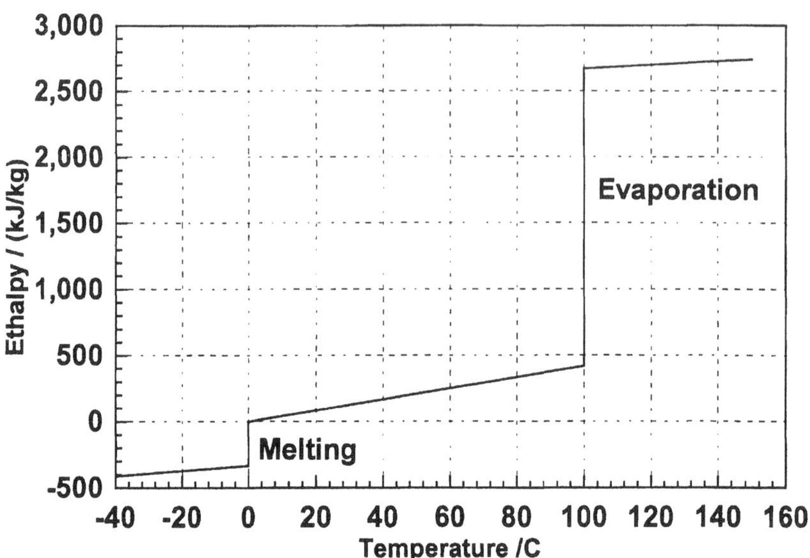

FIGURE 4.6 Energy (enthalpy) changes in water from a subcooled solid to steam.

associated with intermolecular forces as the fluid is compressed into a smaller volume. These are negligibly small for near-ideal gases (zero for ideal gases) and liquids and solids are essentially incompressible implying that the volume change is negligibly small for these materials. Thus, the change in internal energy of a substance between two temperatures without a phase change is given by integration of

$$d\hat{E}_I = \left(\frac{\partial \hat{E}_I}{\partial T}\right)_{\hat{V}} dT \qquad (4.18)$$
$$= C_v dT$$

where C_v is the specific heat at constant volume.

Two intensive properties can also be used to define the enthalpy of a pure phase (as they would any other intensive property of the system). By convention, taking pressure and temperature, the change in enthalpy, H, during a process is described by

$$d\hat{H} = \left(\frac{\partial \hat{H}}{\partial T}\right)_P dT + \left(\frac{\partial \hat{H}}{\partial P}\right)_T dP \qquad (4.19)$$

Generally, the second term is again quite small in the absence of phase changes. For an ideal gas, for example, the term is identically zero. Remember that $P\hat{V} = RT$ for an ideal gas and thus $\hat{H} = \hat{E}_I + P\hat{V}$ does not change with pressure at constant temperature. Neglecting the second term, the change in enthalpy can be determined, in the absence of phase changes, by integration of

$$d\hat{H} = \left(\frac{\partial \hat{H}}{\partial T}\right)_P dT \qquad (4.20)$$
$$= C_p dT$$

or, for a transition between temperatures T_1 and T_2,

$$\hat{H}_2 - \hat{H}_1 = \int_{T_1}^{T_2} C_p dT \tag{4.21}$$

C_p is the specific heat at constant pressure. Note that for an ideal gas, Equations 4.18 and 4.20 with the definition of enthalpy suggest

$$C_p = C_v + R \text{ (for ideal gases)} \tag{4.22}$$

For a constant density fluid or solid, the specific heat at constant pressure and the specific heat at constant volume are equal.

$$C_p \approx C_v \quad \text{(for incompressible phases)} \tag{4.23}$$

Equation 4.21 can be employed to estimate the change in enthalpy during a specified process by assuming a constant heat capacity ($C_p \approx C_{p,avg}$) and integrating to find

$$\hat{H}_2 - \hat{H}_1 = C_{p,avg} \Delta T \tag{4.24}$$

The magnitude of the specific heat depends on the degrees of freedom available to a molecule for the storage of energy. Simple monoatomic gases, that is, gases composed of only one atom per molecule, have few degrees of freedom and exhibit the lowest specific heat. Gases with large, complex molecules exhibit many degrees of freedom and high values of specific heat. At high temperatures, internal vibrations of gas molecules provide additional degrees of freedom and therefore the specific heat of gases tends to increase with temperature. This increase tends to be greatest between 0 and 1000°C with the specific heat approximately constant above that temperature. Table 4.3 illustrates the approximate range of specific heats for monoatomic, linear polyatomic, and nonlinear polyatomic molecules. These estimates work remarkably well. At 0°C, the specific heat of water vapor is about 8 cal·mol^{-1}·°C (compared to a predicted value of 4R = 7.95 cal·mol^{-1}·°C^{-1}) while at 300°C, the specific heat of water vapor is about 13 cal·mol^{-1}·°C^{-1} (compared to a predicted value of 13.8 cal·mol^{-1}·°C^{-1}).

Accurate estimation of enthalpy changes, however, requires use of more precise indications of specific heats and their temperature dependence. The temperature dependence of specific heats are often correlated with an equation of the form

TABLE 4.3
Magnitude of Specific Heats

Type of Molecule	Examples	Specific Heat: High Temperature	Specific Heat: Low Temperature
Monoatomic	He, Ne, Xe	5/2 R	5/2 R
Linear polyatomic	H_2,N_2,O_2,CO_2	(3n – 3/2)R	7/2 R
Nonlinear polyatomic	H_20, NH_3	(3n – 2)R	4 R

Note: n = number of atoms per molecule and R = 1.9872 cal · mol^{-1} · °C^{-1}.

$$C_p \approx a + bT + cT^2 + \ldots \tag{4.25}$$

Data can be readily fit to an equation of this form using the methods in Chapter 3. The use of such a correlation is illustrated in Example 4.6.

Example 4.6: Use of $C_p(T)$ correlating equation

Determine the heat input (in calories) required to raise 1 mole of N_2 from 25 to 2000°C.

A correlation for the specific heat of nitrogen is given by

$$\frac{C_p(T/\text{K})}{\frac{\text{cal}}{\text{mol}\cdot\text{K}}} = 6.529 + 0.1488 * 10^{-2}\,T - 0.02271 * 10^{-5}\,T^2$$

Thus, the heat input required is given by

$$\Delta\hat{H} = \int_{25+273}^{2000+273} [6.529 + 0.1488 * 10^{-2}\,T - 0.02271 * 10^{-5}\,T^2\,]dT$$

$$= \left[6.529\,T + 0.1488 * 10^{-2}\,\frac{T^2}{2} - 0.02271 * 10^{-5}\,\frac{T^3}{3}\right]_{298}^{2273}$$

$$= 15{,}790 \text{ cal/mol}$$

The approach employed in Example 4.6 can be used to estimate the adiabatic flame temperature of a combustion reaction. An *adiabatic* process is one in which no heat is added or removed from the system. The adiabatic flame temperature is thus the temperature that results if all of the heat generated by the combustion is absorbed by the combustion product gases. For a hydrocarbon fuel, the primary products are carbon dioxide and water vapor. If air is employed as the source of oxygen, nitrogen is an unavoidable diluent of the products. Since the heat generated by the reaction does not change whether air or pure oxygen is used as the oxygen source, the use of air results in a reduced adiabatic flame temperature. Excess oxygen or air also is often used to ensure good combustion without the formation of such things as soot, which are products of incomplete combustion (PICs). The extra oxygen and/or air also serves to dilute the product gases and lowers their adiabatic flame temperature.

The balanced combustion reaction for methane can be written

$$CH_4 + 2(O_2 + 3.76N_2) \rightarrow CO_2 + 2H_2O + 7.52N_2 \tag{4.26}$$

Because the coefficients in the balanced equation represent molecules or moles, it is convenient to evaluate enthalpy changes on a molar basis, specifically per mole of methane burned. Note that combustion with air rather than pure oxygen results in a significant increase in the moles (and volume) of the product gases, 10.52 moles per mole of methane rather than 3. Let us assume that the equations for C_P are of the form

$$C_p = a + b\cdot T + c\cdot T^2 + d\cdot T^3 \tag{4.27}$$

If α_i represents the moles of product i per mole of methane, the enthalpy change of the product gases, which is equal to the heat of combustion of the methane, $\Delta\hat{H}_c$, per mole, is given by

$$\Delta \hat{H}_c = \Delta \hat{H} = \int_{T_i}^{T_f} \sum_{i=1}^{3} \left[\alpha_i \cdot (a_i + b_i \cdot T + c_i \cdot T^2 + d_i \cdot T^3)\right] dT \tag{4.28}$$

T_i is the initial temperature (25°C) and T_f is the final temperature. This can be written

$$\begin{aligned} \Delta \hat{H}_c = \Delta \hat{H} &= \int_{T_i}^{T_f} \left[\Sigma(\alpha_i \cdot a_i) + \Sigma(\alpha_i \cdot b_i) \cdot T + \Sigma(\alpha_i \cdot c_i) \cdot T^2 + \Sigma(\alpha_i \cdot d_i) \cdot T^3\right] dT \\ &= \left[\Sigma(\alpha_i \cdot a_i) \cdot T + \Sigma(\alpha_i \cdot b_i) \cdot \frac{T^2}{2} + \Sigma(\alpha_i \cdot c_i) \cdot \frac{T^3}{3} + \Sigma(\alpha_i \cdot d_i) \cdot \frac{T^4}{4}\right]_{T_i}^{T_f} \\ &= (T_f - T_i) \cdot \Sigma(\alpha_i \cdot a_i) + \frac{T_f^2 - T_i^2}{2} \cdot \Sigma(\alpha_i \cdot b_i) + \frac{T_f^3 - T_i^3}{3} \cdot \Sigma(\alpha_i \cdot c_i) + \frac{T_f^4 - T_i^4}{4} \cdot \Sigma(\alpha_i \cdot d_i) \end{aligned} \tag{4.29}$$

Since the final temperature is unknown and not easily derived from Equation 4.29, an iterative solution is often employed. This process is

1. Guess a final temperature.
2. Calculate the enthalpy change, $\Delta \hat{H}_{calc}$, at the guessed temperature.
3. If $\Delta \hat{H}_{calc} < \Delta \hat{H}_c$, guess a lower temperature and repeat Step 2.
4. If $\Delta \hat{H}_{calc} > \Delta \hat{H}_c$, guess a higher temperature and repeat Step 2.
5. Repeat Steps 2 through 4 until $\Delta \hat{H}_{calc} = \Delta \hat{H}_c$.

The iterations required can be reduced by linear interpolation or extrapolation from the results of the first two guesses to produce the third guessed final temperature. Expanding $\Delta \hat{H}_{calc}(T_3)$ in a Taylor series

$$f(x_n) = f(x_{n-1}) + \frac{\partial f}{\partial x}(x_n - x_{n-1}) + \ldots$$

or (4.30)

$$\Delta \hat{H}_{calc}(T_3) \approx \Delta \hat{H}(T_2) + \frac{\Delta \hat{H}(T_2) - \Delta \hat{H}(T_1)}{T_2 - T_1}(T_3 - T_2)$$

Since $\Delta \hat{H}_{calc}(T_3)$ is desired to be the heat liberated by combustion, $\Delta \hat{H}_c$, the third guess of the final temperature

$$T_3 = T_2 + (T_2 - T_1)\frac{\Delta \hat{H}_c - \Delta \hat{H}(T_2)}{\Delta \hat{H}(T_2) - \Delta \hat{H}(T_1)} \tag{4.31}$$

The process can be repeated as often as necessary until the guessed value of temperature provides a calculated enthalpy change sufficiently close to $\Delta \hat{H}_c$. The technique is employed in Example 4.7.

Coefficients for the correlating equation for specific heats of common combustion gases are shown in Table 4.4 for specific heats in $J \cdot mol^{-1} \cdot °C^{-1}$ using temperatures in °C. These correlating equations are only valid up to a temperature of 3000°C. Attempts to use them outside of this range give unreliable and sometimes ludicrous results. For example, the predicted specific heat of water vapor at 5800°C is -371 $J \cdot mol \cdot K^{-1}$, an impossible result.

TABLE 4.4
Specific Heat Equation Parameters for Common Combustion Gases

Compound	a	b	c	d
Carbon dioxide	36.11	$4.233 \cdot 10^{-2}$	$-2.887 \cdot 10^{-5}$	$7.464 \cdot 10^{-9}$
Water	33.46	$0.688 \cdot 10^{-2}$	$0.7604 \cdot 10^{-5}$	$-3.593 \cdot 10^{-9}$
Nitrogen	29	$0.2199 \cdot 10^{-2}$	$0.5723 \cdot 10^{-5}$	$-2.871 \cdot 10^{-9}$
Oxygen	29.1	$1.158 \cdot 10^{-2}$	$-0.6076 \cdot 10^{-5}$	$1.311 \cdot 10^{-9}$

Note: Coefficients in $C_p = a + bT + cT^2 + dT^3$.

Example 4.7: Calculation of the adiabatic flame temperature

Determine the adiabatic flame temperature for the combustion of methane at 25°C for (1) stoichiometric combustion with air and (2) combustion with 25% excess air.

The heat of combustion of methane at 25°C is 890.360 kJ/mol. The combustion reaction can be written

$$CH_4 + \alpha O_2 + \beta N_2 \rightarrow CO_2 + 2H_2O + (\alpha - 2)O_2 + \beta N_2$$

1. Stoichiometric combustion with air ($\alpha = 2$, $\beta = 7.52$)

   In this case there are 10.52 moles of product gases per mole of methane. The enthalpy calculated by Equation 4.28 required for the product gases to achieve a final temperature of 1500°C is 56.14 kJ or 36% less than the heat of combustion of methane. If a final temperature of 2000°C is guessed, the required enthalpy is 77.25 kJ or still 13.2% less. Using Equation 4.28,

$$T_f \approx 2000°C + (2000°C - 1500°C)\frac{89.036\text{ kJ} - 77.25\text{ kJ}}{77.25\text{ kJ} - 56.14\text{ kJ}}$$

$$\approx 2279°C$$

   The calculated enthalpy with a final temperature of 2279°C is 88.45 kJ. Reapplication of Equation 4.28 using 2000 ($\Delta\hat{H}$ = 77.25 kJ) and 2279°C (88.45 kJ) gives an estimate of final temperature of 2293°C ($\Delta\hat{H} = 88.99$ kJ ~ $\Delta\hat{H}_c$). Thus,

   adiabatic flame temperature ~ 2293°C.

2. Combustion with 25% excess air ($\alpha = 2 \cdot 1.25 = 2.5$, $\beta = 2.5 * 3.76 = 9.4$)

   There is now 12.9 moles of product gases per mole of methane so the adiabatic flame temperature is expected to *decrease*. The guessed temperatures and calculated enthalpies are

   T_f = 1500°C, $\Delta\hat{H}$ = 67.71 kJ
   T_f = 2000°C, $\Delta\hat{H}$ = 93.04 kJ
   By Equation 4.28, T_f ~ 1921°C where $\Delta\hat{H}$ = 89.1 kJ ~ $\Delta\hat{H}_c$

   adiabatic flame temperature ~ 1921°C.

Thus, the addition of 25% excess air would be expected to reduce the flame temperature by 372°C.

The specific heat of liquids tends to be even less temperature dependent than that of gases. The specific heat of water is 1cal·g^{-1}·°C^{-1} or 18 cal·mol^{-1}·°C^{-1}. The specific heat of most organic compounds lies between 25 and 50 cal·mol^{-1}·°C^{-1}. Correlations for the prediction of specific heats of both gases and liquids can be found in Lyman et al. (1990).

Note that only the *changes* in internal energy and enthalpy are defined by the above equations. Values for these quantities must always be referenced to a specified temperature and pressure. In steam tables, the reference condition, the point of zero enthalpy, is typically liquid water at 0°C and a pressure of 0.006 atm (the vapor pressure of liquid water at 0°C).

4.2.2 Latent Heat Transitions

The above discussion of internal energy and enthalpy changes is limited to sensible heat changes without phase changes. Additional energy is required for the transition from solid to liquid state and from liquid to vapor state. The enthalpy changes associated with these transitions are the latent heat of melting, $\Delta\hat{H}_m$, and the latent heat of vaporization, $\Delta\hat{H}_m$, respectively. These latent heats could be on either a mass or molar basis, but because chemical reaction is not involved, there is no advantage to employing molar units as in Section 4.2.1.

During the movement from solid to liquid to vapor, heat must be adsorbed by the substance to break the intermolecular attractive forces that occur in the denser materials. During the movement from vapor to liquid to solid, the same amount of heat must be evolved by the substance. As a substance condenses and freezes, the latent heat changes are referred to as the latent heat of condensation and the latent heat of fusion, respectively. The quantity of heat transferred during these transitions is independent of the direction of the transition (e.g., the magnitude of the latent heat transition from liquid to vapor is identical to the latent heat of the transition from vapor to liquid). With respect to the substance undergoing the transitions, the sign of the heat flow is positive for melting and vaporization (heat absorption by the substance) and negative for condensation and freezing (heat evolution by the substance).

In the hypothetical transition of subcooled ice to superheated vapor, the total enthalpy change is the sum of the enthalpy changes during the individual process steps. If the subcooled solid(s) temperature is initially T_1, the melting temperature of the solid, T_m, the boiling temperature of the liquid (l), T_b, and the final temperature of the vapor (v) T_2, the change in enthalpy is given by,

$$\Delta\hat{H} = \int_{T_1}^{T_m} \left(C_p\right)_s dT + \Delta\hat{H}_m + \int_{T_m}^{T_b} \left(C_p\right)_l dT + \Delta\hat{H}_v + \int_{T_b}^{T_2} \left(C_p\right)_v dT \tag{4.32}$$

The change in enthalpy for water from 0°F (-17.8°C) to 150°C was shown in Figure 4.6. For water, the specific heat of liquid water at 0°C is 1 cal·g^{-1}·°C^{-1}, or 18 cal·mol^{-1}·°C^{-1} while the latent heat of melting and vaporization at the normal melting and boiling points is 1436 cal/mol (6012 J/mol, 334 J/g) and 9717 cal/mol (40,680 J/mol, 2260 J/g), respectively. As shown by these values, phase changes typically require much more energy that sensible heat changes. It is not until the temperature changes are a hundred degrees or more that sensible heat requirements are of a similar magnitude to latent heat requirements. As a result, temperature changes in a single phase fluid are easily accomplished whereas phase changes require large energy input or removal. A practical manifestation of this phenomena is the large diurnal changes in temperature that occur in the atmosphere. Temperature variations in the desert where the moisture content of the air is low can be quite large compared to moist environments where significant amounts of evaporation and condensation accompany temperature changes. Rarely can the nighttime temperature cool below the dew point, or the point where moisture begins to condense. Generally, the most important latent heat of interest to environmental engineers is the heat of vaporization. Lyman et al. (1990) provides data on heats of vaporization and methods for its estimation. Perhaps the simplest correlation for heat of vaporization is the correlation of Kistiakovskii, which requires the normal boiling temperature

$$\frac{\Delta\hat{H}_v(T_b)}{T_b} = (8.75 + 1.987\ \ln T_b) \qquad T_b - \text{K},\ \Delta\hat{H}_v - \text{cal/mol} \tag{4.33}$$

This correlation underestimates the heat of vaporization of slightly polar compounds. The correction should be less than 5% for any hydrocarbon, organic halide, organic sulfide, mercaptans, or ethers. The correction is between 5 and 10% for short organic carbon chain esters, aldehydes, ketones, nitrogen-containing compounds (up to 16% for primary amines), and oxides. Alcohols are the only organic compounds that exhibit significant deviations from the above relationship with single alcohols typically exhibiting a 30% higher heat of vaporization, diols approximately 33% higher, and triols 38% higher. Detailed information can be found in Lyman et al. (1990).

Because the boiling point of many compounds differs significantly from ambient environmental conditions (typically 20 to 25°C), the temperature dependence of the heat of vaporization must sometimes be considered. The Watson correlation relates the heat of vaporization at any temperature to the heat of vaporization at the boiling point (estimated by the Kistiakovskii equation) and the critical temperature. The correlation is

$$\Delta\hat{H}_v(T) = \Delta\hat{H}_v(T_b)\left(\frac{1 - T/T_c}{1 - T_b/T_c}\right)^m \tag{4.34}$$

where m = 0.19 for liquids. The critical temperature is often approximated as 1.5 T_b giving the approximation

$$\Delta\hat{H}_v(T) \approx \Delta\hat{H}_v(T_b)(3 - 2T/T_b)^m \tag{4.35}$$

Grain in Lyman et al. (1990) claims an error of less than 5% with this approximation. The value of m (0.19 for liquids) is different for solids depending on the ratio of the temperature to the substances boiling point temperature. Grain in Lyman et al. (1990) suggests

$$m = \begin{cases} 0.36 & T/T_b > 0.6 \\ 0.80 & 0.5 < T/T_b < 0.6 \\ 1.19 & T/T_b < 0.5 \end{cases} \tag{4.36}$$

Table 4.5 compares the heat of vaporization for a variety of compounds with that estimated by Equation 4.35 and the critical temperature with that estimated from the normal boiling point.

4.3 THERMODYNAMICS OF EQUILIBRIUM

Our introduction to phase behavior and the energetics of phase changes has prepared us to evaluate the thermodynamics of equilibrium. This requires use of the first and second law of thermodynamics

TABLE 4.5
Heat of Vaporization and Critical Temperature Estimation

Compound	T_b (K)	$\Delta\hat{H}_v$ (kcal/mol)	$\Delta\hat{H}_v$ (Eq. 4.35)	% error	T_c (K)	T_c (1.5^*T_b)	% error
Ammonia	239.7	5.581	4.708	−15.6	405.5	359.6	−11.3
Benzene	353.3	7.353	7.209	−1.95	562.6	529.9	−5.81
Carbon tetrachloride	349.9	7.17	7.13	−0.5	556.4	524.9	−5.67
Chlorobenzene	405.3	8.73	8.38	−4.0	632.4	607.9	−3.88
Ethanol	351.7	9.22	7.17	−22.2	516.3	527.6	2.18
Heptane	371.6	7.575	7.62	0.61	540.2	557.4	3.18
Methane	111.7	1.955	2.023	3.50	190.7	167.5	−12.1
Nitrogen	77.3	1.333	1.345	0.90	126.2	116	−8.07
n-Propylbenzene	432.4	9.14	8.99	−1.56	638.7	648.6	1.55
Water	373.2	9.717	7.66	−21.2	647.4	559.7	−13.5

and the definition of additional thermodynamic variables before finally defining equilibrium thermodynamically. Each of these topics will be considered in turn.

4.3.1 First and Second Laws of Thermodynamics

The first law of thermodynamics states that the change in internal energy (ΔE_I, kinetic plus potential energy) of a system undergoing any process or transition is equal to the heat added from the surroundings (δQ) minus the work done on the surroundings by the system (δW).

$$\Delta E_I = \delta Q - \delta W \tag{4.37}$$

This equality is depicted in Figure 4.7(a). The first law of thermodynamics is nothing more than an energy balance, a recognition of the law of conservation of energy which states that energy is neither created nor destroyed. The energy added to a system in the form of heat or work must be reflected in a change in the internal energy content of the system. The different symbols for the change in internal energy (Δ) and change in heat and work (δ) are an effort to reflect the fact that internal energy is a *state* variable while heat and work are *path dependent*. You could continuously add heat or work and remove or use it through a number of different cycles and in different ways, but the internal energy is completely specified by the state of the final system. As indicated previously, only two intensive state variables are necessary to define the state of the system. The final temperature and pressure, for example, define the internal energy of the system, regardless of the steps taken to add or remove heat or work to or from the system.

It is also possible to have an equilibrium system that is flowing under stationary or time-independent conditions. In a flowing system the energy between an entrance and exit of the system is desired, and Equation 4.37 must be modified to reflect the fact that the change in energy in the system may be in terms of potential and kinetic energy as well as internal energy. The system on which an energy balance is required is shown in Figure 4.7(b). Potential energy is the result of changes in elevation with respect to gravity and kinetic energy is the result of changes in velocity between the system inlet and outlet. The potential energy per unit mass of material is given by gz where z represents the height above a reference datum. The kinetic energy per unit mass is given by $u^2/2$ where u represents the velocity. In a flowing system, it is also convenient to separate work done on the flowing fluid, for example, by a pump, and the pressure volume work at the entrance and exit of the system. The addition of fluid at pressure P_1 at the entrance to the system does a

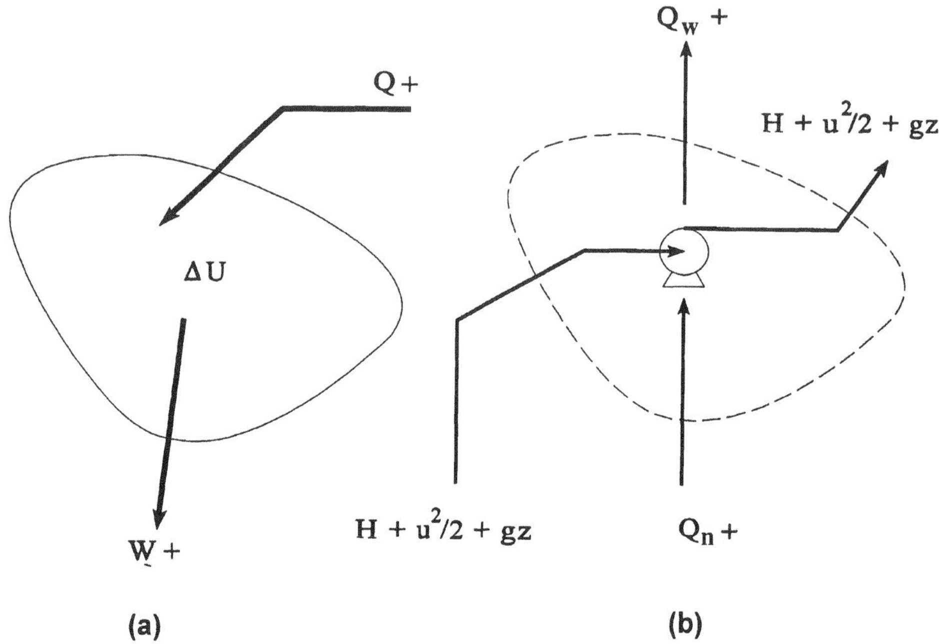

FIGURE 4.7 First law of thermodynamics for a closed (a) and open (b) system.

quantity of pressure-volume work per unit mass on the system equal to $P_1\hat{V}_1$ as indicated previously. Similarly, the loss of fluid at pressure P_2 at the exit does a quantity of work per unit mass equal to $P_2\hat{V}_2$ on the surroundings. Let $\hat{Q}_h$ and $\hat{Q}_w$ represent the flow of heat and work to the system per unit mass, respectively. The energy balance becomes

$$\left.\left(\hat{E}_i + \frac{u^2}{2} + gz\right)\right|_2 - \left.\left(\hat{E}_i + \frac{u^2}{2} + gz\right)\right|_1 = \hat{Q}_h - \hat{Q}_w + P_1\hat{V}_1 - P_2\hat{V}_2$$

$$\left.\left(\hat{E}_i + P\hat{V} + \frac{u^2}{2} + gz\right)\right|_2 - \left.\left(\hat{E}_i + P\hat{V} + \frac{u^2}{2} + gz\right)\right|_1 = \hat{Q}_h - \hat{Q}_w \tag{4.38}$$

$$\left.\left(\hat{H} + \frac{u^2}{2} + gz\right)\right|_2 - \left.\left(\hat{H} + \frac{u^2}{2} + gz\right)\right|_1 = \hat{Q}_h - \hat{Q}_w$$

Thus, when the pressure-volume work is separated from the external work, enthalpy rather than internal energy arises in the energy balance.

Returning to the nonflowing system, let us consider the change in internal energy in a reversible system. The reversible system is one that involves no loss of energy due to friction or a similar mechanism and thus can be continuously repeated forever without additional energy input.

In a reversible system, the heat added to the system is given by $T\Delta S$ where ΔS is the entropy change of the system due to the addition of heat. In real processes, the change in entropy of the system is greater than the heat added. Entropy is a property of a system that identifies its tendency to move toward a completely random or disordered state. All spontaneous processes, that is, processes that are not driven by the external application of energy or work, proceed such that the net entropy or randomness of the universe increases. Using this definition for the heat and with the definition of $\delta W = P\Delta V$ from Equation 4.16, the first law of thermodynamics becomes

$$\Delta E_i = T\Delta S - P\Delta V \tag{4.39}$$

This relation defines the change in internal energy of the system as a function of its temperature, pressure, and change in entropy and volume. It is limited to a *closed* system, or one in which there is no loss of mass by either transport or reaction. Although developed for a reversible system, Equation 4.39 is applicable to real systems if all variables are considered system variables and not indicative of the state of the surroundings. That is, –PΔV refers to the pressure and volume change of the system and TΔS is the temperature and entropy change of the system. These are identical to the heat and work added by or to the surroundings only in a reversible system. Due to irreversible losses such as frictional losses, additional work must be done by the surroundings to transfer this much energy to the system. Substituting the definition of the change in enthalpy of a system gives

$$\begin{aligned}\Delta H &= \Delta E_i + \Delta(PV) \\ &= \Delta E_i + P\Delta V - V\Delta P \\ &= T\Delta S + V\Delta P\end{aligned} \tag{4.40}$$

Note that the expansion of the change in the quantity PV follows the same rules as expansion of a differential (i.e., d(PV) = PdV + VdP) Neither Equation 4.39 or 4.40 are especially convenient to work with in that we normally cannot easily define the entropy of a system. It is much more convenient to define a thermodynamic quantity whose changes depend on the easily measured quantities of temperature and pressure. Such a thermodynamic quantity is the Gibbs' free energy which is defined by

$$G = H - TS \tag{4.41}$$

The change in Gibbs' free energy in a process can be written in the form

$$\begin{aligned}\Delta G &= \Delta H - T\Delta S - S\Delta T \\ &= T\Delta S + V\Delta P - T\Delta S - S\Delta T \\ &= V\Delta P - S\Delta T\end{aligned} \tag{4.42}$$

Thus, at constant temperature and pressure, the Gibbs' free energy change for a closed system is zero. Note that constant temperature and pressure implies that there is no change in the state or condition of the (single phase) system since specification of these two variables was shown by the phase rule to completely specify the system. That is, the system is at equilibrium. This can be evaluated further by rewriting the change in Gibbs' free energy in terms of the heat input and change in entropy.

$$\begin{aligned}\Delta G &= \Delta(H - TS) \\ &= \Delta(E_i + PV - TS) \\ &= \Delta E_i + P\Delta V + V\Delta P - T\Delta S - S\Delta T \\ &= \delta Q - \delta W + P\Delta V - T\Delta S \\ &= \delta Q - T\Delta S\end{aligned} \tag{4.43}$$

For a reversible system the heat added to a system from the surroundings is exactly equal to the temperature times the change in entropy of the system, i.e., $\delta Q = T\Delta S$. Thus, the Gibbs' free energy change is zero for any ideal reversible process that takes place at constant temperature and pressure.

Real processes, however, are irreversible without the input of energy. All real processes can proceed in only one direction without the input of energy required for reversal. The preferred direction occurs spontaneously, that is, without the input of energy from the surroundings. A spontaneous process is one that liberates the energy necessary to drive the process internally, one in which the final Gibbs' free energy is less than the initial. Since the change in Gibbs' free energy is final state minus initial state, this means that the change in Gibbs' free energy for a spontaneous change must be negative.

Alternatively, for any real, irreversible process the change in entropy of the system must be greater than the heat added (otherwise the change in Gibbs' free energy would be positive). This is a statement of the second law of thermodynamics,

$$\Delta S \geq \frac{\delta Q}{T} \tag{4.44}$$

Note that the requirement that the change in Gibbs' free energy for a spontaneous process is negative does not mean that other processes cannot occur. It simply means that the energy to drive the process is not available internally and instead must be provided by the surroundings in the form of heat or work. The process within the system may then exhibit a positive Gibbs' free energy change (i.e., an increase in Gibbs' free energy) or even a decrease in entropy. Because all real processes are irreversible, however, this must be coupled with even more heat or work from the surroundings and a greater increase in the entropy of the surroundings.

This has important implications for the environmental engineer. In particular, it provides an indication of the minimum energy input required to drive a nonspontaneous process. One of the most important environmental engineering concerns is the separation of a waste constituent from a mixture, for example, in the removal of a trace organic contaminant from a wastewater stream. This "ordering" process necessarily decreases the entropy of the effluent stream and thus must necessarily be driven by the introduction of energy from external to the system. The minimum amount of work required to achieve the separation is that equal to the change in Gibbs' free energy for the proposed separation.

$$\delta W_{min} = \Delta G \tag{4.45}$$

In reality, a greater amount of energy input is required to insure that the net change in Gibbs' free energy for the process is negative.

4.3.2 Gibbs' Free Energy and Equilibrium

The Gibbs' free energy relationships described in the previous section are also important in that they provide a means of defining equilibrium. If a hypothetical process is envisioned for which the Gibbs' free energy change is zero, then the process is reversible, which for any real system implies that it is at equilibrium. For any process in which the Gibbs' free energy change is negative, the process will increase the entropy of the system and therefore it is thermodynamically favored. The process will tend to occur spontaneously. The driving force for the process is the maximization of entropy in the universe and the result is the minimization of the Gibbs' free energy. It is this principle that defines both physical (phase) equilibrium and chemical equilibrium.

Our immediate interest is defining physical equilibrium with respect to particular constituents. We must first extend our current definition of Gibbs' free energy to a multicomponent system that can exhibit changes in composition (an open system). The total differential of the Gibbs' free energy can be written

$$dG = \frac{\partial G}{\partial P}dP + \frac{\partial G}{\partial T}dT + \sum \frac{\partial G}{\partial n_i}dn_i$$
$$dG = VdP - SdT + \sum \mu_i dn_i \tag{4.46}$$

As indicated previously, the first two terms are simply the influence on pressure and temperature in a closed system. The last term reflects the Gibbs' free energy change due to composition changes. The term μ_i is the partial molar Gibbs' free energy and n_i is the number of moles of component i in the system. Also μ_i is termed the chemical potential because it indicates the driving force for the transfer of mass between phases.

At constant temperature and pressure, the total Gibbs' free energy of a mixture can be written as simply the sum of contributions from each of its component substances.

$$G = \sum \mu_i n_i \tag{4.47}$$

Again μ_i is the partial molar Gibbs' free energy ($\partial G/\partial N_i$) and the n_i is the number of moles of component i. Note that the change in Gibbs' free energy can be written

$$dG = \sum \mu_i dn_i + \sum n_i \, d\mu_i \tag{4.48}$$

Which after substituting into Equation 4.46, suggests

$$SdT - VdP + \sum n_i \, d\mu_i = 0 \tag{4.49}$$

This is called the Gibbs-Duhem equation and relates the change in partial molar Gibbs' free energy to composition, temperature, and pressure. It is the key starting point in the definition of physical equilibrium. Note that at constant temperature and pressure, Equations 4.48 and 4.49 suggest

$$\sum n_i \, d\mu_i = 0 \qquad \text{Equilibrium at Constant } T, P$$
$$\sum \mu_i \, dn_i = 0 \tag{4.50}$$

That is, both terms in Equation 4.48 are separately zero at equilibrium.

Considering first physical equilibrium, equilibrium will occur when mass transfer between phases results in no net change in the Gibbs' free energy. That is, $\Delta G_I + \Delta G_{II} = 0$ where the subscripts indicate the separate phases. Dividing the change in Gibbs' free energy by the change in moles of component i from phase I, $(\Delta n_i)_I$, and recognizing that this is opposite in sign to the change in moles of i in phase II, the following result is obtained.

$$0 = \Delta G_I + \Delta G_{II} = \frac{\Delta G_I}{(\Delta n_i)_I} + \frac{\Delta G_{II}}{(\Delta n_i)_I} = \left.\frac{\Delta G}{\Delta n_i}\right|_I - \left.\frac{\Delta G}{\Delta n_i}\right|_{II} \tag{4.51}$$

For infinitesimal changes, the last form suggests that the criteria for equilibrium is simply that the partial molar Gibbs' free energies of the distributing component are equal in the two phases,

$$\left(\frac{\partial G}{\partial n_i}\right)_I = \left(\frac{\partial G}{\partial n_i}\right)_{II} \tag{4.52}$$

$$\mu_i|_I = \mu_i|_{II}$$

Because differences in partial molar Gibbs' free energy of a compound result in mass transfer of that compound between the phases, it has come to be called the *chemical potential.* This name emphasizes its importance as the driving force for mass transfer just as electrical potential indicates the driving force for electricity. Use of the chemical potential to define equilibrium requires its specification. Just as with Gibbs' free energy, only differences in chemical potential are important and a reference state is required with which to refer measurements. In addition, it has proven inconvenient to work with chemical potentials directly. It is possible, however, to define the chemical potential with respect to a reference state for an ideal gas and to use this definition to form the basis for defining chemical potential or an equivalent quantity for real liquids and gases.

Consider a system at equilibrium ($\Delta G = 0$) undergoing a constant temperature change in pressure. From Equation 4.49,

$$\sum n_i \, d\mu_i = V \, dP \tag{4.53}$$

By definition of the partial pressure, Equation 4.53 can be used to determine the chemical potential of a single component, A, in an ideal gas.

$$\begin{aligned} d\mu_A &= \frac{V}{n_A} dp_A \\ &= RT \frac{dp_A}{p_A} \end{aligned} \tag{4.54}$$

Integrating between some reference chemical potential and the pressure at that reference state,

$$\mu_A = \mu_A^{ref} + RT \ln\left(\frac{p_A}{p_A^{ref}}\right) \tag{4.55}$$

In the environmental arena, the reference state is almost universally taken to be the *pure component at the same temperature and phase as the mixture at 1 atm pressure.* The ratio p_A/p_A^{ref} then becomes the mole fraction of the component in the vapor phase, $y_A = p_A/P$. The difference in chemical potential between an initial and final composition of an ideal gas (referring both to the same reference chemical potential) is then

$$\mu_A = \mu_A^{ref} + RT \ln(y_A) \tag{4.56}$$

The molar Gibbs' free energy for an ideal gas in the absence of reaction is then

$$\hat{G} = \hat{G}^{ref} + \sum_i y_i RT \ln(y_i)$$
$$\hat{G} = RT \ln\left(\Pi y_i^{y_i}\right) \quad (4.57)$$

where it is recognized that (a ln x) is equal to (ln x^a) and ln x + ln y = ln xy. Πx_i represents the product series $(x_A)(x_B)\ldots(x_N)$. It now becomes possible to estimate the change in Gibbs' free energy for an ideal gas. As indicated previously, this can be used to estimate the minimum energy input required to achieve a certain separation. This is explored in Example 4.8.

Example 4.8: Estimation of minimum work required to achieve a specified separation

Estimate the minimum energy input required to remove 99% of the benzene from an air stream. The inlet mole fraction of benzene is 0.01.

Basis: 1 mole of feed air

benzene $|_1 = 0.01$ moles   air $|_1 = 0.99$ moles

benzene $|_2 = 1\% * (0.01)$ moles   air $|_2 = 0.99$ moles

$$y_{BZ}|_2 = \frac{0.0001}{0.99 + 0.0001} \approx 0.0001 \qquad y_{AIR}|_2 \approx 0.9999$$

The change in Gibbs' free energy is then

$$W_{min} = \Delta\hat{G} = \sum_{i=1}^{2} y_i RT \ln(y_i)$$
$$= \left(8.3145 \frac{\text{J}}{\text{mol}\cdot\text{K}}\right)\cdot(298\ \text{K}) \ln\left(\frac{(0.0001^{0.0001})\cdot(0.9999^{0.9999})}{(0.01^{0.01})\cdot(0.99^{0.99})}\right)$$
$$= 136.2 \frac{\text{J}}{\text{mol}}$$

This is the minimum energy requirement to achieve the desired separation. Depending on the process used to achieve the separation, significantly more energy may be required.

When dealing with gases other than ideal gases, G.N. Lewis defined the fugacity, f, which is nothing more than a component pressure corrected for the nonidealities of the system. For an ideal gas, the fugacity is simply equal to the partial pressure of that component and Equation 4.55 determines the chemical potential. For a non-ideal gas, the chemical potential is given by

$$\mu_A = \mu_A^{ref} + RT \ln\left(\frac{f_A}{f_A^{ref}}\right) \quad (4.58)$$

Where, typically, $f_A^{ref} = p_A^{ref} = P = 1$ atm. The fugacity coefficient is the ratio of the fugacity to the partial pressure.

$$\Phi_A = \frac{f_A}{p_A} \tag{4.59}$$

The fugacity coefficient of an ideal gas is unity, and thus $f_A \sim p_A$.

For a liquid solution, Lewis defined an ideal solution by analogy to the ideal gas using the liquid mole fraction, x_i, rather than the gas mole fraction as in Equation 4.56,

$$\mu_A = \mu_A^{ref} + RT\ln(x_A) \tag{4.60}$$

A nonideal solution is then governed by

$$\mu_A = \mu_A^{ref} + RT\ln(a_A) \tag{4.61}$$

where a_i is the activity of component i in the liquid solution. The activity coefficient is then

$$\gamma_A = \frac{a_A}{x_A} \tag{4.62}$$

The activity coefficient of an ideal solution is also unity. Unfortunately, an ideal liquid solution is encountered much less frequently than an ideal gas. Thus, while the fugacity coefficient is unity for most environmental situations, the activity coefficient is likely to be very much different from one.

In terms of fugacity and activity coefficients, the statement of equilibrium at a vapor-liquid interface can be written

$$\begin{gathered} \mu_A|_I = \mu_A|_{II} \\ \mu_A^{ref}|_I + RT\ln(y_A\Phi_A)|_I = \mu_A^{ref}|_{II} + RT\ln(x_A\gamma_A)|_{II} \end{gathered} \tag{4.63}$$

where, if the choice of the reference states are selected such that

$$\mu_A^{ref}|_I - \mu_A^{ref}|_{II} = RT\ln\left(\frac{f_A^{ref}|_I}{f_A^{ref}|_{II}}\right) \tag{4.64}$$

the statement of equilibrium becomes simply,

$$(y_A\Phi_A f_A^{ref})|_I = (x_A\gamma_A f_A^{ref})|_{II} \tag{4.65}$$

This means that equality of chemical potential is equivalent (with suitable choices of the reference states) to equality of fugacity between the two phases. Equation 4.65 is the actual starting point for the determination of vapor-liquid equilibrium and liquid-liquid equilibrium in routine engineering calculations. In the subsequent sections, this concept will be extended and applied to a variety of important situations. Up until now we have explicitly recognized that a single component, A, is partitioning between the different phases and that the terms in the various equations for chemical potential and fugacity refer only to that component. The explicit designation that the terms refer to a particular component will generally be dropped to allow a simpler system of nomenclature. That is, terms such as γ_A will normally be written simply as γ with the component designation assumed.

Routine calculations using Equation 4.65 require knowledge of the appropriate reference fugacities and the values of the activity and fugacity coefficients. As will be seen, the almost universally employed reference condition for reference, or standard state fugacity, is the *pressure exerted by a pure component at the same temperature, pressure, and phase as the mixture*. Thus, the reference fugacity for a vapor component in the atmosphere is the pressure that would be exerted by the component if the gas phase were composed wholly of that component (i.e., $f_A^{ref} = P = 1$ atm). The reference fugacity for a liquid component is the pressure that would be exerted by a pure liquid phase of that component (i.e., the pure component vapor pressure). To summarize,

$$f_A^{ref}\big|_{gas\ phase} = P \quad \text{(1 atm in the environment)}$$
$$f_A^{ref}\big|_{liquid\ phase} = P_v \quad \text{(the pure component vapor pressure)} \tag{4.66}$$

The fugacity and activity coefficients can often be defined at the concentration or mole fraction at which the pure component pressures (i.e., f_A^{ref}) are exerted in a phase. For a liquid phase, the component will exert the standard state pressure, or vapor pressure, when present at its solubility limit in the water or other solvent. Thus, before turning to the practical applications of fugacity, we must further explore vapor pressure and solubility.

4.4 VAPOR PRESSURE

The phase changes that accompany heating and cooling are closely linked to the concept of vapor pressure. The vapor pressure of a substance is the pressure exerted by a liquid due to the equilibrium between evaporation and condensation processes at the liquid-vapor interface. Under conditions typically of interest to environmental engineers, the vapor pressure is strictly a function of the temperature, which controls the energetics of the molecules attempting to leave the liquid surface by evaporation. At higher temperatures, more energetic molecules can leave the liquid surface and the vapor pressure is higher. At lower temperatures, the molecular energetics are lower and the vapor pressure is lower. Note that the vapor pressure is the pressure that would be exerted by the liquid at a given temperature regardless of the pressure on the system. Water exerts a vapor pressure of 0.006 atm at 0°C. Because this is the pressure that would be exerted by water if it filled the entire vapor space, the vapor pressure is equal to the partial pressure. Thus, in the normal atmosphere exerting 1 atm pressure, the equilibrium mole or volume fraction of water vapor is 0.006 or 0.6% at 0°C. This was illustrated previously in Example 4.5.

When both the liquid and associated vapor state are at equilibrium, both the liquid and the vapor are said to be saturated. Each phase contains as much of the other phase as is possible. If the vapor were to contain any additional water, condensation would occur to cause the system to return to equilibrium. If the liquid contained any more of the "vapor," evaporation would occur to return the system to equilibrium.

This equilibrium state is satisfied as long as the temperature remains constant. If the temperature increases, the energetics of the evaporation process increase driving more water into the vapor phase. As the water concentration in the vapor phase increases, however, the rate of condensation also increases until ultimately a new equilibrium is reached and both phases are again saturated.

Conditions are rarely at equilibrium in the natural environment. At any given instant, a relative saturation can be defined as the ratio of the actual partial pressure to the partial pressure at equilibrium

$$\text{Relative Saturation} = \frac{p_A}{p_A^*} = \frac{p_A}{P_v} \tag{4.67}$$

When applied to water vapor in the atmosphere, the relative saturation is referred to as relative humidity

$$\text{Relative Humidity} = RH = \frac{p_{H_2O}}{p^*_{H_2O}} \tag{4.68}$$

Generally, this ratio is multiplied by 100 and expressed as a percent. The term 100% relative humidity implies that the partial pressure of water vapor is equal to the equilibrium or vapor pressure of water at that temperature. A relative humidity of 100% defines the dew point temperature of the air since the air is then saturated with water and condensation will occur if any further decreases in temperature are experienced.

The rate of increase in vapor pressure with temperature is dependent upon the latent heat required to achieve the vaporization, $\Delta \hat{H}_v$. The relationship between vapor pressure ($P_v = p^*$) and temperature is governed by the Clapeyron equation

$$\frac{dP_v}{dT} = \frac{\Delta \hat{H}_v}{T\left(\hat{V}_v - \hat{V}_l\right)} \tag{4.69}$$

where the subscripts v and l on the molar volumes represent the vapor and liquid phase, respectively. Equation 4.69 recognizes that the temperature dependence of a certain process (in this case, vaporization) is related to the energy required to initiate the process. That is, the more energy required to initiate a process, the more sensitive that process will be to temperature which provides a thermal source for that energy. Generally the molar volume in the gas phase is much larger than that in the liquid phase and given by the ideal gas law. If $\Delta \hat{H}_v$ is assumed approximately independent of temperature, the Clapeyron equations becomes

$$\frac{dP_v}{P_v} = d\ln(P_v) = \frac{\Delta \hat{H}_v dT}{RT^2} \tag{4.70}$$

which can be integrated to

$$\ln P_v = \alpha + \frac{-\Delta \hat{H}_v}{RT} \tag{4.71}$$

where α is a constant of integration. Equation 4.71 indicates that a plot of the logarithm of vapor pressure vs. 1/T (absolute temperature) should yield a curve whose slope is $-\Delta \hat{H}_v / R$ and an intercept to define α. Such a plot is shown in Figure 4.8 for water. The constant α could also be defined from a known vapor pressure at any temperature as long as the heat of vaporization remains essentially constant between the temperature of the known vapor pressure and the temperature of the desired vapor pressure. This is illustrated in Example 4.9.

Example 4.9: Estimation of the vapor pressure of water

Develop an equation for the vapor pressure of water between 0 and 50°C. The vapor pressure and heat of vaporization of water at 25°C are 0.0317 bar (0.03129 atm) and 2442.5 kJ/mol, respectively.

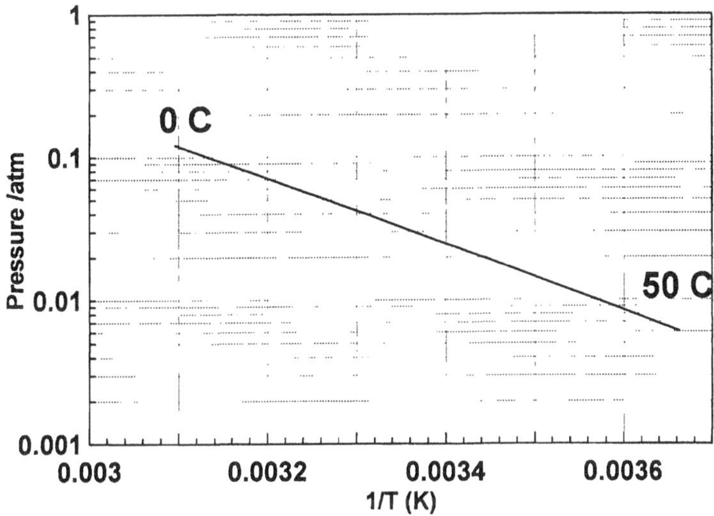

FIGURE 4.8 Logarithm of vapor pressure vs. inverse of temperature.

Assume that the heat of vaporization at 25°C is a good estimate of the heat of vaporization over the entire temperature range 0 to 50°C. The vapor pressure at 25°C can then be used to estimate the constant α.

$$\alpha = \ln\left[\frac{P_v(25°\text{C})}{\text{atm}}\right] + \frac{\Delta\hat{H}_v(25°\text{C})}{R(298\text{ K})} = 14.279$$

Note that 14.279 is the numerical value if the vapor pressure is given in atmospheres. A different value would be determined for a different pressure unit. $\Delta\hat{H}_v / RT$ is dimensionless if a consistent set of units are employed and therefore no specific units need be specified for these quantities. The equation for vapor pressure of water is then given by

$$\ln\left[\frac{P_v(T)}{\text{atm}}\right] = 14.279 - \frac{2442.5\text{ kJ/mol}}{RT}$$

A comparison of predicted and calculated water vapor pressures is tabulated below.

Temperature °C	Vapor Pressure (atm)	Predicted Vapor Pressure (atm)	% error
0	0.00603	0.00616	–0.86
10	0.01211	0.01222	0.43
20	0.02310	0.02312	1.21
30	0.04186	0.04193	1.09
40	0.07285	0.07324	0.76
50	0.12180	0.12357	–0.14

The normal boiling temperature, T_b, is the temperature at which the liquid vapor pressure equals 1 atm. At that point the equilibrium partial pressure of the evaporating compound is equal to the ambient pressure (unless pressurized) and all of the liquid will ultimately move into the vapor state. Thus, knowledge of the compound's boiling point provides the necessary vapor pressure and temperature to determine the constant A.

Use of Equation 4.71 also requires determination of the heat of vaporization of the compound, as discussed previously. Equation 4.71 assumes that the heat of vaporization is independent of temperature. The Clapeyron equation could also be integrated using Equation 4.33 for the temperature dependence of the heat of vaporization. An intermediate course would be to average the heat of vaporization over the temperature range of interest and employ that as a constant value in Equation 4.71.

The temperature dependence of the heat of vaporization is much more important if the compound is a solid at the temperature for which the vapor pressure is desired. As indicated by Equation 4.36, the coefficient m increases for solids well below the substances boiling point.

Under conditions when the temperature dependence of the heat of vaporization must be considered, Grain (1982) integrated the Clapeyron equation using Equations 4.33 and 4.34 and developed the following approximate relation for vapor pressure as a function of temperature,

$$\ln P_v \approx \frac{\Delta\hat{H}_v(T_b)}{RT_b}\left[1-\frac{(3-2T/T_b)^m}{T/T_b}-2m(3-2T/T_b)^{m-1}\ln T/T_b\right] \tag{4.72}$$

This relationship provides a good means of estimating the vapor pressure of a compound when data does not exist or is of questionable value for that compound. Its use is illustrated in Example 4.10.

Example 4.10:Vapor pressure of trichlorobiphenyl

Estimate the vapor pressure of the solid trichlorobiphenyl at 20 and 25°C. The normal boiling point is 315°C.

The heat of vaporization of trichlorobiphenyl can be estimated via

$$\Delta\hat{H}_b(T_b) = T_b\cdot\left[8.75+1.9872\cdot\left(\frac{T_b}{K}\right)\right]\cdot\frac{\text{cal}}{\text{mol}\cdot\text{K}} = 12.6\,\frac{\text{kcal}}{\text{mol}}$$

The ratio of T at 25°C to T_b is [298/(315 + 273)] = 0.507. This suggests that m = 0.8. At 20°C, however, the ratio T/T_b = 0.498 and m = 1.19. Evaluating the vapor pressures by substitution of these values into Equation 4.69.

P_v (20°C) = 1.15 $(10)^{-8}$ atm
P_v (25°C) = 1.33 $(10)^{-7}$ atm

It is very unlikely that the vapor pressure varies this significantly over the 5°C change in temperature. Instead the crude approximation of m = 0.8 at 25°C and m = 1.19 at 20°C is driving the difference. The lesson of this example is that a correlation such as Equation 4.72 should not be relied upon to give much more than order of magnitude estimates of vapor pressure for such low vapor pressure compounds.

4.5 SOLUBILITY

Vapor pressure indicates the potential for a pure liquid compound to equilibrate between the liquid and the gas phase. Solubility is the potential for a pure solid or liquid compound to partition into a liquid phase. As with gaseous equilibrium, the dissolution process is a dynamic equilibrium between molecules of the dissolving compound attempting to leave the solvent phase and molecules attempting to return to the solvent phase. The solubility can be defined as the amount of a substance (the *solute*) that will dissolve in a *solvent* phase (typically water) without forming a separate phase. The solubility of a compound in water is of special importance to us and we will give that the symbol S_w ($\tilde{S}_w$ on a molar basis).

The introduction of additional solute beyond the solubility limit will result in the formation of two phases — pure solvent and pure solute. The formation of two phases if the solute is present above the solubility level occurs regardless of whether the solute is solid or liquid. Thus, a liquid mixture can separate into two distinct phases in the same liquid state. A gaseous mixture cannot separate into distinct phases without a phase change to another state, either liquid or solid. The addition of water vapor to a saturated air phase results in the condensation of some water vapor to liquid water. The addition of water to a liquid phase already saturated with water, however, would result in the formation of a second liquid phase. Also unlike an ideal gas phase, the presence of other substances in a liquid phase can strongly influence the solubility of the solute. Such chemically specific behavior is not observed in ideal gases because of the lack of significant interactions between the gaseous molecules. A further complication in substances dissolving in liquids such as water is that the liquid may ionize or otherwise react with the solvent resulting in a completely different substance in the dissolved state. These effects will be considered in multicomponent partitioning later in this chapter.

Fortunately, the complex interactions between liquid solvents and nonreactive dissolving solutes can often be broken into two distinct classes, near-ideal solutions in which the solute is miscible (soluble) in all proportions and solutions in which the solute is miscible only in trace amounts. Organic solutes in other organic solvents are approximately described by the first class of compounds. Mixtures of these compounds are liquid mixtures that correspond most closely to ideal gaseous mixtures. Since they are miscible in all proportions, these solutes will not be considered in this section.

Often the solvent of interest is water. Water is a polar compound and the individual water molecules form a comparatively structured liquid with which many nonpolar organic molecules are unable to coordinate. The solubility of such compounds is low. Appendix A contains data on the water solubility of a variety of important chemical compounds. The *hydrophobicity* of a compound is a measure of its compatibility with water. Note that any measure of *hydrophobicity* should be correlated with water solubility. The octanol-water partition coefficient, for example, is a measure of the relative solubility of a substance in water and organic phases that indicates hydrophobicity and correlates with water solubility. The definition of the octanol-water partition coefficient is

$$K_{ow} = \frac{\text{Concentration in octanol phase}}{\text{Concentration in water phase}} \tag{4.73}$$

Hydrophyllic compounds will exhibit low concentrations in the octanol phase compared to the water phase and K_{ow} will be very small. *Hydrophobic* compounds, however, will exhibit high concentrations in the octanol phase compared to those in the water phase and the K_{ow} will be very large. Experimental observations with organic compounds have exhibited K_{ow} over the range of at least 10^{-3} to 10^7, or 10 orders of magnitude. Because of this large range, K_{ow} is often given in log units. The range of values of log K_{ow} is then from -3 to 7 (i.e., $10^{-3} \leq K_{ow} \leq 10^7$). Leo and Hansch present data on K_{ow} as well as methods for its estimation from the structure of a compound. Leo and Hansch also have correlated K_{ow} with various environmental parameters, including solubility. Lyman et al. (1990) presents the following correlation between solubility and K_{ow} based on 156 compounds. The correlation has a correlation coefficient of 0.874 suggesting that 87.4% of the variance in the data is described by the fitted equation.

$$\log \frac{1}{\tilde{S}_w} = 1.339 \log K_{ow}^{0.978} \qquad \text{(Solubility in mol/L)} \tag{4.74}$$

The theoretical basis of Equation 4.74 suggests that the slope of a curve of solubility vs. log K_{ow} should be -1, in close agreement with the experimental correlation.

Equation 4.74 is limited to the temperature in the vicinity of 25°C. The effect of temperature on solubility can be related to the energy changes associated with the dissolution. For a liquid dissolution into another liquid, the heat evolved is just the heat of solution, $\Delta\hat{H}_s$, and by analogy to Equation 4.69, the temperature dependence of solubility can be written

$$\ln x_w^* = \frac{\Delta\hat{H}_s}{RT^2}\,dT \tag{4.75}$$

where x_w^* is the mole fraction solubility in water given by the molar solubility in mol/L divided by the solution molar density mol/L, $(\tilde{S}/\tilde{\rho}_s)$. If $\Delta\hat{H}_s$ is independent of temperature (valid for small changes in temperature) and if the molar volume of the solution is assumed constant, this can be written

$$\ln \tilde{S}_w = \alpha - \frac{\Delta\hat{H}_s}{RT} \tag{4.76}$$

for some constant of integration A. For most liquids, $\Delta\hat{H}_s$ is small and can be either positive or negative, resulting in solubilities that can either increase or decrease with temperature. As indicated by Schwarzenbach et al. (1993), in near ambient temperatures the solubility of dichloromethane decreases with temperature, the solubility of trichlorethylene increases with temperature, and the solubility of benzene decreases with temperature below 15°C and increases with temperature above 20°C. In general, the solubility of liquids is only a weak function of temperature near ambient temperatures and its effect is often neglected. Note, however, that the heat of solution of compounds that react with water can be large. Most organic compounds of environmental interest do not react with water, are sparingly soluble, and do not exhibit significant heats of solution.

The solubility of compounds that are solids or gases at the temperatures for which solubility is desired are much more sensitive to temperature. The enthalpy change upon dissolution includes both the heat of solution and the latent heat of the phase change. Because the latent heats associated with phase changes are large, the total heats of solution are large and a strong temperature dependence of solubility results. For gases, the enthalpy of the phase change is the heat of condensation, i.e., the negative of the heat of vaporization, ΔH_v, and the temperature dependence is approximately given by (with α again a constant of integration)

$$\ln \tilde{S}_w \approx \alpha - \frac{-\Delta\hat{H}_v}{RT} \tag{4.77}$$

For solids, the enthalpy of the phase change is the heat of melting, and the temperature dependence is approximately given by

$$\ln \tilde{S}_w = \alpha - \frac{\Delta\hat{H}_m}{RT} \tag{4.78}$$

Because the melting associated with dissolution of solids requires the addition of heat while the condensation associated with dissolution of gases requires its liberation, the solubility of solids tends to increase with temperature and the solubility of gases tends to decrease with temperature. The heat of vaporization of many hydrocarbons is of the order of 0.35 kJ/g suggesting that a 10°C increase in temperature in the vicinity of 25°C will cause a decrease in solubility of about 20% for a gas of molecular weight 50. The heat of melting of many hydrocarbons is 3 to 10 times less resulting in a correspondingly smaller effect on solubility for solids. For the detailed estimation of solubility at temperatures other than ambient, the reader is referred to Lyman et al. (1990).

As indicated previously, the influence of cosolutes, or other compounds in the water, can significantly influence the solubility of a substance. The presence of dissolved salts or minerals leads to a salting out effect, or decreases in the solubility of a substance. The relationship between solubility and salt concentration has been given by

$$\log \frac{\tilde{S}_w}{\tilde{S}_{ws}} = \alpha(\tilde{S}_w)_s \tag{4.79}$$

Here, $\tilde{S}_{ws}$ is the solubility of the compound in the presence of salt and $(\tilde{S}_w)_s$ is the molar concentration of the salt. Seawater is approximately 0.5 *M* (molar, i.e., mol/L). For aromatic and polynuclear aromatic compounds, the value of α is typically between 0.2 and 0.3. Thus, the solubility of many compounds in seawater is about 80% of that which would be observed in freshwater.

Note that the molar solubility in any of the previous equations could be replaced with the mass solubility, S_w, by multiplying by the molecular weight of the solute compound.

Dissolved or dispersed organic matter can produce the opposite of the salting out effect, that is, it can increase solubility. Water soluble organic compounds, such as methanol, could be present at high concentrations effectively destroying the structure of the water matrix and allowing more organic solute molecules to be dissolved. Methanol or other low molecular weight water-miscible organic compounds must be present at concentrations exceeding 5 to 10% for this effect to be important.

Potentially of more significance are small organic carbon-containing particles that might be suspended in the water. Hydrophobic organic contaminants can be sorbed onto these particles, effectively increasing the total amount of contaminant that can be contained in the water phase. Fine particulate matter is normally differentiated into a dissolved, or *colloidal*, fraction and a suspended fraction which contains the larger particles. The dissolved fraction is defined as those particles that pass a 0.45-μm filter. This operational definition is arbitrary and largely based on the difficulty of separating smaller particles by filtration. The total contaminant loading in filtered water samples is generally referred to as the dissolved concentration, regardless of whether the contaminant is sorbed onto the colloidal particulate matter or truly dissolved. For the purposes of this section, however, it should be recognized that it is generally assumed that only the truly dissolved fraction is directly available for partitioning into another phase and is governed by the equilibrium partitioning relationships that will be developed. Only a portion of a hydrophobic contaminant that may be present in the water phase may be available for partitioning into another phase or for adsorption by living organisms coming into contact with the water. The fraction of a contaminant that is available for uptake by a living organism is its *bioavailable fraction*, and it is generally only this portion that is important in assessing the effects of the contaminant on the organism.

Much of the colloidal particulate matter is large molecular weight humic and fulvic acids that are decomposition products of natural organic material. Sediment and soil porewater may contain as much as 10 to 200 mg/L of dissolved organic carbon. Large molecular weight organic molecules are particularly important if the molecule has substituent chemical groups that are hydrophyllic, such as polar OH (alcohols), O (aldehydes, esters, ethers, and ketones), and COOH groups (carboxylic acids). Under some conditions the organic molecules can form micelles, loose structures that arrange themselves such that hydrophyllic groups are exposed to the water and the remaining hydrophobic portion of the molecule collects near the center. A depiction of a micelle like structure is shown in Figure 4.9. Because of the hydrophobic nature of the region at the center of the micelle, hydrophobic contaminants tend to collect there, greatly raising the capacity of the water for the hydrophobic material. Because sorption to the large molecular weight organic colloidal material

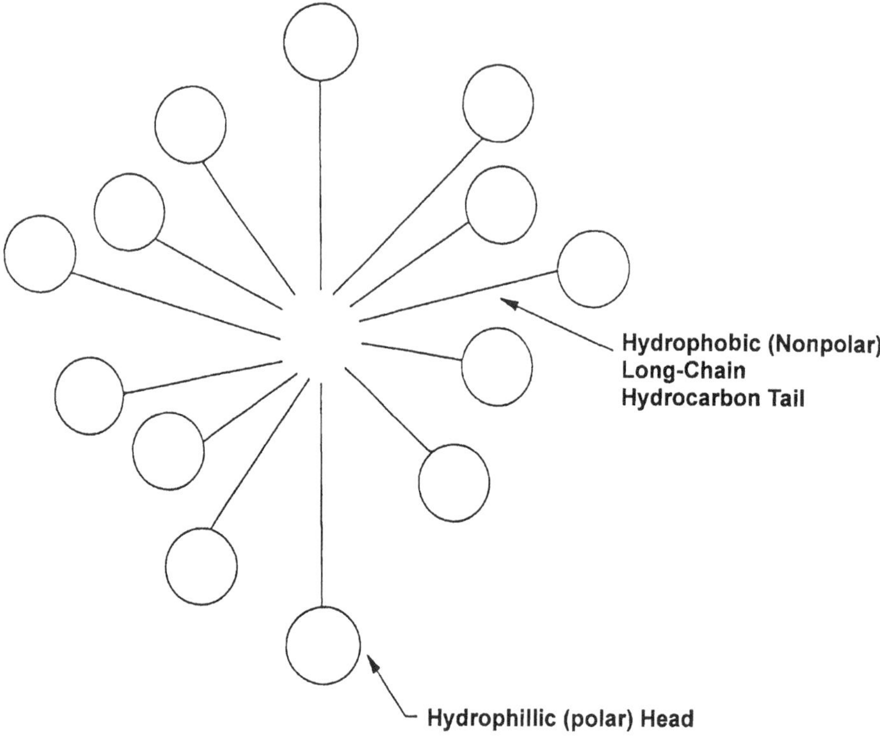

FIGURE 4.9 Crude depiction of micelle structure.

and micelles is similar to sorption to solids in many ways, we will discuss this effect further later in this chapter.

Because of their combined hydrophobic-hydrophyllic nature, these colloidal compounds will also tend to collect at surfaces between water and a separate organic phase with the hydrophyllic portion preferentially oriented toward the water and the hydrophobic portion preferentially oriented toward the organic phase. These compounds are called surface active agents, or *surfactants.* Surfactants can influence the properties of the fluid-fluid interface, for example, greatly reducing its interfacial tension, in addition to influencing the solubility of hydrophobic compounds. It is possible for surfactants to change the interfacial tension at a water-oil interface, for example, by orders of magnitude with only a few percent of the surfactant in the water. It is this process which is employed to enhance oil recovery and soil remediation processes with surfactants.

4.6 FLUID–FLUID PARTITIONING

We are now in a position to define the equilibrium composition of two fluids in contact with each other. As indicated previously, the criteria for physical equilibrium is equality of fugacities of a component across the interface between the two phases. For example, at a vapor-liquid interface, equality of fugacities for a particular component can be written,

$$\begin{aligned} f|_v &= f|_l \\ (y\Phi f^{ref})|_v &= (x\gamma f^{ref})|_l \end{aligned} \tag{4.80}$$

The left-hand side of each equation represents a vapor phase with a mole fraction, y, and fugacity coefficient, Π, for the partitioning component. The right-hand side represents a liquid phase with a mole fraction, x, and an activity coefficient, γ, for the partitioning component. If both phases were liquid, y and Π should also be taken as a liquid phase mole fraction and activity coefficient, respectively. Note that as indicated previously, it is assumed that each term in Equation 4.80 applies to a particular component and the explicit designation via subscript A is dropped. Let us consider the application of Equation 4.80 to a variety of situations important to the environmental engineer.

4.6.1 Ideal Solutions

Ideal solutions were defined earlier by Equations 4.56 and 4.60. In an ideal solution, both the fugacity coefficient in the vapor phase and the activity coefficient in the liquid phase are unity. Thus, the criteria for equilibrium becomes

$$(y f^{ref})|_v = (x f^{ref})|_l \tag{4.81}$$

The reference or saturation fugacities refer to the pressure that would be exerted by the component in a pure state at the *same temperature, pressure, and phase as the mixture*. In a vapor state, a pure phase would exert the total pressure, P, for example, 1 atm in the natural environment. In the liquid state, a pure phase would exert the pure component vapor pressure, P_v. Thus,

$$\begin{aligned} f^{ref} &= P_v \quad \text{(liquid state)} \\ f^{ref} &= P \quad \text{(vapor state)} \end{aligned} \tag{4.82}$$

Physical equilibrium with respect to a component A at a vapor-liquid interface between ideal solutions can now be written

$$\begin{aligned} yP &= p_A = xP_v \\ \tilde{C}_v &= \frac{p_A}{RT} = x\frac{P_v}{RT} \end{aligned} \tag{4.83}$$

Thus, the partial pressure of a component above an *ideal solution* is simply equal to the pure component vapor pressure for that component times the mole fraction of that component in the liquid phase. This is termed Raoult's Law.

It is often convenient to consider equilibrium at an interface in terms of a *partition* or *distribution coefficient*, which is simply the ratio of the component concentration in each of the two phases. If K_{vl} represents the ratio of the concentration in the vapor phase to that in the liquid phase, this becomes for an ideal solution governed by Equation 4.83

$$K_{vl} = \frac{\tilde{C}_v}{\tilde{C}_l} = \frac{xP_v}{RT\tilde{C}_l} = \frac{P_v}{RT\tilde{\rho}_l} \tag{4.84}$$

This equation recognizes that the molar concentration of the particular contaminant in the liquid, $\tilde{C}_l$, is equal to the mole fraction times the molar density of the liquid, $x\tilde{\rho}_l$. Because both the molar vapor and liquid concentrations can be multiplied by the molecular weight of the partitioning compound, the partition coefficient is identical if based upon molar concentrations (e.g., mol/L) or mass concentrations (g/L).

Note that the partition coefficient defined by Equation 4.84 is not dependent upon the concentration of the contaminant in either the vapor or liquid phase. The advantage of employing a distribution coefficient is that the ratio of the concentration between the two phases is generally constant (or approximated as a constant) and independent of the composition of the phases. Thus, a distribution coefficient for a particular compound between two phases can be tabulated. An illustration of Equation 4.84 and Raoult's Law of partitioning can be found in Example 4.11.

Example 4.11: Raoult's Law of partitioning

Estimate the partial pressure and air-liquid partition coefficient for benzene in 1% (by weight) mixture with gasoline. Assume that the gasoline properties are equivalent to pure octane.

The molecular weight of octane is 114 g/mol while that of benzene is 78 g/mol. Therefore, 1% by weight is equivalent to a mole fraction, x_B, of

$$x_B = \frac{\dfrac{1\text{ g}}{78\text{ g/mol}}}{\dfrac{1\text{ g}}{78\text{ g/mol}} + \dfrac{99\text{ g}}{114\text{ g/mol}}} = 0.0146$$

using a basis of 100 g of the "gasoline" mixture. The density of octane and benzene, respectively, is 0.7 g/cm^3 and 8.79 g/cm^3. The mixture mass density is then

$$\rho_M = (0.01)\cdot(0.879\text{ g/cm}^3) + (0.99)\cdot(0.703\text{ g/cm}^3) = 0.705\text{ g/cm}^3$$

Note that as indicated previously, the density of the gasoline is effectively that of pure octane, since octane represents 99% of the mixture. The pure component vapor pressure of benzene is 0.125 atm at 25°C and the equilibrium partial pressure above the mixture is

$$p_B = x_B P_v = (0.0146)\cdot(0.125\text{ atm}) = 0.00182\text{ atm}$$

The concentration of benzene vapors in the air is then

$$\tilde{C}_a = \frac{p_B}{RT} = \frac{0.0182\text{ atm}}{0.08205\dfrac{\text{atm}\cdot\text{L}}{\text{mol}\cdot\text{K}}\cdot 298\text{ K}}$$

$$= 0.0744\text{ mol/m}^3 = 5.8\text{ g/m}^3$$

The partition coefficient between the air, a, and the gasoline or nonaqueous phase, n, is thus

$$K_{an} = \frac{\tilde{C}_a}{\tilde{C}_n} = \frac{5.8\text{ g/m}^3}{(0.01)\cdot(0.705\text{ g/m}^3)} = 0.000823$$

As indicated previously, an ideal gas is one in which the fugacity coefficient is equal to unity. Liquid phases, however, are rarely ideal and governed by the simple equilibrium relationships indicated by Raoult's Law in Equation 4.83 or 4.84. The most common example of liquid solutions that are near ideal include mixtures of a homologous series of hydrocarbons or other organic compounds. A homologous series implies a group of compounds of similar size and structure. For, example, commercially available butane (C_4H_{10}) often contains small amounts of propane (C_3H_8) and pentane (C_5H_{12}). These compounds, all examples of organic alkanes, form a comparatively simple homologous series and they would be expected to form a very nearly ideal solution. Oils and waxes might contain a broader distribution of compounds, but they also tend to be dominated by straight and broken chain alkanes and are very nearly ideal solutions.

Much more complicated solutions are gasoline and most lubricating and fuel oils. These substances contain compounds from a number of different homologous series, including alkanes, aromatics, and trace amounts of other types of organic compounds. Because all of these compounds are nonpolar organics, the assumption of ideality is generally a good one. It should be recognized, however, that this assumption can be quite wrong for some trace constituents. Water, a polar compound, behaves quite nonideally in oils. Its activity coefficient would be very much different from one and the partial pressure of water above an oil containing trace amounts of water would *not* be expected to be simply the vapor pressure of water times the mole fraction of water in the oil.

Typically of more interest is the equilibrium of a constituent such as benzene between an oil or gasoline and an adjacent water or air phase as illustrated in Example 4.11. Benzene is essentially nonpolar as is most of the other oil or gasoline components, and it is soluble in all proportions in the nonaqueous phase. These are key indicators that its activity coefficient in the nonaqueous phase should be near one and the solution is ideal. The activity of the coefficient of benzene in the water is not unity and the mixture is not an ideal solution. Let us now consider such a solution.

4.6.2 Nonideal Solutions

Although gas and hydrocarbon liquid phases are generally ideal or near ideal, mixtures of water and nonpolar organic compounds exhibit behavior far from ideal. Hydrophobic organic compounds exhibit very low solubilities in water, a key indicator of nonideal solution behavior. The solubilities of many organic compounds are so small that the solvent matrix in which they reside is effectively water, regardless of the concentration of the organic compound. That is, at any concentration between zero and the solubility of the organic compound, each molecule of the organic sees a "sea" of water molecules. Compounds with a large solubility typically show a concentration influence on activity coefficient because of the changing nature of the solvent liquid as more of the solute is incorporated. Hydrophobic organic compounds in water, however, show no concentration influence, since at all concentrations (below the solubility) the nature of the solvent phase does not change and the activity coefficient of the compound in water at a given temperature is effectively a constant. Moreover, the value of that constant can be easily determined from measurements of water solubility of the compound.

This can be demonstrated by recognizing that a water solution saturated with respect to a component (that is, the component is present at its solubility) exerts the pure compound vapor pressure of that component. Certainly the addition of even one additional molecule of solute to an already saturated water solution forms a separate pure phase which would exert the pure compound vapor pressure. The vapor pressure of benzene at 25°C is 0.125 atm. Thus, the equilibrium partial pressure of benzene above a saturated water solution (i.e., one containing 1780 g/m^3 of benzene) is 0.125 atm. This puts an entirely different perspective on the importance of trace quantities of contaminants in the environment. A benzene saturated water solution contains less than 0.2% benzene but the air above it will, at equilibrium, contain the same amount of benzene vapors as if a pure benzene phase were present.

Because the equilibrium partial pressure of a compound above a saturated solution is just the pure component vapor pressure and the standard state or reference fugacity of a liquid phase is also the pure component vapor pressure, equality of fugacities then suggests

$$y_i P = x\gamma f^{ref}|_w$$
$$P_v = x^*\gamma P_v|_w \qquad (4.85)$$
$$\gamma = \frac{1}{x_w^*}$$

Thus, the activity coefficient, γ, of a compound is simply equal to the inverse of the mole fraction of the compound at equilibrium with respect to the water, x_w^*. For a sparingly soluble compound such as a nonpolar organic compound in water, the activity coefficient is a very large number. For such compounds, each individual solute molecule sees essentially a "sea of water molecules" surrounding it. As the concentration of the solute is decreased further, there is no significant difference in the interaction of that compound with its surrounding water and the activity coefficient is expected to remain constant. This is termed the activity coefficient at infinite dilution.

The equilibrium partial pressure of a compound above a water solution at any concentration of the hydrophobic solute, A, can then be determined by

$$p_A = \frac{x}{x_w^*} P_v \qquad (4.86)$$

In practice, this is essentially no more complicated than the ideal solution except that the solubility of the compound in water is required. A compound present at some fraction of the solubility will exert that fraction of the pure component vapor pressure. That is, benzene present at a water concentration of half-saturation, or 890 g/m³, will exert, at equilibrium, a partial pressure of half the pure component vapor pressure, or 0.0625 atm. This is valid regardless of whether the water concentration of the solute is measured in mg/L or mol/m³ as long as the solubility is measured in the identical units. These concepts are illustrated in Example 4.12 for benzene in water.

Example 4.12: Partial pressure above a partially saturated benzene-water solution

Estimate the activity coefficient and the partial pressure of benzene above a water-benzene solution containing a benzene mole fraction of 0.00015. The solubility of benzene in water is 1780 g/m³.

Since the solubility of benzene in water is low (<0.2%), we will assume that the effect of benzene on the bulk water properties can be neglected and that the activity coefficient is a constant independent of benzene concentration.

$$x_B^* \approx \frac{\dfrac{S}{MW_B}}{\dfrac{\rho_w}{MW_W}} = \frac{\dfrac{1780\ \text{g/m}^3}{18\ \text{g/mol}}}{\dfrac{10^6\ \text{g/m}^3}{18\ \text{g/m}^3}} = 0.000411$$

Inclusion of the moles of benzene in the total number of moles in the denominator does not change the answer. The activity coefficient of benzene in water is then

$$\gamma = \frac{1}{x_B^*} = 2440$$

and the partial pressure of benzene over water containing 0.00015 mole fraction benzene is

$$p_B = x \gamma P_v = 0.0458 \text{ atm}$$

Recognizing that the ratio of the vapor pressure and the water solubility is a constant (at a given temperature), Equation 4.86 is often rewritten as

$$p_A = H_{aw} x_A \tag{4.87}$$

This is known as Henry's Law, and H_{aw} is known as the Henry's Law constant at the air-water interface. Henry's constant is simply the ratio of a compound's vapor pressure to its water solubility and can be found in a variety of different forms depending on the units used to quantify vapor pressure or solubility. In terms of a partition coefficient given by a ratio of the concentrations in each phase,

$$K_{aw} = \frac{C_a}{C_w} = \frac{\tilde{C}_a}{\tilde{C}_w} = \frac{P_v}{RT\tilde{S}_w} \tag{4.88}$$

where P_v /RT is the molar air concentration of the contaminant at its pure compound vapor pressure and $\tilde{S}_w$ is the molar solubility of the contaminant in water.

The student is cautioned that Henry's Law constant or air-water partition coefficient is never dimensionless. However, if the vapor phase concentration at equilibrium and the water solubility in equilibrium are both defined in the same units as in Equation 4.88, this coefficient will *appear* dimensionless. Unfortunately, this is true whether the concentration units are mole fractions or mass concentrations, and it is easy to err by using a Henry's Law constant in a manner inconsistent with its definition. Example 4.13 illustrates the difference between "dimensionless" Henry's constants for benzene. We will attempt to avoid this problem by using the term "air-water partition coefficient" when the ratio of molar or mass concentrations is implied and explicitly defining the units of the Henry's constant in any other case.

Example 4.13: "Dimensionless" Henry's constant

Estimate the value of Henry's constant for benzene in water given the information in Example 4.11.

Henry's constant is simply the ratio of the air to water concentration of benzene.

$$K_{aw} = \frac{\tilde{C}_a}{\tilde{C}_w} = \frac{\frac{P_v}{RT}}{\tilde{S}_w} = \frac{0.0625 \text{ atm}}{0.08205 \frac{\text{atm} \cdot \text{m}^3}{\text{mol} \cdot \text{L}} \cdot (298 \text{ K}) \cdot (1780 \text{ g/mol})} = 0.224$$

We will use the symbol K_{aw} to represent the ratio of concentrations in air and water. Another common form of Henry's Law is the ratio of the mole fraction in the air to that in the water.

$$H_{aw} = \frac{y}{x} = \frac{P_v / P}{\tilde{C}_w / \tilde{\rho}_w} = \frac{0.125}{0.000411} = 304$$

As you can see the use of both definitions of Henry's constant without an indication of the units can cause significant errors. In this book, the ratio of concentrations will be denoted K_{ow} while H will be used either to denote the dimensionless ratio of mole fractions, or, if dimensions are explicitly identified, any other form of Henry's constant. Since other authors refer to either of the dimensionless forms as H or Henry's constant, the student should always be wary of the units associated with a dimensionless Henry's Law constant.

Note that the air-water partition coefficient can be large if either the vapor pressure of the dissolving compound is large or if the solubility of the compound is small. Thus, dissolved gases tend to exhibit very large air-water partition coefficients constants as do moderate vapor pressure and marginally water soluble compounds such as benzene. Perhaps surprisingly, a number of very low vapor pressure compounds such as certain polynuclear aromatics and polychlorinated biphenyls also exhibit large air-water partition constants because of their equally low solubility in water. As will be seen in numerous examples throughout this text, a compound that has a low vapor pressure does not necessarily mean that it will not vaporize in the environment.

The activity coefficient defined by Equation 4.85 also can be used to evaluate liquid-liquid equilibrium. The presence of two liquid phases suggests that the phase constituents are themselves not mutually miscible, at least in all proportions. The most common situation is a water and an oily or other nonaqueous phase liquid (NAPL) placed in contact with each other. A component distributing between these phases could form nonideal solutions with both phases. The equilibrium relationship is then defined by

$$(x\gamma f^{ref})|_w = (x\gamma f^{ref})|_n \tag{4.89}$$

for the two phases, water (w) and NAPL (n). Recognizing, however, that the saturation fugacities are both equal to the vapor pressure of component at the system temperature,

$$(x\gamma)|_w = (x\gamma)|_n \tag{4.90}$$

Often, a hydrophobic organic compound might form a near-ideal solution with the organic phase while the water phase will exhibit significant deviations from ideality. In such a situation, equilibrium is defined by

$$x_n = x_w\,\gamma_w \tag{4.91}$$

The distribution coefficient, K_{nw}, is simply the ratio,

$$K_{nw} = \frac{C_n}{C_w} = \frac{\tilde{C}_n}{\tilde{C}_w} = \frac{x_n\tilde{\rho}_n}{x_w\tilde{\rho}_w} = \gamma_w\frac{\tilde{\rho}_n}{\tilde{\rho}_w} \tag{4.92}$$

or the activity coefficient in the water times the ratio of the molar densities of the nonaqueous and water phases. Example 4.14 illustrates the evaluation of liquid-liquid partitioning.

Example 4.14: Partitioning of benzene between an organic and aqueous phase

Estimate the concentration of benzene in water that would be in equilibrium with a gasoline composed of 1% benzene and 99% octane.

As calculated in the previous examples, the mole fraction of benzene in this "gasoline" is 0.0146 and the activity coefficient in water is 2440. Noting that the pure phases are both liquids and therefore the reference fugacities are identical and that γ_n is approximately unity gives,

$$x_w = \frac{x_n}{\gamma_w} = \frac{(0.0146)}{(2440)} = 5.98 \cdot 10^{-6}$$

This can be converted to a concentration, recognizing that it is sufficiently small that the water constitutes effectively all of the bulk aqueous phase

$$C_w = x_w \left[\frac{MW_B}{MW_{H_2O} \cdot 1/\rho_w} \right]$$

$$= \left(5.98 \cdot 10^{-6} \frac{\text{mol}_B}{\text{mol}_{H_2O}} \right) \left[\frac{78 \text{ g/mol}_B}{(18 \text{ g/mol}_{H_2O}) \cdot (0.001 \text{ m}^3/\text{kg})} \right]$$

$$= 0.026 \text{ kg/m}^3$$

Since the concentration in the gasoline phase is (0.01) (0.705 g/cm^3) = 7.05 kg/m^3, the organic water partition coefficient is

$$\frac{C_n}{C_w} = \frac{7.05 \text{ kg/m}^3}{0.026 \text{ kg/m}^3} = 271$$

Note that this is identical to the ratio of the air-water to air-organic phase partition coefficients

$$K_{nw} = \frac{K_{aw}}{K_{an}} = \frac{0.224}{0.000823} = 271$$

A particularly useful partition coefficient between a nonaqueous and aqueous phases is the octanol-water partition coefficient. This is the partition coefficient of a contaminant between *n*-octanol (octanol with the OH grouping attached to an end carbon) and water,

$$K_{ow} = \frac{C_o}{C_w} \tag{4.93}$$

The partitioning between the essentially nonpolar octanol and water provides a good indication of the hydrophobicity of the contaminant. As such this partition coefficient is an excellent indicator of sorption onto soil organic carbon and into the organic fraction of living organisms.

Note that distribution coefficients can always be combined to define other distribution coefficients as illustrated in the example. The rules of distribution coefficient math are

$$\begin{aligned} K_{ij} \cdot K_{jk} &= K_{ik} \\ \frac{K_{ij}}{K_{kj}} &= K_{ij} \cdot K_{jk} = K_{ik} \end{aligned} \tag{4.94}$$

Only the truly dissolved fraction tends to partition according to the rules outlined above. As indicated previously, a liquid phase can contain a greater quantity of the contaminant than predicted by these partitioning relationships, however, if additional contaminant is sorbed onto a colloidal or suspended particulate phase.

The simple picture of linear partitioning between phases breaks down when the solute moving into solution disappears as a result of reaction. The dissolution of sulfur dioxide, for example, does not appear to follow Henry's Law of a linear increase in vapor pressure with dissolved sulfur dioxide concentration. Actually, the partitioning of physically dissolved sulfur dioxide does follow Henry's Law. Physically dissolved sulfur dioxide in water disappears according to the following reactions.

$$(SO_2)_g + H_2O \rightarrow (SO_2)_{aq}$$
$$(SO_2)_{aq} + H_2O \rightarrow H^+ + HSO_3^- \tag{4.95}$$
$$HSO_3^- + H_2O \rightarrow H^+ + H_2SO_4^{-2}$$

As the dissolved sulfur dioxide reacts, additional gaseous sulfur dioxide is absorbed to maintain equilibrium of the first reaction. The total sulfur dioxide adsorbed, however, varies nonlinearly with the gaseous sulfur dioxide available due to the varying extent of the subsequent reaction. Carbon dioxide undergoes a similar process of sorption and reaction and similarly appears to exhibit nonlinear partitioning between vapor and liquid phases. The net effect of reactions of the dissolved species is similar to the effect of further sorption of hydrophobic organics on colloidal organic matter. The removal of the truly dissolved fraction by reaction causes additional dissolution just as does removal of the truly dissolved hydrophobic organic compound by sorption onto a colloidal phase.

Compounds that are more soluble in water or form nonideal organic phases also are not described by linear partitioning. As indicated earlier, in this case the activity coefficients are a function of the concentration and therefore the partition or distribution coefficient is also a function of concentration. For most dissolved gases and hydrophobic compounds, the assumption of constant activity coefficient with water or air is a good one and linear partitioning between the truly dissolved contaminants of each phase is valid. As indicated above, reactions or further sorption in the water phase may need to be explicitly evaluated to account for contaminants that undergo other processes after dissolution.

4.6.3 Ionic Solutions

Among the compounds that react in water and therefore do not partition linearly are ionic species. It is generally convenient in ionic solutions to express concentration in molality, or moles per kilogram of solvent (typically water), or for the dilute solutions of interest here, molarity, mol/L of solution, $\tilde{M}$. The activity is then defined by

$$a_A = \tilde{M}_A \gamma_A \tag{4.96}$$

where $\tilde{M}_A$ is the molarity and γ_A is the activity coefficient of component A. The activity coefficient of ions in solution cannot be estimated by any of the methods discussed above. The activity coefficient γ_A is a function of the ionic strength of the solution, I_s, given by

$$I_s = \frac{1}{2}\sum \frac{n_i}{V} z_i^2 \tag{4.97}$$

where z_i is the valence or number of charges that an ion carries and n_i/V is the number concentration of ions. For dilute solutions of ions in water, their activity coefficient can be determined by the extended Debye-Huckel Equation, which at 25°C in water, takes the form

$$\log \gamma_A = \frac{-0.51(z^2)\sqrt{I_s}}{1+(3.281)\,a\sqrt{I_s}} \tag{4.98}$$

Here z^2 is again the valence or number of charges that an ion carries. The term a is a size parameter which is given in Table 4.6 for many common ions in water. The numerical coefficients in Equation 4.99 (i.e., -0.51 and 3.281) are weak functions of temperature. The equation is also only valid for dilute solutions of ionic species or electrolytes, normally less than about 0.1 mol/L. Note that at these dilute ionic strengths the activity of water (x_w γ_w) is effectively unity.

TABLE 4.6
Size Parameter in Debye-Huckel Equation

a	Ion
2.5	NH_4^+
3.0	K^+, Cl^-, NO_3^-
3.5	OH^-, HS^-, MnO_4^-, F^-
4.0	SO_4^{2-}, PO_4^{3-}, HPO_4^{2-}
4.25	Na^+, HCO_3^-, $H_2PO_4^-$, HSO_3^-
4.5	CO_3^{2-}, SO_3^{2-}
5	Sr^{2+}, Ba^{2+}, S^{2-}
6	Ca^{2+}, Fe^{2+}, Mn^{2+}
8	Mg^{2+}
9	H^+, Al^{3+}, Fe^{3+}

The activity coefficient of monovalent ions such as Na^+ and Cl^- decrease with increasing ionic strength but typically fall in the range of 0.8 to 1.0 for $I \leq 0.1$ mol/kg. The activity coefficient of divalent ions such as Ca^{2+}, Mg^{2+}, and SO_4^{2-}, however, are typically of order 0.5 or less at I = 0.1 mol/kg. Note for comparison that the activity coefficient of uncharged solutes such as physically dissolved gases and organic compounds typically have an activity coefficient much greater than one.

4.7 FLUID–SOLID PARTITIONING

No analysis of contaminant partitioning between media would be complete without consideration of the fluid-solid interface. Many common pollutants tend to sorb strongly to soil and sediments and this influences both their ultimate fate, biological accessibility, and their rate of movement in the environment. Unfortunately, the processes of sorption onto or into solid media are much more complicated than between two fluids. In particular, it is difficult to define the activity of a compound in the solid media.

Note that I have referred to *sorption* onto or into a solid phase. This nomenclature emphasizes that it is often unclear whether the component partitioning into a solid phase is *adsorbing*, implying a surface sorption, or *absorbing*, implying a solid solution process or sorption within the solid volume. These processes are shown in Figure 4.10. Neither a conceptual model of soil as a porous structure with exposed surfaces to which *adsorption* can occur nor the model of soil as an open solid structure to which contaminants can *absorb* adequately describe the wide range of experimental observations. As a result, very little in way of practical guidance as to equilibrium partitioning between soils exists, especially between metal species and soils. There are loosely defined model structures, however, that range from simple linear equilibrium concepts to more complicated

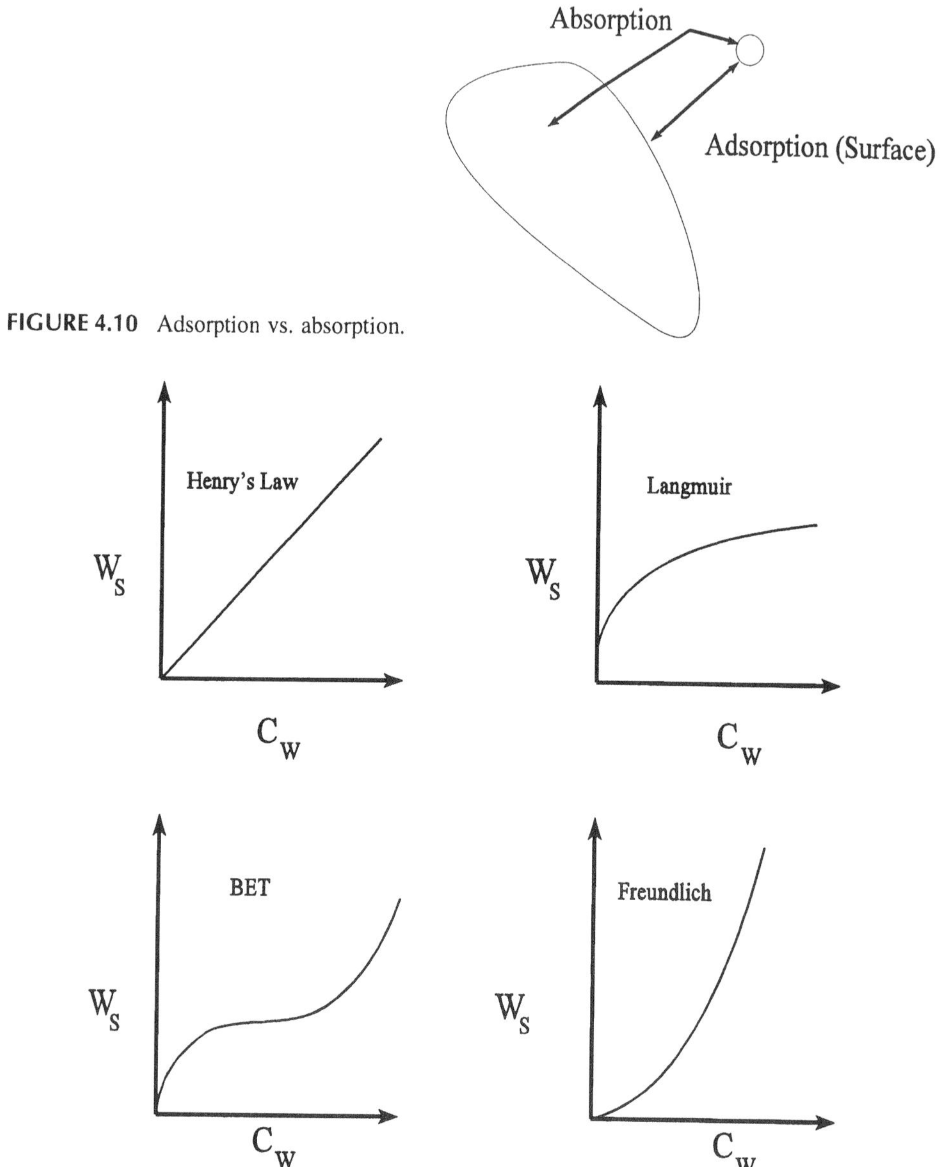

FIGURE 4.10 Adsorption vs. absorption.

FIGURE 4.11 Types of solid sorption isotherms.

nonlinear relationships between sorbed and fluid phase concentration. Figure 4.11 shows examples of the shapes of the linear, Langmuiur, Brauner-Emmett-Teller (BET), and Freundlich isotherms. An isotherm is the measured equilibrium sorption as a function of fluid phase concentration at a given temperature. Each of these isotherm shapes will be discussed in turn. Although it is possible to derive the shape of these isotherms on the basis of a conceptual model of the solid-fluid partitioning process, such a derivation is not nearly as convincing as their ability to adequately correlate specific data.

The discussion of the fluid-solid partitioning will be limited to only the most important phases. That is, the partitioning between soil and water, air and soil, and the three-phase soil-water-air system will be considered. Direct sorption onto a solid phase from an organic phase is of limited environmental interest. The miscibility of a hydrophobic organic in an organic phase is such that significant amounts of solid phase sorption is unlikely to occur. Capillary entrapment of an organic phase is often mistakenly described as a sorption phenomena and will be described later. Considering only sorption onto a solid phase from a water phase, let us consider each of the most important sorption isotherms.

4.7.1 Linear Partitioning

The simplest partitioning relationship between component concentration in the fluid phase and the sorbed phase concentration is linear. It is the fluid-solid equivalent of Henry's Law for fluid-fluid equilibrium. It is rarely completely valid but it is often assumed to be valid because of its simplicity and the lack of sufficient data to support more complicated models. Linear partitioning takes on the following form for sorption of a contaminent i onto a solid phase from water.

$$K_{sw} = \frac{W_s}{C_w} = \frac{[\text{mg/kg}]}{[\text{mg/L}]} = [\text{L/kg}] \tag{4.99}$$

Note that K_{sw} defined in this manner has units of L/kg. Linear partitioning implies that this ratio is a constant, independent, for example, of the concentration in the water phase. The simplicity of the form of Equation 4.99 means that a single measurement of solid and adjacent water concentrations can be used to estimate K_{sw} ($=W_s/C_w$). Reliable estimates of K_{sw}, however, depend upon several measurements to ensure a linear relationship between sorbed and dissolved concentrations and to identify if there is a portion of the contaminant that does not take part in partitioning.

This relationship has several implicit limitations. In general, the linear relationship between sorbed and fluid phase concentrations is approximately valid in the limit of low concentrations of nonreacting and nonionizing components. It has also been shown to be approximately valid for some metals and ionizing species under specific conditions. At high sorbed concentrations, the model tends to break down and, most importantly, cannot be applied to conditions in which it would predict water phase concentrations *exceeding* the solubility of the partitioning compound. The solid phase concentration corresponding to the solubility of component i in the adjacent water phase is termed the *critical loading*.

$$(W_s)_{crit} = K_{sw} S_w \tag{4.100}$$

In reality, the model is rarely accurate at water concentrations approaching the water solubility of the partitioning compound.

The validity of the linear partitioning model is based in empiricism. Do the experimental measurements of sorbed quantity vs. adjacent water concentration suggest a linear relationship? Is there inadequate data to justify a more sophisticated model? Is the use for which the isotherm is to be put insensitive to the form of the assumed relationship between the sorbed and dissolved amount? If the answer to one or more of these questions is yes, then a linear sorption isotherm assumption is appropriate. Note that use of a linear partition coefficient outside of the range of the model's applicability may result in serious errors in the sorbed amounts.

At low concentrations of a partitioning contaminant in solids and water, however, the model has proven exceedingly useful. Although not generally applicable to metals, linear partition coefficients have been measured for various metals. The range in measurements is large as a result of the complex chemistry associated with metals at solid surfaces. The metals and elemental species contained within a solid sample may be chemically fixed, exchangeable, or dissolved in the water fraction in the pores of the solid. The dissolved fraction is, of course, mobile with the water. The chemically fixed fraction is normally not mobile nor does it participate in the partitioning with the adjacent water unless there is a change in the chemical state of the solid, for example, by oxidation or acid treatment. The remainder of the elemental species or the metal is only loosely held in an exchangeable form. In some situations, a large fraction of the metal on a soil or sediment is exchangeable while in others, only a very small fraction is exchangeable. Only the exchangeable metals fraction can be considered for modeling with the linear partition coefficient.

Linear sorption coefficients have proven more useful for the sorption of organic species. This is because organic materials are primarily sorbed into the organic fraction of the soil or sediment matrix. Their interaction with soils is dominated by hydrophobic interactions or their exclusion by the water matrix. Experimental measurements of the soil-water partition coefficient, K_{sw}, have indicated that this quantity often depends linearly on the organic carbon fraction of the soil or sediment. In an attempt to remove this dependence on the soil characteristics in the sorption measurements, an organic carbon based sorption coefficient has been defined

$$K_{oc} = \frac{K_{sw}}{\omega_{oc}} \tag{4.101}$$

where ω_{oc} is the weight fraction of organic carbon in the soil. Note that this organic carbon is typically not related to the sorbing compound but instead represents the partial decomposition products of organic debris that has fallen on the soil. It is the sorbed form of much of the organic substances found in colloidal and dissolved form in the pore waters.

Measurements have often (not always) indicated that K_{oc} is approximately constant for a given compound in various soils. Lyman et al. (1990) reports that K_{oc} estimated in this manner might vary with a coefficient of variation of 10 to 140% for a given compound in different soils. Often organic matter rather than organic carbon content of a soil is known. A commonly used estimate of the ratio of organic matter to organic carbon content, $\omega_{om}/\omega_{oc} \sim 1.724$. Equation 4.101 is often assumed to apply if the fraction organic carbon is greater than about 0.1%. At organic carbon contents less than this amount, direct sorption onto mineral surfaces that is not governed by the compounds hydrobicity is likely dominant.

Note that by assuming that the sorbed concentration of a contaminant is always in direct proportion to the concentration in the adjacent water that the linear model assumes reversibility. That is, it is assumed that the same partition coefficient is observed during sorption as in desorption. Often this has not been found to be the case and desorption is incomplete. This is shown in Figure 4.12. The desorption may also be consistent with a linear relationship between the solid and water but the slope, or desorption distribution coefficient, may not be equal to that observed during sorption. In addition, some portion of the sorbed contaminant may not desorb at all. This is crucial in defining a "clean" soil in that contaminants that cannot desorb from a solid surface may not migrate or be available for uptake by organisms. Thus, a reliable determination of an irreversibly sorbed quantity of contaminant may influence the level of hazard associated with the contaminant and correspondingly modify strategies designed to manage that hazard. The reliable estimate of an irreversibly sorbed fraction, however, requires more information than is often available. In the absence of this information, it is generally assumed that whatever can be sorbed can subsequently desorb. This assumption is generally protective of the environment in that it assumes that all of the contaminants are available for migration or assimilation by organisms.

Under the assumption of linear, reversible sorption of a hydrophobic organic compound, only K_{oc} is needed to characterize the partitioning behavior. As indicated above, K_{oc} is a measure of the compound's hydrophobicity and therefore should correlate with other measures of hydrophobicity

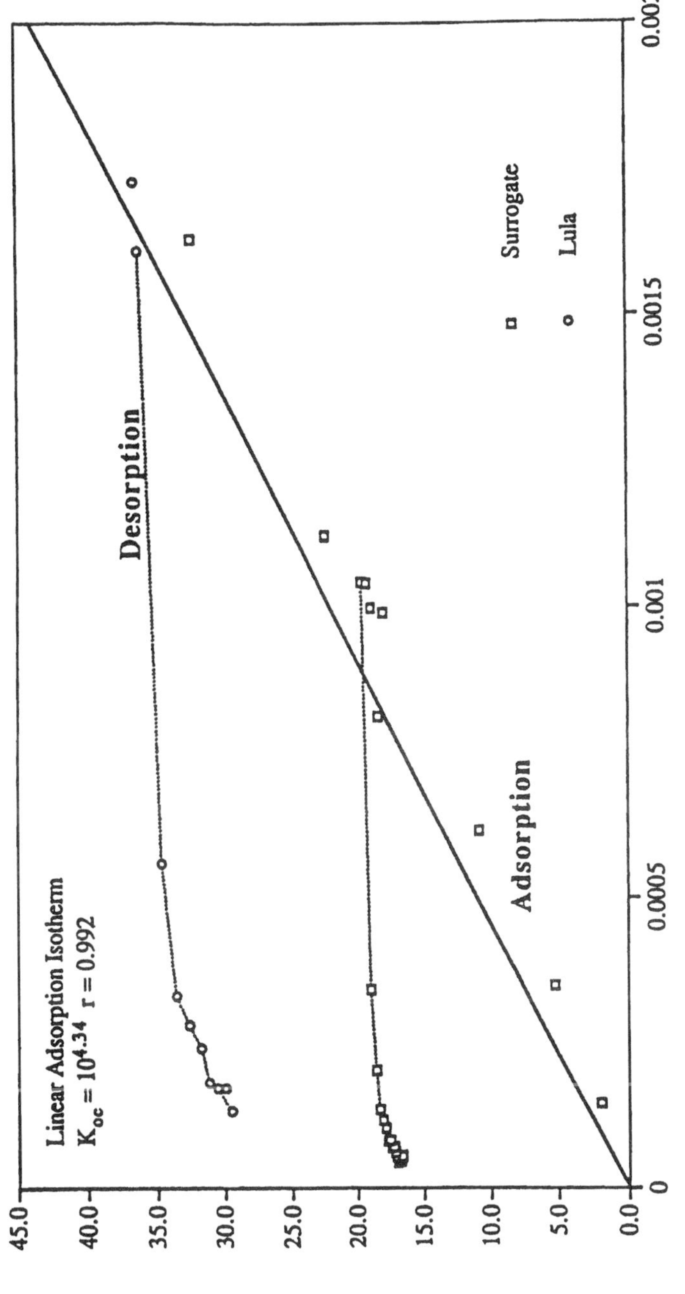

FIGURE 4.12 Hysteresis in adsorption-desorption phenomena. (From Mason Tomson, Rice University. With permission.)

including the octanol-water partition coefficient, K_{ow}, or water solubility, S_w. Lyman et al. (1990) provides a number of example correlations.

Karickhoff et al. (1979) correlated sorption data for ten aromatic and polynuclear aromatic hydrocarbons and found the relationship

$$\log K_{oc} = \log K_{ow} - 0.21 \qquad r^2 = 1.00 \tag{4.102}$$

r^2, the regression correlation coefficient, was found to be 1.00 for this fit. Kenaga and Goring (as referenced in Lyman, 1990) correlated K_{oc} with the water solubility of 106 compounds (in g/m^3) getting the relationship,

$$\log K_{oc} = -0.55 \log S_w + 3.64 \qquad r^2 = 0.71 \tag{4.103}$$

These relationships are potentially subject to large error in a particular application, but, in the absence of site specific experimental data, provide commonly used estimates.

The concept of linear partitioning is illustrated in Example 4.15.

Example 4.15: Estimation of the organic-carbon based partition coefficient

Compare various predicted values of the organic-carbon based partition coefficient for naphthalene to the experimentally observed value of about 83 L/g in this soil. Estimate the equilibrium sorbed concentration of naphthaleneon, a soil containing 4% organic carbon in contact with a saturated water solution.

$$\log K_{ow} = 3.36 \qquad S_w = 30 \text{ g/m}^3 (230\ \mu\text{mol/L})$$

Employing the correlation of Kenaga and Goring (solubility in g/m^3), values of K_{oc} and K_{sw} (4% organic carbon) are

$$I: \qquad \log K_{oc} = -0.55 \log S_w + 3.64 = 2.83$$

$$K_{oc} = 10^{2.83} = 676 \text{ L/kg}$$

$$K_{sw} = K_{oc} \cdot f_{oc} = 676 \text{ L/kg} \cdot 0.04 = 27 \text{ L/kg}$$

Employing a correlation of Chiou et al. (as referenced in Lyman, 1990) (solubility in µmol/L)

$$II: \qquad \log K_{oc} = -0.557 \log \tilde{S}_w + 4.277 = 2.91$$

$$K_{sw} = 10^{2.91} \cdot 0.04 = 33 \text{ L/kg}$$

Employing the correlation of Karickhoff

$$III: \qquad \log K_{oc} = \log K_{ow} - 0.21 = 3.15$$

$$K_{sw} = 10^{3.15} \text{ L/kg} \cdot 0.04 = 57 \text{ L/kg}$$

The variation indicates the uncertainty associated with estimation of the sorption coefficient. The sorbed quantity with a 30 mg/L dissolved concentration is then

$$W_s = K_{sw} C_w$$

$$I: \quad = (27\ \text{L/kg}) \cdot (30\ \text{mg/L}) = 810\ \text{mg/kg}$$

$$II: \quad = (33\ \text{L/kg}) \cdot (30\ \text{mg/L}) = 990\ \text{mg/kg}$$

$$III: \quad = (57\ \text{L/kg}) \cdot (30\ \text{mg/L}) = 1710\ \text{mg/kg}$$

All of which suggests that the capacity of the solid for the compound is greater than the water.

4.7.2 Langmuir Isotherm

If data on a compound partitioning to a particular solid from water is collected, a linear relationship between sorbed and water concentration is often observed at low concentrations. At concentrations close to the water solubility of the compound, however, the sorbed concentration often tends to increase more slowly than the water concentration. This gives rise to the Langmuir isotherm which also is shown in Figure 4.11.

The Langmuir isotherm can be derived by assuming that a finite number of sorption sites in the solid phase exists and that the rate of sorption is proportional to the sites remaining. The Langmuir isotherm has the general form

$$W_s = \alpha_1 \frac{C_w}{1 + \alpha_2 C_w} \tag{4.104}$$

Note that the partition coefficient, $K_{sw} = W_s/C_w$, is no longer constant. Equation 4.104 can be rewritten in a linearized form that makes it convenient to identify the constants from experimental data,

$$\frac{1}{W_s} = \frac{1}{\alpha_1 C_w} + \frac{\alpha_2}{\alpha_1} \tag{4.105}$$

This equation suggests that a plot of $1/W_s$ vs. $1/C_w$ should be a straight line with slope $1/\alpha_1$ and intercept α_2/α_1. Of course, the occurrence of a straight line depends on the degree of the data's adherence to the Langmuir isotherm. In addition, remember that manipulating the fitting equations results in implicit assumptions about the accuracy of that data. Fitting Equation 4.105 via the conventional least squares approach to concentration data assumes that the uncertainty in each of the reciprocal concentrations is the same. This means that the relative uncertainty in low concentrations (large reciprocal values) is implicitly assumed to be less than at high concentrations (small reciprocal values). Thus, a fit of Equation 4.105 by conventional least squares will generally fit lower concentration data measurements at the expense of the fit at higher concentration measurements. This is illustrated in Example 4.16 and in Figures 4.13 and 4.14.

Example 4.16: Fitting of data to the Langmuir isotherm

Develop an isotherm equation for the water and solid concentration data in the table below. Assume that the data should be well represented by a Langmuir isotherm described by

$$W_{act}(C_w) = \frac{40 C_w}{1 + 0.1 C_w}$$

The collected data (W_s vs. C_w) and the linearized form of the data ($1/W_s$ and $1/C_w$) are

C_w (mg/L)	$1/C_w$ (L/mg)	W_s (mg/kg)	$1/W_s$ (kg/mg)	W_{fit} (mg/kg)	W_{act} (mg/kg)
1	1	21.8	0.0459	21.9	36.4
6	0.167	116	0.00863	112	150
11	0.0909	208	0.00480	178	210
16	0.0625	259	0.00386	230	246
21	0.0476	277	0.00361	271	271
31	0.0323	306	0.00326	332	302
36	0.0278	319	0.00314	355	313
41	0.0244	318	0.00315	375	321
46	0.0217	337	0.00296	392	329

The difference between the observed and actual (or expected) soil loading measurements is the result of instrument error (normally distributed with a standard deviation of 20 mg/L). The fit to the linearized data gives

$$\text{slope} = \frac{1}{\alpha_1} = 0.04414 \qquad (\alpha_1 = 22.7 \qquad -43\% \text{ error})$$

$$\text{intercept} = \frac{\alpha_2}{\alpha_1} = 0.00159 \qquad (\alpha_2 = 0.0360 \qquad -64\% \text{ error})$$

Although the fit curve adequately represents the data (See Figure 4.13), the error is caused by the linearization implicitly forcing a good fit to the small values of concentration (greatest values of $1/W_s$) which unfortunately often exhibit the largest error. This is clearly seen in Figure 4.14 in which the large W_s values are seen as a cluster of data that either curve fits well. Note that the intercept gives the saturation value in the Langmuir isotherm as $C_w \to \infty$, as 629 mg/kg rather than 400 mg/L in the actual curve. These errors occur despite the fact that the error in all but the first two measurements are less than 10%.

4.7.3 Brauner–Emmett–Teller (BET) isotherm

The Brauner-Emmett-Teller, or BET, isotherm can be thought of as a generalization of the Langmuir isotherm. In the Langmuir isotherm, a finite number of sorption sites exists. No additional sorption is possible once these sites are filled. The saturation concentration when this occurs is equal to the ratio of A/B in Equation 4.104. In the BET isotherm, this is conceptualized as monolayer coverage of the sorbing material. In the BET isotherm, additional sorption is possible once a monolayer has been sorbed. The characteristic shape of the BET isotherm is as shown in Figure 4.11. At low to medium fluid phase concentrations, the Langmuir isotherm is reproduced. At high concentrations, however, the sorbed amount increases dramatically as multilayer sorption is observed.

Many liquid systems are adequately described by the Langmuir isotherm over the entire range of water concentrations from zero to the solubility limit. Gas-solid systems, however, often give sorption isotherms that can be modeled by the BET isotherm. Its mathematical form is given by

$$W_s = \frac{W_s^0 \, \alpha \dfrac{p_A}{P_v}}{\left(1 - \dfrac{p_A}{P_v}\right)\left[1 + (\alpha - 1)\dfrac{p_A}{P_v}\right]} \tag{4.106}$$

Here p_A is the partial pressure of component A in the gas phase, P_v, is the component vapor pressure and W_s^0 and α are BET constants. W_s^0 is the sorbed amount for monolayer coverage of

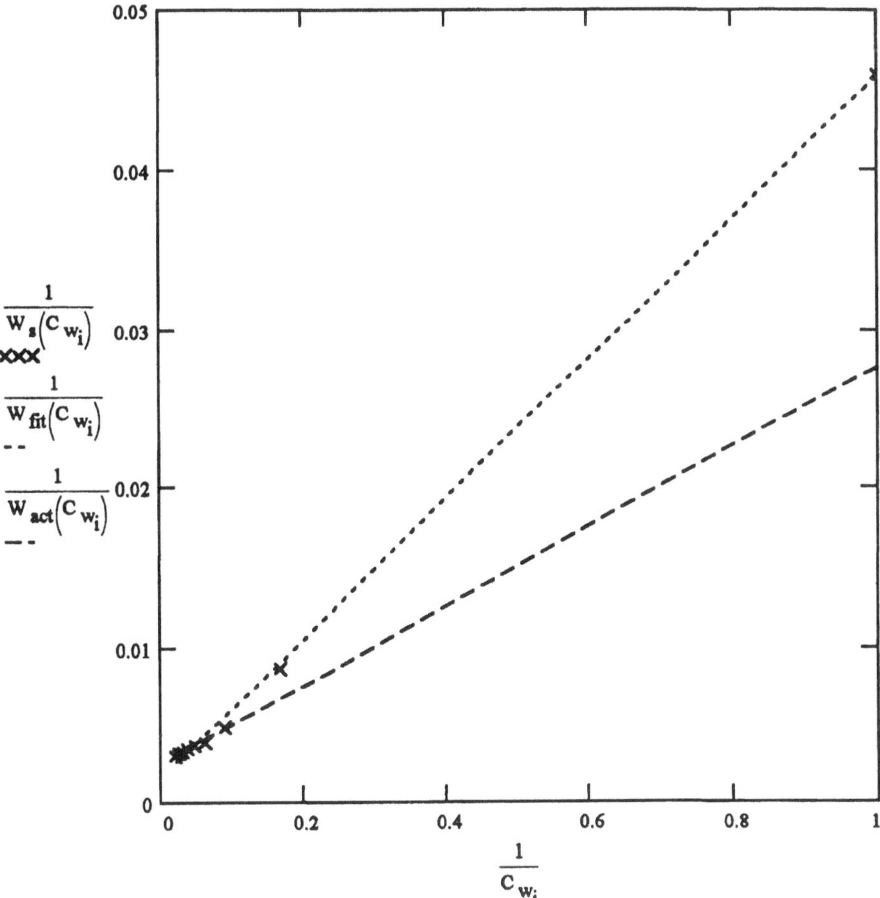

FIGURE 4.13 Linearized Langmuir isotherm curve fit to sorption data.

the surface. For the purpose of estimating these constants, the BET equation can be recast in the linearized form

$$\frac{1}{W_s}\left[\frac{p_A / P_v}{1 - p_A / P_v}\right] = \frac{1}{\alpha W_s^0} + \frac{p_A / P_v(\alpha - 1)}{\alpha W_s^0} \tag{4.107}$$

In this case, a plot of the left-hand side term vs. p_A/P_v gives an intercept of $1/(\alpha W_s^0)$ and a slope of $(\alpha - 1)/(\alpha W_s^0)$. As with the Langmuir approach, linearization influences the way the errors are assumed to be distributed among the individual data points. In general, a direct nonlinear fit to the data is the appropriate choice.

As indicated above, the BET isotherm is especially useful for sorption of gases onto dry soil or solids. The full form is often needed to accurately represent data on completely dry solids exposed to high vapor concentrations. This is perhaps most likely to occur in a packed bed adsorption system used for gas cleanup. In such a system, moisture is strictly controlled and more concentrated vapors generally result in a potentially more efficient cleanup.

In soils the full complexity of the BET isotherm is often not necessary. The BET isotherm is equivalent to linear or Langmuir-type partitioning at low and intermediate sorbed concentrations. For example, for low vapor partial pressures relative to the pure component vapor pressure of a compound, the BET isotherm reduces to

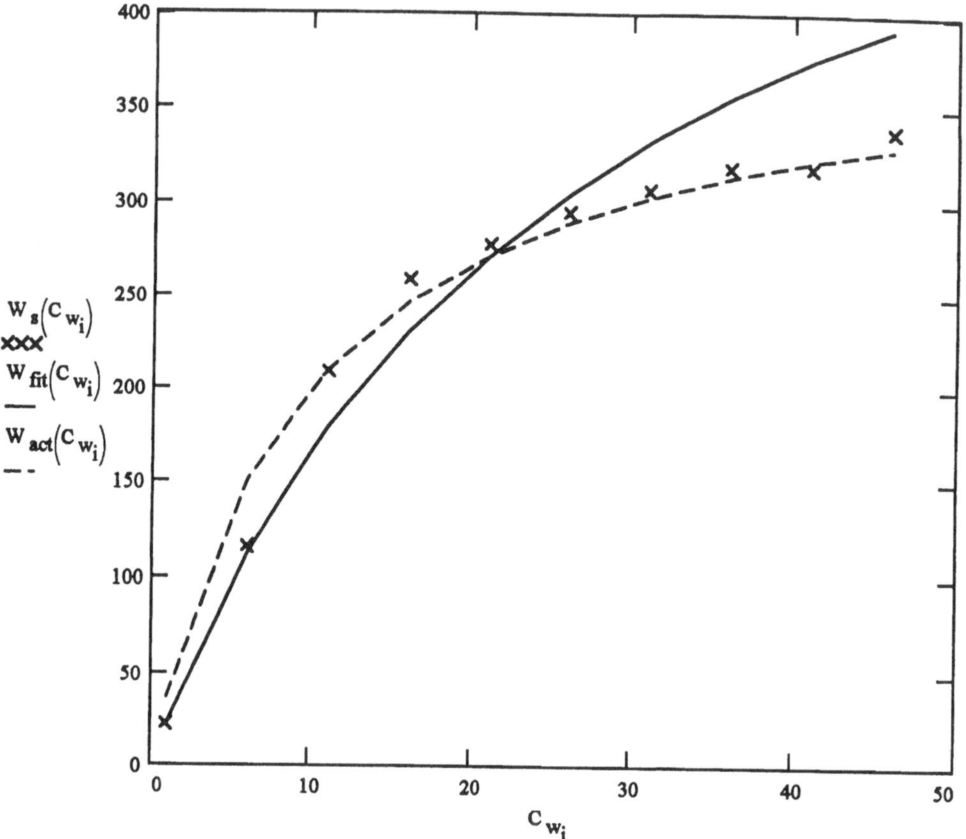

FIGURE 4.14 Evaluation of the linearized Langmuir isotherm fit to sorption data.

$$W_s \approx W_s^0 \alpha \frac{P_A}{P_v} \tag{4.108}$$

In addition, the moisture content of moist soils is generally in excess of 10% (volume water per volume soil). Even in very dry environments, the moisture content just a few centimeters below the surface is generally 10% or more. Under such conditions, the water tends to wet the entire soil surface and sorption of vapors generally proceeds by first sorption into the water and then partitioning at the soil-water interface. Linear or Langmuir-type sorption tends to be observed for at least reversible sorption of trace hydrophobic organics from vapors onto soils.

4.7.4 Freundlich Isotherm

The Freundlich isotherm is a general, empirical power-law adsorption isotherm model. It has the form

$$W_s = \alpha C_w^{1/n} \tag{4.109}$$

where α and n are constants to be determined from experimental data. W_s is the as fraction of the component of interest on the solid phase (e.g., mg/kg) and C_w is the mass concentration in the adjacent water phase (e.g., mg/L). The Freundlich isotherm is generally considered to be purely empirical in nature, although it can be derived from the Langmuir isotherm (discussed below) assuming that there is more than one type of sorption site.

Lyman et al. (1990) reports that the bulk of the experimental data suggest that n lies in the range of 0.7 to 1.1. Rao and Davidson (as referenced in Lyman) found the mean value of 1/n with 26 chemicals to be 0.87 with a coefficient of variation of 15%. The process of fitting the isotherm to data can be simplified by taking the logarithm of both sides of Equation 4.110

$$\ln W_s = \ln \alpha + \frac{1}{n} \ln C_w \tag{4.110}$$

A plot of ln W_s vs. ln C_w should then yield a straight line with slope given by 1/n and intercept (i.e., where ln C_w equals zero) equal to ln α. Although the Freundlich isotherm can be fit to almost any data over a range of concentrations, the nonlinear power often makes it somewhat difficult to apply. In the absence of specific data to the contrary or when 1/n ~ 1, the power is often assumed to be equal to 1. The data of Example 4.16 is fit to the Freundlich isotherm in Example 4.17. Again, the fit curve provides a representation of the data that is adequate for most purposes despite the fact that there is no physical significance to the fit or the fitted parameters.

Example 4.17: Fitting soil loading data to the Freundlich isotherm

Fit the data of Example 4.15 to the Freudlich isotherm

$$W_s = \alpha K \, C_w^n$$

$$\ln W_s = \ln \alpha + n \ln C_w$$

A plot of the W_s vs. C_w data on a logarithmic scale is shown in Figure 4.15. The fitted line which is of slope n and intercept ln α using y = ln W_s and x = ln C_w is also shown in the figure. The calculated values are

$$\text{intercept} = 3.35(\alpha = 28.5) \qquad n = 0.704$$

$$W_s = 28.5\, C_w^{0.704}$$

The resulting fit is shown in Figure 4.16. Note that the shape of the data and the Freundlich isotherm are not in agreement, but it can still be useful in that the fitted curve exhibits an average error of about 16%. The Freundlich isotherm is commonly used because of its simple form yet ability to fit a large range of data.

4.8 EQUILIBRIUM MODELS

In the preceding sections the equilibrium partitioning of a particular compound between any two of the main environmental interfaces was evaluated. This allows us to define the ratio of concentrations between the two contacting phases at equilibrium, but it does not allow us to define the concentrations themselves. In order to define the absolute concentrations we must also consider the volumes or masses of the respective phases. If we consider a spill of some finite volume of an environmental contaminant, the concentration that is ultimately observed depends on the volumes of air, water, or soil over which the contaminant is mixed. To translate the partitioning ratios defined above to expected concentrations we need to turn to an environmental model.

In an environmental model, volumes or masses of the various phases of interest are selected or known. In a closed vessel, for example, a unit operation in a waste treatment system, this can be

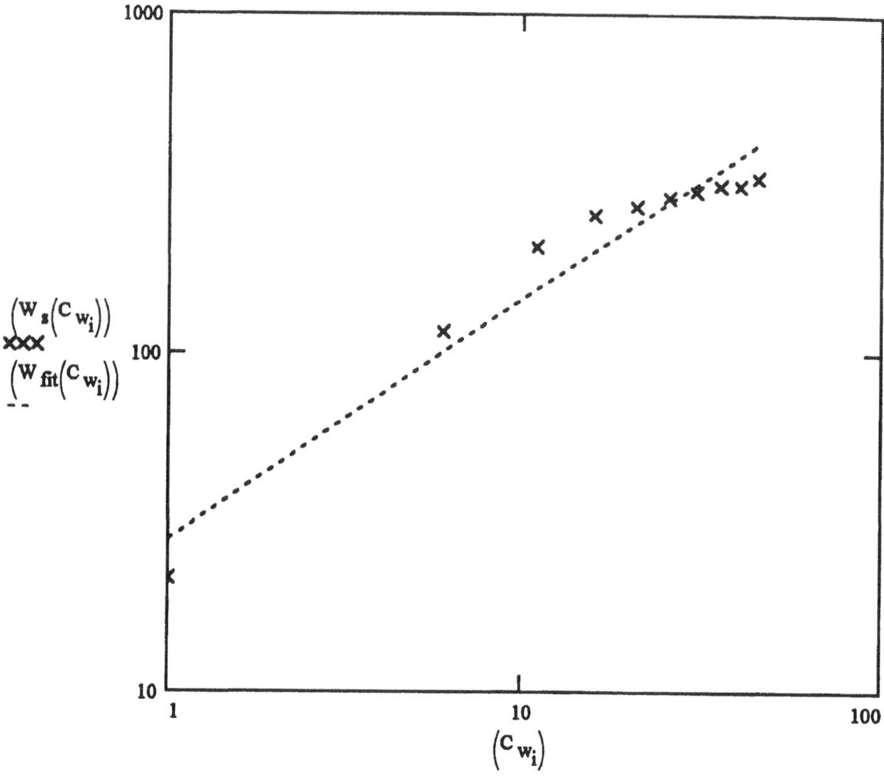

FIGURE 4.15 Linearized Freundlich isotherm fit to sorption data.

easily accomplished. In the open environment, it is much more difficult to identify mixing volumes, or the volumes over which the concentration is uniform. In general, a contaminant is always distributed nonuniformly in space and equilibrium is not obtained. In specific instances, however, it is possible to select phase volumes that provide an indication of either actual or potential concentrations of the environmental contaminant. It is for this purpose that equilibrium models are useful.

Let us first consider a closed vessel containing air and water in known proportions, V_a and V_w, respectively. The concentration of contaminant times these volumes gives the mass in each of these phases. It is known, for a hydrophobic contaminant, that Henry's law defines the ratio of the concentrations between the air and water phase and that the sum of the contaminant mass in each of the phases gives the total mass of contaminant introduced to the system.

$$K_{aw} = \frac{C_a}{C_w} \qquad M = C_a V_a + C_w V_w \tag{4.111}$$

Thus, for this simple system, an equilibrium model is simply two equations in the two unknowns, C_a and C_w. Substituting for C_w in the second equation from the definition of Henry's law gives

$$C_a = \frac{M}{V_a + \frac{V_w}{K_{aw}}} \qquad C_w = \frac{C_a}{K_{aw}} \tag{4.112}$$

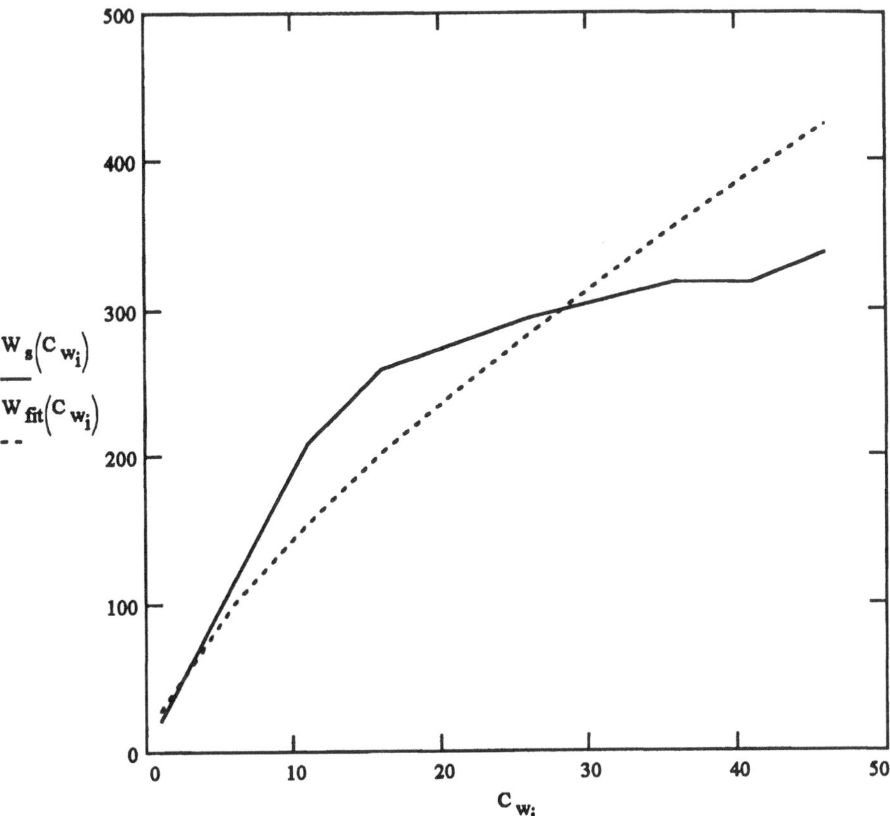

FIGURE 4.16 Evaluation of the linearized Freundlich isotherm fit to sorption data.

Let us now consider an open air, water, soil system in which the contaminant is known to be mixed over a given volume of each phase. This is a more artificial example in that continued contaminant migration will continuously increase the volume of the containing phases and dilute the concentrations. An equilibrium model may still be useful for such a system, however, in that the proportions of the various phases may not vary greatly or the calculation may only be needed to provide an indication of which phase represents the dominant repository or "sink" for the compound. If the volume of the air and water systems are again V_a and V_w, respectively, and the mass of the soil phase is its bulk density, ρ_b, times its volume, V_s, the equilibrium model is now

$$M = C_a V_a + C_w V_w + W_s \rho_b V_s$$

$$K_{aw} = \frac{C_a}{C_w} \qquad K_{sw} = \frac{W_s}{C_w} \tag{4.113}$$

Solving for C_a gives

$$C_a = \frac{M}{V_a + \dfrac{V_w}{K_{aw}} + \dfrac{K_{sw}}{K_{aw}} \rho V_s} \tag{4.114}$$

and C_w and W_s can be determined by the equilibrium partitioning relationships. The equations also can be solved for the water concentration

$$C_w = \frac{M}{K_{aw}V_a + V_w + K_{sw}\rho_b V_s} \tag{4.115}$$

This form would be required to evaluate the distribution of a soil and water system ($V_a = 0$).

Finally extending this to the four-phase system — air, water, soil, and a nonaqueous organic phase — the equilibrium model is

$$M = C_a V_a + C_w V_w + C_n V_n + W_s \rho_b V$$

$$K_{aw} = \frac{C_a}{C_w} \qquad K_{sw} = \frac{W_s}{C_w} \qquad K_{nw} = \frac{C_n}{C_w} \tag{4.116}$$

and the concentration of water is

$$C_w = \frac{M}{K_{aw}V_a + V_w + K_{nw}V_n + K_{sw}\rho_b V_s} \tag{4.117}$$

Note that the air concentration (if present) is simply the water concentration times Henry's law constant, K_{aw}.

We can further extend this to a five-phase system that includes dissolved or suspended particulate matter in the water.

$$M = C_a V_a + C_w V_w + C_c \rho_c V_w + C_n V_n + W_s \rho_b V_s$$

$$K_{aw} = \frac{C_a}{C_w} \qquad K_{sw} = \frac{W_s}{C_w} \qquad K_{nw} = \frac{C_n}{C_w} \qquad K_{cw} = \frac{C_c}{C_w} \tag{4.118}$$

The concentration in the water in the five-phase system is given by

$$C_w = \frac{M}{K_{aw}V_a + V_w + K_{aw}\rho_c V_w + K_{nw}V_n + K_{sw}\rho_b V_s} \tag{4.119}$$

Note that K_{cw} is often approximated by K_{oc}, the organic carbon-based partition coefficient for the particular organic contaminant. If the nonaqueous phase is oil and grease, the K_{nw} also may be well-approximated by K_{oc}. In either case, the K_{oc} should be related to the partitioning of the hydrophobic contaminant to the hydrophobic phase. For nonhydrophobic contaminants, including metals and ions, the partition coefficients in these equations must, in general, be experimentally determined. Note that all of the equations assume linear partitioning which may not be valid although it is computationally convenient.

The use of an equilibrium model is further illustrated in Examples 4.18 and 4.19.

Example 4.18: Equilibrium modeling in an air-water system

Evaluate the distribution of the six compounds below in a system (e.g., a closed vessel) that contains 90% air and 10% water (by volume). The molecular weight, vapor pressure, water solubility, and calculated vapor pressures are shown below

Compound	MW	Vapor Pressure P_v atm	Water solubility S_w mg/L	K_{aw} $P_v/(RTS_w)$
Benzene	78	0.125	1780	0.224
Trichloroethylene	99	0.008	1100	0.0294
Chlorobenzene	112	0.0016	470	0.0156
Naphthalene	130	1.14×10^{-4}	34	0.178
Aroclor 1254	260	5.3×10^{-7}	0.24	0.0234
DDT	350	2×10^{-10}	0.005	0.00057

Let us employ a basis of 1 g of compound (M) and 1 liter total volume with 90% (0.9 L) air and 10% water (0.1 L). The concentration and mass in air and water for each compound is

$$C_w = \frac{M}{K_{aw}V_a + V_w} \qquad M_w = C_w V_w$$

$$C_a = K_{aw} C_w \qquad M_a = C_a V_w$$

The calculated distributions (ratio of M_w and M_a to M) are then

Benzene	33.2% water	66.8% air
Trichloroethylene	79.1% water	20.9% air
Chlorobenzene	87.7% water	12.3% air
Naphthalene	86.2% water	13.8% air
Aroclor 1254	82.5% water	17.5% air
pp-DDT	99.5% water	0.50% air

Note that a significant fraction of each compound except DDT is present in the air phase. Trichloroethylene and Aroclor 1254 have a similar distribution between air and water despite the vapor pressures differing by more than 4 orders of magnitude. The low solubility of Aroclor 1254 in water contributes to a surprising amount in the air phase.

Example 4.19: Equilibrium modeling of a soil/water/colloidal system

Repeat the previous example with soil containing 50% by volume water with 100 mg/L dissolved organic carbon. The organic carbon based partition coefficients are listed below. The soil fraction is 1% organic carbon and the soil grains have a density of 2.5 kg/L.

	Benzene	Trichloroethylene	Chlorobenzene	Naphthalene	Aroclor 1254	pp-DDT
K_{oc}, L/kg	87	120	810	1300	5100	250000

The fraction of each compound in the water, colloidal, and soil phases are given by

$$\frac{C_w V_w}{M} = f_w = \left(V_w + K_{oc}\rho_c V_w + K_{oc} f_{oc}\rho_b V_s\right)^{-1}$$

$$f_c = f_w K_{oc}\rho_c \qquad f_s = f_w K_{oc} f_{oc}\rho_b \frac{V_s}{V_w}$$

The bulk density is the dry density of the soil or 1.25 kg/L, since the void fraction or porosity is 50%. The mass fractions for benzene are

$$f_w = (1\text{ L} + 87\text{ L/kg} \cdot 100\text{ mg/L} \cdot 1\text{ L} + 87\text{ L/kg} \cdot 0.01 \cdot 1.25\text{ kg/L} \cdot 1\text{ L})^{-1} = 0.477$$

$$f_c = 0.477 \cdot 87\text{ L/kg} \cdot 100\text{ mg/L} = 0.00415$$

$$f_s = 0.477 \cdot 87\text{ L/kg} \cdot 0.01 \cdot 1.25\text{ kg/L} \cdot \frac{1\text{L}}{1\text{L}} = 0.519$$

Summarizing each of the compounds:

	Benzene	Trichloroethylene	Chlorobenzene	Naphthalene	Aroclor 1254	pp-DDT
f_w	0.477	0.389	0.0892	0.0575	0.0153	0.00032
f_c	0.00415	0.00478	0.00723	0.00748	0.00781	0.00793
f_w	0.519	0.597	0.904	0.935	0.977	0.992

Note that the more hydrophobic compounds are found primarily sorbed to the soil while benzene is distributed approximately 50% to the water and 50% to the soil. Note also that there are no large differences in the fraction of the mass sorbed onto colloidal organic matter in the water. The fraction of the compound in the mobile water phase is essentially the sum of the colloidal fraction and the truly dissolved fraction. Thus, the mobile fraction of DDT is about 0.00825 (0.825%) with 100 mg/L dissolved organic colloids in the water but only 0.00032 (0.032%) without colloids. The mobile water is carrying more than 25 times more DDT as a result of the colloids.

4.8.1 Fugacity Modeling

An alternative and convenient approach to environmental models is through the use of fugacity as developed by Mackay (1991). To use this approach we must again compartmentalize the environmental media to which our contaminant is exposed, that is define the volumes of the air, water, soil, and other phases in which our contaminant can reside. Through recognition that the fugacity of the contaminant in each of these phases must be the same at equilibrium, it is possible to determine the concentration and total mass of contaminant in each phase.

If linear partitioning is assumed at each of the environmental interfaces, the fugacity of a contaminant within any phase is proportional to a measure of the concentration of the contaminant in that phase. If mole fraction is employed as a measure of concentration, for example, the fugacity of a contaminant in water is given by $f_w = (\gamma_w f_w^0)\, x_w$. The constant of proportionality between fugacity and mole fraction is $\gamma_w f_w^0$. It is generally more convenient to work in concentration units such as mass per volume. Because fugacity has units of pressure, the concentration corresponding to a fugacity f_i is f_i/RT and the proportionality between fugacity (expressed as a concentration) and contaminant concentration can be written

$$\frac{f_i}{RT} = Z_i C_i \tag{4.120}$$

where Z_i represents the "fugacity capacity" of phase i. The Z_i can be defined uniquely for each environmental phase.

Air — The fugacity of a contaminant, A, in air is simply the partial pressure, p_A, since air is assumed to be an ideal gas. Therefore,

$$Z_a = \frac{p_A / RT}{f_A / RT} = 1 \tag{4.121}$$

By making the fugacity capacity a ratio of concentrations, air is automatically unity and the fugacity capacity of all other phases is relative to that of air.

Water — For the case of a hydrophobic organic compound with $C_a = K_{aw} C_w$ the fugacity capacity is

$$Z_w = \frac{C_w}{f_A / RT} = \frac{C_w}{p_A / RT} = \frac{C_w}{C_a} = \frac{1}{K_{aw}} \tag{4.122}$$

In general, the fugacity capacity is the inverse of the concentration ratio partition coefficient between the phase of interest and air.

Soil — The fugacity capacity in soil is given by the inverse of the air-soil partition coefficient

$$Z_s = \frac{C_w}{f_A / RT} = \frac{\rho_b W_s}{C_a} = \frac{\rho_b K_{sw} C_w}{C_a} = \frac{\rho_b K_{sw}}{K_{aw}} = \frac{\rho_b}{K_{as}} \tag{4.123}$$

Here the bulk density arises simply because the traditional unit of concentration in the solid phase is mass-per-mass of dry soil and the bulk density places this in mass per volume concentration units. Note that the air-soil partition coefficient is the ratio of the air-water partition coefficient and the soil-water partition coefficient.

NAPL — An oily or other nonaqueous phase liquid, if assumed to be an ideal solution, has a fugacity capacity of

$$Z_n = \frac{C_n}{C_a} = \frac{K_{nw} C_w}{C_a} = \frac{K_{nw}}{K_{aw}} = \frac{1}{K_{an}} \tag{4.124}$$

The convenience of this approach is that the distribution of a particular contaminant among the various phases can be determined by dividing the total mass of contaminant by the sum of the volume times the fugacity capacity for each phase. This provides the concentration in the air phase directly.

$$C_a = \frac{f_a}{RT} = \frac{M}{\Sigma V_i Z_i} \tag{4.125}$$

The concentrations in any other phase j become simply the concentration in air times the fugacity capacity or $C_a Z_i$. The values for each phase are given by the relationships

$$\begin{aligned} &\text{Water:} && C_w = \frac{C_a}{K_{aw}} \\ &\text{Soil:} && W_s = \frac{C_a}{K_{as}} (C_s = \rho_b W_s) \\ &\text{NAPL:} && C_n = \frac{C_a}{K_{an}} \end{aligned} \tag{4.126}$$

and the mass in each of these phases is simply the concentration times the volume of that phase. These equations are identical to those developed in the preceding section. As shown in that section, this approach could have been developed without any reference to fugacity, but it is normally attributed to Mackay (1991) and referred to as a fugacity model.

4.9 CHEMICAL EQUILIBRIUM

The discussion of equilibrium thus far has been limited to nonreactive substances that retain their identity throughout the physical processes influencing them. For a single species, just two properties (e.g., temperature and pressure) define its state. Equilibrium in a reactive system, however, involves additional constituents and additional degrees of freedom. The Gibbs' phase rule must be modified for chemically reacting systems to

$$\text{degrees of freedom} = \text{components} - \text{phases} - \text{independent reactions} + 2 \quad (4.127)$$

Consider, for example, the combustion of a mixture of hydrogen and methane to the products carbon monoxide, carbon dioxide, and water vapor.

$$\begin{aligned} CH_4 + \frac{3}{2}O_2 &\Rightarrow CO + 2H_2O \\ CO + \frac{1}{2}O_2 &\Rightarrow CO_2 \end{aligned} \quad (4.128)$$

As shown in Equation 4.28, two independent reaction equations exist that describe the system. There are a total of five components and if the reaction is conducted entirely in the gaseous state, one phase. The degrees of freedom of the system are 5 – 1 – 2 + 2 = 4. Thus the number of intensive properties that can be arbitrarily specified are 4, for example, temperature, pressure, and the mole fraction of any two of the compounds. At equilibrium, the mole fraction of the remaining three compounds as well as enthalpy, internal energy, specific volume, etc., cannot be independently specified but their state is fixed by the four specified conditions.

The relationship between the specified mole fractions and those remaining can be determined by again recognizing that the equilibrium state is one that minimizes the Gibbs' free energy of the system. Processes occur spontaneously only if the Gibbs' free energy change during the process is negative. If the Gibbs' free energy change is positive the process does not occur and when the Gibbs' free energy change of a process is zero it is then at equilibrium.

Consider the following chemical reaction,

$$\nu_A A + \nu_B B \Leftrightarrow \nu_c C + \nu_D D \quad (4.129)$$

The stoichiometric coefficients, ν_i, are the moles of component i required to complete the reaction per ν_A moles of component A. For a given change in the number of moles of component A, dn_A, the stoichiometry suggests that the change in the number of moles of the other feed and product substances is given by

$$\begin{aligned} dn_B &= \frac{\nu_B}{\nu_A} dn_A \\ dn_C &= -\frac{\nu_C}{\nu_A} dn_A \\ dn_D &= -\frac{\nu_D}{\nu_A} dn_A \end{aligned} \quad (4.130)$$

At a given temperature and pressure, chemical equilibrium is achieved if

$$\sum \mu_i dn_i = 0 \tag{4.131}$$

or, substituting from Equation 4.130 and incorporating the minus sign into the stoichiometric coefficients (i.e., ν_C, $\nu_D >0$, ν_A, $\nu_B <0$),

$$\sum \mu_i \frac{\nu_i}{\nu_A} dn_A = 0 \tag{4.132}$$

Multiplying through by dn_A/γ_A,

$$\sum \mu_i \nu_i = 0 \tag{4.133}$$

The chemical potential is given by

$$\mu_i = \mu_i^{ref} + RT\ln(a_i) \tag{4.134}$$

where a_i is the activity of component i, f_i/f_i^{ref}. For an ideal solution a_i would be equal to x_i, the mole fraction, whereas for a nonideal solution it would be equal to $x_i\gamma_i$, where γ_i is the activity coefficient. For an ideal gas, a_i is equal to the mole fraction y_i. The reference chemical potential is chosen such that

$$\mu_i^{ref} = RT\ln(f_i^{ref}) \tag{4.135}$$

Equation 4.135 can be rewritten

$$\sum \nu_i \mu_i^{ref} + RT\sum \ln(a_i^{\nu_i}) = 0 \tag{4.136}$$

The first term is defined as the standard Gibbs' free energy change for the reaction and is here given the symbol, ΔG_r. Standard tables give the Gibbs' free energy of formation (ΔG_f) relative to the free energy of the elements that compose the substance. The standard Gibbs' free energy for a given reaction is then the sum of the free energy of formation of each product or reactant times its stoichiometric coefficient

$$\Delta G_r = \sum \nu_i \Delta G_{fi} = \left(\sum \nu_i \Delta G_{fi}\right)\Big|_{products} - \left(\sum |\nu_i| \Delta G_{fi}\right)\Big|_{reactants} \tag{4.137}$$

The minus sign is moved outside of the term for the reactants and the absolute value for the stoichiometric coefficient is used to emphasize that the reactants are subtracted from the products. As indicated previously, the minus signs in Equation 4.137 are usually incorporated within the stoichiometric coefficient. The second term in Equation 4.137 can be written RT ln $\Pi(a_i^{\nu})$, where Π represents the product series,

$$\Pi(a_i^{\nu_i}) = (a_A^{-\nu_A})(a_B^{-\nu_B})(a_C^{\nu_C})(a_D^{\nu_D}) = \frac{(a_C^{\nu_C})(a_D^{\nu_D})}{(a_A^{\nu_A})(a_B^{\nu_B})} \tag{4.138}$$

Thus, the statement of equilibrium can be written

$$\ln \Pi\left(a_i^{\nu_i}\right) = \frac{-\Delta \hat{G}_r}{RT} = \ln K_r \tag{4.139}$$

where K_r is an equilibrium constant. K_r is thus given by

$$K_r = \frac{\left(a_C^{\nu_C}\right)\left(a_D^{\nu_D}\right)}{\left(a_A^{\nu_A}\right)\left(a_B^{\nu_B}\right)} = \Pi\left(a_i^{\nu_i}\right) \tag{4.140}$$

and its relation to temperature is give by

$$\ln K_r = \frac{-\Delta \hat{G}_r}{RT} \tag{4.141}$$

Practical use of this equation requires values for the standard Gibbs' free energies of reaction. Tabulations of these quantities (see Appendix A) exist and approximate estimation methods are available. Example 4.20 illustrates the determination of $\Delta \hat{G}_r$ and the reaction equilibrium constant K_r. Note that at room temperature, the reaction in the example is pushed strongly toward completion. This is commonly true of strong exothermic reactions such as combustion reactions. As the temperature increases, however, the entropy term ($T\Delta \hat{S}$) increases in importance. The Gibbs' free energy is a function of temperature primarily as a result of this term. The enthalpy and entropy change of the reaction also are functions of temperature, but they tend to be much less important than the effect of the explicit reference to temperature in the definition of the Gibbs' free energy. Assuming that the changes in enthalpy and entropy of the reaction are essentially independent of temperature, it is possible to estimate the temperature at which the change in Gibbs' free energy is zero and the reaction is no longer considered to be spontaneous. For methane combustion according to the example, this occurs at about 3660 K.

$$\begin{aligned} \Delta \hat{G}_r &= \Delta \hat{H}_r - T\Delta \hat{S}_r \\ 0 &= (-890.3 \text{ kJ/mol}) - T\left(-0.243 \frac{\text{kJ}}{\text{mol} \cdot \text{K}}\right) \\ T &\approx \frac{-890.3 \text{ kJ/mol}}{-0.243 \dfrac{\text{kJ}}{\text{mol} \cdot \text{K}}} = 3660 \text{ K} \end{aligned} \tag{4.142}$$

At that temperature the Gibbs' free energy is about zero and the reaction equilibrium constant is essentially unity. This emphasizes that even at combustor temperatures (1000 to 2000 K), the reaction is spontaneous.

Example 4.20: Estimation of the Gibbs' free energy change and reaction equilibrium constant

Estimate the Gibbs' free energy change and the reaction equilibrium constant for the combustion of methane at 25°C.

$$CH_4(g) + 2O_2(g) \rightarrow 2H_2O(l) + CO_2(g)$$

Property @ 25°C	CH_4 (g)	O_2 (g)	H_2O (L)	CO_2 (g)
ΔH_f, $kJ \cdot mol^{-1}$	–74.8	0	–285.8	–393.5
S, $kJ \cdot mol^{-1} \cdot K^{-1}$	0.1863	0.2051	0.06991	0.2137

The net change in enthalpy of the reaction is then

$$\Delta \hat{H}_r = (2)\cdot(-285.8) + (1)\cdot(-393.5) - (1)\cdot(-74.8) - (2)\cdot(0)$$

$$= -890.3 \frac{\text{kJ}}{\text{mol of } CH_4}$$

The net change in entropy per unit temperature is

$$\Delta \hat{S}_r = (2)\cdot(0.06991) + (1)\cdot(0.2137) - (1)\cdot(0.1863) - (2)\cdot(0.2051)$$

$$= 0.243 \frac{\text{kJ}}{\text{mol of } CH_4 \cdot K}$$

Thus, the Gibbs' free energy change is given by

$$\Delta \hat{G}_r = \Delta \hat{H}_r - T\Delta \hat{S}_r$$

$$= (-890.3 \text{ kJ/mol}) - (298 \text{ K})\left(-0.243 \frac{\text{kJ}}{\text{mol} \cdot \text{K}}\right)$$

$$= -817.9 \text{ kJ/mol}$$

Finally, the reaction equilibrium constant at 25°C is

$$K_r = e^{\frac{\Delta \hat{G}_r}{RT}} = 2.53 \cdot 10^{156}$$

The very large value of the equilibrium constant indicates that the reaction products would be essentially the only constituents of an equilibrium system. This is consistent with the large negative value of the Gibbs' free energy, which suggests that the reaction occurs spontaneously at room temperature (although an ignition source may be necessary to start the process).

Because the equilibrium constant depends on the activity in the reacting phase, Equation 4.140 takes on very different forms in gases and liquid phases. Let us examine each of these in turn as well as the temperature dependence of the equilibrium constant.

Gases — For gases, the activity of an ideal gas is the partial pressure, yP. The equilibrium constant is then given by

$$K_r = \Pi\left(a_i^{\nu_i}\right) = P^{\Sigma \nu_i} \Pi\left(y_i^{\nu_i}\right) \tag{4.143}$$

Solving for the mole fraction term,

$$\frac{K_r}{P^{\Sigma v_i}} = \Pi\left(y_i^{v_i}\right) \tag{4.144}$$

For the reaction given by Equation 4.129, Equation 4.144 becomes

$$\frac{K_r}{P^{v_C+v_D-v_A-v_B}} = \left(\frac{y_C^{v_C} y_D^{v_D}}{y_A^{v_A} y_B^{v_B}}\right) \tag{4.145}$$

The term on the right is the ratio of the products to the reactants. If $\Sigma v_i = 0$, the equilibrium mole fractions are independent of pressure. If $\Sigma v_i < 0$, which means that the number of moles of reactants are more than the number of moles of products, the amount of products tends to increase with pressure. If $\Sigma v_i > 0$, which means that the number of moles of products is larger than the number of moles of reactants, then the amount of products tends to decrease with pressure. This is termed Le Chatelier's principle of pressure disturbances to equilibrium. High pressure tends to push the reaction toward the direction with the fewest number of molecules. Note that K_r is not a function of pressure but that the pressure, still influences the distribution of the constituents at equilibrium.

Use of Equation 4.145 is relatively straightforward once it is realized that the mole fractions of the individual reactants and products are generally related by the stoichiometry of the reaction. For example, consider the reaction

$$A + 2B \rightarrow C \tag{4.146}$$

Let A_0, B_0, and C_0 represent the initial number of moles of the respective species. Let us define a quantity χ as the extent of reaction or the fraction of A reacted. The moles of A reacted is then $A_0 \cdot \chi$ and the moles of A at any time is $A_0 \cdot (1 - \chi)$. The moles and mole fractions for all species can be summarized.

Species	**Initial**	**Final**	**Mole Fraction**
A	A_0	$A_0 (1 - \chi)$	$\frac{A_0(1-\chi)}{A_o + B_0 + C_0 - 2A_0\chi}$
B	B_0	$B_0 - 2A_0 \cdot \chi$	$\frac{B_0 - A_0\chi}{A_o + B_0 + C_0 - 2A_0\chi}$
C	C_0	$C_0 + A_0 \cdot \chi$	$\frac{C_0 + A_0\chi}{A_o + B_0 + C_0 - 2A_0\chi}$
Total	$A_0 + B_0 + C_0$	$A_0 + B_0 + C_0 - 2A_0 \cdot \chi$	1

Assuming that the reaction occurs in the gas phase and is conducted under ideal conditions (high temperature and low pressure), the equilibrium constant can be written in terms of χ as

$$K_r = P^{-1-2+1}\left[\frac{A_0(1-\chi)}{A_o+B_0+C_0-2A_0\chi}\right]^{-1}\left[\frac{B_0-2A_0\chi}{A_o+B_0+C_0-2A_0\chi}\right]^{-2}\left[\frac{C_0+A_0\chi}{A_o+B_0+C_0-2A_0\chi}\right]^{1}$$
$$=\left[\frac{A_o+B_0+C_0-2A_0\chi}{P}\right]^2\frac{(C_0+A_0\chi)}{[A_o(1-\chi)](B_0-2A_0\chi)^2} \tag{4.147}$$

This can, in principle, be solved for χ (likely by trial and error or by a nonlinear equation solver) for any value of the equilibrium constant. The mole fraction of any particular constituent, for example, the product C, can then be determined by reference to the table above. Equation 4.147 can be simplified significantly for particular values of the initial concentrations and pressure. For example, at 1 atm total pressure and $A_0 = B_0$ and $C_0 = 0$, Equation 4.147 becomes

$$K_r = \frac{\chi(1-\chi)}{(1-2\chi)^2}$$
$$(4K_r+1)\chi^2-(4K_r+1)\chi+K_{rxn}=0 \tag{4.148}$$
$$\chi=\frac{-(4K_r+1)\pm\sqrt{(4K_r+1)^2-4(4K_r+1)K_r}}{2(4K_r+1)}$$

This is illustrated in Example 4.21 for a particular reaction.

Example 4.21: Equilibrium gas composition

Consider the formation of SO_3 in the exhaust gases of an ore smelter.

$$SO_2+\frac{1}{2}O_2\rightarrow SO_3$$

The exhaust gases are 10% SO_2, 10% O_2 and 80% nitrogen at 1 atm pressure. Estimate the equilibrium SO_3 formed by the above reaction at 500 and 600°C from the following data.

	SO_2	O_2	SO_3
ΔH_f, kJ/mol	−296.83	0	−395.72
S, kJ·mol^{-1}·K^{-1}	0.2482	0.2051	0.2568

The net changes in enthalpy and entropy for the reaction are

$$\Delta H_r = (1)(-395.72)-(1)(-296.83) = -98.89 \text{ kJ/mol}$$
$$\Delta S_r = (1)(0.2568)-(1)(0.2482)-(1/2)(0.2051) = -0.094 \text{ kJ/mol}\cdot\text{K}$$

The equilibrium constants at 500 and 600°C (773 and 873 K) are then

$$K_r = e^{\frac{-\Delta G_r(T)}{RT}} = \begin{cases}59.42 & \text{at } 500°\text{C}\\ 10.21 & \text{at } 600°\text{C}\end{cases}$$

Evaluating the mole fractions as a function of reaction extent, χ, on a basis of 10 moles of gas.

	Initial moles	Final moles	Mole fraction
SO_2	1	$1-\chi$	$\frac{1-\chi}{10-\chi/2}$
O_2	1	$1-\chi/2$	$\frac{1-\chi/2}{10-\chi/2}$
SO_3	0	χ	$\frac{\chi}{10-\chi/2}$
N_2	8	8	$\frac{8}{10-\chi/2}$
Total	10	$10-\chi/2$	1

Thus, the equilibrium constant is given by

$$K_r = \frac{\dfrac{\chi}{10-\chi/2}}{\dfrac{1-\chi}{10-\chi/2}\left(\dfrac{1-\chi}{10-\chi/2}\right)^{1/2}}$$

Solving (by trial and error) for χ and the mole fraction of SO_3

$\chi = 0.873$ $\quad y_{SO_3} = 0.091$ $\quad$ 500°C

$\chi = 0.658$ $\quad y_{SO_3} = 0.068$ $\quad$ 600°C

The amount of SO_3 increases as the temperature decreases, which should be expected given that the negativity of the Gibbs' free energy increases as the temperature decreases. Due to the negative change in entropy of the reaction, this term drives the Gibbs' free energy more positive at higher temperatures until at about 780 K when the Gibbs' free energy change is zero. At higher temperatures, the reaction does not occur spontaneously. It should be emphasized, however, that these estimates of temperature assume that the enthalpy and entropy of the reaction are not functions of temperature, an approximation that is approximately correct as long as phase changes do not occur. The temperature dependence of the Gibbs' free energy is discussed in more detail below.

Liquids — In liquids, the effect of pressure is generally negligible under conditions of interest to environmental engineers. Under these conditions, the equilibrium constant becomes

$$K_r = \left(\Pi \gamma_i^{\nu_i}\right)\left(\Pi x_i^{\nu_i}\right) \tag{4.149}$$

or for the reaction of Equation 4.127,

$$K_r = \left(\frac{\gamma_C^{\nu_C}\,\gamma_D^{\nu_D}}{\gamma_A^{\nu_A}\,\gamma_B^{\nu_B}}\right)\left(\frac{x_C^{\nu_C}\,x_D^{\nu_D}}{x_A^{\nu_A}\,x_B^{\nu_B}}\right) \tag{4.150}$$

In order to simplify this further, the ideality, or nonideality, of the various constituents must be specified in addition to the interrelationships of the x_i.

Temperature dependence — The temperature dependence of the Gibbs' free energy can be approximated in many cases by assuming that the enthalpy and entropy change of the reaction are independent of temperature. The temperature dependence is then solely the result of the increasing importance of the entropy change of the reaction at higher temperature (from $T\Delta\hat{S}$). If the entropy change is negative, the Gibbs' free energy gets more positive as the temperature is increased while the opposite is true if the temperature is decreased.

The actual temperature dependence of the Gibbs' free energy also depends on the effect of temperature on the enthalpy of reaction. From the definition of the Gibbs' free energy, it can be shown that the actual temperature dependence of the equilibrium constant follows the relationship

$$\frac{d(\ln K_r)}{dT} = \frac{\Delta\hat{H}_r}{RT^2} \tag{4.151}$$

where ΔH_r is the standard heat of reaction. As indicated previously for an *endothermic* reaction, heat is absorbed by the reaction, the heat of reaction is positive, and the equilibrium constant increases with temperature. That is, more of the products of the reaction can be produced as the temperature rises. For an *exothermic* reaction, heat is evolved by the reaction, the heat of reaction is negative, and the equilibrium constant decreases with temperature. That is, the equilibrium is shifted more toward the reactants and less of the products can be observed at higher temperatures.

Note that use of Equation 4.151 depends on knowing ΔH_r and its dependence on temperature. The heat of formation of various substances can be found in tables, typically at 25°C (see Appendix A). The net heat of reaction can then be determined from

$$\Delta\hat{H}_r = \sum \nu_i \Delta\hat{H}_{fi} = \left(\sum \nu_i \Delta\hat{H}_f\right)_i\Big|_{products} - \left(\sum |\nu_i| \Delta\hat{H}_f\right)_i\Big|_{reactants} \tag{4.152}$$

Equation 4.151 can then be used to estimate the heat of reaction at different temperatures. The approach was previously discussed in Section 4.1.5.

4.10 ACID DISSOCIATION CONSTANT

A particularly important chemical equilibrium constant is the acid dissociation constant for an acidic or basic chemical. An acidic chemical, HA, is one that tends to ionize and give up a proton, H^+, in an aqueous environment.

$$HA \xleftrightarrow{H_2O} H^+ + A^- \tag{4.153}$$

The equilibrium constant that describes the relative amounts of the dissociated acid in solution is here defined as K_a to emphasize that it refers to the acid dissociation

$$K_a = \frac{(a_{H^+})(a_{A^-})}{(a_{HA})} \tag{4.154}$$

As indicated previously, the activities are generally written as $\tilde{M}_i\gamma_i$ where $\tilde{M}_i$ is the molarity of substance I and γ_i is the activity coefficient of that substance in the solution. The activity coefficients of the ionic species are approximately unity and the activity of the undissociated species is often incorporated into the definition of the equilibrium constant giving,

$$K_a = \frac{(\tilde{M}_{H^+})(\tilde{M}_{A^-})}{(\tilde{M}_{HA})} \tag{4.155}$$

Taking the logarithm (base 10) of both sides and defining pH = –log $\tilde{M}_{H^+}$ and $pK_a = -\log K_a$, this becomes

$$pK_a = pH - \log\frac{\tilde{M}_{A^-}}{\tilde{M}_{HA}} \tag{4.156}$$

Thus, for a given value of equilibrium constant (which specifies pK_a), Equation 4.156 relates the ratio of the dissociated and parent species to pH. The concentration of the dissociated species is the same as the concentration as the undissociated substance HA when the pH of the solution is equal to the pK_a. Phenol, which has a pK_a of 10 or a dissociation constant of 10^{-10}, will be 50% dissociated when the pH of the aqueous solution is 10. The specific proportion of the dissociated species is also determined at other values of pH by Equation 4.156. At neutral pH, pH = 7, pK_a – pH = 3 and the ratio of dissociated to undissociated phenol is 10^{-3}. Thus at neutral pH, phenol is expected to be only 0.1% dissociated.

The significance of this for environmental purposes is that the ionized species tends to be essentially nonsorbing onto soils and sediments. The nondissociated species sorbs primarily by the hydrophobic interactions discussed previously. Thus, for example, the neutral form of pentachlorophenol, with a pK_a of 4.75, tends to sorb strongly to soils. If it is assumed that the ionic species is completely soluble in water, at pH = pK_a = 4.75, only 50% of the pentachlorophenol partitions to soil organic carbon in this manner. At a pH > 7, comparatively little of the pentachlorophenol will sorb to the soil or sediment. This is illustrated in Example 4.22.

Example 4.22: Sorption of pentachlorophenol as a function of pH

Estimate the organic carbon based partition coefficient for pentachlorophenol in soil assuming that K_{oc} for the nonionized pentachlorophenol is 10^5 L/kg and the ionized forms are assumed to be negligibly sorbing.

The acid dissociation constant for pentachlorophenol is $1.8 \cdot 10^{-5}$. The pK_a = 4.75. The molar ratio of the ionized to unionized form is then given by

$$\frac{\tilde{M}_{A^-}}{\tilde{M}_{HA}} = 10^{pK_a - pH}$$

The estimated organic carbon based partition coefficient is then the neutral concentration weighted fraction of the organic carbon based partition coefficient (since the ionized species does not contribute to sorption)

$$K_{oc_{avg}} = \frac{\tilde{M}_{HA}}{\tilde{M}_A + \tilde{M}_{HA}} K_{oc}$$

$$= \left(1 + \frac{\tilde{M}_{A^-}}{\tilde{M}_{HA}}\right)^{-1} K_{oc}$$

$$= (1 + 10^{pK_a - pH}) K_{oc}$$

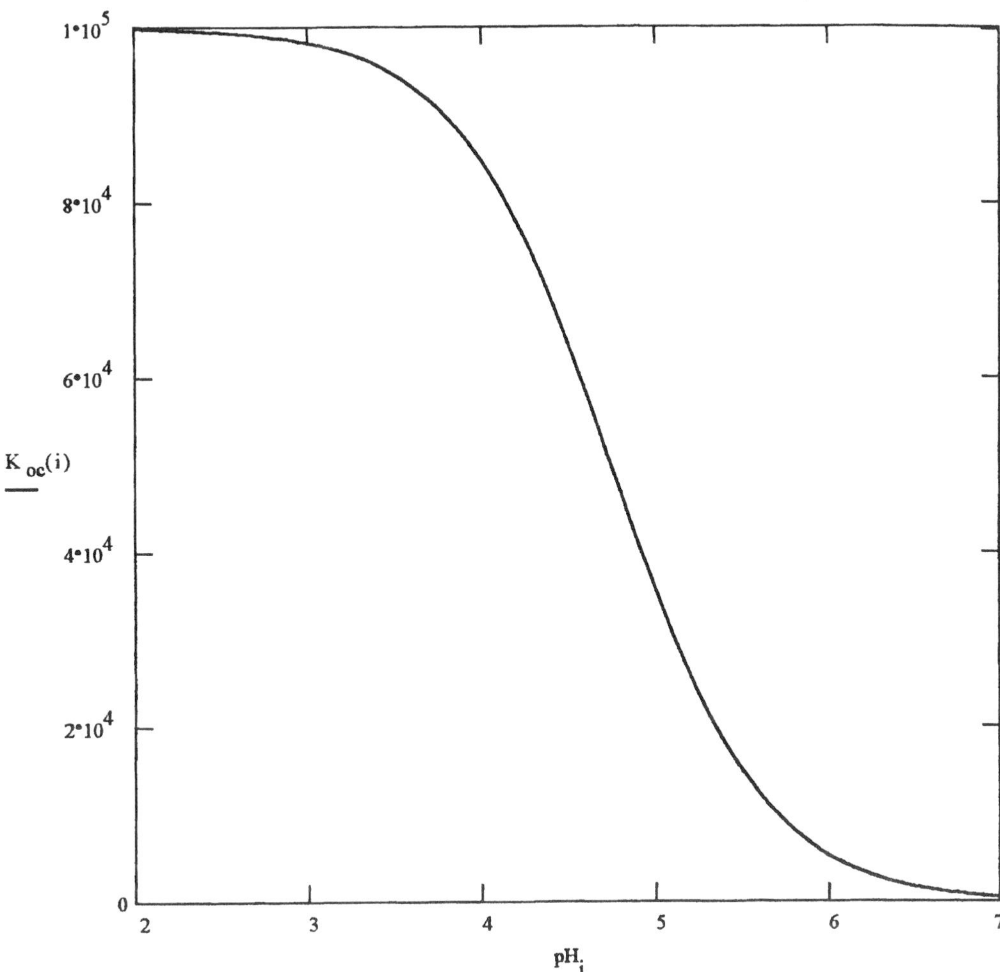

FIGURE 4.17 Ionization of pentachlorophenol and influence on partition coefficient (assuming nonsorbing ionic species).

A plot of the effective organic carbon based partition coefficient as a function of pH is shown in Figure 4.17. Although the sorption coefficient of the ionic species was neglected in this analysis, it should be recognized that some sorption does occur. Ionized pentachlorophenol will still be retarded in its motion through soils as a result of sorption. This sorption is negligible only relative to the sorption of the parent compound.

This situation is more complicated when the substance is a *polyprotic acid*, that is, the acid dissociates into more than one proton or H^+ ion. For example, carbonic acid (H_2CO_3), dissociates in this fashion

$$\begin{aligned} H_2CO_3 &\Leftrightarrow H^+ + HCO_3^- \\ HCO_3^- &\Leftrightarrow H^+ + CO_3^{2-} \end{aligned} \tag{4.157}$$

At 25°C, the equilibrium constants that relate the various species activities are given by

$$K_{H_2CO_3} = \frac{\left(a_{H^+}\right)\left(a_{HCO_3^-}\right)}{\left(a_{H_2CO_3}\right)} = 10^{-6.35} \tag{4.158}$$

$$K_{HCO_3^-} = \frac{\left(a_{H^+}\right)\left(a_{CO_3^{2-}}\right)}{\left(a_{HCO_3^-}\right)} = 10^{-10.33}$$

Note that the assumption of ideal solutions used to relate activity to mole fraction in Equation 4.156 is less appropriate for ions with a valence or charge of greater than one (or less than minus one). At an ionic strength of 0.1 mol/L, the activity coefficient of HCO_3^- is about 0.8 while the activity coefficient of CO_3^{2-} is about 0.4. Despite this, let us assume that the activities of the various species are simply given by the molar concentrations. The equilibrium constants are then (at 25°C)

$$\begin{aligned} K_{H_2CO_3} &\approx \frac{\left(\tilde{M}_{H^+}\right)\left(\tilde{M}_{HCO_3^-}\right)}{\left(\tilde{M}_{H_2CO_3}\right)} = 10^{-6.35} \\ K_{HCO_3^-} &\approx \frac{\left(\tilde{M}_{H^+}\right)\left(\tilde{M}_{CO_3^{2-}}\right)}{\left(\tilde{M}_{HCO_3^-}\right)} = 10^{-10.33} \end{aligned} \tag{4.159}$$

Four species are represented in the two equations of 4.159. Two additional equations must be introduced in order to complete the system of equations. The hydrogen ion concentration may be set by the pH or, if unknown, may be related to

$$H_2O \Leftrightarrow H^+ + OH^-$$

$$K_{diss} = \frac{\left(\tilde{M}_{H^+}\right)\left(\tilde{M}_{OH^-}\right)}{\tilde{M}_{H_2O}} = 1.82 * 10^{-16} \text{ mol/L} \tag{4.160}$$

$$K_w = \left(\tilde{M}_{H^+}\right)\left(\tilde{M}_{OH^-}\right) = 1 * 10^{-14} \text{ mol}^2/\text{L}^2$$

Where K_w has incorporated the molar concentration of water (0.0182 mol/L at 25°C) into K_{diss}. This has now added the molar concentration of OH^- as a variable, but an additional equation can be found from the requirement that the system remain electrically neutral. The system began electrically neutral and any dissociation must maintain this electroneutrality. That is, the sum of the individual ionic species concentrations must be zero.

$$\tilde{M}_{H^+} + \tilde{M}_{OH^-} + \tilde{M}_{HCO_3^-} + 2\left(\tilde{M}_{CO_3^{-2}}\right) = 0 \tag{4.161}$$

Note that twice the molar concentration of CO_3^{-2} is included since its charge is twice that of the other species.

In addition, the concentration of one or more of the other species may be specified by other reactions. In ground water systems, limestone (calcium carbonate, $CaCO_3$) may control the concentration of the carbonate ion through the reaction

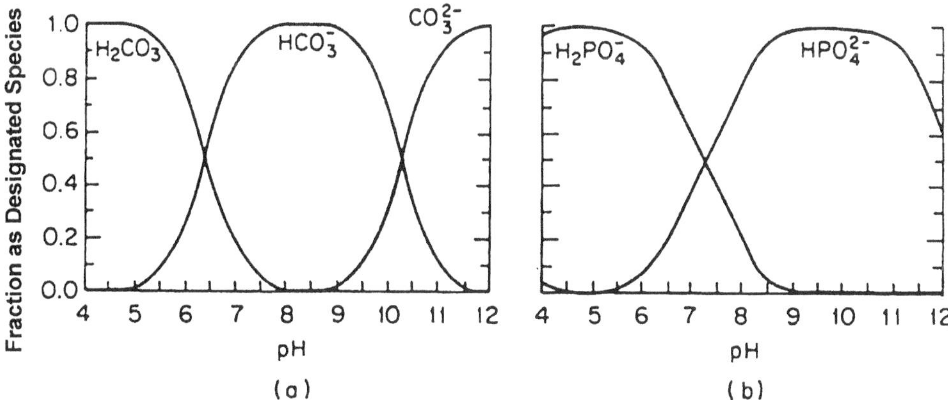

FIGURE 4.18 Distribution of the major species of (a) dissolved inorganic carbon and (b) inorganic phosphorous in water 25°C. (From Freeze, R.A. and J.A. Cherry (1979) *Groundwater*, Prentice-Hall, Upper Saddle River, NJ. With permission.)

$$CaCO_3 \Leftrightarrow Ca^{+2} + CO_3^{-2} \tag{4.162}$$

As we shall see in the next section, the equilibrium in this reaction is governed by a solubility product constant, K_{sp}, which for $CaCO_3$ is $4.57{*}10^{-9}$ mol²/L² at 25°C. Note that the presence of the additional ions affects the electroneutrality balance.

In air or exposed systems, the concentration of the H_2CO_3 is defined by physical equilibrium with the carbon dioxide concentration in the air through a Henry's Law constant or air-water partition coefficient. The Henry's constant (H) for carbon dioxide dissolution into water at 25°C is 29.46 atm·L/mol. With an air concentration of carbon dioxide of 350 ppm or $350 \cdot 10^{-6}$ atm, the concentration of H_2CO_3 is then

$$\tilde{M}_{H_2CO_3} = \frac{p_{CO_2}}{H} = \frac{350 \cdot 10^{-6} \text{ atm}}{29.46 \, \dfrac{\text{atm} \cdot \text{L}}{\text{mol}}} = 1.18 \cdot 10^{-5} \text{ mol/L} \tag{4.163}$$

Which combined with Equations 4.158 defines the concentrations of all species in water. This is illustrated in Figure 4.18. For comparison, the inorganic phosphorous equilibrium is also shown in Figure 4.18.

The sum of the concentrations of H_2CO_3, HCO_3^-, and CO_3^{2-} is essentially the total inorganic carbon in groundwater or surface waters in the environment. Note that the activity and therefore the concentration of undissociated H_2CO_3 increases as the H^+ ion activity (or concentration) increases. Thus, as the pH of the water decreases or the water becomes more acidic, less of the carbonic acid is dissociated. Under normal conditions essentially all of the carbonate present in natural waters is in the HCO_3^- form. Under acidic conditions, however, most of the carbonate will be present as carbonic acid.

4.11 MINERAL DISSOLUTION REACTIONS

Reaction equilibrium concepts can also be applied to mineral dissolution reactions. For example, ordinary table salt, sodium chloride, dissolves in water via the reaction

$$\text{NaCl} \Leftrightarrow \text{Na}^+ + \text{Cl}^- \tag{4.164}$$

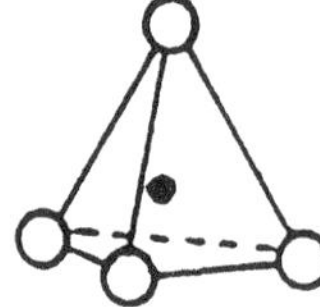
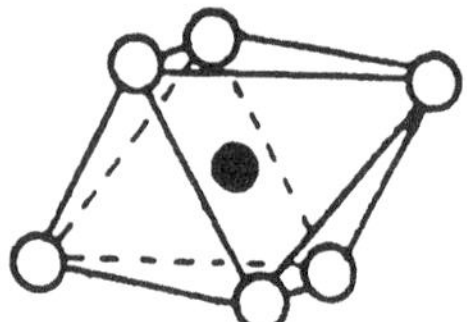

FIGURE 4.19 The basic structural units of aluminosilicate clay minerals: a tetrahedron of oxygen atoms surrounding a silicon ion (left) and an octahedron of oxygens or hydroxyls enclosing an aluminum ion. (From Hillel, D. (1980) *Fundamentals of Soil Physics,* Academic Press, New York. With permission.)

The equilibrium concentration of dissolved sodium chloride is determined by the reaction equilibrium constant,

$$K_r = \frac{(a_{Na^+})(a_{Cl^-})}{(a_{NaCl})} \tag{4.165}$$

Because the activity coefficients vary little at low ionic strength (especially monovalent ions), this is often written in terms of a solubility product constant,

$$K_{sp} = (\tilde{M}_{Na^+})(\tilde{M}_{Cl^-}) \tag{4.166}$$

Solubility product constants for a variety of minerals of interest are available to environmental engineers. Note that the activities in the equilibrium constant and the solubility product are raised to the power of the stoichiometric coefficient. The definition of the solubility product for calcium phosphate $Ca_3(PO_4)_2$ is given by

$$(K_{sp})_{Ca_3(PO_4)_2} = (\tilde{M}_{Ca^{2-}})^3(\tilde{M}_{PO_4^{3-}})^2 \tag{4.167}$$

4.12 ION EXCHANGE AND ADSORPTION

Ionic species can also "dissolve" into a solid state. The solid-liquid partitioning discussed previously was focused on hydrophobic interactions in which nonpolar organic molecules could sorb into the solid phase and, in particular, the organic fraction of the solid phase. The humus, or fully degraded organic fraction of a soil typically retains a small negative charge that can attract or repel ions in solution. More importantly, fine mineral particles, or clays, in soils also retain an electrical charge.

Clay particles are soil particles less than 2 μm in size which are composed primarily of silicates and aluminum oxides. Clay minerals have a crystalline structure in which a silicon or aluminum cation (positive valences) are surrounded in a regular structure by oxygen or hydroxyl (OH) groups. These structures are of two basic types, a tetrahedron of oxygen atoms with silicon (valence +4) at its center and an octahedron of oxygen or hydroxyl groups with aluminum (valence +3) at its center (see Figure 4.19). The structures are stacked in layers with the oxygen shared between layers. Imperfections in the cell lattice which lead to aluminum replacing silicon in the tetrahedryl structure and magnesium (valence +2) replacing aluminum in the octahedryl structure cause a residual negative charge that can attract polar and ionic molecules.

The structures of kaolinite and montmorillonite are shown in Figure 4.20. Kaolinite (a 1:1 mineral) has the general formula, $Al_4Si_4O_{10}(OH)_8$. Adjacent tetrahedryl and octahedryl layers share oxygen forming a very tightly held crystal lattice. Water molecules and ions do not tend to penetrate

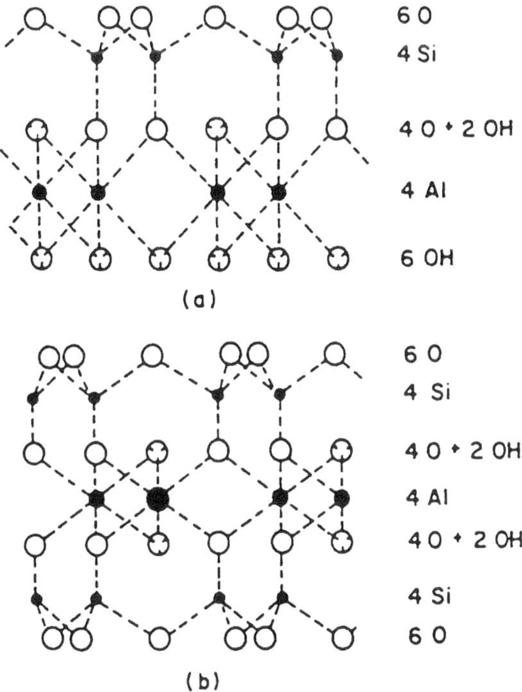

FIGURE 4.20 Schematic representation of the structure of alumnosilicate minerals: (a) kaolinite and (b) montmorillonite. (From Hillel, D. (1980) *Fundamentals of Soil Physics,* Academic Press, New York. With permission.)

between the tightly held crystal sheets of layers and adsorption occurs only on the external surface. In montmorillonite, which has the general formula, $Al_{3.5}Mg_{0.5}Si_8O_{20}(OH)_4$, each octahedryl layer shares oxygen with two tetrahedryl layers (a 2:1 mineral). The layers of montmorillonite are much more loosely connected allowing water and ion penetration between layers, dramatically increasing the available adsorption area. The sorption of water also causes significant swelling to occur in montmorillonite when compared to kaolininte. Upon drying the resulting shrinkage results in cracking. The clay mineral, illite, which has the general formula $Al_4Si_7AlO_{20}(OH)_4K_{0.8}$, has properties intermediate between those of kaolinite and montmorillonite.

The residual negative charge at the surface of these clays is normally balanced by cations or positive charges adjacent to the surface. The addition of water, however, solubilizes some of these ions and separates the negative charge at the surface from the balancing positive ions, creating the electrical *double layer* shown in Figure 4.21. The composition of the double layer can change, for example, by the replacement of monovalent ions with a smaller number of divalent ions. This process is termed ion exchange. The total number of exchangeable ions per unit mass or surface area of the clay particles is the ion exchange capacity. Because the clays typically have a residual negative charge, the exchangeable ions are cations and the ion exchange capacity is called the *cation exchange capacity* (CEC). The cation exchange capacity is normally measured in milliequivalents per 100 gm of soils (mEq/100 gm) and among common clays varies from 3 to 15 mEq/100 g in kaolinite to 80 to 100 mEq/100 g in montmorillonite. The equivalent weight of an ion is its molecular weight divided by its valence or charge. Thus,

milliequivalent = millimoles if valence is 1 (e.g., Na^+ and NH_4^+)
milliequivalent = 2 × millimoles if valence is 2 (Mg^{2+}, Ca^{2+})
milliequivalent = 3 × millimoles if valence is 3 (Al^{3+})

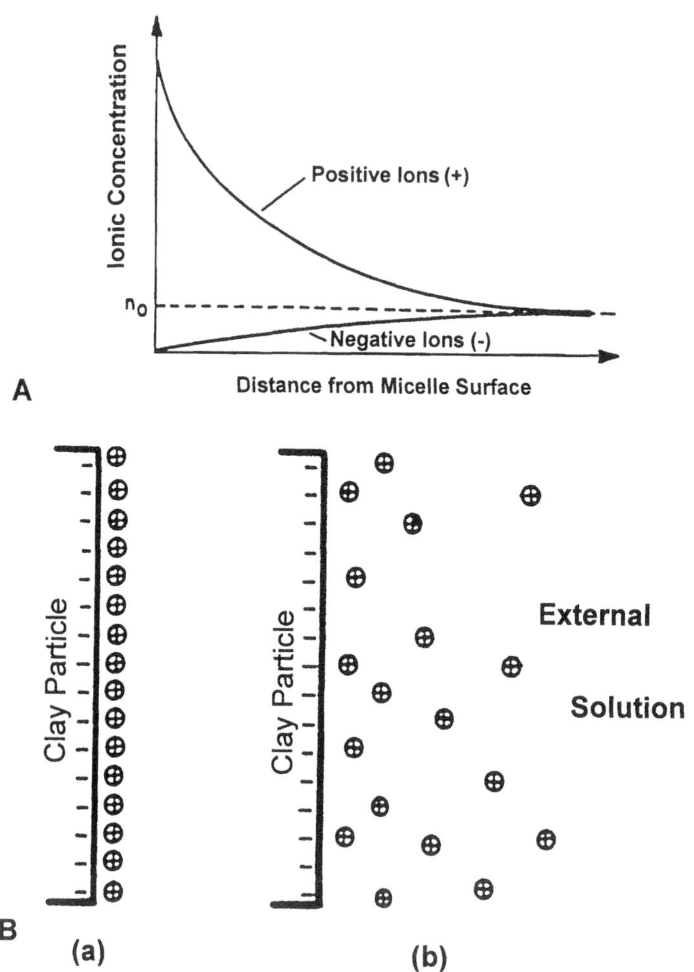

FIGURE 4.21 Double layer phenomena in clay soils. (A) The distribution of positive and negative ions in solution with distance from the surface of a clay micelle bearing net negative charge. Here n_0 is the ionic concentration in the bulk solution outside the electrical double layer. (B) Formation of a diffuse double layer in a hydrated micelle showing (a) the dry and (b) the hydrate state. (From Hillel, D. (1980) *Fundamentals of Soil Physics,* Academic Press, New York. With permission.)

The CEC refers only to reversibly sorbed cations. In general, higher valence substances are more readily attracted to the residual negative charge at the clay surface and will preferentially be sorbed. For ions with a given valence, the smaller ions will tend to be sorbed more strongly. Here, size refers to the size of the hydrated or water-associated form of the ion. Among monovalent and divalent ions, the order of preference (strongest sorbed to weakest and most easily exchangeable) is given by

$Cs^+ > Rb^+ > K^+ > Na^+ > Li^+$ Monovalent (Lithium most exchangeable)

$Ba^{2+} > Sr^{2+} > Ca^{2+} > Mg^{2+}$ Divalent (Magnesium most exchangeable)

Among the most important ions of different valences the order of preference can be written

$$Al^{3+} > Ca^{2+} > Mg^{2+} > NH_4^+ > K^+ > H^+ > Na^+ > Li^+$$

It is important to recognize, however, that complete exchange of one ion for another rarely occurs. Chemical equilibrium concepts can be used to identify the proportion of any ion that might be exchanged. For example, the exchange reaction of calcium for sodium can be written

$$2(Na^{+})_{ad} + Ca^{2+} \Leftrightarrow 2Na^{+} + (Ca^{2+})_{ad} \tag{4.168}$$

This exchange is important in water softeners and in montmorillonite clays. Sodium exchange of calcium in montmorillonite clays will result in spreading of the clay layers and a reduced resistance to flow. The equilibrium constant for this reaction is given by

$$K_{Na-Ca} = \frac{(a_{Ca^{2+}})_{ad}(a_{Na^{+}})^{2}}{(a_{Ca^{2+}})(a_{Na^{+}})^{2}_{ad}} \tag{4.169}$$

This equilibrium constant has come to be termed a *selectivity coefficient* in that it gives an indication of the distribution of the exchangeable ions between the adsorbed state and free state. The activity coefficients of the compounds in the adsorbed state are rarely known and it has been common to assume these equal to one and take the activity of the adsorbed species, A^+, as proportional to its prevalence, that is, as equal to the molar ratio of the cationic exchanging species, $\tilde{f}_{A^+}$.

The selectivity coefficient for sodium and calcium now becomes

$$K_{Na-Ca} = \frac{\left(\tilde{f}_{Ca^{2+}}\right)_{ad}\left(\gamma_{Na^{+}}\tilde{M}_{Na^{+}}\right)^{2}}{\left(\gamma_{Ca^{2+}}\tilde{M}_{Ca^{2+}}\right)\left(\tilde{f}_{Na^{+}}\right)^{2}_{ad}} \tag{4.170}$$

If it is further assumed that the activity coefficients of the ions in solution are near unity (only true for low ionic strengths for divalent ions), then Equation 4.172 becomes

$$K_{Na-Ca} = \frac{\left(\tilde{M}_{Na^{+}}\right)^{2}\left(\tilde{f}_{Ca^{2+}}\right)_{ad}}{\left(\tilde{M}_{Ca^{2+}}\right)\left(\tilde{f}_{Na^{+}}\right)^{2}_{ad}} \tag{4.171}$$

Levy and Hillel (1968) report this selectivity coefficient to have a value of about one-fourth. If the ionic strength is 0.1 molar (0.1 mol/L), the fraction of the exchangeable capacity that is sodium is about 10%. If the ionic strength is 0.001 molar, however, sodium constitutes only about 1% of the exchangeable capacity.

4.13 OXIDATION–REDUCTION REACTIONS

Another important class of reactions for which equilibrium calculations might be desired are oxidation-reduction reactions. An oxidation-reduction reaction is one that involves the transfer of electrons. For example, in the oxidation of iron from the +2 to the +3 oxidation state,

$$O_2 + 4Fe^{2-} + 4H^{+} \Leftrightarrow 4Fe^{3+} + 2H_2O \tag{4.172}$$

the oxidation half-reaction is

$$4Fe^{2+} \Leftrightarrow 4Fe^{3+} + 4e^{-} \tag{4.173}$$

That is, for every mole of iron oxidized, one electron is given off. The reduction half-reaction is

$$O_2 + 4H^+ + 4e^- \Leftrightarrow 2H_2O \tag{4.174}$$

Note that the number of electrons released in the oxidation half-reaction exactly equals that required in the reduction half-reaction. This expresses the concept of electrical neutrality in which the total sum of the positive and negative charges must cancel each other out. If $\tilde{M}_{i^+}$ represents the molarity of the positive or cationic species and $\tilde{M}_{i^-}$ represents the molarity of the negative or anionic species, and z^+ and z^- represent the numbers of electrons transferred by each half-reaction, electrical neutrality can be stated by

$$\sum z_i^+ \tilde{M}_{i^+} = \sum z_i^- \tilde{M}_{i^-} \tag{4.175}$$

The most important ionic species in natural waters include sodium (Na^+), magnesium (Mg^{2+}), calcium (Ca^{2+}), chlorine (Cl^-), carbonate (HCO_3^-), and sulfate (SO_4^{2-}). Thus, electroneutrality suggests

$$\tilde{M}_{Na^+} + 2\tilde{M}_{Mg^{2+}} + 2\tilde{M}_{Ca^{2+}} = \tilde{M}_{Cl^-} + \tilde{M}_{HCO_3^-} + 2\tilde{M}_{SO_4^{2-}} \tag{4.176}$$

Returning to the iron oxidation reaction, the number of electrons required (in this case one per mole of Fe^{2+}) is a measure of oxidizing or reducing potential of the solution in which the reaction is taking place. The equilibrium constant for the iron half-reaction can be written

$$K_{Fe^{2+}} = \frac{\tilde{M}_{Fe^{3+}} \tilde{M}_{e^-}}{\tilde{M}_{Fe^{2+}}} = 10^{-12.53} \tag{4.177}$$

Taking the logarithm of both sides, the pE can be defined as –log of the concentration of the electrons, $\tilde{M}_{e^-}$

$$\log \frac{\tilde{M}_{Fe^{3+}}}{\tilde{M}_{Fe^{2+}}} + 12.53 = -\log \tilde{M}_{e^-} = pE \tag{4.178}$$

The term pE provides an indication of the redox condition of the system. A related quantity is the redox potential, Eh, which is defined, for half-reactions that involve the transfer of only a single electron, by

$$pE = 16.9\ Eh \tag{4.179}$$

Both pE and Eh are functions of pH.

Many of these oxidation-reduction reactions are *catalyzed* by microorganisms. A catalyst is a substance which accelerates the rate of a reaction but is not required for the reaction to occur. A catalyst is neither a reactant nor a product of a reaction. It cannot speed up or otherwise change a reaction that is not thermodynamically favored. The redox potential in soils is an important indicator of the type of microorganisms and the oxidation-reduction reactions that are occurring.

4.14 CLOSURE

Although this chapter has provided an introduction to physical and chemical equilibrium, most of the environment is not described by these concepts. Equilibrium does provide, however, the direction toward which the environment is moving. In some cases a portion of the environment may be at a state of near equilibrium. In addition, the study of equilibrium is important in that the deviation from equilibrium helps us define the rate of transport and fate processes that are working to produce equilibrium in any system. The fundamentals of assessing the rate of these transport and fate processes in the environment and in environmental control systems will be discussed in the next chapter.

PROBLEMS

1. Determine the numerical value of the ideal gas constant in $cal \cdot mol^{-1} \cdot K^{-1}$ and $Btu \cdot lbmol^{-1} \cdot {}^{\circ}R^{-1}$.

2. What is the density of ammonia at 30°C and 2.7 atm pressure?

3. Determine the mass of air in a volume of 1 L at 25°C and 20 atm pressure assuming ideal gas behavior and real gas behavior by applying the compressibility factor chart.

4. At what temperature do you begin to see significant deviations from the ideal gas law for carbon dioxide at atmospheric pressure?

5. Estimate the mass of hydrogen in a vessel at a temperature of 25°C under a pressure of 10^7 Pa.

6. Estimate the volume of stack gas emissions from the stoichiometric combustion of natural gas (per m^3 of natural gas burned at standard conditions). Estimate under stack conditions of 200°C and 1 atm pressure and at ambient conditions of 25°C and 1 atm pressure.

7. Estimate the partial pressure of carbon dioxide, nitrogen, and water vapor in the stack exit under the conditions of Problem 6.

8. What is the partial pressure and density in g/m^3 of water vapor in air at 35°C with a relative humidity of 50%? Refer to Figure 3.2 for the vapor pressure of water as a function of temperature.

9. At what temperature is the relative humidity of the air in Problem 8 equal to 100%? This is referred to as the dew point of air.

10. Estimate the change in enthalpy for the following processes: (a) water temperature change from 25 to 75°C, (b) water temperature change from 25 to 125°C at 1 atm pressure, (c) air temperature change from 25 to 1500°C.

11. Calculate the adiabatic flame temperature of methane as a function of excess air. Use 5, 10, 15, 20, and 25% excess air. Plot the result. Can you explain the reasons for the observed relationship?

12. Calculate the adiabatic flame temperature for a hydrocarbon liquid with a carbon to hydrogen ratio of 1:2.

13. Calculate the adiabatic flame temperature for a coal with a carbon to hydrogen ratio of 1:1.

14. Equation 4.33 should not be applicable to polar compounds. What is the error in attempting to use the equation in estimating the heat of vaporization of water?

15. Compare the density of methane at 125°C and 50 atm pressure using the critical temperature predicted by 1.5 T_b to that estimated using the actual critical temperature. Repeat for hexane.

16. Examine several textbooks on thermodynamics. What sign convention is used for work in the first law of thermodynamics?

17. A stack gas at 200°C contains 1000 ppm of sulfur dioxide. What is the minimum energy required to separate this sulfur dioxide from the remainder of the exhaust gases? What is the minimum energy at 25°C?

18. Benzene has a boiling point of 80.1°C and a heat of vaporization of 30.77 kJ/mol. Develop an equation for vapor pressure as a function of temperature and estimate the vapor pressure of benzene at 25°C. How does this estimate compare to the measured vapor pressure of 0.125 atm at 25°C?

19. Hexane has a vapor pressure of 0.205 atm at 25°C and boils at 69°C. Estimate the heat of vaporization and compare to the experimental value of 28.85 kJ/mol.

20. Derive Equation 4.70.

21. Estimate the vapor pressure of chrysene via Equation 4.70. Montgomery and Welkom (1990), gives two values for the chrysene vapor pressure, 6.3 (10^{-9}) mmHg at 25°C and 6.3 (10^{-7}) mmHg at 20°C. Which do you believe is more accurate?

22. Estimate the solubility of pyrene, 1,2 dichlorobiphenyl, benzene, and hexane in water using Equation 4.72.

23. Estimate the solubility of *o*-xylene at 40°C if $\Delta\hat{H}_v = 0.385$ kJ/gm and its solubility at 25°C is 175 mg/L.

24. The solubilities of several organic compounds in freshwater and seawater are listed below. Determine the constant x in Equation 4.77 for each of these compounds.

Compound	Freshwater Solubility	Seawater Solubility (mg/L)
n-Pentene	38.5	27.6
Tetradecane	0.0022	0.0017
Benzene	1780	1391
Toluene	515	402

25. Estimate the equilibrium partial pressure of toluene above a gasoline containing 15% toluene by weight. The vapor pressure of toluene at 25°C is 0.0374 atm.

26. Estimate the activity coefficient of toluene in water from its solubility of 515 mg/L. Is the activity coefficient of toluene in seawater greater or less than this value?

27. Sketch the equilibrium partial pressure of toluene above a water solution over the entire dissolved concentration range of toluene.

28. Estimate the value of Henry's constant for toluene. Express as a dimensionless concentration ratio, dimensionless mole fraction ratio, and atm·m^3/mol.

29. Estimate the equilibrium concentration of toluene in a water solution in contact with a gasoline containing 15% gasoline.

30. Sketch the equilibrium concentration of toluene in water as a function of toluene concentration in a contacting ideal hydrocarbon mixture.

31. Estimate the air-gasoline partition coefficient of toluene from the air-water and water-gasoline partition coefficients. Compare to Raoult's Law.

32. Estimate the solubility of pyrene, 1,2 dichlorbiphenyl, benzene, and hexane in water using Equation 4.103. Compare to the results of Problem 22.

33. Estimate K_{oc} for pyrene, 1,2 dichlorobiphenyl, benzene, and hexane from Equation 4.102. Estimate K_{sw} for these same four compounds assuming that a soil is 2% organic carbon.

34. Using the K_{sw} values from Problem 33, what is the mass of pyrene, 1,2 dichlorobiphenyl, benzene, and hexane on a 2% organic carbon soil when the compounds are present at their solubility limit in water.

35. Consider the following soil-water equilibrium data. Fit the data to a linear isotherm and determine K_{sw}. If the soil contains 4% organic carbon, estimate K_{oc} for the sorbing compound.

36. Derive the Langmuir isotherm assuming that a finite concentration of sorption sites exist on the solid and that the rate of adsorption is proportional to the concentration of available sites and the fluid concentration of the contaminant. The rate of desoporption is assumed proportional to the concentration of filled sites.

37. Consider the following soil-water equilibrium data. Fit the data to a Langmuir isotherm.

38. Consider the following soil-vapor equilibrium data. Fit the data to a BET isotherm.

39. Refit the data in Problem 37 to a Freundlich isotherm.

40. Refit the data in Problem 38 to a Freundlich isotherm.

41. Consider a gasoline spill in which the gasoline contains 15% toluene. What is the mass distribution (% of total) in each phase in the soil. The soil contains 60% solids with an organic carbon percentage of 2%, 15% vapor space, 10% residual water, and the remainder residual gasoline.

42. Neglecting evaporation define the equilibrium distribution and concentrations of pyrene in each phase of a lake system. The phases include water, suspended and dissolved organic carbon, and sediment.The lake is 20 m in diameter and 2 m deep. Worms and other sediment-dwelling animals mix the upper 5 cm of sediment which contains 5% organic carbon. The bulk density of the sediment is 1 g/cm^3 with a porosity of 60%. The water contains 50 mg/L of dissolved and suspended organic carbon.

43. What is the equilibrium vapor pressure of pyrene above the water surface in Problem 42? Would you expect that evaporation is important?

44. Repeat Problem 41 employing the fugacity approach of Mackay (1991).

45. Repeat Problem 42 employing the fugacity approach of Mackay (1991).

46. Estimate the chemical equilibrium constant for ethane combustion.

47. Would the combustion of ethane be more or less complete at high pressures? That is, does high pressure drive ethane combustion equilibrium toward the right side of the equation or the left?

48. Would the combustion of ethane be more or less complete at high temperatures?

49. Estimate the equilibrium fraction of ethane in a combustion with air at 1000 K and 12000 K.

50. Estimate the sorption coefficient of trichlorophenol as a function of pH.

51. Estimate the redox potential in a system governed by iron oxidation when (a) Fe^{2+} = 100 Fe^{3+}, (b) when Fe^{2+} = Fe^{3+}, (c) when Fe^{2+} = 0.01 Fe^{3+}.

REFERENCES

Freeze, R.A. and J.A. Cherry (1979) *Groundwater*, Prentice-Hall, Upper Saddle River, NJ, 604 pp.

Hillel, D. (1980) *Soil Physics*, Academic Press, New York.

Himmelblau, D.M. (1974) *Basic Principles and Calculations in Chemical Engineering*, 3rd ed., Prentice Hall, New York.

Karickhoff, S.W., D.S. Brown, and T.A. Scott (1979) Sorption of Hydrophobic Pollutants in Natural Sediments, *Water Res.*, 13, 241–248.

Leo, A. and C. Hansch (1971) Linear Free Energy Relationships Between Partitioning Solvent Systems, *J. Organ. Chem.*, 36, 1539–44.

Levy, R. and Hillel, D. (1968) Thermodynamic Equilibrium Constants of Na-Ca Exchange in some Israeli Soils, *Soil Sci.*, 106, 393–398.

Lyman, W.J., W.F. Reehl, and D.H. Rosenblatt, Ed. (1990) *Handbook of Chemical Property Estimation Methods,* American Chemical Society, Washington, D.C.

Masters, G.M. (1990) *Introduction to Environmental Science and Engineering*, Prentice-Hall, New York.

Mackay, D. (1991) *Multimedia Environmental Models. The Fugacity Approach.* Lewis Publishers, Chelsea, MI.

Montgomery, J.H. and L.M. Welkom (1990) *Groundwater Chemicals Desk Reference*, Lewis Publishers, Chelsea, MI.

Rao, P.S.C. and J.M. Davidson (1980) Estimation of Pesticide Retention and Transformation Parameters Required in Nonpoint Source Pollution Models, in *Environmental Impact of Nonpoint Source Pollution*, M.R. Overcash and J.M. Davidson, Eds., Ann Arbor Science Publishers, Inc., Ann Arbor, MI.

Reible, D.D. and T.H. Illangasekare (1989) Subsurface Processes of Non-Aqueous Phase Liquids, in *Intermedia Pollutant Transport: Modeling and Field Measurements*, D. Allen, Y. Cohen, and I. Kaplan, Eds., Plenum Press, New York.

Schwarzenbach, R.P., P.M. Gschwend, and D.M. Imboden (1993) *Environmental Organic Chemistry*, Wiley-Interscience, New York.

Tomson, M., personal communication, 1997.

USEPA (1985) Superfund Public Health Evaluation Manual, Office of Emergency and Remedial Response, Washington, D.C.

Valsaraj, K.T. (1995) *Elements of Environmental Engineering: Thermodynamics and Kinetics*, CRC Press, Boca Raton, FL, 649 pp.

5 Rate Processes

5.1 INTRODUCTION

The procedures discussed in the previous chapter allow determination of the equilibrium concentrations in the various environmental media. Unfortunately, global equilibrium is rarely satisfied in the environment. Equilibrium is likely to be satisfied only locally, for example, at the interface between two phases. Even then, the assumption of equilibrium at an interface is largely the result of assuming that an interface has no volume but simply represents a plane separating the two phases. If it has no volume, then it has no place for accumulation of material and therefore rapidly achieves steady, or equilibrium conditions. Any region of finite volume requires time to achieve equilibrium and often that time is very long. During that time, reaction, mass transfer, or other processes push a system toward equilibrium.

The equilibrium state provides the direction toward which transient, or time-dependent processes progress. Often, the rate of progress toward equilibrium is assumed to be proportional to the difference between the current state of a system, say S, and the equilibrium state, say S^*. This can be written

$$R = k(S - S^*) \tag{5.1}$$

where R represents the rate of progress toward equilibrium and k is a proportionality constant or rate constant. The difference between the current state and the equilibrium state of the system is termed the driving force for the process. Because the rate of the process is proportional to the first power of the driving force, Equation 5.1 expresses the *linear driving force model.*

Among the transient processes that are modeled to behave this way are mass transfer at an interface and many chemical reactions. For a chemical reaction, S would represent the chemical concentration at any time (e.g., in mol/m^3) and S^*, the equilibrium concentration that would be achieved at infinite time. K would be a reaction rate constant, typically with units of $time^{-1}$ (e.g., s^{-1}) and R would be the reaction rate in $mol \cdot m^{-3} \cdot s^{-1}$.

If we consider the simple reaction, A ⇔ B, and assume that the reaction rate is proportional to the concentration of the reactant, the rate of disappearance of A by the forward reaction is given by

$$(R_A)_f = k_f C_A \tag{5.2}$$

where k_f is the forward reaction rate constant. Using equivalent assumptions, the rate of formation of A by the reverse (backward) reaction is given by

$$(R_A)_b = k_b C_B \tag{5.3}$$

The net rate of reaction is then

$$R_A = k_f C_A - k_b C_B \tag{5.4}$$

At equilibrium, the net rate of reaction is 0 and $k_f\ C_A^* = k_b\ C_B$ where C_A^* is the equilibrium concentration of A. Note that this emphasizes that equilibrium does not mean that the rate of either the forward or reverse reaction is zero but only the *net* rate of reaction is zero. Substituting for C_B in Equation 5.4, these assumptions have now allowed us to write the reaction rate at any time as

$$R_A = k_f(C_A - C_A^*) \tag{5.5}$$

Equation 5.5 is of the same form as Equation 5.1, illustrating the linear driving force law as applied to chemical reactions. For many reactions, it is often assumed that the equilibrium concentration (C_A^*) is either zero or sufficiently small that it is negligible compared to the actual or current concentration (C_A). Under these conditions, a linear driving force model of the reaction is simply

$$R_A \approx k_f C_A \tag{5.6}$$

This is the rate equation for a first order, irreversible reaction. Remember that R_A as written here is the rate of *disappearance* of A. This could just as easily have been written as the rate of formation of A and a negative sign would appear in Equation 5.6.

For mass transfer processes, R in Equation 5.1 is usually written as an expression for chemical flux, q_m. A *flux* is simply the quantity per unit area per time as shown in Figure 5.1. A fluid velocity can be thought of as a volumetric flux, i.e., the volume of fluid moving per unit area per time.

$$\begin{aligned} \text{Volumetric Flux} &= q \\ &= \frac{\{\text{volume}\}}{\{\text{area}\}\{\text{time}\}} \\ &= \frac{\{\text{length}\}}{\{\text{time}\}} = \text{Velocity (U)} \end{aligned} \tag{5.7}$$

We will employ the capital U to represent an average velocity. U = q in a free fluid but in a porous media, such as soil, the fluid volume is excluded from all but the pore volume and the velocity is greater in these regions than the average over the total area that q represents. If ε represents the ratio of the void or pore volume to the total volume, or porosity, $U = q/\varepsilon$.

The volumetric flux can be converted to a mass flux, q_m, by multiplying by the density of the fluid,

$$\begin{aligned} \text{Mass Flux} &= q_m = U\rho \\ &= \left(\frac{\text{volume}}{\text{area}\cdot\text{time}}\right)\left(\frac{\text{mass}}{\text{volume}}\right) \\ &= \frac{\text{mass}}{\text{area}\cdot\text{time}} \end{aligned} \tag{5.8}$$

In order to evaluate the flux of a particular component or contaminant, for example, in mass per unit area per time, you must multiply the volumetric fluxes by the *concentration* of the contaminant in that volume. For example, the mass flux of a particular contaminant, q_A, carried by a fluid moving at velocity, U, is the concentration of the contaminant, C_A, multiplied by the velocity.

$$\begin{aligned} \text{Mass Flux}\,|_A &= q_A = U\,CA \\ &= \left(\frac{\text{m}}{\text{s}}\right)\left(\frac{\text{g}}{\text{m}^3}\right) = \left(\frac{\text{g}}{\text{m}^2\cdot\text{s}}\right) \end{aligned} \tag{5.9}$$

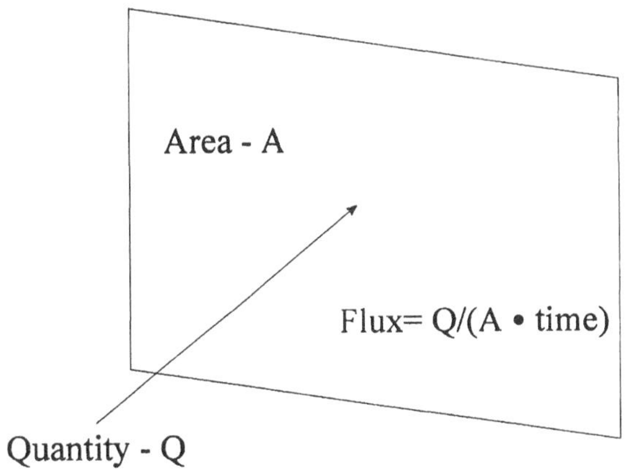

FIGURE 5.1 Depiction of flux and plane through which it is defined.

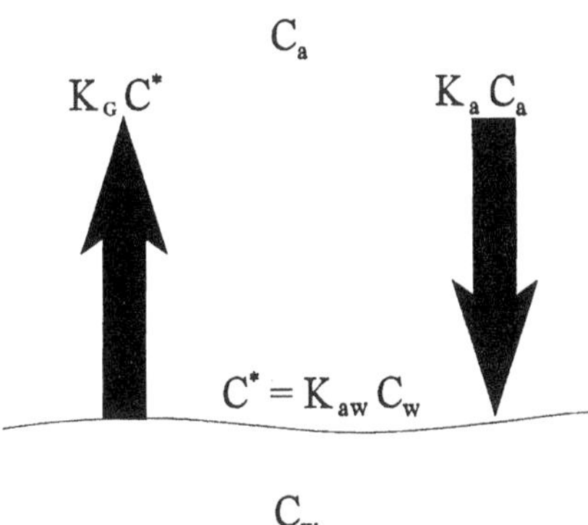

FIGURE 5.2 Depiction of net mass transfer of a contaminant above an air-water interface.

Note that both Equations 5.8 and 5.9 are of the form of the linear driving force model with the fluid velocity, U, equivalent to K and the local concentration representing the driving force.

Near an interface such as an air-water interface or fluid-solid interface, the *net* motion or velocity of the bulk fluid must go to zero. This simply means that any motion toward the interface must be matched by an equal and opposite motion away from the interface, that is $U_{to} = U_{from}$. Although there is no net motion, there may still be contaminant transport if the concentration carried by these motions differs.

Let us consider the evaporation of a chemical, A, at an air-water interface. Let us write the magnitude of the air motion to and from the interface as k_a ($=U_{to} = U_{from}$). The concentration of a contaminant carried to the interface is the bulk air concentration, C_a. The concentration of contaminant carried away from the interface is the interfacial concentration, which as we shall see is given by the concentration that would be in equilibrium with the surface waters, C_a^*. If the equilibrium is governed by Henry's Law, with an air-water partition coefficient K_{aw}, $C_a' = K_{aw} C_w$. As shown in Figure 5.2, the net flux of material out of the water, q_m or, specifically for component A, q_A, is the difference between the fluxes into and out of the water. Because we desire the flux out of the water (the evaporation), let us form the difference in flux from water to air, subscripted wa, minus the flux from air to water, subscripted "aw."

$$q_A = (q_A)_{wa} - (q_A)_{aw} = k_a C_a^* - k_a C_a = k_a (C_a^* - C_a) \tag{5.10}$$

Equation 5.10 is developed in Example 1. Equation 5.10 is just the linear driving force model of Equation 5.1. The flux in Equation 5.10 is positive when the higher concentrations are at the interface and transport is away from the interface. We could have just as easily employed the opposite sign convention. In either case, the transport serves to equalize the concentration within the air.

Example 5.1: Rate of evaporation of water from a lake

Estimate the rate of vaporization of water from a lake with a surface area of 1 ha. The effective mass transfer coefficient (k_a) between the water and the air is 10 m/h. Both the lake and the air temperature is 25°C and the relative humidity in the air is 30%.

The flux of water to the air is given by

$$(q_m)_{wa} = k_a C^*$$

where C^* is the concentration of water vapor in the air near the air-water interface. This is simply the air concentration of water vapor that is in equilibrium with the body of water, or the concentration that corresponds to the vapor pressure of liquid water at 25°C. The concentration very near the interface may depend upon how rapidly water is supplied to the interface, a complication that it unnecessary here since the water is essentially pure and it is *uniformly mixed* throughout the liquid. The flux of water from the air to the water is given by

$$(q_m)_{aw} = k_a C_a$$

At equilibrium, of course, these two fluxes are equal and the water vapor concentration in the air corresponds to the vapor pressure of water. If the system is not at equilibrium, the *net* flux from the water to the air is given by

$$q_m = k_a(C^* - C_a)$$

The equilibrium water vapor concentration of air at 25°C is 0.023 kg/m^3. Air at 30% relative humidity (30% saturation) contains 0.0069 kg/m^3. The rate of evaporation of water is thus

$$q_m = 10 \text{ m/h } (0.023 \text{ kg/m}^3 - 0.0069 \text{ kg/m}^3)$$

$$= 0.161 \frac{\text{kg}}{\text{m}^2 \cdot \text{h}}$$

Here the positive sign simply indicates that the water is evaporating rather than condensing from the air. The total rate of evaporation is the flux times the area of 10^4 m^2 or 1610 kg/h.

5.2 MATERIAL BALANCES

5.2.1 Basic Concepts

In order to assess the effects and implications of fate and transport processes such as chemical reaction and mass transfer it is necessary to develop a model that links them with the dynamics of a system. The basis for such a model is a *material or mass balance* first expressed by French scientist Lavossier. In any system not subject to nuclear reactions, mass is neither created nor destroyed and thus an accounting of the flows into and out of the system and transfers and chemical reactions within the system provide a model of the volumes, concentrations, or mass of material within the system. A material balance is the basic tool for modeling system behavior. The material balance can be used to determine mass flows consistent with inputs and outputs from the remainder of the system and define the dynamics of a system. The material balance will be used as a fundamental tool throughout the rest of this book.

To develop the material balance, let us first consider the simple system of Figure 5.3. The system has a surface area A through which fluid passes with influent velocity U_{in} and effluent velocity U_{eff}. Q_{in} is the influent volumetric flowrate, the product of $A{\cdot}U_{in}$, while Q_{out} is the effluent volumetric flowrate, the product of $A{\cdot}U_{out}$. The material or mass balance can be written

$$\{\text{Accumulation}\} = \{\text{Rate of Gain}\} - \{\text{Rate of Loss}\} \tag{5.11}$$

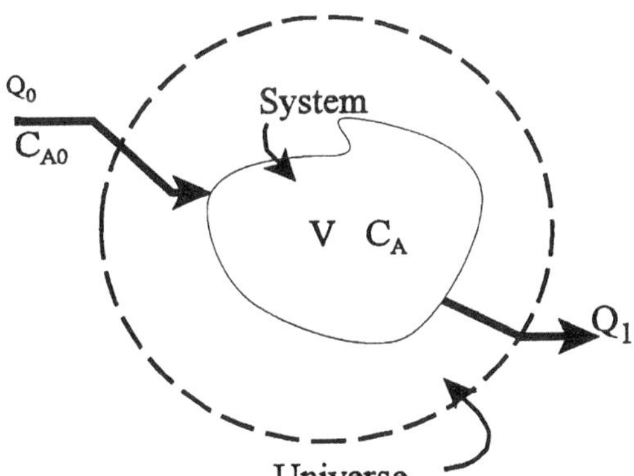

FIGURE 5.3 System for evaluation of simplified material balance. The only operative processes are flow in and out of the system.

The accumulation in Equation 5.11 can always be written quite simply as the time derivative of the quantity for which a balance is being written. In a balance on mass, the rate of accumulation is the time derivative of the mass in the system. For a fluid system this is often written as the fluid density times the volume, ρ V, or for a particular contaminant, the contaminant concentration times the volume. The accumulation term for a mass balance on a particular component A is given by

$$\{\text{Accumulation}\} = \frac{\partial}{\partial t} M_A = \frac{\partial}{\partial t} C_A V \tag{5.12}$$

In the example shown in Figure 5.3, the process that increases mass is just flow in,

$$\{\text{Rate of Gain}\} = Q_0 C_{A0} \tag{5.13}$$

The only process that decreases mass is the output by the effluent fluid. If the concentration within the system is everywhere uniform, then the interior concentration is the same as that measured in the effluent,

$$\{\text{Rate of Loss}\} = Q_1 C_{A1} = Q_1 C_A \tag{5.14}$$

This expression for the rate of loss recognizes that if the contents of the system are homogeneous and *well mixed*, then the effluent properties such as density are the same as that within the system. Thus, the full mass balance can be written,

$$\frac{\partial}{\partial t} C_A V = Q_1 C_{A0} - Q_0 C_A \tag{5.15}$$

Equation 5.15 simply states that if the mass flow into a system (Q_{in} C_{A0}) is different from the mass flow out (Q_{out} C_A), then there must be a change in either the fluid density or system volume with time.

In general it is possible to write a component mass balance for each component in a system. Note that since an overall mass balance is simply the sum of the individual component balances, it is not possible to write component balances *plus* a separate independent overall balance. *For an N component system, it is only possible to write N independent material balance equations.* Typically an overall balance will be coupled with component balances for all but one component. The total number of material balances that can be written remains one overall balance plus N–1 component balances for a total of N.

At long times, the system approaches equilibrium and the changes with time go to zero. Under these *steady-state* conditions the material balance of Equation 5.15 takes on an especially simple form, in particular,

$$(C_A)_{ss} = C_A(t \to \infty) = \frac{Q_{in}}{Q_{out}} C_{A0} \tag{5.16}$$

Let us now consider the transient equation and examine two specific cases. In the first case, let us assume that the concentration or density of a component remains constant and the volume of fluid changes, while in the second case let us assume that the total fluid volume remains constant and the concentration changes. The first case might apply to a material balance on a lake undergoing evaporation with the volumetric flow out, Q_{out}, being given by the product of the mass transfer coefficient and the evaporative area, k A. In that case the concentration of water in the lake is just the density of pure water and this density remains constant throughout the evaporation. The second case might apply to a lake with balanced influent and effluent streams (giving rise to an essentially constant water level) but which contain different amounts of a trace contaminant, for example, dissolved oxygen.

In the first case, the concentration or density of the component is assumed constant and the time change of the volume of the system is governed by

$$\frac{\partial V}{\partial t} = Q_{in} \frac{C_{A0}}{C_A} - Q_{out} \tag{5.17}$$

which can be integrated directly to give,

$$V = \left(Q_{in} \frac{C_{A0}}{C_A} - Q_{out} \right) t + V(0) \tag{5.18}$$

Here V(0) is the initial volume within the system. Equation 5.18 suggests the volume increases or decreases linearly with time unless Equation 5.16 is satisfied, in which case the volume of the system never changes.

In the second case, the volume is assumed constant and the change of fluid concentration with time in the system is governed by,

$$\frac{\partial C_A}{\partial t} + \frac{Q_{out}}{V} C_A = \frac{Q_{in}}{V} C_{A0} \tag{5.19}$$

Equation 5.19 is of the form,

$$\frac{\partial y}{\partial x} + ky = s \tag{5.20}$$

where k represents a first order process constant and s a source term. For constant k and s, this has the general solution

$$y = \frac{s}{k} + \alpha e^{-kx} \tag{5.21}$$

where α is some constant of integration whose value depends upon the specific conditions of the problem. The solution to Equation 5.19 may now be written

$$C_A = \frac{Q_{in}}{Q_{out}} C_{A0} + \alpha \exp\left[-\frac{Q_{out}}{V} t \right] \tag{5.22}$$

As $t \to \infty$, the exponential term dies away and the steady-state solution (Equation 5.16) is recovered. At $t = 0$, there is some concentration $C_A(0)$ which can be used to evaluate the constant of integration, α. Evaluation of Equation 5.22 at $t = 0$ gives,

$$\alpha = C_A(0) - \frac{Q_{in}}{Q_{out}} C_{A0} = C_A(0) - C_A(\infty) \tag{5.23}$$

This allows Equation 5.22 to be written

$$\frac{C_A(t) - C_A(\infty)}{C_A(0) - C_A(\infty)} = \exp\left[-\frac{Q_{out}}{V} t \right] \tag{5.24}$$

Note that both terms in Equation 5.24 are dimensionless and can be written in terms of a dimensionless concentration (θ), given by the left-hand side of Equation 5.24 and a dimensionless time (τ), given by the argument of the exponential. With this substitution, Equation 5.24 takes on the form,

$$\theta(\tau) = e^{-\tau} \tag{5.25}$$

This is consistent with our previous assertion that all dimensionally consistent equations can be written in dimensionless form. The term τ represents the ratio of actual time, t, to V/Q_{out}, which is the average time required for fluid parcels or molecules to move through the system. V/Q_{out} is also known as the *mean residence time or a characteristic time* of the system. Characteristic times are indicators of the time required to achieve the final or steady conditions and their definition is somewhat arbitrary. Common definitions of characteristic times include

$\tau_{1/2}$ — Time required for a process to proceed to 50% of completion (half-life)
$\tau_{1/e}$ — Time required for a process to proceed to within 1/e or 37% of completion
τ_{90} — Time required for a process to proceed to 90% of completion

Characteristic times also are useful as indicators of the dominant process or processes. For processes that occur simultaneously, or in parallel, the dominant process is the one exhibiting the largest rate or, alternatively, the shortest characteristic time. Processes that have much longer characteristic times or much slower rates can often be neglected without loss of accuracy of the system model. This is reversed if the processes are sequential and the completion of one must occur before another. Under such conditions, the slowest process with the longest characteristic is the *rate-determining step* for the overall process.

In the present case, the dynamics of the system are controlled by the mean residence time. As shown in Figure 5.4, the dimensionless concentration is within 37% (e^{-1}) of its final value by $\tau = 1$, or by the time that the average fluid parcel passes through the system. Figure 5.4 also shows that plotting the logarithm of dimensionless concentration vs. the dimensionless time yields a straight line.

The overall mass balance for a single inflow-outflow system without mass transfer or reaction processes are quite simple as shown by the above discussion. This is especially true for steady systems which are governed by algebraic equations. Even in steady form, however, material balances

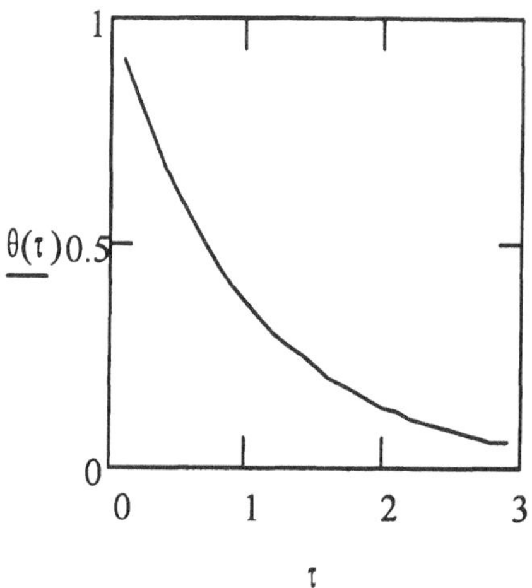

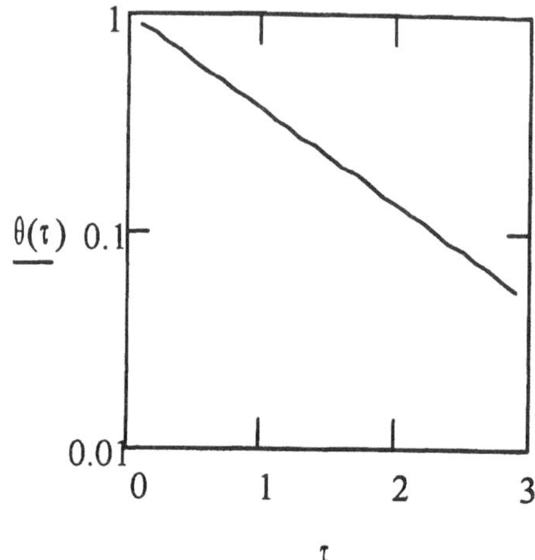

FIGURE 5.4 Plot of dimensionless density, $\theta(\tau)$ vs. dimensionless time, τ, in normal and linearized coordinates.

on multicomponent systems, and systems with multiple inflow and outflow paths and reactive components can be complicated. In some environmental applications, of course, transient material balances are required to describe the physical system and in such conditions differential and not algebraic equations are required. In addition, under many conditions spatial variability is required to fully describe a system and then partial differential equations are generally required. These extensions of the concept of a material balance will each be examined in turn.

5.2.2 Steady–State Material Balance

Under steady conditions, the material balance reduces to simply

$$\{\text{Input}\} = \{\text{Output}\} \tag{5.26}$$

For a single influent and effluent stream, this gives rise to Equation 5.16. For multiple inputs and outputs of a single component in a steady, nonreactive system, Equation 5.16 becomes

$$\sum_{i=1}^{O}(Q_I\, C_{AI})_i = \sum_{i=1}^{I}(Q_0\, C_{A0})_i \tag{5.27}$$

where I and O represent the number of input and output streams and Q_0, C_{A0} and $Q_I C_{AI}$ represent the inlet and outlet flowrates and concentrations of component A in each stream, respectively. Any unknown flowrate or concentration may be determined from Equation 5.27. Remember that an equation equivalent to Equation 5.27 may be written for all of the components of a system or, more commonly, for all but one of the components of a system and for the total or overall mass. As recognized previously, Equation 5.27 may be extended to systems that gain or lose component mass via mass transfer processes or reactions by recognizing that the volumetric flowrate Q is equivalent mathematically to K A for mass transfer processes and to k_r V for first order chemical reactions.

It is appropriate at this point to formalize the process for the preparation of material balances. This will assist in the application of the approach to more complicated problems. The suggested procedure for writing and solving material balances is

1. **Sketch the system to be modeled** — Although the sketch need not be to scale it should indicate the inputs and outputs from the system and allow room for you to label the system variables.
2. **Choose a calculation basis** — This is the amount or flowrate and the units to be used as a basis for all calculations. All quantities should be converted to consistent units before substitution into a material balance. If flows are given, the actual flows may be used otherwise any convenient basis may be used, for example, the material balance may be solved on a per kg of input or per kg/s of input basis.
3. **Label the known flows, concentrations, and fate processes on the sketch.**
4. **Choose algebraic variables for unknown flows and concentrations and label these on the sketch.**
5. **Define the system boundaries and indicate these on the sketch** —The system boundaries should be chosen such that they assist in the solution of the problem. That is, it should isolate wherever possible the unknown variables that you are attempting to use the material balance to determine. More than one system may be defined like a single all-encompassing system to allow isolation of an external flow or process and smaller subsystems to allow isolation of internal flows or processes such as recycle streams. These systems will be illustrated in the examples.
6. **Write the material balance equations** — Care should be taken to include all processes that involve the exchange of material across the defined system boundaries. A more convenient definition of the system may be apparent as a result of this process, in which case the system should be redefined and the material balance equations reformulated. The system should be defined so that any particular material balance equation has the least number of unknowns.
7. **Solve the material balance equations for any desired unknown quantity** — If only a single unknown occurs in a steady-state material balance equation, the algebraic equation can be solved directly for that unknown. Otherwise methods for the solution of simultaneous equations, or, in the case of transient material balances, differential equations must be employed.

Let us explore the approach outlined above in several examples.

Example 5.2: Mass balance of a simple mixer

How much water must be added to dilute a concentrated sulfuric acid solution containing 50% by weight H_2SO_4 to a solution containing 15% by weight H_2SO_4?

A sketch of the system is shown in the Figure 5.5. Because the total flows are not specified, a basis must be chosen. Because all concentrations are given on a weight basis, let us make our calculations on the basis of 100 g/s of produced 15% sulfuric acid. On this basis the exit stream will contain 15 g/s of H_2SO_4 and 85 g/s of water. A material balance on the mass of H_2SO_4, using x to denote the flowrate of concentrated sulfuric acid, gives

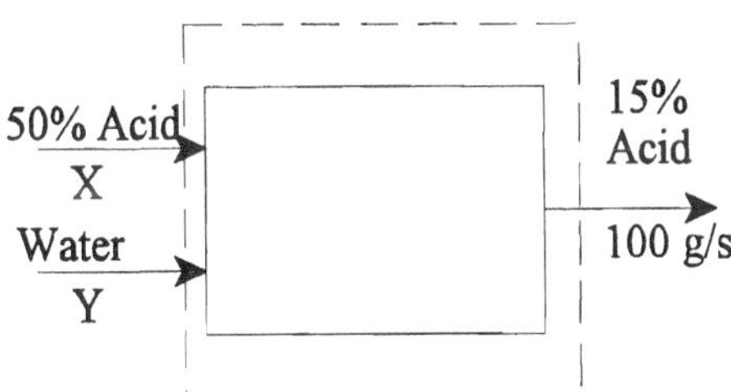

FIGURE 5.5 A simple mixing system for Example 5.2.

$$(x)\,(0.5\ \text{g/g}) = (100\ \text{g/s})\,(0.15\ \text{g/g})$$

$$x = 30\ \text{g/s concentrated } H_2SO_4$$

An overall balance, using y to represent the flowrate of water required gives

$$x + y = 100\ \text{g/s}$$

$$y = 100\ \text{g/s} - 30\ \text{g/s}$$

$$y = 70\ \text{g/s water}$$

If a water component balance were written **rather** than an overall balance

$$(1 - 0.5)\,(\text{g/g})x + (1)y = (1 - 0.5)\ \text{g/s}\,(100\ \text{g/s})$$

$$y = 85\ \text{g/s} - (0.5\ \text{g/g})(30\ \text{g/s})$$

$$y = 70\ \text{g/s water}$$

As expected, no new information is gathered from writing all component balances as well as an overall balance.

Example 5.3: Mass balance on a gas absorber

A gas absorber employs a liquid (for example, water) to absorb contaminants from a gas stream. In Figure 5.6, an absorber for the removal of sulfur dioxide (SO_2) from stack gases is shown. The inlet gases contain $y_0 = 0.001$ SO_2 (mole fraction) and 99% removal of the SO_2 is required. Determine the exit concentration of SO_2 and the required water flowrate. Assume that the exit water (flowrate L_0 and SO_2 mole fraction x_0) is in equilibrium with the inlet gas given by $y^* = 33\,x$ where y and x represent SO_2 mole fractions.

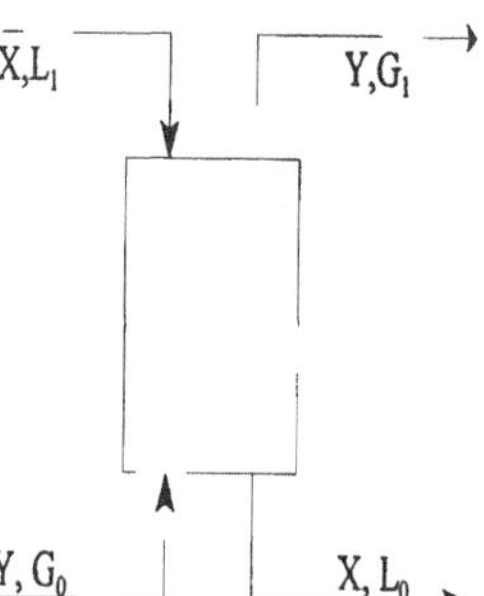

FIGURE 5.6 A gas absorber for Example 5.3.

Basis: 100 mol/s of inlet stack gases

The inlet moles of SO_2 is then 0.1 mol/s, of which 99% or 0.099 mol/s must be removed. The moles in the exit gas stream (assuming no evaporation of the water into the gas) is 99.901 mol/s or effectively identical to the 100 mol/s entering. The exit SO_2 mole fraction is

$$y_1 = \frac{(1 - 0.99) * y_0 * G_0}{G_1}$$

$$\approx 0.01 * y_0 = 1(10)^{-5}$$

Here G_0 and G_1 represent the inlet and outlet flowrates of the gas, respectively, and since the transfer of SO_2 does not change the flowrate appreciably, they are effectively equal. The exit water is in equilibrium with the entering gas or $x_0 = y_0/33 \sim 3.03\ (10)^{-5}$. A mass balance on SO_2 gives

$$y_0 G_0 = y_1 G_1 + x_0 L_0$$

Solving for the exit water flow

$$L_0 = \frac{(y_0 G_0 - y_1 G_1)}{x_0}$$

$$= 3,267 \text{ mol/s or } 58.8 \text{ kg/s (per 100 mol/s of gas feed)}$$

The amount of fresh SO_2 free water feed is

$$L_1 = L_0 - x_0 * L_0 = 3.267 \text{ mol/s}$$

which, again because of the dilute nature of the constituent being removed, is effectively the same as the exit water solution.

Example 5.4: Sludge balance around an activated sludge treatment system

Consider a reactor/clarifier of the activated sludge wastewater treatment system in the Figure 5.7. Microbes degrade organic materials in the reactor producing biological solids (biomass) which must be separated in a clarifier. The effluent water from the clarifier contains almost no solids. The solids leave in a thick slurry or sludge, some of which is recycled to the reactor, hence the name activated sludge. The volume of water to be treated, Q_0, is 1440 m^3/day. Biological solids are produced at the rate of 100 kg/day, requiring a wastage stream of flowrate Q_S The design solids concentration in the reactor, C_S, is 1 kg/m^3 and the concentrated sludges from the clarifier and in the recycle stream contain 10 kg solids/m^3 (C_{Sr}). The reactor basin surface area is 1000 m^2 and its volume is 1500 m^2. The clarifier and piping have negligible volume by comparison. Estimate the wastage rate, Q_S, and the recycle rate, Q_r.

Overall sludge balance

$$Q_m = Q_S C_{Sr}$$

$$Q_S = \frac{Q_m}{C_{Sr}} = \frac{100 \text{ kg/day}}{10 \text{ kg/m}^3} = 10 \text{ m}^3\text{/day}$$

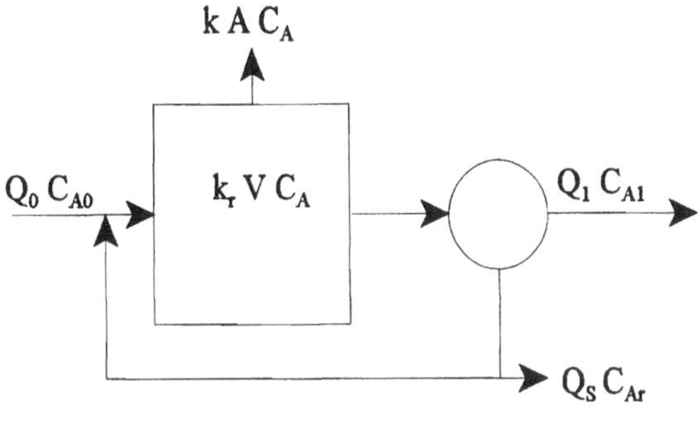

FIGURE 5.7 The activated sludge wastewater treatment system evaluated in Examples 5.4 and 5.5.

Sludge balance around clarifier

$$(Q_0 + Q_r)C_S = Q_S C_{St} + Q_r C_{Sr}$$

or, solving for the recycle rate

$$Q_r = \frac{Q_0 C_S - Q_w C_{Sr}}{C_{Sr} - C_S}$$

$$= \frac{(1440\ \text{m}^3/\text{day})(1\ \text{kg/m}^3) - (10\ \text{m}^3/\text{day})(10\ \text{kg/m}^3)}{10\ \text{kg/m}^3 - 1\ \text{kg/m}^3}$$

$$= 149\ \text{m}^3/\text{day}$$

Example 5.5: Benzene balance in activated sludge treatment system

Benzene, 1 mg/L, is contained within the feed to the wastewater treatment system of Example 5.4. Develop a balance on benzene to estimate the concentration of benzene in the effluent. Assume that benzene is partially vaporized from the bioreactor with a flux given by k C_A where C_A is the concentration in the reactor and k in the aerated basin is about 5 cm/h. The biodegradation rate (per unit volume) of the benzene is approximated by k_r C_A where k_r = 2 day^{-1}. Due to sorption on the biological solids, laboratory tests have suggested that the benzene concentration in the wasted and recycled biomass streams are approximately double that found in the effluent.

Water balance

$$H_2O \text{ in} = H_2O \text{ out} + H_2O \text{ with solids}$$

$$Q_0 = Q_1 + Q_S$$

$$Q_1 = Q_0 - Q_S = 1430\ \text{m}^3/\text{day}$$

Benzene balance

$$\text{Bz in} = \text{Bz out} + \text{Bz in solids} + \text{Bz evaporated} + \text{Bz degraded}$$

$$Q_0 C_{A0} = Q_1 C_A + Q_S 2C_A + kAC_A + k_r V C_A$$

$$C_A = \frac{Q_0 C_{A0}}{Q_0 + 2 * Q_S + kA + k_r V}$$

Substituting, the concentration of benzene in the reactor and in the effluent water is 0.254 mg/L, that is, 74.6% is removed in the waste treatment system. The proportion of the benzene removed by each of the various mechanisms can be determined by taking the ratio of the loss by that mechanism over the total loss. The proportion vaporized is, for example,

$$\%\ \text{vaporized} = \frac{kA}{Q_1 + 2Q_S + kA + k_r V} * 100 = 21\%$$

The proportion biodegraded is 53%, while the proportion lost with the waste sludge is 0.4%.

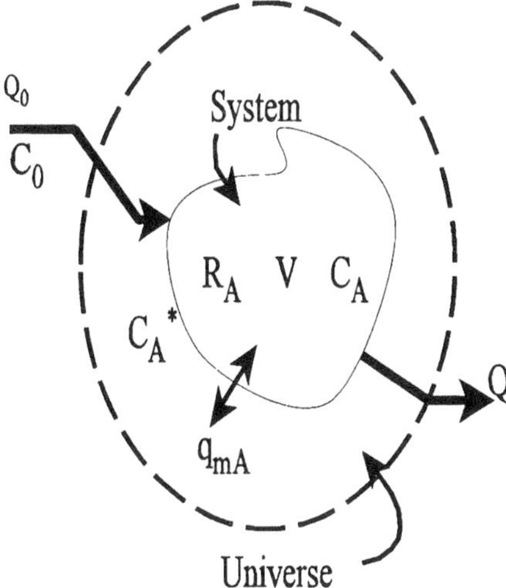

FIGURE 5.8 System and processes influencing system dynamics. Basis for the mass balance development.

5.2.3 Dynamic Balance on a Stirred System

Let us now return to the transient situation for which the material balance was first introduced. Let us consider the uniformly mixed system shown in Figure 5.8. The vessel within the system is of arbitrary shape and has a volume V in which a compound A has a rate of disappearance by reaction R_A V and a rate of disappearance by mass transfer through an area A given by q_m A. The vessel is assumed sufficiently well mixed that all properties, including concentration, are uniform within the vessel and the exit concentration is identical to that within. This characteristic defines a completely stirred reactor or vessel. This system also is termed a continuous flow stirred tank reactor (CSTR). The dynamic material balance on a CSTR is essentially a generalization of Equation 5.15. Using the notation of Figure 5.8,

$$\frac{\partial}{\partial t} C_A V = Q_0 C_0 - Q_1 C_A - R_A V - q_m A \tag{5.28}$$

In order to simplify Equation 5.28 further, it is necessary to define the reaction rate (per unit volume) and the mass transfer flux. As employed previously, let us assume a first order or linear reaction rate model, $R_A = k_r C_A$, and a linear driving force mass transfer model, $q_m = k\,(C_A - C_A^*)$. Equation 5.28 now takes on the form

$$\frac{\partial}{\partial t} C_A V = Q_0 C_0 - Q_1 C_A - k_r C_A V - k(C_A - C_A^*)A \tag{5.29}$$

or, if we assume a constant volume system,

$$\frac{\partial C_A}{\partial t} + \left(\frac{Q_1}{V} = \frac{kA}{V} + k_r \right) C_A = \frac{Q_0}{V} C_0 + \frac{kA}{V} C_A^* \tag{5.30}$$

Equation 5.30 is a general mass balance on a uniformly mixed system or CSTR.

Under steady conditions, the time derivative is zero and there is no change in system properties including the concentration of component A with time. The mass balance then reduces to the algebraic form

$$C_A(\infty) = (C_A)_{ss}\left[\frac{\frac{Q_0}{V}C_0 + \frac{kA}{V}C_A^*}{\left(\frac{Q_1}{V} + \frac{kA}{V} + k_r\right)}\right] \tag{5.31}$$

where the subscript "ss" refers to steady-state conditions. Equation 5.31 gives the steady-state or equilibrium condition for the system.

The transient Equation 5.30 is of the same form as Equation 5.19 and 5.20. If the initial concentration in the system is $C_A(0)$, then the solution is similar to Equation 5.24 and given by

$$\frac{C_A(t) - C_A(\infty)}{C_A(0) - C_A(\infty)} = \exp\left[-\left(\frac{Q_{out}}{V} + \frac{kA}{V} + k_r\right)t\right] \tag{5.32}$$

$C_A(\infty)$ is given by Equation 5.31. Let us examine a continuous flow system without reaction and a no-flow system with reaction in Example 5.6.

Example 5.6: Dynamic balances on stirred vessels

Consider a stirred vessel as in Figure 5.8 with a volume of 1000 L. Estimate the concentration of a component A in the vessel as a function of time under each of the following situations.

Situation 1 — No reaction or mass transfer. Initial concentration, $C_A(0)$, of 100 mg/L. Flowrate in and out of the vessel of 100 L/min. No feed of component A.

Under these conditions, the governing equation becomes

$$V\frac{dC_A}{dt} = -Q_1C_A$$

Since there is no feed of component A ($C_0 = 0$), the steady concentration is zero from Equation 5.31. The concentration as a function of time in the vessel is then given by

$$C_A(t) = C_A(0)\exp\left[-\frac{Q_1}{V}t\right] = (100 \text{ mg/L})\exp\left[-\frac{t}{10 \text{ min}}\right]$$

The mean residence time of fluid in the vessel is 10 min and this is the time required to achieve 37% of the initial concentration, or 37 mg/L. In 30 min the concentration is 5 mg/L in the vessel and in 46 min the concentration is 1 mg/L.

Situation 2 — First order (linear) reaction in the vessel with a decay rate constant of 0.1 min^{-1} with no feed or effluent. Initial concentration as in Situation 1. This is termed a batch reactor.

Under these conditions, the governing equation becomes

$$V\frac{dC_A}{dt} = -k_rC_AV$$

Again, without feed of the component of interest, the steady concentration is zero. The transient concentration within the vessel is given by exactly the same equation as above since the rate constant is 1/(10 min),

$$C_A(t) = C_A(0)\exp[-k_r t] = (100 \text{ mg/L}) \exp\left[-\frac{t}{10 \text{ min}}\right]$$

Note that each of the terms within the parentheses in the argument of the exponential in Equation 5.32 has units of a rate, that is, inverse time. The time scale, or characteristic time, required for each of these processes is the inverse of the rate and as indicated previously, long characteristic times indicate a slow process. The characteristic time, τ_c, of the system with flow, mass transfer, and reaction, is given by

$$\tau_c = \left(\frac{Q_{out}}{V} + \frac{kA}{V} + k_r\right)^{-1} \tag{5.33}$$

The first term in the argument characterizes the rate of change of concentration by effluent removal, the second by mass transfer to a second phase, and the third by disappearance due to reaction. Because these processes are competing, the longer characteristic time processes are generally of lesser importance since the fate of the contaminant is largely determined by the faster processes. Equation 5.32 can now be written in terms of dimensionless concentration, θ, and dimensionless time, τ,

$$\theta(t) = e^{-t/\tau_c} = e^{-\tau}$$

where (5.34)

$$\theta(t) = \frac{C_A(t) - C_A(\infty)}{C_A(0) - C_A(\infty)} = \begin{cases} 1/e = 0.37 & \text{at } \tau = 1 \ (t = \tau_c) \\ 1/2 & \text{at } \tau = 0.693 \ (t = \tau_{1/2} = 0.693\tau_c) \end{cases}$$

Equation 5.34 also shows the relationship between θ and two common definitions of characteristic time, the 1/e time, and the half-life. By further evaluation of Equation 5.34, it can be seen that θ will be within 5% of its final value (θ = 0.05) in three characteristic times (τ = 3), and within 1% of its final value (θ = 0.01) in 4.6 characteristic times (τ = 4.6). This was illustrated in Example 5.6 in which the concentration of the component of interest under the conditions of Situation 1 was 5 mg/L after 30 min (5% after 3τ) and 1 mg/L after 46 min (1% after 4.6τ).

Example 5.7 examines the characteristic times for the mechanisms resulting in benzene loss from the wastewater treatment system of Example 5.5. Half-lives, as an alternative indicator of the rate of various processes, are also included in Example 5.7. The application of a dynamic mass balance is presented for a stirred reaction vessel in Example 5.8.

Example 5.7: Characteristic times and half-lives for the processes of Example 5.5

Determine the characteristic times of the individual benzene loss processes in Example 5.5.

Examining each of the processes in turn using the 1/e time as the characteristic time scale,

Effluent $$\tau_Q = \frac{V}{Q_1} = \frac{1500 \text{ m}^3}{1430 \text{ m}^3/\text{day}} = 1.05 \text{ days}$$

Mass Transfer $$\tau_t = \frac{V}{kA} = \frac{1500 \text{ m}^3}{(1.2 \text{ m/day})(1000 \text{ m}^2)} = 1.25 \text{ days}$$

Reaction $$\tau_r = \frac{1}{k_r} = \frac{1}{2\ \text{day}^{-1}} = 0.5\ \text{days}$$

Although the biodegradation in the reactor is the fastest of the processes, no process is clearly dominant in this case since all are of similar magnitude.

Alternatively, using the half-life of each process as the characteristic time scale.

Effluent $$\tau_Q = 0.693\frac{V}{Q_1} = 0.693\frac{1500\ \text{m}^3}{1430\ \text{m}^3/\text{day}} = 0.728\ \text{days}$$

Mass Transfer $$\tau_t = 0.693\frac{V}{kA} = 0.693\frac{1500\ \text{m}^3}{(1.2\ \text{m/day})(1000\ \text{m}^2)} = 0.866\ \text{days}$$

Reaction $$\tau_r = \frac{0.693}{k_r} = \frac{0.693}{2\ \text{day}^{-1}} = 0.347\ \text{days}$$

Example 5.8: Material balance on stirred reactor with first-order irreversible reaction

Consider a 1000-L mixed reactor with an initial concentration of a component A of zero to which is fed a stream of 100 L/min containing 100 mg/L of the component. A first-order, irreversible reaction with rate constant of 0.1 min^{-1} occurs within the vessel.

The mass balance equation is given by

$$V\frac{dC_A}{dt} = Q_0 C_{A0} - Q_1 C_A - k_r C_A V$$

Setting the inlet and outlet flowrates equal and denoting by Q, the steady-state concentration is

$$C_A(\infty) = \frac{C_{A0}}{1+\frac{k_r V}{Q}} = \frac{100\ \text{mg/L}}{1+\frac{0.1\ \text{min}^{-1}\cdot 1000\ \text{L}}{100\ \text{L/min}}} = 50\ \text{mg/L}$$

This also can be written in terms of the characteristic time for the reaction, $\tau_r = 1/k_r = 10$ min and the residence time in the reactor, $\tau_Q = V/Q = 10$ min. The result is

$$C_A(\infty) = \frac{C_{A0}}{1+\frac{\tau_Q}{\tau_r}} = 50\ \text{mg/L}$$

and the transient from the initial concentration to this steady level is described by

$$C_A(t) = C_A(\infty)\left(1-e^{-\frac{Q_{out}t}{V}-k_r t}\right) = 50\ \text{mg/L}\left(1-e^{-\frac{100\ \text{L/min}\ t}{1000\ \text{L}}-0.1\ \text{min}^{-1}t}\right)$$

$$= 50\ \text{mg/L}\left(1-e^{-\frac{2t}{10\ \text{min}}}\right)$$

5.2.4 Balance on an Unstirred System

A variety of natural systems and processes can be treated as well mixed over large regions if not the entire system. The stirred vessel or CSTR represents one extreme assumption about the flow and mixing behavior in a system. An alternative extreme is to assume no mixing. This is most often represented by flow in a pipe or long channel when material is carried down its length without significant mixing in the axial direction, that is, in the direction of flow. A natural system that might be modeled this way is a river which may be uniformly mixed between the banks but is essentially unmixed along its length. At any given point, the flux of a component A downstream is UC_A where U is the stream velocity and C_A is the component concentration. If a process such as reaction or mass transfer occurs, the concentration changes as the fluid moves downstream. Evaluation of this variation in system properties with distance may be required to understand and describe the system.

For example, oxygen may be depleted as fluid moves downstream due to the presence of oxygen demanding wastes. Mass transfer from the atmosphere, however, will tend to increase the oxygen. The balance between these two competing processes determines the oxygen available to fish and other aquatic animals in the stream. A material balance-based model could be used to predict or simulate the variation in oxygen levels in the stream, in particular as a function of distance from the source of the oxygen-depleting contaminants.

A material balance on an axially unmixed system can be developed with the aid of Figure 5.9. Because the concentration of a component may depend on distance downstream, it becomes necessary to develop a material balance on only a small *differential* length in which system properties can be taken as essentially uniform. The material balance is essentially identical to that of a well-mixed system over this small distance. In Figure 5.9, the only input to this differential segment is the advective transport, $UwhC_A(z)$, where $UC_A(z)$ represents the advective flux at position z while wh is the cross-sectional area of the flow. The outputs include advection at the end of the differential segment, $UwhC_A(z + dz)$, a first-order irreversible reaction, $k_rC_A(z)wh\,dz$, and mass transfer, $kw\,dz(C_A(z) - C_A^*)$. Note that the volume of the differential element is $wh\,dz$, the area through which the advective flux is operative (the cross-sectional area of the stream in the direction of flow) is wh, while the area through which the mass transfer to the atmosphere is operative (surface area of the element) is $w\,dz$. The accumulation in the element is simply the time rate of change of mass (C_AV) in the system, or $\partial/\partial t\; C_Awh\,dz$. The mass balance then becomes

$$\frac{\partial}{\partial t} C_A wh\,dz = UwhC_A(z) - UwhC_A(z+dz) - kwdz(C_A - C_A^*) - k_rC_Awh\,dz \tag{5.35}$$

Dividing by the volume of the element, $(wh)\,dz$, which is assumed constant, and rearranging

$$\frac{\partial C_A}{\partial t} + \frac{UC_A(z+dz) - UC_A(z)}{dz} = -\frac{k}{h}(C_A - C_A^*) - k_rC_A \tag{5.36}$$

The second term on the left-hand side can be recognized as simply the definition of the derivative of (UC_A) with respect to z in the limit of $dz \to 0$, or

$$\frac{\partial C_A}{\partial t} + \frac{\partial}{\partial z}UC_A = -\frac{k}{h}(C_A - C_A^*) - k_rC_A \tag{5.37}$$

which governs the concentration of A as a function of z and t.

Although it is possible to evaluate the dynamic response of the system, for many problems of interest the concentrations and mass transfer coefficients are essentially independent of time (but

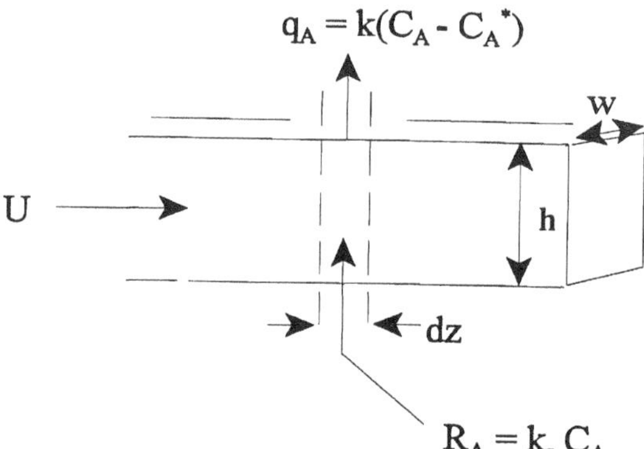

FIGURE 5.9 Definition sketch for an axially unmixed system.

may remain dependent upon distance, z). Under these conditions and assuming constant velocity, Equation 5.37 reduces to

$$\frac{\partial C_A}{\partial z} = -\frac{k}{Uh}(C_A - C_A^*) - \frac{k_r}{U} C_A \tag{5.38}$$

This equation is identical in form to the unsteady-state material balance on a uniformly mixed system. This correspondence can be made more clear by recognizing that without mixing along the z coordinate, a fluid parcel will arrive at a point z from the origin in a time, $t_L = z/U$. Equation 5.38 can then be written

$$\frac{\partial C_A}{\partial \frac{z}{U}} = \frac{\partial C_A}{\partial t_L} = -\frac{k}{h}(C_A - C_A^*) - k_r C_A \tag{5.39}$$

which is identical to the stirred vessel material balance assuming no flow in or out of the vessel. That is, the unmixed material balance is equivalent to a completely mixed balance on a volume moving at the fluid velocity. The subscript L on time in Equation 5.39 is to emphasize that this time is measured from the origin *along the travel path*, sometimes called the Lagrangian time scale after mathematician Lagrange.

The solution to Equation 5.39 can be written identically to the equivalent problem in a well-stirred system with z/U replacing time and setting both inlet and outlet flows to zero. The solution to Equation 5.39 subject to an initial concentration of C_0 at z = 0 is given by

$$\frac{C_A(z) - C_A(\infty)}{C_A(0) - C_A(\infty)} = \exp\left[-\left(\frac{kA}{V} + k_r\right)\frac{z}{U}\right] \tag{5.40}$$

where 1/h is recognized as the ratio of area to volume, A/V and $C_A(\infty)$ is now given by

$$C_A(\infty) = \left[\frac{\frac{kA}{V} C_A^*}{\left(\frac{kA}{V} + k_r\right)}\right] \tag{5.41}$$

Example 5.9 compares an axially unmixed reactor to the fully mixed reactor of Example 5.8.

Example 5.9: Material balance on axially unmixed reactor with first-order irreversible reaction

Consider a 1000-L axially unmixed reactor and an initial concentration of a component A of zero to which is fed a stream of 100 L/min containing 100 mg/L of the component. A first-order, irreversible reaction with rate constant of 0.1 min^{-1} occurs within the vessel.

The steady-state mass balance equation is given by

$$U\frac{dC_A}{dz} = \frac{dC_A}{d(z/U)} = -k_r C_A$$

This suggests that the concentration variation along the length of the reactor is given by

$$C_A = C_{A0}e^{-k_r\frac{z}{U}}$$

Multiplying the argument of the exponential by the cross-sectional area of the reactor the concentration at the reactor exit can be written,

$$C_{A1} = C_{A0}e^{-k_r\frac{V}{Q}} = C_{A0}e^{-\frac{\tau_Q}{\tau_r}} = 36.8 \text{ mg/L}$$

Note that the amount of component A reacted has increased 26.4% over the mixed reactor. For reactions of positive order (that is, those in which the rate increases as the concentration of reactant increases), an axially unmixed reactor is always more effective than a completely stirred reactor. In addition, any real reactor operates between these two extremes. The steady effluent concentration from any real reactor subject to the kinetics of these examples will be between 36.8 and 50 mg/L.

All of the examples above assumed that the reactions were governed by first-order irreversible kinetics. Other fate processes and kinetic forms also exist. Other forms for the transport processes that appear in the mass balance models also exist. Let us now make a more detailed analysis of these processes.

5.3 FATE PROCESSES — SORPTION AND REACTION

The first-order irreversible reactions employed in the above examples could describe chemical reactions or irreversible sorption reactions onto solids. They are specific examples of a much larger class of reactions that include reversible reactions or sorption and nonlinear reactions. Let us examine the kinetics of some of these other systems.

5.3.1 Irreversible Reactions — Kinetic Models

The kinetics of sorption or reaction are defined by the form of the rate equation and the value of the rate constants that arise. In general, both the form and the rate constants of a reaction are defined by experimental measurements. Often one or more steps may control the overall rate of reaction and be referred to as the rate-limiting step. The stoichiometry of this rate-limiting step, which may not correspond to the stoichiometry of the overall reaction, typically controls the form of the kinetic relationship that governs the overall reaction. In particular, consider the reaction

$$A + B \rightarrow C \tag{5.42}$$

If this were the fundamental, rate-limiting step in an overall reaction, it would be expected that the rate of the reaction would be proportional to the probability of a collision between a molecule of A and a molecule of B. The probability of a molecule of A being present in a particular location is proportional to its concentration, C_A, and similarly the probability of a molecule B being in a particular location is proportional to its concentration, C_B. The probability of both being in the same location at the same time is then proportional to the product of the concentrations of A and B. This is a statement of the *principle of mass action* which implies in this case that the rate of formation of compound C is given by

$$R_C = k_r C_A C_B \tag{5.43}$$

where k_r, the reaction rate constant, is the proportionality constant between the probability of collision to the observed rate of reaction. For the purposes of illustration, all reactions considered in this section will be assumed to represent stoichiometry identical to the fundamental rate-limiting step to allow us to write rate relationships.

If we consider a reaction that involves, again as a fundamental, rate-determining step, the reaction of two molecules of A to form compound C, then the principle of mass action suggests

$$\begin{gathered} A + A \rightarrow C \\ R_C = k_r C_A^2 \\ R_A = -2k_r C_A^2 \end{gathered} \tag{5.44}$$

The factor of −2 indicates that A is disappearing (hence a negative rate of formation) and that two moles of A disappear for every one molecule of C formed. The dependence upon the second power of the concentration of A gives rise to its description as a second order model.

Equation 5.43 also describes a second order kinetic model but one that is first order in each of the individual reactants. The rates of formation of the product or disappearance of the reactants in such a reaction are

$$\begin{gathered} A + B \rightarrow C \\ R_C = k_r C_A C_B \\ R_A = -k_r C_A C_B \end{gathered} \tag{5.45}$$

In situations when either A or B is present is large excess, the concentration of the excess component will not change significantly as a result of the reaction and then Equation 5.45 is reduced to a first order reaction system. For example, if component B is present in large excess,

$$\begin{gathered} A + B\,(excess) \rightarrow C \\ R_A = -(k_r C_B) C_A \\ R_A = -k_r C_A \end{gathered} \tag{5.46}$$

Many reactions occur via multiple steps. The simplest example of this is the reaction

$$A \xrightarrow{k_1} B \xrightarrow{k_2} C \tag{5.47}$$

which gives rise to the rates

$$\begin{aligned} R_A &= -k_1 C_A \\ R_B &= k_1 C_A - k_2 C_B \\ R_C &= k_2 C_B \end{aligned} \tag{5.48}$$

Each of these kinetic models describe different concentration behaviors with time. The variations in behavior can be used to differentiate between kinetic mechanisms and to estimate the kinetic rate constants. Perhaps the simplest way to illustrate the different behaviors is to employ a batch reactor as illustrated in Example 5.8. Batch reactors that do not have continuous inflow and outflow are often used to evaluate the kinetic behavior of chemical reactions and to determine kinetic constants.

The general mass balance on a component A in a well-mixed batch reactor subject to a rate equation of the form $R_A = -k_r f(C_A)$ is given by

$$\begin{aligned} \frac{dC_A V}{dt} &= R_A V \\ \frac{dC_A}{dt} &= R_A = -k_r f(C_A) \\ \int_{C_{A0}}^{C_A} \frac{dC_A}{f(C_A)} &= \int_0^t -k_r dt \end{aligned} \tag{5.49}$$

Here C_{A0} represents the initial concentration of component A in the system. Integration of the last of Equation 5.49 then provides the specific concentration vs. time behavior. If the reaction is first-order, that is, $R_A = -k_r C_A$, then

$$\begin{aligned} \int_{C_{A0}}^{C_A} \frac{dC_A}{C_A} &= -\int_0^t k_r dt \\ \ln \frac{C_A}{C_{A0}} &= -k_r t \\ C_A &= C_{A0} e^{-k_r t} \end{aligned} \tag{5.50}$$

If we consider instead a second order reaction, with $R_A = -2k_r C_A^2$,

$$\begin{aligned} \int_{C_{A0}}^{C_A} \frac{dC_A}{C_A^2} &= -\int_0^t 2k_r dt \\ \frac{1}{C_{A0}} - \frac{1}{C_A} &= -2k_r t \\ C_A &= \left(\frac{1}{C_{A0}} + 2k_r t \right)^{-1} \end{aligned} \tag{5.51}$$

Let us now consider the formation of a component B from a first order reaction with $R_B = k_r C_A$ with C_A given by Equation 5.50.

$$\int_{C_{B0}}^{C_B} dC_B = \int_0^t (k_r C_A) dt$$

$$C_B - C_{B0} = \int_0^t k_r C_{A0} e^{-k_r t} \qquad (5.52)$$

$$C_B = C_{B0} + C_{A0}(1 - e^{-k_r t})$$

Now let us consider the multistep reaction of Equation 5.47. The concentration of component A at any time is again given by the last of Equations 5.50. The rate of formation of component B is

$$\frac{dC_B}{dt} = -k_2 C_B + k_1 C_A \qquad (5.53)$$

This first order linear equation can always be integrated by multiplying by the exponential of the integral of the coefficient of C_B. In this case, the equation should be multiplied by the *integrating factor*

$$\text{Integrating Factor} = e^{\int k_2 dt} = e^{k_2 t} \qquad (5.54)$$

Multiplying Equation 5.53 by this integrating factor gives

$$e^{k_2 t} \frac{dC_B}{dt} + e^{k_2 t} k_2 C_B = \frac{d}{dt}\left(C_B e^{k_2 t}\right) = k_1 C_{A0} e^{(k_2 - k_1)t} \qquad (5.55)$$

and integrating from zero time to time t,

$$C_B e^{k_2 t} - C_{B0} = \frac{k_1}{k_2 - k_1} C_{A0}\left(e^{(k_2 - k_1)t} - 1\right)$$

$$C_B - C_{B0} e^{-k_2 t} = \frac{k_1}{k_2 - k_1} C_{A0}\left(e^{-k_1 t} - e^{-k_2 t}\right) \qquad (5.56)$$

This equation reduces to Equation 5.52 when the rate of disappearance of component B is negligible ($k_2 \rightarrow 0$).

5.3.2 Estimation of Reaction Rate Parameters

The use of any of the just discussed kinetic models depends on the ability to estimate the rate constants. This can be accomplished using the methods of Chapter 3 to fit any data to a particular model. Normally such data would include either concentration data at various times or direct information on reaction rate as a function of concentration.

If the data is in the form of measured concentrations vs. time, the integrated form of a particular form of the reaction rate can be compared to the data and the suitability of that reaction rate model can be assessed by the quality of the fit. For example, if a first order reaction is postulated for a particular system, the concentration behavior with time should be adequately described by the model of Equation 5.50. Such a model can be linearized to allow a convenient linear regression to determine the kinetic constants by taking the logarithm of both sides.

$$\begin{aligned} C_A &= C_{A0}e^{-k_r t} \\ \ln C_A &= \ln C_{A0} - k_r t \end{aligned} \tag{5.57}$$

A plot of the logarithm of concentration vs. time should thus provide a straight line whose slope is the reaction rate constant. Alternatively a linear regression between the logarithm of concentration and time will yield the logarithm of C_{A0} and k_r. The integrated second order rate expression, Equation 5.51, is already in a linearized form in that a plot of $1/C_A$ vs. t should yield a straight line or, alternatively, a linear regression between $1/C_A$ and t would yield a best-fit estimate of the rate constant (slope) and C_{A0} (inverse of intercept).

If the data is direct measurements of reaction rate as a function of concentration, the rate equation can be inferred directly from the data. For example, if the reaction is a simple nth order reaction of the form $R_A = -k_r C_A^n$, the logarithm can again be used to linearize the function,

$$\begin{aligned} R_A &= k_r C_A^n \\ \ln R_A &= \ln k_r + n \ln C_A \end{aligned} \tag{5.58}$$

A plot of the logarithm of the rate vs. the logarithm of the concentration should provide a straight line whose slope is given by n, the reaction order. In addition, the intercept of such a plot should give the logarithm of the reaction rate constant. It should be emphasized, however, that the use of the logarithm to linearize the model forms necessarily biases the fit of the data, as discussed in Chapter 3.

The selection of any kinetic model is somewhat ambiguous unless the precision of the data is such that other models can be excluded. Often the error in the data is such that it is nearly impossible to differentiate between reactions of different order. It should be recognized that a kinetic model that is fit to data is consistent with that data, but this does not necessarily prove the assumptions used in development of the kinetic model, for example, the stoichiometry of the fundamental rate-determining step.

It should also be emphasized that everything done thus far assumes that the temperature is constant. Reaction rate constants are strong functions of temperature. A commonly used rule is that a 10°C increase in temperature will approximately double a reaction rate. The actual temperature dependence, however, depends on the activation energy required to initiate a reaction. In particular, reaction kinetic constants have generally been found to follow the *Arrhenius relationship* for the effects of temperature,

$$k_r = A e^{-\frac{E_a}{RT}} \tag{5.59}$$

where E_a is the activation energy and A is a pre-exponential factor. R and T are the ideal gas constant and the absolute temperature, respectively. If reaction rate constant information is known at several temperatures it is possible to fit this data to Equation 5.59 to determine A and E_a, for example, by linearization with the logarithms or plotting directly ln k_r vs. 1/T. Often the activation energy for a chemical reaction is of the order of 80 to 100,000 J/mol which gives rise to the rough approximation that near 300°C, the reaction rate constant doubles with each 10°C rise in temperature.

In general, the description of concentration vs. time in a non-isothermal reactor is significantly more complicated than that described above in that an energy balance must also be calculated to determine the temperature as a function of time. In general, the energy and material balance must be solved simultaneously.

5.3.3 Reversible Reactions

Each of the kinetic models evaluated above assumed that the reactions are irreversible. All reactions are, in general, reversible, although conditions may be such that the reverse reactions may be neglected. If we consider the reversible equivalent of Equation 5.42

$$A + B \Leftrightarrow C \tag{5.60}$$

the material balance on a batch reactor becomes

$$\begin{aligned} \frac{dC_A}{dt} &= R_A \\ &= -k_f C_A C_B + k_b C_C \end{aligned} \tag{5.61}$$

k_f and k_b represent the forward and backward reaction rate constant, respectively. Solution of Equation 5.61 proceeds as before. That is, C_B and C_c must be rewritten in terms of C_A and the Equation 5.61 integrated. The concentrations for both reactants and products can be written in terms of a conversion, χ, which represents the extent of the reaction. If no A has reacted, $\chi = 0$, while if all of the A has reacted, $\chi = 1$. Given the stoichiometry of the equation and the initial concentrations of A and B given by C_{A0} and C_{B0}, respectively, the concentrations of A, B, and C at any time are given by

$$\begin{aligned} C_A &= C_{A0}(1-\chi) \\ C_B &= C_{B0} - (C_{A0}\chi) \\ C_C &= C_{A0}\chi \end{aligned} \tag{5.62}$$

This formulation assumes that component A is present in excess and that there is no component C present initially. If B is present in excess, then χ should be assigned to the extent of the reaction of component B. Using Equations 5.62, Equation 5.61 can be rewritten

$$\begin{aligned} d\frac{C_{A0}(1-\chi)}{dt} &= -k_f[C_{A0}(1-\chi)][C_{B0} - C_{A0}\chi] + k_b C_{A0}\chi \\ -\frac{d\chi}{dt} &= k_f(1-\chi)C_{B0} - C_{A0}\chi) + k_b\chi \end{aligned} \tag{5.63}$$

The last equation is a differential equation in the single unknown variable χ. Use of the equation, however, still requires solution of the resulting differential equation. This and more complicated kinetic models are left to the reader.

For our purposes, the only reversible reactions of interest are those that are sufficiently fast that equilibrium can be assumed. Let us consider the reversible first order reaction,

$$A \Leftrightarrow B$$
$$R_A = -k_f C_A + k_b C_b \tag{5.64}$$

At equilibrium, the net rate, R_A, is zero and the ratio of the concentration of B to A is given by the equilibrium constant,

$$\frac{C_B}{C_A} = \frac{k_f}{k_b} = K_r \tag{5.65}$$

This is identical to the equilibrium sorption processes discussed in the previous chapter. Let us write the solid-water sorption reaction as

$$A_w \Leftrightarrow A_s \tag{5.66}$$

where the subscripts refer to the water and sorbed phase, respectively. If we consider the rate of sorption as k_s Cw and the rate of desorption as k_d W_s, then the net rate of sorption is given by

$$R_A = k_s C_w - k_d W_s \tag{5.67}$$

Since we are applying local equilibrium, $R_A = 0$ and,

$$\frac{W_s}{C_W} = \frac{k_s}{k_d} = K_{sw} \tag{5.68}$$

Here, the true meaning of the equilibrium sorption constant becomes apparent as the ratio of the inherent rate of sorption to the rate of desorption. As indicated in Chapter 4, K_{sw} is approximately given by the product of the organic carbon based partition coefficient, K_{oc} and the fraction organic carbon, ω_{oc} for a hydrophobic organic contaminant.

5.3.4 Reversible Processes and System Dynamics

As indicated in the preceding section, the balance between reversible reactions or sorption defines the ratio of the concentration of a component in one form or phase relative to another form or phase. If a system is controlled such that only that portion of the contaminant in one form or phase is changed, the system must move via mass transfer or reaction to restore the equilibrium balance. Thus, any attempt to remove the component in one form or phase is slowed by the counter effect of the equilibrium attempting to restore that component in that form or phase.

Let us examine the implications of Equation 5.68 for material balances. Although here we have treated sorption as though it were a reversible reaction, it is important to recognize that physical sorption does not change the identity of the sorbing compound. The total amount of the contaminant does not change during the sorption process and at any time is the sum of the sorbed and unsorbed quantity. In addition, physical adsorption does not generally require a large activation energy and an examination of the Arrhenius relationship (Equation 5.59) then suggests that physical adsorption is not as temperature sensitive as chemical reactions.

Consider a particular component in an aqueous slurry in a stirred vessel as shown in Figure 5.10. The stirred vessel contains water (volume fraction, φ_w), suspended solids (volume fraction, φ_s, and a vapor or air space (volume fraction, φ_a) each containing their respective concentration

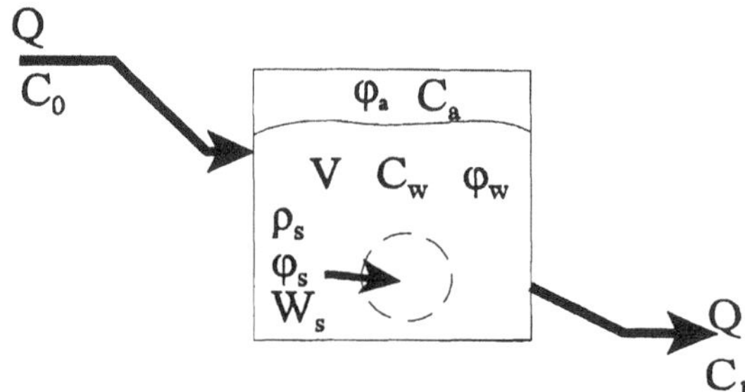

FIGURE 5.10 System with retardation by sorption onto solid slurry within a stirred system.

of contaminant, C_w, W_s, and C_a. Since only a single component is of interest, the subscript component identifier (e.g., A) will be dropped and instead the subscript will be used to denote the phase.

Let us consider the situation where the effluent draws from the aqueous phase and the vapor space remains trapped within the vessel. Let us also assume that the effluent is filtered retaining the solid phase within the vessel. Only a contaminant dissolved in the aqueous phase, at concentration C_w (mg /L water), is mobile. The system contains a slurry of solid particles, each having a density, ρ_s, (e.g., mg solids/L solids) and are present in a volume fraction φ_s (L solids/L total volume). The concentration of the solid particles within the vessel is $\varphi_s\rho_s$ (mg solids/L volume). The distribution coefficient of A between the solids and the water is given by $W_s/C_w = K_{sw}$ (mg contaminant per kg per mg of contaminant per L – L/kg) while between the vapors and the water by $C_a/C_w = K_{aw}$. Using φ_a and φ_w to represent the ratio of the air and water volumes to the total vessel volume, respectively, the total mass of A, both sorbed and unsorbed, in the system can thus be written,

$$\begin{aligned} \text{dissolved} &= \varphi_w C_w V \\ \text{sorbed} &= \varphi_s \rho_s K_{sw} C_w V \\ \text{vapors} &= \varphi_a K_{aw} C_w V \\ \text{total} &= [\varphi_w + \varphi_s \rho_s K_{sw} + \varphi_a K_{aw}] C_w V \\ &= \qquad R_f \qquad C_w V \end{aligned} \tag{5.69}$$

Note that we will use φ_i to represent the ratio of the volume of phase i to the *total volume.* The bracketed term in Equation 5.69 is called a *retardation factor.* A retardation factor can be defined as the ratio of the *total* concentration in a system to that in the mobile or nonstationary phase,

$$\begin{aligned} R_f &= \frac{\text{Total concentration in system}}{\text{Concentration in mobile phase}} \\ &= \frac{\frac{1}{V}(M_s + M_a + M_w)}{M_w / V} \\ &= \frac{[\varphi_w + \varphi_s \rho_s K_{sw} + \varphi_a K_{aw}] C_w}{C_w} \\ &= [\varphi_w + \varphi_s \rho_s K_{sw} + \varphi_a K_{aw}] \end{aligned} \tag{5.70}$$

Remember that the effluent or mobile phase is assumed to be simply the aqueous phase. The effect of storage of contaminant in the solid or vapor phase is that the loss of contaminant from the system in the effluent aqueous phase is slowed due to the accumulation in the phases retained within the system. This retards the dynamics of the system, hence the name retardation factor. To see this, let us write a material balance on the system in Figure 5.10.

$$\frac{d(\text{Mass}_A)}{dt} = \frac{d(R_f C_w V)}{dt} = QC_0 - QC_w \tag{5.71}$$

Assuming that the retardation factor, $R_{f.}$ and the volume, V, are constant, and writing $V/Q = \tau_Q$, the mean residence time of water in the system, this can be rewritten,

$$\frac{dC_w}{dt} = \frac{1}{R_f \tau_Q}(C_0 - C_w) \tag{5.72}$$

This has the solution, assuming none of the component initially within the system,

$$C_w = C_0\left[1 - \exp\left(-\frac{t}{R_f \tau_Q}\right)\right] \tag{5.73}$$

Note that without the partitioning into an immobile solid or vapor phase, $R_f \rightarrow 1$, and the time required to reach steady state is simply a function of $\tau_{Q.}$ which is the mean residence time based on the flow through the system. At $t = \tau_Q$, the concentration in the system is already about 67% of the feed concentration. With partitioning into the immobile phases, however, $R_f > 1$, and the time required to achieve steady conditions can be much longer. In particular, a time $R_f\ \tau_Q$ is required to achieve 67% of the feed concentration. Because K_{sw} can be very large (e.g., exceeding 1000 for many hydrophobic compounds), the retardation factor, R_f, also can be very large implying that times much longer than the mean residence time of water in the system are required to achieve concentrations approaching steady conditions. This is illustrated in Example 5.10.

Example 5.10: The effect of sorption-related retardation on stirred vessel dynamics

Consider the dynamics of a stirred vessel of volume 1000 L initially devoid of a contaminant A to which is introduced a continuous flow of 100 L/h containing 100 mg/L of benzene. Determine the concentration in the water within the vessel assuming (1) no sorption onto a solid phase and (2) sorption onto a 5% slurry of activated carbon (by volume). The partition coefficient of the benzene to the carbon particles is 82 L/kg (82 cm^3/g). The density of the carbon particles is 2.5 gm/cm^3.

1. The mean residence time for water in the vessel is τ_Q = 1h and the concentration vs. time in the vessel is governed by

$$C_w(t) = C_0(1 - e^{-t/\tau_Q}) = 100 \text{ mg/L} (1 - e^{-t/h})$$

2. The retardation factor is $\varphi_w + (1 - \varphi_w)\ \rho_s\ K_{sw} = 0.95 + (0.05)(2.5\ gm/cm^3)(82\ cm^3/g) = 11.2$. The concentration vs. time behavior is governed by

$$C_w(t) = C_0\left(1 - e^{-\frac{t}{R_f \tau_Q}}\right) = 100 \text{ mg/L}\left(1 - e^{-\frac{t}{11.2\ h}}\right)$$

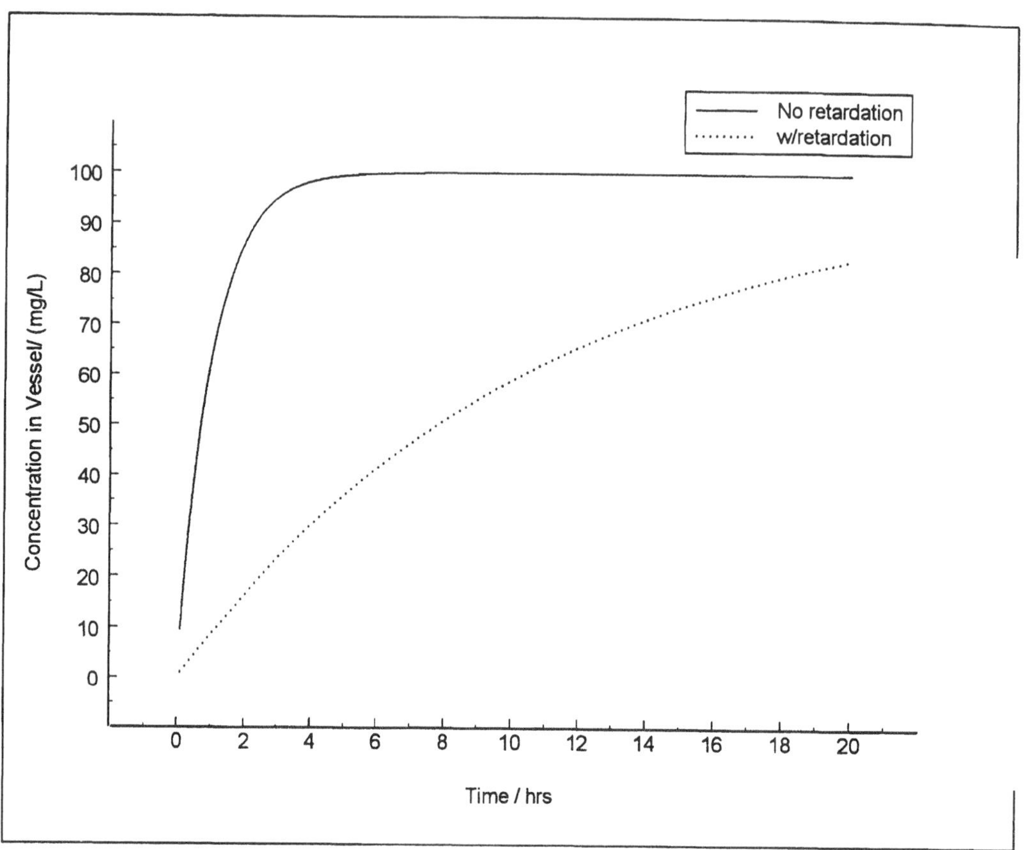

FIGURE 5.11 Contaminant concentration vs. time in vessel with and without retardation by partitioning into phases retained in vessel.

The concentration vs. time behavior predicted by these models is depicted in Figure 5.11. The sorption onto the solid phase in part 1delays the attainment of steady conditions in the vessel by a factor of 11.2.

It should be emphasized that the retardation of the vessel dynamics in Example 5.10 is a result of the partitioning into immobile phases. If the solids and vapors were removed from the vessel with the aqueous phase, no delay in the dynamics of the vessel would be observed. Retardation is especially important in flow in adsorbers or reactors that are packed with solids. Similarly, sorption-related retardation controls the migration of contaminants in groundwater systems. Complete analyses of these systems require analysis of the specific mechanisms of transport, including advection and diffusion. Each will be examined in turn.

5.4 TRANSPORT PROCESSES

5.4.1 Advection

The only transport processes considered in the material balances thus far are advective type processes, transport via a mean velocity or bulk mixing motion. The effluent from the stirred systems or the flow in the axially unmixed systems were examples of advective transport. As the winds blow or the lake currents flow, the fluid properties are likewise transported.

The fluid velocity is by definition a flux of the fluid's volume. That is, it represents the volume of fluid transported per unit area per time, or volumetric flux. The density of the particular property in that volume times the volumetric flux or velocity of the fluid thus defines the advective flux of that property.

$$(q_m)_{adv} = U(C) \qquad \text{mass}$$
$$(q_h)_{adv} = U(\rho C_p T) \qquad \text{heat} \qquad (5.74)$$
$$(\tau_{xz})_{adv} = U(\rho U) \qquad \text{momentum}$$

Here, each of the terms in parentheses in Equation 5.74 represent the "density" of the respective quantity transported (heat, mass, or momentum per unit volume). Note that the direction of the velocity U defines the direction of the flux so again, there are, in general, three components of the fluxes, one each in the x, y, and z directions.

5.4.2 Diffusion — Fick's First Law

Although advection can lead to rapid movement of a fluid property, it does not represent the only mechanism for transport. Molecules travel at a speed related to the energy level defined by their temperature. They travel in straight lines until they collide with another molecule. As a result of the collisions, the molecules exchange properties such as temperature and momentum. The resulting random motion also leads to material mixing and transport. This process is termed molecular diffusion.

If we consider the random motion of individual molecules, any movement by one molecule is offset, on average, by a corresponding motion in the opposite direction by another molecule. If the molecules are identical and exhibit the same chemical and physical properties, the balanced motion causes no change in the average chemical and physical properties of the medium surrounding the molecules. If the molecules have different properties, then the motion of one type of molecule results in transport of this property that is not offset by the corresponding return motion of other molecules with different properties. This is illustrated in Figure 5.12. In 5.12a, the property Ψ being carried by the molecules is already uniform throughout the system. In Figure 5.12b, however, a gradient in the value of the property Ψ exists. Molecular motion upward would cause transport of excess Ψ that would not be offset by molecules traveling downward. If the z axis is taken as positive upward, note that the flux in that direction is the result of a *negative* gradient of the property Ψ. This simply recognizes that the property moves from a region of greater Ψ to a region of lesser Ψ. That is, the transport always leads to the equilibrium state of uniform distribution of the property Ψ. To do otherwise without the input of energy into the system would be to defy the second law of thermodynamics. The system would separate rather than mix and the increase in order and decrease in randomness would cause a net decrease in entropy of the system.

By the reasoning above, the rate of movement of Ψ, or its flux, q_Ψ, should be proportional to the negative of the gradient in the property or

$$q_\Psi \sim -\frac{\partial \Psi}{\partial z} \qquad (5.75)$$

The constant of proportionality is a measure of the random motions of the molecules and their ability to transfer the particular property, Ψ. In diffusion of mass or moles of a substance, the constant of proportional is termed the mass or molar diffusivity. The statement of Equation 5.75 for mass transport is termed *Fick's First Law of Diffusion* which can be written,

$$q_m = -D\frac{\partial C}{\partial z} \qquad (5.76)$$

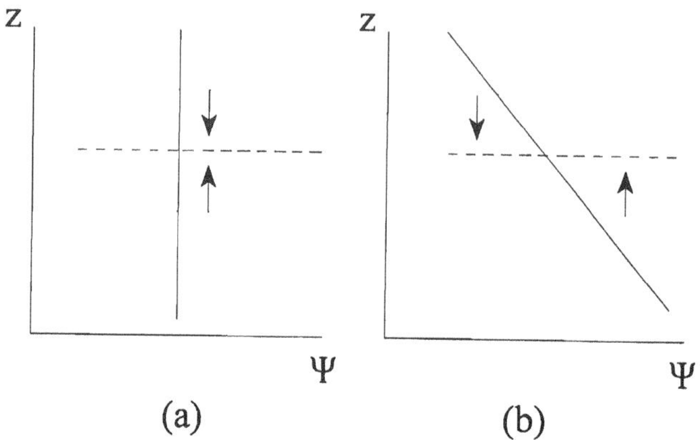

FIGURE 5.12 Depiction of zero gradient of a property (a) and a nonzero gradient (b), which results in diffusive transport due to random molecular motions.

where D is the molecular diffusivity or diffusion coefficient. It has units of {length}^{2}/{time} giving the flux in the required units of mass or moles per unit area per time. Note that Equation 5.76 defines only the diffusive flux in the z direction. In general, there are diffusive mass fluxes in each of the three coordinate directions. These will be nonzero as long as the concentration gradient is nonzero in each of these directions. The diffusivity is normally independent of direction unless conditions such as temperature change in each direction.

The physical meaning of the diffusivity can be seen by considering the displacement of molecules about an initial condition wherein all molecules are located on a particular plane (e.g., $z = 0$). Due to the random motions of these molecules, over time they will slowly spread in both the $z > 0$ and $z < 0$ directions. At any given time, the mean square displacement of the molecules (the average value of z^2) is related to the diffusivity by

$$D = \frac{\overline{z^2}}{2t} \tag{5.77}$$

From this interpretation of the magnitude of the diffusivity, it is possible to get an indication of the time required for diffusion to result in transport over a characteristic distance Z_{c_1}.

$$z_c \approx \sqrt{2Dt} \tag{5.78}$$

Here z_c represents a characteristic distance for diffusion in a time t. It does not represent the actual distance of transport of any particular molecule, but it is a measure of the average distance moved (in the root mean squared sense) in a time t. Example 5.11 examines the significance of this in air and water where a typical value of the mass diffusivity is of the order of 0.1 cm^2/s and 1×10^{-5} cm^2/s, respectively. Note that gas phase diffusivities are typically of the order of 0.1 cm^2/s, regardless of the compound and the gas through which it is diffusing. Liquid phase diffusivities are typically of the order of 1×10^{-5} cm^2/s, again regardless of the compound or the liquid solvent. Diffusivities are always known to a high degree of precision compared to most environmental parameters.

Example 5.11: Characteristic diffusion distances in air and water

Estimation of characteristic diffusion distances in air and water as estimated by Equation 5.78

$$z_c = \sqrt{2Dt}$$

Diffusivity in air ~ 0.1 cm^2/s

Diffusivity in water ~ $1x10^{-5}$ cm^2/s

In 1 s, characteristic diffusion distances

in air	~ 0.447 cm
in water	~ 0.00447 cm

In 1 h, characteristic diffusion distances

in air	~ 26.8 cm
in water	~ 0.268 cm

In 1 year, characteristic diffusion distances

in air	~ 2510 cm
in water	~ 25.1 cm

Although the approximate value of diffusivity of any vapor molecule in air (or other gas) is 0.1 cm^2/s and that of any molecule in water (or other liquid) is 10^{-5} cm^2/s, it is possible to predict these values more accurately. The Fuller, Schettler, and Giddings model of gaseous diffusivities of a compound A in a gas phase, g, can be written

$$D_g = \frac{0.001T^{7/4}(1/Mw_g + 1/Mw_A)^{1/2}}{P\left[\left(\Sigma V_i\right)^{1/3} + 2.72\right]^2} \tag{5.79}$$

Here, Mw_A represents the molecular weight of the diffusing compound while Mw_g is the molecular weight of the gas through which it is diffusing. ΣV_i represents the sum of "diffusion volumes," some of which are tabulated in Table 5.1. For other compounds a group contribution method can be used to estimate the diffusion volumes. Example 5.12 illustrates the calculation of the diffusivity for several important compounds.

Example 5.12: Estimation of gas phase diffusivities

Estimate the diffusivity of

1. Sulfur dioxide (SO_2) in air at 25°C and 1 atm pressure

$$(D_a)_{SO_2} \approx \frac{0.001(298)^{7/4}(1/29 + 1/64)^{1/2}}{1\left[(41.1 + 20.1)^{1/3} + 2.72\right]^2}$$

$$\approx 0.108 \text{ cm}^2/\text{s}$$

$$\text{Actual} \approx 0.103 \text{ cm}^2/\text{s}$$

2. Benzene (C_6H_6) in air at 25°C and 1 atm pressure

TABLE 5.1
Volumes for the Estimation of Molecular Diffusivities

Gas	V_{diff}	Group Contribution Volume Increments
H_2	7.07	C = 16.5
		H = 1.98
N_2	17.9	O = 5.48
O_2	16.6	N = 5.69
		Cl = 19.5
Air	20.1	S = 17.0
CO	18.9	Aromatic rings
CO_2	26.9	Subtract 20.2
NH_3	14.9	Example — CO_2
Cl_2	37.7	$V_{diff} \sim 16.5 + 2(5.48)$
SO_2	41.1	~ 27.46 2.3% error

$$(D_a)_{C_6H_6} \approx \frac{0.001(298)^{7/4}(1/29+1/78)^{1/2}}{1\left[(6(16.5)+6(1.98)-20.2+20.1)^{1/3}+2.72\right]^2}$$

$$\approx 0.082 \text{ cm}^2/\text{s}$$

$$\text{Actual} \approx 0.088 \text{ cm}^2/\text{s}$$

Liquid diffusivities of a compound A through a liquid phase, l, can be estimated with the relation

$$D_l = 7.4(10)^{-8}\frac{(\Phi Mw_l)^{1/2}T}{\mu_l \hat{V}_A^{0.6}} \tag{5.80}$$

Here, $\hat{V}_A$ is the molar volume at the normal boiling point of the diffusing solute in cm³/mol. Guidance for estimating this quantity can be found in Table 5.2. Other properties, the viscosity, μ (in centipoise) and the molecular weight, Mw, refer to the solvent liquid phase. Φ is an *association parameter* that is near unity for a nonpolar solvent and is 2.6 for water. Example 5.13 illustrates the calculation of liquid diffusivity.

Example 5.13: Estimation of diffusivities in liquids

Estimate the diffusivity of

1. Oxygen in water at 25°C
   For water, Φ = 2.6, μ = 1 cp at T = 298 K, and Mw = 18. From 5.2 for oxygen, $\hat{V} = 25.6$

$$D_w = \frac{7.4(10)^{-8}\sqrt{(2.6)(18)}(298)}{(1)(25.6)^{0.6}}$$

$$= 2.16(10)^{-5}\text{cm}^2/\text{s} \quad (8.3\% \text{ error})$$

TABLE 5.2
Molar Volumes of Selected Compounds and Groups

Solute	$\hat{V}$ (cm^3/mol)	Group Contribution Volume Increments		
H_2S	32.9	C = 14.8	H = 3.7	S = 25.6
N_2	31.2	Hg = 19.0	Pb = 48.0	Sn = 42.3
O_2	25.6	Cl	Terminal (RCl)	21.6
Air	29.9		Medial (R-CHCl-R)	24.6
CO	30.7	N	Double bonded	15.6
CO_2	34.0		Primary amines (RNH_2)	11
NH_3	25.8	O	Normal	7.4
Cl_2	48.4		(Except esters, ethers, acids, and rings)	
SO_2	44.8		With S, P, N	8.3
			Six-membered ring	−15.0
			Naphthalene ring	−30.0
			Anthracene ring	−47.5

2. Benzene in water at 25°C
   All parameters are identical except benzene (C_6H_6) $\hat{V} = 6*14.8+6*3.7 = 111$

$$D_l = \frac{7.4(10)^{-8}\sqrt{(2.6)(18)(298)}}{(1)(111)^{0.6}}$$

$$= 8.941(10)^{-6} \text{ cm}^2/\text{s}$$

3. Benzene in a viscous oil (μ = 1000 centipoise, Mw ~200)
   All parameters as in (2) except that μ = 1000 cp, Φ ~1, Mw ~200

$$D_l = \frac{7.4(10)^{-8}\sqrt{(1)(200)(298)}}{(1000)(111)^{0.6}}$$

$$= 2.98(10)^{-8} \text{ cm}^2/\text{s}$$

Note that the diffusivity is considerably smaller in the more viscous liquid.

5.4.3 Film Theory of Mass Transfer

There is a close physical correspondence between mass transfer as described by a mass transfer coefficient and a diffusion process. Remember that diffusion is the only transport process in a fluid in the absence of bulk motion such as winds in the atmosphere or currents in the sea. Very close to the surface of the Earth, experience tells us that the winds do decrease in intensity due to the friction of the Earth's surface. Thus, the mass transfer associated with the evaporation of water from the lake in Example 5.1 could be conceptualized as a diffusion process in the *stagnant* layer of air close to the lake surface. Mixing processes either in the lake or in the overlying air are assumed to be sufficiently rapid that any water molecules are immediately mixed over the entire volume. Within the stagnant layer at the air-water interface, though, the movement is controlled by relatively slow diffusion. In addition, the stagnant layer is usually so small that it is assumed

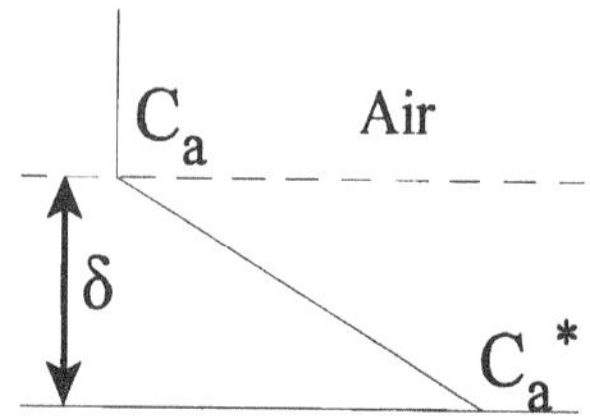

FIGURE 5.13 Depiction of film theory of mass transfer.

that there is no accumulation in the film, that is, the film is at steady state. The concept of mass transfer being controlled by diffusion in a thin film is referred to as the *film theory of mass transfer.*

Figure 5.13 shows a film theory conceptualization of the water evaporation in Example 5.1. The concentration of water vapor in the air at the water surface is the concentration corresponding to the undiluted vapor pressure of water, C_a^*. The concentration in the air well away from the water surface is C_a. The difference in concentration between the two regions is the driving force for mass transfer. The evaporative flux employing a mass transfer coefficient for the air film is

$$q_w = k_{af}(C_a^* - C_a) \tag{5.81}$$

The evaporative flux assuming diffusion in the thin stagnant film at the surface is

$$q_w = -D_a \frac{\partial C}{\partial z} \tag{5.82}$$

Because there is no accumulation or transients in the thin stagnant film, both the flux, and by Equation 5.82, the concentration gradient, are constant. If the concentration gradient is a constant it can be replaced by

$$\frac{\partial C}{\partial z} \approx \frac{\Delta C}{\Delta z} \approx \frac{C_a^* - C_a}{\delta} \tag{5.83}$$

Substituting into Equation 5.82 and comparing to Equation 5.81 suggests that the relationship between the mass transfer coefficient and the diffusivity in the thin film is

$$k_{af} = \frac{D_a}{\delta} \tag{5.84}$$

Although this model of mass transfer near an interface is at best a crude simplification of the actual phenomena, this approximation has proven very useful and its shortcomings do not significantly limit its ability to be used to make practical mass transfer calculations. There are other models of mass transfer and we will employ these models if empirical evidence provides them support in a particular situation. In the absence of such verification, we will continue to focus on the film theory represented by Equation 5.84.

One useful characteristic of any mass transfer theory is that it relates the mass transfer coefficient to molecular properties. Specifically, the film theory relates the mass transfer coefficient to the diffusivity, which is dependent, albeit weakly, on the identity of the compound, and the film thickness, which should be a function of the flow behavior and the particular physical situation of interest. We will discuss the dependence of the film thickness on the flow behavior subsequently but let us focus here on the effect of diffusivity. Because the film theory suggests that the mass transfer coefficient is proportional to diffusivity, by Equation 5.79 the mass transfer coefficient should be inversely proportional to the square root of molecular weight of the diffusing compound. This provides a mean to correct an empirical measurement of a mass transfer coefficient for one compound to allow its application to another by multiplying the measured coefficient by the ratio of the square root of the molecular weights, M, of the two compounds. If the two compounds are A and B, the mass transfer coefficients are related by

$$k_A = k_B \sqrt{\frac{Mw_B}{Mw_A}} \tag{5.85}$$

This might be used, for example, to apply a measured mass transfer coefficient for oxygen in a stream to some volatile organic species. While often used, this relationship is likely to overestimate the effect of the molecular weight of a particular species on the mass transfer coefficient. Remember that experiments tend to indicate that mass transfer coefficients are proportional to the $^1/_2$ or $^2/_3$ power of diffusivity which means that the square root in Equation 5.85 should be replaced by $^1/_4$ or $^1/_3$ power, respectively. A further complication arises when it is recognized that there are mass transfer resistances on both sides of any interface. Any attempt to employ Equation 5.85 must be limited to a single side of an interface, in a single film as in the air-side film depicted in Figure 5.13. Its use is illustrated in Example 5.14.

Example 5.14: Estimation of mass transfer coefficient by correction for compound properties

Measurements of the rate of mass transfer of carbon dioxide across the air-water interface of oceans have suggested that a value of about 20 cm/h is typical. Oxygen is a gas whose mass transfer across the air-water interface should be governed by the same processes as carbon dioxide. Estimate the mass transfer coefficient for oxygen uptake by the ocean.

$$k_{O_2} = k_{CO_2} \sqrt{\frac{Mw_{CO_2}}{Mw_{O_2}}}$$

$$= 20 \text{ cm/h} \sqrt{\frac{44}{32}}$$

$$= 23.5 \text{ cm/h}$$

As we have indicated, the correction for different compounds is generally small.

5.4.4 Two-Film Theory

The film theory provides a means to extend the above discussion to both sides of an interface. In general, a contaminant moving from the bulk phase of one fluid to the bulk phase of an adjoining fluid must face mass transfer resistances in both phases. If we conceptualize these mass transfer resistances as being confined to thin stagnant layers in each phase, the film theory of mass transfer can be applied to both phases. This is depicted in Figure 5.14 at an air-water interface. A contaminant evaporating from the water must first diffuse through the stagnant layer on the water-side of the air-water interface before diffusing through the stagnant layer on the air side to the bulk air phase. In developing this conceptual model of transport at an fluid-fluid interface, it is assumed that

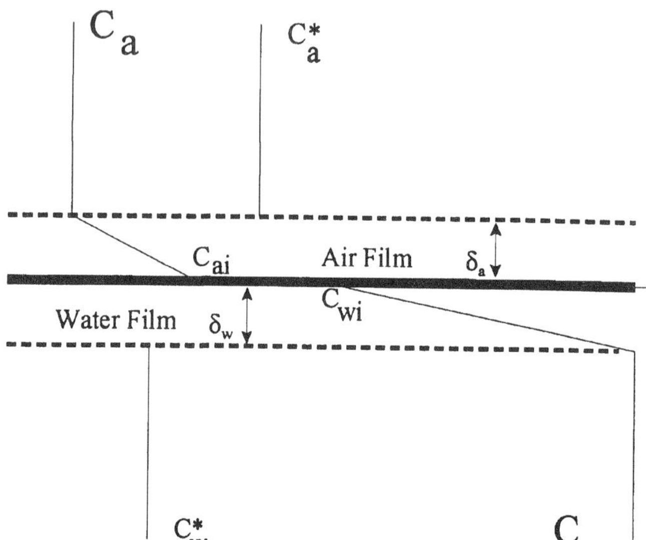

FIGURE 5.14 Depiction of two-film theory at an air-water interface.

- The bulk fluid phases are well mixed and exhibit a uniform concentration of contaminant
- The stagnant layers on each side of the interface are sufficiently thin that quasi-steady state diffusion is exhibited in each layer
- The interface also exhibits steady conditions and is therefore in equilibrium with both fluid phases

The assumption that the stagnant layers and the interface are *quasi-steady* implies that changes in concentration in the bulk fluids occur slowly enough that at any instant there is effectively steady-state transport through the stagnant layer. The characteristic time for transport in the thin layers ($\sim\delta^2/D$) is much faster than the characteristic time of the concentration changes in the bulk phases.

Thus, on the water-side of an air-water interface, the effective mass transfer coefficient, k_{wf}, through the thin stagnant film of thickness δ_w is related to the mass transfer flux by

$$k_{wf} = \frac{D_w}{\delta_w} \qquad q_m = k_{wf}(C_w - C_{wi}) \tag{5.86}$$

Here, C_{wi}, represents the concentration of the contaminant at the interface. The flux is written as positive for a flux out of the water. Similarly, on the air-side of an air-water interface, the effective film mass transfer coefficient, k_{af}, through the thin film of thickness δ_a, is related to the mass transfer flux, q_m, by

$$k_{af} = \frac{D_a}{\delta_a} \qquad q_m = k_{af}(C_{ai} - C_a) \tag{5.87}$$

The flux is written to be consistent with the sign of the water flux, that is, a positive flux represents a flux out of the water and into the air. Note that the flux out of the water must equal the flux into the air (by assumption of no accumulation at the interface) and, therefore, the flux calculated by either 5.86 or 5.87 should be the same.

Use of Equations 5.86 and 5.87 requires estimation of the individual film coefficients, k_{af} and k_{wf}, and the interfacial concentrations on each side of the interface. The values of the mass transfer coefficients depend primarily on the intensity of mixing in each of the individual phases. Experiments or correlations derived from experiments are required to estimate these parameters. These experiments can be conducted with one compound to determine film coefficients that can be used for other compounds, perhaps by correcting with the ratio of diffusivities of the two compounds as suggested by the film theory.

In the open ocean, reasonable estimates of the film coefficients are

$$\begin{aligned} k_{af} &\approx 3000 \text{ cm/h} \quad \text{based on water evaporation} \\ k_{wf} &\approx 20 \text{ cm/h} \quad \text{based on carbon dioxide transport} \end{aligned} \qquad \text{(Open Ocean)} \tag{5.88}$$

In small lakes and impoundments under essentially calm winds, the water-side mass transfer coefficients may be as low as 1 cm/h while the air-side mass transfer coefficients may be as small as 500 cm/h. Under artificially stirred conditions, a wide range of values may be observed depending on the energy input for mixing.

The primary difficulty in applying 5.86 or 5.87 to estimate mass transfer flux is the need to know the interfacial concentrations C_{aI} or C_{wI}. It is much more convenient to write the mass transfer in the form of Equation 5.10 which involves the *overall concentration difference*, that is, the difference between the concentration in the phase and the concentration that would be in the phase if it were in equilibrium with the opposite phase. This could be written based on either the air or water phase concentrations and either should give the same flux.

$$q_m = k_w(C_w - C_w^*) = k_a(C_a^* - C_a) \tag{5.89}$$

Here the ordering of the concentrations is defined such that the flux is positive if the transport of material is *into* the air phase and k_w and k_a are the overall water and air-side mass transfer coefficients, respectively. The utility of this formulation is that more fundamental film coefficients can be used to estimate the overall coefficients by recognizing that the flux must be the same regardless of the method used to calculate it. Thus,

$$\begin{aligned} q_m &= k_w(C_w - C_w^*) \\ &= k_a(C_a^* - C_a) \\ &= k_{wf}(C_w - C_{wI}) \\ &= k_{af}(C_{aI} - C_a) \end{aligned} \tag{5.90}$$

The film and the overall mass transfer coefficients can be related by adding and subtracting the interfacial concentration,

$$\begin{aligned} \frac{a_m}{k_a} &= C_a^* - C_a \\ &= C_a^* - C_{ai} + C_{aI} - C_a \\ &= (K_{aw}C_w - K_{aw}C_{wI}) + (C_{ai} - C_a) \\ &= \quad k_{aw}\frac{q_m}{k} \quad + \quad \frac{q_m}{k} \end{aligned} \tag{5.91}$$

Thus, the relationship between the "overall" mass transfer coefficient and the individual film coefficients is given by dividing through by the flux, q_m,

$$\frac{1}{k_a} = \frac{K_{aw}}{k_{wf}} + \frac{1}{k_{af}} \tag{5.92}$$

A similar development could be followed for the overall water-side coefficient to give the following relationship with the film coefficients

$$\frac{1}{k_w} = \frac{1}{k_{wf}} + \frac{1}{K_{aw}k_{af}} \tag{5.93}$$

In general, the overall mass transfer coefficient is a function of mass transfer resistances on both the air and water sides and dependent not only upon the mixing characteristics in each film but, through the air-water partition coefficient, the individual compound also. For volatile compounds, as defined by large air-water partition constants, Equations 5.92 and 5.93 suggest

$$\text{If } K_{aw} \gg \frac{k_{wf}}{k_{af}} \qquad k_w \approx k_{wf} \qquad k_a \approx \frac{k_{wf}}{K_{aw}} \tag{5.94}$$

This represents *water-side controlled mass transfer* since the air-side mass transfer coefficient is negligible. Note that since $k_{wf} << k_{af}$ in the natural environment (e.g., Equation 5.88), that even a compound for which K_{aw} ~0.01 tends to be water-side controlled. A large number of sparingly soluble hydrophobic compounds, including most gases and the volatile hydrocarbon and chlorinated hydrocarbon solvents, exhibits sufficiently high air-water partition coefficients to meet this criteria. Because carbon dioxide is a sparingly soluble gas that fits this criteria, k_{wf} in Equation 5.88 was estimated directly from measurements of carbon dioxide mass transfer across the air-sea interface.

For compounds exhibiting very low volatility, Equations 5.91 and 5.93 become

$$\text{As } K_{aw} \to 0 \qquad k_w \approx K_{aw}k_{af} \qquad k_a \approx k_{af} \tag{5.95}$$

Under these conditions, the mass transfer is said to be *air-side controlled.* Low Henry's Law constant or air-water partition coefficient compounds would include high solubility compounds or compounds exhibiting extremely low vapor pressure.

It should be recognized that under artificially stirred conditions, a low or high air-water partition coefficient does not necessarily guarantee an air or water-side controlled process. Mechanical mixing of one phase can always be used to increase the film coefficient in that phase to the point that it no longer controls mass transport. It also should be noted that there are no mass transfer resistances in an essentially pure phase. That is, the pure component is always present at the interface and in the "film" regardless of the rate of mixing of that phase. Thus, measurements of water evaporation from a lake or the open ocean provide a direct measurement of the air-side mass transfer coefficient.

The overall mass transfer coefficients for a variety of air and water-side controlled compounds are illustrated in Example 5.15.

Example 5.15: Air and water mass transfer coefficients and controlling resistances

Estimate the overall gas and liquid phase mass transfer coefficients for the listed compounds given the vapor pressure (Pv), solubility (S), and molecular weight (Mw) data and employing the open ocean mass transfer coefficients for carbon dioxide (k_{af} = 3000 cm/h) and water (k_{wf} = 20 cm/h). The ideal gas constant = 0.082 L·atm/mol·K.

The overall mass transfer coefficients are estimated from

$$\frac{1}{k_w} = \frac{1}{k_{wf}} + \frac{1}{K_{aw}k_{af}} \qquad \frac{1}{k_a} = \frac{K_{aw}}{k_{wf}} + \frac{1}{k_{af}}$$

where the Henry's constant and the individual mass transfer coefficients are estimated by

$$K_{aw} = \frac{Pv\ Mw}{RTS} \qquad k_{af} = 3000 \text{ cm/h}\sqrt{\frac{18}{Mw}} \qquad k_{wf} = 20 \text{ cm/h}\sqrt{\frac{44}{Mw}}$$

Compound	Mw	Pv (atm)	S (mg/L)	K_{aw}	k_{af} (cm/h)	k_{wf} (cm/h)	k_a (cm/h)	k_w (cm/h)
1,2-Dichloroethane	99	0.24	5500	0.176	1280	13.3	71.4	12.6
Benzene	78	0.125	1780	0.224	1440	15.0	64.2	14.4
Xylene	106	0.00871	175	0.215	1240	12.9	57.1	12.3
Naphthalene	128	1.14e – 4	34.4	0.017	1120	11.7	423	7.3
Hexachloro-benzene	285	1.4e – 8	0.005	0.033	750	7.9	183	6.0
Benzo(a)pyrene	252	7e – 12	0.004	1.8e – 5	800	8.4	800	0.014
2,3,7,8-TCDD Dioxin	322	1.8e – 12	0.0193	1.2e – 6	710	7.4	709	0.00087

The first five compounds are effectively water-side controlled since $k_w \sim k_{wf}$, while the last two compounds are air-side controlled since $k_a = k_{af}$. A mass transfer of a compound for which $K_{aw} \sim 0.01$ will be equally controlled by air and water side resistances. For $K_{aw} >> 0.01$, the transport is water-side dominated and for $K_{aw} << 0.01$, the transport is air-side dominated.

Air vs. water-side control of mass transfer processes has important implications. For example, the transport of oxygen and other nonreactive gases to water is controlled by water-side processes under natural mixing conditions. Thus the turbulence and mixing characteristics of a stream, which define the value of k_w, control the rate of oxygen transport into the stream. Changes in wind speed or increased turbulence in the wind have essentially no effect.

In addition, as shown in Example 5.15, Equation 5.85 only applies to the individual film mass transfer coefficients. Because it does not incorporate the effect of the air-water partition coefficient, it cannot be used to estimate overall mass transfer coefficients for two different compounds unless these compounds are controlled by the same film-side coefficient.

A further implication of water-side controlled mass transfer can be seen if we consider a hydrophobic compound that is evaporating from a body of water and the background air concentration is essentially zero. Because there is a negligible quantity of the contaminant in the air, the concentration in the water will be zero at equilibrium, i.e., $C_w^* = 0$. The relationship for the flux from the water then reduces to

$$\begin{aligned} q_m &= k_w(C_w - C_w^*) \\ &\approx k_{wf}C_w \end{aligned} \tag{5.96}$$

That is, under these conditions, the rate of evaporation depends only upon the rate of mixing of the water (k_{wf}) and the concentration of the compound in the water. It depends neither on the air-side mass transfer coefficient or, surprisingly, the vapor pressure or Henry's constant for the compound. *The rate of vaporization at natural air-water interfaces is, for many important pollutants, independent of the volatility of the components.* This counterintuitive phenomenon does, of course, depend on the compound having a sufficiently high volatility that $K_{aw} >> k_{wf}/k_{af}$, but once the air-water partition coefficient is sufficiently large, further increases do not increase the rate of vaporization. This is illustrated in Example 5.16.

Example 5.16: Estimation of the rate of contaminant evaporative losses from the ocean

The upper layer of the ocean tends to be relatively well mixed. Assuming that measurements within the well-mixed region were made of each of the compounds in Example 5.15, determine (1) the fraction of the total mass transfer resistance posed by the water, (2) the fraction of each compound left after 30 days, and, (3) the time required to reduce the water concentration to 50% of its initial value. Assume that no fate mechanisms exist other than evaporation for the compounds and that the atmospheric concentration is effectively negligible. Use the mass transfer coefficients of Example 5.15.

1. The fraction of the total mass transfer resistance posed by the water-side is given by

$$f_{H_2O} = \frac{k_w}{k_{wf}}$$

2/3. The fraction of the total mass after 30 days and the half-life can be estimated from a material balance, where H_m is the depth of the well-mixed region, 100 m.

$$\frac{d(C_w V)}{dt} = -k_w A C_w$$

$$\frac{dC_w}{dt} = -\frac{k_w}{H_m} C_W$$

$$\int_{C_0}^{C_w} \frac{dC_w}{C_w} = \int_0^t -\frac{k_w}{H_m} dt$$

$$\ln \frac{C_w}{C_0} = -\frac{k_w t}{H_m}$$

$$\frac{C_w}{C_0} = e^{\frac{-R_w (t \text{ or } 30 \text{ day})}{H_m}}$$

$$f_{30} = \frac{-R_w (t \text{ or } 30 \text{ day})}{H_m} \qquad t_{1/2} = \frac{H_m}{k_w} \ln(0.5)$$

The compounds are ordered by vapor pressure, or pure compound volatility. In this example the rates of evaporation of 1,2-dichloroethane, benzene, xylene, naphthalene, and hexachlorbenzene are all similar, but their vapor pressures vary over seven orders of magnitude and their Henry's Law constants vary over an order of magnitude. The mass transfer of all of these compounds are dominated by water-side mass transfer resistances.

Compound	Mw	Pv (atm)	K_{aw}	k_w (cm/h)	f_{H_2O} (%)	f_{30} (%)	$t_{1/2}$ (days)
1,2-Dichloroethane	99	0.24	0.176	12.6	94.4	40.4	22.9
Benzene	78	0.125	0.224	14.6	95.5	35.6	20.1
Xylene	106	0.00871	0.215	12.3	95.4	41.3	23.5
Naphthalene	128	1.14e – 4	0.017	7.3	62.4	59	39.5
Hexachlorobenzene	285	1.4e – 8	0.033	6.0	75.8	65.1	48.5
Benzo(a)pyrene	252	7e – 12	1.8e – 5	0.014	0.2	99.9	55 year
2,3,7,8-TCDD Dioxin	322	1.8e – 12	1.2e – 6	0.00087	~0	100	910 year

The rates of evaporation of benzo(a)pyrene and dioxin are considerably smaller and controlled by air-side mass transfer resistances.

Although we have chosen to develop the two-film theory from the air-water interface, it is a useful concept at any two-fluid interface. Although the air-water interface is most often of interest to environmental engineers, the approach could be used in identical form at air-oil or oil-water interfaces by using the relevant mass transfer film coefficient and K_{aw} is replaced with the appropriate partition coefficient between the two phases.

5.4.5 Diffusion of Heat

Molecules carry other properties including temperature or energy content and momentum as they move randomly in space. When a gradient exists, these properties can "diffuse" by random molecular motion just as in the case of mass.

The equivalent to Fick's Law of Diffusion for the diffusion of heat is called Fourier's Law of heat conduction. It can be written

$$q_h = -k_T \frac{\partial T}{\partial z} \tag{5.97}$$

where q_h is the thermal flux (energy per area per time), T is temperature, and k_T is the thermal conductivity. The units of the thermal conductivity are

$$k_T - \frac{\{\text{energy}\}}{\{\text{length}\}\{\text{time}\}\{\text{temperature}\}} \tag{5.98}$$

Typical units for thermal conductivity might be $J{\cdot}m^{-1}{\cdot}s^{-1}{\cdot}{}^{\circ}C^{-1}$. As with mass diffusion, it is possible to have temperature variations in each of the three coordinate directions and thus the flux will differ in each of these directions.

In mass diffusion the property being transferred is concentration or mass per unit volume. It is also possible to rewrite Fourier's Law of Heat Conduction in an analogous form. The energy content of a material undergoing only sensible heat changes is given by the change in enthalpy, which on a per unit mass basis is given by $\int C_p dT$. On a per unit volume basis analogous to the concentration employed in mass diffusion, the change in enthalpy per unit mass must be multiplied by the fluid density. Relative to an arbitrarily defined reference temperature of 0, therefore, and assuming a constant specific heat between the reference temperature and the fluid temperature T, the energy content per unit volume is given by $\rho c_p T$. If we write α_T as the thermal diffusivity in units of length^2/time, then Fourier's Law of Heat Conduction can be written

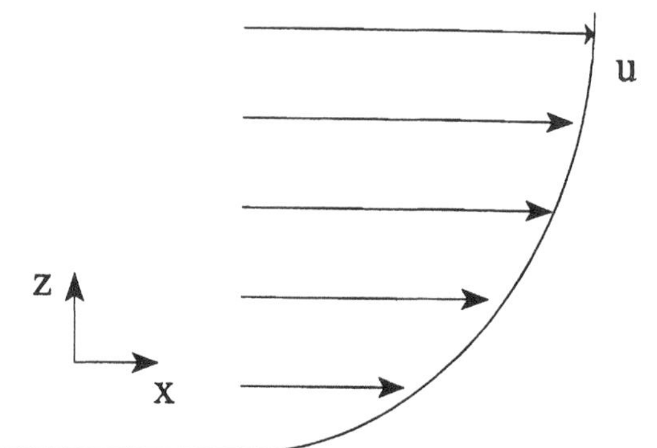

FIGURE 5.15 Depiction of momentum transfer as a result of a gradient in velocity and random molecular motion. Molecules at height carry momentum downward to regions of lesser average momentum as a result of random motion downward.

$$q_h = -\alpha_T \frac{\partial(\rho C_p T)}{\partial z} = -\alpha_T \rho C_p \frac{\partial T}{\partial z} \tag{5.99}$$

Again, the negative sign indicates that the heat flux is toward the regions of lesser temperature or energy content. Equivalence of the two forms of Fourier's Law requires

$$\alpha_T = \frac{k_T}{\rho c_p} = \frac{\{\text{energy}\}\cdot\{\text{length}\}^{-1}\cdot\{\text{time}\}^{-1}\cdot\{\text{temperature}\}^{-1}}{\{\text{energy}\}\cdot\{\text{volume}\}^{-1}\cdot\{\text{temperature}\}^{-1}} = \frac{\{\text{length}\}^2}{\{\text{time}\}} \tag{5.100}$$

5.4.6 Diffusion of Momentum

Similarly, momentum can be transported by a diffusion process. Momentum is the mass of an object times its velocity. Because the moving objects of interest to an environmental engineer are typically fluids (e.g., air or water), it is more convenient to speak of the momentum per unit volume, density times velocity. Thus, the momentum per unit volume of water moving at velocity U is given by $\rho_w U$. Newton's first law states that the time rate of change of momentum of an object is equal to the force applied. Thus, momentum flux, or the transport of momentum per unit time per unit area, has units of force per unit area, or pressure.

The effects of momentum transport are illustrated in Figure 5.15 in which friction slows the horizontal velocity profile at a fluid boundary. The rate of transport of this frictional force through the fluid is the vertical flux of horizontal momentum. The net momentum flux also is termed a shear stress to emphasize its relationship to the frictional force. We will denote the shear stress due to transport of horizontal or x-momentum in the vertical, or z, direction as τ_{xz}. Note that two directional subscripts are required, one indicating the direction of the momentum and the other indicating the direction of transport. Unlike energy content or mass concentration, momentum is a vector quantity, exhibiting both magnitude and direction. As a result, we can speak of quantities such as the transport of *horizontal momentum in the vertical direction.*

The equivalent to Fick's first law for momentum transport is termed Newton's Law of Viscosity and can be written, for a vertical (z) shear stress of horizontal (x) momentum,

$$\tau_{xz} = -\rho \nu \frac{\partial u}{\partial z} \tag{5.101}$$

where $\rho\ \partial u/\partial z$ represents the gradient in horizontal momentum and ν represents the momentum diffusivity. The momentum diffusivity, also called the kinematic viscosity, has units of length²/time as do all diffusivities. The product of the density and momentum diffusivity is the absolute viscosity and is usually given the symbol μ. Shear stress has units of pressure or force per unit area and the velocity gradient has units of inverse time. Therefore, the viscosity, μ, has units of force·time per unit area. Because force is the product of mass and acceleration, the units of viscosity can also be written

$$\mu = \rho \nu = \frac{\tau_{xz}}{-\frac{\partial u}{\partial z}} = \frac{\text{force} \cdot \text{time}}{\text{area}} = \frac{\text{mass}}{\text{length} \cdot \text{tim}} \tag{5.102}$$

$$\text{e.g.,} \quad 1 \text{ poise} = 1 \text{ g/cm} \cdot \text{s}$$

1 centipoise (cp) is 0.01 poise. The viscosity of water is approximately 1 cp.

Again, the negative sign in Equations 5.101 and 5.102 indicates that the momentum flux is toward the regions of lesser velocity or momentum. Many authors will choose to write Newton's Law of Viscosity without the negative sign, which simply means that sign convention on the shear forces within the fluid are taken as opposite of the sense implied here.

5.4.7 Total Flux

The advective (Equation 5.74) and diffusive fluxes for mass (Equation 5.76), heat (Equation 5.99), and momentum (Equation 5.101) can be combined to define the total local flux due to these processes. The total flux is simply given by the sum of the separate fluxes, or, in the z-direction

$$\begin{aligned} q_m &= (q_m)_{adv} + (q_m)_{diff} = U_z C - D\frac{\partial C}{\partial z} \\ q_h &= (q_h)_{adv} + (q_h)_{diff} = \rho C_{pU_z} T - \rho C_p \alpha_T \frac{\partial T}{\partial z} \\ \tau_{xz} &= (\tau_{xz})_{adv} + (\tau_{xz})_{diff} = \rho U_z^2 - \rho \nu \frac{\partial U_z}{\partial z} \end{aligned} \tag{5.103}$$

While strictly valid only in the presence of advection and molecular diffusion, the form of the fluxes given by Equation 5.103 are often used to represent processes other than molecular diffusion or advection. Almost all natural processes that result in the transport of momentum, heat, or mass can be approximated as either an advective or diffusive-like processes. Although these approximations as advective or diffusive may be imperfect, they have proven exceedingly useful for environmental applications.

5.5 TURBULENT TRANSPORT

5.5.1 Turbulent Fluctuations

A process that is often modeled as a diffusive-like process is heat, mass, and momentum transport via turbulence. Turbulence is the effectively random motion that characterizes flow in the atmosphere, lakes, streams, and process vessels and pipes. Gusts of wind and the swirling, eddying motion of air around buildings are some of the most common examples of turbulence in the environment. The rapid mixing associated with these eddies causes rapid mixing and results in our ability to model some turbulent systems as uniformly mixed. For example, a stream might be considered well mixed across its width while a lake might be considered well mixed throughout its volume as a result of turbulence.

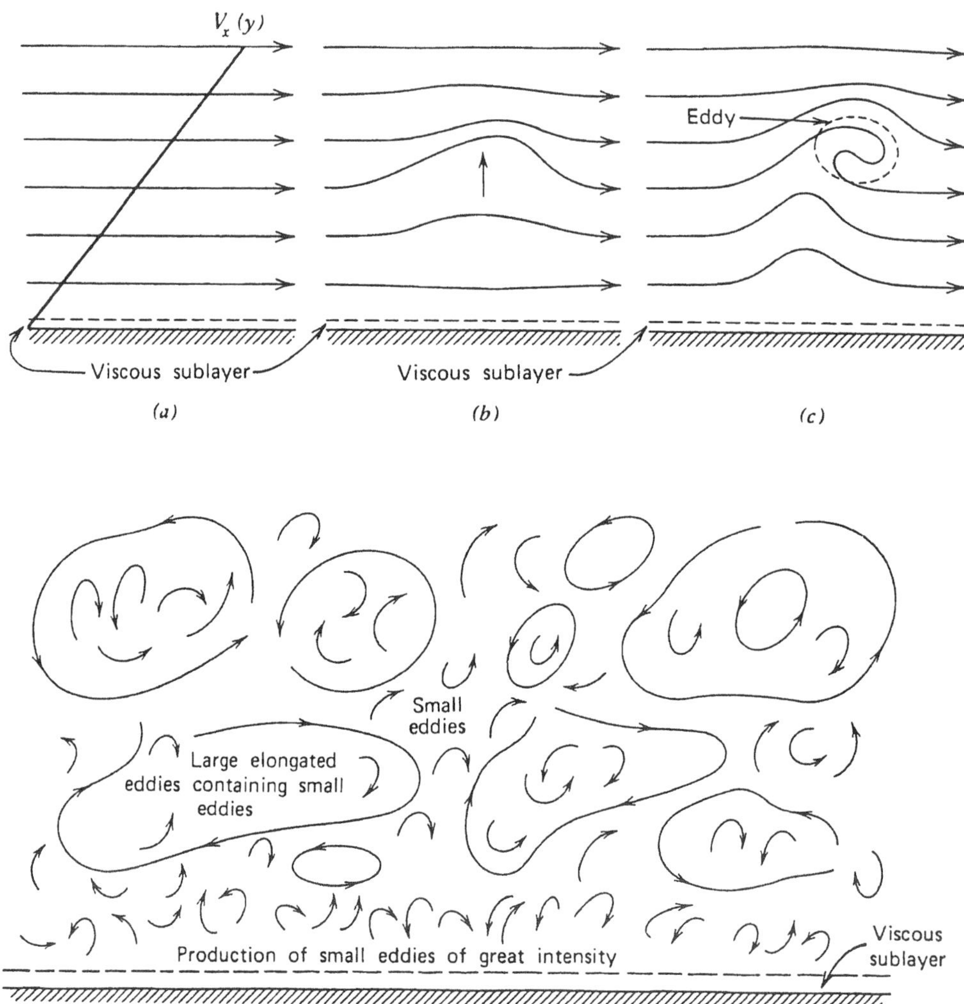

FIGURE 5.16 Turbulent eddies above an air-soil interface. (From Davies, J.T. (1972) *Turbulence Phenomena*, Academic Press, New York, and Thibodeaux, L.J. (1996) *Environmental Chemodynamics, Movement of Chemicals in Air, Water, and Soil*, John Wiley & Sons, New York. With permission.)

Turbulence is a characteristic of the flow and not the fluid that is flowing. It is characterized by mixing events in the form of eddies, typically eddies with many different sizes as shown in Figure 5.16. At low speeds, flows are controlled by viscous effects and momentum is transported solely by diffusion as described by Equation 5.103. At high speeds this description breaks down and small eddies are formed which speed momentum transport. Mass and energy also can be transported by the mixing associated with these eddies.

Almost all environmental flows as well as almost all flows in pollution control equipment are turbulent flows. Typically, the *only* flows of environmental interest that are not turbulent occur in subsurface soils or sediments or in packed beds. Even in these situations the randomness of the medium gives rise to mixing processes that are often modeled in a manner similar to that of turbulence.

If we were to continuously measure the velocity of a fluid in a turbulent flow, we might find the time record shown in Figure 5.17 which shows a continuous measurement of wind velocity. The seemingly random fluctuations can be characterized as a local perturbation about a mean which

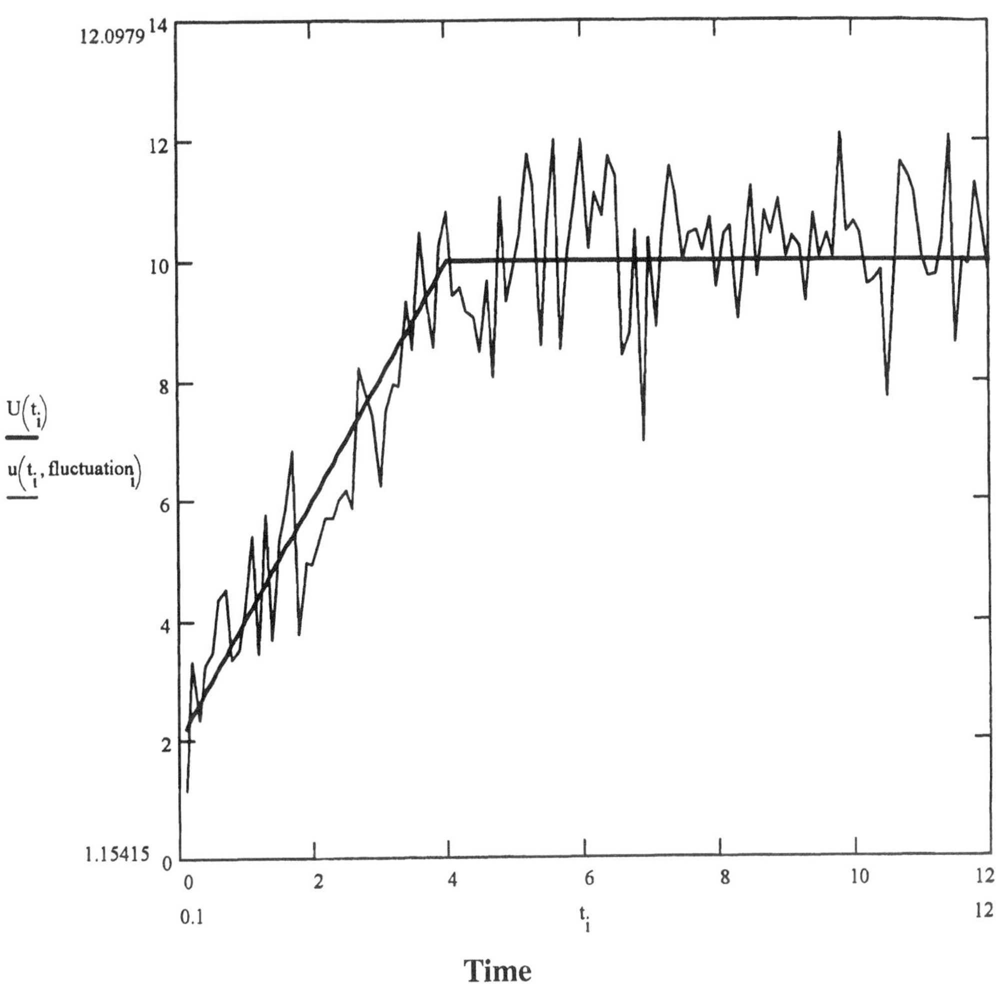

FIGURE 5.17 Time-averaged and fluctuating component of the wind.

is depicted in the figure by a horizontal solid line. That is, the velocity at any instant in time can be written as the sum of a mean velocity and a perturbation,

$$u(t) = U + u' \tag{5.104}$$

u′ is the velocity fluctuation that we feel as a wind gust or meander or as equivalent motions in bodies of water. U, the mean velocity is given by,

$$U = \frac{1}{\tau}\int_0^{\tau} u(t)\,dt \tag{5.105}$$

Here τ is the averaging time. There is, unfortunately, no obvious definition for the averaging time that defines the mean. The time record shown in Figure 5.17 is a superposition of turbulent eddies of varying frequencies. If 12 h were chosen as an averaging time for the data in Figure 5.17, the steady increase in wind speed over the first 4 h of the data would not be recognized. Conversely, a very short averaging time would capture all of the fluctuations in the record and not be very useful. Our interest is generally only in the mean value of wind speed, concentration or temperature and sometimes it is difficult to select an appropriate averaging time to define that mean.

Turbulence also causes fluctuations in temperature and concentration. Equations similar to Equations 5.104 and 5.105 can be written for each quantity.

$$c(t) = C + C' \qquad C = \int_0^\tau C(t)\,dt$$
$$T(t) = T + T' \qquad T = \int_0^\tau T(t)\,dt \tag{5.106}$$

Although our primary interest is typically the mean wind speed, concentration, or temperature, it is not possible for us to disregard the perturbations because of the mixing and transport they provide. Let us evaluate this quantitatively using Equations 5.104 and 5.106 substituted into the first of Equation 5.103.

$$\begin{aligned} q_m &= (q_m)_{adv} + (q_m)_{diff} = (U_z + u_z')(C + C') - D\frac{\partial(C + C')}{\partial z} \\ &= U_z C + u_z' C + U_z C' + u_z' C' - D\frac{\partial C}{\partial z} \end{aligned} \tag{5.107}$$

The first and last terms are generally nonzero and identical to the terms already identified in Equation 5.103. The average value of the second term is given by

$$\frac{1}{\tau}\int_0^\tau u_z' C\,dt = \frac{C}{\tau}\int_0^\tau u_z'\,dt = 0 \tag{5.108}$$

because, by definition, the average of the *perturbations about the average* is zero. Because the average of the concentration perturbations about the mean are also zero, $U_z C' = 0$. Of the terms containing perturbation terms, only the average value of $u_z' C'$ is potentially nonzero. Thus, the nonzero terms of the average flux are

$$q_m = (q_m)_{adv} + (q_m)_{turb} + (q_m)_{diff} = U_z C + u_z' C' - D\frac{\partial C}{\partial z} \tag{5.109}$$

where each term represents an average value. The second term in Equation 5.109 is the additional transport and mixing due to turbulence.

To demonstrate the nonzero value of the average value of the product of the perturbations, let us examine Figure 5.18. This figure shows a mean concentration profile that decreases with height as if there were a source of contaminant at z = 0 that was being dispersed and diluted as you moved up to positions of greater z. Consider a parcel of fluid as shown by the rectangular box initially at the center position in the figure. A vertical velocity fluctuation ($u_z' > 0$) causing the parcel to move upward a distance ℓ would result in an increase in the concentration at the new level, i.e., a positive

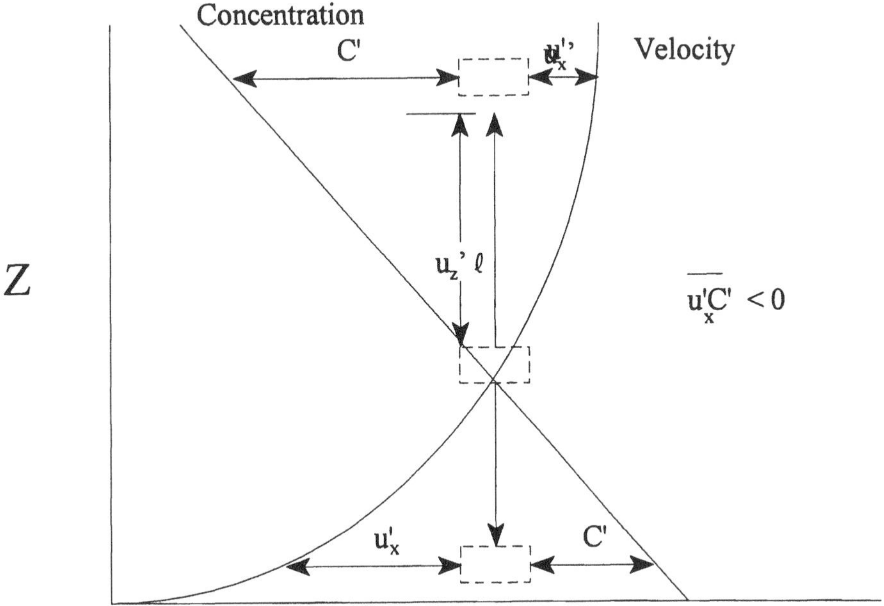

FIGURE 5.18 Conceptualization of turbulent transport.

concentration fluctuation. If the average local concentration gradient were given by ∂C/∂z then the concentration fluctuation is given by

$$C' = -1\frac{\partial C}{\partial z} \tag{5.110}$$

Then the turbulent transport is given by

$$\begin{aligned}(q_m)_{turb} &= u_z'C' = -u_z'1\frac{\partial C}{\partial z}\\ &= -D_t\frac{\partial C}{\partial z}\end{aligned} \tag{5.111}$$

This implies that the mass transport caused by the essentially random turbulent motions is of the same form as diffusion with an effective diffusivity, D_t, given by $u_z'\ell$ Thus the magnitude of the turbulent velocity fluctuations times the length over which they act serves as an effective material diffusivity (and by analogy, momentum and energy diffusivity). This analogy to molecular diffusion is by no means exact but it provides a useful and often employed conceptual model for turbulence.

Note that since the concentration gradient in Equation 5.111 is negative (decreases with height) the turbulent flux is positive (i.e., material is transported upward) in the example of Figure 5.18. The vertical fluctuations that give rise to the transport are equally likely to be positive (upward) and negative (downward). A negative vertical velocity fluctuation is also shown in Figure 5.18. In this case the concentration fluctuation at the new height is negative as is the vertical velocity fluctuation again the product u_z' C'>0. Both negative and positive velocity fluctuations give rise to a product of the velocity and concentration fluctuation of the same sign and the average value of the product is nonzero. In both cases the material flux is positive and the turbulent transport is always upward in the situation shown in Figure 5.18. If the concentration profile in Figure 5.18 were to increase with height, the signs of the turbulent transport term and the flux would change but its magnitude would still be nonzero. Although this is a crude physical picture of turbulence, it is quite useful and will be explored further in the next section.

5.5.2 Mixing Length Theory

Practical use of Equation 5.111 requires specification of u_z' and ℓ For this let us again turn to Figure 5.18 and focus on the mean velocity profile. Note that an upward fluctuation over distance 1 also gives rise to a negative fluctuation in horizontal velocity since the fluid is carrying lower velocity fluid to the new level. The magnitude of the vertical fluctuation is related to the eddy size and the mean horizontal velocity gradient by

$$u_z' = -u_x' = -1\frac{\partial U_x}{\partial z} \tag{5.112}$$

The transport associated with turbulence is then given by substituting into 5.111.

$$u_z'C' = -u_x'C' = -1^2\left|\frac{\partial U_x}{\partial z}\right|\frac{\partial C}{\partial z} \tag{5.113}$$

And now our effective turbulent diffusion coefficient has been related to the size of the eddies and the mean horizontal velocity gradient

$$D_t = 1^2\left|\frac{\partial U_x}{\partial z}\right| \tag{5.114}$$

The absolute value on the right-hand side of Equation 5.114 recognizes that, for the diffusion model to be valid, the material transport always moves from regions of high concentrations to low concentrations, that is, it is defined by the sign of the concentration gradient regardless of the sign of the velocity gradient. This model remains an imperfect but often useful conceptualization of turbulent transport processes. The total flux of mass, heat, or momentum in the vertical direction is now the sum of the advective, diffusive, and turbulent diffusive flux, or

$$\begin{aligned} q_m &= U_zC - (D + D_t)\frac{\partial C}{\partial z} \\ q_h &= \rho C_p U_z T - \rho C_p(\alpha_T + \alpha_t)\frac{\partial T}{\partial z} \\ \tau_{xz} &= \rho U_z U_x - \rho(\nu + \nu_t)\frac{\partial U_z}{\partial z} \end{aligned} \tag{5.115}$$

where, in this simple model,

$$D_t \approx \alpha_t \approx \nu_t = 1^2\left|\frac{\partial U_x}{\partial z}\right| \tag{5.116}$$

It remains impossible to use Equation 5.116 unless we can specify the eddy size, ℓ This also hasbeen termed the mixing length and Equation 5.116 is a statement of the mixing length theory first developed by Prandtl. Equation 5.116 also implies that the turbulent diffusivities for heat, mass, and momentum transport are proportional to the mean horizontal velocity gradient or shear. This is a crude simplification, even in a simple shear flow such as that shown in Figure 5.14, but it is in these flows that the Prandtl mixing length theory is most useful. The generation of turbulence and turbulent transport as a result of velocity shear is referred to as the mechanical generation of turbulence.

5.5.3 The Logarithmic Velocity Profile Law

Let us examine the implications of Equation 5.115 on velocities in the absence of the complicating factor of significant heating or cooling effects. There is no general theory or model for the mixing length to complement Equation 5.115. Very near an interface, however, rather it be a pipe wall, the surface of the Earth or water body, or the sediment-water interface, a mixing length proportional to the distance from the interface has been observed. Clearly the size of the turbulent eddies at a particular height z above the surface of the ground cannot exceed the distance to the ground. Observations have suggested in fact, that the mixing length is related to distance from an interface by a relation of the form,

$$l = \kappa z \tag{5.117}$$

κ is the von Karman constant and is approximately 0.4.

Let us examine further the flow of fluid near an environmental interface. The friction at the surface slows the fluid near the interface and results in momentum transfer from the fluid. In turn the momentum transferred from the fluid is provided by transferring momentum from deeper into the fluid. The frictional force per unit area or shear stress, τ, is approximately constant near the interface in each of these situations. If we denote this approximately constant surface shear flux as τ_0, its relation to the velocity profile is given by

$$\begin{aligned} \tau_0 &= -\rho\, l^2 \left|\frac{\partial U}{\partial z}\right| \frac{\partial U}{\partial z} \\ &= -\rho\, \kappa^2 z^2 \left|\frac{\partial U}{\partial z}\right| \frac{\partial U}{\partial z} \end{aligned} \tag{5.118}$$

This can be written

$$\begin{aligned} \frac{\tau_0}{\rho}\frac{1}{(\kappa z)^2} &= -\left|\frac{dU}{dz}\right| \frac{dU}{dz} = \left(\frac{dU}{dz}\right)^2 \\ \frac{dU}{dz} &= \sqrt{\frac{\tau_0}{\rho}}\frac{1}{\kappa z} = \frac{u_*}{\kappa z} \\ \int_0^U dU &= \frac{u_*}{\kappa}\int_{z_0}^{z} \frac{dz}{z} \\ U &= \frac{u_*}{\kappa} \ln \frac{z}{z_0} \end{aligned} \tag{5.119}$$

$u_* = \sqrt{t_0/\rho}$ has the dimensions of velocity and is called the friction velocity due to its relationship with the surface shear stress, τ_0. The applicability of this turbulent model extends from some height z_0 where the turbulence is effectively negligible to some height z in the fluid where the momentum flux is still approximately constant. The roughness height, z_0, is related but not equal to the height of the objects that roughen the surface including buildings, trees, etc. The final relation in Equation 5.119 is termed the logarithmic law and it accurately describes the shape of the velocity profile in the air above the ground, in a stream above the sediment, and even near the wall in a duct or pipe.

We can now estimate the effective momentum diffusivity as a result of turbulence above an environmental interface through Equation 5.116.

$$\begin{aligned} \nu_t &= 1^2\left|\frac{\partial U_x}{\partial z}\right| \\ &= \kappa^2 z^2 \frac{\partial}{\partial z}\frac{u_*}{\kappa}\ln\left[\frac{z}{z_0}\right] \\ &= \kappa^2 z^2 \frac{u_*}{\kappa z} \\ &= \kappa u_* z \end{aligned} \tag{5.120}$$

Table 5.3 summarizes values of the roughness height and typical ranges of the friction velocity under various environmental conditions. Such a table can be useful in providing rough guidance as to values of roughness height in a particular situation, but it is generally necessary to measure values of friction velocity. This can be done by measuring the velocity in the air or stream of interest at a particular height above the ground or stream bottom and then solving for the friction velocity from the last of Equations 5.119. In the atmosphere winds are usually measured at a height of 10 m and the use of this height, the velocity at that height, and an estimated value of surface roughness allows calculation of a friction velocity. If velocity is known at two heights it is not necessary to estimate the roughness height and it also can be calculated from Equation 5.119. These approaches to estimating friction velocity and roughness height are illustrated in Example 5.17.

TABLE 5.3
Typical Values of Friction Velocity and Surface Roughness

Condition	z_0 (cm)	v_* (m/s)
Winds		
Above mud, snow, sea, or desert	10^{-3}–0.01	0.16–0.27
Above lawn grass (5 cm)	1–2	0.43
Above grass to 60 cm	4–9	0.60
Open area, occasional trees and buildings	10	—
Fully grown root crops	14	1.75
Water currents		
Deep sea	2	0.001
Continental shelf	—	0.01
Riverbeds	3 –90	—

Example 5.17: Estimation of logarithm velocity profile parameters

1. A wind velocity of 3 m/s is measured at a height of 10 m above the ground. There is open terrain with occasional buildings and trees with an estimated roughness height, z_0, of about 0.1 m. Estimate the friction velocity. Solving the logarithmic velocity profile for u_*.

$$u_* = \frac{\kappa u}{\ln\left(\dfrac{z}{z_0}\right)} = \frac{(0.4)(3\text{ m/s})}{\ln\left(\dfrac{10\text{ m}}{0.1\text{ m}}\right)} = 0.261\text{ m/s}$$

2. A second measurement of wind velocity at an elevation of 5 m is 2.5 m/s. Estimate both friction velocity and roughness height. Forming the ratio of the formula for v_* at each height and solving for the roughness height gives

$$z_0 = \exp\left[\frac{u_{x1}\ln(z_2) - u_{x2}\ln(z_1)}{(u_{x1} - u_{x2})}\right] = \exp\left[\frac{2.5\ln(5) - 3 1\ln(10)}{(2.5 - 3)}\right] = 0.156\text{ m}$$

Using this value of z_0, v_* can then be evaluated as in part 1.

$$u_* = \frac{\kappa u}{\ln\left(\dfrac{z}{z_0}\right)} = \frac{(0.4)(3\text{ m/s})}{\ln\left(\dfrac{10\text{ m}}{0.156\text{ m}}\right)} = 0.289\text{ m/s}$$

The effective turbulent diffusion coefficient at for example 10 m above the surface is then given by

$$\begin{aligned} v_t &= \kappa u_* z \\ &= (0.4)(0.289\text{ m/s})(10\text{ m}) \\ &= 1.16\text{ m}^2/\text{s} \end{aligned}$$

In the absence of significant surface heating or cooling, this quantity also provides a good estimate of D_t and α_t. (See Section 5.5.4.)

5.5.4 Heat, Mass, and Momentum Transfer Analogies

The assumption of equality of the eddy diffusivity, conductivity, and viscosity in Equation 5.116 and in Example 5.17 is dependent upon the molecular diffusivities, conductivities, and viscosities being approximately the same. Although the theory does not explicitly contain these quantities, their effect can be thought of as a complicated influence upon the appropriate mixing lengths for heat, mass, or momentum transfer. To understand their influence we will turn to empirical observations of the effects of different values of the molecular transport coefficients. The ratio of the momentum diffusivity or kinematic viscosity of the fluid, $\nu = \mu/\rho$, to the molecular diffusivity of the component in the fluid, D, is called the Schmidt number, N_{Sc}, while the ratio of the momentum diffusivity to the thermal conductivity, k_T, is the Prandtl number, N_{Pr}.

$$N_{Sc} = \frac{\nu}{D} = \frac{\mu}{\rho D} \qquad N_{Pr} = \frac{\nu}{\alpha_T} = \frac{\mu C_p}{k_T} \tag{5.121}$$

Equation 5.116 and equality of turbulent diffusivities in heat, mass, and momentum would apply when $N_{Sc} = N_{Pr} = 1$. For compounds in air, this is generally a very good assumption. In air, therefore, the equivalence of turbulent heat, mass, or momentum transfer is fairly good. In water, however, many compounds exhibit a Schmidt number of the order of 1000 or more. This is especially important near an environmental interface such as the sediment-water interface or the air-soil interface in that the velocities approach zero near these interfaces. Under these conditions, the effective diffusion coefficient very near the interface is strongly influenced by molecular diffusion and the equivalence of momentum and mass transfer is weakened. Empirical evidence suggests that the ratios of the turbulent heat, mass, and momentum diffusivities, or turbulent Schmidt and Prandtl numbers near a motionless, solid interface are given by normal (or molecular) Schmidt and Prandtl numbers to the 2/3 power. This is sometimes expressed as a relationship between the turbulent Schmidt (ν_t/D_t) and Prandtl (ν_t/α_t) numbers and the molecular quantities defined by Equation 5.121.

$$(N_{Sc})_t = \frac{\nu_t}{D_t} \approx N_{Sc}^{2/3} \qquad (N_{Pr})_t = \frac{\nu_t}{\alpha_t} \approx N_{Pr}^{2/3} \tag{5.122}$$

Near an interface

Thus, benzene in air with $N_{Sc} = 1.76$ would exhibit a turbulent Schmidt number of about 1.45 or the turbulent momentum diffusivity would be expected to be about 1.45 times the turbulent mass diffusivity near an air-soil interface. For benzene in water, $N_{Sc} = 1025$, and the turbulent momentum diffusivity near a sediment-water interface would be expected to be more than 100 times larger than the turbulent mass diffusivity. Both estimates are valid only as long as the turbulence is driven and controlled by velocity shear or gradients. Often buoyancy forces and density stratification is an equal, if not more, important mechanism for the production (or consumption) of turbulence and this analogy is less valid. This is especially important in the atmosphere due to solar heating or nighttime cooling of the ground surface. The small correction to atmospheric turbulent transfer coefficients posed by Equation 5.122 is often neglected by comparison to this larger effect of solar heating and cooling.

5.5.5 The Effect of Stratification on Turbulent Transport

The logarithmic velocity profile law and the estimation of a mixing length as proportional to height above a surface is generally only valid near a boundary such as a pipe wall or near the soil surface. The logarithmic velocity profile implicitly assumes that the only source of turbulence is friction at a surface. Another, often more important source of turbulence, is due to buoyancy. Buoyancy is motion driven by density variations in a fluid in the presence of gravity. Lighter, less dense fluid will tend to rise and be replaced near a surface by heavier, more dense fluid. Because the fluids tend to conserve their original properties during this movement, transport of heat, mass, and momentum can occur.

Let us consider a shallow body of fresh water. If the temperature was uniform as a function of depth, the water would all be of constant density and there would be no tendency for transport to occur via buoyancy. Warming of the surface waters can lead to a warm body of fluid that will tend to remain at the surface. The motion of wind or currents that might cause fluid to mix downward will be resisted by this buoyancy force, slowing the rate of heat, mass, or momentum transfer downward. If the surface of the water were to be cooled by winter conditions to the point that the surface water is cooler, and therefore, more dense than the fluid below, the surface waters will tend to fall downward. Thus, any current or wind motion that would cause transport downward would be enhanced and rapid transfer of heat, mass, or momentum would occur.

Note that this discussion is limited to freshwaters since salinity can also affect water density. In addition, water below 4°C increases in density as temperature increases. Very cold water responds to heating or cooling in the opposite fashion to that described above. Although temperature is a good indicator of the density profile in waters (or in the atmosphere), it is important to recognize that it is density variations that define buoyancy.

Let us examine the effect of temperature and density variations in fluids in detail. The uniform temperature condition is shown in Figure 5.19a. A small parcel of fluid that would be perturbed upward or downward as a result of any turbulent motion will always exhibit the same temperature and density as its surroundings and buoyancy will not tend to cause it to either return to its original position or continue its motion away from that position. This is referred to as a neutrally stable condition, since the *temperature or density stratification* neither enhances nor dampens the turbulent motion.

In Figure 5.19b, the fluid above is less dense than the fluid below. An upward perturbation of a fluid parcel as a result of turbulence results in its movement into fluid which is less dense. It would thus be heavier than its surroundings and tend to return to its original position. A downward perturbation of the same fluid parcel would result in a movement into fluid which is more dense. It would then tend to again rise and return to its original position. The tendency to return to its original position reflects the *stable stratification* of the original system. A stable environment slows and dampens turbulent mixing.

In Figure 5.19c, the fluid above is more dense than the fluid below. An upward perturbation of a fluid parcel will now cause movement into a fluid of even greater density and the parcel will tend to continue to rise. Conversely, a downward movement will bring the parcel into contact with fluid of lesser density and it will continue to fall. When perturbations or movements lead to further movement and a larger deviation from the original state, the system is referred to as *unstable*. In many lake systems, cooling at the surface in late fall results in heavier fluid above than below, or an unstable stratification, and this causes the entire lake to turnover and become completely mixed in a matter of minutes.

Often neutral, stable, or unstable stratification does not apply to a body of water or to the atmosphere but only to a small region within the fluid. In the atmosphere, the near surface layer of the atmosphere is often unstable and relatively well mixed while above 1 to 2000 m, the atmosphere might exhibit a temperature that increases (density decreases) with height. This layer is stably stratified and very poorly mixed. Temperature normally decreases with height in the atmosphere, and an increase with height is termed a temperature inversion. In a lake, the near-surface environment might be relatively well mixed due to friction at the air-water interface and the resulting wind-driven currents. Because the surface tends to be warmer than the deeper waters below, there is often a poorly mixed region that represents the transition from the warmer water above to the cooler water below. This is termed a thermocline and like a temperature inversion in the atmosphere, it is a layer that is stable and significantly limits vertical transport of heat, mass, or momentum.

We have indicated that turbulence results from both friction at a surface (velocity shear or mechanical generation of turbulence) and density variations in a gravitational field (buoyancy generation of turbulence). Friction and velocity shear, however, always result in the generation of turbulence while buoyancy forces can either generate or consume turbulence. The relative importance of these two processes is indicated by the Richardson number,

$$N_{Ri} = \frac{-\dfrac{g}{\rho}\dfrac{d\rho}{dz}}{\left(\dfrac{dU}{dz}\right)^2} \tag{5.123}$$

Stratified Environments

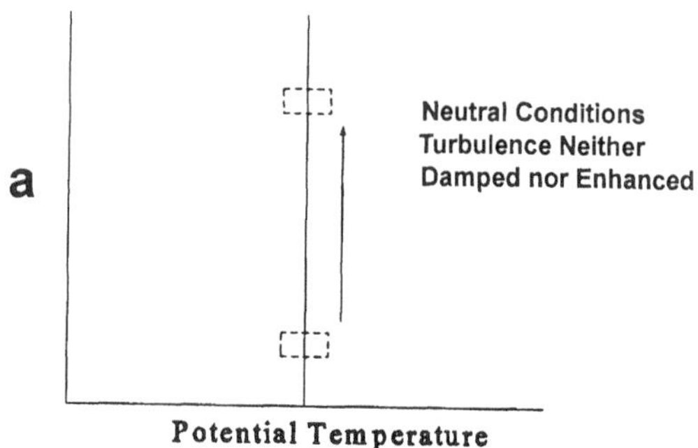

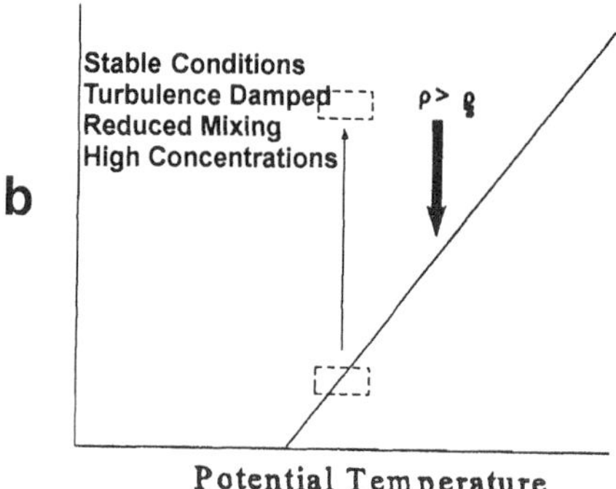

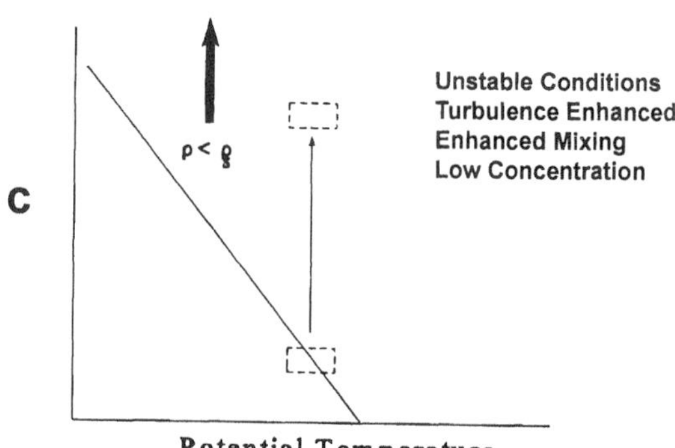

FIGURE 5.19 Depiction of stability in a density stratified environment: (a) neutrally stable conditions, (b) stable conditions, and (c) unstable conditions.

The Richardson number is the ratio of the consumption of turbulent energy by buoyancy forces to the generation of turbulent energy by velocity shear. If buoyancy forces increase turbulence, the Richardson number is thus negative while positive values occur when stable stratification prevails. The Richardson number is a good indicator of the strength of turbulence and the mixing that results.

The Richardson number is also commonly written in terms of temperature due to its ease of measurement,

$$N_{Ri} \approx \frac{\beta_T g \frac{dT}{dz}}{\left(\frac{dU}{dz}\right)^2} \tag{5.124}$$

$$\text{where } \beta_T = -\frac{1}{\rho}\frac{d\rho}{dT} \begin{cases} \approx 1/T \text{ in air} \\ \approx 1.1\,(10)^{-4}\,°C^{-1} \text{ in water} \end{cases}$$

The reader is cautioned, however, that saline water or water below 4°C is not described by the value of β_T shown above. In either case, Equation 5.124 would lead to misleading interpretations of density stratification. Both Equations 5.123 and 5.124 also are only applicable to low and essentially constant pressures. The change of pressure with height in the atmosphere and its effect on density and stability will be discussed in the next chapter.

5.6 MICROSCOPIC MASS, ENERGY, AND MOMENTUM BALANCES

5.6.1 Development of Microscopic Balances

In our material balances, we have assumed that temperature and concentration are uniform over some region in space. In the previous section, we developed models of heat, mass, and momentum transport that are associated with continuous gradients in these quantities. Let us extend our notion of a material balance to such a system.

Consider the one-dimensional motion of a contaminant as a result of advective and diffusive fluxes in the arbitrary differential volume of Figure 5.20. A material balance around the volume can be written, as before,

$$\{\text{Accumulation}\} = \{\text{Input}\} - \{\text{Output}\} \tag{5.125}$$

The accumulation term is, as before, the time rate of change of the total mass of contaminant in the volume. If C_T represents the total concentration of the contaminant in the system, it can be written in terms of the fluid phase concentration as $C_T = R_f C_f$, where R_f is the retardation factor defined previously as the ratio of the total concentration to the concentration in the mobile, or fluid, phase. As a result, the accumulation term can be written,

$$\{\text{Accumulation}\} = \frac{\partial}{\partial t}[C_T A\,\Delta x] = \frac{\partial}{\partial t}[R_f C_f A\,\Delta x] \tag{5.126}$$

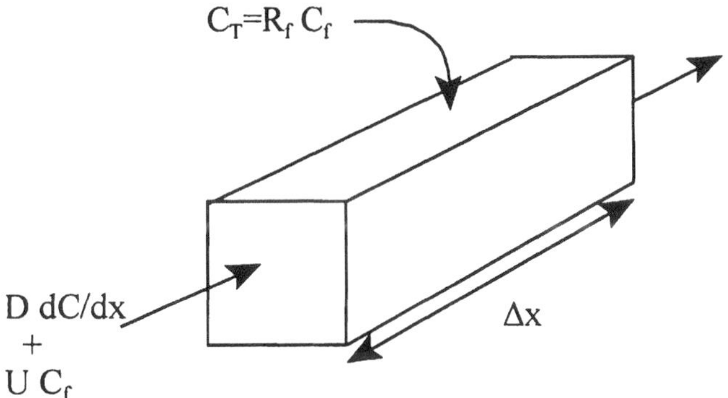

FIGURE 5.20 Control volume for development of differential mass balance.

At the upstream edge of the arbitrary volume (defined as the face perpendicular to the direction of flow and pointed toward the direction of flow) the transport of contaminant by advection at velocity U and diffusion with diffusivity D through the area A is given by

$$\{\text{Input}\} = \left[UC_f - D\frac{\partial C_f}{\partial x} \right]_x A \tag{5.127}$$

Similarly, at the downstream edge,

$$\{\text{Output}\} = \left[UC_f - D\frac{\partial C_f}{\partial x} \right]_{x+\Delta x} A \tag{5.128}$$

Combining these separate terms and dividing by the area,

$$\frac{\partial R_f C_f}{\partial t} = \frac{1}{\Delta x}\left\{ \left[UC_f - D\frac{\partial C_f}{\partial x} \right]_x + \left[UC_f - D\frac{\partial C_f}{\partial x} \right]_{x+\Delta x} \right\} \tag{5.129}$$

In the limit of $\Delta x \to 0$, the right-hand side is the negative of the derivative with respect to x

$$\frac{\partial R_f C_f}{\partial t} = -\frac{\partial}{\partial x} UC_f + \frac{\partial}{\partial x} D \frac{\partial C_f}{\partial x} \tag{5.130}$$

Commonly both the velocity and diffusivity are assumed to be independent of distance, giving rise to the *advection-diffusion equation.*

$$\frac{\partial C_f}{\partial t} + U\frac{\partial C_f}{\partial x} = D\frac{\partial^2 C_f}{\partial x^2} \tag{5.131}$$

This equation is often used to describe the dynamics of a contaminant in the environment. Solutions of this equation describe concentrations as a function of time and space, $C_f(t,x)$. A three-dimensional version can also be derived if advection and diffusion occur in more than the single direction. Note that the diffusivity in Equation 5.131 can be either the molecular diffusivity or a turbulent diffusivity.

Example 5.18: The Lagrangian form of the advection-diffusion equation

Equation 5.131 is sometimes referred to as the Eulerian advection-diffusion equation. This refers to the fact that it describes transport relative to a fixed location. An alternative is to view the transport relative to the moving frame of reference given by the location of a fluid parcel moving with the fluid (Lagrangian form). Write the Lagrangian form of Equation 5.131.

The movement of the frame of reference is simply given by the velocity of the fluid parcel in each of the coordinate direction. The time derivative of concentration experienced by such a moving fluid parcel is given by the chain rule of partial differentiation in the single dimension for which Equation 5.131 is written

$$\frac{dC_f}{dt} = \frac{\partial C_f}{\partial t} + \frac{\partial x}{\partial t}\frac{\partial C_f}{\partial x}$$

$$= \frac{\partial C_f}{\partial t} + U\frac{\partial C_f}{\partial x}$$

Thus, Equation 5.131 could be written in Lagrangian form as

$$\frac{dC_f}{dt} = D\frac{\partial^2 C_f}{\partial x^2}$$

which describes the change in time of the fluid concentration in a reference frame moving with the fluid. The total time derivative on the left-hand side is sometimes referred to as the substantial derivative to recognize that it is the total derivative in a reference frame moving with the fluid. The partial derivative may be thought of as the time rate of change detected by an observer fixed on the side of stream, while the substantial derivative is that detected by an observer sitting in a boat drifting with the motion of the stream.

Example 5.19: Momentum transport equation

An equation similar to Equation 5.131 also may be written for the transport of momentum. Develop this equation for the vertical transport of horizontal (x) momentum. Recognize that external forces represent a source of momentum whereas Equation 5.131 does not include a material source term.

Following the development leading to Equation 5.131, the accumulation of horizontal momentum due to transport in the vertical direction is governed by

$$\frac{\partial \rho U_x}{\partial t} = -\frac{\partial P}{\partial x} - \frac{\partial \tau_{xz}}{\partial_z}$$

$$= -\frac{\partial P}{\partial x} - \frac{\partial}{\partial z}\left[\rho U_z U_x - \rho(\nu + \nu_t)\frac{\partial U_x}{\partial z}\right]$$

$$\frac{\partial U_x}{\partial t} + U_z\frac{\partial U_x}{\partial z} = -\frac{1}{\rho}\frac{\partial P}{\partial x} + (\nu + \nu_t)\frac{\partial^2 U_x}{\partial z^2}$$

Remember that the definition of the flux and thus all terms in this equation represent mean or time-averaged quantities. Generally the time-averaged vertical velocity and molecular diffusion of momentum is negligible compared to turbulent transport. Thus, the vertical transport of horizontal momentum is normally described by

$$\frac{\partial U_x}{\partial t} = -\frac{1}{\rho}\frac{\partial P}{\partial x} + \nu_t\frac{\partial^2 U_x}{\partial z^2}$$

Note that momentum is a vector equation and a balance analogous to this can be written in each of the three coordinate directions. Furthermore, in multiple dimensions advection and turbulent diffusive transport of each component of momentum in these other directions must also be considered.

Finally note that an energy balance based on the equation for energy fluxes also can be derived that is identical in form to Equation 5.131 or mass. Proof is deferred to the problems.

5.6.2 Error Function Solution to the Microscopic Material Balance

Deriving a solution to Equation 5.131 requires specification of an initial condition in time and two conditions on each of the directions in space. Different solutions exist for different boundary and initial conditions. We will summarize a few of the solutions that are widely used to describe the environmental behavior of contaminants.

One of the most commonly used solutions to Equation 5.131 is one that corresponds to diffusion in a large zone of initially uniform concentration exposed to a zone of different concentration. This has been used to describe contaminant transport from initially uniformly contaminated soils and sediments. Typically a different concentration exists at the ground surface or at the sediment-water interface and the solution to the equation subject to these conditions describes the migration of the contaminant between the interface and the bulk soil or sediment. The form of the equation and initial and boundary conditions that describe this problem are

$$\begin{aligned} \frac{\partial R_f C_f}{\partial t} &= D\frac{\partial^2 C_f}{\partial z^2} \\ C = C_0 &\quad \text{at} \quad t = 0 \\ C = C_f(0) &\quad \text{at} \quad z = 0 \\ C \to C_0 &\quad \text{at} \quad z \to \infty \end{aligned} \tag{5.132}$$

The advective component in 5.131 is not included since only diffusion or a diffusion-like process is assumed to apply. The first condition suggests that initially the concentration is uniform for all $z > 0$ at $C_f = C_0$. The second condition suggests that for all time ($t > 0$), the concentration at $z = 0$ is given by a different concentration, $C_f(0)$. Finally, the third condition indicates that the concentration remains unchanged as you move far away from the interface. Although this implies that the zone of initial contamination is infinite, it really only requires that the model be applied to times such that diffusion has not influenced the entire depth of contamination. As indicated earlier, very long times are often required for diffusion to influence concentrations over even modest distances. The solution to the equations in 5.132 is called the error function solution and is given by

$$C(z,t) = C_f(0) + [C_0 - C_f(0)]\, erf\left(\sqrt{\frac{z^2 R_f}{4Dt}}\right) \tag{5.133}$$

The contaminant flux, in this case just the diffusive flux, is given by,

$$\begin{aligned} q_m &= -D\frac{\partial C_f}{\partial z} \\ &= [C_0 - C_f(0)]\sqrt{\frac{DR_f}{\pi t}} \end{aligned} \tag{5.134}$$

The error function is simply an integral that cannot be directly integrated but for which numerical values have been generated and tabulated. For a value of the argument $\sqrt{z^2 R_f / 4Dt}$ equal zero, the error function is also zero, while at infinite values of the argument, the error function is one. Thus, at $z = 0$, Equation 5.133 reduces to $C(z,t) = C_f(0)$ and at either large z or small time, $C(z,t) \to C_0$,

TABLE 5.4
Values of the Error and Complementary Error Functions

x	erf(x)	erfc(x) = 1 − erf(x)
0	0	1
0.25	0.2763	0.7237
0.5	0.5205	0.48
0.75	0.7112	0.2888
1	0.8427	0.157
1.25	0.9229	0.0771
1.5	0.9661	0.034
2	0.99532	0.005
2.5	0.999593	4.07×10^{-4}
3	0.9999779	2.209×10^{-5}
3.5	0.999999257	7.431×10^{-7}
4	0.9999999845	1.542×10^{-8}
4.5	0.999999999803	1.966×10^{-10}
5	0.99999999999846	1.537×10^{-12}

consistent with the specified boundary conditions. Values of the error function at intermediate values of the argument are given in Table 5.4.

Note that the retardation factor divides the diffusivity in the equation for concentration (Equation 5.133) while it multiplies the diffusivity in the equation for the flux (Equation 5.134). This seems to imply that the flux of contaminant increases when retardation occurs due, for example, to sorption onto a solid phase. This is not true, however, since the concentration in the diffusing phase also decreases as the retardation factor increases.

To understand this, note that it is the fluid phase concentration that is predicted in Equation 5.133. Due to sorption or entrapment on an immobile phase, this concentration decreases relative to the total concentration in the system as a result of $C_f = C_T/R_f$. The net effect is that, for a given total concentration in a system governed by Equation 5.134, retardation slows the contaminant flux according to $q_m \sim (R_f)^{-1/2}$. Example 5.20 illustrates the application of the error function solution.

Example 5.20: Application of the error function solution

Compare the transport rate of trichloroethylene (Mw = 133.5, S = 1100 mg/L) and pyrene (Mw = 202, S_w = 150 μg/L) during sorption onto a sediment from an aqueous solution saturated with respect to each compound solution. The retardation factors for the two compounds are, respectively, 2 and 1500. The effective diffusion coefficient in the soil is 85.7 cm²/year for trichloroethylene and 48.3 cm²/year for pyrene.

The concentration 5 cm into the sediment after 100 years of exposure to the pyrene saturated water is

$$C(z,t) = 150\ \mu g/L + (0\ \mu g/L - 150\ \mu g/L)\,\mathrm{erf}\left[\frac{5\ \mathrm{cm}\cdot\sqrt{1500}}{\sqrt{4\cdot 48.3\ \mathrm{cm}^2/\mathrm{year}\cdot 100\ \mathrm{year}}}\right]$$

$$= 7\ \mu g/L$$

The flux at the surface after 100 years is given by

$$q_m(0, 100\ \text{year}) = [0 - 150\ \mu\text{g/L}]\sqrt{\frac{1500 \cdot 48.4\ \text{cm}^2/\text{year}}{\pi \cdot 100\ \text{year}}}$$

$$= -22.8\ \frac{\text{mg}}{\text{m}^2 \cdot \text{year}}$$

where the minus sign indicates the direction of transport is into the sediment. By comparison, trichloroethylene is essentially saturated at 1050 mg/L, 5 cm below the surface after 100 years. The flux at the sediment surface is 7900 mg·m^{-2}·year^{-1} after 100 years. Concentrations are still about 10% of the saturated level more than 1.5 m deep into the sediment. The more rapid transport is largely the result of the much smaller sorption as measured by the retardation factor.

5.6.3 Solutions to the Advection–Diffusion Equation

Advection is often an important mechanism for transport in fluids. Consider first the equivalent to Equation 5.132 with advection and diffusion/dispersion in a single direction the specific equation and boundary conditions are

$$\begin{gathered}\frac{\partial R_f C_f}{\partial t} + U\frac{\partial C_f}{\partial z} = D\frac{\partial^2 C_f}{\partial z^2}\\ C = C_0 \quad \text{at} \quad t = 0\\ C = C_f(0) \quad \text{at} \quad z = 0\\ C \to C_0 \quad \text{as} \quad z \to \infty\end{gathered} \tag{5.135}$$

The solution to this equation subject to the given initial and boundary conditions is

$$C_f(z,t) = C_0 + \frac{(C_f(0) - C_0)}{2}\left[erfc\left(\frac{R_f z - Ut}{\sqrt{4DR_f t}}\right) + \exp\left(\frac{Uz}{D}\right)erfc\left(\frac{R_f z + Ut}{\sqrt{4DR_f t}}\right)\right] \tag{5.136}$$

Here the erfc is the complementary error function given by erfc(η) = 1 – erf(η). Values are also included in Table 5.4. Example 5.21 illustrates the application of this equation to advection/diffusion in soil.

The flux predicted by this equation at z = 0 also is often of interest. This is given by

$$\begin{aligned}q_m\big|_0 &= UC_f(0) - D\frac{dC_f}{dz}\\ &= UC_f(0) + (C_f(0) - C_0)\left[\sqrt{\frac{DR_f}{\pi t}}\exp\left(-\frac{U^2 t}{4DR_f}\right) - \frac{U}{2}erfc\left(\frac{U}{2}\sqrt{\frac{t}{DR_F}}\right)\right]\end{aligned} \tag{5.137}$$

Note that the velocity and flux are assumed positive in the direction of positive z in Equation 5.137. If we consider a sediment from z = 0 to ∞ and we are concerned with the upward (and outward) velocity and flux, the sign of the entire equation and the velocities would change in Equation 5.137.

Example 5.21: Application of advection-diffusion equation solution

An underground storage tank is leaking pure trichloroethylene into groundwater. The groundwater has an effective velocity of 10 m/year toward a water well 100 m away. Estimate the time required for trichloroethylene to reach the well at a concentration 10 and 90% of saturation. Trichloroethylene is found to exhibit a retardation factor of 1.9 and an effective diffusion coefficient of 10^6 cm²/year (10^2 m²/year) in this soil.

At the source of the contamination, water is in equilibrium with the pure trichloroethylene and contains 1100 mg/L of the contaminant. The concentration at the well 100 m away can be estimated from

$$C_f(100m,t)=\frac{1100\text{ mg/L}}{2}\left[erfc\left(\frac{1.9\cdot 100-10\cdot t}{\sqrt{4\cdot 10^2\cdot 1.9t}}\right)+\exp\left(\frac{10\cdot 100}{10^2}\right)erfc\left(\frac{1.9\cdot 100+10\cdot t}{\sqrt{4\cdot 10^2\cdot 1.9t}}\right)\right]$$

Here, all distance units are in meters and time is in years. Determination of the time required to achieve 10% of the saturation concentration (110 mg/L) is by trial and error.

Trial	C_f(100 m,t)
5 years	1.17 mg/L
15 years	409 mg/L
12 years	218 mg/L
10 years	110 mg/L

It is expected that 10 years will be required to achieve concentrations 10% of saturation at the well.

Another commonly used transport equation occurs when contaminant migration occurs in one direction and diffusion or diffusion-like processes cause migration in all coordinate directions. This might be used to represent a point-like source of pollution in an unbounded atmosphere or body of water. The equation and appropriate conditions for this case are

$$\frac{\partial R_f C_f}{\partial t}+U\frac{\partial C_f}{\partial z}=D_x\frac{\partial^2 C_f}{\partial x^2}+D_y\frac{\partial^2 C_f}{\partial y^2}+D_z\frac{\partial^2 C_f}{\partial z^2}$$

$$C_f\to 0 \quad \text{as} \quad x^2+y^2+z^2\to\infty \tag{5.138}$$

$$M_0=\int_V R_f C_f dV$$

The first condition ensures that the concentration far enough away from the point source of contaminant is effectively 0. The second condition ensures that the mass of material introduced at the source equals the total mass in the entire system at any time. The solution to this equation subject to these conditions is given by

$$C(x,y,z,t)=\frac{M_0 R_f^{3/2}}{8(\pi t)^{3/2}\sqrt{D_x D_y D_z}}\exp\left[-\frac{(R_f x-Ut)^2}{4D_x R_f t}-\frac{R_f y^2}{4D_y t}-\frac{R_f z^2}{4D_z t}\right] \tag{5.139}$$

This has the form of a Gaussian distribution function where the standard deviation in a given direction is given by

$$\sigma_i^2 = 2\,D_i t \tag{5.140}$$

Using this definition of a standard deviation, Equation 5.138 can be written

$$C(x,y,z,t) = \frac{M_0 R_f^{3/2}}{(2\pi)^{3/2}\sigma_x\sigma_y\sigma_z}\exp\left[-\frac{(R_f x - Ut)^2}{2\sigma_x^2 R_f} - \frac{R_f y^2}{2\sigma_y^2} - \frac{R_f z^2}{2\sigma_z^2}\right] \tag{5.141}$$

The standard deviations in Equations 5.140 and 5.141 are measures of the spreading of the contaminant in each of the coordinate directions. As indicated in Chapter 3, Section 3.4.3, σ separates the locations that are 10% of the maximum in a standard Gaussian distribution. Thus, the values of the σ parameters directly provide an indication of the width of the dispersing contaminant. In theory, this means that the spread of a dispersing cloud is proportional to the square root of time through Equation 5.140. In the environment, a spread proportional to time is usually a better description of observations, at least at short times. Despite this inadequacy, Equation 5.141 is often used to describe contaminant dispersion in the environment, sometimes with experimentally observed correlations for σ as a function of distance or time rather than Equation 5.140. An illustration of the use of Equation 5.141 is shown in Example 5.22.

In many situations, the spreading in one direction or another may be limited resulting ultimately in the concentration becoming uniform in that direction. In the atmosphere there usually exists a limit to vertical mixing, or the mixing height. In a lake or ocean it might be the bottom or a barrier to significant transport such as a thermocline, which we will discuss in Chapter 7. If H_m represents the region over which the concentration is well mixed (say in the ith direction), Equation 5.141 must be modified by setting $\sigma_i = H_m \sqrt{R_f / 2\pi}$ and eliminating the corresponding exponential term in that direction. This could be applied in as many directions as there are limitations to mixing.

Example 5.22: Application of point source model

Estimate the concentration at the well in Example 5.21 after 10 years if the trichloroethylene source is an essentially instantaneous release of sufficient size to saturate a volume 3 m in diameter (1.5 m radius). The effective diffusion coefficient in the direction of travel is 100 m²/year while in directions perpendicular to the direction of travel it is approximately 10 m²/year. All other conditions remain as in Example 5.21. Assume that the well is directly in the path of the migrating contaminant from the source (i.e., y = z = 0).

The mass of trichloroethylene released, M_0, is

$$M_0 = (1100\text{ mg/L})\cdot[4.3\pi(1.5\text{ m})^3] = 15.55\text{ kg}$$

The concentration after 10 years at a position 100 m from the source is given by

$$C(10\text{ year}) = \frac{(15.55\cdot 10^6)\cdot(1.9)^{3/2}}{8[\pi(10)]^{3/2}\sqrt{(100)(10)(10)}}\exp\left[-\frac{(1.9\cdot 100 - 10\cdot 10)^2}{4\cdot 100\cdot 1.9\cdot 10} - 0 - 0\right]$$

$$= 0.1\text{ mg/L}$$

Due to the spreading in all directions, the concentration at the well is approximately 1000 times smaller after 10 years than in Example 5.21. Repeated evaluation of this model at other times shows that the concentration at the location of the well never exceeds 0.14 mg/L and that the maximum occurs approximately 14 years after the initial release.

After 10 years, using the listed effective diffusion coefficients, the spread in the trichloroethylene plume (as measured by the distances between points 10% of the maximum) is

$$\text{Longitudinal spread} = 4.3\sigma_x = 4.3\sqrt{2D_x t} = 4.3\sqrt{2(100\ \text{m}^2/\text{year})(10\ \text{year})} = 192\ \text{m}$$

$$\text{Lateral spread} = 4.3\sigma_y = 60.8\ \text{m} = \text{Vertical spread}$$

A final variant of Equation 5.130 solves the problem for a continuous release of material at the origin. Under such conditions, the gradient in concentration in the direction of the flow and therefore the diffusion in that direction, is often negligible. That is, the transport in the direction of the velocity is primarily due to advection. Because the source is continuous and diffusion in the direction of flow is neglected, only a steady solution is sought. As a result, transient sorption that causes retardation is not a factor. Under such conditions, the governing equation is given by

$$U\frac{\partial C_f}{\partial x} = D_y\frac{\partial^2 C_f}{\partial y^2} + D_z\frac{\partial^2 C_f}{\partial z^2}$$

$$C_f \to 0 \quad \text{as} \quad x^2 + y^2 + z^2 \to \infty \tag{5.142}$$

$$Q_m = \int_A UC_f dA$$

Here the final condition requires that the integral of the advective flux over a cross-section is given by the rate of contaminant released.

The solution to Equation 5.137 is given by

$$C(y,z,t) = \frac{Q_m}{4\pi\sqrt{D_y D_z}\,x}\exp\left[-\frac{y^2}{4D_y(x/U)} - \frac{z^2}{4D_z(x/U)}\right] \tag{5.143}$$

and in terms of $\sigma_i = \sqrt{2D_i(x/U)}$,

$$C(y,z,t) = \frac{Q_m}{2\pi\sigma_y\sigma_z U}\exp\left[-\frac{y^2}{2\sigma_y^2} - \frac{z^2}{2\sigma_z^2}\right] \tag{5.144}$$

This is the so-called Gaussian plume model which forms the basis for modeling the concentration field away from a continuous point source of pollution into an essentially infinite volume such as an industrial stack releasing into the atmosphere or an industrial or municipal outfall releasing into a large lake or ocean. As indicated earlier, dispersion in one direction may be limited, and the concentration rapidly becomes uniform in this direction. The same modification applied earlier also applies here (recognizing there is no transient sorption in a steady-state problem so that $R_f = 1$). If H_m represents the mixing region or height over which the contaminant is well mixed, Equation 5.144 becomes

$$C(y,z,t) = \frac{Q_m}{\sqrt{2\pi}\,\sigma_y\,H_m\,U}\exp\left[-\frac{y^2}{2\sigma_y^2}\right] \tag{5.145}$$

If the mixing was limited in the lateral direction rather than vertical, Equation 5.145 would still apply if z, the vertical coordinate, replaces y and H_m is taken to represent the lateral mixing range.

The application of any of the models described above assumes that the transport processes can be modeled as either advective with a constant velocity or diffusive with a constant diffusivity. A constant diffusivity model is termed a Fickian diffusion model. Such a model is a crude, but often useful, approximation to the natural environment and to environmental systems. Specific applications of each of these models to problems of contaminant transport or cleanup in air, water, and soil will be investigated in subsequent chapters.

PROBLEMS

1. Consider a contaminant present in a concentration of 10 mg/L in water subject to an average velocity of 0.5 m/s in a stream 1 m deep and 5 m wide. Estimate mass flux of water downstream, mass flow of water, contaminant flux downstream, and contaminant mass flow.

2. Naphthalene has a vapor pressure of 0.037 kPa at 25°C. If the air-side mass transfer coefficient is 10 m/h, what is the rate of evaporation from a pool of pure naphthalene and a pool of a mixture containing 10 mol% naphthalene, if the mixture is kept completely mixed by mechanical stirring?

3. Write a material balance for the mass of apples on a tree if the average density of an individual apple is ρ_A, its volume, V_A, and if the number of apples that drop from the tree per unit time is proportional to the wind velocity, U.

4. Consider a well-mixed vessel with an influent and effluent stream (both of volumetric flowrate Q) containing 2 components, A and B. A reacts to form B at a rate per unit volume of $k_r C_A$. Show that an overall material balance is simply the sum of a material balance on component A and component B.

5. In the simple mixer of Example 5.2, use material balances to show that recycling 10% of the effluent stream does not influence the product stream. What is the new flowrate out of the mixer (prior to the recycle stream draw)?

6. For the absorber of Example 5.3, determine the exit flowrate and concentration (mole fraction) of SO_2 in the liquid if the bottom water is partially cleaned of SO_2 and recycled. The mole fraction of SO_2 in the recycled feed is 5 (10^{-6}).

7. Assume that inlet SO_2 composition to the absorber in Example 5.3 is 0.25. If 99% of the SO_2 must be removed in the absorber, estimate the change in gas flowrate between the inlet and exit conditions. Assuming that the equilibrium relationship remains the same ($y = 33x$) and that the inlet water is free of SO_2, again calculate the exit flowrate and mole fraction in the water.

8. Consider the activated sludge wastewater treatment system of Example 5.4. If the produced biomass was removed from the reactor rather than from the thickened solids, what volume of wastage is required?

9. In the activated sludge wastewater treatment system of Example 5.5, a sudden injection of phenol due to an upset in an upstream processing plant has effectively killed the microbes and significant biodegradation is no longer occurring. The flowrate to the system has increased to 1800 m^3/day. Assuming that the recycle and wastage rates remain as in Example 5.5, estimate the respective proportions of benzene that evaporate and discharged as effluent. Assume quasi-steady or approximately steady conditions.

10. Assume that the stream in Problem 1 feeds a lake initially 10 m deep and 1 ha in area with no outflow. Determine the change in water depth with time. Determine the contaminant concentration in the lake as a function of time assuming the initial concentration in the lake is zero. Assume that the lake is always well mixed.

11. Determine the contaminant concentration in the lake as a function of time if the effluent from the lake is of the same volumetric flow as the influent. Assume that the lake is well mixed.

12. Determine the contaminant concentration in the lake as a function of time if the contaminant concentration in the influent stream is 10 mg/L for a period of 2 days after which time the concentration in the influent stream is 0. Again assume that the lake is well mixed.

13. Determine the contaminant concentration in the lake as a function of time if, in addition to the conditions of Problem 12, the contaminant is subject to an evaporative loss of k_eC_w where $k_e = 0.5$ day^{-1}.

14. Compare the characteristic times of evaporative losses to dilution in the lake of Problem 13.

15. Consider a batch reactor with no flow in or out with a component A subject to a second order reaction, i.e., the rate of disappearance of A is given by $k_rC_A^2$. Derive an equation for the concentration as a function of time in the reactor with $C_A = C_A(0)$ initially.

16. Consider a continuous, stirred-tank reactor with influent and effluent volumetric flowrates, Q. A component A fed to the reactor at concentration C_{A0} is subject to a second order reaction in the reactor. Derive an equation for the concentration as a function of time in the reactor with $C_A = 0$ initially.

17. Consider a continuous, stirred-tank reactor with influent and effluent volumetric flowrates, Q, of an aqueous solution. A component A fed to the reactor at concentration C_{A0} is subject to a first order reaction in the reactor. The reactor contains an air space of V_a and a liquid space, V_w The air space remains in the reactor and is in equilibrium with the water with air-water partition coefficient, K_{aw}, at all times Derive an equation for the concentration as a function of time in the reactor with $C_A = 0$ initially.

18. Show that the effluent concentration from N continuous, stirred-tank reactors of individual volume V_i in which a first order reaction occurs approaches the effluent concentration from a plug-flow (axially unmixed) reactor of volume N V_i.

19. Determine if a first order rate equation is the most appropriate for the data of Problem 3.33. That is, fit n for the rate equation, kC_A^n.

20. Determine the activation energy and preexponential factor for the reaction described by the following rate data.

Temp (K)	**1100**	**1150**	**1200**	**1250**	**1300**	**1350**	**1400**	**1450**	**1500**
Rate cm^3/(mol·s)	23.5	88.5	126	303	363	932	1330	2110	2890

21. Consider a stirred vessel of volume 1000 L initially devoid of a contaminant A to which is introduced a continuous flow of 100 L/h containing 100 mg/L of 1,2-dichloroethane (EDC). Determine the steady-state concentration in the water within the vessel assuming (1) no sorption onto a solid phase and (2) sorption onto a 10% slurry of soil containing 3% organic carbon.

22. If the effluent from the stirred vessel in Problem 21 is filtered to ensure the retention of particles, estimate the steady-state concentration within the vessel with (1) no sorption and (2) with sorption onto the 10% slurry. A first-order reaction with rate constant 1 day occurs within the vessel.

23. Estimate the characteristic time required to achieve steady state in the stirred vessel of Problem 22 for both without sorption and with sorption onto the soil slurry.

24. Estimate the diffusivity of ammonia vapors in air at 25°C. Calculate the characteristic time to diffuse 1 m.

25. Estimate the diffusivity of naphthalene at 25°C in (1) water, (2) oil with a molecular weight of 100 and viscosity of 1 cp, and (3) oil with a molecular weight of 200 with a viscosity of 1000 cp. Calculate the characteristic times required to diffuse 10 cm in each case.

26. Derive Equation 5.93.

27. Under particular wind conditions over a small lake, the air-side mass transfer coefficient is 1000 cm/h and the water-side mass transfer coefficient is 1 cm/h. For what valves of the air-water partition coefficient would you expect (1) water-side controlling, (2) air-side controlling, or (3) both air- and water-side resistances being important?

28. Evaporation to air caused the following water concentration vs. time observations in a laboratory beaker. Why is the rate of evaporation of the two compounds essentially identical? Estimate the water film mass transfer coefficient in the beaker.

	P_v (mmHg)	S_w (mg/L)	Concentration in Water (mg/L) at Various Times (min)						
			t — min	0	20	40	60	80	100
Dichloroethane	426	19800	C_w (mg/L)	1	0.48	0.21	0.1	0.05	0.03
Trichloroethane	123	1300	C_w (mg/L)	1	0.49	0.23	0.13	0.04	0.02

29. Evaporation to air caused the following water concentration vs. time observations in a laboratory beaker. Why are the rates of evaporation of the two compounds substantially different? Estimate both the air- and water-side mass transfer coefficients in the beaker assuming that chloropropene is effectively water-side controlled.

	P_v (mmHg)	S_w (mg/L)	Concentration in Water (mg/L) at Various Times (min)						
			t — min	0	25	50	75	100	125
Chloropropene	361	3370	C_w (mg/L)	1	0.48	0.27	0.13	0.07	0.05
Tetrachloroethane	6.5	3000	C_w (mg/L)	1	0.75	0.55	0.40	0.29	0.22

30. Plot the shape of the wind velocity profile over lawn grass and compare it to the velocity profile over fully grown root crops.

31. Estimate the roughness height of the surface if a wind velocity of 4 m/s is observed at a height of 10 m above the surface while 3 m/s is observed at 2 m above the surface.

32. Estimate the Schmidt number of ammonia in air and of naphthalene in water.

33. A stably stratified and, therefore, stagnant open-topped, deep vessel protected from the wind contains trichloroethane initially uniformly mixed in water at 100 mg/L. Assuming that diffusion to the essentially trichloroethane free air controls the mass transfer, estimate the flux to air vs. time.

34. Under the conditions of Problem 33, estimate the concentration profile of trichloroethane vs. depth in the vessel after 1, 10, and 100 days.

35. For the conditions of Example 5.21, plot the concentration of trichloroethylene as a function of position after 5, 10, and 15 years.

36. For the conditions of Example 5.22, plot the concentration of trichloroethylene as a function of distance from the source along the centerline of the plume, ($y = z = 0$).

37. For the conditions of Example 5.22, plot the concentration of trichloroethylene as a function of the crossflow coordinate, y, after 10 years, at 10 and 100 m downgradient.

38. Repeat Problem 37, assuming that the trichloroethylene source is continuous at 1 kg/year and the source has been releasing long enough that steady conditions apply at both 10 and 100 m downgradient.

39. Estimate the time required to achieve essentially steady conditions at 10 and 100 m downgradient in Problem 38 assuming that the source is continuous.

40. Consider the spill of 100 kg of a pollutant from a ship in a large open bay under low wind and current conditions. The mixing in the vertical direction is limited by a thermocline at 50 m. The horizontal dispersion of the pollutant is given by the standard deviation in concentration of which a variety of field observations have provided an approximation. (Adapted from data in Csanady, 1973.)

$$\sigma_x \sigma_y = 0.08\, t^{2.3} \quad t \text{ in s, } \sigma_x \sigma_y \text{ in cm}^2$$

How long would you expect it to take to observe significant concentrations of the pollutant at a beach 10 km from the spill location? Assume that significant concentration is 10% of the maximum concentration at any time and that the spread occurs equally in all directions.

REFERENCES

Csanady, G.T. (1973) *Turbulent Diffusion in the Environment*, D. Reidel, Boston.

Felder, R.M. and R.W. Rousseau (1986) *Elementary Principles of Chemical Processes*, John Wiley & Sons, New York, 668 pp.

Himmelblau, D.M. (1974) *Basic Principles and Calculations in Chemical Engineering*, 3rd ed., Prentice Hall, New York.

Russell, T.W.F. and M.M. Denn (1972) *Introduction to Chemical Engineering Analysis*, John Wiley & Sons, New York, 502 pp.

Davies, J.T. (1972) *Turbulence Phenomena*, Academic Press, New York.

Thibodeaux, L.J. (1996) *Environmental Chemodynamics, Movement of Chemicals in Air, Water, and Soil,* John Wiley & Sons, New York, 593 pp.

6 Air Pollution and Its Control

6.1 INTRODUCTION

Our goal in this chapter is to apply the knowledge we have gained in the thermodynamics and transport phenomena of environmental processes to understand and control air pollution. Our interest as an environmental engineer is in both contaminant processes in the atmosphere and in treatment technologies designed to reduce or eliminate the introduction of air pollutants to the atmosphere. Together these processes relate the potential source of air pollutants to the concentrations observed by humans and other receptors at the point of exposure. We will begin with a discussion of atmospheric processes and how they influence transport and dilution of air pollutants. We will then examine specific air pollution treatment technologies.

6.2 ATMOSPHERIC PROCESSES

6.2.1 Global Atmospheric Structure and Mixing

The atmosphere of interest to air pollution studies is a very thin layer on the surface of the essentially spherical Earth. This layer of the atmosphere, the *troposphere*, extends only to a height of about 10 km. The temperature in this layer tends to decrease with height at a rate of about 6.5°C/km. The lower 1 to 2 km of the troposphere constitutes the planetary boundary layer which is directly influenced by the diurnal or daily temperature variations at the surface. The troposphere is the layer of primary interest for air pollutants since the inhabitants of the Earth spend all of their lives within this layer. Even commercial air travelers are limited to the *tropopause*, the region at the top of the troposphere forming the boundary with the next highest layer, the stratosphere.

The *stratosphere* is also influenced by air pollution and, more importantly, the effects can negatively impact the residents of the Earth's surface. The stratosphere is the region of the atmosphere between about 10 and 50 km above the surface. The temperature is approximately constant in the lower reaches of the stratosphere but increases at a rate of 2–3°C/km in the middle and upper portions of the layer. The importance of the stratosphere in air pollution studies is largely the result of the presence of ozone in this layer. Ozone, one of the most important air pollutants at the Earth's surface, serves a beneficial role in the stratosphere by absorbing significant amounts of short-wave (ultraviolet, UV) radiation from the sun, as discussed in Chapter 2. It is the absorption of this radiation which gives rise to the slow increase of temperature with height in the upper reaches of the stratosphere.

The temperature variation with height is an important indicator of density in the atmosphere and, as a result, largely defines the degree of mixing between layers of the atmosphere. Figure 6.1 shows the temperature variation with height in the lower layers of the atmosphere and Table 6.1 shows the temperature and pressure changes with height in the U.S. Standard Atmosphere (1976)* from which the figure is drawn.

Also shown in Figure 6.1 is an indication of the variation of eddy diffusivity with height. As indicated in Chapter 5, the eddy diffusivity is an indicator of the rate of mixing in that it characterizes the intensity of turbulent mixing with an effective diffusion coefficient. In the lower 1 to 2 km of atmosphere that constitutes the planetary boundary layer the eddy diffusivity is a strong function of height, atmospheric conditions, and surface heating. Under midday, sunny conditions the eddy

* U.S. Standard Atmosphere (1976) National Oceanic and Atmospheric Administration, Washington, D.C.

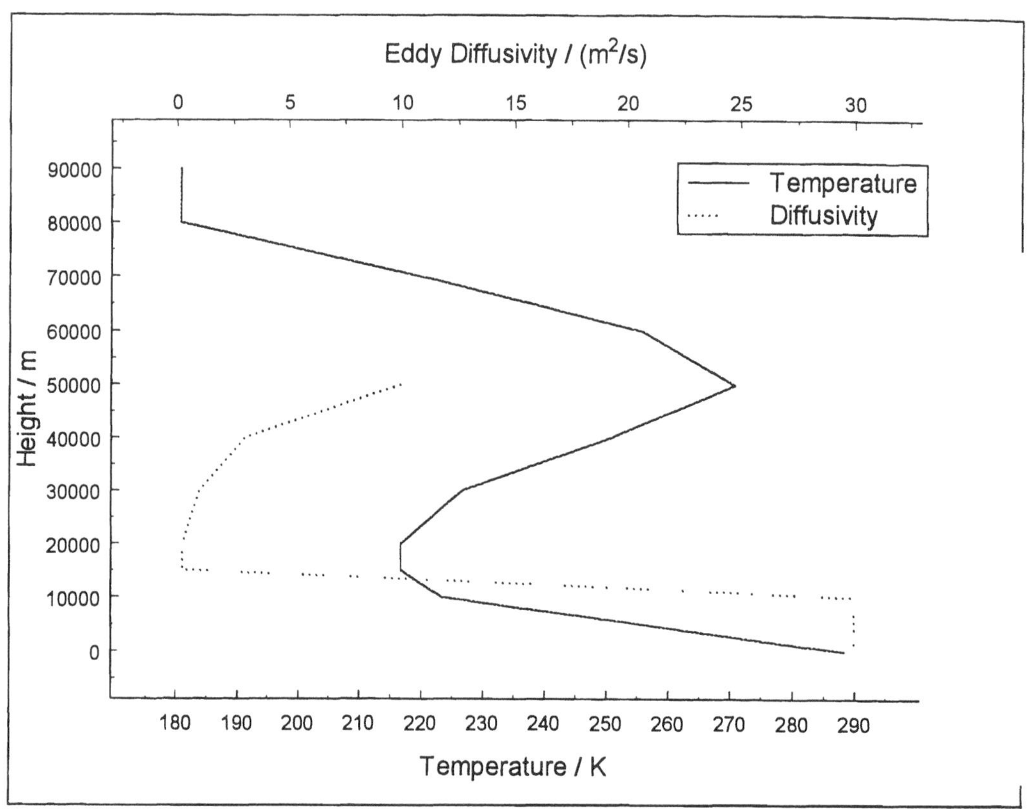

FIGURE 6.1 Temperature and eddy diffusivity profile in the U.S. Standard Atmosphere.

diffusivity can be as much as 1000 m^2/s or more. This can be compared to a molecular diffusion coefficient of about 0.1 cm^2/s (0.1×10^{-4} m^2/s). Above this layer the deep tropospheric eddy diffusivity is approximately constant and of the order of 20 m^2/s (Warneck, 1988). Immediately above the tropopause, the eddy diffusivity drops abruptly to 0.2 to 0.4 m^2/s. As a result surface-emitted pollutants tend to mix relatively rapidly over the first 1 to 2 km, at least during sunny, daytime conditions, somewhat more slowly over the remainder of the troposphere, and only very slowly into the stratosphere.

Figure 6.2 shows vertical profiles of carbon monoxide, hydrogen, and ozone concentration over a location in western France in 1972. The major sources of carbon monoxide are at the Earth's surface and carbon monoxide is found in higher levels in the troposphere. At the tropopause, the rapid decrease in vertical mixing rates results in a sharp concentration gradient to the much lower values in the stratosphere. With the exception of the region in the immediate vicinity of the tropopause, both the stratosphere and the troposphere exhibit approximately uniform concentrations. As indicated previously local concentrations near source areas such as cities and in the very lowest 1 to 2 km of the atmosphere would not be well mixed with the rest of the troposphere. Let us assume that the bulk of both layers are well mixed, and develop a model of the atmosphere as shown in Figure 6.3.

Let φ_S represent the volume or molar fraction of a particular contaminant in the stratosphere and φ_T represents the volume or molar fraction of the same contaminant in the troposphere. These are volume or molar ratios representing the volume filled by the contaminant divided by the volume of air in the stratosphere and troposphere, respectively. Then, if Q represents the volume of air exchanged between the stratosphere and the troposphere per unit time, then $Q\varphi_S$ represents the contaminant exchange from the stratosphere to the troposphere and $Q\varphi_T$ represents the contaminant

TABLE 6.1
U.S. Standard Atmosphere

Altitude (m)	Temperature (K)	P/P_0 ($P_0 = 1.013$ Pa)	ρ_a/ρ_0 ($\rho_0 = 1.225$ kg/m^3)
–500	291.4	1.061	1.049
0	288.2	1.000	1.000
500	284.9	0.9421	0.9529
1000	281.7	0.8870	0.9075
1500	278.4	0.8345	0.8638
2000	275.2	0.7846	0.8217
2500	271.9	0.7372	0.7812
3000	268.7	0.6920	0.7423
4000	262.2	0.6085	0.6689
5000	255.7	0.5334	0.6012
6000	249.2	0.4660	0.5389
8000	236.2	0.3519	0.4292
10,000	223.3	0.2615	0.3376
15,000	216.7	0.1195	0.1590
20,000	216.7	0.05457	0.07258
30,000	226.5	0.01181	0.01503
40,000	250.4	0.002834	0.003262
50,000	270.7	0.0007874	0.0008383
60,000	255.8	0.0002217	0.0002497
70,000	219.7	0.00005448	0.00007146
80,000	180.7	0.00001023	0.00001632
90,000	180.7	0.000001622	0.000002588

Adapted from Fox, R.W. and A.T. McDonald (1978) *Introduction to Fluid Mechanics*, 2nd. ed., John Wiley & Sons, New York.

exchange in the opposite direction. The accumulation in the stratosphere is given by the time rate of change of $V_S\varphi_S$. Thus, a simple box model of the atmosphere suggests that the stratospheric and tropospheric contaminant concentrations are related by

$$\frac{d(V_S\varphi_S)}{dt} = Q(\varphi_T - \varphi_S)$$
$$\frac{d(V_T\varphi_T)}{dt} = Q(\varphi_S - \varphi_T) \tag{6.1}$$

or, for a constant stratospheric and tropospheric volume, this can be written

$$\frac{d\varphi_S}{dt} = \frac{(\varphi_T - \varphi_S)}{\tau_S}$$
$$\frac{d\varphi_T}{dt} = \frac{(\varphi_S - \varphi_T)}{\tau_T} \tag{6.2}$$

Here $\tau_i = Q/V_i$ is a measure of the time required to move the contaminant from layer i to the other layer. Note that $\tau_S \neq \tau_T$ since the volumes of the two layers are different. If a contaminant has a

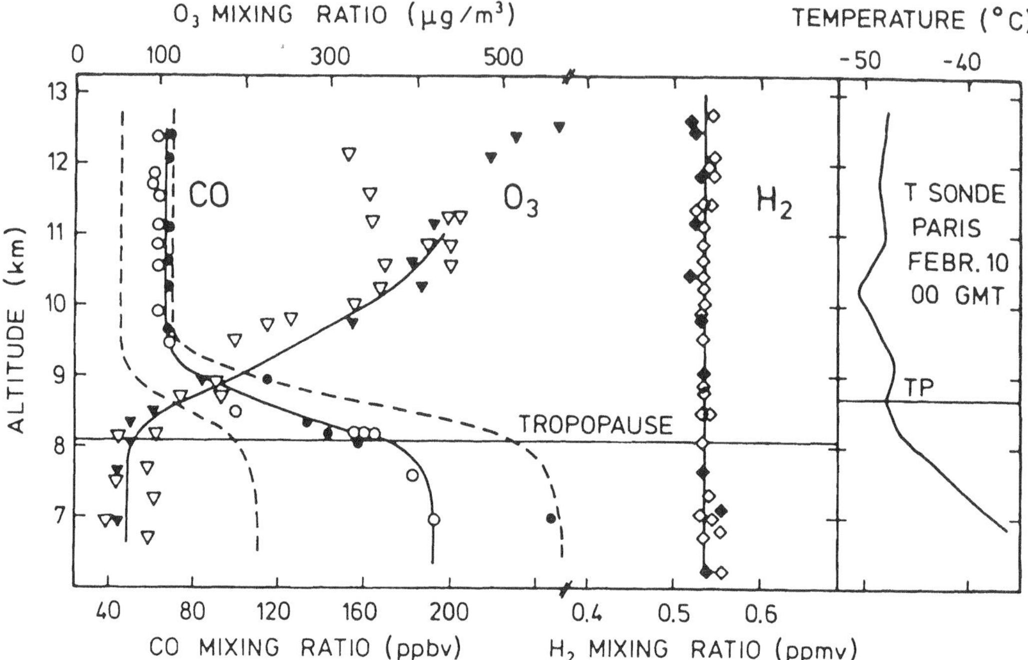

FIGURE 6.2 Vertical profiles of carbon monoxide, ozone, hydrogen, and temperature over western France on February 9–10, 1972. (From Warneck, P. (1988) *Chemistry of the Natural Atmosphere*, Academic Press, New York. With permission.)

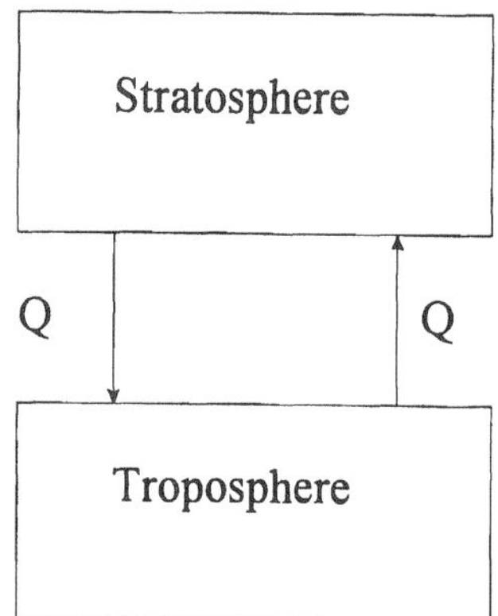

FIGURE 6.3 Conceptual model of the atmosphere assuming that both the stratosphere and troposphere are well mixed with Q defining the flow between the two layers over time.

source in either the stratosphere or the troposphere and if φ_T and φ_S are monitored as a function of time, Equation 6.2 can be used to estimate the layer exchange times.

In particular, let us consider a contaminant released directly into the stratosphere that is rapidly removed by tropospheric processes. Under these conditions, the tropospheric concentration of the contaminant can always be assumed to be negligible and Equation 6.2 becomes

$$\frac{d\varphi_S}{dt} \approx -\frac{\varphi_S}{\tau_S} \tag{6.3}$$

which has the solution, assuming that the initial volume fraction in the stratosphere is given by φ_{S0},

$$\varphi_S = \varphi_{S0} \exp\left(-\frac{t}{\tau_S}\right) \tag{6.4}$$

Strontium-90 was injected directly into the stratosphere during surface testing of atomic weapons. Strontium-90 has a radioactive decay half-life of 28 years so it is essentially inert over the time required to mix throughout the atmosphere. Reiter (1975) reported the total quantity of strontium-90 in the northern and southern stratospheres as a function of time between 1963 and 1967. The strontium-90 monitored during this period was released between 1958 and 1962 and no additional tests were conducted until 1967. Equation 6.4 suggests that if the Strontium-90 is well mixed it should decay exponentially between 1963 and 1967. Equation 6.4 can be written

$$\ln(^{90}Sr) = \ln(^{90}Sr_0) - \left(\frac{1}{\tau_S}\right)t \tag{6.5}$$

Thus, a plot of the logarithm of the amount of strontium in the atmosphere vs. time will yield a straight line with intercept (at $t = 0$) of the initial strontium concentration and with slope of $1/\tau_S$. As shown in Figure 6.4, the logarithm of the strontium vs. time is a straight line. The slope of the total quantity of strontium on a semi-logarithmic scale is about 0.77 year^{-1} which corresponds to a characteristic exchange time of 1.3 years. Warneck (1988) compared this estimate to a variety of other estimates of the characteristic exchange time between the stratosphere and troposphere and all of the estimates were clustered about 1.4 years. Thus atmospheric pollutants injected into the stratosphere such as strontium-90 or gases from a strong volcanic eruption will mix downward to the troposphere with a characteristic exchange time of 1.4 years. As indicated by Equation 6.4, the stratospheric concentration will drop to e^{-1} or 37% of its initial value within this period of time.

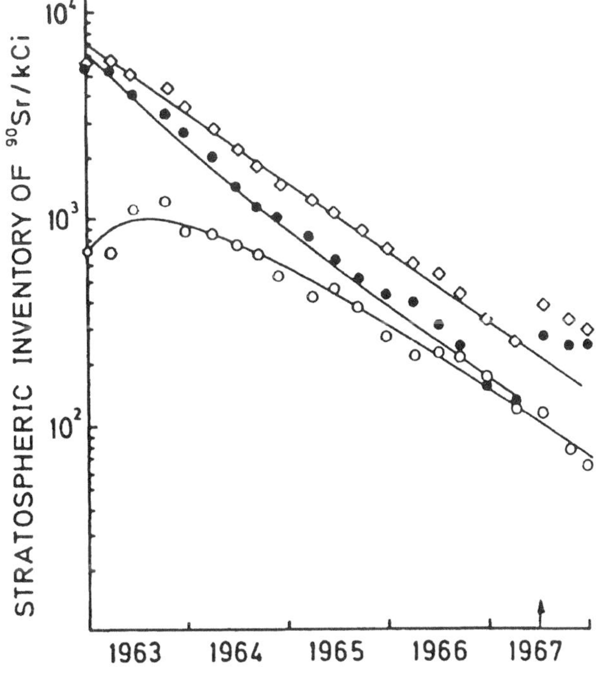

FIGURE 6.4 Inventory of strontium-90 in the stratosphere following the injection by nuclear weapons tests. The arrow indicates a new injection due to Chinese tests in June 1967, ● — northern stratosphere, ○ — southern stratosphere, and ◇ — sum of both. (From Warneck, P. (1988) *Chemistry of the Natural Atmosphere*, Academic Press, New York. With permission.)

Similarly pollutants injected into the troposphere will mix upward to the stratosphere at a rate governed by Equation 6.3. The solution of the equation assuming that there is initially none of the contaminant in the stratosphere is

$$\varphi_S = (\varphi_S)_\infty \left(1 - \exp\left[-\frac{t}{\tau_S}\right]\right) \tag{6.6}$$

where $(\varphi_S)_\infty$ is the ultimate concentration of the contaminant in the stratosphere at long times (i.e., steady-state concentration). Again the characteristic exchange time is given by τ_S and thus after 1.4 years the concentration in the stratosphere reaches 63% of its final value ($1 - e^{-1}$).

Note that the characteristic exchange time of 1.4 years refers only to the response of the stratosphere to contaminants introduced from or lost to the troposphere. The troposphere contains a much greater mass of air and the concentration in the troposphere changes at a much slower rate. For an air contaminant mixing downward into the troposphere, the tropospheric concentration of the contaminant would remain close to a constant over the 1.4 year period despite the loss of 63% of the contaminant in the stratosphere. Similarly, a contaminant mixing upward into the stratosphere would reach 63% of its final value in the upper atmosphere in a time of order 1.4 years, but during that time the tropospheric concentration would remain approximately constant.

Although mixing times between the troposphere and stratosphere are slow compared to times required to mix within the troposphere, they remain small compared to the lifetimes of chlorofluorocarbons. It is for this reason that chlorofluorcarbons have sufficient opportunity to mix to the stratosphere and attack the ozone layer, as discussed in Chapter 2.

6.2.2 Global Circulation

Although the simple well-mixed box model of the troposphere and stratosphere presented above illustrates the rate of exchange between these two layers of the atmosphere, it does not recognize the large spatial variations in source strength and atmospheric concentrations over the face of the Earth. In particular, the bulk of the atmospheric contaminants associated with anthropogenic activities are released over the land masses in the Northern Hemisphere. The atmospheric circulation patterns are such that these contaminants can mix throughout the Northern Hemisphere over a period of weeks to months, but only slowly will they migrate upward to the stratosphere (with a characteristic exchange time of about 1.4 years as demonstrated above) and across the equator to the Southern Hemisphere.

Figure 6.5 shows the mean circulation patterns that are observed in the atmosphere. The major feature is the so-called *Hadley cell* near the equator in both the Northern and Southern Hemispheres that results from the intense solar heating in this region. Heating of the air near the equator decreases its density causing it to rise relative to the cooler air in the more temperate regions. Air moves toward the equator at the surface to fill the void left by this rising air. In addition the air moving aloft moves away from the equator, forced poleward by the continued addition of rising air from the surface. As this air moves poleward it cools, becomes more dense, and falls toward the surface to complete the cycle. This flow directed toward the equator at the surface gives rise to convergence near the equator at what is termed the *interhemispheric tropical convergence zone*. It is this zone that acts as a barrier to rapid transport between the two hemispheres. Airborne material carried toward the equator at the surface is carried aloft before moving into the other hemisphere. This convergence is not always directly over the equator but moves back and forth as the Hadley cells in each hemisphere strengthen or weaken. During June, July, and August, the sun is directed north of the equator and the northern hemispheric Hadley cell is stronger and causes the zone of convergence to move into the Southern Hemisphere. The opposite occurs during January and February.

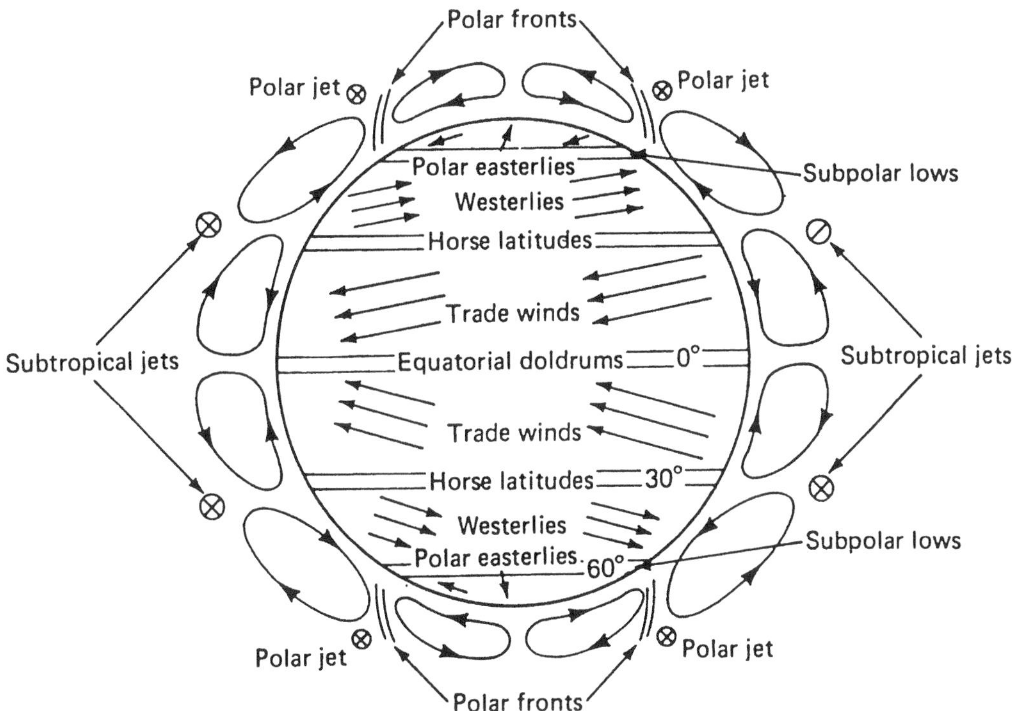

FIGURE 6.5 Mean circulation patterns in the atmosphere. (From Seinfeld, J. (1975) *Air Pollution: Physical and Chemical Fundamentals*, McGraw-Hill, New York. With permission.)

The Hadley cell is also a major factor in climate. Air cools as it rises so that the equatorial air loses much of the moisture it contained at the surface during the upward portion of the cycle. This results in large amounts of rainfall near the equator and deserts near 30° latitude where this dried air returns to the surface.

The limited mixing between the Northern and Southern Hemispheres is illustrated in Figure 6.4, in which Northern Hemisphere strontium levels are much larger than the amount of strontium in the Southern Hemisphere in 1963 and 1964. The atomic weapons testing was conducted in the Northern Hemisphere and the delay in mixing across the equator is graphically illustrated by the southern hemispheric levels.

Because the strontium-90 levels shown in Figure 6.4 are stratospheric levels, this data cannot be used to indicate interhemispheric transport in the troposphere. Warneck (1988) illustrates that carbon dioxide can be used to estimate interhemispheric transport since about 90% of the carbon dioxide emitted to the atmosphere is emitted into the northern troposphere. A model similar to that used for stratospheric-tropospheric transport can be employed, given by

$$\frac{d(V_n\varphi_n)}{dt} = Q(\varphi_n - \varphi_s) + Q_m$$
$$\frac{d(V_s\varphi_s)}{dt} = Q(\varphi_s - \varphi_n) \tag{6.7}$$

Q_m is the carbon dioxide emission rate, here assumed entirely within the Northern Hemisphere, and Q is the volumetric exchange rate between the Northern and Southern Hemispheres. All quantities refer to the troposphere. The subscripts n and s refer to the Northern and Southern

Hemispheres, respectively. Losses of carbon dioxide to the oceans or to the stratosphere are neglected. For constant and equal tropospheric volumes in the northern and Southern Hemispheres, this can be written

$$\frac{d\varphi_n}{dt} = \frac{(\varphi_s - \varphi_n)}{\tau_T} + 2\frac{Q_m}{V_T}$$
$$\frac{d\varphi_s}{dt} = \frac{(\varphi_n - \varphi_s)}{\tau_T} \tag{6.8}$$

where V_T is the volume of the entire troposphere. The term τ_T represents the characteristic exchange rate between the tropospheres (V/Q). The sum and difference of the two equations can be written

$$\frac{d\varphi_n}{dt} + \frac{d\varphi_s}{dt} = 2\frac{Q_m}{V_T}$$
$$\frac{d(\varphi_n - \varphi_s)}{dt} = -\frac{2(\varphi_n - \varphi_s)}{\tau_T} + 2\frac{Q_m}{V_T} \tag{6.9}$$

Estimation of the characteristic exchange time between the two tropospheres then proceeds by use of Figure 6.6. The time derivative or slope of the carbon dioxide concentration with time is essentially identical in the two hemispheres and thus the difference between the two tropospheric concentrations is approximately constant. Under these conditions, the equations in 6.9 can be written

$$2\frac{d\varphi_n}{dt} = 2\frac{Q_m}{V_T}$$
$$0 = \frac{2(\varphi_n - \varphi_s)}{\tau_T} + 2\frac{Q_m}{V_T} \tag{6.10}$$

The average rate of increase of carbon dioxide from Figure 6.6 is 0.64 ppm/year while the average concentration difference between the Northern and Southern Hemispheres is 0.66 ppm. By combining these two equations, the estimated characteristic exchange time between the northern and southern tropospheres is given by

$$\tau_T = \frac{\varphi_n - \varphi_s}{\frac{d\varphi_n}{dt}} = \frac{0.66 \text{ ppm}}{0.64 \text{ ppm/year}} = 1.03 \text{ year} \tag{6.11}$$

Thus, the transport of air pollutants between the northern and southern tropospheres takes place at a slightly faster rate than the transport between the troposphere and the stratosphere. Note that the assumption that the losses from the troposphere to the stratosphere are negligible is valid even though the calculated exchange times are approximately the same. This is again the result of the greater mass in the troposphere which moderates the influence of exchange with the stratosphere.

The assumption that the global atmosphere is composed of distinct well-mixed compartments is clearly a crude approximation. The approximation has proven quite useful, however, in the estimation of global budgets of a variety of air contaminants whose primary source is the Northern Hemisphere. Once the exchange rate between these compartments is accepted on the basis of one atmospheric tracer, it can be applied to, for example, to estimate the source strength of other

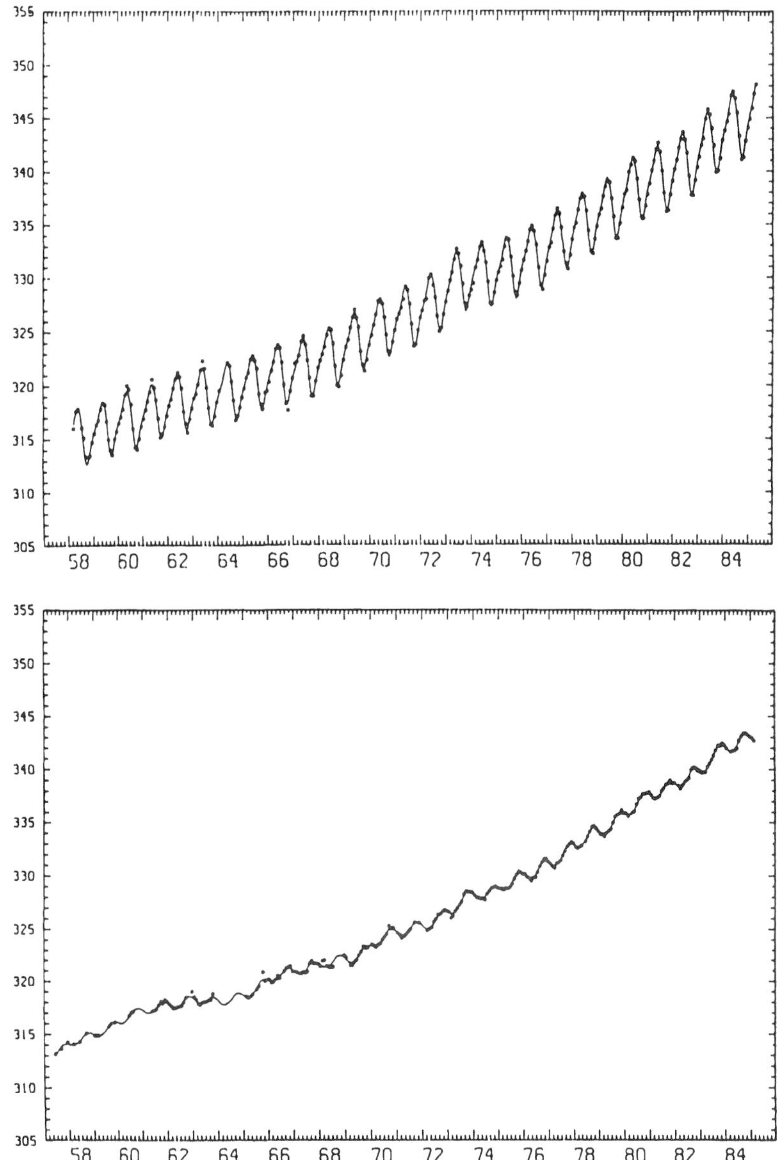

FIGURE 6.6 Carbon dioxide levels vs. time in the Northern and Southern Hemispheres. (From Warneck, P. (1988) *Chemistry of the Natural Atmosphere*, Academic Press, New York. With permission.)

contaminants from observed average concentration levels. Generally the interhemispheric gradient of a particular contaminant is the most useful indication in that through the second of the equations in 6.10, this gradient is related directly to the source strength in the Northern Hemisphere. It is also possible to use this type of analysis for reactive contaminants by modifying these model equations. These extensions are employed in the problems at the end of this chapter.

The ability of this approach to model global atmospheric contaminant budgets depends on atmospheric processes that result in mixing times within the individual hemispheric tropospheres or stratospheres that are less than the characteristic exchange time between the hemispheres or between the troposphere and the stratosphere. As a result of a variety of processes to be discussed below, this approximation is generally valid.

6.2.3 Regional Processes

The preceding section focused on atmospheric processes that occur over global scales, that is tens of thousands of kilometers. Regional processes are those that occur over scales of tens of kilometers to, at most, a few thousand kilometers. Included among these processes are the large-scale frontal systems that dominate weather patterns in most inland areas. In coastal and mountainous areas, the weather patterns are further influenced by local buoyancy driven winds that cause the land-sea breeze circulation and the mountain-valley breezes. Our interest here is not to develop a detailed understanding of local weather patterns but only to summarize the key processes involved and how they influence contaminant transport and dispersion.

Atmospheric flows are driven by pressure gradients which are themselves largely driven by variations in heating of the Earth's surface. The variations that drive regional atmospheric flows are primarily seasonal or due to variations in the albedo and temperature response of the Earth's surface. Exposed land surface temperatures respond relatively quickly to solar heating due to the slow conduction of heat away from the surface. Forested areas and bodies of water respond much more slowly. The resulting variation in surface temperatures are largely responsible for large scale atmospheric motions and weather.

The air motion that responds to these large scale pressure gradients is modified by friction at the surface and Coriolis forces. Coriolis forces cause the apparent motion resulting from the variations in angular momentum on the surface of the rotating Earth. The angular momentum of an object (or parcel of fluid) is the product of the radius of curvature and the velocity of motion on the Earth's surface. Near the poles, the radius from the axis of rotation is less and the angular momentum is therefore smaller than near the equator (see Figure 6.7). Fluid moving toward the equator, or southward, in the Northern Hemisphere encounters fluid of greater angular momentum and its apparent motion is to lag the surrounding air or move to the right. The greater angular momentum of air (or water) moving northward in the Northern Hemisphere again causes a veering motion to the right. The identical effects occur during movement in the opposite direction in the Southern Hemisphere. Because motion always tends to veer right in the Northern Hemisphere and to the left in the Southern Hemisphere, motion of sufficient speed and scale will tend to form a circular motion. Thus, the flow responding to a high-pressure region in the Northern Hemisphere will initially flow outward toward regions of low pressure, but Coriolis forces will cause the flow to rotate in a clockwise (also referred to as *anti-cyclonic*) fashion around the region of high pressure. Motion around low-pressure zones in the Northern Hemisphere is counterclockwise and the system is referred to as a cyclone. Thus, interestingly, flow in the atmosphere (and large ocean circulations) is not along the direction of maximum pressure gradients but instead tends to their perpendicular. This is true except in the near-surface layer where friction tends to reduce the veering in the wind.

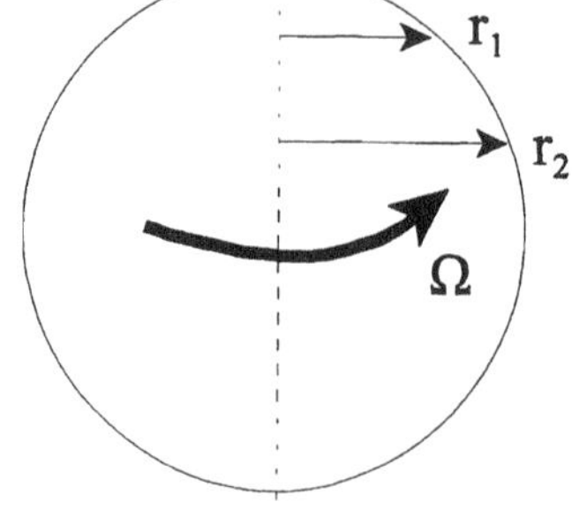

FIGURE 6.7 Diagram illustrating the reduction in angular momentum as one moves poleward on the surface of a rotating sphere.

Mathematically, the horizontal momentum balance presented in Example 5.19 must be revised to account for the additional source term from Coriolis force. This force is proportional and perpendicular to the velocity and in the Northern Hemisphere is given by fU_y in the easterly direction and $-fU_x$ in the northerly direction. The Coriolis parameter is termed f and is given by two times the rate of rotation of the Earth times the sine of the latitude. Thus, the Coriolis parameter (and Coriolis forces) are zero at the equator and about 10^{-4} s^{-1} at mid-latitudes.

Example 6.1: Values of the Coriolis parameter

Estimate the value of the Coriolis parameter as a function of latitude

If Θ represents the latitude and Ω the rate of rotation of the Earth, the Coriolis parameter is given by

$$f = 2\,\Omega \sin \Theta$$

The angular rate of rotation of the Earth is 2π radians per 24 h or 1.45×10^{-4} radians per second. The Coriolis factor is then given by

Latitude	Coriolis Parameter, f (s^{-1})	Latitude	Coriolis Parameter, f (s^{-1})
0°	0	50°	1.11×10^{-4}
10°	2.53×10^{-5}	60°	1.26×10^{-4}
20°	4.97×10^{-5}	70°	1.37×10^{-4}
30°	7.27×10^{-5}	80°	1.43×10^{-4}
40°	9.35×10^{-5}	90°	1.45×10^{-4}

The horizontal momentum balances incorporating Coriolis forces are given by

$$\frac{\partial U_x}{\partial t} = -\frac{1}{\rho}\frac{\partial P}{\partial x} + v_t \frac{\partial^2 U_x}{\partial z^2} + fU_y \tag{6.12}$$

$$\frac{\partial U_y}{\partial t} = -\frac{1}{\rho}\frac{\partial P}{\partial y} + v_t \frac{\partial^2 U_y}{\partial z^2} - fU_x \tag{6.13}$$

Here x and y represent the coordinate directions in the easterly and northerly directions, respectively. In both of these equations it has been assumed that diffusion of horizontal momentum is only important in the vertical direction. Momentum transport is also assumed to be completely governed by turbulent transport modeled with eddy kinematic viscosities.

Often the hydrostatic equation (discussed earlier in Chapter 3) is used to write these in a different form. The hydrostatic equation is given by

$$\frac{\partial P}{\partial z} = -\rho g \tag{6.14}$$

Recognizing that the pressure gradients in Equations 6.12 and 6.13 can be written

$$\begin{aligned} \frac{\partial P}{\partial y} &= -\frac{\partial P}{\partial z}\left(\frac{\partial z}{\partial y}\right)_P & \qquad \frac{\partial P}{\partial x} &= -\frac{\partial P}{\partial z}\left(\frac{\partial z}{\partial x}\right)_P \\ \frac{\partial P}{\partial y} &= \rho g\left(\frac{\partial z}{\partial y}\right)_P & \qquad \frac{\partial P}{\partial x} &= \rho g\left(\frac{\partial z}{\partial x}\right)_P \end{aligned} \tag{6.15}$$

where terms such as $(\partial z/\partial y)_P$ represent the slope of lines of constant pressure (in this case the change in elevation in the y, or easterly, direction. Using this approximation, the horizontal momentum equations become

$$\frac{\partial U_x}{\partial t} = -g\left(\frac{\partial z}{\partial x}\right)_P + v_t \frac{\partial^2 U_x}{\partial z^2} + fU_y$$
$$\frac{\partial U_y}{\partial t} = -g\left(\frac{\partial z}{\partial y}\right)_P + v_t \frac{\partial^2 U_y}{\partial z^2} - fU_x \tag{6.16}$$

These equations describe the horizontal wind field in the presence of atmospheric pressure gradients and Coriolis forces as modified by turbulent diffusivity. Two common simplifications of Equation 6.16 exist, the geostrophic flow equations and the Eckman spiral.

6.2.3.1 Geostrophic Flow

Geostrophic flow represents the atmospheric wind field that would result in the steady-state balance of Coriolis forces and a pressure gradient, i.e., neglecting surface friction and turbulent transport of momentum to the ground. Under these conditions, the time derivatives and the turbulent diffusion terms in Equations 6.16 are neglected and the geostrophic winds in the easterly and northerly direction are respectively,

$$U_{xg} = -\frac{g}{f}\left(\frac{\partial z}{\partial y}\right)_P \qquad U_{yg} = \frac{g}{f}\left(\frac{\partial z}{\partial x}\right)_P \tag{6.17}$$

Note that the easterly geostrophic wind (U_{xg}) and the northerly geostrophic wind (U_{yg}) depend on the slope of the constant pressure surface in their respective directions. That is, the previous assertion that the wind moves perpendicular to the pressure gradients in the absence of surface friction and turbulent diffusion is affirmed.

Equation 6.17 is written in terms of the slope of the constant pressure lines. Upper air data is often defined on maps of constant pressure. For example, a 500 mbar map provides the elevations where the pressure is 500 mbar, approximately half the pressure at the ground. On such a map, lines of constant elevation provide the local slope of the constant pressure lines in much the same way a topographic map provides indications of local slope of the land surface.

6.2.3.2 Eckman Spiral

As you move closer to the ground, the motion of the winds is no longer strictly perpendicular to pressure gradients. This can be seen by solving the steady-state form of the equations in 6.16 but retaining the turbulent diffusion terms that transport momentum to the ground as a result of friction.

$$0 = -g\left(\frac{\partial z}{\partial x}\right)_P + v_t \frac{\partial^2 U_x}{\partial z^2} + fU_y$$
$$0 = -g\left(\frac{\partial z}{\partial y}\right)_P + v_t \frac{\partial^2 U_y}{\partial z^2} - fU_x \tag{6.18}$$

Replacing the terms involving the slope of the constant pressure surface with the geostrophic velocities, this can be written

$$0 = v_t \frac{\partial^2 U_x}{\partial z^2} + f(U_y - U_{yg})$$
$$0 = v_t \frac{\partial^2 U_y}{\partial z^2} - f(U_x - U_{xg}) \tag{6.19}$$

If we choose the x-coordinate such that U_x approaches the geostrophic wind at infinite height, i.e., choose the coordinate system such that $U_{yg} = 0$, then the solution to these equations defines an *Eckman spiral* that can be written

$$U_x(z) = U_{xg}(1 - e^{-az} \cos az)$$
$$U_y(z) = U_{xg}(e^{-az} \sin az) \tag{6.20}$$
$$a = \left(\frac{f}{2v_t}\right)^{1/2}$$

Example 6.2: Direction of the Eckman spiral near the ground

Evaluate the direction of the Eckman spiral winds near the ground's surface.

At large distances from the ground, we know that the Eckman spiral approaches the geostropic wind. In this case, this means that the winds approach U_{xg} in both speed and direction at height. Near the ground, however, $\cos(az) \to 1$ and $\sin(az) \to az$. Then

$$U_x = U_{xg}\, az$$
$$U_z = U_{xg}\, az\, (1 - az)$$

Thus, for small values of az, the velocity in both the eastward and northward directions approach the same value, $U_{xg}az$, and the resultant wind is 1.4 $U_{xg}az$ and directed at 45° between the two components.

As shown in Example 6.2, this velocity profile is shifted 45° from the geostrophic wind at the ground. As the elevation above the ground is increased the predicted velocities in an Eckman spiral increase and shift toward the geostrophic wind. The wind direction predicted by the Eckman spiral is shown in Figure 6.8. The magnitude and direction of the winds slightly overshoot the geostrophic values then match as $z \to \infty$. The depth of the Eckman boundary layer is of order "1/a." As indicated earlier, the value of f in the mid-latitudes is of the order of 10^{-4} s^{-1} while average values of v_t are of the order 50 to 100 m^2/s (but highly variable). This means that the depth of an Eckman boundary layer is of the order of 1000 m. This layer is often termed the *atmospheric boundary layer.*

The Eckman boundary layer equations are only a crude approximation of the behavior in the actual atmosphere due primarily to the assumption of a constant turbulent viscosity. At higher altitudes, the atmosphere typically exhibits a "stable" layer that reduces the turbulent diffusion of mass and momentum and leads to a decoupling between the geostrophic winds and the surface winds. The observed shift between the geostrophic and ground level winds typically varies between 5 and 50° depending upon atmospheric conditions and terrain. Note that this shift in both speed and direction has important consequences for pollutant transport in the atmosphere in that elevated pollutant plumes are transported in a manner different from ground-based emissions. Alternatively, transport of pollutants over long distances results in interactions with the entire surface layer of

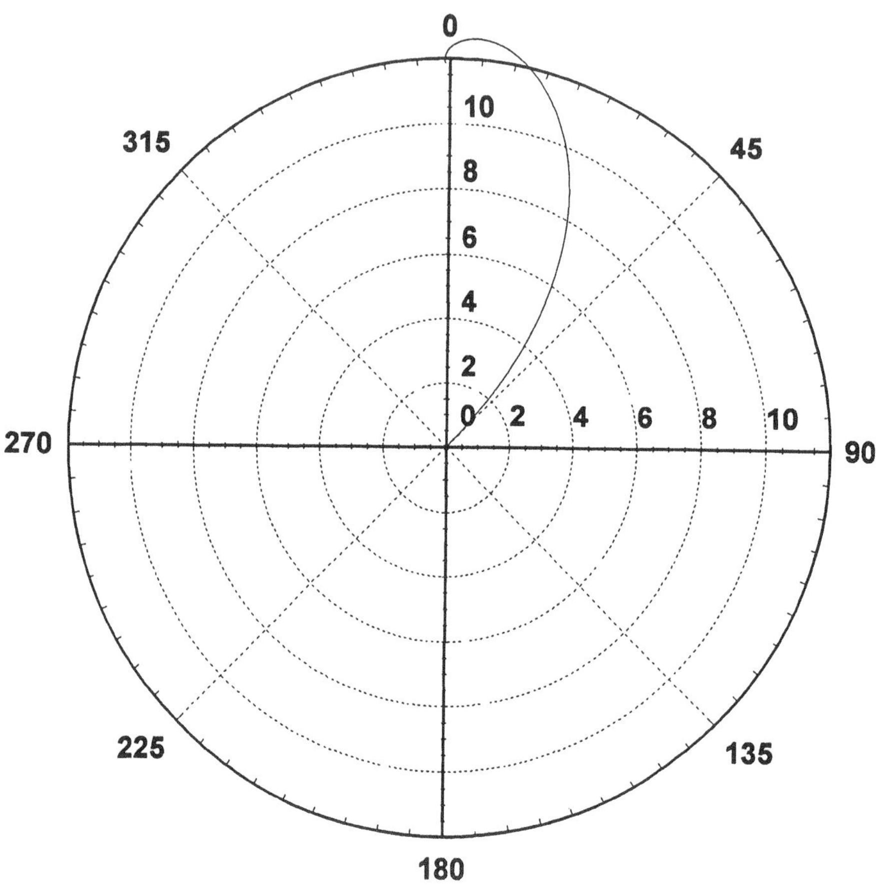

FIGURE 6.8 Polar diagram indicating the 45° shift between the ground and the top of an Eckman boundary layer. The line represents the magnitude and direction of the resultant wind vector as a function of height in the Eckman layer.

the atmosphere and portions of the pollutants are advected in different directions and speeds then the remaining pollutants.

6.2.3.3 Local Winds

The description of winds developed above also neglected the effect of local topography and geography. These features influence the direction and strength of the winds. Terrain tends to channel flow and direct winds along lines of constant elevation. In addition, variations in the albedo and heat capacity of various surfaces result in temperature variations in the lower layers of the atmosphere causing buoyancy driven flows to develop. In a mountain-valley system the upper elevations are heated more rapidly during the day than at lower elevations. The warmer air aloft rises and is replaced by cooler air from below. The result is an upslope flow during the day. At night the more rapid cooling aloft results in a cool, dense layer of air that moves downhill in a downslope flow. A similar process occurs at the land-sea interface. Bodies of water can absorb heat and mix it effectively downward resulting in limited changes in temperature at the surface. The limited thermal conductivity of the land surface causes the heat absorbed to be concentrated near the surface. During the day, the land surface gets much hotter than an adjacent water surface. The warmer air over the land is less dense, rises, and is replaced by the cooler, more dense air over the waterbody. At night, the reverse flow occurs since the cooling over the land is much more rapid.

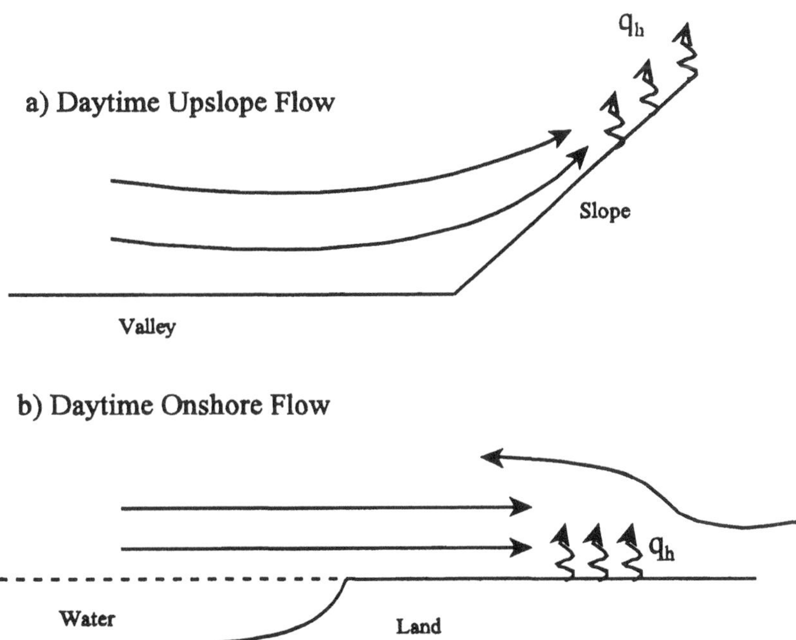

FIGURE 6.9 Depiction of local winds: (a) daytime upslope flow and (b) daytime onshore flow.

Both of these local flows are shown in Figure 6.9. Figure 6.9a shows the mountain-valley breeze during the daytime while Figure 6.9b shows the daytime land-sea breeze.

6.2.4 Observed Structure and Mixing in the Atmospheric Boundary Layer

6.2.4.1 Boundary Layer Temperature Structure

As indicated previously, the Eckman boundary layer is overly simplistic in that it assumes that the turbulent diffusion coefficient is independent of height and it neglects surface heating or cooling effects. The lower layer of the atmosphere is influenced by surface friction and the diurnal variations in temperature on the Earth's surface, while geostrophic winds are typically observed aloft. We defined the height of the planetary boundary layer previously on the basis of the Eckman spiral. More often, the height of the boundary layer is defined by surface heating and cooling and the limited effect of these perturbations at high altitudes.

In order to understand the observed velocity and mixing behavior in the lower atmosphere we must discuss its temperature structure and how this influences turbulent intensity. The diurnal heating and cooling of the Earth's surface gives rise to the typical atmospheric structure depicted in Figure 6.10. At night, a comparatively cool and dense layer of air tends to stagnate in the near surface environment. As day breaks, the land begins to warm causing *thermally driven convection,* the rise of warmed air from the surface due to its buoyancy relative to the air aloft. This warmed air will rise and mix the lower layers of the atmosphere. The transport process that causes the mixing of this layer is referred to as convection in the atmospheric sciences literature to differentiate it from *advective* transport by pressure driven motion such as winds.* The depth of the layer that is mixed is termed the *mixing layer* or *convectively mixed layer* and it will continue to grow unless a layer of warm and therefore less dense air overlies the surface layer.

* Some disciplines do not differentiate between transport driven by buoyancy or pressure gradients and use the terms advection and convection interchangeably. In chemical and mechanical engineering, buoyancy driven flows and transport are referred to as natural or free convection, but the term convection also is used to indicate transport by a forced flow.

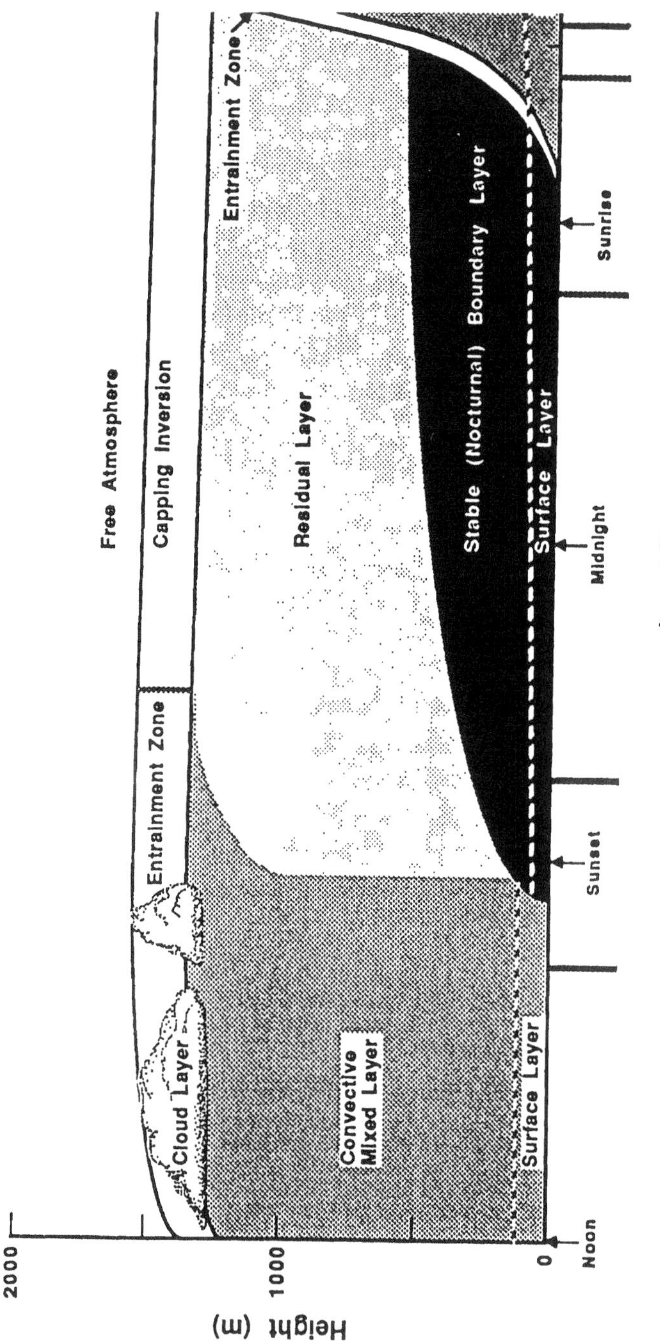

FIGURE 6.10 Diurnal evolution of the atmospheric boundary layer. (From Stull, R.B. (1988) *An Introduction to Boundary Layer Meteorology*, Kluwer Academic Publishers, Boston. With permission.)

Normally the temperature of the atmosphere decreases with altitude but in a layer of warm air aloft it may actually increase, a phenomena termed *temperature inversion.* An elevated inversion layer tends to serve as a cap on the boundary layer in that transfer of heat, mass, and momentum is hindered by the limited turbulent intensity within this layer. As indicated in Chapter 5, a warm layer of fluid overriding a cooler, more dense layer is stable and vertical motions tend to be damped. Because the land surface continues to warm until late afternoon, the penetration of the thermal motions into the atmosphere typically increases during the day, deepening the convectively mixed layer. As evening arrives, the cooling of the Earth's surfaces erodes this mixing layer from below. That is, the coolest and most dense air will again be found at the Earth's surface. Because the coldest air is at the surface the temperature tends to increase with altitude giving rise to a temperature inversion, in this case a *nocturnal inversion.*

The heat transfer that gives rise to the destruction of the elevated inversion and the growth of the mixing layer during the day is an important indicator of the intensity of turbulence in the daytime mixing layer. As indicated in Chapter 5, the vertical flux of heat is given by

$$(q_h)_z = \rho C_p u_z' T' \tag{6.21}$$

We also can speak of the vertical flux of buoyancy which is the source of the energy for the thermal rise. Note that the buoyancy force per unit mass is given by $g(\Delta T/T)$. Thus, the vertical flux of buoyancy is

$$b_z = \frac{g}{T}(u_z' T') \tag{6.22}$$

Note that the growth in the mixing layer height, H_m, is dependent upon the buoyancy flux at the surface, b_{z0}, and the intensity of vertical motion in the convective boundary layer, is usually described by a characteristic convection velocity, w_*. The units of these variables are

$$\begin{aligned} &b_{z0} : (L \cdot t^{-2})(T^{-1})(L \cdot t^{-1})(T) = L^2 \cdot t^{-3} \\ &H_m : (L) \\ &w_* : (L \cdot t^{-1}) \end{aligned} \tag{6.23}$$

Dimensional analysis then suggests that the only possible dimensionless grouping of these variables is $w_*^3/b_{z0}H_m$. This suggests in turn that the characteristic velocity scale for the motion in the convective boundary layer is

$$w_* = (b_{z0} H_m)^{1/3} = \left(\frac{g}{T}(u_z' T')_0 H_m\right)^{1/3} \tag{6.24}$$

w_* is a useful indication of the rise velocity in the unstable convective boundary layer. Most importantly, it suggests that in a time given by H_m/w_*, the convective boundary layer approaches a completely mixed state. If H_m is about 1000 m and w_* is about 2 m/s (a typical midday value under sunny conditions), this means that only about 500 s or less than 10 min is required for the convective boundary layer to become mixed over its entire depth. The reason for its description as a convectively mixed layer becomes clear. This has important implications for pollutant mixing which will be explored in a subsequent section.

Chapter 5 related the intensity of mixing to temperature profiles in a fluid. In particular, a uniform vertical temperature profile was said to be neutrally stable in that vertical motions would neither be enhanced nor damped. The only source of turbulence in such a situation is the friction at the Earth's surface which is reflected in velocity shear in the lower layers of the atmosphere. When cool fluid overlays warmer fluid, however, turbulence and vertical mixing are enhanced (unstable conditions giving rise to the daytime convectively mixed layer described above). When the opposite occurs, that is when warmer fluid overlays cooler, more dense fluid, and mixing was hindered (stable conditions, as might occur during the night).

This simple picture of stability as a function of fluid temperature and density is somewhat more complicated in the actual atmosphere. During the day, for example, as warm air rises from the surface, it encounters lower pressures and expands. This results in work (PΔ-type work) being done on the atmosphere. During the rapid rise of an air parcel such that it cannot exchange heat with its surroundings (adiabatic rise), the only source of the energy required for this expansion must come at the expense of the sensible heat or temperature of the fluid parcel. That is, the temperature of a parcel of air rising from the surface drops typically 6 to 9°C/km of altitude. This makes it difficult to compare temperatures at the surface with temperatures aloft in order to determine atmospheric stability. To overcome this problem, it is more appropriate to evaluate stability on the basis of *potential temperature*, the temperature that the air would exhibit if it were brought to sea level. By comparison, at a common pressure, temperature defines density and buoyancy directly. The simple picture of stability presented in Chapter 5 is then valid when potential temperature is considered, not the actual temperature.

In order to define potential temperature, we must estimate the actual rate of temperature decrease with altitude. Let us consider a dry atmosphere and apply the first law of thermodynamics to a rapidly ascending air parcel that does not exchange heat with its surroundings.

$$\begin{aligned} \delta E_t &= \partial Q - \delta W \\ C_v dT &= 0 - Pd\hat{V} \end{aligned} \tag{6.25}$$

Recognizing that $d(P\hat{V}) = Pd\hat{V} + \hat{V}dP$ and, from the ideal gas law, that $dP\hat{V} = d(RT) = RdT$,

$$\begin{aligned} 0 &= (C_v + R)dT - \hat{V}dP \\ &= C_p dT - \frac{1}{\tilde{\rho}} dP \end{aligned} \tag{6.26}$$

since also from the ideal gas law, $\tilde{\rho} = \dfrac{P}{RT}$,

$$\begin{aligned} C_p dT &= \frac{dP}{P/RT} \\ \frac{dT}{T} &= \frac{R}{C_p}\frac{dP}{P} \\ \frac{dT}{T} &= \frac{C_p - C_v}{C_p}\frac{dP}{P} \\ \ln\frac{T}{T_0} &= \left[1 - \frac{C_v}{C_p}\right]\ln\frac{P}{P_0} \end{aligned} \tag{6.27}$$

$$\frac{T}{T_0} = \left[\frac{P}{P_0}\right]^{1-\frac{C_v}{C_p}}$$

The temperature of dry air decreases as the pressure is decreased. The quantity $1-(C_v/C_p)$ is about 0.288 for dry air. This also can be written in terms of height by the hydrostatic equation,

$$\frac{dP}{dz} = -\hat{\rho}\, g$$
$$-\frac{1}{\hat{\rho}}\, dP = g\, dz \tag{6.28}$$

After dividing through by dz and solving Equation 6.28 for the vertical temperature gradient,

$$\frac{dT}{dz} = -\frac{g}{C_P} = -9.8\frac{°\text{C}}{\text{km}} \tag{6.29}$$

This means that dry air rising from the surface will cool at a rate of 9.8°C per km. This is termed the *dry adiabatic lapse rate*, Γ, in that it refers to adiabatic motions in a dry atmosphere. For moist air, the specific heat is different and, in addition, the latent heat associated with condensation must be taken into account. The net effect is that water-saturated air near the tropics tends to cool at the lower rate of about 6.5°C/km. For intermediate moisture contents, the temperature decrease with height would be between 6.5 and 9.8°C/km.

We are now in a position to define potential temperature, the temperature that air would exhibit if referenced to sea level and 1 atm pressure. The potential temperature, θ, can be calculated from the temperature, T, at altitude z

$$\theta = T + \Gamma z \tag{6.30}$$

or, from the temperature at pressure, P

$$\theta = T\left[\frac{1\ \text{atm}}{P}\right]^{0.288} \tag{6.31}$$

The potential temperature of an air parcel remains constant during a change in elevation. An atmosphere that exhibits a uniform potential temperature would be expected to be neutrally buoyant and neither stable nor unstable conditions would prevail. If the potential temperature were to increase with height, this would mean that warmer, less dense air would override cooler, more dense air and any vertical motions would be damped (stable conditions). Conversely, if the potential temperature were to decrease with height, the cooler, more dense air aloft would tend to mix downward in the unstable conditions. A quantitative parameter that characterizes this phenomenon is the Brunt-Vaisala frequency, f_{BV}, defined by

$$f_{BV} = \left(\frac{g}{\theta_a}\frac{d\theta}{dz}\right)^{1/2} \tag{6.32}$$

This quantity, which has units of radians per second, is a measure of the frequency of oscillations about an equilibrium point that would be exhibited by an air parcel in a stable atmosphere.

Remember that an air parcel would always tend to return to its original position in a stable environment and the rate of that return (or frequency) is dependent upon the strength of the stability.

We now see that in the atmosphere, the temperature changes associated with a rising air parcel mean that the neutrally stable atmospheric condition in which turbulence is neither enhanced nor damped is one in which the background temperature gradient decreases at the same rate as air rising from the surface, that is 9.8°C/km. When the background temperature decreases more slowly or even increases with height the atmosphere is stable since the rising air parcel would cool more rapidly and then become more dense than its surroundings as it rises. Air in which the background temperature decreases more rapidly with height is unstable since the rising air parcel would cool less rapidly than the surroundings and be less dense than its surroundings as it rises.

The relationship of temperature, potential temperature, and stability is depicted in Figure 6.11. Figure 6.11a shows the potential temperature and actual temperature as a function of height in a neutrally stable atmosphere. The actual temperature would decrease at 9.8°C/km (dry air) while the potential temperature is uniform. An air parcel moving up or down in such an atmosphere would always exhibit the same temperature and potential temperature as its surroundings and the vertical motion would not be enhanced or damped.

The temperature and potential temperature profiles observed under stable conditions are shown in Figure 6.11b. The temperature and potential temperature of the surroundings would always be greater than for an upward moving air parcel and less than for a downward moving air parcel. In both cases the vertical motion is stable since there is a tendency to return to the original position.

Figure 6.11c shows the unstable condition when vertical upward motions would cause an air parcel to always be warmer and less dense than its surroundings and therefore continue to rise. Heated surface air will continue to rise as long as the air it encounters is of lesser potential temperature (and therefore more dense than the rising air). Thus, the mixing height at any time is the height at which the air aloft exhibits the same potential temperature as the surface level. Due to turbulent mixing under unstable conditions, essentially all of the air within the mixing layer is effectively at the same potential temperature.

Let us apply the above discussion to a diurnal variation in temperature above the surface. Figure 6.12 shows the actual and the potential temperature profile measured in a hypothetical atmosphere in early morning. The condition shows a nocturnal temperature inversion at the surface which would be poorly mixed because of its stability. Above this layer is a deep layer that is essentially unchanged from temperatures on the previous day and above which is depicted a semi-permanent elevated temperature inversion. As the surface temperature increases due to solar heating, the surface air rises until its potential temperature equals that of its surroundings.

The convective motion causes erosion and ultimately elimination of the surface nocturnal inversion. As the surface temperature continues to rise, the convective process begins to erode the elevated temperature inversion. The mixing height is that region which has been mixed to a constant potential temperature. The maximum mixing height is the height that would be reached by the air warmed to the maximum surface temperature, that is, the height of air with a potential temperature equal to the maximum surface temperature. This approach to estimating the growth of the mixing layer during the day is illustrated in Example 6.3.

Example 6.3: Estimation of maximum mixing height from a morning temperature sounding

Estimate the mixing height over the course of the day given the following 6 a.m. (0600) vertical temperature profile and daytime surface temperatures. The temperature gradient is linear between 200 and 700 m.

The 0600 vertical temperature profile indicates a nocturnal inversion at the surface (temperature increases continuously until at least 200 m). Between 200 and 700 m the average rate of temperature

Stratified Atmospheres

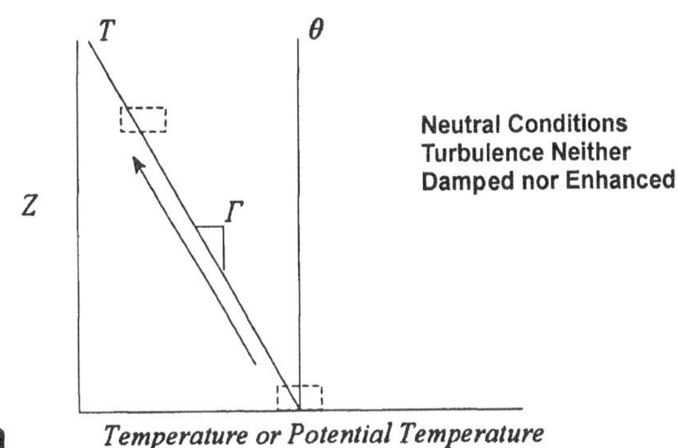

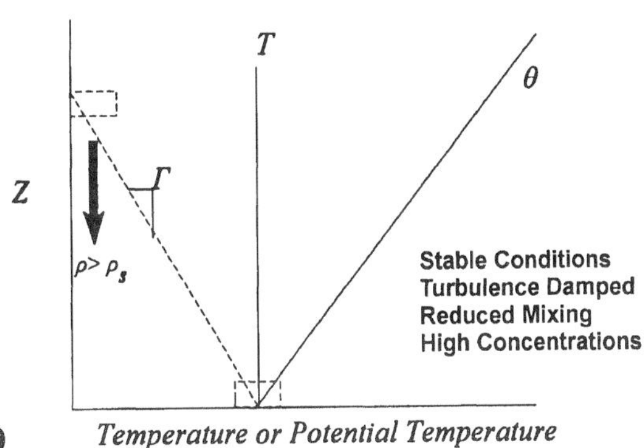

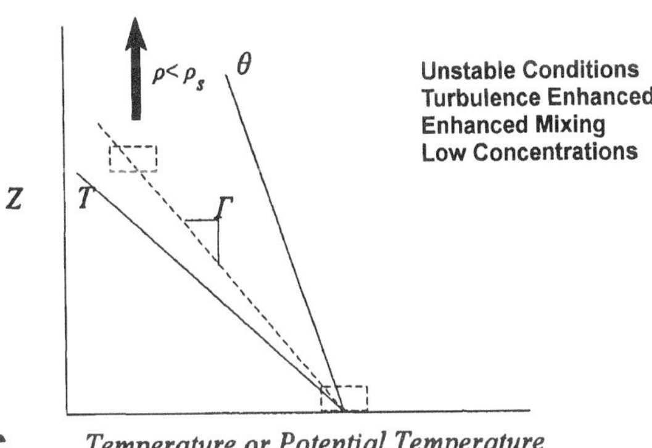

FIGURE 6.11 Temperature, potential temperature, and stability in the atmosphere.

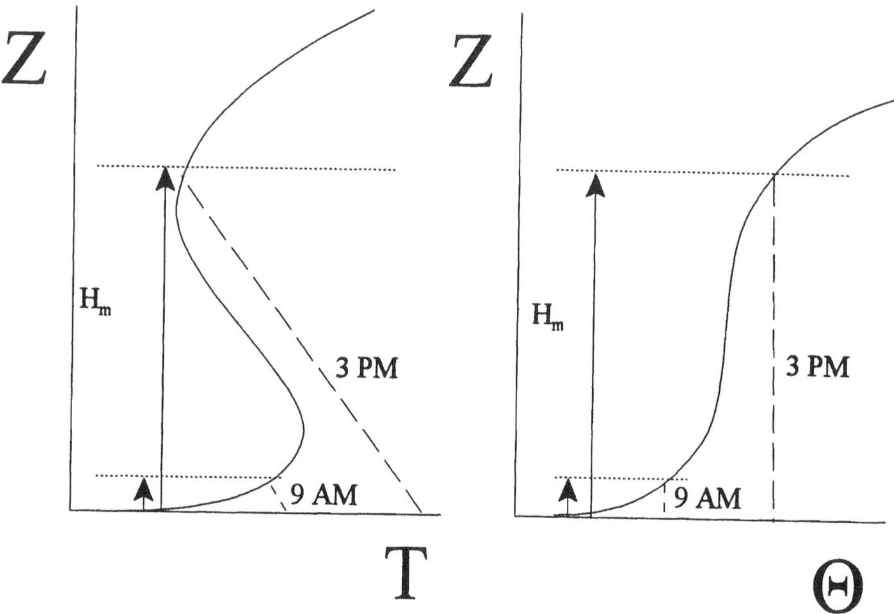

FIGURE 6.12 Hypothetical temperature and potential temperature profile as measured at 6 a.m. and estimated mixing heights at 9 a.m. and 3 p.m. based upon extrapolation from surface temperature.

Elev (m)	T (°C)	Elev (m)	T (°C)	Time	T(z = 0)	Time	T(z = 0)
10	12	700	11	0600	12	1600	24
50	14	750	11	0800	14	1800	23
100	15	800	12	1000	16	2000	22
150	16	850	14	1200	20	2200	20
200	16	900	16	1400	22	2400	18

decrease is 10°C/km, or approximately adiabatic. Above 700 m an elevated temperature inversion is observed. The general shape of the temperature profile is similar to that depicted in Figure 6.12.

There is no well-defined mixing height at 0600 and 0800 and there is no appreciable upward mixing of surface air. By 1000 the mixing height would be expected to be about 100 m since the potential temperature of the air at 100 m (15 + 9.8°C/km × 0.1 km = 16°C) equals the surface temperature. By 1200, the mixing height is about 800 m since the surface temperature is equal to the potential temperature of air at that altitude (12 + (9.8)(0.8) = 20°C). The maximum mixing height is about 875 m at 1600 since the potential temperature at that altitude is approximately 24°C, the same as the maximum surface temperature. During the subsequent evening cooling of the surface it is difficult to define the mixing height and another temperature sounding is needed to clearly define it. Because the cooling occurs at the surface the nocturnal inversion begins to form, reducing mixing into the layers above the surface layer. Note, however, that any pollutants contained within the air aloft may still be returned to the surface layer during the growth of the mixing layer during the subsequent day unless the nighttime winds effectively purge the upper layers.

Note also the rapid increase in mixing height during late morning followed by the very slow growth during late afternoon due to the strong elevated temperature inversion. Although the particular pattern of growth observed varies with the particular meteorological conditions, the depicted behavior is not atypical for many locales subject to a semi-permanent elevated inversion (e.g., the Los Angeles basin).

6.2.4.2 Boundary Layer Velocity Profile

We are now in a position to examine the velocity behavior in the lower atmospheric boundary layer. As indicated previously, the velocity profile near any environmental interface follows a logarithmic velocity profile. In the absence of significant surface heating and cooling, horizontal winds near a relatively flat surface are well approximated by this profile. That is, the wind speed, U, as a function of height, z, is described by

$$U = \frac{u_*}{\kappa} \ln\left(\frac{z}{z_0}\right) \tag{6.33}$$

where u_* is a friction velocity given by the square root of the surface shear stress divided by density, $\sqrt{\tau_0/\rho_a}$, and z_0 is the surface roughness height. κ is the von Karman constant of approximately 0.4. Both u_* and z_0 must be determined from observations, usually by measurements of wind velocity and fitting to the logarithmic profile under conditions for which it is applicable. Also as indicated previously, the effective turbulent diffusion coefficient is given by $\kappa u_* z$ in this surface layer where the logarithmic velocity law is valid.

The picture presented above assumes that friction at the surface and the resulting wind shear (i.e., velocity gradient) is the dominant source of turbulence. This picture must be modified when there is substantial heating and cooling near the Earth's surface. The parameter that defines the importance of heating and cooling effects in comparison to wind shear is the Richardson number (Equation 5.120). The Richardson number, however, is the ratio of the consumption of turbulence by buoyancy to the generation of turbulence by velocity shear and these quantities are dependent upon height above the surface. It is often easier to think of the height when these quantities are approximately equal, the Monin-Obukhov length scale, L_{MO}. We seek the relationship between the thermally generated buoyancy flux and surface friction. Using dimensional analysis

$$\begin{aligned} L_{MO} &= \frac{u_*^3}{\kappa\left(\dfrac{g}{T}\right)\left(-u_z'T'\right)} \\ &= \frac{-\rho C_p u_*^3}{\kappa\left(\dfrac{g}{T}\right)q_h} \end{aligned} \tag{6.34}$$

where q_h is the surface heat flux ($= \rho C_p u_z' T'$). The von Karman constant is included here for consistency with other's definition of L_{MO} and because it arises in a more formal analysis using the turbulent energy balance. Because it is a dimensionless constant, it does not affect our analysis and L_{MO} could be equally well defined without its inclusion.

The Monin-Obukhov length scale takes on small negative values under very unstable conditions (large positive heat flux, air being warmed) and small positive values under very stable conditions (large negative heat flux, air being cooled). Under near neutral conditions, the parameter is infinite. The absolute value of the Monin-Obukhov length scale provides a direct indication of the height of the surface layer in which friction is important and the layer over which the logarithmic velocity profile is approximately valid. In very stable or unstable conditions, L_{MO} may only have a magnitude of 10 m or less, and the thickness of the layer in which the logarithmic velocity profile is valid is similarly small. Under neutral conditions, however, L_{MO} is infinite and the logarithmic velocity profile is uniformly valid throughout the atmospheric boundary layer.*

The relationship between the Richardson number (N_{Ri}) and the Monin-Obukhov length scale is

$$\frac{z}{L_{MO}} = \frac{N_{Ri}}{(1-5N_{Ri})} \qquad \text{stable conditions}$$
$$= N_{Ri} \qquad \text{unstable conditions} \tag{6.35}$$

Note that, as suggested by the stable condition equation of Equation 6.35, the Richardson number is limited to a positive value of about 0.2. At this value, turbulence effectively ceases and $L_{MO} \rightarrow 0$. For Richardson's number near 0 and $L_{MO} \rightarrow \infty$, heating and cooling effects are unimportant and Equation 6.33 is valid as long as the terrain is flat. As indicated previously, in diabatic, or other than neutral conditions, the logarithmic profile is approximately valid only up to height $|L_{MO}|$. Above this height, the velocity profile tends to be approximately constant in the unstable convectively mixed boundary layer. A somewhat more complicated picture arises under stable conditions. Corrections for the velocity profile as a function of Richardson number or Monin-Obukov length scale have been suggested, but for our purposes it is sufficient to analyze the effect of surface heating and cooling more simply through empirical power law formulas and crude stability classification schemes.

6.2.4.3 Empirical Turbulence Classification Schemes

The state of the atmosphere and the magnitude of the Richardson number can be inferred from the degree of cloud cover and solar intensity and the wind speed. Cloud cover during the night reduces surface radiative cooling while during the day, cloud cover reduces solar *insolation.* This means that buoyancy forces are negligible and the Richardson number is near zero which describes neutral atmospheric conditions. During the day with low cloud cover and high solar insolation, surface heating drives unstable conditions, Richardson numbers that are large and negative, and intense turbulence. During the night, low cloud cover and rapid radiative cooling of the Earth's surface causes cold surface conditions and mild turbulence. Conversely, under low wind speed conditions, the gradient in the wind near the surface is also low, the denominator of the Richardson number is small, and the intensity of turbulence is governed entirely by the buoyancy forces. Under high wind speeds, however, the velocity gradient near the surface is typically also high and buoyancy forces are small relative to the denominator of the Richardson number and the generation of turbulence by wind shear.

These semi-empirical indications of the Richardson number give rise to the Pasquill-Gifford stability classifications which characterize the infinity of stability stratifications observed in the actual atmosphere with a small number (6) of conditions from highly unstable (A) through neutral (D) to very stable (F). The relationship between atmospheric conditions and these classifications are shown in Table 6.2.

Figure 6.13 shows a more quantitative relationship between the Pasquill-Gifford stability classification and Monin-Obukhov length scale.

Having defined a turbulence classification scheme it becomes possible to relate a variety of atmospheric properties to this scheme. Two of the most important properties are the velocity profile and the eddy diffusivity variation with height. Under neutral conditions, the velocity profile is logarithmic and the eddy diffusivity linear with height, as described previously. For non-neutral or diabatic conditions it is useful to employ a power law fit to the profiles.

* Note that neutral conditions rarely exist throughout the entire boundary layer. Generally the atmosphere aloft is slightly stable in the absence of significant surface heating or cooling.

TABLE 6.2
Pasquill-Gifford Stability Classification

Wind Speed (10 m–m/s)	Day Incoming Solar Radiation: Strong	Moderate	Weak[a]	Night Cloud Cover: Mostly Overcast[a] (>4/8 Cloud Cover)	Mostly Clear (<3/8 Cloud Cover)
<2	A	A-B	B	E	F
2–3	A-B	B	C	E	F
3–5	B	B-C	D	D	E
5-6	C	C-D	D	D	D
>6	C	D	D	D	D

[a] Neutral or D stability should be employed during overcast conditions, day or night.
From Turner, D.B. (1969) *Workbook of Atmospheric Dispersion Estimates*, HEW, Washington, D.C.

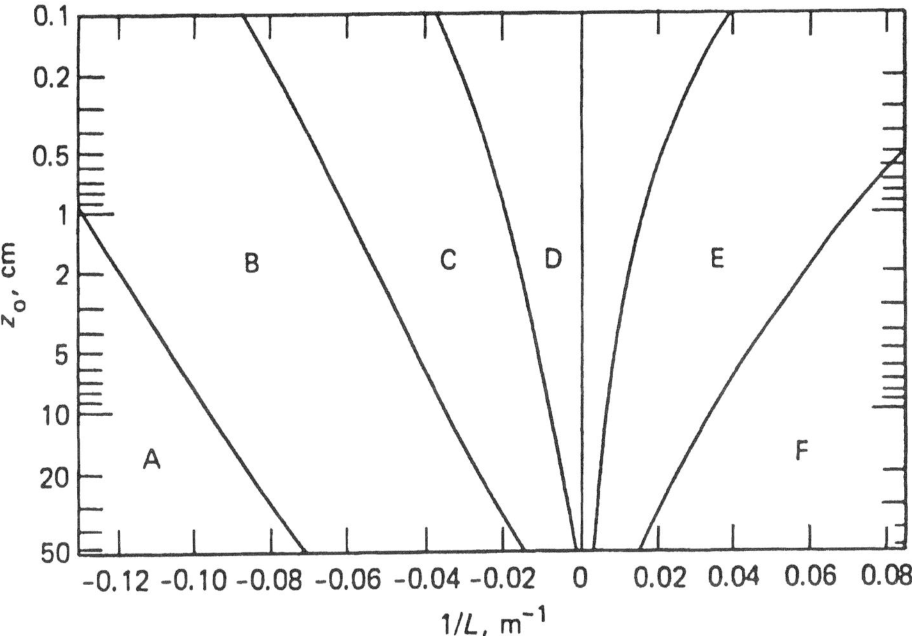

FIGURE 6.13 The relationship between the Monin-Obukhov length scale and Pasquill-Gifford stability class. (After Golden, D. (1972) Relations among stability parameters in the surface layer, *Bound. Layer Meteorol.*, 3:47-58. With permission.)

The power law profile is expressed by the following relationship

$$\frac{U(z)}{U(z_1)} = \left(\frac{z}{z_1}\right)^{\alpha} \tag{6.36}$$

Here U(z) is the wind velocity at height z as estimated from the known velocity $U(z_1)$ at height z_1. In flat, open terrain without significant surface heating effects, α is about $^1/_7$ giving rise to what is called the $^1/_7$ power law. The shape of such a curve is quite close to that predicted by the logarithmic profile of Equation 6.33. Under stable conditions and in urban built-up areas, the velocity profile tends to be more curved ($1/2 > \alpha > 1/7$) while under unstable conditions, the velocity profile tends to be more "flat" ($1/15 < \alpha < 1/7$). Values of the velocity power law exponent as a function of Pasquill-Gifford stability classifications and urban vs. rural conditions can be found in Table 6.3.

The vertical transport of heat, mass, and momentum tends to be governed by an eddy diffusivity that is linear with height in the neutral atmosphere ($\kappa u_* z$ from the logarithmic law). Under *diabatic* conditions (i.e., those other than neutral or adiabatic), a power law profile can also be used to indicate the near-surface behavior of the eddy diffusivity.

$$\frac{D_t(z)}{D_t(z_1)} = \left(\frac{z}{z_1}\right)^{\beta} \tag{6.37}$$

Unlike velocity, the eddy diffusivity profile tends to be more curved under unstable conditions ($\beta > 1$) and flatter under stable conditions ($\beta < 1$). Values of the eddy diffusivity exponent as a function of the Pasquill-Gifford categories and Monin-Obukhov length scale can also be found in Table 6.3.

TABLE 6.3
Typical Values of Atmospheric Boundary Layer Parameters

	L_{MO}–m (z_0 = 0.1 m)		Velocity Exponent (α)		D_t Exponent (β)	$\sigma\theta$
Stability Class	**Lower Bound**	**Upper Bound**	**Urban**	**Rural**	**Rural**	**Degrees**
A	−10	−5	0.15	0.07	1.37	25
B	−30	−10	0.15	0.07	1.32	20
C	−100	−10	0.20	0.10	1.20	15
D	−100	100	0.25	0.15	1.00	10
E	30	100	0.40	0.35	0.80	5
F	10	30	0.60	0.55	0.59	2.5

Velocity exponents from Hanna et al. (1982), *Handbook on Atmospheric Diffusion*, DOE/TIC-11223, U.S. DOE.

The final column in Table 6.3 is a compilation of standard deviations in wind direction as a function of stability classification. The standard deviation in wind direction characterizes wind meandering. In addition to directly indicating the variability in wind direction in the boundary layer, this parameter often controls the crosswind spreading of pollutants.

6.3 POLLUTANT TRANSPORT AND DISPERSION IN THE BOUNDARY LAYER

As an environmental engineer, our interest in winds and the forces that drive them includes their effects on the transport and mixing processes of pollutants. The primary processes of interest are advection by the mean winds and mixing or effective diffusion/dispersion by turbulence. As described above, the wind speed and direction are largely defined by the large or *synoptic* scale pressure variations as modified by local conditions. Vertical profiles in winds and contaminant levels, however, are controlled by the intensity of mixing and turbulence which is in turn controlled by the thermal stratification and stability aloft and the ratio of wind shear and buoyancy forces, or Richardson number, near the surface.

During nighttime conditions, mixing is limited due to the surface cooling driven stability. As a result contaminants released near the ground remain relatively undiluted by mixing with clean air. For pollutants emitted at or very near the ground, it is typically the nighttime conditions that are of the most concern. During the day, as the surface is warmed by the sun, these contaminants are mixed over a deepening mixing layer lowering their concentrations.

The deepening of the surface mixing layer also may cause the incorporation of pollutants from above into the surface layer. Elevated emissions from a smokestack may pose little or no problem during the night due to the lack of mixing to ground level but may *fumigate* to the surface by incorporation into the mixing layer during the day.

The qualitative aspects of the behavior of a single smokestack plume under various stability conditions are depicted in Figure 6.14. Under very unstable conditions (6.14a), strong vertical motions in the atmosphere give rise to the *looping* plume sometime observed. Under such conditions, elevated contaminant releases may be transported directly to the ground and then "bounce" back to the deeper atmosphere. It should be noted that quantitative description of this process requires a very sophisticated mathematical model which is not normally available for routine dispersion calculations.

Under near neutral conditions (6.14b), spreading is nearly equal both upward and downward. Further spreading occurs laterally. It is these conditions, or small deviations therefrom, that can be readily and accurately predicted by simple mathematical models. Spreading in all directions also occurs under stable conditions, but the rate of spreading is considerably smaller than with neutral atmospheric conditions.

When a surface inversion occurs during the night (6.14c), mixing to the ground is hindered as indicated above. Fumigation occurs (6.14d) when growth of the mixing layer incorporates the smokestack plume and transports it to ground level. This can be the result of mixing layer growth with time or as a result of horizontal variations in mixing layer height. Near a coastline, for example, there is little thermal convection over the sea, but heating over the land gives rise to a mixing layer that grows as one moves inland. An elevated stack plume may move inland and become entrained within the mixing layer and fumigate to the ground. Fumigation can result in rapid increases in pollutant concentration at ground level and because it involves atmospheric variations on a very small scale, it can be one of the most difficult processes to identify or predict. Elevated stable layers or inversions can also serve to reduce ground level impacts of a pollutant source by causing *lofting* of the plume (Figure 6.14e) or *entrapment* (Figure 6.14f). Note that either condition can result in a fumigating plume if further growth of the mixing layer occurs.

The transport and dispersion of pollutants from a smokestack can also be significantly influenced by the presence of rough terrain or buildings. As an illustration consider the flow of air around a building of height H_b as shown in Figure 6.15. Flow around any bluff body such as a building results in the development of a wake behind the building. The wake is a zone of poorly oriented flow that effectively acts as a stirred vessel with a volume of 2 to 3 times the building volume. Stacks within 5 building heights ($5H_b$) laterally and up to 2.5 building heights ($2.5H_b$) vertically are likely to be influenced by the wake. The effect is that the emissions from the stack will tend to be mixed over the wake, typically spreading the pollutants over a much larger zone than would be expected for dispersion in open terrain so close to the source.

Let us consider a steady material balance over the wake. The input from the emission source (Q_m) is balanced by advection out of the assumed well-mixed wake, $C_aU(2W_bH_b)$. Here W_b represents the width of the building and the factor of 2 represents the assumption that the cross-sectional area of the wake is double the area of the building. Solving for the air concentration within the wake,

$$C_a = \frac{Q_m}{U(2W_bH_b)} \tag{6.38}$$

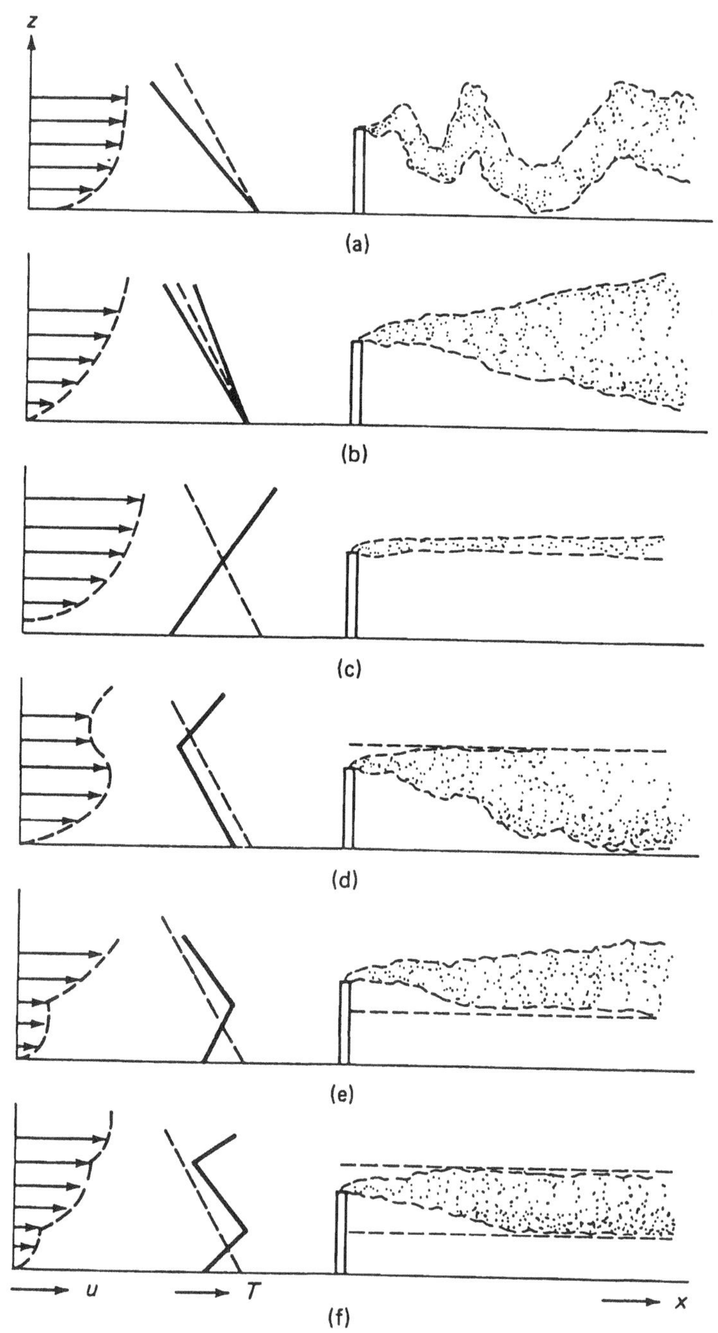

FIGURE 6.14 Plume behavior from an elevated smokestack under various meteorological conditions. (From Wark, K. and C.F. Warner (1981) *Air Pollution: Its Origin and Control*, Harper Collins, New York. With permission.)

In addition to providing rapid spreading of a smokestack plume, the wake can have important implications for people within the wake or within the adjacent building. Unless the architectural design of the building considers the effect of the wake, it is likely that the building ventilation system exhausts into the wake and intakes are positioned to pull from the wake. Thus, the wake concentration often defines the building concentration. Laboratory buildings often exhibit the

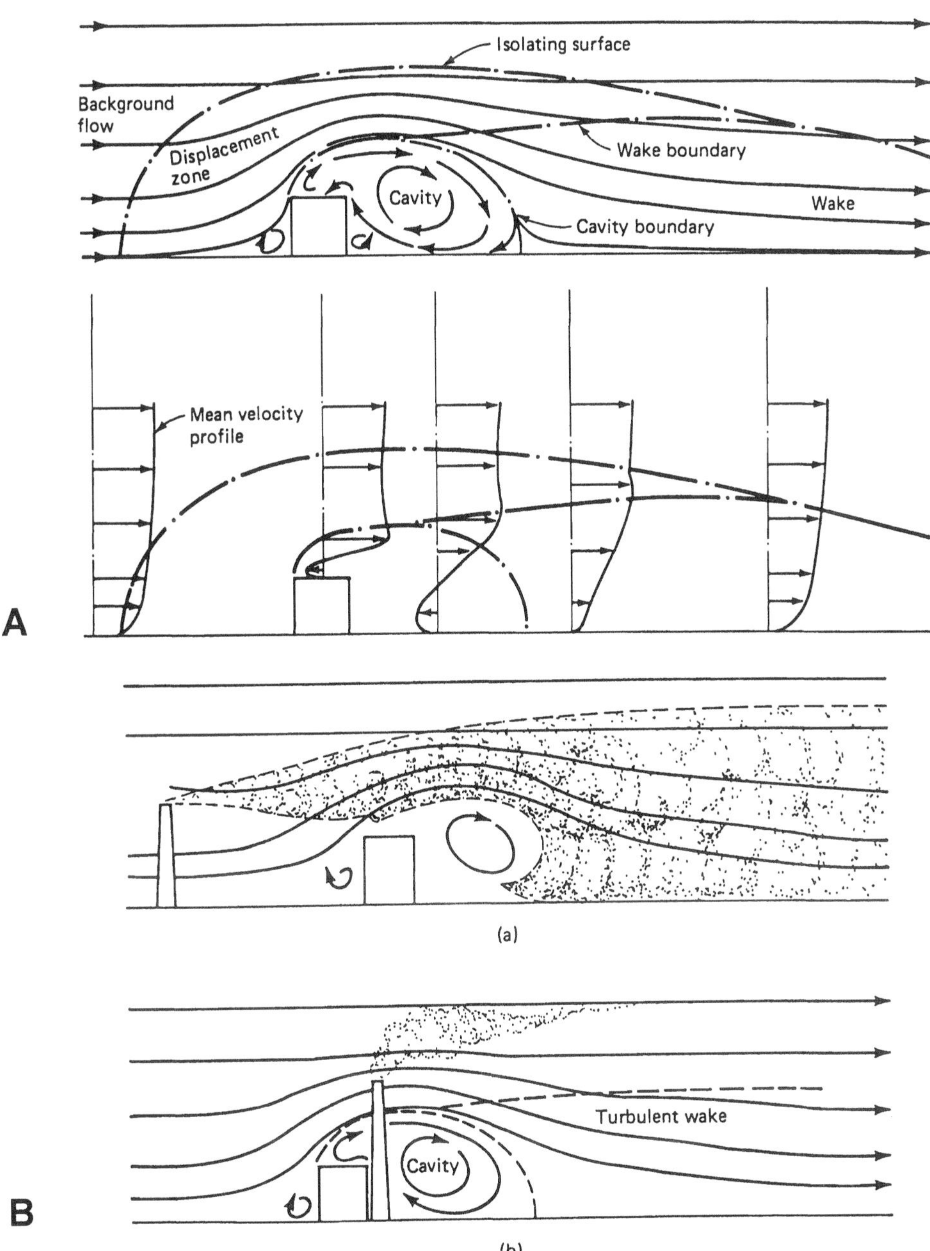

FIGURE 6.15 Characteristics of flow around an isolated building: (A) From Slade, D.H. (1968) Ed. Meteorology and Atomic Energy, Washington, D.C., AEC. (B) From Briggs, G.A. (1969) Plume Rise, AEC Critical Review Series.)

additional problem of fumehoods that can place a building under a slight negative (gauge) pressure further aggravating reentry of the exhausted contaminants. Perhaps the easiest way to eliminate reentry of exhausted building air is to build tall exhaust stacks of the order of 2.5 times the building height or taller. This is rarely practical for office or laboratory buildings wishing to avoid factory-style smokestacks. In such instances, placement of the building intakes toward the prevailing wind direction and avoidance of negative pressures are likely the most effective means of reducing reentry.

Equation 6.38 is an example of stirred tank or box model of air pollution. A similar model can be applied under any conditions that cause pollutant concentrations to be mixed uniformly over some volume. Often pollutant sources in a city are distributed over a wide area and such an assumption may be appropriate. Pollutants from distributed, mobile sources such as automobiles are often modeled very well by such a model. As indicated previously, these emissions would be mixed over the depth of a mixing layer in a matter of minutes under typical conditions and thus the concentration of such pollutants would be expected to be approximately uniform both laterally and vertically. Under these conditions a box model analogous to Equation 6.38 can be applied to a very large area. An urban or large area box model for the prediction of pollutant concentrations under such conditions is discussed below.

6.3.1 Well-Mixed Box Model for Area Sources

Under conditions where pollutants are emitted over a large area and only average area-wide concentrations are required, the pollutants can often be described by a well-mixed box model. The extent of such a box would include the region over which the pollutant concentration is assumed to be approximately uniform. Most commonly this may be the area of a city in which a large number of pollutant sources tend to keep the concentration approximately uniform. The vertical extent of these emissions is assumed to be the mixing height, normally a function of time of day by the methods described in the preceding section.

The model is based upon a pollutant material balance over the box shown in Figure 6.16. The along-wind dimension of the box is the length of the city, L; the cross-wind dimension is its width, W; and its height is the mixing height, H_m. The accumulation of a particular pollutant within the box present in concentration, Ca, is given by

$$\text{Accumulation} = \frac{\partial}{\partial t} C_a(L\,W\,H_m) \tag{6.39}$$

The accumulation is balanced by input from upwind sources that cause a concentration, C_0, at the upwind edge of the box and output from the downwind edge of the box. Pollutant sources within the city are assumed to cause the release of q_m mass of pollutant per unit area per time (e.g., $kg \cdot km^{-2} \cdot h^{-1}$). The advective flux in or out of the box is given by the wind speed, U, times the concentration at the respective edge of the box. From Figure 6.16, C_0 is the concentration at the upwind edge and by the assumption of well mixedness, C_a is the concentration everywhere within the box and at the downwind edge. Thus, the input and output terms are given by

$$\begin{aligned} \text{Input} &= UC_0(WH_m) + q_m(LW) \\ \text{Output} &= UC_a(WH_m) \end{aligned} \tag{6.40}$$

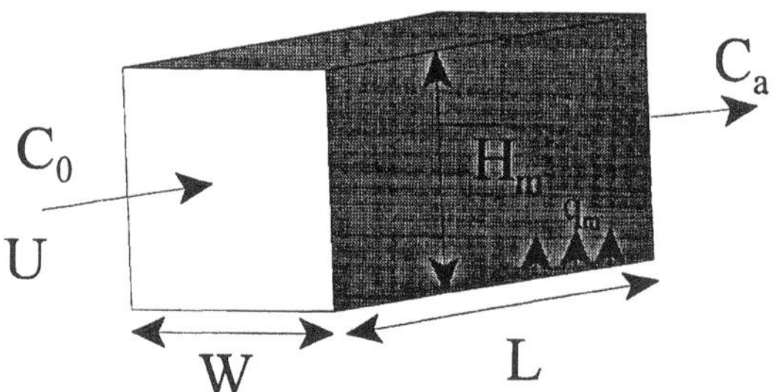

FIGURE 6.16 Conceptualization of well-mixed box air quality model.

and the full material balance for the emitted pollutant is

$$\frac{d}{dt}C_a(L\,W\,H_m) = UC_0WH_m - UC_aWH_m + q_mWL \tag{6.41}$$

Solution of this equation requires specification of the time dependence of the mixing height and an initial condition on the concentration of the pollutant in the box. During the day, the mixing height increases according to the processes described previously. As the mixing height increases, contaminated air from above may also be incorporated into the box at a rate given by the concentration above the box and the rate that this volume is incorporated is given by $(dH_m/dt)WL$. The influence on the growth of the mixing height will be explored in the problems. For now, let us assume that the box is effectively of constant height. Equation 6.41 can then be written

$$\frac{dC_a}{dt} = \frac{UC_0}{L} - \frac{UC_a}{L} + \frac{q_m}{H_m} \tag{6.42}$$

At long times, the solution to this equation can be found by setting the time derivative equal to zero (that is, solve for the steady solution). Solving for the concentration within the box,

$$\begin{aligned} C_a(\infty) = C_{ss} &= C_0 + \frac{q_m}{H_m}\frac{L}{U} \\ &= C_0 + \frac{q_m}{H_m}\tau_Q \end{aligned} \tag{6.43}$$

where τ_Q is the volumetric residence time of air within the box. This solution suggests that the ultimate concentration within the box is equal to the concentration coming from upwind sources modified by emissions introduced into the box. The additional amount added by the emissions is proportional to the time air spends over the source area and inversely proportional to the depth over which these emissions are mixed.

If the initial concentration within the box is negligible and we write the emissions on a total rate basis (i.e., $Q_m = q_mLW$) then the equivalence to Equation 6.38 becomes more clear

$$C_{ss} = \frac{Q_m}{U\,W\,H_m} \tag{6.44}$$

Remember that the factor of 2 in Equation 6.38 was simply to suggest that the wake cross-sectional area was double the building area. In Equation 6.44, the width and height of the well-mixed area are explicitly included.

This solution emphasizes that the two most important influences on the concentrations to which individuals are exposed are the velocity of the wind and the depth of mixing. As these quantities increase, the observed concentrations decrease. Observed concentrations in the atmosphere during high winds and unstable daytime conditions are quite low compared to those observed during nighttime stable, low wind speed conditions. Often residents located near industrial facilities or other pollutant sources are concerned about increases in emissions at night "when no one is watching." This may simply be due to the increased odor or other adverse effects due to reduced mixing and increased concentrations during the night.

If the time-dependence of the concentrations is desired, Equation 6.41 or 6.42 must be solved subject to an initial condition in the "box" over the city. We will assume that the initial concentration within the box is that at the upwind edge of the box, i.e., Ca(0) = C_0. If we assume no growth in the mixing layer depth, the solution to Equation 6.42 is then

$$C_a = C_0 + \frac{Q_m}{H_m W U}\left(1 - e^{-\frac{U}{L}t}\right) \tag{6.45}$$

Note that the steady condition is only reached when time t is very large compared to the residence time in the box, i.e., $t \gg \tau_Q$ (= L/U). Under these conditions, the second term within the parentheses goes to zero and the air concentration approaches the steady-state concentration. If we consider a city (or effective length of source area) of 10 km and a wind speed of 2 m/s, this means that the concentration approaches the steady-state concentration as long as the time is significantly longer than about 1.4 h. Because this is often short compared to the time scales of interest, the assumption of steady state and well-mixedness over the box are often good ones as long as the sources are distributed over nearly the entire box. The box model is illustrated in Example 6.4.

Example 6.4: Use of a box model to predict air quality

Estimate the steady-state concentration and time required to reach 90% of steady state in a box model of SO_2 concentrations in Baton Rouge, LA. Assume well mixedness over a box 10 km by 10 km in the industrial area of the city in which emissions of SO_2 total about 31,320 tons/year (1994). Assume zero SO_2 from upwind sources, a wind speed of 2 m/s, and a constant mixing depth of 500 m.

Estimating the steady-state concentration from Equation 6.44

$$C_{ss} = C_0 + \frac{Q_m}{H_m W U}$$

$$= 0 \text{ mg/m}^3 + \frac{(31320 \text{ tons/year})\left(28.748 \dfrac{\text{mg/s}}{\text{ton/year}}\right)}{(500 \text{ m})(10000 \text{ m})(2 \text{ m/s})}$$

$$= 0.09 \text{ mg/m}^3 = 0.034 \text{ ppm}$$

This is perhaps surprisingly close to the maximum SO_2 concentration of 0.041 ppm in West Baton Rouge Parish and 0.035 ppm in East Baton Rouge Parish detected in 1995. For comparison, the ambient air quality standard is 0.14 ppm. The time required to achieve a fraction, f = 0.9 (90%), of the steady-state value is given by

$$t_f = -\ln(-f + 1)\frac{L}{U}$$

$$t_{0.9} = -\ln(0.1)\frac{10000 \text{ m}}{2 \text{ m/s}}$$

$$= 3.2 \text{ h}$$

6.3.2 Point Source Dispersion Model

The box model fails when we are interested in pollutants very near an individual point source. Under such conditions an alternative model that does not depend upon well-mixedness is necessary.

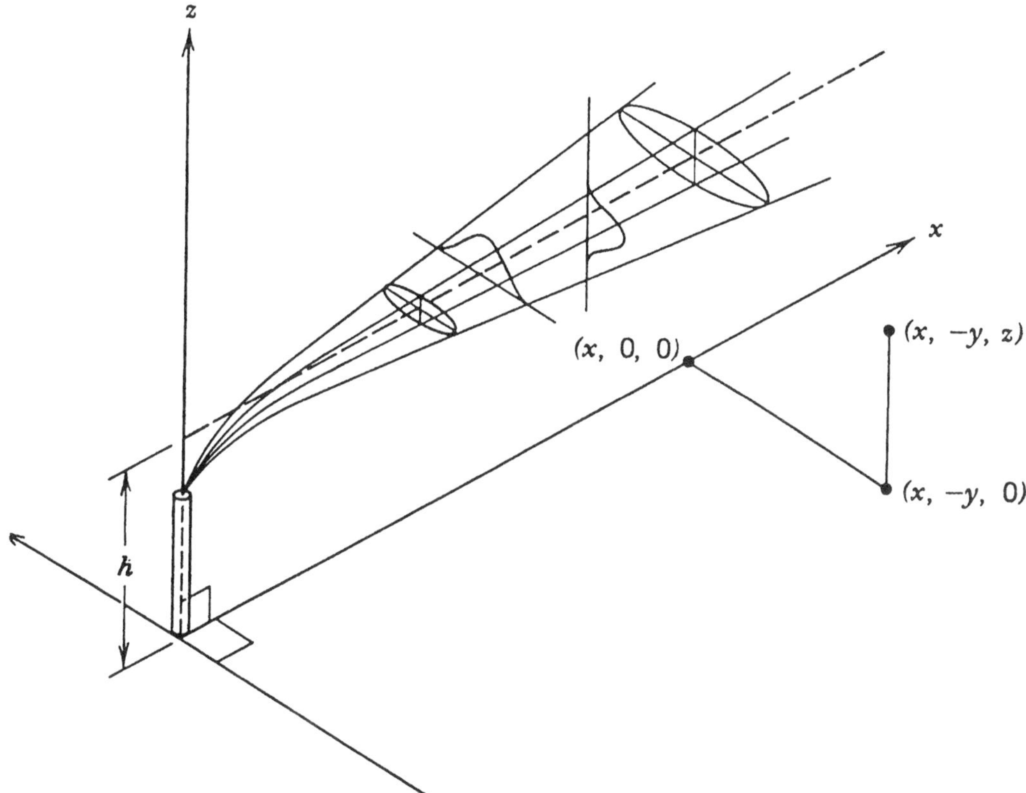

FIGURE 6.17 Idealization of contaminant dispersion from an isolated smokestack plume. (From Wark, K. and C.F. Warner (1981) *Air Pollution: Its Origin and Control*, Harper Collins, New York. With permission.)

Figure 6.17 depicts a pollutant plume exiting from an isolated smokestack. The movement of the plume can be broken into two distinct phases, plume rise due to the buoyancy and momentum of the plume exiting the stack, and transport and dispersion of the plume by the wind and turbulence. Each of these processes will be discussed in turn.

6.3.2.1 Plume Rise

A typical stack emits pollutants from a furnace, boiler, or combustion chamber of an engine. The pollutants are emitted with some velocity and are generally hotter than the surrounding air. Both effects cause the plume to rise to an effective height, H_s, that can be considerably higher than the physical height of the stack. The effective height is the sum of the physical stack height, H_0, and the additional height, ΔH, associated with buoyancy or momentum induced plume rise. As a plume rises it entrains ambient air, reducing the effective velocity or buoyancy. This continues until the plume is effectively indistinguishable (with respect to buoyancy or momentum) from the ambient air. Plume rise under stable atmospheric conditions is the best understood and will be considered here. Plume rise under unstable conditions is heavily influenced by ambient turbulence levels and there is less data available to support predictive models. The relatively rapid vertical mixing often makes the elevation of the emissions less important under unstable conditions.

The vertical momentum, τ_{zz}, carried by the gases exiting at velocity, v_s, from a stack of diameter, d_s, is given by

$$\tau_{zz} = v_s^2 \frac{d_s^2}{4} \frac{\rho_s}{\rho_a} \tag{6.46}$$

A constant π from the cross-sectional area of the stack exit would normally be included in this equation, but this is rarely done and the constant is incorporated into empirical coefficients that are fit to field data. The ratio of densities between the stack and ambient air could also be written as the inverse ratio of absolute temperatures since we are considering an ideal gas. The buoyancy, B, carried by these same gases is given by

$$B = \frac{g}{T_a}(T_s - T_a)v_s \frac{d_s^2}{4} \tag{6.47}$$

Again π has been omitted by convention. The ratio of Equations 6.46 and 6.47 give an indication of the relative magnitude of momentum relative to buoyancy as a mechanism for plume rise.

$$\frac{\tau_{zz}}{B} = \frac{v_s \frac{T_a}{T_s}}{\frac{g}{T_a}(T_s - T_a)} \tag{6.48}$$

This ratio has units of time and indicate the time until buoyancy begins to dominate the plume rise. This is often of the order of 10 s or less. For most large industrial stacks, the plume rise associated with buoyancy is likely to be the most important process.

The rise of a buoyant stack gas plume is a function of the rate of buoyancy release, B, the wind speed, and the atmospheric stability. Using the Brunt-Vaisala frequency (Equation 6.32) to characterize stability, the ultimate plume rise due to buoyancy under stable conditions is approximated by (Hanna et al., 1982)

$$\Delta H = 2.6\left(\frac{B}{Uf_{BV}^2}\right)^{\frac{1}{3}} \tag{6.49}$$

This relationship is applicable only to "bent-over" plumes that are influenced by the horizontal wind speed and is not applicable to neutral or unstable conditions. When the wind speed is less than about 1 m/s, the plume is essentially vertical and the horizontal wind speed no longer affects the rise.

Since the plume is bent over by the effect of the horizontal wind, the ultimate plume rise is realized some distance downwind from the stack. This horizontal distance downwind can be approximated by

$$x \approx 2.1 \frac{U}{f_{BV}} \tag{6.50}$$

which was derived from models of plume trajectory and ultimate plume rise. This equation also is only valid for stable atmospheric conditions. For purposes of dispersion modeling, that is, the prediction of concentrations downwind of the source, this distance is usually neglected and the source is assumed to be located at the stack at an elevation of $H_0 + \Delta H$. The trajectory of the plume may be important, however, in explaining near-field ground level concentrations or interactions with nearby hills or buildings. Equation 6.50 can be used as a check of the distance required to achieve the full plume height. It will at least provide an indication of the uncertainty in estimating distances to downwind locations.

Carpenter et al. (1970) suggest a different formula for the estimation of plume rise that overcomes the limitation to stable conditions of Equations 6.49 and 6.50. The formula is semi-empirical and based on plume rise from Tennessee Valley Authority coal-fired power plants. This formula can be written

$$\Delta H = 114\ \text{m}^{2/3}\left(1.58 - [41.4\ \text{m/K}]\frac{d\theta}{dz}\right)\frac{B^{1/3}}{U} \tag{6.51}$$

Here, the result is in m, the potential temperature gradient is in K/m, B is in m^4/s^3, and U is in m/s. This formula is an empirical fit to data for plume rise in potential temperature gradients from –0.001°C/m to 0.013°C/m. Example 6.5 illustrates the use of these plume rise models.

Example 6.5: Estimation of plume rise

Estimate the plume rise from a 3-m diameter stack, 30 m tall. The stack exit temperature is 200°C and velocity 10 m/s. The ambient air temperature at stack height is 25°C and the background temperature gradient is –0.002°C/m (potential temperature gradient of 0.0078 K/m). The wind speed at 10 m is 2 m/s.

Calculating the buoyancy and momentum flux of the stack exhaust and the Brunt-Vaisala frequency

$$B = \frac{9.8\ \text{m/s}^2}{298\ \text{K}}(473\ \text{K} - 298\ \text{K})(10\ \text{m/s})\frac{(3\ \text{m})^2}{4} = 130\ \text{m}^4/\text{s}^3$$

$$\tau_{zz} = (10\ \text{m/s})^2\frac{(3\ \text{m})^2}{4}\left(\frac{298\ \text{K}}{473\ \text{K}}\right) = 142\ \text{m}^4/\text{s}^2$$

$$f_{BV} = \left(\frac{9.8\ \text{m/s}}{298\ \text{K}}(-0.002\ \text{K/m} + 0.0098\ \text{K/m})\right)^{1/2} = 0.016\ \text{s}^{-1}$$

The ratio τ_{zz}/B is 1.1 s, thus the plume is buoyancy dominated almost immediately. The average wind velocity influencing the plume should be used in the plume rise equations. Let us estimate the winds at 100 m approximate this average. Using the moderately stable exponent of 0.35 from Table 6.2, $U = (100\ \text{m}/10\ \text{m})^{0.35} = 4.5$ m/s. Equation 5.49 suggests a plume rise of

$$\Delta H = 2.6\left(\frac{130\ \text{m}^4/\text{s}^3}{(4.5\ \text{m/s})(0.016\ \text{s}^{-1})^2}\right)^{1/3} = 127\ \text{m}$$

Equation 5.50 indicates that this plume rise would occur about 595 m downwind. Using Equation 5.51 gives an estimate of plume rise that differs by about 29% from this value.

$$\Delta H = 114\ \text{m}^{2/3}[1.58 - (41.4\ \text{m/K})(0.0078\ \text{K/m})]\frac{(130\ \text{m}^4/\text{s}^3)^{1/3}}{4.5\ \text{m/s}} = 163\ \text{m}$$

It is difficult to choose between these two values, but the lower plume rise would generally result in predictions of higher concentrations at ground level and thus a design (e.g., selection of a stack height) based on this plume rise would tend to be more protective of human health. For more neutral or unstable conditions, Equation 5.49 gives invalid results and should not be used.

6.3.2.2 Point Source Dispersion

Upon reaching its ultimate plume height, a plume will then be transported and dispersed by the winds and turbulence. Of primary interest are continuous point sources such as an industrial smokestacks. With a continuous source, the gradient in concentration in the direction of the wind is small. Thus, advection by the mean wind is the primary mechanism for transport in the along-wind direction as opposed to any turbulent mixing process. Under these conditions a modification of the model represented by Equation 5.140 is most applicable to dispersion from stacks and is repeated here as Equation 6.52 with Q_m representing the rate of mass release from the stack (mass/time).

$$C(y,z,t) = \frac{Q_m}{2\pi\sigma_y\sigma_z U}\exp\left[-\frac{y^2}{2\sigma_y^2}-\frac{z^2}{2\sigma_z^2}\right] \tag{6.52}$$

Because the form of this solution is equivalent to the Gaussian probability distribution the model is referred to as the Gaussian dispersion model. The Gaussian dispersion or Gaussian plume model has been widely used for more than 40 years and despite improvements in our knowledge of atmospheric transport and mixing processes remains the basic tool for routine dispersion estimates. Commonly used computer models of stack plumes, such as the EPA Industrial Source Complex Model, still employ the Gaussian plume model as its foundation. Alternatives exist that address problems to which this model is not applicable, but they will not be discussed here.

The dispersion parameters that appear in the Gaussian model, σ_y and σ_z, represent the standard deviations in the crosswind and vertical spread of the concentration distribution, respectively. Note that the equation does not explicitly depend upon travel time or distance in the x-direction. These dependencies appear implicitly in that the standard deviations are a function of travel time (t) or distance (x = Ut) and turbulent intensity or diffusivity. As defined in Chapter 5, the theoretical relationship is

$$\sigma_i = \sqrt{2(D_t)_i t} = \sqrt{2(D_t)_i x / U}\,.$$

In practice, these standard deviations are estimated by empirical correlations which suggest that the standard deviations are approximately proportional to travel time or distance as opposed to its square root. The correlations also relate the standard deviation to measures of turbulent intensity, in particular the Richardson number, Monin-Obukhov length scale, or semi-empirical measures thereof. Equation 6.52 describes the concentration downwind of a source at x = y = z = 0 in an infinite medium subject to advection in the x-direction and diffusive-like processes in the y and z directions. It assumes that the wind velocity is approximately constant and uniform in direction, which based upon our previous discussion is clearly a crude approximation to the atmosphere. This approximation is especially poor in hilly or mountainous terrain or when the winds are controlled by local phenomena such as urban street canyons or the land-sea breeze cycle at a coastline. Even under nearly uniform wind conditions, however, two modifications must be considered to apply the model to dispersion from tall stacks. In the atmosphere, ground level concentrations are generally of primary interest and the boundary condition the ground represents must be included in the model. In addition, we must incorporate the ability to source the emissions at the effective plume rise, i.e., $x = y = 0$, $z = H_s = H_0 + \Delta H$. Let us first consider the latter modification.

In an infinite medium, an elevated source can be incorporated by simply defining a new vertical coordinate, ζ, that is given by $\zeta = z - H_s$. Still, z = 0 represents the ground but $\zeta = 0$ represents the elevation of the source. Replacing the source location in Equation 6.52 with its location relative to the ground,

$$C(x,y,z;H_s) = \frac{Q_m}{2\pi\,\sigma_y\,\sigma_z\,U}\exp\left[-\frac{y^2}{2\sigma_y^2} - \frac{(z-H_S)^2}{2\sigma_y^2}\right] \quad (6.53)$$

where H_s on the left-hand side of Equation 6.53 is used to represent that the solution depends on the value of the parameter H_s, the effective height of the source. Also the downwind distance, x, rather than the travel time, t, is included in the list of variables. The relationship between these quantities is x = Ut so either could be used as the independent variable. Remember that the dependence upon x (or t) is implicit in the standard deviations.

Note that the exponential terms are a maximum (unity) when their argument equals zero. Thus the concentration immediately downwind of the source of a plume ($y = 0$, $z = H_s$, $\zeta = 0$), is given by

$$C(X,0,H_3) = \frac{Q_m}{2\pi\,\sigma_y\,\sigma_z\,U} \quad (6.54)$$

The model predicts a lower concentration at the ground, $C(X,0,H_3)e^{-\frac{H_s^2}{2\sigma_z^2}}$ as would be expected for an elevated source.

The second modification to the basic model recognizes that Equation 6.52 or 6.53 does not recognize the existence of the boundary condition posed by the ground. The model assumes an infinite medium and thus the gaseous pollutants are predicted to pass through the ground. Most gaseous pollutants, however, are reflected at the ground surface and remain in the air above. Most of the plume is reflected even for gaseous pollutants that react with the ground or vegetation since these reactions typically occur at low rates. The solution to this problem for a "linear" model such as the Gaussian plume model is to restore that portion of the plume that the model predicts is "lost" by movement underground. The modification of the model can perhaps best be understood by considering a ground level pollutant source, i.e., H = 0. For such a source the current model predicts concentrations above and below ground that are symmetric. That is, exactly half of the plume is predicted to disperse below ground at the same rate as that above ground. The simplest means of restoring the material to the atmosphere that is incorrectly predicted to be absorbed into the ground is to double the rate of emission. Alternatively, an additional *virtual* or *imaginary source* could be placed at the same location as the actual source and the air concentrations downwind are equal to the sum of the concentrations predicted by the two sources. Thus, the Gaussian plume model for a ground level source with *reflection* at the surface is

$$C(x,y,z;0) = \frac{Q_m}{\pi\,\sigma_y\,\sigma_z\,U}\exp\left[-\frac{y^2}{2\sigma_y^2} - \frac{z^2}{2\sigma_z^2}\right] \quad (6.55)$$

For an elevated source, the source strength cannot simply be doubled to account for material lost to the ground. At any distance downwind, only that portion of the pollution that is predicted to have mixed below ground must be returned to the air. The solution is to add a contribution from a source *below ground*, that is a mirror image of the above ground actual source, as shown in Figure 6.18. If the "atmosphere" below the ground is identical to that above, any pollutants predicted to go below ground from the actual source will be predicted to go above ground from the virtual

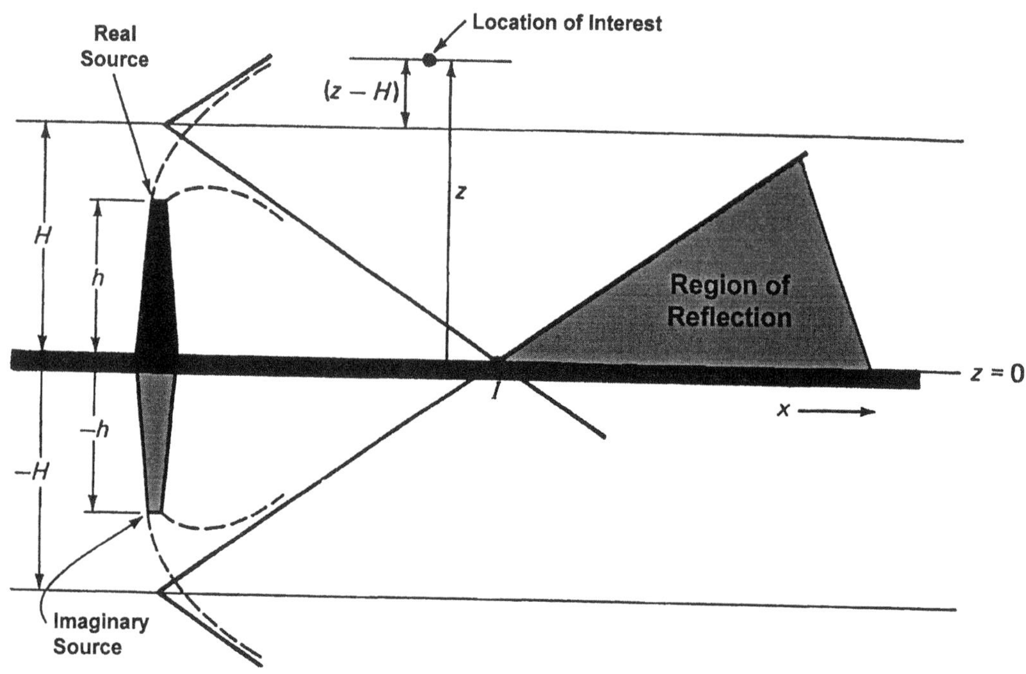

FIGURE 6.18 Definition of a virtual source for prediction of reflection at the ground. (From Wark, K. and C.F. Warner (1981) *Air Pollution: Its Origin and Control*, Harper Collins, New York. With permission.)

source. Because the virtual source is at a height $-H_s$, the coordinate ζ for that source is $z - (-H_s) = z + H_s$. Thus the solution for an elevated source with reflection at the ground is given by

$$C(x,y,z;H_s) = \frac{Q_m}{2\pi\,\sigma_v\,\sigma_z\,U}\exp\left[-\frac{y^2}{2\sigma_y^2}\right]-\left[\frac{(z-H_s)^2}{2\sigma_z^2}+\frac{(z+H_s)^2}{2\sigma_z^2}\right] \tag{6.56}$$

It is this form that is normally referred to as the Gaussian plume model. Note that a similar process could be instituted to allow for reflection at the top of the mixing layer if overlayed by a stable inversion layer. Again the model in its present form would predict loss of pollutant through the stable layer above the mixing layer and a virtual source that is the mirror image of the actual source on the other side of the top of the mixing layer must be added. Unfortunately, this seemingly simple solution to the full problem grows rather tedious when it is recognized that the virtual source above the mixing layer will ultimately predict nonzero concentrations below ground and that a virtual source below the ground will predict nonzero concentrations above the mixing layer. Virtual sources would in turn need to be included to offset these difficulties and the result is an infinite series of real and virtual sources. A Gaussian plume model accounting for these additional sources has been developed but a simpler alternative is normally available. Once you are sufficiently downwind that reflections from both the ground and the top of the mixing layer are important, the concentration within the mixing layer is approximately uniform vertically. When the concentration is assumed vertically uniform, spreading only continues to occur laterally and the Gaussian plume model takes on the form

$$C(x,y;H) = \frac{Q_m}{(2\pi)^{1/2}\,\sigma_y\,H_m\,U}\exp\left[-\frac{y^2}{2\sigma_y^2}\right] \tag{6.57}$$

The distance downwind before this model is valid depends on the rate of spreading in the vertical direction. If we consider a ground level source, the concentration decreases in the vertical according to the Gaussian term, $e^{-(z^2/2\sigma_z^2)}$ from Equation 6.55. This term is unity at z = 0 (the ground or source level) and is 0.1 or 10% of the maximum at about z = 2.15 σ_z. For a pollutant plume, this means that the concentrations at 2.15 σ_z are 10% of the maximum concentration, an often used characterization of the "edge" of the plume. Thus, at a distance downwind, x_L, such that $\sigma_z(x_L) = H_m/2.15$, we would expect a pollutant plume beginning to be influenced significantly by the presence of an overlying stable layer as indicated by the presence of concentrations 10% of the maximum. A commonly used heuristic rule is that by $2x_L$, or double this distance, the pollutant would be entirely mixed in the vertical direction and Equation 6.57 would now apply. For distances $x < x_L$, Equation 6.56 should be usable without regard to the presence of a finite mixing layer. In between, linear interpolation between these two models has been suggested.

6.3.2.3 Parameter Estimation

Regardless of the form of the model, our ability to use it depends on the availability of the dispersion parameters or standard deviation in concentration in the vertical and horizontal directions. The standard deviations are related to the width of the concentration distributions in the respective directions. An often used approach is to consider the effective width of a pollutant plume to be the distances to the points that are 10% of the maximum concentration. In the vertical direction, from a ground-based pollutant source, the spreading is only upward and a measure of the depth of a pollutant plume is the height until the concentration is 10% of the maximum, or 2.15 σ_z as indicated above. For a horizontal plume (or an elevated vertical plume), spreading can occur in both positive and negative directions and the plume is approximated by 2.15 σ in each direction. The total width of a plume is approximated by 4.3 σ.

In principle, therefore, the determination of the dispersion parameters to be used in the Gaussian models is essentially an effort to determine the width of a dispersing plume under various meteorological conditions. Unfortunately, we are attempting to apply a deterministic model to a stochastic or random process. Such an approach has inherent limitations, although we will see that it is possible to develop such a model subject to certain limitations. To illustrate the potential problems let us consider the effect of averaging time on the dispersion of a point source plume.

Figure 6.19 provides a view from above of a pollutant plume from a point source. Each set of solid lines represents an essentially instantaneous view of the plume taken at different times. Due to the random meandering of the wind, the instantaneous plume takes radically different realizations. The concentration profile in the crosswind direction is shown for one of the instantaneous realizations. It is shown as a Gaussian profile but over short time averages such a profile would not necessarily be expected. In some cases, we would like to be able to predict the near instantaneous concentration field, for example, if we are concerned about a dispersing cloud of acutely toxic gases from which a single breath may result in death. In most cases, however, our interest is in a pollutant that poses a potential chronic concern where long-term averages of at least an hour or more are of interest. For such a situation, the individual realizations in Figure 6.19 are not important but only their longer time average. The region within which the hourly averaged concentration exceeds a certain level is shown by the broken line in the figure. The average concentration measured as a function of crosswind location within these bounds is shown by the Gaussian shaped curve in the figure. For the time-averaged plume, the crosswind spread is much greater and the maximum averaged concentration is much less than the short-term averaged plume. This is a general result. The longer the averaging period, the greater the crosswind spread and the lower the maximum concentration. This ambiguity requires that the dispersion parameters are linked not only to travel time but sampling time. This is often quantified by a relationship of the form

$$\frac{\sigma_y(\Delta T_1)}{\sigma_y(\Delta T_2)} = \left(\frac{\Delta T_1}{\Delta T_2}\right)^{\alpha}$$

$$\alpha \approx 0.25 - 0.30 \quad (1\ \text{h} < \Delta T < 100\ \text{h}) \tag{6.58}$$

$$\alpha \approx 0.2 \quad (3\ \text{min} < \Delta T < 1\ \text{h})$$

for different sampling or averaging times ΔT_1 and ΔT_2. The powers given are crude guidelines and values well outside of these ranges have been reported. Because maximum concentrations are generally of the greatest interest, some investigators describe this effect by referring to concentrations that increase as the averaging time decreases. While the maximum concentrations do decrease as the averaging time increases, the concentrations at any particular location may increase or decrease depending upon the particular wind direction during sampling. For the purposes of routine dispersion modeling, the most important consideration is that the dispersion parameters are based upon similar conditions, including averaging time, as the conditions that are being simulated. Most of the data that has been collected to estimate the standard deviations in concentration, σ_y and σ_z, for example, represent sampling times of the order of 10 min. Despite this, they are generally accepted as valid for the prediction of hourly averaged concentrations. Longer time averages can be estimated by formally averaging hourly simulations.

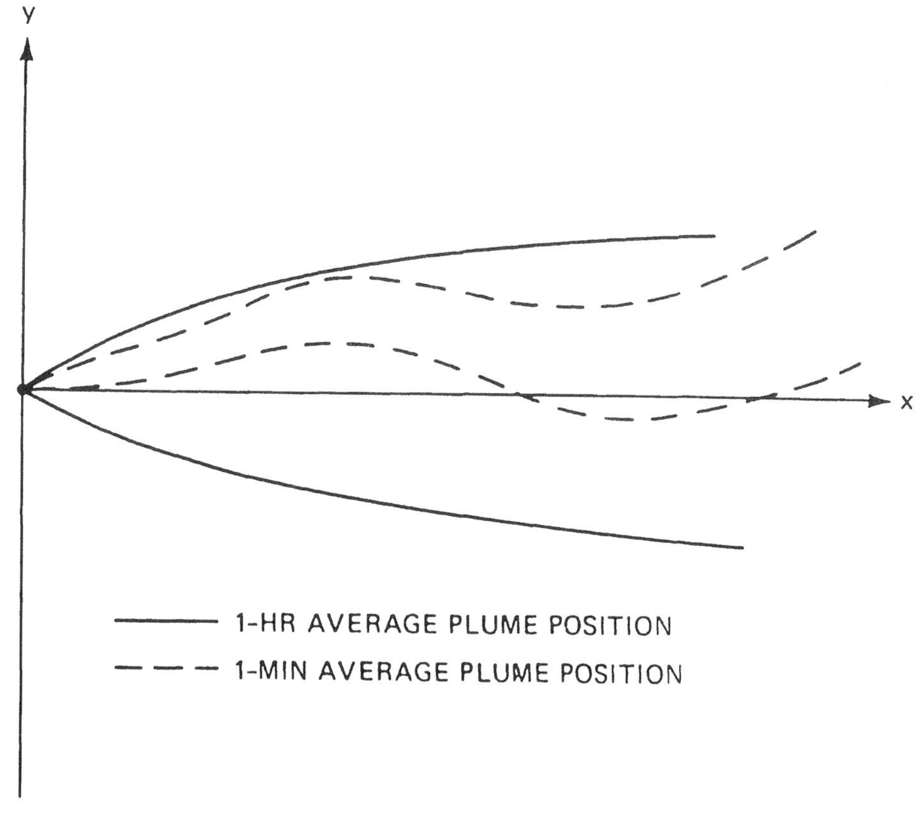

FIGURE 6.19 Depiction of instantaneous vs. time-averaged spreading of a pollutant plume from a point source. (From Hanna, S.R., et al. (1982) *Handbook on Atmospheric Diffusion,* DOE/TIC-11223, Technical Information Center, U.S. DOE.)

Subject to the above caveats, let us provide estimates of values of the dispersion parameters in the Gaussian plume model. Unless stated otherwise, the parameters are applicable to single point sources in nearly flat terrain under steady and uniform wind conditions. The predictions of the model are assumed to represent 10 min to 1 h average concentrations of nonreactive pollutants. As indicated previously, the values of the horizontal and vertical dispersion parameters or standard deviations depend upon the intensity of turbulence which, in turn, depends upon the Richardson number and distance downwind.

$$\sigma_y, \sigma_z = \sigma(x, N_{Ri}) \tag{6.59}$$

Typically, an empirical classification system such as the Pasquill-Gifford turbulence typing scheme is employed to define this relationship.

A long-used relationship between these quantities has been reported by Turner (1969) and is shown in Figures 6.20 and Figure 6.21. Curve fit functions for these curves have also been generated from a number of sources for convenience in making concentration and exposure concentrations using the Gaussian plume model. Table 6.4 includes a particular popular form attributable to Briggs.

The crosswind and vertical dispersion parameters provide a direct indication of plume width as a function of distance downwind under various stability conditions. If we consider a measure of the width of a plume to be the distance to locations that are 10% of the maximum concentration, for example, then the properties of the Gaussian distribution suggest that 4.3 times σ_y at any distance downwind is equal to that width. For a ground level source, spreading is only upward and 2.15 times σ_z is a measure of the plume height. At a distance of 1 km downwind from a ground level point source, for example, σ_y = 75 m and σ_z = 31 m under neutral conditions, suggesting that at this distance downwind this measure of the width of the pollutant plume is 323 m and its height is about 67 m.

The ground-level concentration predictions of the Gaussian dispersion model for an elevated point source have the general shape shown in Figure 6.22. Very close to the elevated source, concentrations are low because there has been insufficient time for the pollutants to mix to ground level. At some distance downwind, with the actual distance dependent upon the stability and degree of vertical mixing in the atmosphere, the ground-level concentration increases rapidly due to mixing downward. The distance to this point must generally be estimated by trial and error using the Gaussian equation. For the particular case of constant σ_y/σ_z, however, which is approximately valid under neutral to slightly unstable atmospheric conditions, the location of the maximum concentration can be determined by the location, x_m, that satisfies

$$\sigma_z^2(x_m) \approx \frac{H_s^2}{2} \tag{6.60}$$

That is, the effective stack height H_s suggests a value of σ_z which would result in the approximate maximum ground level concentration. Using the Briggs' formulation for σ_z allows an estimate of the distance to the maximum ground-level concentration. For class A stability, the distance is $x_m \approx (0.707/0.2)H_s \approx 3.5\,H_s$ while for Class B stability, $x_m \approx 5.9\,H_s$. This can also be applied to the other stability classifications. Under stable conditions, though, the assumption of the ratio of σ_y/σ_z being constant is not as good and the distance downwind to the point of maximum concentration may exceed 10 km, the limit of the applicable range of the Briggs' dispersion formula. Although the distance downwind to the maximum concentration increases as the stability class grows more stable, the dispersion is weaker resulting in higher concentrations at a given distance downwind. These somewhat offsetting effects typically cause the concentration at ground level to

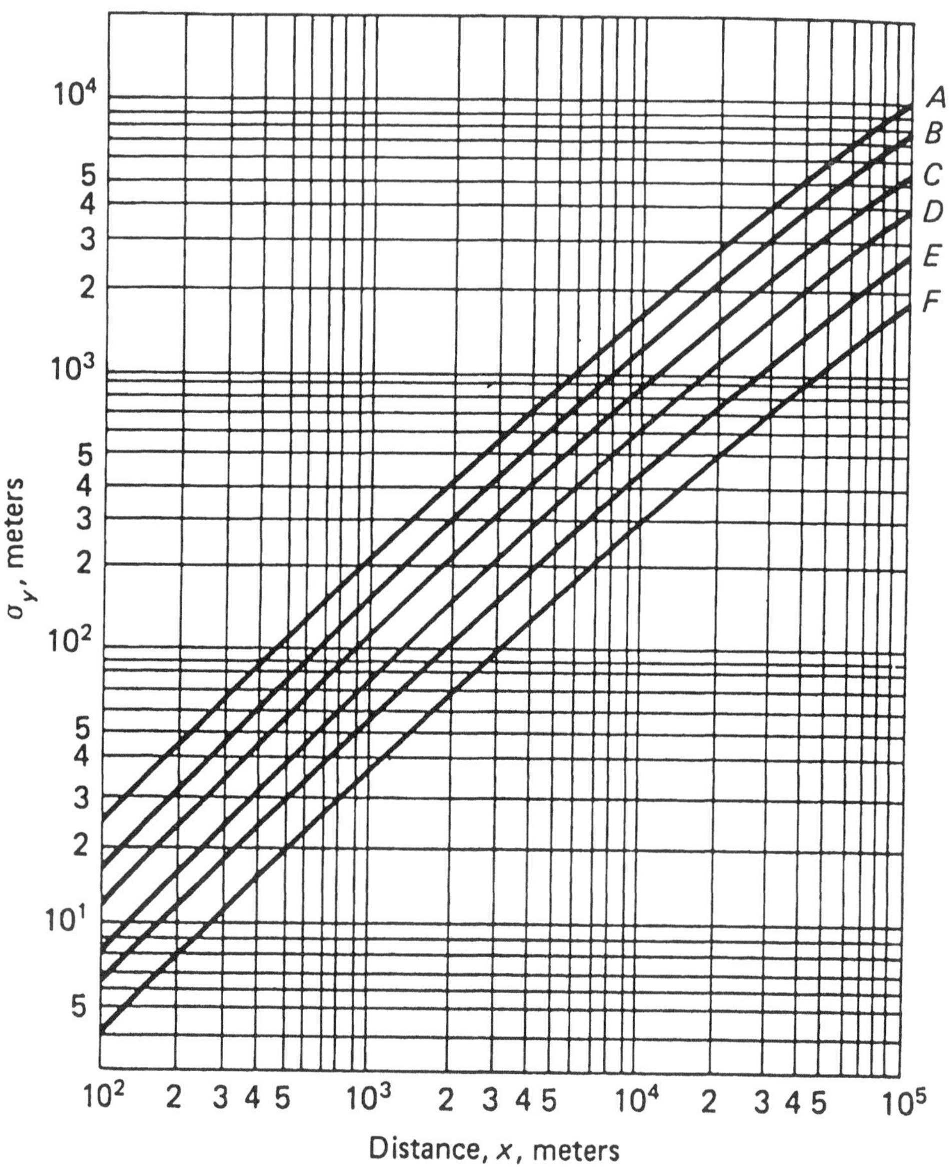

FIGURE 6.20 Crosswind standard deviation in concentration, σ_y, vs. downwind distance for various stability classifications. (From Turner, D.B. (1969) *Workbook of Atmospheric Dispersion Estimates*, Washington, D.C., HEW.)

be maximum under slightly unstable conditions, class B or C. The dispersion is so rapid under A stability conditions that concentrations reaching the ground are low, while under more stable conditions the distance downwind to the maximum concentration is so far that the concentrations are also low.

As one moves downwind from the point of maximum ground level concentration, the concentration decreases approximately in proportion to the square of the distance. Note that theoretically, since $\sigma \sim t^{1/2}$ (or $x^{1/2}$), the product of σ_y and σ_z in the denominator of the Gaussian model should cause the concentration to drop linearly with distance. This deficiency of the model has not precluded its application to the prediction of atmospheric dispersion. In an atmosphere

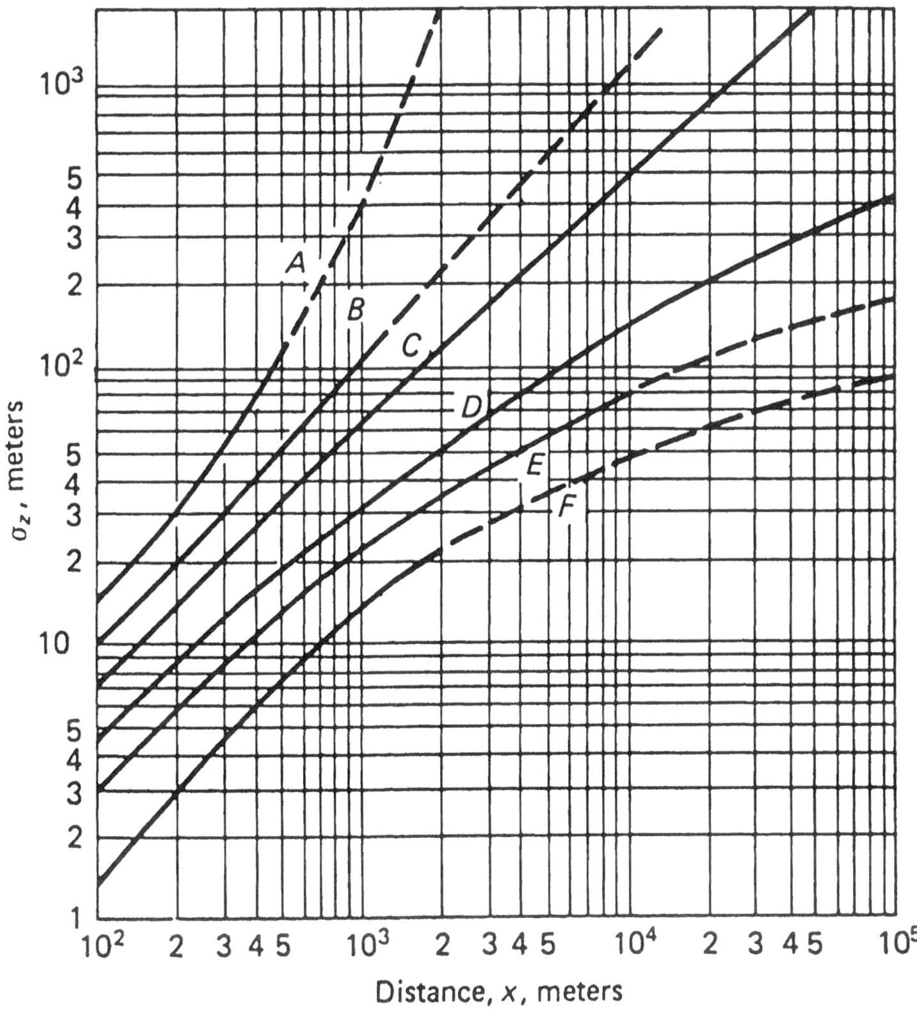

FIGURE 6.21 Vertical standard deviation in concentration, σ_z, vs. downwind distance for various stability classifications. (From Turner, D.B. (1969) *Workbook of Atmospheric Dispersion Estimates*, Washington, D.C., HEW.)

bounded by a finite mixing layer height, H_m, the plume will ultimately become well mixed in the vertical direction and subsequent spreading will only occur in the crosswind direction. Under these conditions, the concentration will decrease only linearly with downwind distance as shown in the later stages of Figure 6.22.

Despite the fact that the curves in Figures 6.20 and 6.21 extend to a distance of 100 km, the reader should be cautioned that use much beyond 10 km is subject to great uncertainty, not only due to the lack of supporting data beyond that distance downwind but also due to the likelihood of significant changes in wind direction or terrain and surface characteristics if the travel exceeds that distance. In reality, most of the data used to generate the curves in Figures 6.20 and 6.21 were collected from a 10-min average, ground-level samples collected less than a kilometer from the source. In most cases, the vertical dispersion parameter, σ_z, was actually inferred from the ground level data by assuming that the vertical profile was Gaussian. To emphasize the greater uncertainty in the vertical dispersion parameter, often the curves are indicated as broken lines beyond approximately 1 km.

TABLE 6.4
Briggs Dispersion Parameter Formulas

Pasquill-Gifford Category	Standard Deviation Crosswind (σ_y, m)	Standard Deviation Vertical (σ_z, m)
	Rural Conditions	
A	$0.22x(1 + 0.0001x)^{-1/2}$	$0.20x$
B	$0.16x(1 + 0.0001x)^{-1/2}$	$0.12x$
C	$0.11x(1 + 0.0001x)^{-1/2}$	$0.08x(1 + 0.0002x)^{-1/2}$
D	$0.08x(1 + 0.0001x)^{-1/2}$	$0.06x(1 + 0.0015x)^{-1/2}$
E	$0.06x(1 + 0.0001x)^{-1/2}$	$0.03x(1 + 0.0003x)^{-1}$
F	$0.04x(1 + 0.0001x)^{-1/2}$	$0.016x(1 + 0.0003x)^{-1}$
	Urban Conditions	
A–B	$0.22x(1 + 0.0001x)^{-1/2}$	$0.24x(1 + 0.0004x)^{-1/2}$
C	$0.22x(1 + 0.0001x)^{-1/2}$	$0.20x$
D	$0.22x(1 + 0.0001x)^{-1/2}$	$0.14x(1 + 0.0003x)^{-1/2}$
E–F	$0.22x(1 + 0.0001x)^{-1/2}$	$0.08x(1 + 0.00015x)^{-1/2}$

Note: For $\sigma_y(x)$ and $\sigma_z(x)$ for $100 < x < 10000$ m.

From Briggs, G.A. (1969), *Plume Rise*, AEC Critical Review Series.

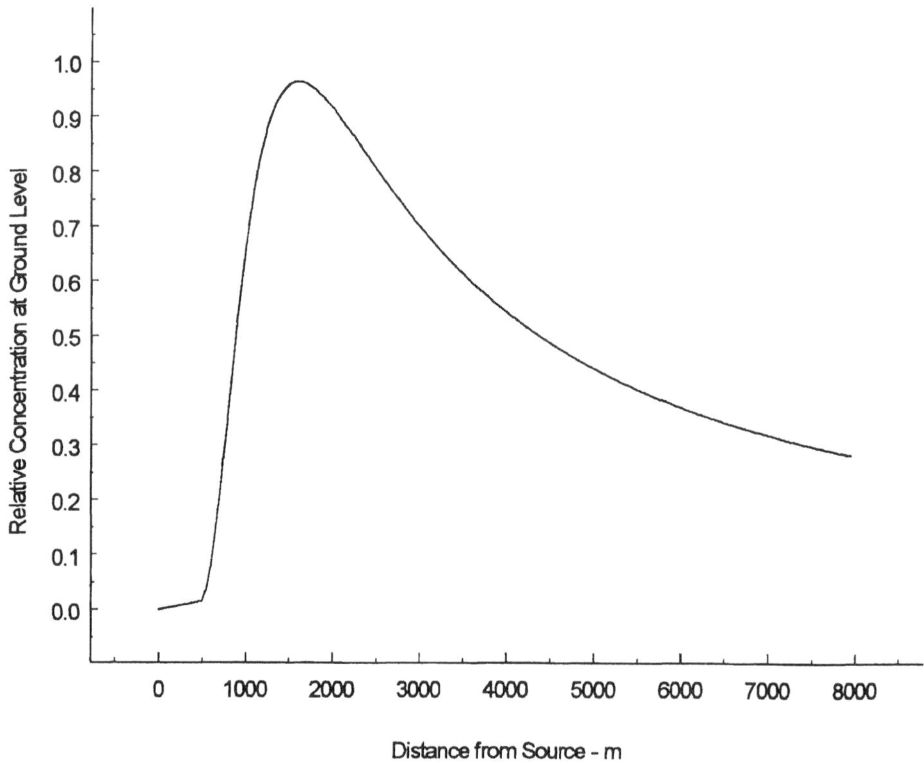

FIGURE 6.22 Typical shape of Gaussian plume model predictions of concentration vs. distance at ground level for an elevated source.

Despite these problems with the supporting data, predictions of hourly averaged pollutant concentrations based upon them are assumed to be as accurate as within a factor of two (under the best conditions). At times this degree of accuracy could best be described as wishful thinking, but the modeling approach and its widespread use ensure at least a good relative prediction of downwind concentrations. That is, the relative effect of changes in wind speed or stack height are probably reproduced well at a given source as is the impact of one source relative to another. The modeling approach thus provides a convenient tool for regulatory purposes even if more sophisticated models may be more appropriate for the purpose of making absolute concentration and exposure predictions. Several illustrations of the application of the Gaussian plume model can be found in the examples below.

Example 6.6: Estimation of maximum ground level concentration downwind of an elevated source

Estimate the maximum ground level concentration of SO_2 and its distance from the source for the stack in Example 6.5 assuming E stability and rural conditions. The SO_2 concentration in the stack is 1000 ppm.

The emission rate of SO_2 is

$$Q_m = C_s v_s \pi \frac{d_s^2}{4}$$

$$= \left(1000 \text{ ppm} \frac{64 \text{ g/mol}}{0.0245 \text{ m}^3\text{/mol} \cdot \dfrac{473 \text{ K}}{298 \text{ K}}} \right) 10 \text{ m/s} \left(\pi \frac{3 \text{ m}^2}{4} \right)$$

$$= 116 \text{ g/s}$$

From Example 6.5, the average wind speed in the plume is 4.5 m/s and the effective stack height is 157 m (30 m + 127 m). Estimating ground level (z = 0) and maximum centerline concentrations (y = 0), substitution into the Gaussian model gives

$$C(x,0,0;H_s) = \frac{Q_m}{\pi \sigma_y \sigma_z U} \exp\left[-\frac{H_s^2}{2\sigma_z^2} \right]$$

Using the Briggs' formulas for Class E, rural conditions at various distances downwind

5000 m	C_a = 0.018 mg/m^3	
10,000 m	C_a = 0.029 mg/m^3	(11,000 m C_a = 0.0293 mg/m^3)
15,000 m	C_a = 0.028 mg/m^3	
20,000 m	C_a = 0.026 mg/m^3	

The maximum concentration about 0.0293 mg/m^3 (0.112 ppm) occurs approximately 11,000 m downwind. This is below the ambient air quality standard of 0.14 ppm, although only slightly.

Example 6.7: Effect of finite mixing height

Estimate the ground-level concentration 10 km downwind of the source in Example 6.6 under B stability conditions when the maximum mixing height is 250 m.

Because a limit to vertical mixing exists, we should first assess the distance downwind in which this mixing height becomes important. Under B stability conditions, $\sigma_z = 0.12x$ and the distance to the location where the mixing height influences the plume is given by

$$\sigma_z(x_L) = 0.12\,x_H = 2.15\,H_m$$

$$x_H = \frac{2.15}{0.12}\,250 \text{ m} = 17.92\ (250 \text{ m})$$

$$x_H = 4.5 \text{ km}$$

This means that by 9 km downwind the concentration should be vertically uniform and the concentration at further distances should be estimated with

$$C_a = \frac{Q_m}{(2\pi)^{1/2}\ \sigma_y\ H_m\ U} \exp\left[-\frac{y^2}{2\sigma_y^2}\right]$$

Along plume centerline, y = 0 and $\sigma_y = 0.16x(1 + 0.0001x)^{-1/2}$. Substituting at x = 10 km,

$$C_a = \frac{116 \text{ g/s}}{(2\pi)^{1/2}(1130 \text{ m})(250 \text{ m})(4.5 \text{ m/s})}$$

$$= 0.037 \text{ mg/m}^3 = 0.014 \text{ ppm}$$

Example 6.8: Application of point source model to area sources

Consider a surface impoundment that is approximately 100 m on a side. The estimated rate of evaporation of a volatile organic compound from this impoundment is 100 g/s. Use the Gaussian plume model to predict concentrations downwind of the impoundment on a clear, sunny day with a wind speed at 10 m of 3 m/s.

The Gaussian plume model is a point source model and thus cannot be applied directly. The two simplest means of addressing this problem is by

1. Dividing the impoundment up into smaller areas that can each be effectively treated as a point source.
2. Using an imaginary or virtual point source, dispersion from which approximates the actual area source.

The first approach is straightforward but tedious. The second approach involves choosing a point upwind of the impoundment such that the plume from the point source spreads to the same width as the impoundment. A measure of the width of a Gaussian plume is 4.3 σ_y. The meteorological conditions suggest B stability and thus $\sigma_y = 0.16x(1 + 0.0001x)^{-1/2}$. A point source plume under these conditions would spread to a total width (4.3 σ_y) of 100 m about 146 m downwind. *Thus, the area impoundment has the approximate appearance of the horizontal spread from a point source located 146 m upwind of the center of the impoundment.* Such an approach will be very inaccurate close to the area source but will provide quite reasonable results at downwind distances much greater than the width of the source.

To illustrate the application of this approach, consider a location 500 m downwind of the center of the impoundment. The concentration at that point using the virtual point source approach is given by evaluating the Gaussian model with σ_y evaluated at a downwind distance of 500 m + 146 m (the distance to the virtual source) while σ_z is evaluated at just the distance downwind (500 m).

$$C_a(x) = \frac{Q_m}{\pi \sigma_y (x + x_v) \sigma_z (x) U}$$

$$= \frac{100 \text{ g/s}}{\pi 0.16(646 \text{ m})(1 + 0.0001(646 \text{ m}))^{-1/2}(0.12)(500 \text{ m})(3 \text{ m/s})}$$

$$= 0.41 \text{ mg/m}^3$$

6.4 PARTICULATE CONTROL PROCESSES

An objective equally important to assessing the fate and transport processes influencing air pollutants is the control technologies necessary to reduce or eliminate significant emissions of these pollutants. The assessment of the transport and dispersion processes outlined in the previous section allows, at least in principle, the identification of the level of emissions that would not result in significant adverse exposures to humans and the ecosystem. Our goal would then be development of control or prevention strategies to ensure emissions remain at or below these levels. In this and the subsequent sections of this chapter we will examine the control technologies and their fundamental design considerations for each of the major air pollutants. The purpose of this discussion is not to develop design equations that the student can use to prepare the final process and equipment design but instead to develop an understanding, through models and basic design equations, of the key parameters that control the effectiveness of a particular pollution control process.

In this section we will focus on the control of airborne particulate matter. The design of control processes generally involves two steps:

1. Identification of a mechanism for collection or elimination of a pollutant on which a control device might be based.
2. Development of an overall material balance that includes macroscopic flow and mixing behavior as well as the microscopic description of the collection or elimination mechanism.

This design/model development process will be illustrated with each of the major control processes for particulate and gaseous pollutants in the coming sections.

6.4.1 Gravity Settling

One of the simplest of all air pollution control devices depends on the tendency of particles to settle in the atmosphere. The buoyancy force acting on an isolated particle is the weight of the particle relative to the fluid that would occupy the same space if the particle were not there. Weight is given by mass times the acceleration of gravity, so the buoyancy force is

$$\text{Buoyancy force} = F_B = m_p g - m_a g = (\rho_p - \rho_a)\pi \frac{d_p^3}{6} g \qquad (6.61)$$

where ρ_p and ρ_a is the density of the particle and air, respectively. This assumes that the particle is essentially spherical with diameter d_p. Typically, the density of air is a factor of a thousand or more smaller than the density of the solid particle so that ρ_a can be neglected. If the particle were falling through water or another liquid, however, the density of the fluid could not be so easily neglected.

The buoyancy force is resisted by the frictional drag of the fluid through which the particle falls. This *drag force*, F_D, is generally described by a dimensionless *drag coefficient*, C_D, defined by normalizing the force with the kinetic energy per unit volume in the fluid displaced by the particle, $\rho_a v_p^2/2$ and the *projected area* of the assumed spherical particle, $\pi d_p^2/4$,

$$C_D = \frac{F_D}{\left(\pi \frac{d_p^2}{4}\right)\left(\rho_a \frac{v_p^2}{2}\right)} \tag{6.62}$$

The projected area is the area of the particle's projection onto a two-dimensional surface. A sphere's projected area is thus that of a circle of the same diameter. The drag coefficient is a function of the diameter of the particle, d_p, the air viscosity, μ_a, and density, ρ_a, and the velocity of the fluid relative to the particle, v_p,

$$C_D = C_D(d_p, \mu_a, \rho_a, v_p) \tag{6.63}$$

Since the drag coefficient depends on four variables in three fundamental dimensions of mass (M), length (L), and time (t) only a single combination exists of these variables into a dimensionless group which should correlate the drag coefficient. Forming that group and applying dimensional analysis, the Reynolds number, N_{Re}, arises.

$$\begin{aligned}
\pi &= \rho_a v_p^\alpha d_p^\beta \mu^\gamma \\
&= \left[\frac{M}{L^3}\right]\left[\frac{L}{t}\right]^\alpha [L]^\beta \left[\frac{M}{L \cdot t}\right]^\gamma \\
M &: 1 + \gamma = 0 \qquad \gamma = -1 \\
t &: -\alpha + (-\gamma) = 0 \qquad \alpha = 1 \\
L &: -3 + \alpha + \beta + (-\gamma) = 0 \qquad \beta = 1 \\
\pi &= \frac{\rho_a v_p d_p}{\mu_a} = N_{Re}
\end{aligned} \tag{6.64}$$

Thus, the drag coefficient is a function of the Reynolds number

$$C_D = C_D\left(\frac{\rho_a v_p D_p}{\mu_a}\right) = C_D(N_{Re}) \tag{6.65}$$

The Reynolds number is a measure of the ratio of inertia forces (advection of momentum) to viscous forces (diffusion of momentum). At low velocities, viscous forces dominate while inertial forces dominate at high velocity. The drag and the drag coefficient are very different in these two regimes.

For slow relative velocities, it is possible to derive the following relationship for the drag coefficient and drag force,

$$\text{Drag coefficient} = C_D = 24\left(\frac{\mu_a}{\rho_a v_p d_p}\right) = \frac{24}{N_{\text{Re}}} \tag{6.66}$$

Substituting into Equation 6.62, the drag force under these conditions is given by

$$\text{Drag force} = F_D = \left(24\frac{\mu_a}{\rho_a v_p d_p}\right)\left(\pi\frac{d_p^2}{4}\right)\left(\rho_a\frac{v_p^2}{2}\right) = 3\pi d_p \mu_a v_p \tag{6.67}$$

Equation 6.67 is termed Stokes' Law after the English mathematician who first derived it. The acceleration of the particle is related to the difference between the drag and buoyancy forces. A steady velocity, the *terminal velocity*, is reached when the forces are balanced. The terminal velocity of the particle relative to the fluid is given by,

$$F_B = F_D$$

$$\left(\rho_p - \rho_a\right)g\pi\frac{d_p^3}{6} = 3\pi d_p \mu_a v_p \tag{6.68}$$

$$v_p = \frac{(\rho_p - \rho_a)g d_p^2}{18\mu_a}$$

Experiments indicate that Equation 6.68 is valid for $N_{\text{Re}} \leq 1$ consistent with Stokes' derivation which assumed that the inertial forces were negligible. Note that the equation suggests that particles that are twice as large will settle due to gravity at four times the rate.

This rapid increase in settling velocity, however, is somewhat offset by changes in drag at high Reynolds number. At higher values of the Reynolds number (i.e., higher velocities or for a less viscous fluid), the terminal velocity of the particle can be written in terms of the drag coefficient.

$$F_B = F_D$$

$$\left(\rho_p - \rho_a\right)g\pi\frac{d_p^3}{6} = C_D\left(\pi\frac{d_p^2}{4}\right)\left(\rho_a\frac{v_p^2}{2}\right) \tag{6.69}$$

$$v_p = \left[\frac{4}{3}\frac{d_p g(\rho_p - \rho_a)}{C_D\rho_a}\right]^{1/2}$$

The drag coefficient in this equation is given by $24/N_{\text{Re}}$ for $N_{\text{Re}} \leq 1$ but for larger values N_{Re} must be correlated with experimental data. A commonly used relationship between drag coefficient and N_{Re} is

$$C_D = 0.22 + \frac{24}{N_{\text{Re}}}\left[1 + 0.15(N_{\text{Re}})^{0.6}\right] \tag{6.70}$$

Because of increased drag, larger particles (which increases the N_{Re}) fall more slowly than if they were governed by Equation 6.68.

One difficulty in applying Equations 6.69 and 6.70 is that they both include the velocity. Typically a velocity is assumed, a drag coefficient is calculated from Equation 6.70, and this drag coefficient is used to calculate the terminal velocity from Equation 6.69. If this calculated velocity agrees well with the assumed value, then the calculated terminal velocity is correct. If the assumed and calculated values disagree, then a new value of velocity must be guessed which is then used to calculate a new drag coefficient (from Equation 6.70) and a new terminal velocity (from Equation 6.69). This iterative process is continued until the velocity calculated by Equation 6.69 agrees with the guess. This process is illustrated in Example 6.9.

In this example the fact that the difference between the initial guess and the calculated velocity indicates whether the new guess of settling velocity should be smaller or larger. Because the calculated value is always closer to the actual settling velocity (that is, the settling velocity consistent with Equations 6.69 and 6.70), the next guess is taken *beyond* the calculated value. This is an example of *overrelaxation*, a commonly used process to speed convergence of an iterative calculation. In the example, the new guess for substitution into Equation 6.70 is assumed to be the original guess plus 1.5 times the difference between the value calculated in Equation 6.69 and the original guess. That is, the new guess is 50% further away from the original guess than the value calculated by Equation 6.69. As shown in Example 6.9, this allows the calculation process to converge to a consistent estimate of settling velocity in only two iterations. The results of calculations for gravity settling velocity as a function of particle size and density are contained within Figure 6.23. The straight portions of these curves indicate the region where Stokes' Law is valid. A flattening of the curve at large particle sizes indicates that the settling velocity of these particles are sufficiently large that the drag coefficient correlation, Equation 6.70, is required.

Figure 6.23 also shows curvature at small particle sizes. This is associated with the reduced drag of small particles that are of nearly the same size as the mean free path between gaseous molecular collisions. The drag on the particles in this size range are generally corrected by the Cunningham correction factor, α_C,

$$F_D = \frac{3\pi\mu d_p v_p}{\alpha_C} \tag{6.71}$$

where the Cunningham correction factor is given in terms of the mean free path of the gas molecules, λ, as

$$\alpha_C = 1 + \frac{2\lambda}{d_p}\left(1.257 + 0.4e^{-1.1d_p/2\lambda}\right)$$

where (6.72)

$$\lambda = \frac{\mu}{0.499\rho_g\sqrt{8RT/\pi M_w}}$$

For particles in air, this factor is approximately given by

$$\alpha_C = 1 + 0.00973\frac{T^{1/2}}{d_p} \quad (\text{T in K, } d_p \text{ in } \mu\text{m}) \tag{6.73}$$

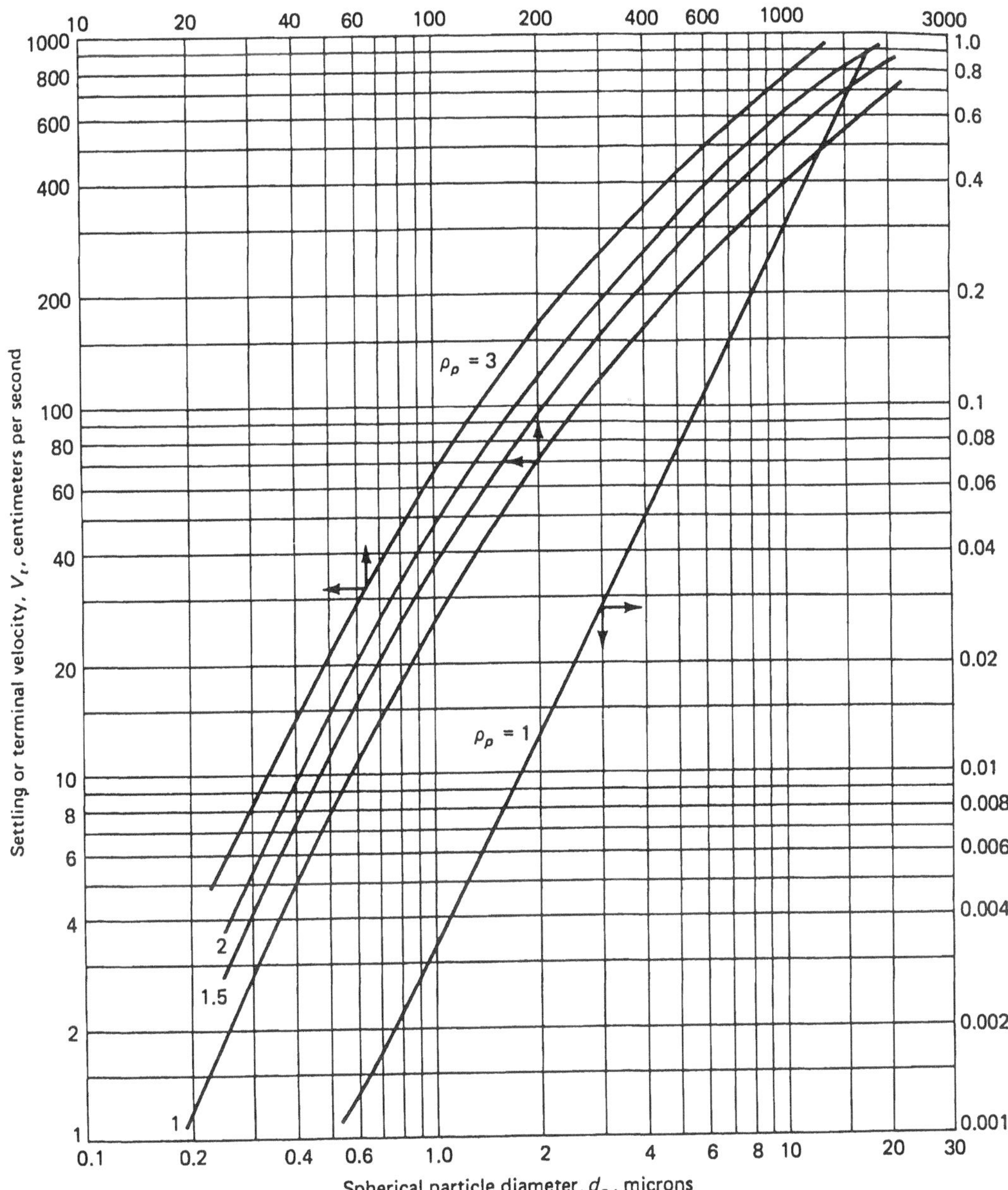

FIGURE 6.23 Stokes' law settling velocities. (From Wark, K. and C.F. Warner (1981) *Air Pollution: Its Origin and Control*, Harper Collins, New York. With permission.)

The Cunningham correction factor is unimportant except for particles of the order of 1 μm in diameter or smaller. The value of the correction factor at ambient temperature and pressures is about 1.05 for 5-μm diameter particles, 1.16 for 1-μm particles, and about 3 for 0.1-μm particles.

Example 6.9: Estimation of gravity settling velocity

Estimate the gravity settling velocity of 1-, 10-, and 100-μm diameter particles (ρ_p = 2500 kg/m^3) in air at 25°C.

The viscosity of air at 25°C and 1 atm pressure is 0.067 kg·m^{-1}·h^{-1}. Assuming Stokes' Law is applicable the settling velocity of the 1-μm particles is given by

$$v_p(1\ \mu\text{m}) = \frac{(2500\ \text{kg/m}^3 - 1.18\ \text{kg/m}^3)(9.8\ \text{m/s}^2)(1\cdot 10^{-6}\ \text{m})^2}{18(1.86\cdot 10^{-5}\ \text{kg}\cdot\text{m}^{-1}\cdot s^{-1})}$$

$$= 7.3\cdot 10^{-5}\ \text{m/s}$$

Calculation of the Reynolds number shows that the Stokes' Law assumption is valid.

$$N_{Re} = \frac{(1.18\ \text{kg/m}^3)(7.3\cdot 10^{-5}\ \text{m/s})(1\cdot 10^{-6}\ \text{m})}{1.86\ \text{kg}\cdot\text{m}^{-1}\cdot s^{-1}}$$

$$= 4.65\cdot 10^{-6}$$

Because of the Cunningham correction factor, the actual settling velocity is

$$v_p(1\ \mu\text{m}) = \frac{7.3 * 10^{-5}\ \text{m/s}}{\alpha_C} = \frac{7.3 * 10^{-5}\ \text{m/s}}{1.16} = 6.3 * (10)^{-5}\ \text{m/s}$$

Repeating the calculation for a 10-μm particle, which requires no Cunningham correction factor, indicates a Stokes' Law settling rate of about 0.73 cm/s and an N_{Re} of about 0.0047. These values remain low but some gravity settling should occur. Repeating once again for the 100-μm particles, the predicted settling rate is 73.1 cm/s but the N_{Re} is about 4.6, using Equations 6.69 and 6.70 starting from the calculated Stokes' Law settling rate.

Assumed v_p (v_{p0})	N_{Re}	C_D From Eq. 6.70	v_p From Eq. 6.69	Δv_p	$v_{p0} - 1.5\Delta v_p$
73.1 cm/s	4.65	7.34	61.3 cm/s	−11.8 cm/s	54.5 cm/s
54.5 cm/s	3.47	9.33	54.4 cm/s	0.1 cm/s	54.3 cm/s
54.3 cm/s	3.45	9.36	54.3 cm/s	0.0 cm/s	—

The 100-μm particle falls at about 54.3 cm/s. Such a particle would be expected to be subject to significant gravity settling, although in the atmosphere such settling might still be significantly deterred by turbulence.

6.4.2 Particle Settling in the Atmosphere

Let us consider particles falling in the atmosphere. Let us focus on particles large enough that they are not influenced by atmospheric mixing processes. As such particles move downwind from the source at elevation H_s, and they continuously fall at their gravity settling velocity, v_p. The time required before the particles will fall to the ground is H_s/v_p. In that time, the horizontal distance traveled is given by $U \cdot t = UH_s/v_p$ where U is the mean wind speed. Alternatively we can consider the trajectory of the particles. In a time, t, the vertical fall distance will be $\Delta z = v_p t$ while the horizontal travel distance will be $x = U t$, or $t = x/U$. Thus, the trajectory from the initial height H_s is given by

$$z = H_s - \Delta z = H_s - v_p \frac{x}{U} \tag{6.74}$$

This is for particles that only settle and are not influenced by turbulent mixing. Let us now consider particles that are mixed by normal atmospheric motions, including turbulence, but that the settling velocity is superimposed on these motions. The reflection condition at the ground is no longer applicable since settling particles presumably remain at the ground surface upon striking it. In addition, the effective height of the plume center is following the trajectory of the settling particles, Equation 6.74. With these two modifications, the Gaussian model could be rewritten.

$$C(x,y,z;H_s)=\frac{Q_m}{2\pi\sigma_y\sigma_z U}\exp\left[-\frac{y^2}{2\sigma_y^2}\right]-\left[\frac{\left\{z-\left(H_s-\frac{v_p x}{U}\right)\right\}^2}{2\sigma_z^2}\right] \tag{6.75}$$

This is called the tilting plume model because the plume centerline tilts downward as the particles settle. This model is a crude approximation of the behavior of settling particles in the actual atmosphere and in particular, is inaccurate for distances downwind such that the centerline has struck the ground.

The model does provide estimates of concentrations as a function of distance downwind. These can be used to estimate the particulate deposition rate as a function of distance downwind. Very near the ground turbulence is expected to be small and vertical motions in the air must effectively cease at the air-soil interface. Due to gravity settling particles are still expected to have a vertical velocity downward. The volumetric flux of particles is v_p and the mass flux of particles is then given by $v_p\ C_a(z = 0)$ where $C_a(z = 0)$ is the concentration of particles in the air at the air-soil interface. From Equation 6.75, the particle deposition rate as a function of distance downwind is given by

$$q_m=v_p C(x,y,0;H_s)=\frac{Q_m v_p}{2\pi\sigma_y\sigma_z U}\exp\left[-\frac{y^2}{2\sigma_y^2}\right]-\left[\frac{\left(H_s-\frac{v_p x}{U}\right)^2}{2\sigma_z^2}\right] \tag{6.76}$$

The maximum deposition rate as a function of downwind distance can be estimated by setting y = 0. This ability of particles to be collected on the ground can also form the basis for a particulate device in a gas stream. This is explored in the next section.

6.4.3 Gravity Settlers

Let us now consider the application of the gravity settling of particulate matter to a pollution control device. Because particles settle relative to the carrier gas stream, this mechanism can be used, in principle, to separate particulate matter from the gas stream. Unless turbulence levels in the device are very low, only particles 100 μm and larger might be expected to settle in a reasonable time. In reality a gravity settler is not an affective particulate control device, but it does provide a convenient model for the evaluation of design equations for particulate control equipment and it does exhibit features important of all such devices.

Consider a simple rectangular chamber. A particulate laden gas stream introduced at one end of the chamber deposits particulate matter according to gravity settling onto the floor of the chamber. It is assumed that once particles deposit on the floor of the chamber the particulate matter is collected and is not reentrained into the gas stream. This is a simple gravity settler.

We have now identified a mechanism for particulate collection, gravity settling, and the geometry of the device where the collection will occur. In order to develop a model of the effectiveness of the device, however, we must still define the flow behavior within the device. The mixing characteristics of the device and the response of the particulate distribution to this mixing is an important factor in understanding the effectiveness of the collector. There are three extremes of mixing that can be considered and the implications of each for the effectiveness of a particulate control device will be considered in turn.

1. No mixing, that is turbulent levels are sufficiently low or the particulates settle at a sufficiently fast rate that their trajectory is defined solely by their horizontal velocity and the settling rate as in Equation 6.74.
2. Rapid mixing perpendicular to the flow direction as a result of turbulent motions but with mixing in the direction of flow minimal and transport in that direction controlled by advection, this is the equivalent to the plug flow or axially unmixed system discussed in Chapter 5.
3. Rapid mixing in all directions within the chamber as a result of intense turbulent motions, the concentration of particulate matter everywhere within the chamber is assumed uniform.

6.4.3.1 Unmixed Collector

In the unmixed model, turbulence does not influence the motion of the particles. The trajectory of a single particle entering the device in Figure 6.24a at height z(0) is simply that given by the equivalent of Equation 6.74

$$z(x) = z(0) - \Delta z = z(0) - v_p \frac{x}{U} \tag{6.77}$$

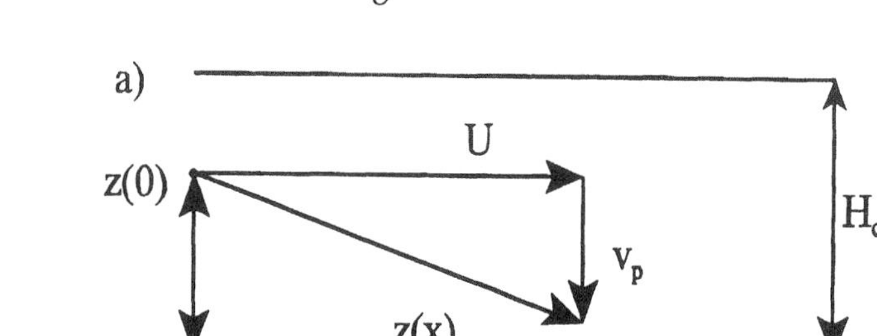

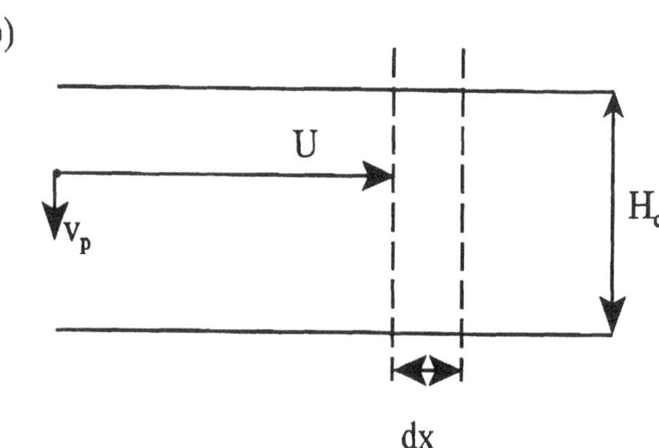

FIGURE 6.24 Schematic of gravity particulate settler/collector: (a) laminar or unmixed collector and (b) plug flow collector.

Collection of the particle depends on whether the change in elevation over the length of the device is sufficient to cause the particle to come in contact with the bottom of the chamber. Any resuspension after contacting the bottom surface is also assumed to be negligible. For a chamber of effective length L, all particles with an elevation at the chamber inlet less than or equal to v_pL/U will have been collected while all particles at a higher elevation will have not yet reached the floor of the chamber. The efficiency of collection, that is the fraction of particles in the inlet gas stream that will be collected, is given by

$$\eta = \frac{C_a(0) - C_a(L)}{C_a(0)} = \frac{V_p P}{UH_c} = \frac{\tau_Q}{\tau_c} \tag{6.78}$$

where this has been written in terms of the volumetric residence time of air in the device, $\tau_Q = L/U$, and a time required for collection by settling, $\tau_c = H_c / v_p$. Note that if $v_pL/U \geq H_c$, or $\tau_Q \geq \tau_c$, then all of the particles will be collected and $\eta = 1$ or 100%. Note also that since the settling velocity is a function of particle size, the efficiency of collection is also a function of particle size, $\eta = \eta(d_p)$.

6.4.3.2 Plug Flow Collector

Let us now consider the second mode of mixing, mixed perpendicular to the flow (z-direction) and unmixed along the axis of flow (x-direction). This is the mixing often attributed to turbulent flow in a long pipe where mixing across the relatively small pipe diameter is assumed but the fluid remains unmixed along the length of the pipe. A material balance for this case was previously discussed in Chapter 5. For the particulate collector of interest here, the material balance should consider three processes over a differential element of length in the collector of Figure 6.24b.

1. Advection of particulate with the inlet carrier gas at x.
2. Advection of particulate with the outlet carrier gas at x + dx.
3. Deposition of particulate along the settler floor at a rate given by V_pC_g where C_g is the uniform concentration in the differential element.

Let us consider only steady flow through the device. The material balance then takes on the form,

$$\text{Inlet} = UC_a(x)H_cW \quad \text{Outlet} = UC_a(x+dx)H_cW \quad \text{Deposition} = v_pC_aWdx$$

$$UC_a(x)H_cW - UC_a(x+dx)H_cW - v_pC_aWdx = 0$$

$$\frac{C_a(x+dx) - C_a(x)}{dx} = -\frac{v_p}{H_cU} \tag{6.79}$$

$$\frac{dC_a}{dx} = -\frac{v_p}{H_cU}$$

Integrating the equation for $C_a(x)$ and setting x = L for the exit concentration,

$$C_a(L) = C_a(0)e^{-\frac{v_p L}{H_c U}} = C_a(0)e^{-\frac{\tau_Q}{\tau_c}} \tag{6.80}$$

where $C_a(0)$ is the inlet particulate concentration in the air stream. Note that when $\tau_Q = \tau_s$, the exit concentration is equal to $e^{-1}C_a(0)$ or 37% of the initial concentration.

The efficiency of collection for any residence time is given by

$$\eta = \frac{C_a(0) - C_a(L)}{C_a(0)} = 1 - e^{-\frac{\tau_Q}{\tau_s}} \tag{6.81}$$

and when $\tau_Q = \tau_c$, the efficiency of collection is 1 – 0.37 = 0.63 or 63% compared to 100% for the completely unmixed settler of the same size. Again, larger particles with a shorter settling time would be collected with more efficiency than smaller particles.

6.4.3.3 Completely Mixed Settler

The final device for consideration is the completely mixed collector which might occur if turbulent levels are sufficiently high to maintain an essentially uniform concentration of particulate matter throughout the device. Deposition and collection will still occur, but the rapid mixing throughout the device means lower concentrations near the collection surface and thus a lower deposition rate v_pC_a. A material balance on the device is

$$\text{Inlet} = UC_a(0)H_cW \quad \text{Outlet} = UC_aH_cW \quad \text{Deposition} = v_pC_aWL$$

$$UC_a(0)H_cW - UC_aH_cW - v_pC_aWL = 0 \tag{6.82}$$

$$C_a = \frac{C_a(0)}{1 + \frac{v_p}{H_c}\frac{L}{U}} = \frac{C_a(0)}{1 + \frac{\tau_Q}{\tau_c}}$$

Now the efficiency of collection is given by

$$\eta = 1 - \frac{1}{1 + \frac{\tau_Q}{\tau_c}} = \frac{1}{\frac{\tau_c}{\tau_Q} + 1} \tag{6.83}$$

This means that when $\tau_Q = \tau_c$, only 50% of the particles are collected compared to 63% with a plug flow collector and 100% with a completely unmixed collector. This behavior is typically true in that mixing will tend to reduce the effectiveness of any process proportional to concentration (such as deposition rate, V_pC_a) because mixing will reduce the effective concentration.

Example 6.10 illustrates the application of the design equations that have been developed for the gravity settler.

Example 6.10: Application of settler design equations

Define the length of settling required in a settler 10 cm high and 20 cm wide with a 1000 m^3/h gas flow containing 1 g/m^3 of each of the particles whose settling velocity was evaluated in Example 6.9. Assume that the design will be such that 100% collection of the 10-μm particles will occur if the settler is completely unmixed and all particles impacting the bottom of the settler are collected. Estimate the efficiency of collection of the remaining particles with each of the three mixing assumptions.

The axial velocity, U, and the length based on 100% collection of the 0.729 cm/s settling velocity of the 10-μm particles are given by

$$U = \frac{(1000\ \text{m}^3/\text{h})}{(0.2\ \text{m})(0.1\ \text{m})} = 13.9\ \text{m/s} \qquad L = \frac{H_c U}{v_p} = \frac{(0.1\ \text{m})(13.9\ \text{m/s})}{0.00729\ \text{m/s}} = 190\ \text{m}$$

This means that a **gravity settler 190 m long is required** to settle the 10-μm particles. By comparison only about 2.6 m is required to settle the 100-μm particles. The efficiency of collection for the other particles and under the various mixing assumptions are

d_p	v_p (m/s)	τ_Q (s)	τ_c (s)	$\eta_{unmixed}$ (τ_Q/τ_c)	$\eta_{plug\ flow}$ $1 - \exp(-\tau_Q/\tau_c)$	η_{mixed} $(1 + \tau_Q/\tau_c)^{-1}$
1 μm	0.000073	13.7	1370	1%	1%	0.99%
10 μm	0.00729	13.7	13.7	100%	63%	50%
100 μm	0.543	13.7	0.18	100%	~100%	98.7%

All of these efficiencies of collection are likely to be overestimates because of incomplete collection of particles striking the bottom surface. It is clear, however, that gravity settling is likely to be a viable particulate control technology only for very large particles.

Note that the evaluation of the gravity settler developed in two steps

1. Identification and quantification of a mechanism for collection (deposition at a rate v_pC_a).
2. Evaluation of a mixing model that allows determination of the concentration C_a (and thus a model of collector effectiveness).

In every case, the ratio of the residence time to the particle collection time is the key parameter in defining the efficiency of the settler. This ratio can also be written in terms of the air volumetric flowrate, $Q_a = UH_cW$, and the volumetric rate of particle removal, $Q_p = v_pWL$

$$\frac{\tau_Q}{\tau_c} = \frac{v_p}{H_c}\frac{L}{U} = \frac{Q_p}{Q_a} \tag{6.84}$$

Although a gravity collector is not an effective particulate control device, these steps will be repeated in the development of design equations for all other devices. A gravity settler is just the simplest and therefore the most convenient system with which to illustrate this process. Gravity settlers also exhibit an efficiency of collection that is a function of particle size. This feature is also reproduced with all other particulate control devices.

Larger particles contain a greater proportion of the total mass of particulate matter in a gas stream. Because the efficiency of collection of larger particle sizes is higher in a gravity settler and most other particulate control devices, the overall or average efficiency on a mass basis is greater than the overall efficiency on a number basis. Reporting mass removal efficiencies is misleading, however, in that it is increasingly being recognized that the fine particles provide the most potential for adverse health effects. For this reason, regulations on particulate emissions specify allowable concentrations or removal efficiencies on the basis of less than the 10-μm particles and less than 2.5-μm particles. Otherwise reported removal efficiencies would be biased by the relatively easy removal of large particles.

6.4.4 Cyclone Separator

A more effective and commonly used particulate separator is a cyclone separator. As with the gravity settler, its analysis proceeds by evaluation of the mechanism whereby particulates can be collected and development of a design equation that incorporates a mixing assumption in order to evaluate the effectiveness of the process. The mechanism of collection in a cyclone separator is centrifugal force. By forcing the inlet gas stream to follow a curved path at high velocity, the centrifugal forces acting upon the particles result in a movement of the particles away from the axis of the gas' rotation, ultimately allowing collection at a wall. It is effectively identical to a gravity settler but by using centrifugal force as an "artificial gravity," a more compact and effective system results. The centrifugal acceleration directed outward is U_t^2/R where U_t is the tangential velocity of the carrier gas stream and R is the radius of curvature.

Figure 6.25 depicts a typical cyclone separator composed of a cylindrical chamber on top of a conical chamber. Flow enters on the side of the cylinder and is forced in a curved path around the outside wall. Particles collide with the wall and fall to the bottom of the separator to be collected. The gas stream eventually flows up and out through the center of the device. The sloping sides of the cyclone separator allow for continuous collection of particles at the bottom, a significant advantage over the gravity separator in that the system need not be shutdown to harvest the collected particles. In addition, the tangential velocity and the radius of curvature can be controlled to maximize the movement of particles toward the wall, whereas in a gravity settler the mechanism of action is not as easily controlled. Let us develop a model of the mechanism of particulate collection and a design equation capable of predicting efficiency of collection as we did with the gravity settler.

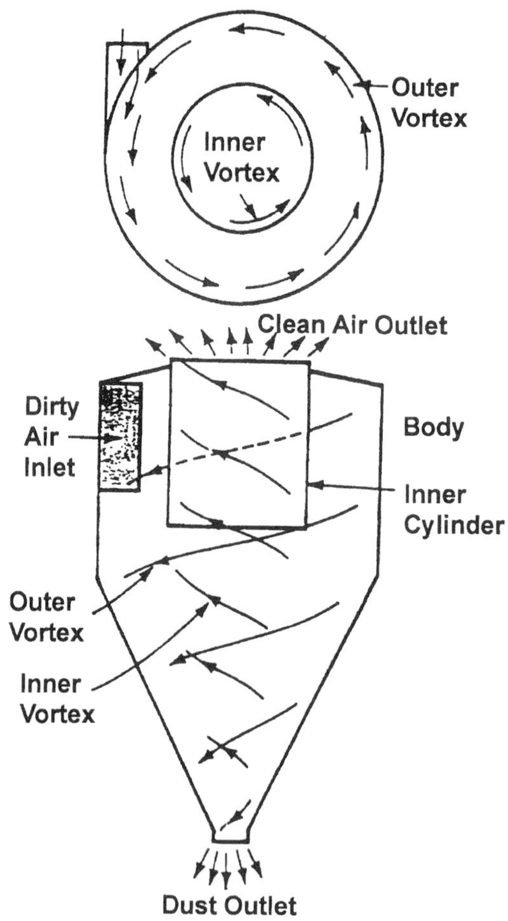

FIGURE 6.25 Centrifugal separator.

In a gravity settler, a particle velocity normal to the direction of gas flow developed as a result of the balance between gravity or buoyancy forces and the drag force. In a cyclone separator, the same result is achieved from a balance between the centrifugal force and drag. The centrifugal force is the particle mass times the centrifugal acceleration (U_t^2/R), or, solving for the particle velocity, both in general and for $N_{Re} \leq 1$,

$$F_C = F_D$$

$$(\rho_p - \rho_a)\frac{U_t^2}{R}\pi\frac{d_p^3}{6} = C_D\left(\pi\frac{d_p^2}{4}\right)\left(\rho_a\frac{v_p^2}{2}\right)$$

$$v_p = \left[\frac{4}{3}\frac{d_p U_t^2(\rho_p - \rho_a)}{C_D R \rho_a}\right]^{1/2} \tag{6.85}$$

$$v_p = \frac{(\rho_p - \rho_a)d_p^2 U_t^2}{18\mu_a R} \qquad \text{for } N_{Re} \leq 1$$

Thus, the collection velocity for a cyclone separator is identical to that of a gravity settler except that the acceleration of gravity is replaced with the centrifugal acceleration.

Development of a design equation for a cyclone separator also proceeds as with the gravity settler. Although we developed a design equation for the gravity settler under unmixed, mixed, and plug flow assumptions, only one assumption is generally applicable to any real system. For the cyclone separator and, indeed, for most particulate control devices they tend to best described by the plug flow or axially unmixed assumption. Other approaches have been employed with the cyclone separator, but this is sufficient for our purpose which is simply to identify the key parameters that influence a particular pollution control device and not attempt to define final design equations.

The flow path in a cyclone separator tends to be described in terms of the number of revolutions made by the dust-laden gas. If H_i is the height of the rectangular inlet and ΔH_e is the height difference between the entrance and the gas exit, an approximation of the number of revolutions made by the gas during passage through the cyclone separator is $N = \Delta H_e/H_i$. This approximation assumes that a volume of inlet air retains its integrity during passage through the separator in keeping with the lack of mixing along the axis of motion. The total length of the collecting path is then $2\pi RN$. The residence time in the device is then $2\pi RN/U_t$. The "settling" time is the ratio of the distance between the inner outlet flow pipe and the outer particulate collection wall, and the collection velocity, i.e., $\Delta R/v_p$. Thus, via the model developed previously for the plug flow assumption, the efficiency of collection of any particular size particle in a cyclone separator is given by

$$\eta = 1 - e^{-\tau_Q/\tau_c}$$

$$= 1 - e^{-\frac{2\pi RNv_p}{U_t\Delta R}}$$

$$= 1 - \exp\left\{-\frac{2\pi N}{\Delta R}\left[\frac{4}{3}\frac{d_p R(\rho_p - \rho_a)}{C_D\rho_a}\right]^{1/2}\right\} \tag{6.86}$$

$$= 1 - \exp\left\{-\frac{2\pi N}{\Delta R}\frac{(\rho_p - \rho_a)d_p^2 U_t}{18\mu_a}\right\} \qquad \text{for } N_{Re} \leq 1$$

As usual for a plug flow collector, the condition that $\tau_c = \tau_Q$ results in an efficiency of collection of 63%.

As with gravity settlers, particle diameter and fluid viscosity are key parameters in the effectiveness of a cyclone separator. Example 6.11 illustrates the application of the design equation for a cyclone separator. As indicated previously, this approach is generally as simple but much more effective than conventional gravity settling. Cyclone separators are widely used for routine solid removal from gas streams from everything from saw mills to chemical plants. They are relatively inexpensive and effective for particulates 5 to 10 μm in diameter and above.

Example 6.11: Application of cyclone separator design equations

Define the efficiency of collection of each of the particle sizes in the gas stream from Example 6.10 in a cyclone separator. Assume the inlet to the separator is 10 cm wide (ΔR) and 20 cm high giving the same entrance velocity of 13.9 m/s as in that example. The gaseous exit piping is 1 m below the entrance and the average radius of curvature in the cyclone is 20 cm.

The centrifugal acceleration in this case is U_t^2/R = 965 m/s², nearly 100 times stronger than the acceleration of gravity. This should allow much greater collection efficiencies and/or a smaller, cheaper control device. The number of turns or revolutions of the gas in the cyclone (N) is 1 m/0.2 m = 5. The collection velocity calculated from Equation 6.85 is shown in the table below. The collection velocity of both the 10- and 100-μm particles was corrected with the drag coefficient, although the correction for the 10-μm particle was less than 10%.

Calculating the efficiency of collection for the cyclone and comparing to the gravity settler,

d_p	v_p (m/s; Stokes)	v_p (m/s)	τ_Q (s)	τ_c (s)	$\eta_{plug\ flow}$ $1 - \exp(-\tau_Q/\tau_c)$	$\eta_{plug\ flow}$ from Ex. 6.10
1 μm	0.0072	0.0072	0.45	13.9	3.2%	0.99%
10 μm	0.719	0.657	0.45	0.15	95%	63%
100 μm	71.9	16.4	0.45	0.006	~100%	~100%

Note that the use of the cyclone separator requires a much smaller (and presumably cheaper) system than the gravity settler and efficiencies are better.

6.4.5 Electrostatic Precipitator

Neither the gravity settler nor the cyclone separator can generally be used to effectively remove small particles (<1 – 10 μm) from a gas stream. Unfortunately, it is precisely these particles that are the primary human health concern. One commonly used process that can be used to collect these small particles is *electrostatic precipitation*. If the particles that require collection are electrically charged or can be charged, exposure to an electric field will result in deflection of their motion. The electric field can replace gravity or centrifugal force in the settlers discussed above and the motion of the particles relative to the gas stream can be used to collect the particles.

The physical configuration of a common type of electrostatic precipitator, a plate and wire type, is shown in Figure 6.26. The system basically involves passing the particulate laden gas stream between electrodes to which is applied a very high voltage, e.g., 50,000 V. This voltage is applied across a gap between the electrodes of a few centimeters, creating electric field intensities of the order of 10,000 V/cm. Under these high electric fields, electrons are discharged from the negatively charged wire electrode and migrate toward the positive electrode (the plate in the configuration shown in Figure 6.26). Molecules in the gas stream are ionized by collisions with the electrons and the negative ions migrate toward the collecting electrode. The subsequent collision of particulate

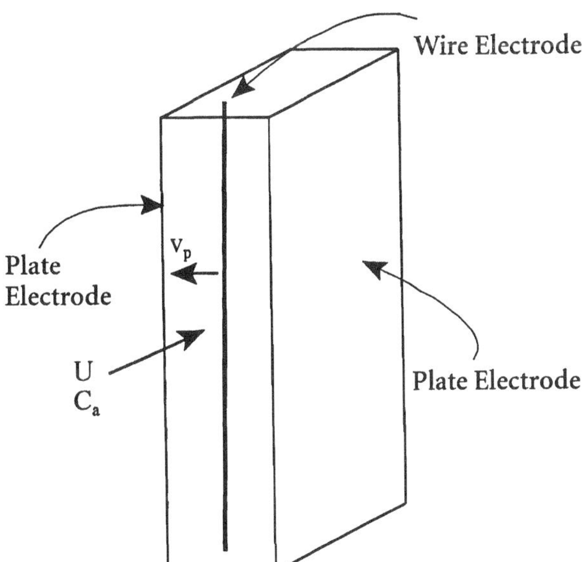

FIGURE 6.26 Plate and wire electrostatic precipitator.

matter with these negatively charged gas molecules results in charging and migration of the particles. The collecting (or precipitating) force, F_p, acting on a particle containing charge q_e is given by

$$F_p = q_e E_p \tag{6.87}$$

where E_p is the electric field intensity associated with the particle collection. The collecting electric field intensity, is in general, different from the local *charging* electric field at the discharge electrode, E_c, despite the fact that the same potential difference is responsible for the formation of both electric fields. The electric field in the vicinity of the discharge electrode is also a function of electrode geometry and the conductivity and composition of the gas stream. Theoretically, it has been shown that the charge that will accumulate on a particle of diameter, d_p, is given by

$$q_e = \pi \kappa_0 \left(\frac{3\kappa_r}{\kappa_r + 2} \right) d_p^2 E_c \tag{6.88}$$

where κ_r is the relative permittivity or dielectric constant of the particles (typically 2 to 8) and κ_0 is the permittivity of free space, 8.854 (10^{-12}) $C \cdot V^{-1} \cdot m^{-1}$.

The force causing migration of the particle to the collecting electrode is again resisted by the hydrodynamic drag. Thus, the steady state or terminal velocity toward the collecting electrode is given by

$$F_p = F_D$$

$$\pi \kappa_0 \left(\frac{3\kappa_r}{\kappa_r + 2} \right) d_p^2 E_c E_p = \frac{3\pi \mu_a d_p v_p}{\alpha_C}$$

or (6.89)

$$v_p = \frac{\alpha_C}{3\mu_a} \kappa_0 \left(\frac{3\kappa_r}{\kappa_r + 2} \right) d_p E_c E_p$$

Note that unlike the gravity settler and cyclone separator, the velocity of the particle toward the collecting electrode depends only linearly on the particle diameter. The effectiveness of electrostatic precipitation for collection of fine particles does not decrease as rapidly as with the other settling equipment. In addition, the correction for noncontinuum effects (α_C) increases for small particles, further improving the effectiveness of collection.

Quantitative evaluation of the efficiency again requires consideration of the flow and mixing characteristics in the device. If we consider the configuration in Figure 6.26, the gap between the electrodes tends to be very small relative to the length of the electrodes. This combined with the fact that the gas flowrates are generally high (and therefore turbulent) suggests that the plug flow assumption is again appropriate. The residence time in the device, τ_r, is its volume V divided by the volumetric flowrate Q while the collection time, τ_c, is the width between the electrodes, W, divided by the particle velocity, v_p.

$$\begin{aligned}\eta &= 1 - e^{-\tau_Q/\tau_c} \\ &= 1 - e^{-\frac{V/Q}{W/v_p}} \\ &= 1 - e^{-\frac{Av_p}{Q}} \\ &= 1 - \exp\left[-\frac{A}{Q}\frac{\alpha_C}{3\mu_a}\kappa_0\left(\frac{3\kappa_r}{\kappa_r+2}\right)d_pE_cE_p\right]\end{aligned} \tag{6.90}$$

A is the area of the collecting electrode. Although derived for a plate and wire type precipitator, the equations are approximately valid for any geometry subject to the assumption of a uniform electric field and plug flow. The relationships for efficiency in the electrostatic precipitator are frequently referred to as the Deutsch equation. Although approximate, it is adequate for our purpose of identifying the key parameters that influence the efficiency of pollution control equipment. Its use is illustrated in Example 6.12.

Example 6.12: Estimation of collection efficiency in an electrostatic precipitator

Estimate the collection efficiency of 0.5-,1,5-, and 10-μm particles in an electrostatic precipitator with a uniform electric field of 5000 V/cm. The gas flowrate to collection area is 3.33 m³/(min·m²). The temperature is 300 K where the viscosity of air is 0.0666 kg/(m·h). The dielectric constant of the particulate in the gas stream is 3.

Let us consider the 0.5-μm particles and then summarize the results for the other particles. The charge on the 0.5-μm particle is calculated by Equation 6.88,

$$q_e = 8.854(10)^{-12}\frac{\mathrm{C}}{\mathrm{V\cdot m}}\left(\frac{3\cdot 3}{3+2}\right)(0.5\cdot 10^{-6}\ \mathrm{m})(5000\ \mathrm{V/m}) = 20(10)^{-19}\ \mathrm{C}$$

This is approximately 12 times the unit charge of $1.6\,(10)^{-19}$ C. If the calculated charge on a particle were less than the unit charge, we would not expect to charge or collect such a particle. The Cunningham correction factor for a 0.5-μm particle at 300 K is about 1.34 from Equation 6.73. Thus, the collection velocity is, combining Equations 6.88 and 6.89

$$v_p = \frac{1.34}{3\cdot 0.0666\ \mathrm{kg/(m\cdot h)}}(20\cdot 10^{-19}\ \mathrm{C})\frac{5000\ \mathrm{V/cm}}{0.5\cdot 10^{-6}\ \mathrm{m}} = 4.8\cdot \mathrm{cm/s}$$

The efficiency of collection is then given by Equation 6.90

$$\eta = 1 - e^{\frac{-0.048\ \text{m/s}}{3.33\ \text{m}^3/(\text{m}^2\ \text{min})}} = 0.58$$

That is, about 58% of the 0.5-μm particles would be expected to be collected, comparing the other size particles. In all cases the correction of Stokes' equation for high Reynolds number is unnecessary.

d_p, μm	q_e (# of unit charges)	α_C	v_p	η
0.5	12	1.34	4.8	0.58
1	50	1.17	8.4	0.78
5	1245	1.03	37	0.99
10	4980	1.02	73	~1

Among the limitations of the Deutsch equation are the previously identified need to have plug flow and uniform electric fields throughout the device. The electric field intensity in the vicinity of the discharge (particle charging) electrode is, in general, different from the collecting electric field intensity applicable throughout the device. This is a function of electrode geometry as well as the conductivity and other properties of the gas stream. An even more important consideration is the deposition of particles at the collecting electrode creating a dust layer that poses an electrical resistance. The resistivity of the dust can vary dramatically depending upon the characteristics of the particulate matter being collected. If the resistivity is less than about 10^4 Ω·cm, that is if the dust is a good conductor of electricity, the charge on the particles will rapidly "bleed" to the collecting electrode. The lack of a residual charge on the particles means that the turbulence in the gas stream will easily re-entrain the collected dust, reducing the overall efficiency. Very resistive dust, greater than about 10^{10} ohm·cm, causes much of the overall voltage drop between electrodes to occur over the dust layer. This reduces the effective electric field intensity in the gas stream, again reducing the overall efficiency of the device. For sufficiently resistive dust layers the voltage drop may be sufficiently high that the layer may serve as a discharge electrode, ionizing the air in the dust layer and effectively reversing the mechanism of collection in the precipitator. This phenomenon is referred to as back ionization or back corona.

Dust resistivity is very important to the effectiveness of electrostatic precipitation. Parameters important to the resistivity of the dust are its temperature and the moisture content in the gas stream, in addition to dust composition. For dry gas streams, the resistivity of the dust layer decreases rapidly as the temperature is increased. Moisture in the gas stream and in the dust layer tend to reduce the magnitude of the resistivity and at low temperatures, change the effect of temperature. At a given moisture content, the resistivity may increase with temperature up to 100 to 200°C. Above these temperatures, the resistivity decreases at rates approaching that of dry dust. Changes in moisture content of several percent are required to achieve significant decreases in resistivity of a dust layer. Temperature changes of 20 to 50°C also are required to causes significant reductions in dust resistivity. The dust resistivity can be influenced more easily by the addition of small amounts of SO_3, NH_3 or related compounds. As low a concentration as 10 to 20 ppm SO_3 can reduce the resistivity of the dust layer by an order of magnitude or more. These have been added in the past to improve dust collection efficiencies, although it is ironic that an air pollutant to be avoided in the exhaust gases can actually contribute to dust collection efficiency. As a result of this phenomenon, power plant switching to lower sulfur fuels to reduce the emissions of oxides of sulfur had the unfortunate offsetting effect of reducing precipitator efficiency.

Another important factor in the resistance posed by the dust layer is its thickness. As the dust layer accumulates, the voltage drop associated with the increasing resistance means that the charge will bleed from freshly deposited dust. This dust will not be held as tightly and is subject to resuspension in the gas stream. Thus regular collection of the deposited particulate matter is important. Often mechanical rapping devices are used for this purpose that simply loosen the dust layer so that it will fall by gravity to a collection bin.

6.4.6 Wet Collectors

In some electrostatic precipitators, water is used to enhance the collection of particulate matter. The sludge that is formed will also flow to the bottom of the device eliminating the need for mechanical rappers to remove a dust layer. This is just one example of a wet collection system in which the suspended particles are removed by contact with a water droplet. The water droplet, being larger, can be easily collected by gravity settling or via a cyclone separator. The most common types of wet collectors in addition to the electrostatic precipitator are shown in Figure 6.27 and include

- Spray towers
- Baffled spray settlers
- Spray cyclones
- Venturi scrubbers

All of the previously discussed particulate collection processes involved the movement of a dust particle to a collection surface. A wet collector and a fabric filter, discussed below, however, place the collection surface in the path of the gas stream. The gas stream will flow around the water droplet or filter fiber, but the particle may still be collected by one of three mechanisms:

1. Direct Interception — The gas flow path which the particle is following may pass sufficiently closely to the water droplet to contact the surface and be collected.
2. Inertial Impaction — The particle may exhibit sufficient inertia to be unable to follow the flow path around the water droplet and will continue on to contact the surface.
3. Brownian Diffusion — The particle may be sufficiently small so that collisions with gas molecules cause it to deviate from the gas stream flow path and potentially contact the surface of the water droplet.

Each of these processes are depicted in Figure 6.28. Large particles, for example those easily collected by simple cyclone separation, tend to be collected by direct interception. Brownian diffusion is applicable only to extremely small particles, of the order of 0.1 μm or less, and this mechanism dominates only for these particle sizes. Thus, the primary mechanism of contact collection of particles in the size range generally of the most interest, 0.1 to 10 μm, is inertial impaction.

The process of intertial impaction can be analyzed by evaluating the trajectories of individual particles located in the gas stream incident upon the collecting surface. Particles in the gas stream in the region upstream of the droplet will be collected depending upon their initial location relative to the droplet. Particles initially located on or near the centerline of the droplet must follow the most curved trajectories and be displaced the greatest distance to avoid contact with the collector. The efficiency of collection of these particles as a result of their inertia is greatest. In the gravity settler, cyclone separator, and electrostatic precipitator the flow was assumed steady and the gas and the particle velocities were assumed to be identical. Here the delayed (i.e., unsteady) motion of the particle as the gas stream is accelerated is responsible for its collision with a water droplet and collection. Evaluation of an individual particle's trajectory with time depends on solving Newton's second law

$$m_p \frac{dv_p}{dt} = \sum F \tag{6.91}$$

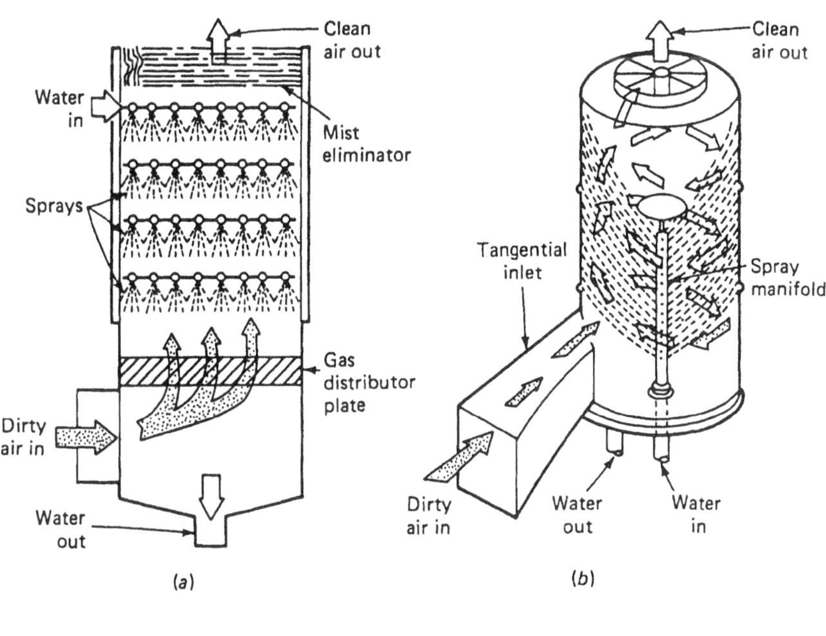

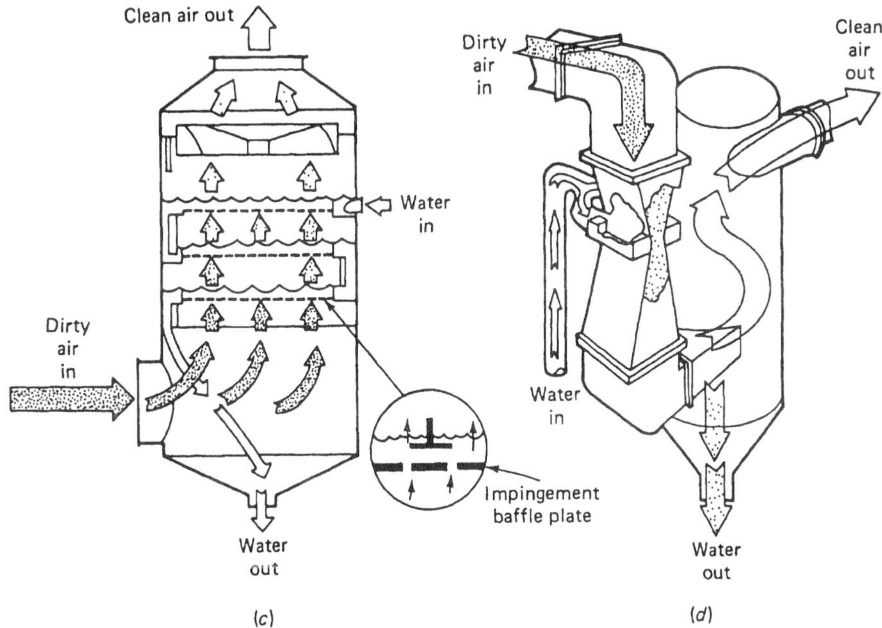

FIGURE 6.27 Wet separators: (a) spray tower, (b) cyclone spray tower, (c) impingement scrubber, and (d) venturi scrubber. (From Seinfeld, J. (1975) *Air Pollution: Physical and Chemical Fundamentals*, McGraw-Hill, New York. With permission.)

The drag force acting on the particle is given by Stokes' Law where the relative velocity of the particle and the gas stream is $v_p - U_a$. Note that these are, in general, vector quantities in that differences in either the direction or speed between the gas and the particle results in a drag force on the particle. Because the particles of interest generally lie between 0.1 and 10 μm, the Cunningham correction factor may be important. If we assume a spherical particle, Equation 6.91 can be written,

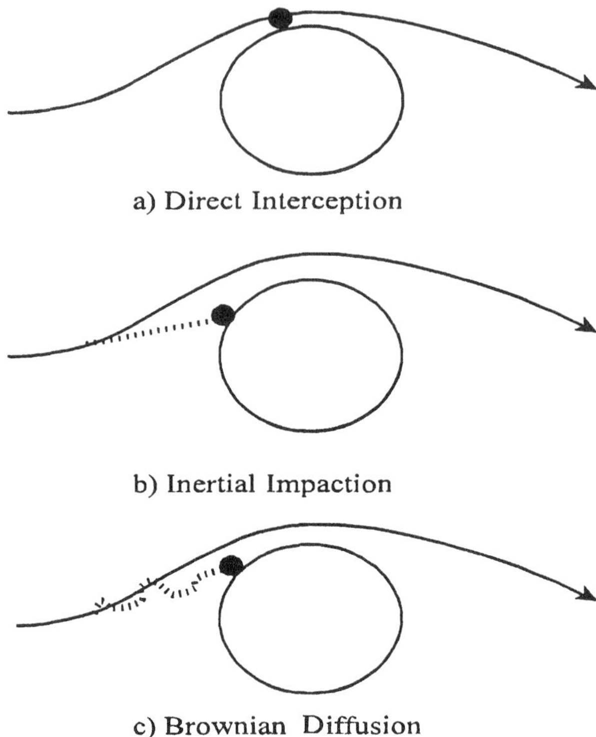

FIGURE 6.28 Mechanisms for collection of particulate matter on collectors such as water droplets or filter fibers.

$$(\rho_p - \rho_a)\frac{d_p^3}{6}\frac{dv_p}{dt} = \frac{3\pi d_p \mu (v_p - U_a)}{\alpha_C} \tag{6.92}$$

The parameters important in controlling the effectiveness of inertial impaction can be identified from this equation if we nondimensionalize it. Let us define the dimensionless velocity and time scales

$$\varepsilon = \frac{v_p}{v_0} \qquad \tau = t\frac{d_c}{v_0} \tag{6.93}$$

where v_0 represents the initial particle velocity (generally presumed to be the same as the air velocity in the direction of particle motion well away from the collecting droplet). The diameter of the collecting droplet, d_c, is the order of the distance over which the velocity changes due to the presence of the droplet. In terms of these variables, Equation 6.92 can then be written

$$\begin{aligned}\frac{dv}{d\tau} &= \frac{18\mu_g d_c}{(\rho_p - \rho_a) d_p^2 \alpha_C v_0}(\varepsilon - U_a / v_0) \\ &= \frac{1}{N_{Imp}}(\varepsilon - U_a / v_0)\end{aligned} \tag{6.94}$$

N_{Imp} is the impaction number. As the impaction number decreases, the effects of particle inertia increases (larger particles) and the ability of the particle to follow the flow path around the collecting

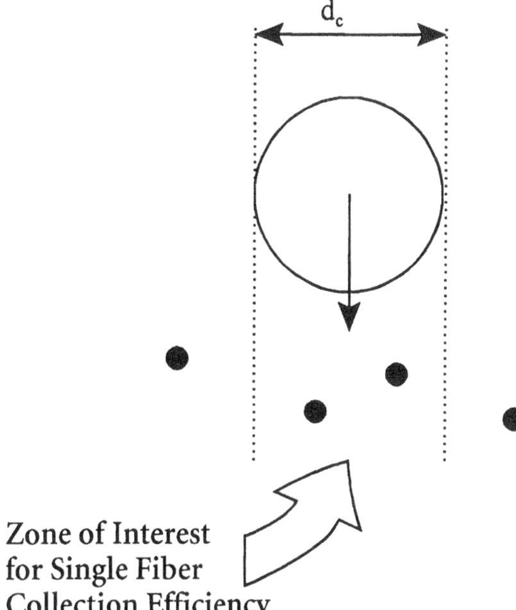

FIGURE 6.29 Depiction of single water droplet collection efficiency showing zone on which it is based.

droplet is reduced, increasing collection efficiency. Experiments have shown that the single droplet collection efficiency, η_0, the fraction of particles immediately upstream of the droplet that are collected, can be approximated by

$$\eta_0 = \left[\frac{2N_{Imp}}{2N_{imp} + 0.7}\right]^2 \tag{6.95}$$

We can now define the efficiency of a wet collector. Consider the wet collector shown in Figure 6.29. A single water droplet of diameter d_c sweeps out a volume $\pi(d_c^2/4)H_c$. This droplet sweeps this volume with efficiency η_0. The number of individual water droplets per unit time is given by the volumetric flowrate of water divided by the volume per droplet, $\pi(d_c^3/6)$. Thus, the volume of air cleaned per unit time is given by

$$Q_c = \text{(volume swept by drop) (single drop efficiency) (number of droplets per time)}$$

$$= \left(\pi\frac{d_c^2}{4}H_c\right) \quad (\eta_0) \quad \left(\frac{Q_w}{\pi\, d_c^3/6}\right) \tag{6.96}$$

where Q_w is the volumetric flowrate of water. The effective velocity of particles to the wall is this volume divided by the area perpendicular to the water droplet flow.

$$v_p = \frac{1}{WL}\frac{3}{2}\eta_0\frac{Q_w}{d_c} \tag{6.97}$$

The ratio of the gas residence time to particulate collection time is the same as the ratio of the volume of air passing through the device per unit time divided by the volume of air cleaned in the device per unit time. The volume of air passing through the device per unit time is simply the volumetric flowrate of air, $Q_a = UH_cW$. The ratio of residence to collection time and the overall efficiency of the device (assuming plug flow) is

$$\frac{\tau_Q}{\tau_c} = \frac{v_p}{H_c}\frac{L}{U} = \frac{3}{2}\eta_0 \frac{H_c}{d_c}\frac{Q_w}{Q_a}$$

$$\eta = 1 - \exp\left[-\frac{3}{2}\eta_0 \frac{H_c}{d_c}\frac{Q_w}{Q_a}\right] \tag{6.98}$$

As with the preceding models of pollution control devices, more sophisticated and detailed design procedures are available. This equation illustrates the basic principles, however, and indicates the key variables that control the efficiency of a wet collector. In particular, the higher the water flowrate or single droplet efficiency the greater the overall efficiency of the collector. A finer water spray also improves collection efficiency, primarily because the greater number of water droplets for a given water flowrate offsets the reduction in volume swept by a single droplet. Note that if the water droplets fall by gravity, as in a spray tower, the settling velocity of fine droplets is reduced relative to larger droplets and an optimum spray diameter exists. Other work has suggested that the optimum gravity settling water droplet diameter is often about 800 μm. Smaller particles settle too slowly to maximize particulate collection efficiency. In a cyclone separator, the optimum size is reduced due to the strong cyclonic motion allowing easier separation of the collecting water droplets. In an electrostatic precipitator, the optimum water droplet size can be reduced further. The use of the Equations 6.98 is demonstrated in Example 6.13.

Example 6.13: Particulate collection in a gravity spray tower

For the particles in Example 6.12 estimate their collection in a gravity spray tower with a water flowrate of 0.001 m^3/m^3 of air flow. The mean water droplet diameter is 1 mm which has a fall velocity of 4 m/s. Temperature and air viscosity are as in Example 6.12. Particle density is 2.5 g/cm^3.

Let us illustrate the calculations with the 5-μm particle which has a Cunningham correction factor of 1.03. From Equation 6.94, the impaction number is

$$N_{Imp} = \frac{(2.5\ \text{g/cm}^3)\,(5\cdot 10^{-6}\ \text{m})^2\,(1.03)\,(4\ \text{m/s})}{18[0.0666\ \text{kg/(m}\cdot\text{h)}]\,(0.001\ \text{m})} = 0.78$$

The single droplet collection efficiency is then approximately 0.48. For a fall height of 3 m with $Q_w = 0.001\ Q_a$, the overall collection efficiency for the 5-μm particle is

$$\eta = 1 - \exp\left[-\frac{3}{2}(0.78)\frac{3\ \text{m}}{0.001\ \text{m}}0.001\right] = 0.882$$

or 88% of this size particle should be collected in such a spray tower. Completing the calculations for the other particle sizes,

d_p, μm	α_C	N_{Imp}	η_0	η
0.5	1.34	0.01	0.00078	0.0035
1	1.17	0.035	0.0083	0.037
5	1.03	0.78	0.48	0.88
10	1.02	3.05	0.81	0.97

6.4.7 Filter Collectors

The final particulate collection device that will be considered here is a fabric filter. Fabric filters were one of the earliest means of controlling particulate emissions, but they have become more common in recent years due to improvements in fabric materials, their relative simplicity, and their ability to remove very fine particulate matter in the 0.1 to 1 μm range. Unlike all previous collection systems, the efficiency of a fabric filter system tends to increase with use, that is as a dust layer collects in the filter the total collecting surface and the overall efficiency of collection increases. This occurs at the expense of the pressure drop through the collection system, however, in that increased dust loadings reduce the void space available for flow and decrease the permeability of the filter. As indicated earlier, the efficiency of almost any device can be related to the pressure drop or other form of energy input into the system. We remain limited by the thermodynamic limit of energy input into a system required to achieve a certain separation as discussed in Chapter 4. Ultimately, of course, the pressure drop through a fabric filter exceeds acceptable limits for economic operation of the gas cleaning and the filters must be replaced or cleaned. Often a high pressure or high flow, short duration backflow is introduced to the filters to remove the particles. Because this backflow is of low volume with a high particulate concentration, collection and disposal of the particles is generally easier.

To evaluate the particulate collection by a fiber filter let us consider the single collector in Figure 6.29 as the cross section of a fiber and part of a larger bed of particles. The single fiber efficiency is defined as it was for the water droplets, i.e., η_0 is the proportion of particles collected that were initially in the projected area of the fiber. The volumetric rate of air passage in the zone upstream of the fiber is given by the interstitial velocity of the air in the fiber bed, q/ε, multiplied by the projected area of the fiber (length, ℓ per unit volume and height, d_c, the diameter of the collecting fiber). The quantity q is the volumetric flowrate divided by the total bed cross-sectional area, also called the superficial or Darcy velocity. The volumetric rate of air cleaned of particles is then

$$Q_c = \eta_0 \,\mathrm{l} d_c \frac{q}{\varepsilon} \tag{6.99}$$

where η_0 is the single fiber collection efficiency. The length of fiber per unit volume of bed, ℓ is the solid fraction $(1 - \varepsilon)$ divided by the cross-sectional area of the fibers, $\pi(d_c^2/4)$. Thus, the key parameter defining the overall collection efficiency, the ratio of the residence time to the collection time of particulate matter, is given by

$$\frac{\tau_Q}{\tau_c} = \frac{Q_c}{Q_a} = \frac{4\eta_0 \dfrac{(1-\varepsilon)}{\pi d_c^2} d_c \dfrac{q}{\varepsilon}}{q/L} = \frac{4}{\pi}\eta_0 \frac{(1-\varepsilon)}{\varepsilon}\frac{L}{d_c} \tag{6.100}$$

and the overall efficiency of particle collection in the fabric filter is given by

$$\eta = 1 - e^{-\tau_Q/\tau_c} = 1 - \exp\left[-\frac{4}{\pi}\eta_0 \frac{(1-\varepsilon)}{\varepsilon}\frac{L}{d_c}\right] \tag{6.101}$$

Again for the particles of primary interest between 0.1 and 10 μm, the dominant mechanism is inertial impaction. The single fiber collection efficiency, η_0 can be approximated as with the wet collector. The application of Equation 6.101 is illustrated in Example 6.14.

Example 6.14: Collection of particulates with a fabric filter

Under the same conditions as Examples 6.12 and 6.13, estimate the efficiency of collection of 0.5-, 1, 5-, and 10-μm particles in a fabric filter composed of fibers 1 mm in diameter. The filter has a porosity of 90% and is 3 cm long. The velocity of the particulate laden gas stream in the bed (q/φ) is 4 m/s.

This calculation proceeds identically with the calculation for the spray tower in Example 6.13. Only the calculation of the overall efficiency differs. The results are contained in the table below. For the 5-μm particle, the overall efficiency is given by

$$\eta = 1 - \exp\left[-\frac{4}{\pi}(0.475)\frac{(1-0.9)}{0.9}\frac{0.03\text{ m}}{0.001\text{ m}}\right] = 0.867$$

Note that the 3-cm filter bed is as effective as the 3-m gravity spray tower at collecting the 5-μm particles. This is also true of the remaining particle sizes. The pressure drop per unit length and, therefore, the energy requirements of a filter bed are greater than a spray tower and filters require replacement or regeneration. In a spray tower, the water flow continuously removes the particulate matter from the system.

d_p, μm	α_C	N_{Imp}	η_0	η
0.5	1.34	0.01	0.00078	0.0033
1	1.17	0.035	0.0083	0.035
5	1.03	0.78	0.48	0.87
10	1.02	3.05	0.81	0.97

6.5 GASEOUS CONTROL PROCESSES

The processes appropriate to control gaseous emissions depend on the characteristics of the emission and the desired degree of control. In general, process changes to avoid emission of the pollutant is always to be preferred and we will illustrate this through process changes to control oxides of nitrogen. Oxides of nitrogen also provide a good example of a pollutant that can be further controlled by reactive treatment processes. We will then discuss two processes that are applicable to a wide range of gaseous pollutants — absorption, commonly used to control oxides of sulfur, and adsorption, commonly used to control volatile hydrocarbons. As with particulate control processes, simplified design equations will be developed that are intended to illustrate the important factors influencing their behavior.

6.5.1 Process Changes to Control Gaseous Emissions

Oxides of nitrogen are an important component in the formation of ozone and photochemical smog. The primary sources of the pollutant are mobile and stationary fossil fuel combustion sources in which nitrogen in the feed air combines with excess oxygen under the high temperature conditions. Greater than stoichiometric air-fuel ratios for the purpose of ensuring complete combustion worsen the situation by increasing the amount of oxygen available. High temperatures, also desirable for combustion efficiency, also speed the formation of oxides of nitrogen. Typically 90+% of the immediate emissions are in the form of nitric oxide, NO, but in the atmosphere the NO rapidly forms nitrogen dioxide, NO_2. Neither of these gaseous compounds are easily removed from the exhaust gases by physical separations and control of oxides of nitrogen has instead focused on process changes to achieve control at the source and more sophisticated reactive means of control (e.g., catalytic converters in automobiles). In this section we will focus on process changes to achieve control and defer discussion of effluent reactor methods to the next section. Note that source control, i.e., pollution prevention, is always a valued goal and oxides of nitrogen simply represent an excellent example of the approach.

As indicated above, the primary causes of nitrogen oxides in the effluent from a fossil fuel combustor are the combination of high temperatures and the presence of both oxygen and nitrogen. Elimination of any of these factors can dramatically reduce the formation of nitrogen oxides. Let us examine each of these in turn.

Sources of nitrogen in a combustor include the fuel and the air which provides the oxygen source. For most fuels, the nitrogen content is small and the portion of nitrogen oxides formed from fuel nitrogen is likewise small. As with sulfur, coal generally contains more fuel nitrogen than oils and an insignificant amount is normally present in natural gas. Regardless of the source of fuel, however, the primary source of nitrogen in the combustor is in the air normally employed as the oxygen source. Of course, pure oxygen could always be used as the combustion oxygen, eliminating the formation of nitrogen oxides and providing a more efficient combustion system. The cost of producing the required volumes of pure oxygen is prohibitive and this is not a viable solution.

Oxygen is, of course, the source of the oxidant in the combustion reactions. In order to ensure complete combustion, it is normally necessary to provide greater than stoichiometric quantities of oxygen. Incomplete combustion increases the amount of organic compounds and carbon oxides in the exhaust, both of which are generally more serious concerns than nitrogen oxides. Excess air to provide this extra oxygen means that there is always significant quantities of oxygen and nitrogen in the combustion chamber and in the exhaust air. Improvements in the design of the combustion chamber and in the fuel-air mixers/burners can reduce the requirements for excess oxygen but not eliminate it entirely.

One of the most effective means of reducing oxygen levels is to stage the combustion. An initial combustion can be conducted under fuel-rich conditions. This may reduce the maximum temperatures observed in the fuel-air mixing region, but it definitely reduces the amount of free oxygen available for nitrogen oxide formation. After some of the heat has been removed from the system (e.g., in the heating of process water to make steam for power generation), the balance of the required air can be introduced into the system to complete the combustion. This has the effect of reducing the maximum temperatures achieved in the combustor but, more importantly, it means that oxygen is present in lower concentrations at the hottest points in the combustor, at the fuel-air mixer/burners. In some cases, this approach could be implemented on industrial burners by simply feeding a fuel-rich mixture to the burners and opening vents in the combustion chamber to allow additional air to be brought in under natural draft conditions for the second stage. The location of these vents above the burners led to the use of the term "overfire air" for this approach to the control of oxides of nitrogen.

Direct control of reaction temperature is the third option for reducing oxides of nitrogen emissions from a combustion chamber. One approach is to ensure good mixing at the burner to avoid the formation of "hot spots" that lead to local zones of high nitrogen oxide production. Improvements in burner design have significantly reduced this problem. Overall decreases in temperature can be obtained by diluting the combustion gases. As indicated in our discussions of the adiabatic flame temperature, the maximum temperature in a combustor is largely defined by the volume of exhaust gases that must be heated by the energy released by the reaction. The effect on the formation of oxides of nitrogen can be illustrated by the fact that the amount of these pollutants initially increases as the feed air is increased above stoichiometric due to the presence of additional oxygen. As the excess air is increased further, however, the decrease in combustion temperature associated with the additional exhaust gases offsets the availability of additional oxygen. This is illustrated qualitatively in Figure 6.30. This effect can also be achieved by the introduction of inert compounds into the combustion chamber. Because they are inert, the compounds do not take part in the reaction but do dilute the exhaust and reduce the flame temperatures. Steam injection has been used for this purpose. The difficulty of such an approach is that the overall efficiency of the combustion system is dramatically reduced by such an approach.

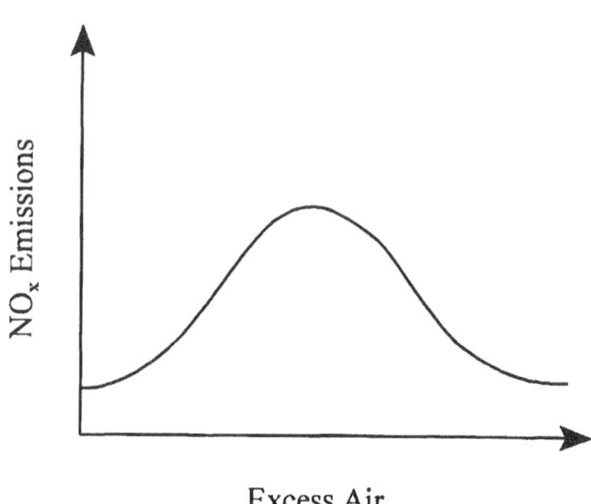

FIGURE 6.30 General behavior of oxides of nitrogen emitted from a combustor as a function of excess air.

An alternative that is not as damaging to the overall efficiency of the combustion system is the use of exhaust gas recirculation. This has been used in automobiles and in large combustion systems. Recirculation of the exhaust gases provides an additional diluent at the combustion point, but it reduces efficiency only as a result of the energy losses in the recycle loop and any energy requirements associated with the exhaust injection.

It has been possible to reduce nitrogen oxide formation by a factor of 2 to 5 by these methods. In many cases, as with overfire air and with improved burner design, these improvements have had a minimal impact or even improved combustor efficiency. Reductions in nitrogen oxide emissions beyond this point, however, generally require exhaust gas treatment of some sort. As indicated previously, simple physical separations are unable to achieve these reductions economically and more reactive schemes have been developed. These are illustrated in the next section.

6.5.2 Reactive Processes

Two important examples of treatment of gaseous effluents are considered in this section, the use of combustion, or afterburners, to remove combustible contaminants such as organic compounds, and the use of catalytic reaction systems to control oxides of nitrogen. Let us consider each in turn. We will provide only a qualitative discussion. Development of design equations for these reactive systems is beyond the scope of this text.

6.5.2.1 Afterburners

A gas stream contaminated with combustible material can be treated by injection into a flame. This is the basis of a flare at a chemical plant which is designed to avoid release of combustible compounds during process upsets. A small flame might also be used for routine elimination of combustible materials from a continuous source. Generally additional fuel must be injected to initiate the combustion. Exhaust gases containing less than about 2% organic vapors will not burn without the addition of fuel. Because the primary purpose of this afterburning operation is the "polishing" of a gas stream, it is generally conducted with relatively long residence times, 0.2 to 0.8 s, at moderate temperatures, 750 to 1000 K, to avoid the formation of oxides of nitrogen at higher temperatures. An important consideration is the hydrogen to carbon ratio of the fuel with a ratio of 1:3 or higher (e.g., 1:2 or 1:1) being best to avoid the generation of smoke and soot associated with partially burned hydrocarbons. Steam injection into the combustion zone can assist in the avoidance of soot in that the additional gas enhances mixing, encouraging complete combustion. Although the injection of steam reduces the efficiency of the combustion process, this is not a difficulty in that the primary objective in this case is compound destruction and not extracting energy from the system.

6.5.2.2 Catalytic Reaction Systems

The residence time required to achieve destruction of the pollutant can be reduced dramatically by the use of a catalytic system. The catalyst is not used up during the reaction but speeds its progress. Typically a solid catalyst adsorbs the reactants and the reaction proceeds more rapidly in the sorbed state. Note that catalysts can speed the rate of a reaction toward equilibrium but they cannot change the composition at equilibrium. That is, if a reaction is not favored thermodynamically, a catalyst will not change that. In addition, catalytic activity is reduced by the presence of contaminants in the gas stream such as sulfur which "poisons" many catalysts at concentrations as low as 1 ppm. Noble metals such as platinum and palladium are often used as catalysts. Due to the expense of these metals, increased catalyst activity is sought by maximizing the exposed surface area by dispersing the metal on an inert support such as alumina.

Catalytic systems can also be used for other than combustion-type reactions. Direct decomposition of NO into N_2 and O_2 can be accomplished with catalysis. Large scale use of this process in commercial fossil fuel combustors is not feasible. An alternative approach which has been used is selective catalytic reduction. In selective catalytic reduction, hydrogen, carbon monoxide, or ammonia can be used to form nitrogen from oxides of nitrogen. For example, with carbon monoxide one set of possible reduction reactions involving NO and NO_2 is

$$\begin{aligned} CO + 2NO &\rightarrow CO_2 + N_2 \\ 4CO + 2NO_2 &\rightarrow 4CO_2 + N_2 \end{aligned} \tag{6.102}$$

With ammonia, the corresponding reactions might be

$$\begin{aligned} 4NH_3 + 6NO &\rightarrow 6H_2O + 5N_2 \\ 8NH_3 + 6NO_2 &\rightarrow 12H_2O + 7N_2 \end{aligned} \tag{6.103}$$

Platinum group catalysts are commonly used for these reactions but these are catalysts that are easily poisoned by sulfur.

Nonselective reduction can also occur in the presence of hydrocarbons, for example, with methane

$$\begin{aligned} CH_4 + 4NO_2 &\rightarrow CO_2 + 2H_2O + 4NO \\ CH_4 + 4NO &\rightarrow CO_2 + 2H_2O + 2N_2 \end{aligned} \tag{6.104}$$

Noble metal catalysts also can be used for these reactions. The reactions are highly exothermic and also require a minimum temperature to initiate. These reactions are the basis of carbon monoxide, hydrocarbon, and oxides of nitrogen control in a modern catalytic converter system in automobiles.

6.5.3 Sorption Processes — Adsorption

Although the processes described above can be quite effective in particular circumstances, the most common separation methods for the control of gaseous emissions are based on contacting the gas with a phase which preferentially sorbs the contaminant. If the contacting phase is a liquid in which the sorbing contaminant is ultimately dissolved into the bulk phase, this process is called *absorption*. If the contacting phase is a solid in which the sorbing contaminant is ultimately assumed to physically sorb to the surface of the solid, this process is called *adsorption*. Although seemingly a straightforward definition, the distinction between the two processes is not always clear. Surface sorption followed by diffusion and deep penetration into the solid matrix, for example, might be termed an absorption process. Here, we will refer to all sorption onto solid surfaces as adsorption and all sorption into liquid phases as absorption. In this section let us focus on adsorption.

An adsorption process involves the physical binding of a contaminant molecule (the *adsorbate*) on the surface of a solid *adsorbent*. The binding is the result of intermolecular attractive forces (van der Waals forces). Physical binding, or *physisorption*, in this manner is characterized by easy reversibility and rapid rate. Solid catalyzed reactions usually involve *chemisorption*, attachment to surfaces involving chemical bonding. Chemisorption involves chemical changes on the solid surface and is not generally reversible. One of the clearest differentiations between the two types of surface binding is the latent heat associated with the sorption process with physisorption releasing 2 to 20 kJ/mol and chemisorption releasing 80 to 100 kJ/mol.

Common adsorbents include activated carbon, alumina, and silica gel. Among the characteristics of a good adsorbent are a porous structure that provides a significant internal porosity (i.e., within the particles) and large surface area. Table 6.5 provides some typical physical properties of these adsorbents. Note that the physical properties are heavily dependent upon the preparation procedures for the adsorbents. Activated carbon is dried, carbonized by heating in the absence of air, and then exposed to a controlled oxidation. Activated alumina is prepared by heating alumina trihydrate in air. Silica gel is formed by treating sodium silicate with a strong acid. The solid is then washed, dried, and heated for activation.

TABLE 6.5
Physical Properties of Common Adsorbents

Adsorbent	Internal Porosity, ε (%)	Bulk Density, ρ_b (g/cm³)	Surface Area, a_s (m²/g)
Activated carbon	55–75	0.2–0.5	600–1400
Activated alumina	30–40	0.7–0.9	200–300
Silica gel	70	0.4	300

Adsorption depends on the adsorbate being preferentially adsorbed onto the adsorbent relative to the bulk constituents of the gas stream. A surface area of *hundreds of square meters per gram* implies that these adsorbents have a large capacity for sorption of contaminants. Activated carbon, for example, can adsorb 30 to 80 g of pure carbon tetrachloride per 100 g of adsorbent. When exposed to a mixture, the capacity decreases according to the equilibrium isotherm for the particular adsorbate and adsorbent. As discussed in Chapter 4, the equilibrium partitioning between a gas and a dry solid cannot generally be described by a simple linear isotherm and experimental measurements of sorbed vs. gas phase concentrations are generally required. If W_s^* represents the mass of contaminant that can be sorbed onto a solid at a given inlet gas phase concentration, then the maximum capacity of a bed of adsorbent of cross-sectional, A; length, L; and bulk density, ρ_b; is given by

$$M_{ads} = \rho_b W_s^* A\, L \tag{6.105}$$

When the capacity of the sorption bed is reached, the adsorbent must be regenerated or replaced before gas cleaning can continue. In practice multiple beds of adsorbent are often used and one bed is replaced or regenerated while another is being used for gas cleaning. Regeneration involves changing temperature or pressure on the adsorbent bed to change the equilibrium toward release of the contaminant back into the gas stream. This can sometimes be done in place by heating or depressurizing the adsorbent bed and driving off the contaminant. Sometimes the adsorbent must be removed and regenerated off-site under more severe conditions then can be achieved in-place. Regeneration of activated carbon, for example, essentially requires a repeat of the same steps involved in the production of the adsorbent initially.

In an ideal adsorbent bed, the entire bed will become saturated with the contaminant before regeneration/replacement is necessary. The time between regeneration or replacement of the bed is then given by its mass capacity defined by Equation 6.105 divided by the rate at which contaminants are fed to the system. If the volumetric flowrate of contaminated air to the bed is Q_a and the contaminant concentration at the inlet is $C_a(0)$, then the time before regeneration/replacement is necessary is given by

$$\tau_{ads} = \frac{M_{ads}}{(Q_m)_{feed}} = \frac{\rho_b W_s^* A L}{Q_a C_a(0)} \tag{6.106}$$

Thus, the zone of fully utilized or saturated adsorbent grows with a velocity, V_{ads}, given by

$$V_{ads} = \frac{L}{\tau_Q} = \frac{Q_a C_a(0)}{\rho_b W_s^* A} \tag{6.107}$$

If use is made of the empirical Freundlich isotherm for generality,

$$W_s^* = \left(\frac{C_a}{\alpha}\right)^{1/\beta} \tag{6.108}$$

then the velocity with which the saturated region grows with time can be written

$$\begin{aligned} V_{ads} &= \frac{Q_a C_a(0)}{\rho_b A}\left(\frac{C_a(0)}{\alpha}\right)^{-1/\beta} \\ &= \frac{Q_a \alpha^{1/\beta}}{\rho_b A}\left[C_a(0)\right]^{(1-1/\beta)} \end{aligned} \tag{6.109}$$

In reality the bed cannot be operated for as long as the time defined by Equation 6.106 before regeneration/replacement is necessary. An active adsorption zone exists in which the adsorbent is not saturated and whose presence reduces the time before regeneration is necessary. The following is an analysis of that zone following the approach of Wark and Warner (1981).

A plot of the effluent concentration from a packed bed adsorber, or *breakthrough curve*, as a function of time is shown in Figure 6.31. Immediately after introduction of a contaminated gas stream into a fresh adsorbent bed, the effluent concentration is effectively zero. This assumes that the adsorbent is chosen such that very little contaminant remains in the gas phase. After some time, however, the concentration of contaminant in the effluent stream is measurable and continues to increase as long as feed to the system is continued. When the concentration in the effluent exceeds some arbitrarily defined low concentration, the *breakpoint* of the adsorbent bed has been said to be reached. The breakpoint is usually defined to be the maximum contaminant concentration that is allowable in the effluent. When the effluent concentration exceeds the breakpoint, shutdown and regeneration/replacement of the adsorbent bed is necessary. If the adsorbent bed is not regenerated, the concentration in the bed continues to increase until the *endpoint* is reached. At this concentration,

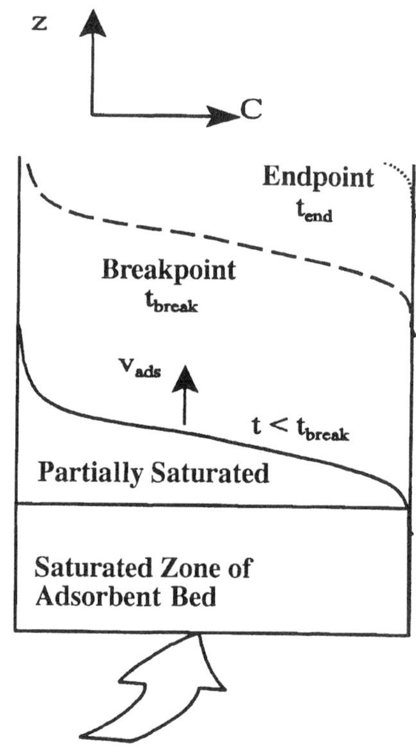

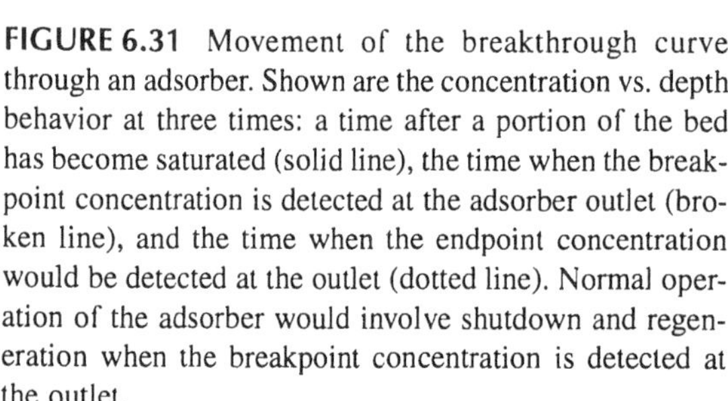

FIGURE 6.31 Movement of the breakthrough curve through an adsorber. Shown are the concentration vs. depth behavior at three times: a time after a portion of the bed has become saturated (solid line), the time when the breakpoint concentration is detected at the adsorber outlet (broken line), and the time when the endpoint concentration would be detected at the outlet (dotted line). Normal operation of the adsorber would involve shutdown and regeneration when the breakpoint concentration is detected at the outlet.

the effluent concentration is essentially identical to that in the influent and the bed is ineffective at removing the contaminant from the gas stream. For convenience the endpoint is often arbitrarily specified as 95 or 99% of the influent concentration.

If instead of considering the effluent concentration as a function of time we consider the concentration profile within the adsorbent bed at a particular time, the plot in Figure 6.31 results. There is a zone within the adsorbent bed that is effectively saturated with respect to the adsorbing contaminant. The gas phase concentration in this zone is identical to that in the influent since no further adsorption can take place. Above (or downstream of) this zone is a zone where active adsorption is taking place. The concentration in the gas stream in this active adsorption zone changes from the influent concentration (effectively the endpoint concentration) to zero (effectively the breakpoint concentration). The measurement of the breakpoint and endpoint concentrations in the effluent from the adsorption system correspond to the movement of the beginning and end of the active adsorption zone to the end of the bed. Because shutdown of the bed must occur when the breakpoint concentration is observed in the outlet, the upper (or final) portion of the bed is only partially saturated in the active adsorption zone. For this reason, Equation 6.106 overestimates the operating time of an adsorbent bed before regeneration or replacement is required.

In order to maximize the operating time of the adsorber, it is desirable to keep the active adsorption zone as short as possible. Its length is the result of mass transfer limitations and diffusion/dispersion in the bed. If the rate of adsorption is slow, a finite time is required to reduce the gas phase concentration to the breakpoint when fresh adsorbent is encountered in the active adsorption zone. Alternatively, if there are mixing processes in the bed or nonuniform flow, the high influent concentration extends further in some areas of the bed. The average concentration at any point within the active adsorption zone then reflects fluids with varying degrees of adsorption. Both mass transfer and dispersion effects result in the same basic profile, an extended transition from the endpoint concentration at the top of the saturated zone to the breakpoint concentration at the downstream end of the active adsorption zone. Dispersion can be minimized by ensuring a uniform packing and flow in the adsorbent bed and by avoiding large adsorbent particles. Let us assume that there is negligible dispersion and that the rate of adsorption is governed by the local deviation from the equilibrium concentration. In this situation, the width and concentration profile of the active adsorption zone is always constantly reflecting only the time required to saturate the bed when the contaminated gas stream is exposed to fresh adsorbent. The total adsorbed mass in this moving active adsorption zone is also constant. Our picture of adsorption in a finite bed is one in which the active adsorption zone moves through the bed at a constant speed governed by the growing zone of saturated adsorbent, given by Equation 6.109.

Note that since the speed and profile in the active adsorption zone is everywhere constant, Equation 6.107 also applies to anywhere within this zone where the concentration in the air phase is C_a and the local concentration on the solid is W_s,

$$V_{ads} = \frac{Q_a C_a}{\rho_b W_s A} \tag{6.110}$$

Thus, the relationship between the gas phase concentration and the solid concentration can be written, with the help of Equation 6.109, as

$$\begin{aligned} W_s &= \frac{Q_a C_a}{\rho_b V_{ad} A} \\ &= C_a \alpha^{-1/\beta} [C_a(0)]^{(1/\beta - 1)} \end{aligned} \tag{6.111}$$

The rate of mass transfer per unit time in an element of length dz in the adsorbent bed is given by

$$dQ_m = k_a a_v (C_a - C_a^*) A\, dz \tag{6.112}$$

Here k_a is the overall gas phase mass transfer coefficient (or a lumped parameter mimicking the effects of dispersion), a_v is the surface area of the adsorbent per unit volume of bed ($a_s\rho_b$), and C_a^* is the concentration in equilibrium with the local solid phase concentration. From Equation 6.111, C_a^* is given by

$$\begin{aligned} C_a^* &= \alpha W_s^\beta \\ &= \alpha \left\{ C_a \alpha^{-1/\beta} [C_a(0)]^{(1/\beta - 1)} \right\}^\beta \\ &= C_a^\beta C_a(0)^{\beta - 1} \end{aligned} \tag{6.113}$$

A material balance over the interval dz then gives

$$\begin{aligned} & Q_a C_a\big|_z - Q_a C_a\big|_{z+dz} - k_a a_v (C_a - C_a^*) A\, dz = 0 \\ & \frac{dC_a}{dz} = -\frac{k_a a_v}{q} \left[C_a - C_a^\beta C_a(0)^{1-\beta} \right] \end{aligned} \tag{6.114}$$

where q = U/A, the superficial velocity in the adsorber.

Let us solve this equation for the thickness of the active adsorption zone, δ_{ads}, by integrating between the concentrations that span this zone, the breakpoint, and the endpoint concentrations (C_b and C_e, respectively).

$$\int_{C_e}^{C_b} \frac{dC_a}{\left[C_a - C_a^{\beta} C_a(0)^{1-\beta}\right]} = -\int_0^{\delta} \frac{k_a a_v}{q} dz \tag{6.115}$$

This integral is not bounded if the endpoint is taken as the inlet concentration and the breakpoint as 0. As indicated earlier, the endpoint concentration is defined arbitrarily as one essentially identical to the inlet concentration while the breakpoint concentration is normally set by the allowable concentration in the effluent from the system. Let us arbitrarily assume an endpoint equal to 99% and a breakpoint equal to 1% of the inlet concentration. Integrating Equation 6.115 and solving for the zone width, δ, gives

$$\delta = \frac{q}{k_a a_v}\left[4.595 + \frac{1}{\beta - 1}\ln\frac{1-(0.01)^{\beta-1}}{1-(0.99)^{\beta-1}}\right] \tag{6.116}$$

If we instead assumed an endpoint equal to 99.9% and a breakpoint equation to 0.1% of the inlet concentration, the constant in Equation 6.116 is replaced with 6.907 and 0.001 and 0.999 replace 0.01 and 0.99, respectively.

When the active adsorption zone reaches the outlet of the adsorber, the effluent concentration is just equal to the breakpoint concentration and the adsorber must be shutdown and regenerated. If the mass of contaminant in the active adsorption zone is neglected, this means that the time until regeneration is given by

$$\tau_{ads} = \frac{M_{ads}}{(q_m)_{feed}} = \frac{\rho_b W_s^* A(L-\delta)}{Q_a C_a(0)} = \frac{(L-\delta)}{V_{ads}} \tag{6.117}$$

Because the adsorption zone does contain some mass, however, the time until regeneration is required is actually somewhere between the times given by Equations 6.117 and 6.106. The use of these equations is illustrated in Example 6.15.

Example 6.15: Design of an activated carbon adsorber

Activated carbon with a bulk density of 400 kg/m^3 is to be used to separate a contaminant in an adsorber 6.65 m long and 1.5 m in diameter. The equilibrium sorption of the contaminant is given by $\alpha = 10$ kg/m^3 and $\beta = 2$ in Equation 6.108. The mass transfer coefficient in the bed is $ka_v = 24$ s^{-1}. The feed flowrate is 1 m^3/s of air at 25°C containing 0.1% (by weight) contaminant. Estimate the maximum adsorption time (time to bed saturation), the width of the active adsorption zone, and the time until regeneration is required neglecting the contaminant adsorbed in the active adsorption zone.

The density of air at 25°C is 1.18 kg/m^3. The inlet concentration of contaminant is 1.18 g/m^3 and the equilibrium solid phase concentration from Equation 6.108

$$W_s^* = \left(\frac{1.18 \cdot 10^{-3}\ \text{kg/m}^3}{10\ \text{kg/m}^3}\right)^{1/2} = 0.011\ \text{kg/kg}$$

From Equation 6.106, the adsorption time until complete saturation of the bed is

$$\tau_{ads} = \frac{(400\ \text{kg/m}^3)\ (0.011\ \text{kg/kg})\ (1.77\ \text{m}^2)\ (6.65\ \text{m})}{(1\ \text{m}^3/\text{s})\ (1.18\cdot 10^{-3}\ \text{kg/m}^3)} = 43200\ \text{s} = 12\ \text{h}$$

The velocity of the adsorption zone from Equation 6.110 is then

$$V_a = \frac{(1\ \text{m}^3/\text{s})\ (1.18\cdot 10^{-3}\ \text{kg/m}^3)}{(400\ \text{kg/m}^3)\ (0.011\ \text{kg/kg})\ (1.77\ \text{m}^2)} = 0.55\ \text{m/h}$$

The width of the adsorption zone from Equation 6.116 is

$$\delta = \frac{(1\ \text{m}^3/\text{s})}{(1.77\ \text{m}^2)(24\ \text{s}^{-1})}\left[4.595 + \frac{1}{2-1}\cdot \ln\left(\frac{1-0.01^{2-1}}{1-0.99^{2-1}}\right)\right] = 0.22\ \text{m}$$

Finally, the time until all but the active adsorption zone is calculated from Equation 6.117

$$\tau_{ads} = \frac{L-\delta}{V_{ads}} = \frac{6.65\ \text{m} - 0.22\ \text{m}}{0.55\ \text{m/h}} = 11.6\ \text{h}$$

Note that the width of the active adsorption zone is essentially negligible (in this case).

The analysis of the regeneration process can often be conducted in exactly the same manner in that only the direction of mass transport changes when a clean regenerator gas stream is fed to a desorbing bed. Rates of desorption and equilibrium quantities, however, are generally different for desorption compared to adsorption. This is by design in order to minimize the time and volume of fluid required to regenerate the adsorbent.

6.5.4 Sorption Processes — Absorption

Certain gaseous pollutants also can be removed from an air stream via absorption into a liquid stream. Sulfur dioxide, for example, has long been removed from a gas stream in this manner. When absorbed into water, sulfur dioxide undergoes the series of reactions,

$$\begin{aligned} &SO_2(g) + H_2O \rightarrow SO_2(aq) \\ &SO_2(aq) + H_2O \rightarrow HSO_3^- + H^+ \\ &HSO_3^- + H_2O \rightarrow H_2SO_4^{-2} + 2H^- \end{aligned} \tag{6.118}$$

Although sulfur dioxide is only partially soluble, the presence of the additional reactions continuously drives the equilibrium to the right by removing the physical dissolved SO_2. These reactions can be enhanced further by the presence of a base that reduces the concentration of the hydrogen ion, further encouraging movement of the reactions toward the right and further reducing the concentration of SO_2 in the physically sorbed state. Sulfur dioxide can be effectively removed from a gas stream by exposure of the stream to a basic liquid phase in an absorption system. A wide variety of other gases can also be removed by absorption into a liquid, although large scale application requires an absorbing liquid that is inexpensive or one that is easily recycled. For these reasons, water or aqueous solutions are by far the most common absorbing phase. The basic design of an absorption system is the subject of this section.

The mass transfer between the solute contaminant in the air stream and the absorbing liquid is dependent upon the characteristics of the phases as well as the mixing characteristics in the system. Remember that the overall mass transfer coefficient across an air-liquid interface can be described by the two-film theory as a combination of mass transfer resistances in both phases.

$$\frac{1}{k_a} = \frac{1}{k_{af}} + \frac{K_{al}}{k_{lf}}$$
$$\frac{1}{k_l} = \frac{1}{k_{lf}} + \frac{1}{K_{al}k_{af}} \tag{6.119}$$

The individual film coefficients, k_{lf} and k_{af}, are defined only by the mixing characteristics in their respective fluid phases. For contaminants exhibiting a large air-liquid partition coefficient for the absorbing solvent, only mixing in the liquid phase is important while the opposite is true for low volatility contaminants. The contacting between the air and liquid phases and the mixing within phases in an absorber are generally enhanced by the presence of trays or packing within the system. The more common packed tower is shown in Figure 6.32. The purpose of the packing is solely to maximize mixing and contacting within and between phases.

Generally an absorption tower or column is designed to maximize mixing across the tower and minimize axial dispersion. As a result, the plug flow model is once again appropriate. Using the notation in Figure 6.32, a material balance on the air side on an element of length dz can be written identically to that of an adsorber.

$$Q_a C_a\big|_z - Q_a C_a\big|_{dz} - k_a a_v (C_a - C_a^*) A\, dz = 0$$
$$\frac{dC_a}{dz} = -\frac{k_a a_v}{q}\left(C_a - C_a^*\right) \tag{6.120}$$

As before, q is the superficial or Darcy velocity in the adsorber column (Q_a/A) and a_v is the interfacial area per unit volume of adsorber. Generally a_v must be measured as a combined factor with the mass transfer coefficient for a particular system. That is, experimental measurements or tabulated data provide estimates of the product of the film or individual phase mass transfer coefficients and the combined overall coefficient can be estimated from

$$\frac{1}{k_a a_v} = \frac{1}{k_{af} a_v} + \frac{K_{aw}}{k_{lf} a_v} \tag{6.121}$$

Note also that mass transfer between phases changes the mass of each phase and, in general, the velocities. In most environmental applications the concentration of the absorbing solute is sufficiently small that this can be neglected and herein we will only consider effectively constant volumetric flowrates and velocities of both phases.

Integration of Equation 6.120 requires specification of C_a', the gas phase concentration that would be in equilibrium with the contaminant in the absorbing liquid. Unlike the adsorber the absorbing liquid can, in principle, be provided to the system at any rate. In the limit of a high volume of absorbing liquid feed (or an ideal absorber that can effectively capture an infinite amount of the contaminant), the equilibrium concentration is $C_a^* \sim 0$. Under these conditions, the contaminant concentration at the exit of the absorber, at z = L, is governed by

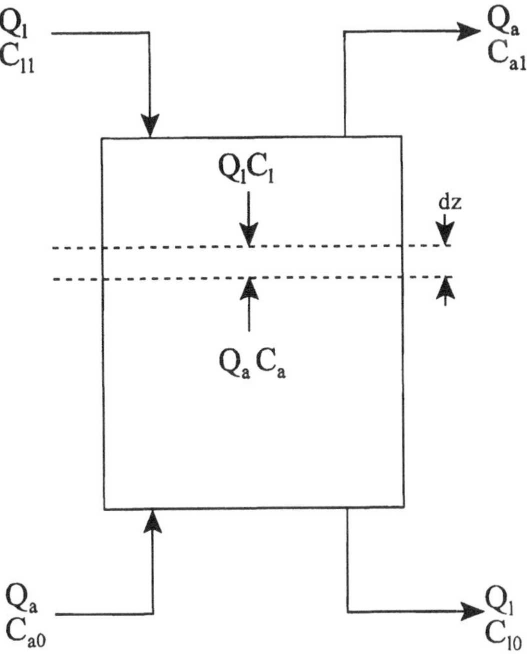

FIGURE 6.32 Notation for absorber material balance.

$$C_a(L) = C_a(0)e^{\frac{-k_a a_v}{q}L} \tag{6.122}$$

For any reasonable liquid flow, however, the equilibrium concentration is not zero and the concentration as a function of absorbing column length is defined by integrating Equation 6.120.

$$\int_{C_{a0}}^{C_{al}} -\frac{q}{k_a a_v}\frac{dC_a}{C_a - C_a^*} = \int_0^L dz \tag{6.123}$$

The quantity $q/k_a a_v$ is often approximately constant and has units of length. It is the ratio of the speed with which the gas moves through absorber to the intrinsic rate at which mass transfer to the liquid phase occurs. It is related to the height required to achieve a given separation and has come to be known as the *height of a mass transfer transfer unit* (HTU). The remaining integral is dimensionless and related to the degree of difficulty of the separation. It has come to be known as the *number of mass transfer units* (NTU). Using this nomenclature, Equation 6.123 can be written

$$\begin{aligned} L &= \int_{C_{al}}^{C_{a0}} \frac{q}{k_a a_v}\frac{dC_a}{C_a - C_a^*} \\ &= \frac{q}{k_a a_v}\int_{C_{al}}^{C_{a0}} \frac{dC_a}{C_a - C_a^*} \\ &= (HTU)\ (NTU) \end{aligned} \tag{6.124}$$

In order to complete the integration calculating the number of transfer units (NTU), the difference between the gas phase concentration and the gas phase concentration that would be in equilibrium with the liquid phase must be known everywhere in the absorption column. This depends upon the ratio of the liquid to gas flowrate. In the limit of very high liquid flowrate, when C_a^* is approximately zero, the number of transfer units is simply $\ln(C_{al}/C_{a0})$ and Equation 6.122 is reproduced. At low liquid flowrates, C_a^* approaches C_a within the column. It is not possible for absorption to continue if $C_a^* \geq C_a$ anywhere in the system. Thus, the minimum liquid flowrate that can be employed to achieve a given separation of the contaminant is that which would result in equilibrium between the gas and liquid phases at the exit of the system.

This can be illustrated graphically in a plot of the gas vs. liquid concentrations in which both the equilibrium curve and the actual concentrations in an absorption tower are plotted. The actual concentrations in both the gas and liquid phases when plotted on such a manner are referred to as the *operating lines*. The equations for the operating lines can be defined by making an overall material balance on the bottom portion of the column in Figure 6.33. A balance on the rate of influent and effluent contaminant through the control volume reduces, under steady conditions, to

$$Q_a C_{a0} + Q_l C_l = Q_a C_a + Q_l C_{l0} \tag{6.125}$$

where C_a and C_l are the air and liquid phase concentrations, respectively. Solving for the gas phase concentration exiting the control volume

$$C_a = C_{a0} + \frac{Q_l}{Q_a}(C_l - C_{l0}) \tag{6.126}$$

This is an equation of a straight line between C_a and C_l with a slope given by the ratio of the liquid to gas flowrates, Q_l/Q_a. This assumes that the mass transfer from one phase to another does not significantly alter the flow rates, otherwise the ratio Q_l/Q_a is not constant and the equation is curved. The operating line connects the concentrations at the bottom of the absorber where the entering gas concentration is C_{a0} and the exiting liquid concentration is C_{l0} to the gas and liquid concentrations anywhere in the absorber. This is shown in Figure 6.33. If the control volume includes the entire absorber the operating line relates the concentrations at the bottom to those at the top.

$$C_{a1} = C_{a0} + \frac{Q_l}{Q_a}(C_{ll} - C_{l0}) \tag{6.127}$$

At very high flowrates, the operating line is effectively vertical as shown in Figure 6.33 and any degree of removal of the absorbing contaminant can be achieved by specifying an absorber of sufficient length as defined by Equation 6.122. Although this minimizes the number of transfer units and, thus, the size of the absorber, it generally requires unacceptably high flowrates of liquid. Remember that the contaminant must generally be removed from the liquid before disposal. High liquid flowrates entail not only large pumping costs but large liquid treatment costs as well.

Conversely, the minimum flowrate is one in which the operating line would intersect the equilibrium curve as shown in Figure 6.33. The effluent liquid at the bottom of the tower is then in equilibrium with the influent gas. Any lower flowrate would mean that the effluent water carried more than an equilibrium amount of contaminant. There is a minimum slope to the operating curve that depends upon the inlet concentration and the equilibrium. This slope corresponds to the minimum liquid to gas flowrate that will achieve the desired separation. Even at the minimum flowrate, the driving force for mass transfer (C_a-C_a^*) goes to zero as would the rate requiring, in effect, an absorber of infinite length to achieve the desired separation.

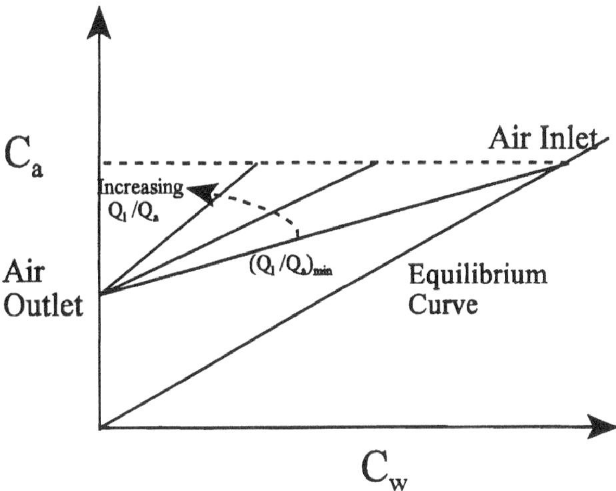

FIGURE 6.33 Operating lines in an absorber, i.e., the air and corresponding water phase concentrations everywhere in the absorber. Straight equilibrium and operating lines (dilute solution) are assumed. The inlet water is assumed to contain none of the contaminant. The minimum liquid-to-gas flowrate curve is the operating line that intersects the equilibrium curve at the air inlet conditions.

The optimum condition is a liquid to gas flowrate that is intermediate between these two extremes. As indicated above an essentially infinite column is required if the minimum liquid to gas flowrate is employed and the number of transfer units and therefore the length of the absorption column decreases as the liquid to gas flowrate is increased. Definition of the optimum liquid-to-gas flowrate is beyond the scope of this text, but it is often a small multiple of between one and two times the minimum required flowrate. Once the actual liquid-to-gas flowrate is known, the operating line (either graphically or Equation 6.127) specifies the actual effluent concentration in the liquid and through Equation 6.126, the relationship between the gas and liquid concentrations everywhere in the absorption column. Thus, the integral in Equation 6.124 can then be evaluated since the integrand involves the actual air-phase concentration (at each point in the column) and the concentration in the air phase that would be in equilibrium with the liquid-phase concentration (at that same point in the column). If $C_a(z)$ and $C_l(z)$ represent the air- and water-phase concentrations at a particular point in the column,

$$C_a - C_a^* = C_a(z) - K_{aw} C_l(z) \tag{6.128}$$

With the operating line equation, this difference can be calculated at every point in the column and the integral in Equation 6.124 evaluated to estimate the total height of the column required to achieve the desired separation. This process is illustrated in Example 6.16.

Example 6.16: Design of an SO_2 absorption column

Estimate the total required height of an absorber designed to remove 95% of the SO_2 in an air stream initially containing 10% SO_2 using clean water as the absorbing liquid. The air-water partition coefficient of SO_2 in water can be crudely approximated as 0.02 for the purposes of this problem (due to the reactions of SO_2 in water, it actually varies between about 0.03 and 0.01 over the concentration range in this problem). The air and water film mass transfer coefficients, $k_{af}a_v$ and $k_{wf}a_v$, are 735 h^{-1} and 11.25 h^{-1}, respectively, in this packing at these conditions. The air flowrate is 0.117 $m^3/s/m^2$ of column cross-sectional area. The column temperature is 25°C.

The height of a transfer unit (HTU) can be calculated by using Equation 6.121 to estimate the overall mass transfer coefficient

$$k_a = \left(\frac{1}{k_{af}a_v} + \frac{K_{aw}}{k_{wf}a_v}\right)^{-1} = \left(\frac{1}{735\ \mathrm{h}^{-1}} + \frac{0.02}{11.25\ \mathrm{h}^{-1}}\right)^{-1} = 316\ \mathrm{h}^{-1}$$

The mass transfer resistances in each of phases are nearly equal. The HTU is then

$$HTU = \frac{q}{k_a a_v} = \frac{(0.117\ \mathrm{m/s})(3600\ \mathrm{s/h})}{316\ \mathrm{h}^{-1}} = 1.34\ \mathrm{m}$$

The concentrations of SO_2 in the inlet (0.10 mole fraction) and outlet (95% removal ~0.005 mole fraction) at 30°C are

$$C_a = \frac{p_{SO_2} \cdot MW_{SO_2}}{RT} \begin{cases} C_{a0} = \dfrac{(0.1\ \mathrm{atm})\ (64\ \mathrm{g/mol})}{24.5\ \mathrm{L \cdot atm/mol}} = 0.261\ \mathrm{kg/m^3} \\ C_{al} = \dfrac{(0.005\ \mathrm{atm})\ (64\ \mathrm{g/mol})}{24.5\ \mathrm{L \cdot atm/mol}} = 0.013\ \mathrm{kg/m^3} \end{cases}$$

The minimum liquid-to-gas (water to air in this case) flowrate, Q_w/Q_a is then given by Equation 6.127 with the $C_{w0} = C_{a0}/K_{aw}$ or 12.9 kg/m³.

$$\left(\frac{Q_w}{Q_a}\right)_{min} = \frac{C_{al} - C_{a0}}{C_{a0}/K_{aw}} = \frac{(0.261\ \mathrm{kg/m^3} - 0.013\ \mathrm{kg/m^3}}{12.9\ \mathrm{kg/m^3}} = 0.019$$

Note that the inlet water, C_{wl}, was assumed to be free of SO_2.

Let us assume that the optimum liquid-to-gas flowrate is 50% greater than the minimum. The liquid-to-gas flowrate and resulting water concentration in the column effluent is then

$$\left(\frac{Q_w}{Q_a}\right)_{act} = (1.5)(0.019) = 0.029$$

$$C_{w0} = \frac{(C_{a0} - C_{al})}{(Q_w/Q_a)_{act}} = 8.6\ \mathrm{kg/m^3}$$

As expected, the effluent water contains less SO_2 than with the minimum liquid-to-gas flow. The operating line and C_a^* then become

$$C_w(z) = C_{w0} + \frac{C_a(z) - C_{a0}}{(Q_w/Q_a)_{act}} \qquad C_a^*(z) = K_{aw}C_w(z)$$

The number of transfer units (NTU) then becomes,

$$NTU = \int_{C_{al}}^{C_{a0}} \frac{dC_a(z)}{C_a(z) - C_a^*(z)} \approx 6.4$$

The integration can be accomplished by choosing values of C_a between C_{al} and C_{a0}, calculating C_a^* corresponding to those values of C_a. Thus, values of the integrand can be calculated at particular points between the limits of integration and the approximate value of the integral determined by numerical integration.

The total height of absorber required is then L = HTU·NTU = 8.5 m.

Note that if a less stringent separation were required, for example, to only 1% SO_2 in the effluent air, then the NTU would become approximately 4.6 and the total height required is only 6.2 m. The required height of the column is reduced as the degree of separation required is decreased.

6.6 SUMMARY

The above discussion has illustrated some of the technological approaches to controlling both particulate and gaseous air pollutants as well as the atmospheric processes of air pollutants after emission. In every case semi-quantitative design tools were developed that at least illustrate the key factors influencing the design of such systems and the parameters that influence their effectiveness. We will continue this approach for both water pollutants and soil contaminants in the next two chapters.

In most cases, individual sources of air pollutants have a measurable impact only locally and models were discussed that can predict that impact, at least under simple meteorological and terrain conditions. More sophisticated and presumably more accurate models exist, but they are beyond the scope of this text.

A wide variety of pollution control devices exist to limit the atmospheric emissions. Particulate emissions are generally the easiest to control due to their distinct physical properties when compared to air. Significant reduction in particulates can be accomplished cheaply with simple filtration or centrifugal separation. More sophisticated tools such as electrostatic precipitation may be necessary for very fine particles and other special situations. Generally, the smaller the particulate matter the more difficult it is to achieve capture and control.

Gaseous emissions are more difficult to control due to their similarity to the air stream in which they are transported. Controls are often required to be more contaminant and process specific. A few examples of particular processes were summarized and two widely applicable approaches were discussed in more detail. Adsorption onto a solid phase works well for nonpolar organic contaminants that sorb strongly to particular adsorbents. Absorption into a liquid phase is quite useful for gaseous components that adsorb or react with the liquid phase. Adsorbents and absorbents must be either readily capable of regeneration or inexpensive for the process to be economically viable.

PROBLEMS

1. Compare the reported density at 1000 m elevation in the U.S. Standard Atmosphere to that predicted by the ideal gas law given the reported temperature and pressure. What could be the cause of any discrepancy?

2. Estimate the mass of air contained within the troposphere and compare it to a similar thickness of the stratosphere.

3. Rewrite Equation 6.1 for a contaminant that undergoes degradation reactions described by a rate $k_S\varphi_S$ in the stratosphere and $k_T\varphi_T$ in the troposphere. Show the form of Equation 6.4 under these conditions if $k_T \gg k_S$.

4. Develop material balance equations for the northern and southern hemispheric stratospheres and use the strontium-90 data in Figure 6.4 to estimate the rate of exchange between them.

5. It has been estimated that 10^{10} kg of particulate matter was injected into the southern troposphere by the explosion of the volcano Krakatoa in the East Indies in 1883 (Seinfeld, 1975). Assuming that this is fine particulate matter that does not settle or move into the stratosphere, estimate the concentration in the northern troposphere as a function of time. Compare to ambient air quality standards for particulate matter.

6. Assume that the particles deposit to the Earth at a rate of 1.1×10^{-5} g/(m^2·h), estimate the tropospheric concentrations.

7. Determine the slopes of a constant pressure surface required to generate a geostrophic wind of hurricane force (i.e., 75 mph).

8. Derive the Eckman spiral equations from Equation 6.19.

9. Integrate the Eckman spiral velocity profile to determine the magnitude and direction of the mean velocity over the boundary layer.

10. Estimate the convective velocity scale, w_*, for a heat flux of 100 W/m^2 and a typical midday mixing height of 1000 m. What is the characteristic time for mixing over the mixing height? For a horizontal wind velocity of 2 m/s, how far downwind from a pollutant source would the pollutants be expected to be well mixed over the depth of the mixing layer?

11. Derive the temperature lapse rate for air containing ω kg of water vapor per kg of dry air. As the air rises, some of the water vapor condenses releasing heat, $\Delta \hat{H}_v d\omega$. Compare the dry and wet adiabatic lapse rate if the surface air is 30°C to the dry and wet adiabatic lapse rate if the surtace air is 0°C.

12. What is the magnitude of the Brunt-Vaisala frequency in a environment in which the potential temperature **increases** 1°C/100 m of height?

13. In Example 6.3, how hot would the surface temperature need to be before the mixing depth extends beyond 900 m? If the air at the surface were saturated with water what would the maximum mixing height be?

14. It is conceivable that a source of heat at the surface could eliminate elevated inversions and allow deep mixing and dilution of surface air pollutants. Estimate the amount of energy that must be provided to mix air to a height of 900 m using the morning temperature sounding of Example 6.3. Is it feasible to provide this energy, for example, from the combustion of fossil fuels?

15. In an area with an average roughness height of about 0.1 m, the wind velocity at 10 m elevation is 2 m/s. The surface heat flux is about 100 W/m^2. What is the Monin-Obukhov length scale and a qualitative characterization of the stability of the atmosphere? What would be the expected Pasquill-Gifford stability classification? What would be the expected velocity at 2 m and 30 m above the ground?

16. A chemist has a small lecture bottle containing 1 kg of hydrogen sulfide (H_2S) that he no longer needs. Recognizing that this compound is quite dangerous and detectable by a distinctive rotten egg odor at concentrations well below 100 ppb and can be dangerous at concentrations in the ppm range, the chemist decides to slowly release the hydrogen sulfide in a fumehood at a rate of about 100 g/min. The fumehood exhaust is a surface mounted vent, however, allowing the vapors to be trapped within the wake of the building which is 10 m high and 30 m long. In a 2 m/s wind speed, what would be the maximum expected concentration in the wake of the building. Assuming that this concentration is drawn into the building intake vents, would everyone in the building be expected to smell the hydrogen sulfide? (This should be a caution to anyone who assumes that fumehoods safely remove contaminants from a laboratory.)

17. Consider a heavily populated and largely unregulated downtown area 1 km on a side in the developing world. During morning rush hour traffic emissions of carbon monoxide within this area are expected to be as great as 10 kg/s. On a cool morning when the wind speeds are low, 2 m/s, and the mixing height small (~50 m), estimate the concentration of carbon monoxide within the area. Compare to the ambient aır quality standard in the U.S. for carbon monoxide of 9 ppm. If the afternoon mixing height is 500 m and the wind speed 4 m/s, estimate the lower concentrations expected at that time.

18. Under the conditions of Example 6.4, estimate the fractional contribution of upwind sources to the concentration of SO_2 over Baton Rouge if the upwind concentration is of the order of 0.03 mg/m^3. Estimate the steady-state concentration over the city.

19. In addition to upwind contamination as in Problem 18, the air above the mixing height also contains SO_2 at 0.03 mg/m^3. If the mixing height is growing 100 m/h, estimate the concentration after 5 h assuming that the initial concentration is the steady-state concentration without upwind sources calculated in Example 6.4.

20. Compare Equations 6.49 and 6.51 under stable conditions. Are their predictions of plume rise for a particular stack condition similar?

21. Estimate the plume rise from a 1-m diameter stack, 50 m tall. The stack exit temperature is 250°C and velocity 10 m/s. The ambient air temperature at stack height is 25°C and the background temperature gradient is –0.002°C/m (potential temperature gradient of 0.0078°C/m). The wind speed at 10 m is 2 m/s. Estimate plume rise for the following background temperature (not potential temperature) gradients: –0.002°C/m (potential temperature gradient is 0.0078°C/m), –0.0098°C/m, and +0.01°C/m.

22. A power plant releases 1000 g/s of SO_2 with an effective stack height of 200 m. On a clear day with a wind speed of 2 m/s at 10 m, estimate the concentration as a function of distance downwind.

23. Repeat Problem 22 on a cloudy day all other conditions being the same.

24. Repeat Problem 22 with an elevated temperature inversion at 500 m.

25. On a day with conditions similar to those in Problem 22, residents downwind have detected an odor that is later attributed to SO_2 which has an odor threshold of about 1300 mg/m^3. Estimate the minimum emission rate of SO_2 from the stack on that day. The effective stack height is again 200 m. Note that odor detection might be associated with short term average concentrations (e.g., 3 min averages).

26. On a clear, cool night a 10,000-gallon tank sulfuric acid fails allowing the liquid to fill a 30 m × 30 m diked area around the tank. It is very difficult to estimate how much sulfuric acid is likely to give rise to vapors or fumes, but an employee located 200 m downwind detected heavy, choking acid fumes. Knowing that concentrations of 10 mg/m^3 of sulfuric acid fumes can be definitely unpleasant, estimate the air release flux of sulfuric acid vapors in g/(m^2·s). The wind speed is about 2 m/s.

27. A highway might be considered a "line-source" of pollution. Integrate Equation 6.52 assuming that the wind is perpendicular to the highway to derive an equation for the concentration as a function of distance from the highway.

28. Use the model developed in Problem 27 to determine the CO concentration as a function of distance from the roadway assuming that a busy road has about 10,000 automobiles per hour, each releasing an average about 3 g/mi CO. The wind speed is 2 m/s on a cloudy morning.

29. Repeat Problem 28 but instead of using a line-source dispersion model, assume that the buildings adjacent to the street cause the CO to be mixed in the "street canyon." The separation between the buildings on each side of the street is 50 m and their height averages 10 m. The wind speed is again 2 m/s.

30. If it is assumed that the ratio of σ_y/σ_z is a constant (good under neutral to unstable conditions), show that the maximum concentration occurs at $2^{-1/2}H_s$ where H_s is the effective stack height. Using this estimate of the location of the maximum, show that the concentration at the maximum is given by

$$C_{max} = \frac{0.1171 Q_m}{U\sigma_y\sigma_z}$$

31. A failure in the particulate collection system in the power plant stack in Problem 22 results in the release of 10 kg/s of particulate matter with an average aerodynamic particle diameter of 20 μm and a particle density of 2 g/cm^3. Estimate the concentration of particulate matter at ground level as function of distance downwind.

32. Repeat Problem 31 if the particulate matter were 100-μm particles. What is the N_{Re} of the settling velocity of these larger particles?

33. Derive a formula for the overall efficiency, η_o, of two identical particulate collection devices in series if the efficiency of each of them individually is $\eta(d_p)$.

34. Derive a formula for the overall efficiency, η_o, of two identical particulate collection devices in parallel if the efficiency of each of them individually is $\eta(d_p)$.

35. Determine the collection efficiency of a cyclone separator for the collection of 10-μm particles with a density of 2 g/cm^3. The cyclone has three effective turns and an inlet width of 10 cm. The inlet gas velocity is 10 m/s. The temperature of the air stream is 75°C where the viscosity of air is about 0.75 g/(cm·h).

36. Test the formulas developed in Problem 33 by consideration of two identical cyclone separators as described in Problem 35 in series removing a stream containing only 10-μm particles.

37. Test the formulas developed in Problem 34 by consideration of two identical cyclone separators as described in Problem 35 in parallel removing a stream containing only 10-μm particles.

38. Consider a particle size distribution that is log normal with a geometric mean of 4 μm (log mean = 1.39) and a standard deviation in ln d_p of 0.542 μm. Approximately what fraction of the particle mass is greater than 4 μm or 6μm in diameter?

39. Determine the electric field strength required in V/cm to remove 90% of the 1-μm particles from a gas stream. A collection area of 100 m^2 exists and the feed flowrate is 500 m^3/min. The particle density is 2 g/cm^3 and their dielectric constant is 2.5. The temperature of the gas stream is 300 K.

40. Determine the single fiber/droplet collector collection efficiency as a function of particle size for a 500-μm collector in air at 25°C. The particles all have a density of 2 g/cm^3 and their initial speed relative to the collector is 10 m/s.

41. Estimate the collection of 10-μm particles in a gravity spray tower with a water flowrate of 0.002 m^3/m^3 of air flow. The mean water droplet diameter is 800 μm. Temperature is 300 K and the air viscosity is 0.666 g/(cm·h). Particle density is 2.5 g/cm^3.

42. How long must a fabric filter be to collect 90% of 1-μm particles with a density of 2 g/cm^3? The filter has a porosity of 90% and is composed of fibers 1 mm in diameter. The velocity of the particulate laden gas stream in the bed (q/ε) is 4 m/s. Temperature is 300 K and the air viscosity is 0.666 g/(cm·h).

43. Consider a linear gas-solid equilibrium $C_a = \alpha W_s$. For this case, rederive the velocity of the adsorption zone and the width of the active adsorption zone, δ.

44. Activated carbon with a bulk density of 400 kg/m^3 is to be used to separate a contaminant in an adsorber 10 m long and 1.5 m in diameter. The equilibrium sorption of the contaminant is given by α = 50 kg/m^3 and β = 1.5 in Equation 6.108. The mass transfer coefficient in the bed, ka_v = 50 s^{-1}. The feed flowrate is 1 m^3/s of air at 25°C containing 0.2% (by weight) contaminant. Estimate the maximum adsorption time (time to bed saturation), the width of the active adsorption zone, and the time until regeneration is required neglecting the contaminant adsorbed in the active adsorption zone.

45. Determine the length of an activated carbon adsorber required to adsorb a contaminant for which the equilibrium adsorption is given by a Freundlich isotherm with α = 20 kg/m^3 and β = 1.5. The adsorbent has a bulk density of 400 kg/m^3. The equilibrium sorption of the contaminant is given in Equation 6.108. The mass transfer coefficient in the bed, $k_a a_v$ = 20 s^{-1}. The feed flowrate is 1 m^3/(m^2·s) of air at 25°C containing 0.1% (by weight) contaminant. It is desired to operate for 12 h between regenerations.

46. Acetone is present in an air stream at a mole fraction of 0.1. An activated carbon adsorption bed is available for this purpose. The bed has a bulk density of 450 kg/m^3, is 5 m thick, and has a cross-sectional area of 5 m^2. The total gas flow is 1 kg/s. α = 15 kg/m^3 and β = 2.2. $k_a a_v$ = 20 s^{-1}. The bed and air stream temperature is 25°C. Determine the time to breakthrough. How appropriate is it to assume that the active adsorption zone is of negligible length?

47. Estimate the total required height of an absorber designed to remove 99% of the SO_2 in an air stream initially containing 1% SO_2 using clean water as the absorbing liquid. The air-water partition coefficient of SO_2 in water can be crudely approximated as 0.03 for the purposes of this problem. The air and water film mass transfer coefficients, $k_{af} a_v$ and $k_{wf} a_v$, are 735 h^{-1} and 11.25 h^{-1}, respectively in this packing at these conditions. The air flowrate is 0.117 m^3/s per m^2 of column cross-sectional area. The column temperature is 25°C.

48. The following data for the equilibrium of ammonia and water at 30°C is available from Wark and Warner (1981).

C_a^- kg/100 kg	1	1.5	3.5	5.0	9.0	19.0	30.0
p_a – mmHg	10	15	35	50	100	250	450

It is proposed to use water to collect ammonia from a gas stream. of 10 m^3/min. If the inlet ammonia content in the gas stream is 20% by volume and the inlet water is ammonia free, what is the minimum water flowrate needed to remove 90% of the ammonia?

49. The acetone in a contaminated air stream in Problem 46 also may be removed by contact with water. The equilibrium relationship between the gas and liquid phase mole fractions (y and x, respectively) and a correlation for the height of a transfer unit are given by

$$y = -0.33xe^{1.95(1-x)^2} \qquad HTU = 3.3G^{0.33}L^{-0.33} \quad \text{(meters)}$$

Here G and L are the mass flows in kg/($m^2 \cdot h$) to the tower. (The source of the data is McCabe and Smith, 1956.)

Estimate the minimum water flowrate to remove 90% of the acetone. What is the required height of the absorption tower if the actual water flowrate is 50% more than the minimum?

REFERENCES

Briggs, G.A. (1969) *Plume Rise*, AEC Critical Review Series, Washington, D.C..

Carpenter, S.B. (1970) Principal plume dispersion models, TVA Power Plants, 63rd Annual Meeting, Air Pollution Control Association, June.

Golden, D. (1972) Relations among stability parameters in the surface layer, *Bound. Layer Meteorol.*, 3:47-58.

Gifford, F.A. (1976) Turbulent diffusion typing schemes: A review, *Nucl. Saf.*, *17*(1): 71.

Fox, R.W. and A.T. McDonald (1978) *Introduction to Fluid Mechanics*, 2nd. ed., John Wiley & Sons, New York.

Hanna, S.R., G.A. Briggs, R. P. Hosker (1982) *Handbook on Atmospheric Diffusion*, DOE/TIC-11223, Technical Information Center, U.S. DOE.

McCabe, W.L. and J. Smith (1956) *Unit Operations in Chemical Engineering*, McGraw-Hill, New York.

Reiter, E.R. (1975), Stratospheric-tropospheric exchange processes, *Rev. Geophys. Space Phys.*, *13*, 459–474.

Seinfeld, J. (1975) *Air Pollution: Physical and Chemical Fundamentals*, McGraw-Hill, New York, 523 pp.

Slade, D.H. (1968), Ed. *Meteorology and Atomic Energy*, AEC, Washington, D.C.

Stull, R.B. (1988) *An Introduction to Boundary Layer Meteorology*, Kluwer Academic Publishers, Boston, MA.

Turner, D.B. (1969) *Workbook of Atmospheric Dispersion Estimates*, HEW, Washington, D.C.

Wark, K. and C.F. Warner (1981) *Air Pollution: Its Origin and Control*, Harper Collins, New York, 526 pp.

Warneck, P. (1988) *Chemistry of the Natural Atmosphere*, Academic Press, New York, 757 pp.

7 Water Pollution and Its Control

7.1 INTRODUCTION

Our focus in this chapter is on the processes and control of pollutants in surface waters and in the sediments with which they are associated. The oceans cover about 70% of the earth but while the oceans constitute the largest volume of surface waters, they will not be our primary focus. Most identified environmental problems of surface water are associated with rivers and lakes of finite volume and assimilation capacity located near human activities. Even in the ocean our primary interest is only the upper layers, since mixing between the surface layer and the layers below is limited. Even in this layer, however, the volume of water is large and the assimilation capacity great. This is not to say that the oceans are not a potentially significant environmental problem. There has simply not been a global problem identified that has been demonstrated to be as significant as the atmospheric concerns of global warming and ozone depletion. Much of our discussion of environmental processes will focus on rivers and lakes and the behavior of contaminants.

Contaminated sediments associated with surface waters also tend to be a more localized problem. Sediment contaminants that are observed in measurable and potentially significant concentrations were often placed in the sediments as a result of past industrial activity or accidents. Many of these introduced contaminants may be spread throughout the biosphere by this time and are present at low concentrations often with unknown potential or actual effects. Those that remain with the sediments are largely immobile, in sediments that are themselves largely immobile. The volume of the sediments that are contaminated tends to be large while the concentrations found in these sediments tend to be lower than might be expected to be an immediate risk to human health or the ecosystem. Our discussion of these problems will focus on an effort to identify and quantify, in so far as possible, the processes that lead to exposure and risk to human and ecological health from these contaminants.

Finally, we shall evaluate the common water treatment technologies available for the prevention of surface water pollution. In this material we shall take the same approach as in the previous chapter of developing approximate design equations for those technologies that illustrate the fundamental principles and key process variables but may not be the current basis for a detailed design.

7.2 SURFACE WATER PROCESSES

7.2.1 Ocean and Lake Circulations

As in the atmosphere, solar heating and the differences in temperature both laterally and vertically largely define the ocean circulation. Also as with large scale atmospheric motions, Coriolis forces play a major factor in development of the large scale circulations that form the world's currents. Ocean circulations are also driven by salinity gradients for which there is no counterpart in the atmosphere. Figure 7.1 displays the major surface currents in the oceans. Perhaps the most prominent features are the large clockwise circulations in the northern hemispheric oceans and the opposing counterclockwise circulations in the southern hemispheric oceans. The western portion of these circulations in the Northern Hemisphere, the Gulf Stream in the Atlantic and the Kuroshio current in the western Pacific, are well-defined currents that bring warm water from the equator northward and have a profound effect on the climate in the upper latitudes. Similar flows are

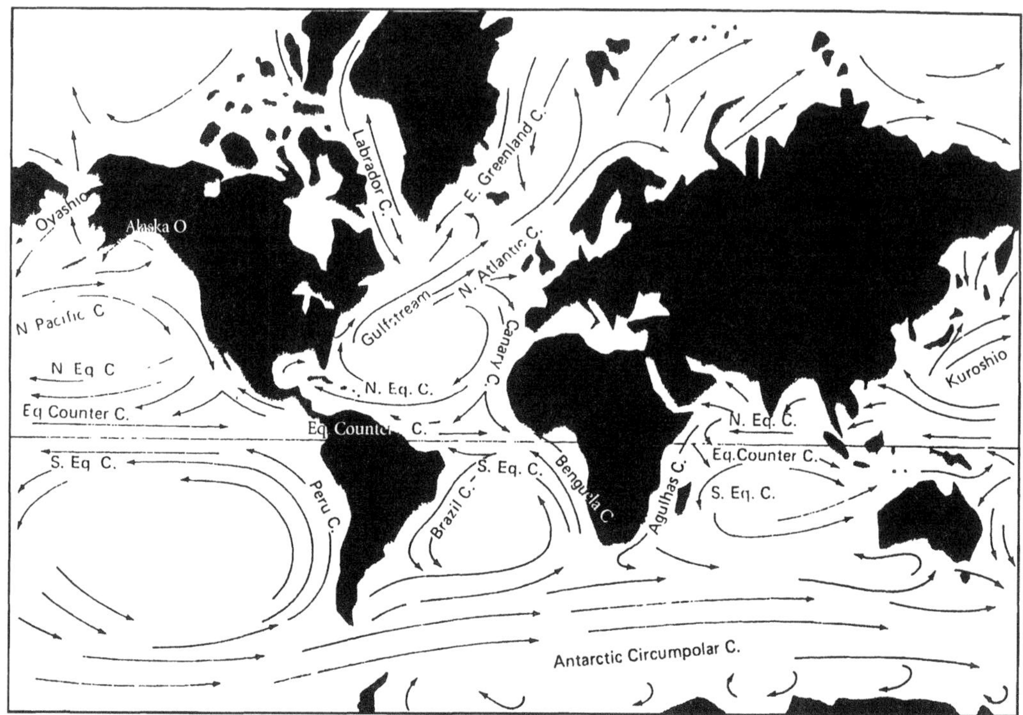

FIGURE 7.1 Major surface currents of the world's oceans. (From Knauss, J.A. (1978) *Introduction to Physical Oceanography*, Prentice-Hall, Upper Saddle River, NJ. With permission.)

observed in the Southern Hemisphere and are also directed poleward on the western edge of the oceans. The direction of these circulations agrees with the tendency of flows to veer right in the Northern Hemisphere and to the left in the Southern Hemisphere as a result of Coriolis forces. To a first approximation all major ocean currents are *geostrophic*, that is they represent a balance between Coriolis forces and the driving force for the current such as pressure gradient, density gradient, or sea surface slope.

Generally of more interest to the environmental engineer than the global circulation is local wind-driven flows. If there is a flat water surface with no horizontal pressure gradients or slope to the water surface, the only significant driving force for flow is the wind. Wind causes both surface currents and waves in the sea as a result of friction at the interface. Typically the surface current produced by a given wind is of the order of 2 to 3% of the wind velocity (as measured at a height of 5 to 10 m). Thus a 10 m/s wind should give rise to a surface water current of about 0.3 m/s. This can also be written in terms of shear stress or friction. The friction at the sea surface due to a wind of speed U (as measured at a height of 10 m) is given approximately by

$$\tau_{xz}(\text{dyn/cm}^2) \approx 0.02\,U^2 \qquad U \text{ in m/s} \tag{7.1}$$

This has important consequences for mixing within a lake, the exchange of gases at the air-water interface, and for the release of volatile chemicals from water. The energy imparted to the water by the wind is often the primary mechanism for mixing in lakes and impoundments.

Because of Coriolis forces, the direction of the wind-induced current is not in the same direction as the surface winds but is instead directed 45° toward the right in the Northern Hemisphere (that is, toward the west as a result of a wind from the north). This can be derived by assuming a constant eddy viscosity with depth as was done with the atmospheric Eckman spiral. Because the Eckman spiral suggests that the surface winds are shifted 45° from the geostophic winds, a wind-generated current is expected to be in the direction of the geostrophic winds aloft and *perpendicular* to the atmospheric pressure gradients. In the real atmosphere and ocean systems, of course, rarely do the rigid assumptions of the Eckman spiral apply. Wind-driven currents are normally observed to be shifted 20 to 40° from the wind.

One consequence of the shift between wind motions and the resulting currents is that a flow along the shoreline will tend to cause a current directed perpendicular to the shore as shown in Figure 7.2. Thus, winds from the direction of the poles that predominate along the Peruvian and California coasts cause a surface flow that is directed offshore. To replace this water, cooler nutrient rich water rises from the ocean floor. This *upwelling* gives rise to the rich fisheries as well as cold water off these coasts. Note that this motion is superimposed upon the global circulation which also encourages cooler water (but not upwelling) to be found along these coasts.

Motion within the deeper layers of a lake or ocean is resisted by friction at the sediment-water interface. Conceptually this interface is equivalent to the air-soil interface and the processes of heat, mass, and momentum transfer at this interface are described in a similar manner. Because solar heating of the sediment-water interface is seldom important, however, the currents near the bottom are described by the neutrally stable logarithmic law. This means that near the sediment-water interface the velocity varies logarithmically with depth while the eddy conductivity, diffusivity, or viscosity vary linearly with depth. In streams, the relatively high velocities and shallow depths eliminate significant stratification and the logarithmic law is generally a good approximation over their entire depth, except at the surface of slow moving streams when wind-driven currents may again become important. Flow in streams is explored in more detail in the section below.

7.2.2 Flow in Streams

Flow in streams is generally a balance between the gravity flow due to the slope of the water surface (which generally mirrors the slope of the sediment bed) and friction at the sediment surface. The direction is, of course, downslope. Figure 7.3 shows a cross section of a sloping stream. The force driving the stream downslope in the interval dx is given by

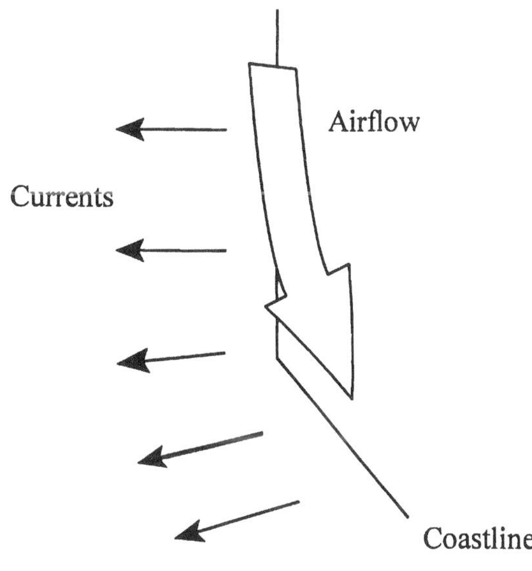

FIGURE 7.2 Offshore current resulting from along shore winds.

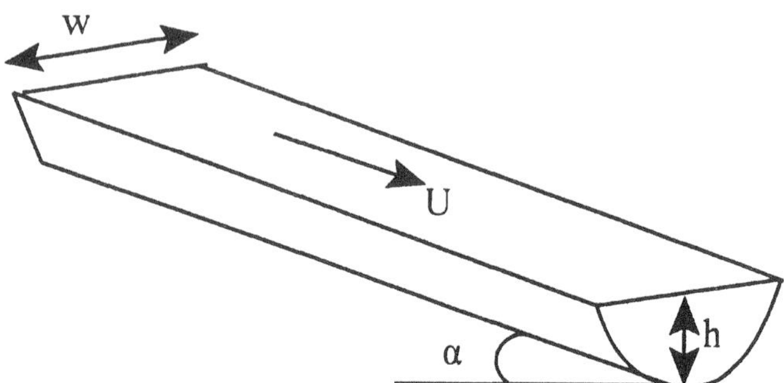

FIGURE 7.3 Cross section of a flowing stream.

$$F_x = \rho_w g \sin\alpha (h\,w) dx \tag{7.2}$$

where $\rho_w g$ is the force per unit volume acting on the water if it were aligned with gravity, sin α provides the component of gravity acting in the direction of the river slope and h·w·dx is the volume of the interval (h is the depth of the stream). Note that for small angles the sin α ~ α (in radians). That is, sin α represents the actual slope of the stream and is often given the symbol s.

The opposing frictional force, F_f, is given by the surface shear stress, τ_0, which is a force per unit area times the area of the stream bottom, the source of the friction,

$$F_f = \tau_0\, w\, dx \tag{7.3}$$

Remember that the logarithmic velocity profile should be valid over nearly the entire depth of a stream due to the lack of heating or cooling effects in the underlying sediment. The shear stress at the sediment bed is then related to the friction velocity in the logarithmic profile by $\tau_0 = \rho_w u_*^2$. Setting Equations 7.2 and 7.3 equal gives a relation for the bottom surface friction or the friction velocity as a function of stream depth and slope.

$$F_f = F_x$$

$$\tau_0\, w\, dx = \rho_w\, g\, s\, w\, h\, dx \tag{7.4}$$

$$\tau_0 = \rho_w\, g\, s\, h \quad \text{or} \quad u_*^2 = g\, s\, h$$

where s = sin α.

As indicated previously, the estimation of the surface shear stress or friction velocity is dependent upon measurement of the velocity profile. The flow of water in streams and channels has long been studied and there exists a means to relate velocity in a stream to routinely available quantities such as stream slope.

$$U = \frac{r_H^{2/3} s^{1/2}}{\alpha_M} \tag{7.5}$$

This is termed Manning's formula. The term r_H is the hydraulic radius of the wetted surface of the stream and is given by the ratio of the cross-sectional area of the flow to the wetted perimeter.

$$r_H = \frac{\text{cross-sectional area of flow}}{\text{wetted perimeter}} \tag{7.6}$$

TABLE 7.1
Manning's Roughness Coefficient in $s/m^{1/3}$

Situation	Roughness Coefficient (α_M – $s/m^{1/3}$)
Alluvial channels — sand bed without vegetation	
Plane bed	0.014–0.020
Ripples	0.018–0.028
Dunes	0.018–0.035
Natural stream channels — earthen bed with vegetation	
Clean, straight banks	0.025–0.030
Winding stream, some pools and shoals	0.035–0.040
Winding stream with stony bed	0.040–0.055
Sluggish reaches, deep pools, significant vegetation	0.070
Canals and ditches	
Earthen canals, straight and uniform	0.023
Winding sluggish canals	0.025
Dredged earth canals	0.028
Pipe	0.010–0.017 (0.013 typical)

$$= \frac{\pi \frac{r^2}{2}}{\pi r} = \frac{r}{2} \qquad \text{for a stream of circular cross section}$$

For a rectangular channel of depth h and width w, $r_H = wh/(w + 2h)$. Note that for a wide channel, w >> h, and the hydraulic radius is simply the depth of the stream, h. The coefficient α_M depends on the roughness of the streambed. Note that α_M is not dimensionless. Generally, the hydraulic radius is in meters and the calculated velocity is desired in m/s. Then α_M must have units of $s/m^{1/3}$. Table 7.1 includes values of the roughness coefficient in these units under a variety of conditions. Note that Equation 7.5 is also applicable to gravity flow in engineered pipes, canals, and channels as well as natural streams, and values of the roughness coefficient for these situations are also included in Table 7.1.

Equation 7.4 and 7.5 can be combined to relate the friction velocity to common stream parameters,

$$u_* = \frac{U\alpha_M}{r_H^{2/3}}\sqrt{gh} \tag{7.7}$$

Example 7.1: Estimation of stream velocity

Employ Equation 7.5 to relate flow and stream geometry under the following conditions:

- The Mississippi River has an average slope of about 1 m per 10 km of length. Estimate the velocity of the river under conditions of an average depth of about 8 m and where the bottom is rippled sand.
- Estimate the required slope of an unlined drainage ditch 1 m deep and 2 m wide designed to remove 1 m^3/s of water.

1. Rippled sand would have a roughness coefficient of about 0.025 s/m$^{1/3}$. Because the Mississippi River is wide, $r_H \sim h$ where h is the depth of 8 m. Thus, the velocity of the river under these conditions is

$$U = \frac{r_H^{2/3} s^{1/2}}{\alpha_M} \frac{(8\text{ m})(1\text{ m}/10000\text{ m})}{(0.025\text{ s/m}^{1/3})} = 1.55\text{ m/s}$$

This represents relatively low flow conditions on the southern Mississippi River.

2. A ditch 1 m deep and 2 m wide has a hydraulic radius of

$$r_H = \frac{wh}{w+2h} = \frac{(2\text{ m})(1\text{ m})}{2\text{ m} + 2(1\text{ m})} = 0.5\text{ m}$$

Straight, earthen banks would suggest a roughness coefficient of about 0.023 s/m$^{1/3}$. The required velocity is Q/(h·w) or 0.5 m/s suggesting a required slope of

$$s = \left(\frac{U\alpha_M}{r_H^{2/3}}\right)^2 = \left(\frac{(0.5\text{ m/s})(0.023\text{ s/m}^{1/3})}{(0.5\text{ m})^{2/3}}\right)^2 = 3.33(10)^{-4}$$

This represents a slope of about 33 cm per km.

7.2.3 Surface Water Runoff

A major source of water as well as contaminants for streams and other surface waters is direct runoff from rainfall. Water that falls as rain can initially either infiltrate into the soil, evaporate, or flow along the surface of the Earth in the form of runoff. Over a longer period, *transpiration*, the release of moisture by plants after its absorption into root systems, may occur. Here we will consider only runoff due to its effect on the *quantity and quality* of surface waters. This *nonpoint source* of pollution of surface waters is extremely difficult to control because it is distributed and not confined to a single controllable outfall as is an industrial wastewater discharge. Runoff control requires an active effort at agricultural fields, construction sites, paved surfaces, and other areas where normal infiltration patterns have been disrupted. Undisturbed vegetated surfaces tend to have a much greater capacity for infiltration of rainwater.

Surface runoff is normally estimated via an empirical approach developed by the U.S. Soil Conservation Service. The volume of runoff, V_r, is approximated as a function of the drainage area, A, of the basin exposed to rainfall, H_r (for example, in inches or centimeters of rainfall), in which the soil's storage capacity for the rainfall is S_r (in equivalent inches or centimeters of rainfall). The relationship given by Mockus, Soil Conservation Service (1972), can be written,

$$V_r = A\frac{(H_r - 0.2S_r)^2}{(H_r + 0.8S_r)} \tag{7.8}$$

In the empirical approach of the Soil Conservation Service, S_r is correlated with the characteristics of the soil through a runoff curve number, CN, the relationship between the two being given by

$$S_r = \frac{1000}{CN} - 10 \qquad S_r \text{ in inches} \tag{7.9}$$

Curve number 100 corresponds to no significant capacity for water infiltration into the soil and all of it becomes runoff while small curve numbers imply significant infiltration capacity. The curve numbers were based on rainfall in inches and thus use of SI or other systems of units requires correction of S_r *after estimation of* S_r *from the curve number.* The curve numbers are based on soil characteristics and the moisture contained in the soil prior to the rainfall. The Soil Conservation Service defines four soil groups,

A. Low runoff potential soils, well-drained, deep layers of sands or gravels, infiltration rate of 8 to 12 mm/h
B. Less deep, sandy soils or less, infiltration rate of 4 to 8 mm/h
C. Shallow soils and silty or clayey soils, infiltration rate of 1 to 4 mm/h
D. High runoff potential soils, poorly drained, shallow soils or swelling clays, infiltration rates of 0 to 1 mm/h

These categories must be modified by the initial moisture content in the soil according to the following conditions:

I. Dry soils, less than 0.5 in. of rain in previous 5 days
II. Average conditions, 0.5 to 1.1 in. of rain in the previous 5 days
III. Wet conditions, over 1.1 in. of rain in previous 5 days

Table 7.2 defines curve numbers according to these categories under selected conditions and land uses. More complete guidance on curve numbers and runoff from various land uses exists, but Table 7.2 provides a preliminary indication of runoff curves.

TABLE 7.2
Representative Soil Conservation Service Runoff Curve Numbers (CN)

	Site Type and Initial Moisture Conditions											
	Overall Site — Farm			Roads/Paving			Meadow			Forests		
Soil	I	II	III	I	II	III	I	II	III	I	II	III
A	39	59	77	55	74	88	15	30	50	26	45	65
B	55	74	88	68	84	93	38	58	76	46	66	82
C	66	82	92	78	90	96	52	71	86	59	77	89
D	72	86	94	81	92	97	60	78	90	67	83	93

Because runoff produced is a nonlinear function of the soil storage and curve number, the runoff volume should be calculated for each land use area separately and added. Use of an average curve number will in generally not be correct. This is illustrated in Example 7.2.

Example 7.2: Estimation of runoff volume

Consider a 10 ha area composed of 25% cultivated land, 25% paved, 25% open pasture, and 25% forested which is exposed to 10 cm of rainfall after a 0.75″ of rainfall 5 days previously. The entire area is covered by silty soils except for the pasture which is sand and gravel. Estimate the volume of runoff water.

The silty soils suggest a type C soil group in the area except for type A in the pasture. This suggests runoff curve numbers of 82 for the cultivated soil, 90 for the paved area, 30 for the open pasture, and

77 for the forested portion of the catchment. The associated soil storage capacity and volume of runoff generated on each portion of the catchment is then

Area	CN	S_r (in)	S_r (cm)	V_r (m^3)
Cultivated land	82	2.2	5.6	1365
Paved area	90	1.1	2.8	1816
Open pasture	30	23.3	59.3	15
Forest	77	3.0	7.6	1119

The total runoff generated is therefore about 4300 m^3 due to the 10-m rainfall. If the average curve number of 80 had been used, the (incorrectly) predicted runoff volume would only be about 3230 m^3.

The importance of runoff is not just associated with the volume of water that it adds to surface waters. In addition, runoff tends to carry sediment from the catchment area and the contaminants contained in those sediments. The sediment carried by the runoff is not just a function of the total volume of runoff, V_r, but also the peak rate of runoff. Haith (1980) estimated the peak runoff rate with the formula

$$Q_r = \frac{V_r}{T_r}\frac{H_r}{H_r - 0.2S_r} \tag{7.10}$$

where in addition to the terms defined previously, T_r is the duration of the runoff, generally taken as the duration of the storm event. This volumetric flow of runoff could be converted to a velocity of runoff by dividing by the depth of flow and width of the drainage basin. Equation 7.5 could then be used to estimate the depth of the runoff flow as a function of the Manning roughness coefficient and drainage basin slope

$$\begin{gathered}\frac{Q_r}{wh} = \frac{r_H^{2/3} s^{1/2}}{\alpha_M} \approx \frac{h^{2/3} s^{1/2}}{\alpha_M} \\ h \approx \left[\frac{Q_r \alpha_M}{w\, s^{1/2}}\right]\end{gathered} \tag{7.11}$$

This form of the equation assumes that $h \ll w$, i.e., that the hydraulic radius of the runoff is given by its depth. We shall see in a subsequent section that the ability of the flow to erode soil and sediment depends on the friction velocity or applied shear stress associated with flow. From Equation 7.4, assuming a logarithmic velocity profile in the runoff, the friction velocity is given by $u_* = gsh$, i.e., the product of the gravitational constant, the local slope, and the depth of the runoff.

7.2.4 Stratification and Stability in Temperate, Shallow Freshwater Systems

In surface waters as in the atmosphere the density stratification often defines the rate of vertical mixing of pollutants. In the atmosphere, the air density and therefore stratification is a function of only temperature and pressure (or altitude). In surface waters, however, these factors are further complicated by salinity and, in cold water, by water density that can increase as the temperature increases. Before considering these complications, let us discuss the stability of a shallow column (<100 m) of warm (>4°C) freshwater.

In a temperate, shallow freshwater system, the evaluation of stability and the tendency for turbulence and vertical motions to be enhanced or damped in such a system is essentially equivalent to the simple picture presented in Chapter 5. If the temperature is uniform with depth under these conditions, the density is also uniform and the column of water is neutrally stable. If, however, the temperature increases as you near the surface the presence of the lighter, less dense water above causes stable stratification and turbulence and mixing to be damped. If the temperature decreases as you near the surface, the cooler, more dense water at the surface will cause the column to *overturn* and mix effectively over its entire depth. Normal seasonal temperature changes in the mid-latitudes will cause moderately deep lakes to pass through each of these conditions over the course of a year.

Figure 7.4 displays the temperature profile that would be expected in such a lake during the spring, summer, and fall. During the summer, warming of the surface layers of the lake will cause a stable region to form at the surface. Its density relative to the water below will tend to eliminate mixing downward. This warming trend will continue during the summer resulting in the development of a strongly stable layer. Wind-driven circulation in the upper few meters of the lake will likely keep this region relatively well mixed despite the tendency to form a stable layer. Below the wind mixed region, however, the temperature decreases rapidly due to the lack of mixing and heat, mass, or momentum transport through the stable layer. This layer is called a *thermocline*, referring to a relatively thin, well-defined layer where the temperature decreases as one goes deeper into the lake. This region acts analogously to an elevated temperature inversion in the atmosphere.

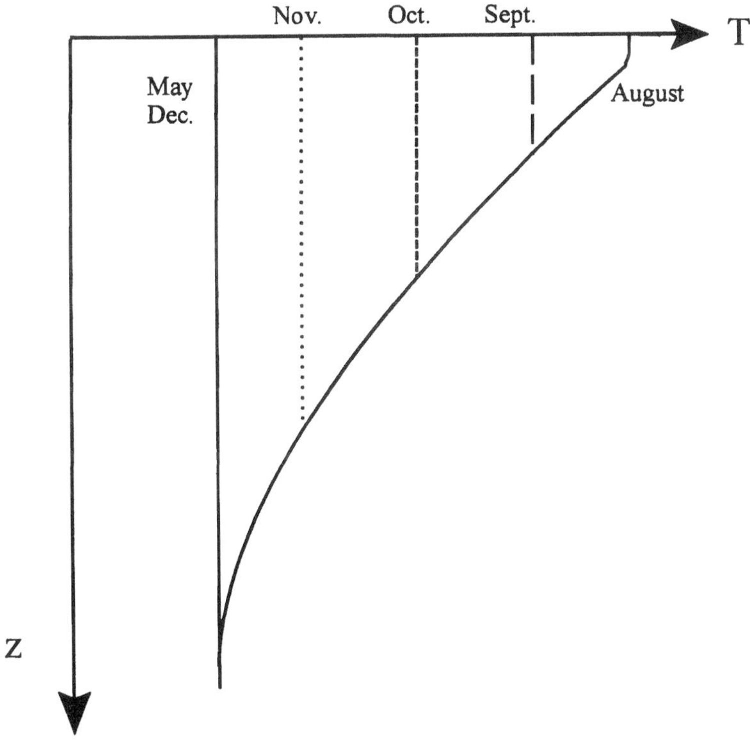

FIGURE 7.4 Temperature profiles in a mid-latitude lake. A strong thermocline in late summer gives way to a growing mixed-layer at the surface during the fall until the surface layer cools to the point that it is lower in temperature than the deeper water causing the lake to overturn, mixing the entire lake to a uniform temperature.

This process will continue until the surface waters begin to cool with the onset of fall and winter. Any cooling of the surface water during this time is translated almost immediately into mixing of a portion of the upper layer and erosion of part of the thermocline. Cooling and the subsequent fall of the cooled, dense water is equivalent to the erosion of the elevated temperature inversion in the atmosphere by rising thermals from a warmed ground. The cooled water will fall until its temperature (and density) equals that of the surroundings. This process will continue, eroding and weakening the thermocline until the surface cooling has resulted in the formation of water at or below the temperature of the water below the thermocline. The lake will then immediately overturn due to the settling of this cold water and mix throughout its depth.

The presence of stratification in lakes can be quantified by examination of the numerator of the Richardson number, which represents the consumption of turbulent energy by buoyancy forces. This quantity is the square of the Brunt-Vaisala frequency,

$$f_{BV}^2 = \frac{g}{T}\frac{dT}{dz} \tag{7.12}$$

When this quantity is positive and real values of the Brunt-Vaisala frequency exist, the stratification is stable and its strength is proportional to the temperature gradient.

The process of stratification formation and destruction is the central feature of the environmental dynamics of lakes. During periods of strong stratification and limited mixing of lakes, degradation of pollutants in the sediments can cause depletion of oxygen in the lower levels of the lake. If this occurs long enough and the pollutant loading is high enough, the bottom of the lake can be depleted of so much oxygen that sediment dwelling organisms at the base of the food chain may die with higher organisms to follow. Even for higher organisms, the higher temperatures at the upper layers of the lake may mean that insufficient oxygen is available in either layer.

Temperature stratification is a permanent feature of the oceans. Figure 7.5 displays the vertical temperature and salinity profile in the ocean. In the tropics and mid-latitudes the water is significantly warmer at the surface. This warmer water is generally less dense and therefore the oceans are stably stratified and turbulent mixing between the surface and deeper layers is reduced. This is demonstrated by the lack of mixing implied by the rapid changes in salinity at the thermocline in Figure 7.5. The relatively uniform levels of salinity in the upper layer suggest that wind-driven or other surface currents keep this layer well mixed. The uniform temperature profiles provide no significant resistance to turbulent mixing.

There is a seasonal strengthening or weakening of the thermocline in the near surface layers but because the seasonal variations in water temperature are small compared to those overland, the presence of a thermocline at depth is a permanent feature of the stratification in the ocean and provides an effective decoupling between contaminants and other characteristics of the upper layers from those below. For many environmental concerns, it is the upper layer of the oceans that is of primary importance.

7.2.5 Other Effects on Stability of Surface Waters

The above discussion was largely limited to relatively shallow freshwater or uniform salinity systems above 4°C. It is important to recognize that the density gradient controls the stability of a column of fluid and not the temperature. In seawater, salinity gradients also can have a strong effect on density, while in very cold waters, the dependence of temperature on density changes. Below 4°C, freshwater density decreases as the temperature decreases. This means that cooler water (below 4°C) will be less dense and tend to rise through or float on top of slightly warmer water. Thus, in the upper mid-latitudes during cold conditions, the coldest water at the surface is the least dense and the water column is stably stratified. Continued cooling will result in ice formation,

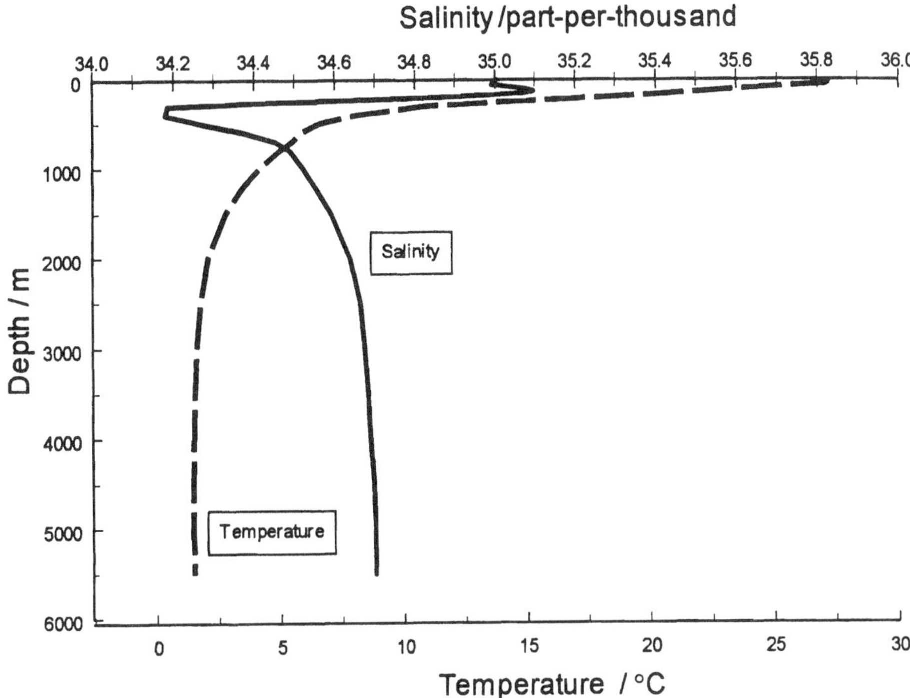

FIGURE 7.5 Temperature and salinity profiles in the ocean. (Adapted from Knauss, J.A. (1978) *Introduction to Physical Oceanography*, Prentice-Hall, NY.)

which due to its lesser density will also remain at the surface. It is the stability associated with the more dense water depth which keeps many streams from freezing completely during the winter.

The median salinity of the ocean is about 34.69 g/kg of seawater, about 85% of which is sodium chloride. The presence of the salt depresses the freezing point of water to about –2°C and changes the temperature at which the density of water is a maximum to about –4°C. The presence of salt modifies the influence of temperature on stability. In addition, saltwater is significantly more dense than freshwater at between 1.02 and 1.03 g/cm^3. Major ocean currents are driven by small variations in salinity. For example, the salinity of the Mediterranean Sea is about 38 g/kg compared to 35 to 36 g/kg in the eastern north Atlantic Ocean. The water in the Mediterranean Sea is more dense despite being warmer. The greater density of the Mediterranean Sea causes a current directed outward (westward) and down at Gibraltar.

Vertically variations in salinity can also influence stability. A column of water with higher saline content at depth would be stably stratified. The salinity gradient strengthens the stability of the upper layer in the ocean depicted in Figure 7.5. Often, since higher temperatures are associated with increased evaporation, they may also exhibit higher salinity. Water exhibiting an apparently stable temperature profile may be destabilized by variations in salinity.

For very cold and saline waters, the quantitative picture of stability presented by Equation 7.12 must be modified. Now it is recognized that it is not solely an increasing temperature with height (or decreasing temperature with depth) that controls the stratification and whether significant consumption of turbulent energy will occur. The consumption of turbulent energy by buoyancy effects is better correlated with the actual density gradient, regardless of whether that density gradient is the result of temperature variations or salinity. Under such conditions, the Brunt-Vasaila frequency can be written,

$$f_{BV}^2 = \frac{g}{\rho_w}\frac{d\rho_w}{dz} \tag{7.13}$$

Positive real values of the Brunt-Vasaila frequency are again the case for stably stratified conditions.

The interpretation of the effects of temperature and salinity on stability also must be modified in very deep water. Under the extreme pressures at depth in the ocean, water is slightly compressible and changes in depth force a change in temperature in much the same way that the temperature decreases when air rises in the atmosphere. It is convenient to define a potential temperature for the comparison of temperatures at different depths in the ocean. The potential temperature, by analogy to the potential temperature in the atmosphere, is the temperature that would be exhibited by a parcel of water if brought to the surface adiabatically, or without exchanging heat with its surroundings. The conversion between actual and potential temperature is a function of salinity, depth, and temperature so it is somewhat more complicated than in the atmospheric case. The difference is small and for shallow water systems such as lakes or above the thermocline in the ocean, is generally unimportant to the environmental engineer. Water 10°C at a depth of 1000 m containing about 35 g/kg salt will exhibit a temperature of about 9.87°C if raised to the surface. Thus, the difference between the actual and potential temperature of the water at this depth is only 0.13°C.

The development of a thermocline in rivers and streams is usually discouraged by the turbulence and mixing within a stream. Saltwater intrusions into rivers at the coast are quite common. A dense wedge of saline water will move upriver a distance depending on the velocity of the downriver flow. Because this saline water is more dense than the overriding flow above, there tends to be little mixing between the two fluids. The flow of a dense fluid such as a saline intrusion into the less dense freshwater is an example of a gravity or density current. If the friction at the bottom of the stream is neglected, often a surprisingly good assumption, a gravity current will move horizontally with a velocity given by the following relationship

$$N_{Fr} = 0.5\frac{U}{\sqrt{g\frac{\Delta\rho}{\rho}h}} \tag{7.14}$$

N_{Fr} is called a Froude number, the ratio of the inertial to gravitational forces. U is the velocity of the saline intrusion, $\Delta\rho$ is the density difference between the saline and fresh waters, and h is the thickness of the saline wedge. To a first approximation, the density current will be expected to continue to progress upriver as long as U is greater than the current in the stream directed in the opposite direction. Because the river bottom is sloped upward the limit of upriver progression of a saline wedge cannot extend beyond sea level.

7.3 CONTAMINANT PROCESSES IN SURFACE WATERS

The influence of surface waters on contaminants can be divided into either fate processes such as chemical degradation or equilibrium phase partitioning or transport processes such as mixing and vaporization. Equilibrium phase partitioning recognizes that surface waters come in contact with sediment and air and also contain significant quantities of suspended and dissolved or colloidal sediment particles. The ultimate fate of a chemical is significantly influenced by its partitioning into these phases. Equilibrium concentrations also define the state to which the system is proceeding, although dynamic calculations are needed to define the rate of progress toward this state.

The ultimate fate process affecting many chemicals in the natural environment is degradation or chemical fixation in the sediment. Focusing specifically in the water phase, chemical degradation can occur as a result of a variety of processes including photodegradation (degradation driven by light, especially sunlight), hydrolysis, and biological processes. Biological processes include the microbial processes of bacteria as well as processes of larger animals and plants. In the discussion of these processes, let us focus on the hydrophobic organic contaminants because they engage in both a variety of degradation processes and partition to all environmental phases.

7.3.1 Phase Partitioning in Surface Waters

Partitioning in water has been discussed previously in Chapter 4 but a short review here is in order. Hydrophobic organics tend to sorb to the organic matter in suspended/dissolved sediment particles and the settled sediment vaporize to the atmosphere and accumulate in fish and other organisms. Surface waters are normally composed of at least the five phases — water, biota (or animal life), suspended organic matter (in suspended or dissolved/colloidal sediment particles), settled sediment, and the air as depicted in Figure 7.6. Normally these phases are not at equilibrium but we will defer the analysis of the dynamics of the mass transfer processes until a subsequent section. As indicated in Chapter 4, the partitioning of a compound among these phases is controlled by a series of partition coefficients, which for hydrophobic organic compounds is controlled by the hydrophobicity of the compound as measured, for example, by the organic carbon based partition coefficient or the octanol-water partition coefficient.

Often sediments were contaminated by past uncontrolled pollutant discharges to surface waters. Subsequent control of those discharges have led to an improvement in water quality and the development of a situation where the sediment now poses a source of pollution to the water. Let us therefore start with the sediment. At equilibrium, the concentration of a pollutant in the overlying water will approach the concentration in the pore waters. That is the concentration in the overlying water will approach that given by the sediment-water partition coefficient,

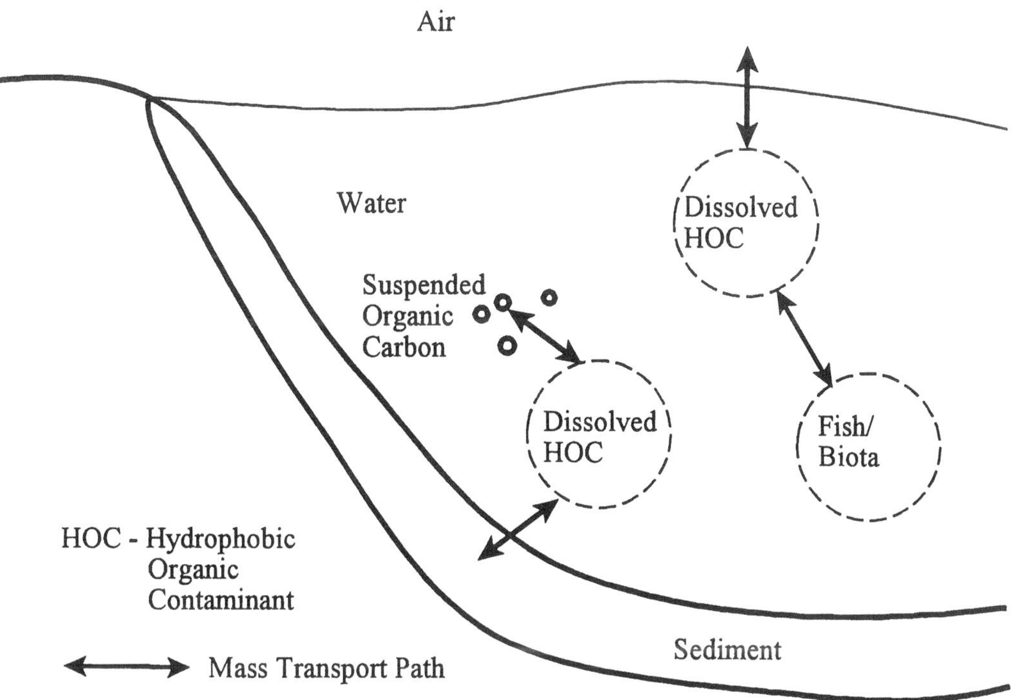

FIGURE 7.6 Depiction of five phases into which hydrophobic organic contaminants can partition.

$$\frac{W_s}{C_w} = K_{sw} \approx K_{oc}\omega_{oc} \tag{7.15}$$

Here W_s and C_w are the concentrations in the sediment (mg/kg) and water (mg/L), respectively. K_{oc} is the organic carbon based partition coefficient and ω_{oc} is the organic carbon fraction in the sediment. C_w is the concentration of the hydrophobic organic that is truly dissolved in water. The total concentration in the water phase is the sum of the truly dissolved and the quantity associated with the dissolved (doc) and suspended organic matter (som). For a hydrophobic organic the partitioning to both of these phases is often approximated by assuming that its sorbing characteristics are equivalent to the organic matter in the sediment, i.e., $K_{doc} \sim K_{soc} \sim K_{oc}$. If ρ_{doc} represents the mass concentration of dissolved organic carbon and ρ_{soc} represents the equivalent quantity for suspended organic matter, this means that the ratio of the total water concentration to the dissolved concentration is given by

$$\begin{aligned} \frac{C_T}{C_w} &= 1+\rho_{doc}K_{doc}+\rho_{soc}K_{soc} \\ &\approx 1+\rho_{doc}K_{oc}+\rho_{soc}K_{oc} \\ &\approx 1+\rho_{oc}K_{oc} \end{aligned} \tag{7.16}$$

where ρ_{oc} is the total mass of organic carbon suspended or dissolved in the water. Although this is only an approximate relationship, it is the best way to estimate the ratio of the total amount of contaminant in the water to that which is truly dissolved and available for partitioning to the sediment and the air in the absence of specific on-site experimental data. It should be noted that the suspended organic carbon can generally be filtered from water samples to eliminate the quantity of contaminant associated with that phase. Separation of dissolved organic carbon from the truly dissolved fraction is rarely attempted, however, and so-called "dissolved" concentrations of a contaminant in water actually represent the sum of the dissolved and that associated with colloidal organic matter (i.e., C_w $(1 + \rho_{doc}\ K_{doc})$).

The partioning of truly dissolved contaminant in the water and the biota follows a similar approach employing the bioaccumulation factor or biota-water partition coefficient, $K_{bw} = W_b/C_w$. The bioaccumulation factor for a particular organism depends upon the available organic fraction in the organism, best measured by the organic lipid content, ω_L. The fraction lipid can vary widely but many fish and benthic animals have a lipid content of the order of 4 to 5%. K_{bw} might be written $K_{Lw}\,\omega_L$ where the subscript L represents the lipid content of the organism. If the partitioning to the lipid content of an organism is essentially identical to the partitioning to the organic carbon of the soil, K_{Lw} might be approximated by the organic carbon based partition coefficient, K_{oc}. The ratio of K_{Lw} to K_{oc} defines what is called a biota-sediment accumulation factor (often abbreviated BSAF).

$$BSAF = \frac{K_{Lw}}{K_{oc}} = \frac{\dfrac{K_{bw}}{\omega_L}}{\dfrac{K_{sw}}{\omega_{oc}}} = \frac{\dfrac{W_b}{C_w\omega_L}}{\dfrac{W_s}{C_w\omega_{oc}}} = \frac{W_b}{\omega_L}\frac{\omega_{oc}}{W_s} \tag{7.17}$$

Thus, the BSAF is the mass loading or concentration in the biota (mg/kg biota) normalized by the lipid fraction to the mass loading in the sediment (mg/kg sediment) normalized by the organic carbon fraction. If the BSAF is unity, the partitioning of the contaminant to the organism lipids and to the sediment organic carbon are the same.

$$\frac{W_b}{C_w} = K_{bw} = K_{Lw}\omega_L \approx K_{oc}\omega_L \tag{7.18}$$

The BSAF ratio is often found to be near unity (perhaps 0.5 to 5) for a system *at equilibrium.* A measured BSAF, for example, between a fish living in the overlying water and the sediment, may be used to indicate the deviation of the system from equilibrium. A measured value in one location or at one time also may be used to predict the concentration in biota at another location or time based on the sediment concentration. Unless both systems are at equilibrium, however, such an extrapolation must be employed very cautiously since differences in the rate of processes at the different locations or times may give rise to different deviations from equilibrium and therefore different measured BSAFs.

Finally, the partitioning of the contaminant to the air is controlled by the Henry's Law constant or air-water partition coefficient for hydrophobic organic compounds. The ratio of the air concentration to the truly dissolved concentration is then given by

$$\frac{C_a}{C_w} = K_{aw} \tag{7.19}$$

It should be emphasized that all of these relationships are at equilibrium. While no actual system is at equilibrium, the equilibrium estimates may provide a good indication of the distribution of a contaminant in a system. Equilibrium calculations certainly provide an indication as to the direction and rate of mass transfer.

The goal of evaluating equilibrium conditions in such a system is defining the distribution of a contaminant between the various phases, or compartments. In order to fully define the system it is necessary to know the ratios of the concentrations (partition coefficients) and the volumes of the various phases. In addition, either the total mass of the contaminant in the system or the concentration in a single phase must be known. For example, if the equilibrium water concentration is desired in an air, water, or sediment system, Equation 4.115 is applicable, here repeated as Equation 7.20.

$$C_w = \frac{M}{K_{aw}V_a + V_w + K_{sw}\rho_b V_s} \tag{7.20}$$

If we were equally interested in the four phase system that includes air, water, sediment and biota, Equation 4.117 could be modified to give

$$C_w = \frac{M}{K_{aw}V_a + V_w + K_{sw}\rho_b V_s + K_{bw}M_b} \tag{7.21}$$

where M_b is the mass of biota in the system. For a hydrophobic organic compound, $K_{sw} \sim K_{oc}\omega_{oc}$ where K_{oc} is the organic carbon based partition coefficient and ω_{oc} is the organic carbon fraction in the sediment, as indicated above.

A compartmental model in a lake-sediment system is explored in Example 7.3 in preparation for a discussion of transport in such a system.

Example 7.3: Partitioning to sediments and bioaccumulation in a lake

Consider a lake 10 ha in surface area and 10 m deep with an approximately uniform temperature of 10°C. The upper 10 cm of sediment is kept in near equilibrium with the water by the action of organisms that populate the sediment-water interface. The bulk density of the sediment is 1 g/cm^3 and its organic carbon fraction is 1%.

- Estimate the equilibrium distribution of 100 kg of pyrene ($K_{oc} = 10^5$ L/kg) in the lake and sediment. Neglect the mass of pyrene in fish and plants.
- Estimate the equilibrium concentration of pyrene in fish and in sediment-dwelling organisms assuming each is 5% lipid and the BSAF ~4.

1. The volume of the lake is 10^6 m^3 while the sediment is 10^4 m^3. Using the concentration in the water as a basis the mass in the water and sediment and their contaminant fractions in each are given by

$$M_w = C_w V_w$$

$$M_s = C_w \rho_b K_{oc} \omega_{oc} V_s$$

$$f_w = \frac{V_w}{V_w + \rho_b K_{oc} \omega_{oc} V_s} \qquad f_s = \frac{\rho_b K_{oc} \omega_{oc} V_s}{V_w + \rho_b K_{oc} \omega_{oc} V_s}$$

Substituting from the problem statement with the given volumes, $\rho_b = 1$ gm/cm^3, $K_{oc} = 10^5$ L/kg, $\omega_{oc} = 0.01$, gives the fraction 0.091 (9.1%) of pyrene in the water and 0.909 (90.9%) in the sediment. The concentrations in the respective phases are then

$$C_w = \frac{(100 \text{ kg})(0.091)}{10^6 \text{ m}^3} = 9.09 \text{ µg/L} \qquad W_s = \frac{(100 \text{ kg})(0.909)}{(10^4 \text{ m}^3)(1 \text{ kg/m}^3)} = 9.09 \text{ mg/kg}$$

Note that 9.09 µg/L is commonly referred to as 9.09 ppb and that 9.09 mg/kg is commonly referred to as 9.09 ppm. Note that they have completely different bases, however.

2. The biota at the sediment-water interface are assumed to be in equilibrium with the sediment and the biota in the water are assumed to be in equilibrium with the water.

$$(W_b)_{sed} = BSAF\, W_s \frac{\omega_L}{\omega_{oc}} = 182 \text{ mg/kg} \qquad (W_b)_w = BSAF\, C_w K_{oc} \omega_L = 182 \text{ mg/kg}$$

So despite how we calculate it, the concentrations of pyrene in the sediment-dwelling organisms and in the fish are identical since the BASF and lipid content was assumed identical. This is simply because the entire system is in equilibrium so that the fish in the water are, in essence, in equilibrium with the sediment just like the sediment-dwelling organisms.

7.3.2 Turbulent Mixing and Transport in Streams, Lakes, and Oceans

Although equilibrium calculations indicate the direction of mass transport, the rate of progress toward the equilibrium state is dependent upon mixing and transport processes. The mixing of pollutants and other species within streams, lakes, and the ocean is characterized by the effective or turbulent diffusivity, D_t. At the air-water interface similar processes control exchange between the air and the water. In surface waters like the atmosphere, D_t is controlled by the mechanical generation of turbulence by velocity shear and the consumption or generation of turbulence by buoyancy forces as defined by vertical density gradients. Near the water surface, wind-driven or other currents cause effective mixing over a finite depth without significant resistance by buoyancy forces. In streams, the entire depth is generally influenced by the velocity shear associated with the friction of the sediment bed. In the thermocline of lakes and in the ocean, however, the effect of buoyancy dominates fluid motion except very near the sediment-water interface. Let us examine the processes in each of these regimes.

7.3.2.1 Transport Across the Air–Water Interface

In Chapter 5, we defined the mass transfer processes near a fluid-fluid interface in the environment as a combination of air- and water-side processes through the *two-film theory*. In the open ocean or on large lakes, an estimate of the air- and water-side film mass transfer coefficients are given by

$$\begin{aligned} K_{af} &\approx 3000 \text{ cm/h} \\ K_{wf} &\approx 20 \text{ cm/h} \end{aligned} \tag{7.22}$$

In general, these coefficients depend upon the air and water velocities near the air-water interface. In forced systems, such as in aerated impoundments or ponds, both the air- and water-side coefficients can be of the order of 10,000 cm/h, significantly larger than under any unforced natural conditions. For now, let us consider only natural systems. As indicated in Chapter 5, the vast majority of volatile organic contaminants of interest are hydrophobic compounds that are rarely controlled by the air-side coefficient in natural systems. We will not attempt to refine estimates of that coefficient here. The water-side coefficient is often important and we should provide additional information about its magnitude.

There are two conditions that are normally of interest in natural systems. In relatively quiescent lakes, the water-side mass transfer coefficient is largely determined by the wind speed. The wind speed forces the surface water motion as a result of surface friction. Under such conditions, the water-side mass transfer coefficient is approximately given by Thibodeaux (1996),

$$\frac{K_{wf}}{\text{cm/h}} = \begin{cases} 0.094\left(\dfrac{U}{\text{m/s}}\right)^2 & U \geq 3.5 \text{ m/s measured at 10 m} \\ 1 \text{ cm/h} & U < 3.5 \text{ m/s} \end{cases} \tag{7.23}$$

By this relationship, the water-side mass transfer coefficient does not approach that of the open ocean until the wind speed is of the order of 15 m/s. This is a reflection of the absence of current and wave motion in small impoundments and lakes which significantly enhances the rate of vaporization in the ocean.

The second limit that is of interest in natural systems occurs in streams when the water-side mass transfer coefficient is a function of the stream current. Experiments and theory have suggested that for a stream of depth, h, the film mass transfer coefficient is related to stream velocity, U, and water diffusivity, D_w, by (Thibodeaux, 1996)

$$\frac{K_{wf}}{\text{cm/h}} \approx \left(\frac{D_w U}{\text{h}}\right)^{1/2} \tag{7.24}$$

Not surprisingly, water currents mix the water more effectively than the winds over the surface. Equation 7.24 suggests that a 1 m/s water current in a 3 m deep river will result in a water-side mass transfer coefficient of about 6.6 cm/h for a compound with a water diffusivity of (10^{-5}) cm^2/s. Equation 7.23 suggests that a similar wind speed would give rise to only a 1 cm/h water-side mass transfer coefficient at the air-water interface in a quiescent pond. Examples 7.4 and 7.5 illustrate the estimation of transport across the air-water interface.

Example 7.4: Rate of water evaporation from a lake

Estimate the rate of evaporation of water from a lake at 20°C when the relative humidity of the air is 30%.

The vapor pressure of water at 20°C is 17.5 mmHg or 0.023 atm. Air saturated with water vapor at 20°C contains 2.3% water (molar or volume basis). This is the amount of water present right at the air-water interface. At 30% relative humidity, the air well away from the lake surface contains 0.3 (2.3%) = 0.69% water. The concentrations in air at the lake surface and in the bulk air are then

$$C_a = RH\frac{P_v}{RT} = 0.3\frac{0.023 \text{ atm}}{0.0822\dfrac{\text{atm}\cdot\text{L}}{\text{mol}\cdot\text{K}}(293\text{ K})} = 5.17 \text{ mg/L}$$

$$C_a^* = \frac{P_v}{RT} = \frac{0.023 \text{ atm}}{0.0822\dfrac{\text{atm}\cdot\text{L}}{\text{mol}\cdot\text{K}}(293\text{ K})} = 17.25 \text{ mg/L}$$

There is no concentration gradient of water in the lake. Therefore there are no mass transfer resistances in the water film. Assuming that the air film mass transfer coefficient of Equation 7.22 applies, the rate (or flux) of water via evaporation is given by

$$q_m = k_{af}(C_a^* - C_a) = 3000 \text{ cm/h } (17.25 \text{ mg/L} - 5.17 \text{ mg/L}) = 362\frac{\text{gm}}{\text{m}^2\cdot\text{h}}$$

Example 7.5: Rate of chemical transport across the air-water interface

Estimate the rate of (1) uptake of oxygen into a body of water initially devoid of oxygen and (2) rate of evaporation into clean air of 1,2-dichloroethane (EDC) present at a concentration of 9 mg/L (9 $\mu g/cm^3$) in the water. Assume the film mass transfer coefficients of Equation 7.22 apply. The solubility of oxygen from air in water at the stream temperature of 20°C is approximately 9 mg/L. At 20°C, the vapor pressure of EDC is 0.24 atm, its water solubility is 5500 mg/L, and its molecular weight is 99.

1. Noting that the partial pressure of oxygen (Mw = 32 g/mol) in air is 0.21 atm, the air-water partition coefficient at 20°C is

$$K_{aw} = \frac{P_{O_2}MW_{O_2}}{RTC_w^*} = \frac{(0.21 \text{ atm})(32 \text{ g/mol})}{\left(8.22\cdot 10^{-5}\dfrac{\text{atm}\cdot\text{m}^3}{\text{mol}\cdot\text{K}}\right)(293\text{ K})(9 \text{ g/m}^3)} = 31$$

The overall mass transfer coefficient based upon water concentrations is then

$$k_w = \left(\frac{1}{k_{wf}} + \frac{1}{K_{aw}k_{wf}}\right)^{-1} = \left(\frac{1}{20 \text{ cm/h}} + \frac{1}{(31)(3000 \text{ cm/h})}\right)^{-1} = 19.996 \text{ cm/h}$$

Note that the mass transfer is effectively completely controlled by the water-side mass transfer resistances. If the water contains no oxygen, the flux **into** the water is

$$q_m = k_w C_w^* = (20 \text{ cm/h})(9 \text{ µg/cm}^3) = 180 \frac{\text{µg}}{\text{cm}^2 \cdot \text{h}}$$

2. Similarly, the air-water partition coefficient of EDC is 0.179. The overall mass transfer coefficient is then 19.3 cm/h, or still effectively completely controlled by water-side mass transfer resistances. If the air contains no EDC, the flux **out of** the water is

$$q_m = k_w C_w^* = (19.3 \text{ cm/h})(9 \text{ µg/cm}^3) = 174 \frac{\text{µg}}{\text{cm}^2 \cdot \text{h}}$$

Note that the fluxes are essentially identical even though the vapor pressure and the air-water partition coefficients are quite different. This confirms that mass transfer of compounds with sufficiently high air-water partition coefficients are water-side controlled and, if none of the compound is present in the other phase, a function of only the water concentrations. In this case, the driving water-side concentrations (for oxygen the interfacial water concentration and for EDC the bulk water phase concentration) are equal and therefore the magnitude of the fluxes are identical.

7.3.2.2 Transport Near the Sediment–Water Interface

The turbulence associated with velocity shear generally controls the mixing within the stream and near the sediment-water interface. Because the sediment surface is not directly heated by the sun, convection associated with buoyancy effects does not occur and the Richardson number characterizing the turbulence and mixing behavior is near zero and near neutrally stable conditions apply. The velocity profile is then well approximated by the logarithmic velocity profile and the turbulent momentum diffusivity is then given by

$$\nu_t = \kappa u_* z \tag{7.25}$$

where z is the distance above the bottom. As indicated previously, u_*, the friction velocity, can be estimated from the velocity profile above the bottom or for gravity flow streams by either Equation 7.4 or by evaluation of only the mean velocity and the use of Equation 7.7.

Away from the sediment-water interface, Equation 7.25 also provides a reasonable estimate of the effective diffusivity of mass or heat in the stream, that is $\nu_t \sim D_t$. In streams, this eddy diffusivity tends to cause rapid mixing across the width and depth of a stream. Mixing over similar distances along the length of a stream also would be expected, but generally the transport of contaminants in this direction is controlled by the velocity of the stream. Thus, a stream is typically modeled as a plug flow system with rapid and complete mixing perpendicular to the flow but no mixing in the direction of flow. Transport in the direction of flow is controlled only by advection due to the mean velocity of the stream. In lakes and in the ocean the dynamics of the water body is controlled by the internal density stratification and this is discussed in the next section.

Very near the sediment-water interface the influence of molecular properties (viscosity, diffusivity, and thermal conductivity) cannot be neglected. When the Schmidt ($N_{Sc} = \nu/D$) number is near one, as in gases, Equation 7.25 still gives a good estimate of the turbulent mass and thermal diffusivities. The Schmidt number of most liquids is far greater. The kinematic viscosity of water is about 0.01 cm^2/s while the diffusivity of most chemicals in water is about 10^{-5} cm^2/s giving a Schmidt number of about $0.01/10^{-5} = 10^3$. Thus, the turbulent mass diffusivity in water near the sediment-water interface should be written, as described in Chapter 5,

$$D_t \approx \kappa u_* z N_{sc}^{-2/3} \tag{7.26}$$

In any body of water subject to a significant bottom current, Equation 7.26 may give a reasonable description of the effective diffusivity near the sediment-water interface and may be useful in the estimation of the release of contaminants from that interface. The use of Equation 7.26 is hindered by the fact that the turbulent diffusivity varies as a function of depth above the stream bed. Often, it is more useful to have some indication of an average rate of mixing upward from the sediment bed rather than a description of its variation. Because the eddy diffusivity is linearly dependent on height above the stream bed, the average at any height is half the value at that height. The mass transfer coefficient or ratio of eddy diffusivity to height is then given by

$$\begin{aligned}(k_{wf})_{avg} &= \frac{1}{z}\frac{D_t(z)}{2} \\ &= \frac{1}{2}\kappa u_* N_{Sc}^{-2/3}\end{aligned} \tag{7.27}$$

Experiments indicate that over a smooth sediment bed the observed values of the mass transfer coefficient might be smaller by about a factor of 2.

If we use the exact analogy suggested by the film theory, the mass transfer coefficient is the ratio of the molecular diffusivity to a film thickness in which motion is effectively negligible. From Equation 7.27, this suggests

$$\begin{aligned}\frac{D_w}{\delta_{bl}} &\approx k_{wf} \approx \frac{1}{2}\kappa u_* N_{Sc}^{-2/3} \\ \delta_{bl} &\approx \frac{2}{\kappa}\frac{\nu_w}{u_* N_{Sc}^{1/3}} \approx \frac{5\nu_w}{u_* N_{Sc}^{1/3}}\end{aligned} \tag{7.28}$$

Here, δ_{bl}, is the thickness of a *diffusive sublayer* or boundary layer in which molecular diffusion is important. Again, experiments suggest that this layer may extend farther, especially near a rough sediment bed. Although approximate, the equation provides a rough indication of the thickness of the diffusive sublayer and its dependence on flow and chemical properties. Values of the transport parameters near the sediment-water interface are illustrated in Example 7.6.

Example 7.6: Estimation of the bottom mass transfer coefficient

Estimate the bottom mass transfer coefficient and diffusive sublayer thickness in the conditions of the Mississippi river employed in Example 7.1

The relevant parameters of the Mississippi River employed in Example 7.1 include: U ~ 1.6 m/s, α_M = 0.025 s/m$^{1/3}$, and r_H = h = 8 m. Based upon this information, the friction velocity characterizing the surface shear stress is

$$u_* = \sqrt{\frac{\tau_0}{\rho_w}} = \frac{U\alpha_M}{r_H^{2/3}}\sqrt{gh} = 0.089 \text{ m/s}$$

The kinematic viscosity of water is about 0.01 cm²/s while the molecular diffusivity of a contaminant in water is of the order of 10^{-5} cm²/s. Thus, the Schmidt number, ν_w/D_w = 1000. The water film mass transfer coefficient at the sediment-water interface is then

$$k_{wf} = \frac{1}{2}\kappa u_* N_{Sc}^{-2/3} = 1.77\ 10^{-4}\ \text{m/s} = 64\ \text{cm/h}$$

although, as indicated in the text, this may be an overestimate of the actual mass transfer coefficient at the sediment-water interface.

The diffusive sublayer thickness is

$$\delta_{bl} = 5\frac{\nu_w}{u_* N_{Sc}^{1/3}} = 5.645\ \mu\text{m}$$

As indicated in the text, the thickness of this layer is exceedingly small.

7.3.2.3 Mixing in Lakes and Oceans

Unlike streams, the predominant characteristic of mixing in lakes and in the ocean is the density stratification of the body of water. In lakes the seasonal temperature variations may lead to a well-developed thermocline and strong stratification in later summer and early fall. In the ocean, seasonal variations also play a role in development of a shallow thermocline but a permanent thermocline also exists in the temperate latitudes and near the equator. In general, mixing through the thermocline is still controlled by the Richardson number. Because the magnitude of the currents and the velocity shear tends to be small in the vicinity of the thermocline, the mixing process and the magnitude of the eddy or turbulent diffusivity tends to be controlled solely by the density stratification.

Because only the consumption of turbulent energy in the thermocline is important, the effective diffusivity (either α_t or D_t) within this region can be correlated with the Brunt-Vasaila frequency or an equivalent quantity, the normalized density gradient. Figure 7.7 from Koh and Fan (1970) illustrates a correlation of the effective thermal diffusivity as a function of the normalized density gradient. Although the data is certainly scattered, almost all of the data are estimated within an order of magnitude by the relation

$$\alpha_t \approx D_t \approx \frac{10^{-8}\ \text{m/s}}{\left|\dfrac{1}{\rho_w}\dfrac{d\rho_w}{dz}\right|} \tag{7.29}$$

Equation 7.29 is only valid under stable stratification when the normalized density gradient is negative if z is taken as positive upward and positive if z is taken as positive downward from the surface. The absolute value allows equation 7.29 to be valid regardless of the sign convention on the vertical coordinate. Under unstable conditions, the body of water will overturn and mix throughout its depth and Equation 7.29 is not valid. If the normalized density gradient were given in units of m^{-1}, the calculated turbulent diffusivities would be in m^2/s (= 10^{-4} cm^2/s).

The denominator also can be written in terms of temperature gradient if temperature is the primary indicator of density as in freshwater lakes. In temperate, freshwater systems, Equation 7.29 can be written

$$\alpha_t \approx D_t \approx \frac{10^{-8}\ \text{m/s}}{\left|\dfrac{1}{\rho_w}\dfrac{\partial\rho_w}{\partial T}\dfrac{\partial T}{\partial z}\right|} = \frac{10^{-8}\ \text{m/s}}{\left|-\beta_T\dfrac{\partial T}{\partial z}\right|} \tag{7.30}$$

Here β_T is the coefficient of thermal expansion of water, approximately $1.1(10^{-4})$/°C at 15°C.

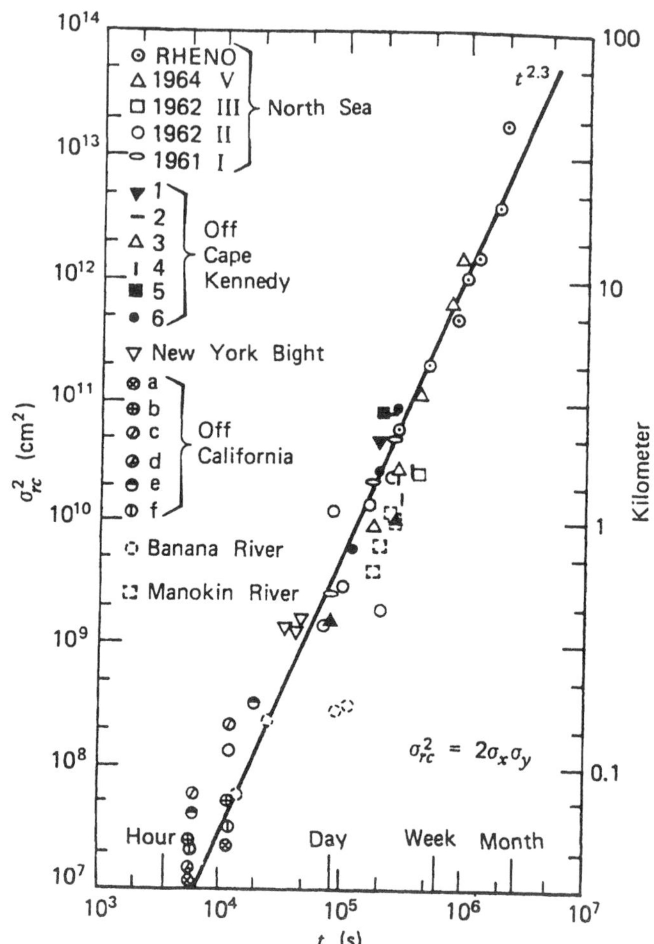

FIGURE 7.7 Diffusion diagram for variance vs. diffusion time $\sigma^2_{rc} = 2\sigma_x \sigma_y$. (From Koh, C.Y. and L. N. Fan (1970) Water Pollution Control Research Series, Washington, D.C. With permission.)

Let us examine the magnitudes of the mixing coefficients that would be estimated by Equation 7.30. If the temperature gradient in the water body is of the order of 1°C/m increasing upward, the predicted eddy diffusivity is about 0.9 cm²/s. While this is considerably larger than molecular diffusion rates, it is at least an order of magnitude smaller than the predicted eddy diffusivity in the absence of a thermocline. The significantly faster rate of mixing within the regions above and below the thermocline often causes a lake to be divided into two relatively well-mixed regions, an upper *epilimnion* and a lower *hypolimnion*, separated by a thermocline through which the rate of transport is relatively slow. This picture of a lake is shown in Figure 7.8. A commonly used model of contaminant dynamics in a lake is as two weakly coupled well-mixed stirred tanks. If C_e and C_h represent the well-mixed concentration in each of the regions, then the flux between the two regions is given by

$$q_m = \frac{D_t}{\delta_T}(C_h - C_e) \tag{7.31}$$

where D_t is the eddy diffusion coefficient in the thermocline and δ_T is its thickness. The sign of the flux is selected here to be positive for transport *from* the hypolimnion *to* the epilimnion ($C_h > C_e$).

The quantity D_t/δ_T in Equation 7.31 represents an effective mass transfer coefficient through the thermocline. For a 5-m thick thermocline with $D_t \sim 0.9$ cm²/s, which corresponds to a thermo-

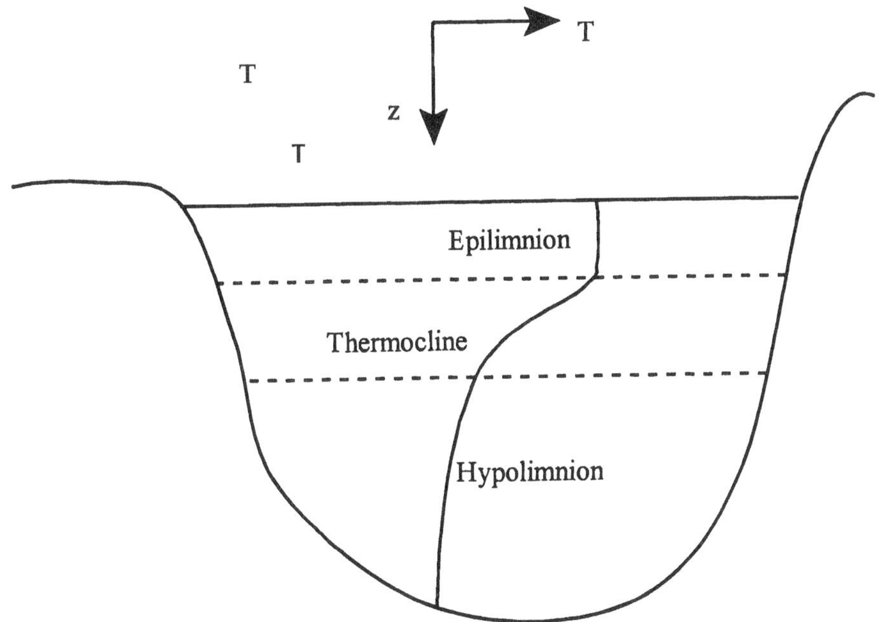

FIGURE 7.8 Depiction of temperature profile in a lake identifying relatively well-mixed upper and lower layers.

cline temperature gradient of 1°C/m as indicated above, this suggests an effective mass transfer coefficient of about 6.5 cm/h. This value can be compared to the water film mass transfer coefficients at the air-water and sediment-water interface which are similar in magnitude. This lends further support to the idea of a lake as two relatively well-mixed regions with mass transfer resistances at the boundaries of the lake and through the thermocline in its interior.

7.3.3 Chemical and Biological Degradation Processes in Surface Waters

As indicated previously, the processes of primary importance for chemical degradation of hydrophobic organic compounds in surface waters include photodegradation, ionization, hydrolysis, and microbial transformation.

Photodegradation is the light-induced reaction of these compounds. It's likely to be of importance only near the surface of waters due to the attenuation of sunlight with depth, especially in sediment laden streams or lakes. Many organic compounds are photosensitive, however, just as these same organics take part in photoinduced reactions in the atmosphere.

Hydrolysis is important in an even greater group of compounds. Hydrolysis reactions have the general form

$$R-X+H_2O \rightarrow R-OH+H^{+}+X^{-} \tag{7.32}$$

Here R represents some hydrocarbon chain and X is hydrogen or some other constituent group that can be replaced by the hydroxyl group (OH) during hydrolysis. The addition of the hydroxyl group tends to make the organic compound more soluble and less sorbing than its parent R-X compound. Whether a compound undergoes hydrolysis and at what rate is very compound dependent. For example, dichloromethane has a hydrolysis half-life of the order of 700 years, while trichloromethylbenzene hydrolyzes in less than a minute (half-life of 19 s). The hydrolysis reaction is also pH (= $-\log \tilde{M}_{H+}$) dependent as indicated by the presence of the hydrogen ion as a product. Many inorganic compounds also hydrolyze. As indicated previously, for example, sulfur dioxide reacts in water forming sulfite (HSO_3^-) and bisulfite ion (HSO_4^{-2}).

Ionization of organic and inorganic compounds is also an important process for particular compounds. Acids and bases dissociate to varying degrees in water dependent upon their strength. Organic acids and bases tend to be weak and only partially dissociate, and this dissociation is a strong function of pH. An organic alcohol and carboxylic acid dissociate, for example, according to the reactions

$$\begin{aligned} R-OH &\Leftrightarrow R-O^- + H^+ \\ R-COOH &\Leftrightarrow R-COO^- + H^+ \end{aligned} \tag{7.33}$$

As indicated previously, the fraction of a weak acid or base that is dissociated is related to the pH through the pK_a. When the pH is given by the pK_a of a particular acid or base, the compound will be observed half in the dissociated state and half in the parent state (at equilibrium). When the pH is 2 units greater than the pK_a, the concentration of the dissociated species is 100 times greater than the neutral parent species. Alternatively, when the pH is 2 units less than the pK_a, the concentration of the dissociated species is 100 times less than the neutral parent species.

The most important of the chemical pathways that affect a wide range of organic and inorganic compounds are microbially mediated oxidation-reduction and degradation reactions. A variety of microorganisms live in surface waters but bacteria are the most important. Bacteria range in size from 0.5 to 3 μm and are characterized by a loosely structured nuclear area, i.e., they have no nucleus defined by a nuclear membrane. To survive and grow, bacteria require energy and a carbon source, usually referred to as a substrate. Energy is provided by reduction-oxidation (redox) reactions, which involve the transfer of electrons between reactive species. Microbes employ enzymes to mediate the transport of electrons from an electron donor to an acceptor. Enzymes enhance the rate of the reaction without being consumed, that is they catalyze the reactions. This proceeds according to the following general reaction

$$\text{Enzyme} + \text{Substrate} \Leftrightarrow \text{Enzyme Substrate Complex} \rightarrow \text{Enzyme} + \text{Products} \tag{7.34}$$

The products can be additional biomass or degraded substrate. The free energy change that accompanies these chemical reactions is the energy that is available for use by the organism. Some of this energy is stored for latter use by the organism in the form of adenosine triphosphate (ATP). The ATP can later be hydrolyzed releasing energy for the formation of new cell material if a carbon source is available.

Microbes are most commonly classified by their source of energy and carbon and oxygen and temperature conditions under which they thrive.

Classification by carbon source — Autotrophs employ carbon dioxide while heterotrophs employ organic carbon.

Classification by energy source — Phototrophic bacteria employ light while chemotrophs use redox reactions as described above. Chemotrophic heterotrophs employ the redox reactions of organic compounds while chemotrophic autotrophs employ inorganic redox reactions as an energy source.

Classification by optimum oxygen conditions — Aerobic organisms thrive in the presence of dissolved oxygen while anaerobic bacteria live in the absence of dissolved oxygen. Facultative organisms can adapt to either aerobic or anaerobic conditions. Aerobic organisms employ oxygen as the electron acceptor in the redox processes while anaerobic organisms employ nitrate (denitrification), sulfate (sulfate reduction), or carbon dioxide (methanogenesis) as the electron acceptor.

Classification by optimum temperature conditions — Cryophyllic organisms live in cold temperatures as low as –2°C and thrive at 12 to 18°C. Mesophyllic organisms thrive in temperatures of 25 to 40°C, while thermophyllic organisms can live at temperatures up to 75°C and thrive in temperatures of 55 to 65°C.

Bacteria are sensitive to pH with the optimum pH being in the range of 6.5 to 7.5. They generally cannot tolerate pH levels outside of the range of 4 to 9.5. Bacteria can thrive in saline environments, even at levels of salinity much greater than that of seawater. Bacteria also are insensitive to pressure. As indicated by the various classifications, bacteria can survive in a wide range of environmental conditions and using vastly different food and energy sources. A typical bacteria population is composed of many different types of microbes with the dominant members being those that thrive under the particular environmental conditions.

Heterotrophic aerobic bacteria are the most important in surface waters due to their ability to utilize organic waste material as food and energy sources. The basic reactions under these conditions can be written

$$C(HONS) + O_2 \xrightarrow[\text{nutrients}]{\text{bacteria}} CO_2 + NH_3 + C_5H_7NO_2 + \text{misc. products} \tag{7.35}$$

Here C(HONS) represents a carbon source substrate that may contain hydrogen, oxygen, nitrogen, and sulfur. $C_5H_7NO_2$ represents additional cell matter or biomass. Miscellaneous products include partial oxidation products of the biodegradation processes. Although complete reduction of the nitrogen to ammonia is possible, partial reduction to N_2, or nitrite, NO_2, is also observed under certain conditions and with certain bacteria. The removal of nitrate by reduction of the nitrogen to nitrite, nitrogen, or ammonia is a very important aerobic process in soils, where it removes an important plant nutrient (nitrate) and in sediments, where it is an integral part of the degradation of organic matter. Aerobic reactions are relatively rapid but as implied by the discussion above, sensitive to the biological population and environmental conditions. We are, therefore, left without means to predict the rate of these processes in lakes and streams, although some guidance is available in specific wastewater treatment process conditions. A wide variety of aliphatic and aromatic hydrocarbons are biodegradable at significant rates by this process. In addition, the presence of these degradable compounds may cause other compounds to be degraded, a process called cometabolism.

Anaerobic processes also can result in degradation of organic compounds. The reduction of iron (III) coupled with the oxidation of organic compounds is an important cause of high dissolved iron concentrations in ground waters. Sulfate reduction forming, for example, hydrogen sulfide, also is important under anaerobic conditions. Methanogenesis, or the formation of methane from carbon sources in soils and sediment, is another important anaerobic process. Key products of anaerobic reactions are reduced compounds such as hydrogen sulfide and methane. Chlorinated organic compounds also undergo dechlorination under anaerobic conditions and this is an important fate process for such contaminants. The anaerobic processes are not normally observed in surface waters due to the presence of oxygen transfer from the atmosphere. These processes are of primary importance in deep groundwaters or in sediments where oxygen levels may be low.

Although bacteria are generally the most important type of microorganism to the environmental engineer, a variety of other microorganisms exist in surface waters. Bacteria contain no nuclear membrane and this separates them from higher members of the kingdom protista (i.e., generally classed separately from plants and animals). Fungi are multicellular aerobic heterotrophs that have the ability to survive under low nitrogen and pH conditions, making them useful for the treatment of some wastes. They are often classed with bacteria. Algae are photosynthetic, autotrophic protists. They produce foul tastes in water supplies and produce oxygen by photosynthesis.

$$CO_2 + 2H_2O + Light \rightarrow CH_2O(\text{new algae cells}) + O_2 + H_2O \tag{7.36}$$

and at night will reduce oxygen by respiration

$$CH_2O + O_2 \rightarrow CO_2 + H_2O \tag{7.37}$$

As a result of these processes, dissolved oxygen levels in the presence of algae will tend to be high during the day and low at night. The depletion of the oxygen at night may have important consequences as a result of the need of animals for dissolved oxygen by fish and other animals in the water. The usage and release of CO_2 also causes a diurnal variation in pH.

The presence of the nutrients such as nitrogen and phosphorus in surface waters supports rapid growth of algae. Although the algae do not directly employ organic matter as food, the presence of the algae in the surface waters also may significantly influence the partitioning of hydrophobic organic compounds. The algae can serve as sorption sites for hydrophobic organics just as any other suspended organic matter. In the absence of site-specific data, it is again common to assume for hydrophobic organic compounds that the partitioning to the algae is given by the organic carbon based partition coefficient, K_{oc}, times the concentration of algae, i.e., the mass of a hydrophobic contaminant in the algae is given by

$$M_{algae} = \rho_{algae} K_{oc} C_w \tag{7.38}$$

7.3.4 Kinetics of Biological Degradation Processes

As indicated above, bacteria depend upon the utilization of substrates as a source for energy and carbon for cell growth. Chemotrophic heterotrophs utilize organic compounds as substrates resulting in a significant mechanism for the degradation/depletion of such compounds in surface waters. As we shall see, this mechanism is an important means of treating wastewaters as well. In this section, let us examine the dynamics of bacterial growth and substrate utilization that we can use to understand and predict the dynamics of organic contaminants.

Figure 7.9 shows the general dynamics of bacterial growth in a batch system. Four distinct phases are observed.

1. Lag — Time is required for bacteria to acclimate to the available substrates and environmental conditions. During this time little or no growth or substrate utilization is noted.
2. Log-growth — A period of rapid growth which is associated with plentiful substrate and necessary nutrients.
3. Stationary — A period marked by little or no growth associated with limitations in substrate and nutrient concentrations.
4. Endogenous — During this phase, organisms must rely on their own energy storage due to the lack of external food and energy sources which have been fully utilized during the growth phase. The rate of death of the organisms becomes greater than their growth and the *biomass* decreases.

While this picture is applicable to a single organism in a batch system, in both the natural environment and biological wastewater treatment systems, a population composed of a variety of interacting organisms are observed and this leads to much more complicated dynamics. It also is often preferred to operate a wastewater treatment system under steady conditions such that the acclimation phase can be avoided. Under these conditions the rate of growth of organisms can be written

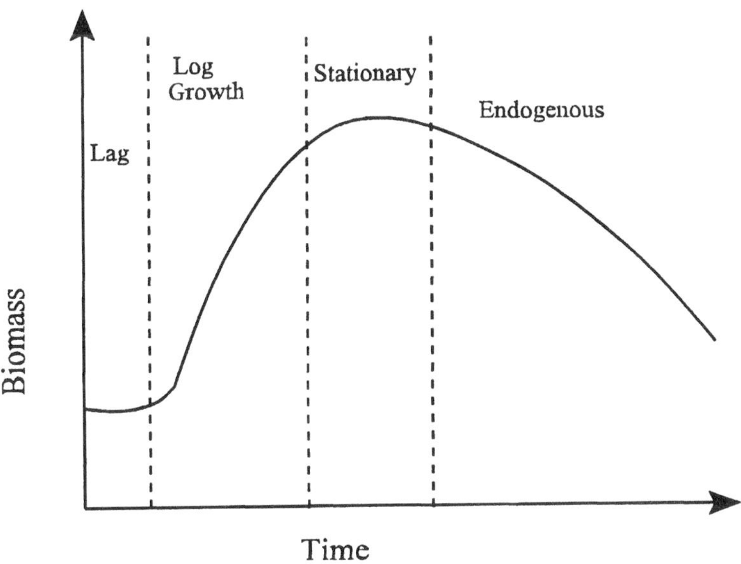

FIGURE 7.9 Typical behavior for microbial growth with time.

$$R_X = k_X X \tag{7.39}$$

where μ represents a growth rate per organism or specific growth rate and has units of inverse time. The term X is the biomass or concentration of microorganisms. In a batch culture if the specific growth rate is a constant, the rate of change of biomass is given by this rate.

$$\frac{dX}{dt} = R_X = k_X X$$

$$\int_{X_0}^{X} \frac{dX}{X} = \int_{0}^{t} k_X dt \tag{7.40}$$

$$\ln \frac{X}{X_0} = k_X t$$

This illustrates what is meant by the logarithmic growth phase.

At high biomass concentrations, there is insufficient food to maintain this rate of growth and the specific growth rate decreases. The specific growth rate under these conditions is often modeled by a relationship of the form (Monod kinetics)

$$k_X = k_m \frac{C_S}{C_{S_{1/2}} + C_S} \tag{7.41}$$

where C_S is the substrate concentration. The units of substrate concentration may be milligrams of contaminant per liter or similar units, but often a nonspecific measure of substrate concentration such as milligrams of oxygen demand per liter is employed. $C_{S_{1/2}}$ is a constant that represents the substrate concentration that corresponds to $k_X = k_m/2$ and k_m is the maximum specific growth rate (i.e., the specific growth rate with a large excess of substrate as food, k_M when $C_S >> C_{S_{1/2}}$). In addition the net rate of biomass growth equals that given by Equation 7.39 minus the rate of death or decay, generally assumed first order, k_dX. Substituting Equation 7.41 into 7.39 and incorporating the decay rate, the net rate of growth of the biomass becomes,

$$R_X = \frac{k_m C_S X}{C_{S_{1/2}} + C_S} - k_d X \tag{7.42}$$

Our primary interest often is not in the biomass but in the substrate, which may be an organic compound which is being degraded. Let us define the quantity Y as the ratio of the mass of biomass formed to the mass of substrate consumed. Then the rate of "formation" of the substrate is the rate of growth of biomass (not the net rate) divided by -Y, or,

$$R_S = -\frac{k_m C_S X}{Y\left(C_{S_{1/2}} + C_S\right)} \tag{7.43}$$

Two limits of this rate relationship are often important. At low substrate concentrations, $C_S <<$ $C_{S_{1/2}}$, the rate of substrate utilization is first order in biomass and substrate concentration.

$$R_S = -\left[\frac{k_m}{Y C_{S_{1/2}}}\right] C_S X \qquad C_S << C_{S_{1/2}} \tag{7.44}$$

At high substrate concentrations, $C_S >>$ $C_{S_{1/2}}$, the rate of substrate utilization is first order in only the biomass concentration.

$$R_S = -\left[\frac{k_m}{Y}\right] X \qquad C_S << C_{S_{1/2}} \tag{7.45}$$

In a stream or lake, the biomass concentration, X, is largely defined by the background loading of substrates and nutrients, which may be relatively constant. Then an additional contaminant added to the system in low concentrations ($C_{S_{1/2}} >> C_S$) implies that the rate of substrate removal is first order in that substrate. In a wastewater treatment system, however, substrate concentrations may be large and the rate of substrate removal is dependent only upon the biomass concentration. An activated sludge wastewater treatment system involves the recycling of effluent biomass in an effort to keep biomass concentrations high for this reason.

Temperature affects the rate of biomass growth and substrate utilization as well. Temperature also can affect the settling rate of biological solids and the rate of transport (e.g., of oxygen) across the air-water interface. As a result, theoretical descriptions of the effect of temperature are not employed and the effect is usually described by an empirical relationship based on the rate at 20°C of the form

$$R(T) = R(20°\text{C})\,\alpha^{(T-20)} \qquad T \text{ in °C} \tag{7.46}$$

α is typically in the range of 1.04 to 1.08.

7.4 MODELING CONTAMINANT DYNAMICS IN SURFACE WATERS

7.4.1 Basic Model Forms

Based on the above discussion of transport processes in lakes and streams, two idealized conceptual models of transport and fate of contaminants in surface waters exist. In the case of a lake, two well-mixed boxes separated by an unmixed thermocline is an appropriate idealized model. In the case of a stream, a laterally mixed, plug flow system seems to be the appropriate idealized model. Any real system can be described by a combination of these models, but let us limit our attention to these idealized lake and stream conditions.

7.4.2 Stream Model

As indicated in the previous section, mixing in streams tends to be relatively rapid due to the turbulence associated with the friction at the sediment-water interface. In the direction of stream travel, this mixing is still generally small compared to the advective transport. Over the comparatively short distances across the width and over the depth of a stream, however, turbulence serves to rapidly mix any constituent. There are instances, such as wide rivers or very near a pollutant outlet, that contaminants may not be considered well mixed across the width and depth of a river, but for many problems facing an environmental engineering, this assumption is a good one. The behavior of the contaminant in the stream can then be modeled as in the plug flow system.

There are two immediate consequences of modeling contaminant behavior in a stream by a plug flow assumption. Both are depicted in Figure 7.10. First, any release of contaminant into the stream is assumed to be immediately diluted by mixing across the width and depth of the stream. If Q_0 represents the volumetric flow of water in the stream before introduction of the effluent from a waste source and C_0 the concentration of a particular contaminant in that stream (again prior to the introduction of the additional waste), then the rate of mass of that contaminant carried by the stream is Q_0C_0. Similarly if Q_e is the volumetric flow of effluent from the waste source and C_e represents the concentration of the contaminant in that effluent, the rate of mass introduced to the stream is Q_eC_e. The resulting concentration in the stream, $C_w(0)$, is then given by

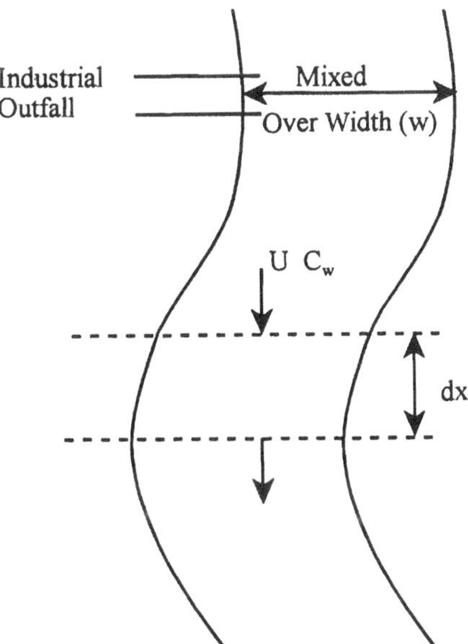

FIGURE 7.10 Differential element for stream material balance assuming complete mixing across the width (and depth) of the stream.

$$Q_0 C_0 + Q_e C_e = (Q_0 + Q_e) C_w(0)$$

$$C_w(0) = \frac{Q_0 C_0 + Q_e C_e}{Q_0 + Q_e} \tag{7.47}$$

This concentration represents the starting concentration in the stream which will then change as a result of stream processes such as evaporation to the air, mass transfer to the sediment, or degradation within the stream. This assumes that the emission of contaminant into the stream is continuous. If the emission is of short duration, some spreading along the length of the stream cannot be neglected, resulting in a reduction in the stream concentration over that expected from a continuous emission.

If the contaminant is subject to some degradation or loss process, the concentration also will change (decrease) with time. As time proceeds, the contaminant will proceed downstream as well. If the stream velocity is U, the location of the fluid initially introduced at x = 0, is given by

$$x = U t_L \tag{7.48}$$

Where t_L is the Lagrangian time, that is, time measured along the path of the fluid parcel. Thus, the second consequence of modeling the transport in the stream by the plug flow assumption is that either the downstream position or time could be used equally well to characterize the period that has transpired since introduction of the contaminant to the stream.

Let us now develop an equation that relates the contaminant concentration to distance (or time) downstream from its point of introduction into a stream. Consider the differential element of length in the stream as shown in Figure 7.10. Advection in and out of the element and losses to the air and sediment can be written

$$\text{Flow in} = U\, C_w\, w\, h\big|_x$$

$$\text{Flow out} = U\, C_w\, w\, h\big|_{x+dx} \tag{7.49}$$

$$\text{Loss} = k_{aw}\left(C_w - \frac{C_a}{K_{aw}}\right) w\,dx + k_{sw}\left(C_w - \frac{W_s}{K_{sw}}\right) w\,dx + k_r C_w\, w\, h\, dx$$

Each term has units of mass of contaminant per time. The first two loss terms represent loss at the air-water interface, with overall mass transfer coefficient, k_{aw}, and at the sediment-water interface, with overall mass transfer coefficient, k_{sw}, respectively. Both coefficients involve the overall driving force based on water concentrations. C_a/K_{aw} and W_s/K_{sw} represent the concentration that would be observed in the water if it were in equilibrium with the air and water, respectively. The final loss term represents a first order degradation reaction appropriate for low substrate concentration in the water. Note that if any of these terms represented gains (for example, transport into the water across the air-water interface) the term would simply be incorporated into the material balance with a change of sign.

Under steady conditions, flow in – flow out = loss. Substituting and dividing by the length of the differential element, dx, and the width and depth of the stream,

$$\frac{UC_w\big|_{x+dx} - UC_w\big|_x}{dx} = -\frac{k_{aw}}{h}\left(C_w - \frac{C_a}{K_{aw}}\right) - \frac{k_{sw}}{h}\left(C_w - \frac{W_s}{K_{sw}}\right) - k_r C_w \tag{7.50}$$

Recognizing that the left-hand side is the derivative (as dx → 0), for constant velocity this becomes

$$U\frac{dC_w}{dx} = \frac{dC_w}{dt_L} = -\frac{k_{aw}}{h}\left(C_w - \frac{C_a}{K_{aw}}\right) - \frac{k_{sw}}{h}\left(C_w - \frac{W_s}{K_{sw}}\right) - k_r C_w \tag{7.51}$$

where it is recognized that the spatial derivative can be written in terms of the time derivative following the fluid motion. This may be written

$$\frac{dC_w}{dt_L} + \left(\frac{k_{aw}}{h} + \frac{k_{sw}}{h} + k_r\right)C_w = \left(\frac{k_{aw}C_a}{hK_{aw}} + \frac{k_{sw}W_s}{hK_{sw}}\right) \tag{7.52}$$

Solving this equation subject to an initial stream concentration, $C_w(0)$, given by Equation 7.47.

$$C_w = C_w(0)\exp^{-\alpha t_L} + \frac{\beta}{\alpha}(1 - e^{-\alpha t_L})$$

$$\text{where } \alpha = \frac{k_{aw}}{h} + \frac{k_{sw}}{h} + k_r \tag{7.53}$$

$$\text{and } \beta = \frac{k_{aw}C_a}{hK_{aw}} + \frac{k_{sw}W_s}{hK_{sw}}$$

which represents the basic governing equation for contaminant concentration in the plug flow stream. Note that $1/\alpha$ represents a characteristic time, τ_c, for the disappearance of the contaminant and that the individual terms represent the characteristic times for the individual processes, evaporation to air, τ_a, sorption onto the sediment, τ_s, and reaction, τ_r

$$\tau_c = \left(\frac{k_{aw}}{h} + \frac{k_{sw}}{h} + k_r\right)^{-1}$$

$$\tau_a = \frac{h}{k_{aw}} \qquad \tau_s = \frac{h}{k_{sw}} \qquad \tau_r = \frac{1}{k_r} \tag{7.54}$$

A comparison of these fate processes for a particular contaminant in a stream is investigated in Example 7.7.

Example 7.7: Comparison of fate processes in a river for 1,2 dichloroethane

Consider the fate of 1,2-dichloroethane in a river with the conditions listed below. Known conditions include

River: U = 1.6 m/s	h = 200 cm	$u_* = 11.2$ cm/s
Water: T = 20°C	$\nu_w = 0.01$ cm²/s	
Compound:	$D_w = 9.9 \cdot 10^{-6}$ cm²/s	$k_r = 0.183$ h⁻¹

Mass transfer sediment-water interface: Let us approximate the k_{sw} by k_{wf}, although it is likely to be much smaller due to sediment-side resistances. $N_{Sc} = 1010$ and, therefore, $k_{wf} = 79.8$ cm/h by the method of Example 6.6. The characteristic time of loss by transport to the sediment is

$$\tau_s = \frac{h}{k_{sw}} = \frac{200 \text{ cm}}{79.8 \text{ cm/h}} = 2.5 \text{ h}$$

Mass transfer air-water interface (assuming water film controlled and using Equation 7.24):

$$\tau_a = \frac{h}{k_{aw}} = \frac{h}{\left(\frac{D_w U}{h}\right)^{0.5}}$$

$$= 200 \text{ cm}\left(\frac{(9.9 \cdot 10^{-6} \text{ cm}^2/\text{s})(160 \text{ cm/s})}{(200 \text{ cm})}\right)^{-0.5}$$

$$= \frac{(200 \text{ cm})}{(10 \text{ cm/h})} = 20 \text{ h}$$

Reaction:

$$\tau_r = \frac{1}{k_r} = \frac{1}{0.183/\text{h}} = 5.5 \text{ h}$$

Clearly transport to the sediment-water interface occurs rapidly. We have not yet evaluated sediment-side processes and these processes are likely to control the migration of contaminants into the sediment. Note that both mass transfer resistances are very sensitive to the stream depth.

Degradation in the water is a very important process assuming that the assigned degradation rate is valid. Reported degradation rates are widely variable and often measured without consideration of other processes (such as evaporation) that are contributing to the apparent loss rate. We also are considering only the parent compound and have not considered the importance or fate of any degradation byproduct.

For the case of a nonvolatile contaminant that does not sorb significantly to the sediment, $\tau_r << \tau_a, \tau_s$, this equation reduces to

$$C_w = C_w(0)e^{-k_r t_L} = C_w(0)e^{-k_r x/U} \tag{7.55}$$

An important application of this model is to the estimation of oxygen levels in a stream. Oxygen is depleted by reaction with oxygen-demanding substrates in the water and at the sediment-water interface and replaced by transport across the air-water interface. Because transport to the sediment-water interface is likely controlled by the slow sediment-side processes, let us neglect the *sediment oxygen demand.* A material balance on dissolved oxygen in a stream reflecting the remaining processes can then be written,

$$U\frac{dC_{O_2}}{dx} = \frac{dC_{O_2}}{dt_L} = \frac{k_{aw}}{h}\left(C_{O_2} - C^*_{O_2}\right) - k_r C_S \tag{7.56}$$

where $C^*_{O_2} = (C_a)_{O_2}/K_{aw}$ and the rate of reaction by microbial processes is recognized to be first order in substrate concentration and largely independent of oxygen concentration in the water. Substituting for the substrate concentration assuming that it is modeled by an equation of the form of 7.55, and rewriting in terms of the oxygen deficit in the stream, $\Delta_{O_2} = C_{O_2} - C^*_{O_2}$

$$\frac{d\Delta_{O_2}}{dt_L} = \frac{k_{aw}}{h}\Delta_{O_2} - k_r C_S(0)e^{-k_r t_L} \tag{7.57}$$

which has the solution, subject to an initial oxygen deficit in the stream of $\Delta_{O2}(0)$

$$\Delta_{O_2} = \frac{k_r C_S(0)}{k_{aw}/h - k_r}\left[e^{-k_r t_L} - e^{-\frac{k_{aw}}{h}t_L}\right] + \Delta_{O_2}(0)e^{-\frac{k_{aw}}{h}t_L} \tag{7.58}$$

The implications of this model for dissolved oxygen concentrations in a stream are explored in Example 7.8.

Example 7.8: Estimation of dissolved oxygen sag curve

Estimate the dissolved oxygen sag curve for the stream in Example 7.7 if the oxygen demand loading is initially 50 mg O_2/L with a consumption rate of 0.25 day^{-1}. The water temperature is 20°C and the initial oxygen concentration is 5 mg/L (deficit of 4 mg/L at this temperature).

Substituting into Equation 7.58, the deficit as a function of time (in hours) is given by

$$\Delta_{O_2}(t_L) = \frac{(0.25)(50\text{ mg/L})}{\frac{(10.1)(24)}{200} - 0.25}\left(\exp\left[-\frac{0.25}{24}t_L\right] - \exp\left[-\frac{10}{200}t_L\right]\right) + 4\text{ mg/L}\exp\left[-\frac{10}{200}t_L\right]$$

This equation shows a maximum deficit at about 30 h of 7.53 mg O_2/L followed by a slow return to oxygen saturated conditions (if no additional oxygen-demanding wastes are introduced to the water). Note that a deficit of 7.53 mg O_2/L would leave only 1.5 mg O_2/L, insufficient for many fish. A plot of the dissolved oxygen concentration in the stream as a function of time after the introduction of the wastes is shown in Figure 7.11.

7.4.3 Lake Model

The lake model (or a two-layer model of the ocean) includes the two well-mixed regions defined in Figure 7.8. We will denote the contaminant concentration in the upper and lower layers as C_e and C_h, respectively. We will assume that the thermocline of thickness, δ_T, contains only a small amount of the contaminant compared to the amounts in the well-mixed regions. Under these conditions, the thermocline is simply an interface of area, A, transport through which in the downward direction is given by D_t/δ ($C_e - C_h$) A. We will assume that volatile transport to and from the atmosphere can occur in the upper compartment, the epilimnion, while transport to and from the sediment can occur in the lower compartment, the hypolimnion. In addition, a first order reaction governed by a reaction rate constant, k_{re}, may occur in the upper compartment, while a first order reaction governed by a reaction rate constant, k_{rh} may occur in the lower compartment. A material balance over the upper region with streams entering and exiting the lake of volumetric flowrate, Q_{in} and Q_{out}, respectively, gives

$$\frac{d}{dt}C_e V_e = Q_{in}C_{in} - Q_{out}C_e - \frac{D_t}{\delta_T}(C_e - C_h)A - k_{aw}\left(C_e - \frac{C_a}{K_{aw}}\right)A - k_{re}C_e V_e \tag{7.59}$$

As usual, the effluent concentration is assumed identical to the concentration within the well-mixed layer. The equivalent balance over the lower region, with no inflow or outflow, is

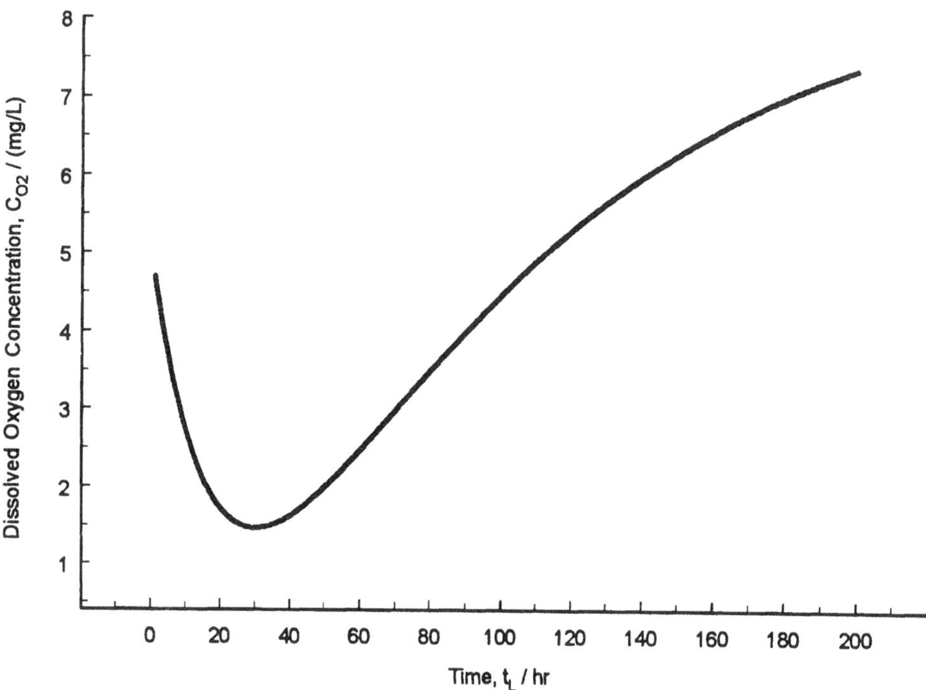

FIGURE 7.11 Dissolved oxygen sag curve for Example 7.8.

$$\frac{d}{dt}C_hV_h = +\frac{D_t}{\delta_T}(C_e - C_h)A - k_{sw}\left(C_h - \frac{W_s}{K_{sw}}\right)A - k_{rh}C_hV_h \tag{7.60}$$

If the volume of each layer is effectively constant and $A/V_e = h_e$ and $A/V_h = h_h$, the depths of the respective layers, then

$$\begin{aligned}\frac{dC_e}{dt} &= \frac{Q_{in}}{V_e}C_{in} - \frac{Q_{out}}{V_e}C_e - \frac{D_t}{h_e\delta_T}(C_e - C_h) - \frac{k_{aw}}{h_e}\left(C_e - \frac{C_a}{K_{aw}}\right) - k_{re}C_e \\ \frac{dC_h}{dt} &= \frac{D_t}{h_h\delta_T}(C_e - C_h) - \frac{k_{sw}}{h_h}\left(C_h - \frac{W_s}{K_{sw}}\right) - k_{rh}C_h\end{aligned} \tag{7.61}$$

In general, these equations must be solved simultaneously to describe the dynamics of each region of the lake. Although not difficult, the solution to the full two-chamber model is beyond the scope of this text. Let us explore a subset of the full models in Example 7.9.

Example 7.9: Concentrations in a lake after a surface spill

Consider a lake of area 1 ha with a 2-m thick thermocline with a temperature gradient of 1°C/m separating the epilimnion and hypolimnion which are each 5 m thick. Determine the maximum concentration in the hypolimnion that results from a spill of 100 kg of 1,2-dichloroethane (EDC) to the surface layer. Assume no degradation of the EDC or sorption onto the sediment.

The EDC will quickly disperse over the upper layer giving rise to an initial concentration of

$$C_e = \frac{100 \text{ kg}}{(10^4 \text{ m}^2)(5 \text{ m})} = 2 \text{ mg/L}$$

The only processes influencing this concentration are transport to the atmosphere and into the hypolimnion. The material balances can be written

$$\frac{d}{dt}C_eV_e = -\frac{D_t}{\delta_T}(C_e - C_h)A - k_{aw}\left(C_e - \frac{C_a}{K_{aw}}\right)A$$

and

$$\frac{d}{dt}C_hV_h = +\frac{D_t}{\delta_T}(C_e - C_h)A$$

The overall air-water interface mass transfer coefficient, k_{aw}, was estimated previously as about 10 cm/h based upon water-side concentrations. The effective mass transfer coefficient through the thermocline is given by

$$D_t = \frac{10^{-8} \text{ m/s}}{\left|-\beta_T \frac{\partial T}{\partial z}\right|} = \frac{10^{-8} \text{ m/s}}{1.1 \cdot 10^{-4} \text{ K}^{-1} \cdot 1 \text{ K/m}} = 0.909 \text{ cm}^2\text{/s}$$

An analytical solution for these two equations can be found or the equations integrated numerically. The concentration behavior in each of the layers of the lake are depicted in Figure 7.12 assuming that the bulk air concentration, C_a, is effectively 0. Note that the epilimnion continuously decreases due to evaporation and initially due to transport into the lower layer. The hypolimnion concentration peaks after 30 h under the conditions of this example at about 0.7 mg/L. Note that the assumption of well mixedness within each layer of the lake may be questionable over this short time frame.

7.5 SEDIMENT PROCESSES

7.5.1 Summary of Processes

The sediments beneath a stream or lake are much different in character from the overlying water. Up until now we have characterized transport in sediment as only a component of the overall mass transfer coefficient between sediment and water. We have not identified the processes that define pollutant migration or fate within the sediment nor have we even decided that this approach is appropriate. The use of a "sediment-side" mass transfer coefficient as a component of an overall mass transfer coefficient is, in fact, of extremely limited usefulness in dealing with sediments. A film coefficient is generally only useful when a stagnant film exists near an interface of very little volume which constitutes essentially all of the mass transfer resistance of the phase. The limited volume implies negligible rate of accumulation and, therefore, steady or quasi-steady transport through the film. In sediments, however, the entire layer of sediment poses the resistance to transport and concentration variations must often be considered over this entire layer. Instead of a film coefficient that lumps all mass transfer processes, it is more appropriate to consider the transport processes more explicitly.

There are potentially a large number of processes responsible for transport of contaminants in sediments. A few of these are depicted in Figure 7.13.

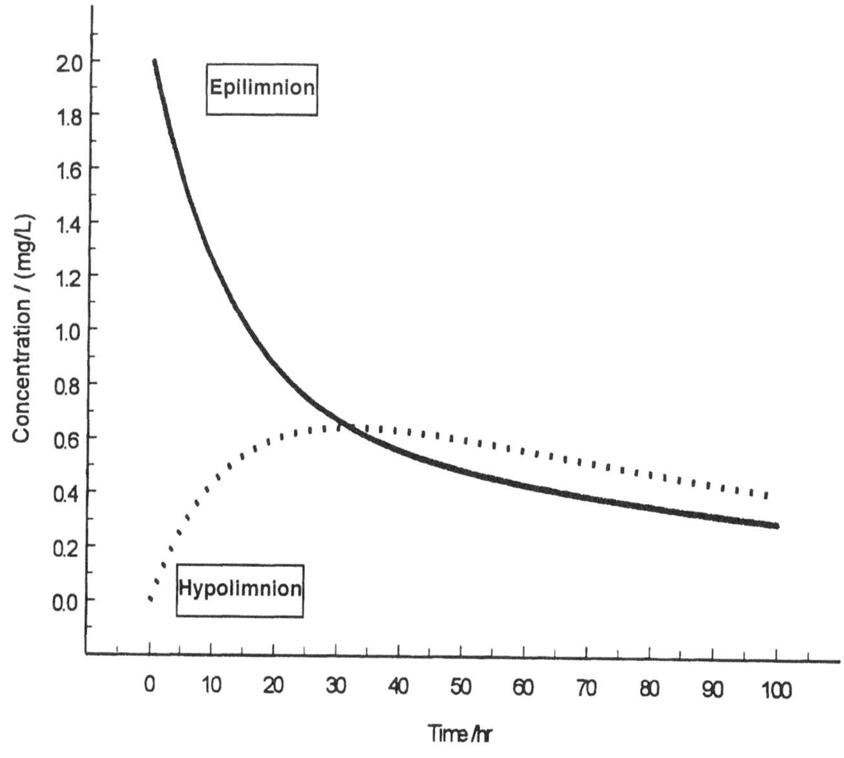

FIGURE 7.12 Concentration vs. time in a lake under the conditions of Example 7.9.

Particles that are buried deep within the sediment tend to be exposed to a low oxygen, anaerobic environment that influences the type, rate, and effectiveness of biological and chemical fate processes in the sediment. Erosion at the bottom of a stream results in the migration downstream of contaminated sediment, and subsequent deposition in a slow moving portion of the stream can serve to cause accumulation of contaminants at a particular location. Similarly, as excess rainfall causes erosion on a hillslope, the sediment carried by the runoff carries with it a significant fraction of contaminants present on the hillslope.

Organisms that live in or near the sediment-water interface also disturb the sediment, exposing fresh sediment and potentially destabilizing the sediment. Bioturbation is the sediment movement associated with the normal activities of organisms that live at or near the sediment-water interface. Such activity might include burrowing, direct processing of sediment for food, or simple mixing processes associated with organism movement, for example, disturbance of the sediment by walking across its surface. These activities expose sediment particles and the contaminants they may contain increasing the opportunity for release of these contaminants to the overlying water. Bioturbation of soils and exposed sediments also can enhance erosion and runoff.

In addition to these processes that can move both sediments and the contaminants they contain, transport processes that affect the water and waterborne contaminants in the pore spaces of the sediment include advection and diffusion. Each of these processes will be discussed in this section.

7.5.2 Sediment Structure

Sediments whether exposed on dry land, regularly moistened in wetlands, or submerged beneath a body of surface water are composed of solid particles that have settled, perhaps over geologic time scales, to reside in a generally stable environment. The particles may have aggregated as a result of cohesive forces or remained separate as a noncohesive sandy bed. The basic characterization of

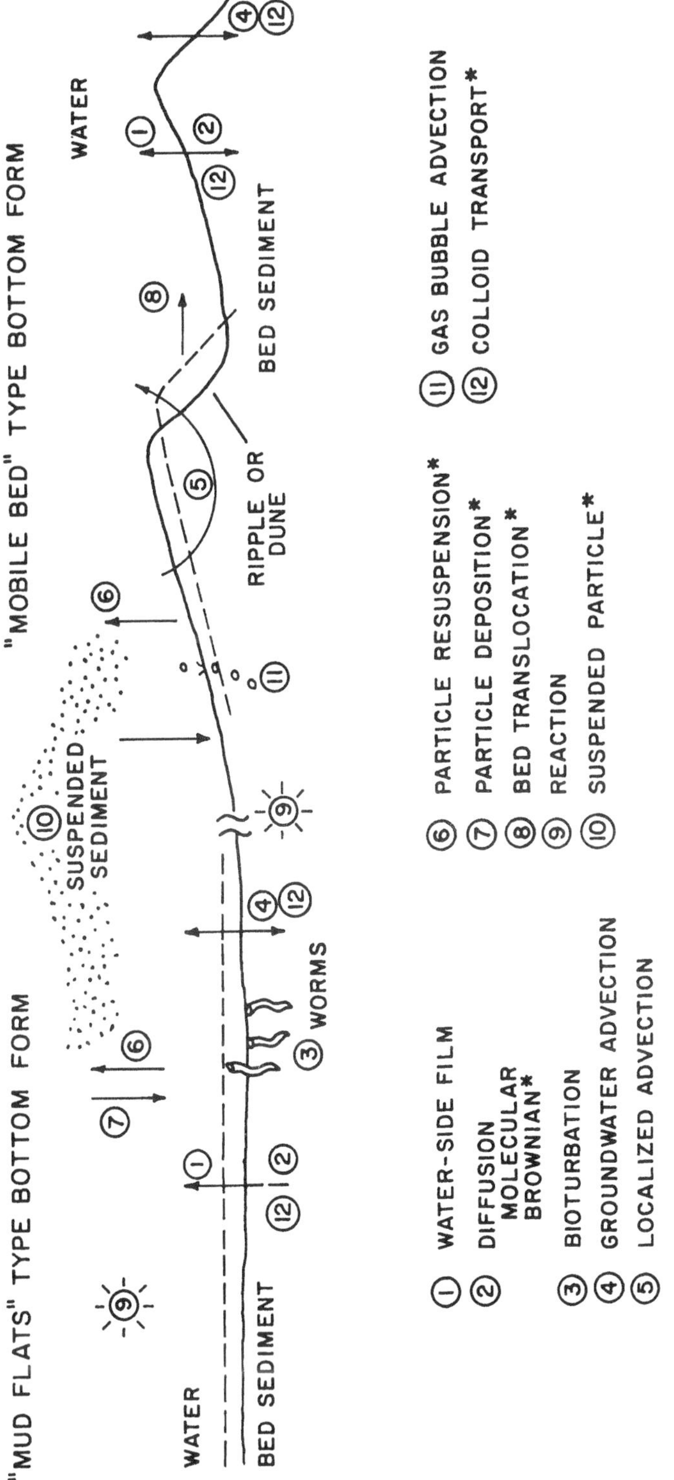

*Denotes mechanisms tied to fine particle behavior

FIGURE 7.13 Processes at the sediment-water interface.

sediments is by particle size. The largest particles in sediments, *sand*, are those particles that range in size from 20 μm (International Soil Science Society Classification) or 50 μm (U.S. Department of Agriculture classification) to 2000 μm (2 mm). *Silt* are the particles that range in size from 2 μm to the lower limit of the sand range and *clay* is everything smaller than 2 μm. Although the classification is based simply on size, the classifications also are significant indicators of the chemical and physical properties of the sediment. Sand is made up largely of quartz and other minerals and exhibits a relatively small surface area per unit mass of the particles. Clay, however, is generally composed of aluminosilicates and exhibits very large surface areas per unit mass, several hundred square meters per gram of clay being typical. Aluminosilicates are crystalline structures composed primarily of silicon, oxygen, hydroxyl groups (OH), and aluminum or magnesium. The crystalline structures form layers of thin sheets of clay *lamellae* as discussed in Chapter 4.

The size characterization also provides a significant indication of the effect of sediment grains on contaminants. Most of the organic matter in the sediments are associated with the finer particles of clay. *Humus* is the stable organic matter that remains after degradation of organic litter from animals or plants. Humus can bind and stabilize clay aggregates by serving as a cementing agent. It is this organic matter that serves as the primary sorption site for hydrophobic organics and also contributes significantly to the cation exchange capacity of the sediment. Sand contains very little of this material and hydrophobics tend to sorb only weakly to its minerals. The primary site of sorption of both hydrophobic organics and exchangeable inorganic species is the organic matter in the clay fraction. A fraction of the organic matter in the sediment degrades further and can dissolve in the pore waters as colloidal matter, the high molecular weight organic material that can serve to increase the suspended fraction of organic contaminants.

Of course our primary interest is not in the solid particles but in the void or pore spaces between particles where the mobile fluid, either or air or water, resides. The most basic characterization of the void space of a soil or sediment is its volume, the voidage or porosity of a sediment. The porosity as we have used it previously is defined simply as the ratio of the volume of void space to the total volume,

$$\varepsilon = \frac{V_v}{V_t} \tag{7.62}$$

The bulk or dry density of a soil or sediment, ρ_b, can be related to the porosity through the solid grain density, ρ_s,

$$\varepsilon = \frac{\rho_s - \rho_b}{\rho_s} \tag{7.63}$$

In soils, the bulk density varies little. In sediments, both the porosity and the bulk density can vary significantly due to the wide variations of water content observed. Loose, fluffy sediments composed of fine particles may be 60 to 80% moisture (or exhibit porosities of 60 to 80% and the resulting bulk density can be 0.5 to 1 g/cm^3. Sandy sediments, however, have a relatively narrow range of porosities (35 to 45%) and bulk densities (1.35 to 1.65 g/cm^3).

Another particle size related influence on contaminants is the size exclusion effect of small pores. Pore sizes tend to be of the order of the smallest particle size since these smaller particles fill the larger pores. As a result pore sizes in sandy sediments may be a 100 μm or larger while the majority of the pore sizes in clays may be only about 0.1 μm. Although this poses no hindrance to individual contaminant molecules, colloidal organic matter may be of this size or larger limiting its ability to penetrate the smaller pores. Similarly, bacteria are generally 0.5 to 3 μm in size and again would be limited to the larger pores.

The processes that lead to sediment and contaminant transport and availability in sediments are strongly influenced by the vertical stratification of sediments. Due to the stability of many sediments, contaminants placed even decades ago may be limited to a relatively shallow layer near the surface of the sediment. These surface layers often contain significant amounts of organic matter as well due to the degradation of natural organic matter that falls to the sediment bed from the overlying water. For hydrophobic organic contaminants this further increases the sorption onto the largely immobile solid fraction. The surface layer of sediment is exposed as well to the water column where dissolved oxygen is present. The sediments may pose a significant *sediment oxygen demand* associated with the oxygen required for microbes to degrade the organic litter that falls to the sediment bed. This often results in a rapid decrease in oxygen content with depth in the sediment bed. Thus, the upper layers of the sediment are *aerobic*, that is containing excess oxygen, while the deeper layers are *anaerobic*, largely devoid of oxygen. A commonly used indicator of this behavior is the reduction-oxidation potential or *redox potential* that is measured as a function of depth in the sediment. The mobility and reactions of organic and inorganic contaminants are a strong function of redox conditions and the availability of oxygen. Many elemental species, for example, tend to be less soluble and, therefore, less mobile under reducing or low oxygen conditions. These species may be more mobile and available to organisms and the overlying water in the near sediment surface than at depth. Organic matter is often degraded by microorganisms that are present in the pore waters of the sediment. More different types of microorganisms populate sediments in aerobic or oxidizing conditions than under reducing or anaerobic conditions. These organisms influence the chemical characteristics of the sediment they inhabit including, for example, the natural degradation of organic contaminants in the sediment. Hydrocarbons such as alkanes, aromatics, and polyaromatic hydrocarbons tend to degrade naturally under aerobic conditions. These compounds, present in gasoline and petroleum oils, tend to degrade fastest in the near-surface sediments. Chlorinated organics, however, tend to undergo dechlorination reactions mediated by microorganisms living in reducing or anaerobic environments. The deeper sediments may be conducive to degradation of these contaminants.

The depth of a layer of sediment tends to have a significant impact on the processes, including contaminant-influencing processes, that occur there. In addition, dynamic processes that move sediment from depth toward the surface have a significant impact on the chemical fate of the contaminant as well as potentially increasing exposure and the potential for loss to the overlying water. Both sediment dynamics and bioturbation will be examined from the perspective of their impact on sediment movement from depth to the surface.

7.5.3 Sediment Erosion

A major source of contamination of surface waters is associated with the suspension of soil by runoff or the resuspension of sediment from the bottom of a stream or lake. Because many of the contaminants of interest tend to sorb to the solid phase (either soil or sediment particles), the movement of these particles results in the movement of contaminants.

Fundamentally, resuspension of a soil or sediment grain as a result of water flow occurs when the frictional force of the fluid acting on the grain exceeds the resisting force of gravity and the cohesive forces associated with the surrounding soil grains. This is depicted in Figure 7.14. The total forces resisting soil grain resuspension are generally assumed to be proportional to the weight of the soil grain relative to the weight of the water it displaces.

$$\begin{aligned} F_s &= \alpha_1 (m_p g - m_w g) \\ &= \alpha_1 (\rho_p - \rho_w) g \pi \frac{d_p^3}{6} \end{aligned} \tag{7.64}$$

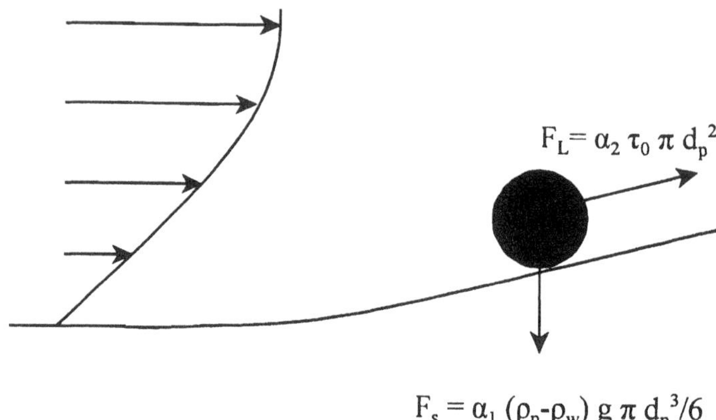

FIGURE 7.14 Balance of forces on an isolated grain of sediment as a result of fluid motion over the sediment bed. α_1 and α_2 are functions of angle of repose and flow and sediment conditions.

Here α_1 is a function of the angle of repose of the soil grain, the cohesive forces associated with adjacent particles, and the nonsphericity of the sediment grains. The buoyancy force acting on the soil grain also can be thought of as the force that serves to restore the particle to the sediment layer upon resuspension. The force that serves to resuspend the soil grain is the lift force acting on the particle. The source of this lift is ultimately the frictional or drag force associated with the passage of water over the sediment and this force is simply the shear stress at the surface of the sediment times the exposed area of the particle.

$$F_L = \alpha_2 \tau_0 \pi d_p^2 \tag{7.65}$$

Here, α_2, is a coefficient depending upon the degree of nonsphericity of the sediment grain and the effectiveness of the drag force for generating lift. Resuspension of the soil grains will begin to occur when these two forces are equal. Equating Equations 7.64 and 7.65 and combining all coefficients into a single coefficient, α_{Sh} $(= \alpha_1/(6\alpha_2))$, the *critical shear stress*, τ_{0c}, or *critical friction velocity*, u^*_c, to *just* achieve mobilization of particles of diameter, d_p, is given by

$$\tau_{0c} = \alpha_{Sh}(\rho_p - \rho_w) g d_p$$

or (7.66)

$$u_{*c}^2 = \alpha_{Sh} \frac{\rho_p - \rho_w}{\rho_w} g d_p$$

$$u_{*c} = \alpha_{Sh}^{1/2} \sqrt{\frac{\rho_p - \rho_w}{\rho_w} g d_p}$$

In the last two equations of 7.66, u_* is recognized as τ_0/ρ_w. The quantity α_{Sh}, the ratio of the square of the friction velocity to the relative buoyancy of the soil particle is also called a *Shield's parameter*. In general the Shield's parameter would depend on the flow and sediment characteristics. Experiments suggest that a value of 0.06 ± 0.03 ($\alpha^{1/2} = 0.25 \pm 0.17$) is appropriate for noncohesive granular sediments (e.g., sands). Correlations of the Shield's parameter exist with the onset and concentration of resuspended sediment. These correlations are quite useful with noncohesive sandy sediments and soils but are of limited usefulness in cohesive, clay-type sediments. Understanding resuspension of cohesive sediments remains an active area of research.

For our purposes, we are only interested in how flows give rise to particle resuspension. From 7.66, the threshold of particle mobilization (in noncohesive sandy soils and sediments) occurs when $u_* > u^*_c$, or

$$\text{Resuspension if} \quad u_* > u_{*c}$$

$$\text{stream} \quad \frac{U\alpha_M h^{1/2}}{r_H^{2/3}\sqrt{(\Delta\rho/\rho)d_p}} > \alpha_{Sh} \approx 0.25 \tag{7.67}$$

$$\text{runoff} \quad \frac{s^{0.7}}{\sqrt{(\Delta\rho/\rho)d_p}}\left[\frac{Q_r n}{w}\right]^{3/5} > \alpha_{Sh} \approx 0.25$$

where the equations for stream and runoff are written in terms generally most useful for each medium.

The primary limitations of Equation 7.67 are associated with the assumption of noncohesive sediments and the presumption that a single particle size characterizes the sediment. Cohesive sediments, or sediments stabilized by vegetation, are significantly less likely to be eroded by runoff or stream flow. An illustration of the effect of vegetation can be inferred from empirical relationships used to calculate soil losses by erosive processes. The universal soil loss equation (Mills et al., (1982)) is one such approach that includes a cover factor generally given the symbol C. Bare ground corresponds to a cover factor of unity while for soil 40% covered by grass, C ~ 0.1, and for soil covered 95 to 100% with grass, C ~ 0.003. Thus, the soil resuspended by runoff over 40 and 100% ground-covered soils are only 10 and 0.3%, respectively, of that resuspended by runoff over bare ground.

The effect of particle diameter is that fine particles are more easily resuspended than coarse particles. As a crude approximation, particles in the upper 1 cm of soil or sediment smaller than the diameter corresponding to the critical friction velocity might be considered available for resuspension. As an alternative, the calculations might employ a particle diameter smaller than 65% (or larger than 35%) of the sediment particles (d_{35}). If the particles of diameter d_{35} would be expected to be resuspended, significant resuspension of essentially all of the soil or sediment might be expected. The critical particle diameter that might be resuspended under some representative conditions is explored in Example 7.10.

Example 7.10: Threshold of particle motion in the Mississippi River

Estimate the critical particle size in the sandy bottom of the Mississippi River using the conditions of Example 7.1 assuming a particle density of 2.5 g/cm^3.

Solving Equation 7.67 for the critical particle diameter

$$d_{pc} = \frac{U^2\alpha_m^2 h}{r_H^{2/3}\Delta\rho/\rho_w(0.06)}$$

$$= \frac{(1.6\ \text{m/s})(0.025\ \text{s/m}^{1/3})(8\ \text{m})}{(8\ \text{m})^{4/3}(1.5)(0.06)}$$

$$= 8.5\cdot 10^{-3}\ \text{m} = 8.5\ \text{mm}$$

Although particles this size and smaller may be subject to mobilization, resuspension of such a large particle is unlikely. Bed load transport by movement along the surface of the sediment and not complete resuspension is more likely. As indicated in the text any cohesiveness in the sediment will also significantly reduce the likelihood of mobilization of the sediment.

7.5.4 Bioturbation

Another mechanism for particle migration in sediments is bioturbation, the sediment reworking associated with the normal life cycle activities of organisms that populate the upper layers of soil and sediment. Examples of bioturbation influencing the soil or sediment include everything from human agricultural practices or dredging in rivers to the activities of organisms that live in the upper layers of soil or sediment and process sediment for food. Normally we think of bioturbation as the uncontrolled action of animals in the natural environment and do not include human activities in these processes. Regardless of the organism responsible, the processes of movement along the surface, burrowing into the soil or sediment for food or shelter, and feeding on sediment to extract organic matter all involve particle movement. Because most contaminants of interest are found primarily sorbed to the particle fraction, these sediment reworking activities also result in significant contaminant movement.

A wide variety of organisms live and interact with the upper layers of sediment. Figure 7.15 shows a few of these organisms. The activities of bioturbating organisms can generally be divided into three different groups.

Foraging — Many animals that live in or near sediments assimilate organic matter from the sediment. During the process of foraging for this food, the sediment is disturbed and with it the associated contaminants. Surface-dwelling organisms, such as amphipods, tend to limit the sediment movement and their influence on contaminants to the upper surface layers of the sediment. Some fish and larger bottom-dwelling organisms, however, more actively mix the sediment and their influence penetrates deeply. In-dwelling organisms such as worms tend to forage within the sediment and their sediment reworking activities extend far deeper than the surface dwellers.

Burrowing — In-dwelling organisms must burrow into the sediment causing further sediment movement. This might entail removal of the sediment from the location of burrow by placement at the surface or simply displacement in and around the original location. Clearly the importance of this sediment reworking as a contaminant migration mechanism depends on the disposition of the sediment removed from the burrows. In addition to the direct sediment reworking associated with the burrowing process, burrowing leaves significant secondary porosity in the sediment allowing more rapid water exchange between the sediment and the overlying water. Burrowing might be the result of foraging for food or a burrow may be used to provide shelter for a stage in the organisms life cycle. For example, flies may spend a portion of their lives as larvae dwelling in sediment burrows.

Conveyor-Belt Sediment Processing — Certain organisms ingest and process the sediment directly. Some of these organisms ingest sediment at depth and defecate at the surface carrying sediment and associated contaminants to the surface where it can be released to the overlying water. This process is referred to as *conveyor belt bioturbation*. It has the most potential for sediment and associated contaminant transport of the bioturbating mechanism. Some of the most common organisms that populate contaminated sediment interact with sediment in this fashion. Deposit feeding worms, for example, often ingest sediment at the bottom of burrows 3 to 10 cm deep in the sediment but expose their posterior to the overlying water to aid oxygen transport.

The net effect of bioturbation is significant sediment destabilization and movement. Reworking rates of the upper few 5 to 10 cm of sediment are often of the order of 0.03 to 3 cm/year. That is, sediment equivalent to 67% ($1 - e^{-1}$) of the sediment in a layer 0.03 to 3 cm thick is reworked in a given year. While slow compared to sediment erosion under high flow conditions, they are significant in a stable sediment otherwise subjected only to porewater transport processes such as diffusion and advection. As indicated previously and in more detail below, porewater processes

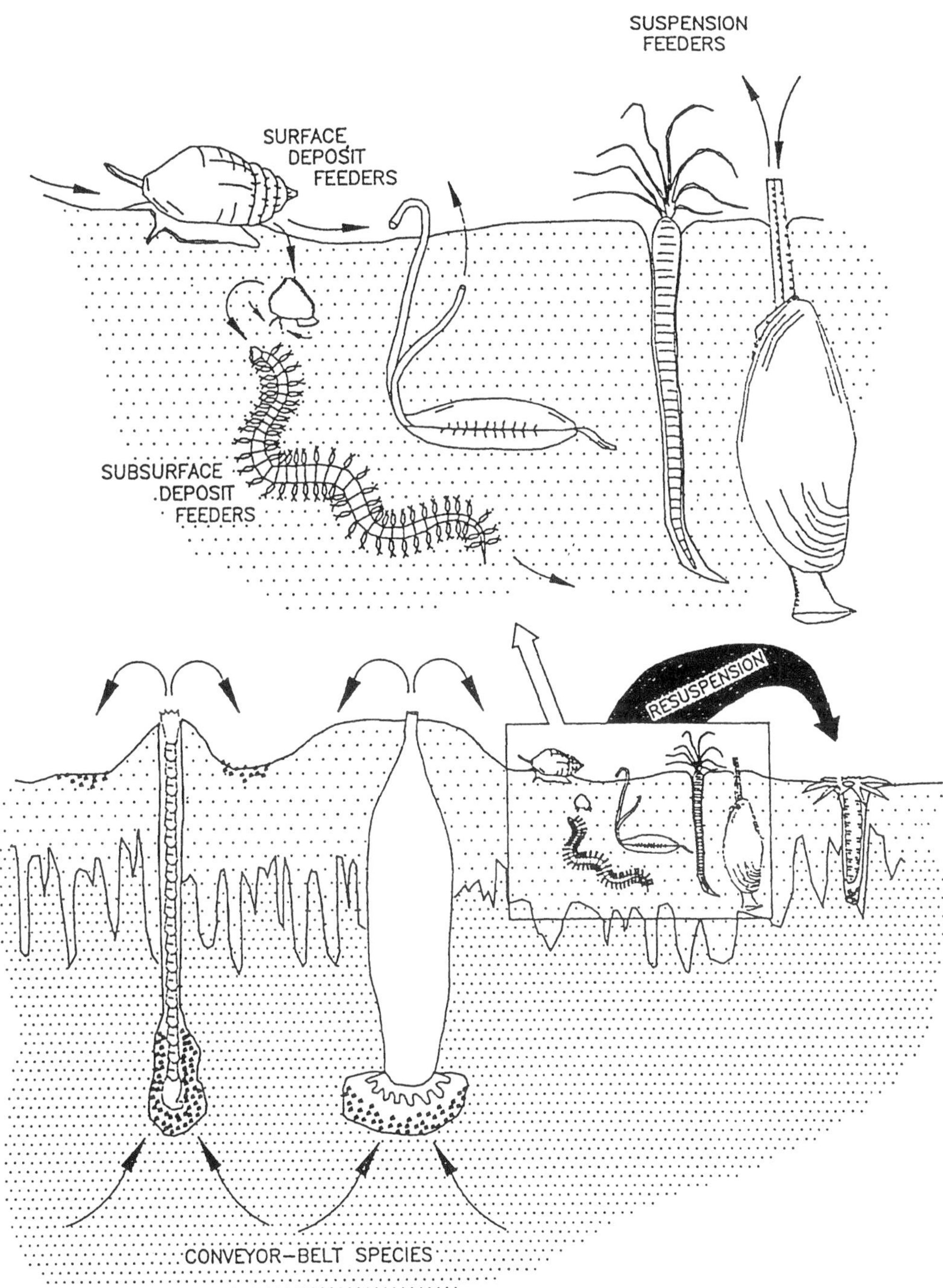

FIGURE 7.15 Feeding types of benthic organisms. (From Rhoads, D.C. (1974) *Oceanography and Marine Biology — Annual Review*. With permission.)

influence only that fraction of the contaminant not sorbed to the sediment, which is only a very small fraction of the total sediment loading for most sediment contaminants including hydrophobic organics. Sediment reworking during bioturbation affects both aqueous phase and sorbed contaminants. The influences of these processes on contaminant transport will be evaluated in the next section.

7.5.5 Contaminant Transport by Sediment Relocation

Due to the loading of contaminants onto particles, sediment movement, when it occurs, is the most important contaminant transport process. Typically the sediment loading of a contaminant, W_s, is the mass of contaminant per mass of dry sediment. The concentration in the associated porewater is the mass of contaminant per volume of porewater. Thus, $\rho_b W_s$, where ρ_b is the bulk density, represents the total contaminant loading per unit volume of sediment, and εC_w represents the contaminant loading in the porewater per unit volume of sediment. Thus,

$$\rho_b W_s = C_w(\varepsilon + \rho_b K_d)$$

or (7.68)

$$\frac{\rho_b W_s}{C_w} = \varepsilon + \rho_b K_d = R_f$$

The ratio of the mass of contaminant mobile during sediment erosion or movement by bioturbation to the mass of contaminant mobile in porewater processes is just given by the retardation factor. Most transport processes are directly proportional to the contaminant concentration, and this means that the mass transported by a sediment transport process is the retardation factor times the mass transported by a porewater transport process of equivalent intrinsic rate. That is, if bioturbation is reworking the upper layers of sediment with a reworking coefficient of k_b (perhaps 1 cm/year), the contaminant flux is effectively

$$q_m = k_b \rho_b W_s \tag{7.69}$$

The flux due to the equivalent porewater process is

$$q_m = k_w C_w \tag{7.70}$$

or the ratio between the two is simply the simply k_b/k_w R_f. Even if k_b is identical in magnitude to k_w, the contaminant flux would be a factor R_f times larger for the sediment transport process.

7.5.6 Contaminant Transport by Porewater Processes

Although slower, sediments in many areas of concern are essentially stable and not subject to significant erosion. Neither bioturbation nor erosion may be an issue at depths of 10 cm or more into the sediment. Transport under these conditions is via porewater processes. Transport by these processes involves a multistep process as shown in Figure 7.16. These steps include:

1. Desorption of a contaminant from a sediment particle.
2. Transport within the porewater away from the immediate vicinity of the site of desorption. Among the available transport processes within the porewater are advection and diffusion and both processes must operate within the limited pore volume and tortuous path provided by the space between sediment particles.
3. Potential sorption onto vacant sorption sites on sediment particles along the transport path, slowing the net rate of transport.
4. Release to the sediment surface and the overlying water, resisted by any mass transfer resistances posed by the overlying water.

Diffusion in soils and sediments is a very different process than in the atmosphere or in bodies of surface water. It is not controlled by turbulence and the random motion of fluid eddies but by

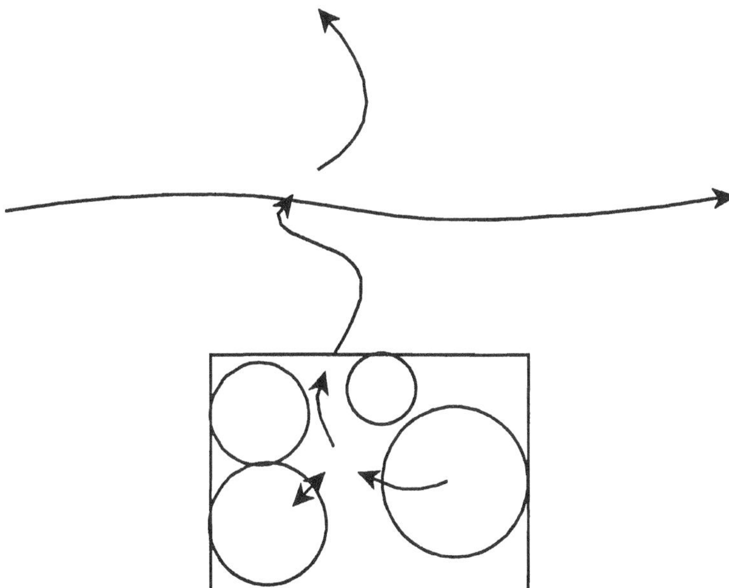

FIGURE 7.16 Depiction of bed sediment contaminant release processes including desorption from a sediment particle, transport within the pore space between sediment particles (with sorption/desorption onto other sediment particles), and ultimate migration to the surface and the overlying water.

the random orientation of the diffusion path in the porous media. The effective diffusion coefficient in the porous and tortuous (characterized by ε and τ, respectively) path of the sediments, D_{sw}, can often be approximated by the model of Millington and Quirk (1961) as

$$\begin{aligned} D_{sw} &= D_w \frac{\varepsilon}{\tau} \\ &\approx D_w \varepsilon^{4/3} \end{aligned} \tag{7.71}$$

This may not be an appropriate model for sediments of high clay content but we will assume it here without experimental data to the contrary. The size of molecules (although not necessarily colloidal particles) would be expected to be very much smaller than the size of pores thus indicating the effects of steric hindrances are negligible. The range of porosity values observed in sediments is typically 25 to 75%. Using the lower value and a molecular diffusion coefficient typical of medium to high molecular weight hydrocarbons of about $0.5(10^{-5})$ cm^2/s, the effective diffusion coefficient by Equation 7.71 is of the order of $0.08\ (10^{-5})$ cm^2/s. The characteristic time to diffuse through the upper 10 cm of sediment is distance^2/time ~1.2 (10^8) s or a little under 4 years. Remember, however, that only the amount of contaminant present in the porewater is available for diffusion. As described in Chapter 5, the diffusion of contaminant in the upper sediment layers is governed by

$$\begin{aligned} \frac{\partial \rho_b W_s}{\partial t} &= D_{sw} \frac{\partial^2 C_{sw}}{\partial z^2} \\ \frac{\partial C_w(\varepsilon + \rho_b K_{sw})}{\partial t} &= D_{sw} \frac{\partial^2 C_{sw}}{\partial z^2} \end{aligned} \tag{7.72}$$

or

$$\frac{\partial C_w}{\partial t} = \frac{D_{sw}}{\varepsilon + \rho_b K_{sw}} \frac{\partial^2 C_{sw}}{\partial z^2}$$

Thus, the effective diffusion coefficient that describes the change in the porewater concentration (and through the equilibrium partition coefficient the change in the sediment concentration) is given by

$$\begin{aligned} D_{eff} &= \frac{D_{sw}}{\varepsilon + \rho_b K_{sw}} \\ &\approx \frac{D_w \varepsilon^{4/3}}{\varepsilon + \rho_b K_{sw}} \\ &\approx \frac{D_w \varepsilon^{1/3}}{R_f} \end{aligned} \tag{7.73}$$

For a compound such as pyrene that might have a retardation factor of the order of 1000 in a typical sediment, this means that the characteristic time to diffuse 10 cm to the surface of the sediment is not 4 years but nearly 4000 years. Thus, the porewater diffusion process is very slow. In a stable sediment without significant advection, this is likely the most important transport mechanism below the level of bioturbation. Contaminant lifetimes in sediments can be very long indeed.

In those conditions where diffusion is applicable over the entire depth of sediment (e.g., a stable sediment sufficiently contaminated that benthic organisms cannot populate the sediment and bioturbation is unimportant), the concentration as a function of time in the sediment and the flux to the overlying water can be estimated by the error function solution to the diffusion equation discussed in Chapter 5 (Equations 5.133 and 5.134).

Even when bioturbation is important, the concentration and flux of contaminants from the sediment might be described by a variant of the diffusion model. Just as transport due to random eddies in the atmosphere might be modeled with a turbulent diffusion coefficient, the random effects of the bioturbating organisms also might be modeled with an effective biodiffusion coefficient. It should be emphasized that this is a useful tool for describing the effects of bioturbation, but it is of limited predictive value since we have no guidance for making other than very crude estimates of the magnitude of the effective bioturbation diffusion coefficient. Modeling bioturbation as a diffusion phenomenon is fundamentally different from porewater diffusion since it results in particle movement as well as porewater movement. Equations 7.72 and 7.73 become

$$\begin{aligned} \frac{\partial \rho_b W_s}{\partial t} &= D_b \frac{\partial^2 \rho_b W_s}{\partial z^2} \\ D_{eff} &= D_b \end{aligned} \tag{7.74}$$

That is, the effective diffusion coefficient for bioturbation is not reduced by the retardation factor. Predictions of concentration or flux from the sediment by bioturbation are made by substituting $\rho_b W_s$ for C_w, D_b for D, and $R_f = 1$ in the error function solution Equations 5.133 and 5.134.

These descriptions of contaminant migration must be modified if advective processes are important. In particular Equation 5.136 might be used to deal with the semi-infinite sediment case. More complicated initial and boundary conditions than those of Equation 5.136 must generally be dealt with via a numerical model. The advective flux can be estimated as well by recognizing that it is equal to q C_w, where q is the volume of fluid expressed from the sediment per unit area per time and C_w is the concentration of the contaminant in the porewater. Example 7.11 compares the diffusive, advective, and bioturbative fluxes of contaminants in a particular situation.

Example 7.11: Release flux from sediment by diffusion, advection, and bioturbation

Estimate the flux out of a sediment bed containing 100 mg/kg pyrene by (1) diffusion, (2) advection (with q = Q/A = 1 m/year) and (3) bioturbation (with D_b = 1 cm²/year). Treat each process independently as though it were the only operative process.

Sediment conditions:	$\rho_b = 1$ g/cm³	$\varepsilon = 0.4$	$\omega_{oc} = 0.03$
Contaminant conditions:	$W_s = 100$ mg/kg	$D_w = 5.2 \cdot 10^{-6}$ cm²/s	$K_{oc} = 10^5$ L/kg

1. **Diffusion** — The concentration driving diffusion is the pore water concentration given by

$$C_f = C_w = \frac{W_s}{K_{sw}} = \frac{100 \text{ mg/kg}}{(0.03)(10^5 \text{ L/kg})} = 0.033 \text{ mg/L} = 33 \text{ mg/m}^3$$

The diffusive flux is then given by

$$(q_m)_{diff} = C_w \sqrt{\frac{D_w \varepsilon^{4/3} R_f}{\pi t}}$$

where $R_f = \varepsilon + \rho_b K_{sw} = 3000$. The calculated fluxes are presented in the table below.

2. **Advection** — If the layer of sediment is considered essentially infinitely deep, the advective flux is steady at q C_w = 33 mg/(m²·year). This is significant, but smaller, than the initial diffusive flux. The actual combined flux could be evaluated by Equation 5.137 with U = –q, $C_f(0) = 0$ and $C_0 = C_w$. These results are also included in the table below.
3. **Bioturbation** — Using Equation 5.134 replacing the total concentration for the porewater concentration, $R_f = 1$ and D_b for $D_w\varepsilon^{4/3}$.

$$(q_m)_{bio} = \rho_b W_s \sqrt{\frac{D_b}{\pi t}}$$

Results also are included in the table below.

Table of fluxes for Example 7.11

Time (year)	$(q_m)_{diff}$ mg/(m²·year)	$(q_m)_{adv/diff}$ mg/(m²·year)	$(q_m)_{bio}$ mg/(m²·year)
1	71.6	89.5	1780
10	22.7	43.1	564
20	16.0	37.9	399
50	10.1	34.5	252
100	7.2	33.6	178
200	5.1	33.4	126

Initially diffusion is as high as advection but falls off rapidly while advection ultimately reaches the steady-state value (because an infinite layer of contaminated sediment was assumed) of about 33 mg/(m^2·year). Bioturbation, however, stays high for considerably longer and with the assumed parameter values clearly dominates the flux from the sediment. Although the flux from these mechanisms can vary significantly depending on the particular conditions, the trend with bioturbation >> advection or diffusion is often true.

Release from the sediment into the adjacent porewaters is often assumed to be rapid compared to the slow rate of movement through the pores to the overlying water. Because organic matter and metal contaminants are generally associated with the clay and silt fraction of the sediment, an upper bound to the particle size containing significant quantities of contaminants is 100 μm, a much smaller dimension than the 1 to 10 cm of interest in terms of transport to the overlying water. The assumption of rapid release from sediment particles may be inappropriate during bioturbation or when the sediment is subject to erosive processes. Under these conditions, the rate of transport of sediment to the surface may be sufficiently rapid that desorption and equilibration with the overlying water may not occur before burial by subsequent transport. It is important to recognize that some of the contaminants are often to sorb irreversibly to the sediment and thus do not desorb at all or at very slow rates from sediment. This can have a profound impact on bioavailability of the contaminants to organisms and is an area of much current research. It is beyond the scope of this text to deal with this issue, but it is important to recognize that only a fraction of the contaminants in sediments or soils may actually be readily available for uptake and exposure to organisms.

7.6 PRIMARY WATER TREATMENT

Treatment for the removal of contaminants that may redistribute naturally in the environment by the processes described above generally means wastewater treatment before introduction of these contaminants into the surface waters. Wastewater treatment is generally classified into one of three categories depending upon the order and degree of sophistication of the treatment. These are titled quite simply primary, secondary, and tertiary treatment. Let us consider each in turn and provide an indication of the basic mechanisms on which they are based.

Primary wastewater treatment is the term reserved for physical separation of solids from a wastewater stream. This includes simple sieving of larger particles and debris to sedimentation or settling of particulate matter. Since many contaminants and biological oxygen-demanding components are associated with the particulate matter, this can provide a significant reduction in the pollutant loading in the effluent from the wastewater treatment system. Due to its simplicity, primary treatment by particle separation is universally used in modern treatment systems. In times past, it was often the only treatment method employed giving rise to many of the surface water and sediment contamination issues that we still face today. Let us focus on the more complicated of the primary treatment processes, sedimentation.

7.6.1 Fundamentals of Sedimentation

Sedimentation proceeds by the same mechanism and process as gravity settling of particulate matter in air streams. Considering a single spherical particle of particle diameter, d_p, in water, the terminal settling velocity is given as before by

$$
\begin{aligned}
v_p &= \frac{g(\rho_p - \rho_w)d_p^2}{18\,\mu_w} \qquad N_{\mathrm{Re}} < 1 \\
&= \left[\frac{4}{3}\frac{g(\rho_p - \rho_w)d_p}{C_D\rho_w}\right]^{1/2} \qquad N_{\mathrm{Re}} > 1
\end{aligned}
\tag{7.75}
$$

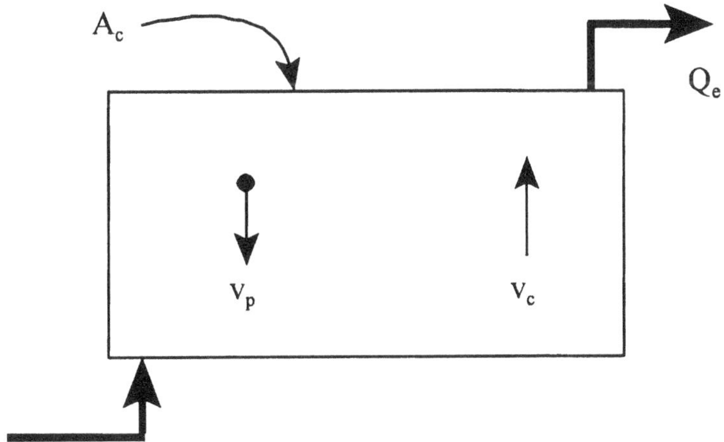

FIGURE 7.17 Schematic indicating fundamentals of clarifier operation.

where

$$C_D = \frac{24}{N_{Re}} + \frac{3}{\sqrt{N_{Re}}} + 0.34$$

$$N_{Re} = \frac{\rho v_p d_p}{\mu_w}$$

The primary differences between gravity settling in water vs. air is that the density of the surrounding fluid is significant. As a result particles of a given size tend to settle more slowly in water. The Reynolds number for a given size particle is often not very different despite the slower velocity because the kinematic viscosity of water (μ/ρ) is about 20 times smaller in water than in air. Thus, 60-μm particles of density 2.5 g/cm^3 in 25°C air fall at a rate of about 26 cm/s and exhibit a Reynolds number of about 1 while the same particles fall in water at a velocity of only 0.33 cm/s exhibiting a Reynolds number of about 0.22.

The speed of settling can be increased dramatically by the use of hydrocyclones. These are the wastewater equivalents of cyclone separators for air streams and essentially allow an artificial gravity much larger than occurs at the surface of the Earth. In dilute water systems, the analysis of particle collection is identical to that discussed for cyclone separators in Chapter 6.

The basic design approach for a *clarifier*, or process unit employing sedimentation to remove solids from a water stream, is to ensure that the upflow associated with the removal of water is less than the settling velocity of the smallest particle desired to be removed. Although turbulence will still cause carryover of some of these particles, a significant fraction of these particles and an increasing fraction of larger particles (due to the increased settling velocity of larger particles) will be retained within the clarifier. This is illustrated in Figure 7.17. Using the notation in the figure the basic design criteria can be written

$$v_c < v_p \quad \text{or} \quad \frac{Q_e}{A_c} < v_p \tag{7.76}$$

where v_c and A_c are the vertical upflow velocity and cross-sectional area of the clarifier, respectively, and Q_e is the volumetric effluent flow from the clarifier.

This approach suggests that all particles equal to or larger than the particles that meet the criteria in Equation 7.76 will be collected (in the absence of practical flow considerations such as turbulence) and that all smaller particles will pass through the clarifier. A real clarifier, however, will have an inlet section over which particles may be distributed over depth and an outlet section in which Equation 7.76 should apply. Between these regions, the clarifier is designed to have very low velocities and essentially all particles will tend to settle. Clarification of the water with respect to a certain particle will occur as long as the particle settles before entering the outlet section. Clarification depends upon the initial height of the particle upon entrance to the settling section and the settling velocity in a manner equivalent to the gravity settling chamber for particulates in an air stream as discussed in Chapter 6. The efficiency of collection of a given particle size can be written in terms of the residence time (τ_Q) and the settling or collection time (τ_c) in the clarifier.

$$\eta(d_p) = \frac{\tau_Q}{\tau_c} = \left(\frac{A_c H_c}{Q_e}\right)\left(\frac{v_p(d_p)}{H_c}\right) = \frac{A_c v_p(d_p)}{Q_e} \tag{7.77}$$

where, here A_c and h_c, the cross-sectional area and depth of the clarifier should only refer to that portion of the clarifier outside of the active inlet and exit regions. Equation 7.77 is consistent with the basic design criteria presented by Equation 7.76, that is, complete collection of particles that settle at a rate equal to or greater than the vertical upflow velocity is observed. The efficiency of collection of particles in a clarifier using Equation 7.77 is illustrated in Example 7.12.

Example 7.12: Design of a clarifier

Consider a clarifier 5 m in diameter and 1 m high to which is fed 10 m^3/min of water containing a uniform distribution of biological solids with effective diameters between 100 and 500 μm in size. Estimate the collection efficiency of discrete particles as a function of particle size if the density of the settling solids is 1.1 g/cm^3.

The residence time of water in the clarifier is given by

$$\tau_Q = \frac{Q}{A_c H_c} = \frac{(10\ \text{m}^3/\text{min})}{(\pi(5\ \text{m})^2/4)\ (1\ \text{m})} = 1.96\ \text{min}$$

The calculation of settling velocities proceeds as before with checking of the Reynolds number to ensure appropriate selection of the drag coefficient, C_D. The settling velocity of a 100-μm solid is 0.033 m/min, for which the Reynolds number is very much less than unity (Stokesian particle). The efficiency of collection of this size solids is then

$$\eta = v_p \frac{A_c}{Q} = 0.033\ \text{m/min}\ \frac{\pi(5\ \text{m})^2/4}{10\ \text{m}^3/\text{min}} = 0.063$$

The table below summarizes the settling velocity, Reynolds number (N_{Re}), and efficiency of other particle sizes.

Particle Diameter (d_p, μm)	Stokes' Settling Velocity, (v_S cm/min)	Actual Settling Velocity, (v_p cm/min)	Reynolds Number N_{Re}	Efficiency of Collection η
100	0.033	0.033	0.05	0.063
200	0.133	0.122	0.41	0.24
300	0.299	0.258	1.29	0.506
400	0.532	0.425	2.83	0.835
500	0.831	0.614	5.10	1.00

Assuming that all particles are present in the same proportion and spherical, the overall or average number-based, η_n, and mass-based, η_m, efficiencies are

$$\eta_n = \frac{\Sigma \eta(d_p) d_p}{\Sigma d_p} = 0.694 \qquad \eta_m = \frac{\Sigma \eta(d_p) \pi \frac{d_p^3}{6}}{\Sigma \pi \frac{d_p^3}{6}} = 0.863$$

As with a gravity settling device for air pollutants, the largest particles are more easily collected in a clarifier. These particles also represent the vast majority of the mass of particles in the stream undergoing clarification. In all particulate control devices where the larger particles are preferentially collected, the mass-based efficiency is always greater than the number-based efficiency. This is illustrated in Example 7.12 where both the number- and mass-based efficiencies are defined and calculated. In the air pollution context we did not discuss this behavior in detail because it is increasingly recognized that the concentration of fine particles most directly relates to the adverse health effects of breathing particulate matter. The primary focus of particulate and solids removal from a water effluent is the reduction of the oxygen demand and the concentration of particulate toxic contaminants. These may or may not be associated with specific sizes of particulate matter. If the solids to be removed in the clarifier are primarily flocs of biological solids, for example, the contaminant loading might be essentially independent of particle size. Regardless, the real measure of effectiveness of a clarifier is the reduction in oxygen demand and toxic contaminants in the effluent and not the removal efficiency of the particulate matter itself.

7.6.2 Flocculation

As indicated previously, the above approach assumes that the particle concentration is small enough that the settling velocity is not influenced by the presence of other particles. At medium particle concentrations, *flocculation* or coalescence of particles, may occur and the effective particle diameter increases. This will increase the rate of settling by Equation 7.75. Sometimes alum (aluminum sulfate), lime, or ferrous sulfate are added to increase flocculation. Alum is one of the most common and acts by forming aluminum hydroxide in water that contains calcium or magnesium bicarbonate. For example, with calcium bicarbonate

$$\underset{\text{Aluminum sulfate}}{AL_2(SO_4)_3 \cdot 18H_2O} + \underset{\text{Calcium bicarbonate}}{3Ca(HCO_3)_2} \rightarrow 3CaSO_4 + \underset{\text{Aluminum hydroxide}}{2Al(OH)_3} + 6CO_2 + 18H_2O \tag{7.78}$$

The insoluble aluminum hydroxide that is formed is a gelatinous floc that collects other particles as it settles.

Polyelectrolytes, or long chain polymeric compounds that have ionizable functional groups, also are used to enhance flocculation to improve settling qualities. Solids often exhibit residual surface charges that would normally hinder their coalescence and flocculation. Electrolytes partially neutralize these charges to enhance the coalescence process to produce larger and more rapidly settling particles. The settling characteristics of flocculants generally cannot be predicted mathematically. The assumption of a spherical (or approximately spherical) particle of uniform density that is central to Equation 7.75 is no longer valid. Instead, experimental settling tests are required to estimate settling velocities. These experiments simply measure the percent settled as a function of time. The measurements provide a direct indication of an appropriate collection time, τ_c required to achieve a given clarification of the water. An indication of the efficiency of the clarifier might then be given by the ratio of the residence time to this collection time, $\eta \sim \tau_Q/\tau_c$, as in Equation 7.77.

7.6.3 Hindered Settling

Although this criteria is sufficient for design of clarifiers with dilute solids concentrations, it will not apply to a real clarifier unless the solids can be removed before they accumulate. The particles that settle in the clarifier will continue to settle forming a region of high particle concentration at the bottom of the unit. This region is termed a *thickener.*

At high particle concentrations the presence of the other particles will hinder further settling. Water is displaced as the solids settle and the motion of this water serves to slow further settling. The particles tend to settle without significant relative motion in this regime giving rise to an alternative description as *zone* settling. The rate of settling in the hindered or zone settling region is a strong function of particle concentration that must also be measured experimentally. In zone settling the water immediately above the settling blanket is generally clear and sharply defined and the experiments involve measurement of this interface as a function of time. Measurement of the concentration in the settling blanket at any time with the slope of the interface vs. time curve allows development of a settling velocity vs. concentration curve.

As the settling continues still further, the solids concentration is sufficiently high that a sludge blanket is said to form on the bottom of the settler/clarifier. As further settling occurs, it is the result of compression and dewatering of the sludge blanket. The rate of settling continues to slow and again experiments are required to predict the rate of settling with time.

In an experiment, the migration of a the clean water/solids laden water interface with time provides the settling velocity vs. time directly. By repeating this at different solids concentrations it is possible to plot a curve of velocity vs. solids concentration. Finally the product of the velocity and concentration provides the solids flux. This process and the general shape of the resulting curves are shown in Figure 7.18. Note that this flux represents the maximum flux at which the sludges can be withdrawn from a settler since it is not possible to withdraw solids faster than they are supplied by settling. For each concentration, therefore, there is a limiting solids flux which, given a total mass of solids that must be removed, defines the required area of the thickener region. The larger of the required areas for clarification and thickening then defines the required area of the clarifier/thickener combination.

Analysis of a primary settler/clarifier requires an estimation of the residence time (or alternatively, cross-sectional area) of a clarifier section to ensure removal of a given particle size in the effluent water **and** an estimation of the removal rate of solids to avoid their accumulation in the settler/clarifier. A full thickener/clarifier is shown schematically in Figure 7.19. A water balance indicates that the total volume flow into the system is balanced by the effluent water (the overflow) and the underflow containing solids as well as residual water. With the overflow, there are essentially no residual solids (at least by design), while the underflow contains a high concentration of solids.

At any point in the thickener, the total flow of solids is the sum of the advective component, $Q_u C_w$, and the settling component, $v_s A_t C_w$, where A_t is the cross-sectional area of the thickening section of the settler. The terms v_s and C_w are the velocity and concentration of solids in the settling sludge. The flow of solids downward must be equal to the solids that are desired to be removed in the thickener, Q_m.

$$Q_m = Q_u C_w + A_t v_s C_w$$

or (7.79)

$$A_t = \frac{Q_m - Q_u C_w}{v_s C_w}$$

Recognizing that the settling velocity is a function of the solid concentration, C_w, through the experimental data, Equation 7.79 suggests that a different area is required at every point in the thickener. If settling velocity as a function of concentration is measured (as described above), the

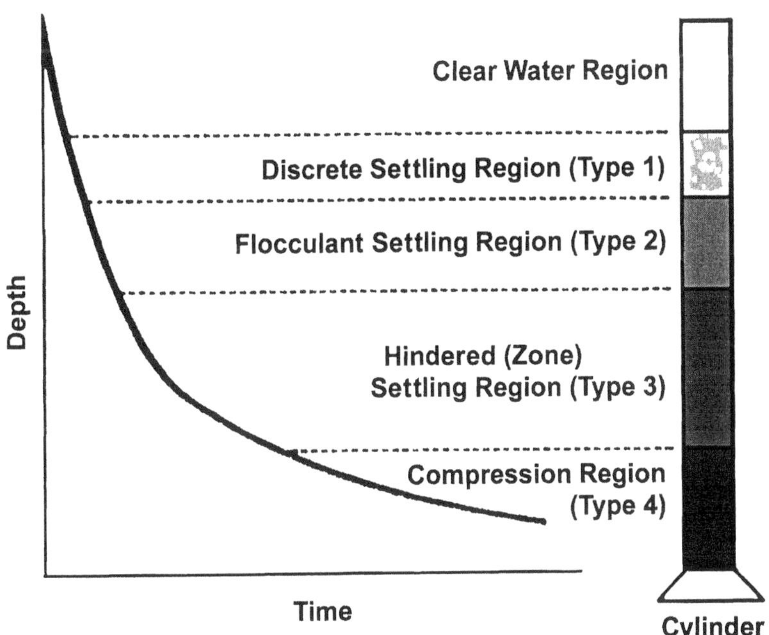

FIGURE 7.18 Schematic of settling regions for activated sludge. (Metcalf and Eddy Staff (1996) *Wastewater Engineering,* 3rd ed., McGraw-Hill, New York. With permission.)

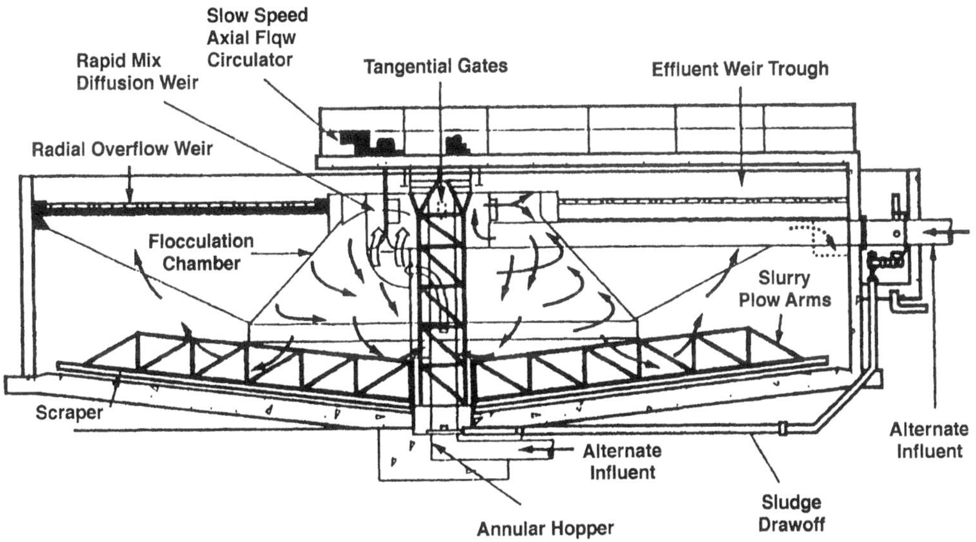

FIGURE 7.19 Typical flocculator-clarifier used in wastewater treatment. (From Metcalf and Eddy Staff (1996) *Wastewater Engineering,* 3rd ed., McGraw-Hill, New York. With permission.)

substitution of these values into Equation 7.73 allows estimation of the area required at every concentration between the inlet and exit (underflow) of the thickener section. The largest calculated area defines a limiting condition to achieve the desired solids separation and this should be used to size the thickener. Because the clarifier/thickener is often combined into a single vessel, the larger of either the clarifier area (from Equation 7.76) or thickener area (from Equation 7.79) should be used to size the vessel. This process is illustrated in Example 7.13.

Example 7.13: Determination of required area for a clarifier/thickener

Consider the initial settling rate vs. solids concentration contained in the table below. Use this data to define the required surface area of a clarifier-thickener combination fed 5 m^3/min of water containing 2 kg/m^3 of solids with an underflow 10% of that flowrate. The entire solids loading of Q_m = 10 kg/min is designed to be removed with the underflow.

Clarifier — From the table, the initial settling velocity of the 2000 mg/L solution is 11.5 cm/min. This can be used to define the required area of the clarifier section.

$$A_c = \frac{Q}{v_s(2000\ \text{mg/L})} = \frac{5\ \text{m}^3/\text{min}}{0.115\ \text{m/min}} = 43.5\ \text{m}^2$$

Thickener — The product of the settling velocity and solids concentration gives the solids settling flux at that concentration and is included in the table. The required area of the thickener can be calculated for each concentration. For example, the required thickener area at the inlet solids concentration is

$$A_t(2\ \text{kg/m}^3) = \frac{Q_m - Q_u C_w}{v_s C_w} = \frac{(10\ \text{kg/min}) - (0.5\ \text{m}^3/\text{min})(2\ \text{kg/m}^3)}{(0.115\ \text{m/min})(2\ \text{kg/m}^3)} = 39.2\ \text{m}^2$$

The required areas at the other concentrations up to the underflow concentration of 20 kg/m^3 are included in the table. The maximum required area is 53.2 m^2 which is therefore the needed area of the thickener. Because it is larger than the required clarifier area, a settler with this area will accomplish both the clarification and thickening. This area corresponds to a diameter of 8.2 m.

C_w (kg/m³)	v_s (cm/min)	q_m (kg/m²·min)	A_t, m²
2	11.5	0.230	39.2
4	6.58	0.263	30.3
6	3.78	0.227	30.9
8	2.17	0.173	34.6
10	1.24	0.124	40.2
12	0.713	0.086	46.7
14	0.409	0.057	52.3
16	0.235	0.038	53.2
18	0.135	0.024	41.2
20	0.077	0.015	0.0

7.7 SECONDARY WASTEWATER TREATMENT

Secondary wastewater treatment generally refers to biological treatment processes that are used to remove organic contaminants from industrial and municipal wastewaters. As indicated previously, bacteria have the capability of using organic contaminants as both food and energy sources and it is this capability that is utilized in this process.

A variety of biological processes are in use to treat wastewater streams. By far the most important are those that employ aerobic bacteria due to the generally faster rate of reactions and the wider range of organic substrates that are utilized by such organisms. Anaerobic systems are used to treat high-strength organic wastes and to treat sludges. The reactions occurring in a composting pile are an example of anaerobic degradation processes. Aerobic processes remain the most important in natural surface waters and in treatment of municipal and industrial wastewaters. Both aerobic and anaerobic treatment processes can be divided into *suspended growth* and *attached growth* processes. In a suspended growth process the microorganisms grow and degrade organic substrates while suspended in solution. Common aerobic processes that fall into this category include the activated sludge process and treatment ponds and lagoons. In an attached growth process, the microorganisms are stabilized by attachment onto a solid medium. A slime layer grows on the surface of the solid support and the microbial reactions occur in that slime layer. The solid supports can be designed to afford good contacting between the wastewater and the slime layer. Due to the attachment between the slime layer and the solid support, a problem of suspended growth processes called *washout* can be better avoided. Some of the difficulties with attached growth processes are associated with control of the hydraulics in the system to avoid *sloughing*, the loss of the slime layer and the equivalent of washout, and *flow channeling* which results in efficient use of portions of the support bed of solids. Examples of attached growth biological processes include trickling filters and rotating biological contactors. Due to their wide usage and comparative simplicity, let us focus on suspended growth processes.

7.7.1 Aerated Lagoon

The simplest of these is a stabilization pond in which solids are allowed to settle to the bottom and natural biological degradation reactions are allowed to occur. If the loading of oxygen-demanding wastes in the feed water to such a pond is too high, however, such a pond will turn anaerobic. Under anaerobic conditions, the generation of reduced byproducts such as hydrogen sulfide will result in objectionable odors (hydrogen sulfide is often described as the odor of rotten eggs). To maintain aerobic conditions it is generally necessary to aerate the pond with fans that stir the surface of the water or by bubbling air through the water, producing an *aerated lagoon*. The process of aeration also tends to enhance mixing in the lagoon, suggesting that a reasonable model of the system is as a well-stirred vessel. Let us consider the simple process flow diagram represented by Figure 7.20. Because biomass is neither input nor removed from the lagoon, the organism death and settling to the bottom of the lagoon must be in balance with the rate of growth to give a constant biomass concentration. A material balance on a degradable organic substrate in the water can be written

$$\frac{dC_S}{dt} = QC_{S_0} - QC_S + R_S V \tag{7.80}$$

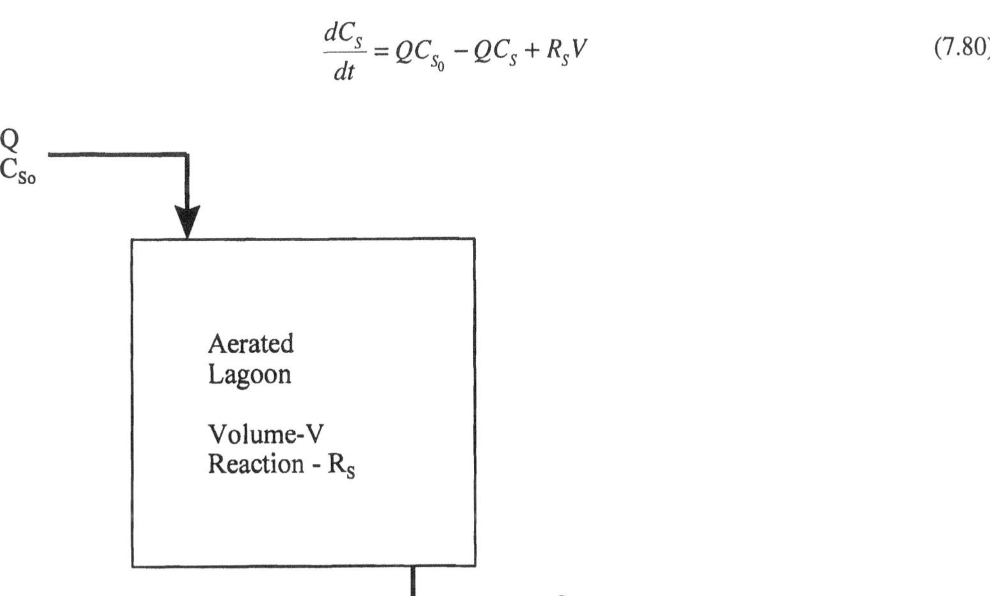

FIGURE 7.20 Schematic of an aerated lagoon for wastewater treatment.

where R_S is the net rate of formation of substrate. That is, since the substrate degrades, R_S is negative. Under steady operating conditions this gives for the effluent substrate concentration (which by the assumption of a stirred vessel is identical to the substrate concentration in the lagoon)

$$C_S = C_{S_0} + R_S \frac{V}{Q} = C_{S_0} + R_S \tau_Q \tag{7.81}$$

where τ_Q is the hydraulic residence time in the lagoon. The reaction rate R_S is given by Equation 7.43. Substitution of this relationship for R_S gives a quadratic equation for the effluent substrate concentration. If the lagoon is effective, it might be expected that the substrate concentration in the effluent is low, i.e., $C_S << C_{S1/2}$ and

$$R_s \approx -\frac{k_m C_S X}{Y C_{S_{1/2}}} \tag{7.82}$$

Under these conditions, the effluent substrate concentration is given by

$$C_S = \frac{C_{S_0}}{1 + \frac{k_m X}{Y C_{S_{1/2}}} \tau_Q} \tag{7.83}$$

7.7.2 Activated Sludge System

An activated sludge system is the most common wastewater treatment system. The process follows the basic process flow diagram presented in Figure 7.21. The name refers to the fact that since acclimated microbes are continuously recycled back to the biological reactor, the microbial population is "activated." This recycling is accomplished by employing a clarifier/thickener that separates the biological solids from the effluent water. The process works very well for a wastewater stream of essentially unvarying composition. Process upsets, for example, by a pulse input of different or more concentrated wastewater requires a new period of acclimation and lagging growth of the organisms. The large volume of the aerated lagoon that serves as the reactor helps damp changes in feed composition or conditions.

An activated sludge wastewater treatment system can be designed in a very similar manner to the simple aerated lagoon. That is, the fundamental basis is a material balance on the reactor, in this case for both the substrate and the biological solids that are fed to the thickener/clarifier. Let us consider a well-mixed, aerated lagoon of volume V as shown in Figure 7.21. For simplicity it is often assumed that the reaction occurs only in the aerated lagoon and that the thickener/clarifier serves only to separate solids from the effluent water. Some of the these solids are recycled to the reactor at volumetric flowrate Q_r, while the remainder is "wasted" in the form of sludge. The maintenance of high growth conditions in the reactor requires removal of some of the produced biomass. This sludge is normally dewatered and then applied to soil or otherwise treated. The influent wastewater flow is Q (volume/time) which is generally very close to the same volumetric flow in the effluent ($Q \sim Q_e$). The volumetric flow of sludge exiting after thickening ($Q_u - Q_r = Q_t$) also contains water but the volume is generally small compared to the wastewater influent or effluent. In addition, the biodegradable substrate of interest is associated with the water and only small amounts (compared to the total amount charged to the treatment system) are found in the waste sludge. Both the water and substrate content of the waste sludge significantly impact its subsequent treatment and disposal, but the amounts are small compared to the influent/effluent water streams. An overall material balance on the biodegradable substrate can then be written

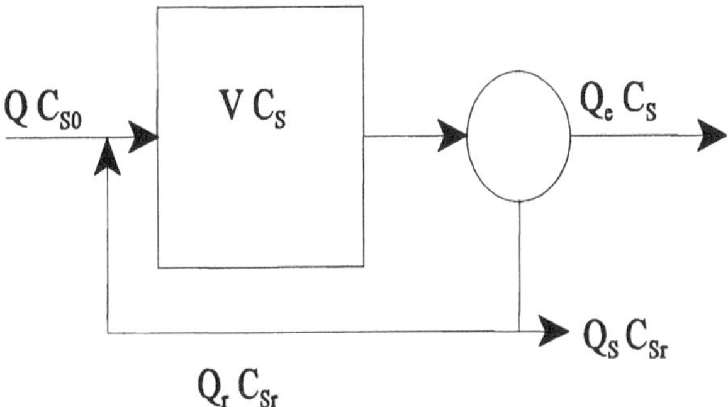

FIGURE 7.21 Schematic of an activated sludge wastewater treatment system.

$$V\frac{dC_S}{dt} = QC_{S_0} - Q_eC_S - R_SV$$

where (7.84)

$$R_S = \frac{k_mC_SX}{Y(C_{S_{1/2}} + C_S)}$$

At steady state and setting Q ~ Q_e, and τ_Q = V/Q, the mean hydraulic residence time in the reactor, this can be written

$$C_{S_0} = C_S + \frac{k_mC_SX}{Y(C_{S_{1/2}} + C_S)}\tau_Q$$

or (7.85)

$$\frac{Y(C_{S_0} - C_S)}{\tau_QX} = \frac{k_mC_S}{(C_{S_{1/2}} + C_S)}\tau_Q$$

Correspondingly, an overall material balance on the biological solids can be written

$$V\frac{dX}{dt} = -(Q_tX_t + Q_eX_e) + R_XV$$

where (7.86)

$$R_X = \frac{k_mC_SX}{C_{S_{1/2}} + C_S} - k_dX$$

The biomass concentration in the reactor is usually measured as the solids loading, as the *mixed-liquor suspended solids* (MLSS) or *mixed-liquor volatile suspended solids* (MLVSS). The term mixed-liquor simply refers to the mix of wastewater and biomass that resides in the reactor. Because the recycle stream provides only internal recirculation it represents neither an input nor output on the total system. Equation 7.86 assumes that there is no significant amount of bacterial cells in the influent to the treatment system. Under steady conditions, Equation 7.86 can be written

$$\frac{Q_tX_t + Q_eX_e}{V} = \frac{k_mC_SX}{C_{S_{1/2}} + C_S} - k_dX \quad (7.87)$$

The ratio of the biomass in the reactor, XV, to the flowrate of biomass leaving the system, $Q_tX_t + Q_eX_e$, is called the mean cell residence time, τ_X. The inverse of the mean cell residence time is the fractional rate of cell removal, $\dot{f}_X$. That is, if the mean cell residence time is 10 days, the fraction rate of cell removal is 0.1 or 10% per day. Using this definition, Equation 7.87 can be written

$$\frac{Q_t X_t + Q_e X_e}{VX} = \frac{k_m C_S}{C_{S_{1/2}} + C_S} - k_d$$
$$\frac{1}{\tau_X} + k_d = f_X + k_d = \frac{k_m C_S}{C_{S_{1/2}} + C_S} \tag{7.88}$$

Setting Equations 7.87 and 7.88 equal, the biomass concentration in the reactor is

$$X = \frac{\tau_X}{\tau_Q} \frac{Y(C_{S_0} - C_S)}{(1 + k_d \tau_X)} \tag{7.89}$$

and the substrate concentration is

$$C_S = C_{S_0} - X\left(\frac{1 + k_d \tau_X}{Y}\right)\left(\frac{\tau_Q}{\tau_X}\right) \tag{7.90}$$

The first term in parentheses is the inverse of what is termed the observed yield coefficient

$$Y_{obs} = \frac{Y}{1 + k_d \tau_X} \tag{7.91}$$

X/Y_{obs} represents the maximum amount of substrate that could be utilized by the given biomass concentration. Equation 7.90 could thus be rewritten

$$\Delta_S = (\Delta_S)_{\max} \frac{\tau_Q}{\tau_X}$$

where

$$\Delta_S = C_{S_0} - C_S \tag{7.92}$$

$$(\Delta_S)_{\max} = \frac{X}{Y_{obs}}$$

Note that $(\Delta_S)_{max}$ could be greater than the amount of substrate charged to the activated sludge system since it refers only to the potential utilization by the biomass. Equation 7.92 clearly indicates that conversion efficiency, defined by

$$\text{Efficiency} = \eta = \frac{C_{S_0} - C_S}{C_{S_0}} \tag{7.93}$$

increases as the hydraulic residence time or the biomass concentration increases and decreases as the cell residence time increases. Of course, decreases in cell residence time increases the amount of sludge that must be treated or disposed. In addition, the cell residence time cannot be reduced

TABLE 7.3
Typical Kinetic and Design Parameter Values — Activated Sludge Wastewater Treatment Process

Parameter	Units	Values Range	Values Typical
Maximum cell growth rate, k_m	day^{-1}	1–6	3
Substrate concentration at half growth $C_{S_{1/2}}$	mg/L BOD	25–100	60
	mg/L COD	15–70	40
Substrate utilization coefficient, Y	mg S/mg BOD	0.4–0.8	0.6
Cell death rate, k_d	day^{-1}	0.025–0.075	0.06
Hydraulic residence time, τ_Q	hours	3–5	4
Mean cell residence time, τ_X	days	5–15	10
Ratio of recycle to influent flow, Q_r/Q	—	0.25–1.0	0.5

Note: BOD — biological oxygen demand, COD — chemical oxygen demand.

below a minimum level below which a phenomenon termed washout occurs. Washout occurs when the substrate conversion efficiency is zero, i.e., E = 0 and S = S_0. According to Equation 7.88, this occurs when

$$\frac{1}{(\tau_X)_{\min}} = (\dot{f}_X)_{\max} = \frac{k_m C_{S_0}}{C_{S_{1/2}} + C_{S_0}} - k_d$$

or when

$$C_{S_0} \gg C_{S_{1/2}} \tag{7.94}$$

$$(\tau_X)_{\min} \approx \frac{1}{k_m - k_d}$$

$$(\dot{f}_X)_{\max} = k_m - k_d$$

The substrate concentration charged to the system is often much greater than $C_{S_{1/2}}$, and the maximum fractional rate of removal of biomass cells is often simply the difference between the maximum growth and death rates.

Table 7.3 provides typical values for the kinetic constants that appear in the design equations. Based upon the typical values in that table, washout would not be expected to occur until mean cell residence time of about 10 h is observed (fractional rate of removal of 10%/h). Mean cell residence times are normally in the range of 5 to 15 days, well above the residence times that would be expected to lead to washout of the organisms and loss of treatment efficiency.

An illustration of the use of these equations to analyze activated sludge systems can be found in Example 7.14.

Example 7.14: Design of an activated sludge wastewater treatment system

Determine the volume, sludge wastage rate, and recycle rate of a well-mixed activated sludge reactor designed to convert a 10 m^3/min stream containing 200 mg BOD/L to an effluent containing 20 mg/L. Previous experience suggests that the system will operate best with a 10-day biomass residence time and a biomass concentration of 4000 mg/L in the reactor with a substrate utilization coefficient of 0.6. The biomass death rate is 0.06 day^{-1}. The clarifier/thickener is expected to produce 10000 mg/L sludge in the underflow.

The required volume from Equation 7.89 is

$$V = \tau_X QY \frac{C_{S0} - C_S}{X(1 + k_d \tau_x)}$$

$$= \frac{10 \text{ days}(10 \text{ m}^3/\text{min})(0.6)(200 \text{ mg/L} - 20 \text{ mg/L})}{4000 \text{ mg/L}[1 + (0.06 \text{ day}^{-1})(10 \text{ days})]} 1440 \text{ min/day} = 2430 \text{ m}^3$$

The residence time is then V/Q~4.05 h. From the definition of the mean cell residence time, the sludge wastage rate is given by

$$Q_t = \frac{VX}{\tau_X X_t} - Q\frac{X_e}{X_t} = \frac{(2430 \text{ m})(4000 \text{ mg/L})}{(10 \text{ days})(10000 \text{ mg/L})} = 0.067 \text{ m}^3/\text{min}$$

if it is assumed that the biomass carried over the clarifier, X_e ~0. Because the recycle stream is the only source of biomass feed into the reactor

$$Q_r X_r = Q_r X_t = XQ$$

$$Q_r = \frac{4000 \text{ mg/L}}{10000 \text{ mg/L}} 10 \text{ m}^3/\text{min} = 4 \text{ m}^3/\text{min}$$

In addition to the parameters defined above, wastewater treatment systems are often designed with reference to the specific utilization rate and the food to microorganism ratio. The specific utilization rate, U_s, is the rate of substrate removal divided by the biomass concentration,

$$U_s = \frac{k_m C_S}{Y(C_{S_{1/2}} + C_S)} \tag{7.95}$$

From Equations 7.85, U_s also can be written

$$U_s = \frac{C_{S_0} + C_S}{\tau_Q X} \tag{7.96}$$

Although the specific utilization rate is a useful characterization for a wastewater treatment system, the mean cell residence time, τ_X, is generally a more convenient design parameter in that it does not require evaluation of the active biomass concentration as shown by Equation 7.88. The specific utilization rate using the conditions of Example 7.14 is 0.267 day^{-1}, typical for wastewater treatment systems.

Closely related to the specific utilization rate is the food to microorganism ratio, commonly given the symbol F/M

$$F/M = \frac{C_{S_0}}{\tau_Q X} = \frac{U_s}{(C_{S_0} + C_S)/C_S} = \frac{U_s}{\eta} \tag{7.97}$$

Typically, F/M varies from 0.2 to 0.6 of BOD per MLVSS per day (0.2 to 0.6 mg BOD·mg MLVSS^{-1}·day^{-1}). For the conditions of Example 7.14, the F/M is 0.296 day^{-1}, within this range.

As indicated previously, a wide variety of other biological treatment processes exist. The activated sludge process just discussed represents one of the most common for routine wastewaters. We will defer discussion of other wastewater treatment processes to more specialized texts and courses.

7.8 TERTIARY WASTEWATER TREATMENT

Upon completion of primary and secondary treatment a wastewater stream is often of sufficiently high quality for discharge to surface waters. It is not of sufficient quality to be used as drinking water nor can high strength industrial waters be discharged directly. In either case, tertiary treatment is required to remove residual contaminants from the wastewater stream. A variety of tertiary treatment processes are available depending upon the nature of the contaminant and the degree of separation required. In our discussion we will illustrate these processes in detail with only three processes: filtration, ion exchange, and membrane. While not comprehensive these processes are among the most important. Before discussing these processes, let us discuss adsorption and absorption (or stripping) which are conceptually identical to the processes evaluated for removal of contaminants from a gas stream.

7.8.1 Adsorption

Adsorption of waterborne contaminants onto activated carbon or other adsorbent can be achieved in the same fashion as it was applied to the adsorption of vapors. Equilibrium data provide an estimate of the maximum capacity of an adsorbent bed. An estimate of the time available for sorption between regeneration of the adsorbent is then simply this capacity divided by the rate of feed of the sorbing contaminant to the adsorber. If mass transfer rate or diffusion/dispersion information is known, this estimate can be improved upon by use of the techniques described in the previous chapter.

7.8.2 Stripping

Stripping is the opposite of the absorption process discussed in the last chapter. During stripping, a volatile component of a gas stream can be removed from the water by partitioning into the gas phase. Steam is often used for the stripping process in that the heat provided by steam raises the vapor pressure of the contaminant forcing the equilibrium partitioning toward the vapor phase, improving the degree of separation between the liquid and the stripped component.The design of a stripper is identical to that of an absorber. The mass transfer is simply from liquid to vapor.

7.8.3 Filtration

Perhaps the first example of a tertiary treatment process was to filter the secondary effluent through a bed of sand, gravel, or other solid. This can remove suspended particulate matter, generally arbitrarily defined as solids with a nominal particle diameter greater than 0.45 μm. Dissolved particulate matter, or colloidal matter, with a nominal particle size less than this is generally not captured effectively by filtration. Molecular contaminants also are generally not captured by filtration.

Filtration of contaminants from water and wastewater is not conceptually different from the filtration of particulate matter from gas streams. The mechanisms of collection of suspended solid matter are identical with larger particles collected via direct interception, intermediate size particles being collected by inertial impaction and some collection of "dissolved" and colloidal solids by Brownian diffusion.

Unlike a gas filter, however, the pressure drop across a water filter is generally large even for clean filtration media. In most cases, gravity flow drives the flow through the filter limiting the pressure drop that can be sustained. Pressure drop across the clean media is one of the most important design considerations. Generally a water filter is a bed of sand of either uniform or varying particle sizes or an adsorbent filter material such as charcoal. The filter bed is porous with a typical void fraction in the range of 30 to 50%. The flow of fluids and pressure drop in a porous bed of solids will be discussed in more detail in the next chapter. Here let us simply refer to a commonly used model relating pressure drop to bed material and flow properties. The Carmen-Kozeny equation relates the head loss or pressure drop across a filter bed to the flow velocity, and the bed and particle geometries.

$$\Delta P = \alpha_f \rho_w \frac{L}{\alpha_p d_p} \frac{1-\varepsilon}{\varepsilon^3} q^2 \tag{7.98}$$

Here $q = Q/A$ is the superficial velocity in the bed (volumetric flowrate divided by cross-sectional area), ε is the bed void fraction or porosity, d_p is the nominal diameter of the particles that make up the bed, and L is the length of the bed. The term α_p represents the shape factor for the particles that make up the bed and in a clean bed is usually given the values

$$\begin{aligned} \alpha_p &= 1 && \text{for uniform spherical particles} \\ &= 0.82 && \text{for rounded sand} \\ &= 0.75 && \text{for average sand} \end{aligned} \tag{7.99}$$

α_f is the friction factor often estimated via the Ergun equation

$$\alpha_f = 150\left(\frac{1-\varepsilon}{N_{\text{Re}}}\right) + 1.75$$

where

$$N_{\text{Re}} = \frac{\alpha_p \rho_w d_p}{\mu_w} q \tag{7.100}$$

Equation 7.98 is often written

$$\Delta P = 180 \frac{L\mu_w}{\alpha_p d_p^2} \frac{(1-\varepsilon)^2}{\varepsilon^3} \tag{7.101}$$

An examination of this relationship indicates that the pressure drop is a strong function of particle diameter ($\Delta P \sim 1/d_p^2$) and flowrate (Example 7.15). Thus, the simplest means of controlling the pressure drop is to change the grain size of the particles that make up the bed. Unfortunately, the collection efficiency, especially for small particulate matter, will decrease if the particle diameter is large. In order to simultaneously have relatively small pressure drops across the bed and have high efficiency of contaminant collections, a multimedia filter is often used. In a multimedia filter, the grain size is varied with the larger grains exposed to the raw water feed followed by finer and finer particles. The product water is the effluent from the finest bed particles. For a stratified sand or granular media bed composed of a variety of particle sizes, Equation 7.98 becomes

$$\Delta P = \rho_w \frac{1-\varepsilon}{\varepsilon^3} q^2 \sum_i \left[\frac{\alpha_f(d_p)}{\alpha_p d_p} \right]_i L_i \qquad (7.102)$$

where the sum is taken over all particle sizes in the multilayered filter bed. A multimedia filter bed also has a distinct advantage over single layer beds in that contaminants are collected over the entire depth of bed. In a single media filter bed, particulates tend to collect at the inlet to the bed effectively blocking off the bed long before its full capacity is utilized.

Example 7.15: Pressure drop in a clean filter bed

Estimate the pressure drop of a 1 m^3/min water flow in a sand filter of 0.4-mm particles that is 5 m in diameter and 3 m high.

Assume that the porosity of the bed is about 40% and that the particle shape factor is typical of that of average sand (0.75). With a diameter of 5 m, the bed has an area of 19.6 m^2 and therefore q = Q/A = 0.085 cm/s. The Reynolds number of the flow through the bed is

$$N_{Re} = \frac{d_p q}{\nu_w} = \frac{(0.04\ \text{cm})(0.085\ \text{cm/s})}{0.01\ \text{cm}^2/\text{s}} = 0.339$$

This defines the friction factor

$$\alpha_f = 150 \frac{1-\varepsilon}{N_{Re}} + 1.75 = 150 \frac{(1-0.4)}{0.339} + 1.75 = 268$$

Which in turn defines the pressure drop for Equation 7.98, or using Equation 7.101

$$\Delta P = 180L \frac{\mu_w}{\alpha_p d_p^2} \frac{(1-\varepsilon)^2}{\varepsilon^3} q$$

$$= 180(300\ \text{cm}) \frac{0.001 * \text{kg/(m} \cdot \text{s)}}{(0.75)(0.04\ \text{cm})^2} \frac{(1-0.4)^2}{(0.4)^3} (0.085\ \text{cm/s})$$

$$= 21500\ \text{kg/(m} \cdot \text{s}^2) = 21500\ \text{Pa} = 0.212\ \text{atm}$$

This pressure drop is the equivalent of $\Delta P/(\rho_w g)$ = 2.2 m of water. Because the filter bed is 3 m high, gravity feed of the water flow through the filter could be accomplished when it is clean. As the filter collects solids in the water, the pressure drop will increase and the filter would need to be backwashed for cleaning.

As with the gas filtration systems, a water filter must be cleaned of contaminants on a regular basis. This restores close to the clean water pressure drop calculated by, for example, Equation 7.102, and allows continued water treatment. Cleaning is most often conducted by backwashing the filter with water. The collected contaminants are removed with a small amount of reversed flow water. The backwash velocity should be of sufficient magnitude to fluidize the bed, that is overcome its own weight. The bed will expand and the pressure drop decreases as the void fraction increases. Thus, a high volume of backwash water can be pushed through the filter for a short period to achieve the cleaning of the filter. After backwashing the filter can be returned to contaminant removal service.

7.8.4 Ion Exchange

Ionic contaminants can be removed from water and wastewater via ion exchange. Zeolites, a naturally occurring porous sand and a variety of synthetic materials and cation (positively charged) and anion (negatively charged) exchange resins, are employed as ion exchange media. These media are charged and as a result attract ionic species. A cationic exchange media will be negatively charged and initially this charge will be offset by a particular ionic species. Other ionic species accessible to the media may replace the original ion and it is this process that is termed ion exchange. It is built on the principle that different ions are attracted and held more tightly than others. Thus, ion exchange is the process of replacing an easily displaced ionic species with a contaminant ion that is not easily displaced. Generally, ions exhibiting multiple charges, such as lead (Pb^{+2}) or cadmium (Cd^{+2}) are held more tightly by ion exchange media than singly charged ions. There is also a preference order among identically charged ions, as indicated in Chapter 3, for ion exchange in soils. For cationic species, the preference order is generally

$$Ba^{+2} > Pb^{+2} > Sr^{+2} > Ca^{+2} > Ni^{+2} > Cd^{+2} > Cu^{+2} > Co^{+2} > Zn^{+2} > Mg^{+2} > Ag^{+} > Ca^{+} > K^{+} > NH_4^{+} > Na^{+}$$

and for anionic species the preference order is generally

$$SO_4^{-2} > I^- > NO_3^- > CrO_4^{-2} > Br^- > Cl^-$$

The hydrogen ion (H^{+}) and hydrozxyl ion (OH^-) also can be used to exchange with ions in the water, but their location in the preference list depends upon the acidity of the ion exchange media. This preference lists suggests that a negatively charged cation exchange resin containing sodium (Na^+) can be used to remove lead or cadmium ions effectively from a wastewater. Such a process has long been used to "soften" water and can be used to completely demineralize water. The latter process would require use of a strongly acidic cation exchange resin that replaces the hydrogen ion with any cationic mineral species followed by a strongly basic anionic exchange resin that replaces the hydroxyl ion with any anionic species.

The preference for particular ions (I_1 and I_2) can be expressed quantitatively through consideration of the equilibrium constant for the exchange reaction, as discussed in Chapter 3

$$I_1 + I_2 \cdot S \Leftrightarrow I_2 + I_1 \cdot S$$

$$K_{exch} = \frac{C_{I_2} W_{I_1 S}}{C_{I_1} W_{I_2 S}} \tag{7.103}$$

where C and W represent the fluid and solid phase concentrations, as usual. The exchange equilibrium constant is the ratio of the fluid phase concentrations C_{I_2}/C_{I_1} to the sorbed phase concentrations $W_{I_2 S}/W_{I_1 S}$. It therefore represents a selectivity coefficient and indicates the degree of separation that can be achieved.

The total capacity of exchange media is termed the ion exchange capacity, for example, the cation exchange capacity discussed in Chapter 3. At any time, the total concentration of ions equals the ion exchange capacity which is the ion concentration required to just offset the surface charge on the medium. The selectivity coefficient defines what fraction of the ion exchange capacity will ultimately consist of the contaminant for which removal is desired.

Ion exchange is essentially identical to adsorption. There exists an equilibrium amount of exchange that can occur. In an ideal ion exchange bed, available sites are filled in order and the entire ion exchange bed can be equilibrated prior to shutting the feed down and regenerating or replacing the bed. If W_s^* represents the equilibrium capacity of the ion exchange column at the inlet feed concentration of exchangeable ion $C_w(0)$, then the time between regenerations is given by

$$\tau_{exch} = \frac{\rho_b W_s^* A L}{Q C_w(0)} \tag{7.104}$$

As with an adsorber, however, the actual time between regenerations must be somewhat less than this due to the effects of a finite mass transfer rate of exchange between the ions and dispersion in the bed. The models developed in Chapter 6 remain applicable. The mass transfer and width of the breakthrough curve is also often determined by experiment and these experiments used to determine the size of the final column (e.g., as described in Reynolds and Richards, 1996).

The regeneration of an ion exchange bed involves the renewal of the original ion back on the exchange resin. At first glance this seems like a difficult task since the ion exchange material has already demonstrated the preference for the exchanged ion whose removal is the goal of the treatment system. An examination of Equation 7.103 shows that the equilibrium can be adjusted by control of the concentrations of the various ions. That is, for an ion exchange resin that is to be impregnated with sodium, Na^+, a very strong brine with a salt concentration of 30 to 40% can be introduced to the regenerating ion exchange bed. This extremely high concentration drives the equilibrium selectivity toward the sodium rather than the exchange ion and regeneration can proceed. Upon the conclusion of regeneration, the system can be returned to its original status as an ion exchange treatment system.

7.8.5 Membrane Processes

A membrane is a thin sheet of natural or synthetic material that is preferentially permeable to a particular contaminant or solute. In principle, if a membrane allows only a single contaminant to pass, this contaminant could be completely separated from the water. Although lumped together here, membrane processes constitute a wide variety of materials and processes, including ultrafiltration, dialysis, electrodialysis, and reverse osmosis. We will define these processes below but Figure 7.22 compares their applicability for contaminants of a specific size to sedimentation, centrifugation, and conventional filtration. The differentiation between the various forms of filtration — cloth, micro-, and ultra- — is based upon the pore sizes in the filter media. *Ultrafiltration* is a filtration process with a membrane that exhibits pore sizes much smaller than conventional filters. As a result the pressure drop is large but the degree of separation is high, especially for dissolved, colloidal solids and large molecular weight contaminant molecules that tend to pass unhindered through conventional filters.

The mechanisms for dialysis, electrodialysis, and reverse osmosis are completely different from filtration. In *dialysis*, a concentration gradient between the two sides of a membrane drives solute passage. The membrane has pore sizes that exclude larger ions from passing despite a concentration gradient while allowing smaller ions to pass. This process is depicted in Figure 7.23. The net effect of dialysis is separation of the water and solute that is excluded by the membrane. Dialysis is not used for much large scale water treatment operations, however, because of the slow rate of mass transfer by diffusion through the membrane.

Electrodialysis can speed the process of dialysis for ionic species. By placing a voltage across the membrane, the motion of the ionic species can be increased due to the response to the voltage gradient. This process can be further enhanced through use of multiple layers of membranes that serve as ion exchange membranes which have a net charge and are therefore permeable only to a species of opposite charge. By alternating anion and cation permeable membranes as shown in Figure 7.24, alternating layers concentrated in either the anion or cation can be created. The cation, for example, will progress toward the cathode, passing through the cation-permeable exchange

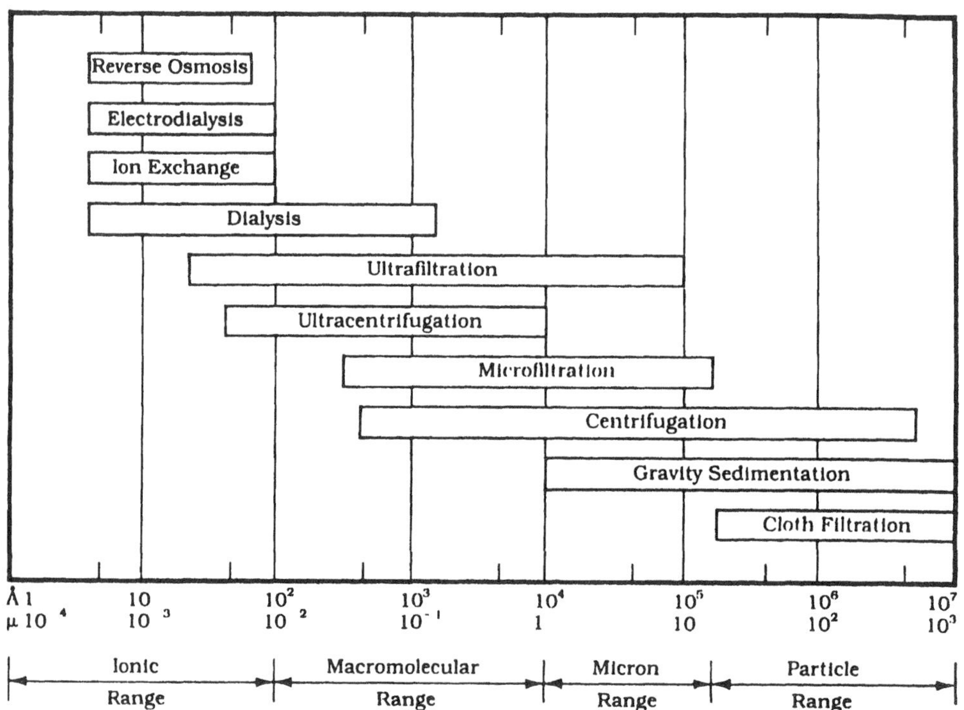

FIGURE 7.22 Particle sizes for which various membrane technologies are effective. (From Lacey, R.E. (1972) *Chemical Engineering*, McGraw-Hill, New York. With permission.)

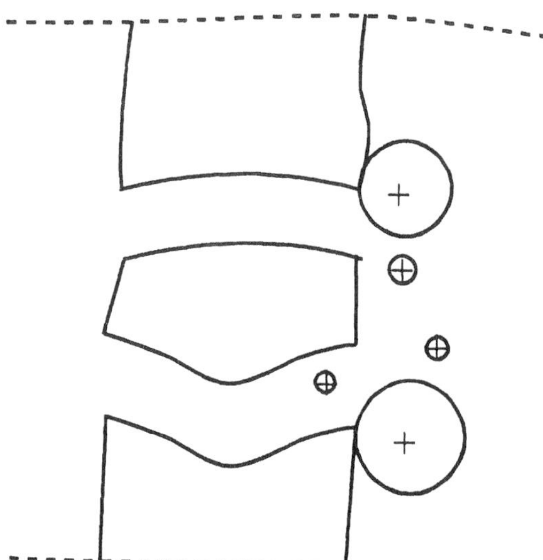

FIGURE 7.23 Cross section of dialysis membrane showing size exclusion of ionic species.

membrane but stopped by the anion-permeable membrane. The cation concentrates on the cathode side of the cation-permeable membranes. Similarly, the anion passes through the anion-permeable membrane in its movement toward the anode but is stopped by the cation-permeable membrane. All ions, both positive and negative, are concentrated in the cathode side of cation-permeable membranes while the water on the anode side of the cation-permeable membranes is depleted of ionic species. Electrodialysis has been used to demineralize wastewaters and also has been used to remove salt from seawater.

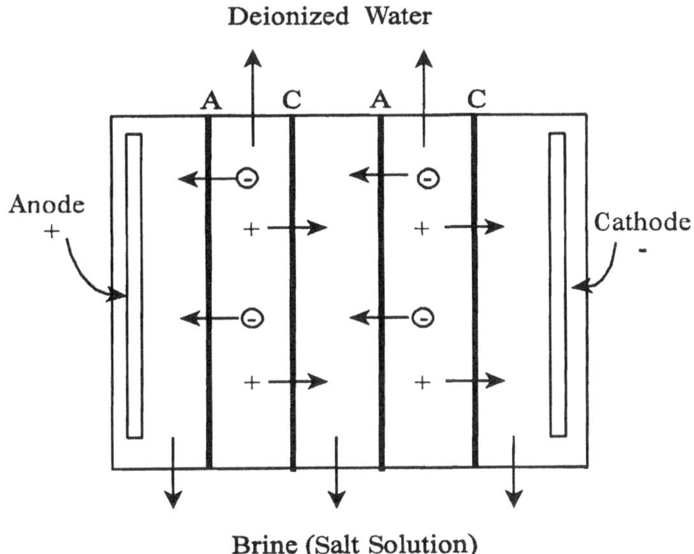

FIGURE 7.24 Depiction of an electrodialysis unit for the desalination of sea water.

One Faraday (F_e) of electricity (96,500 A/s or C) will move a gram equivalent weight of a substance from one electrode to the other. That is, one Faraday will move one mole of a singly charged ion and half of a mole of a doubly charged ion. If there are (2n) membranes in an electrodialysis unit this current moves through n different cells capable of collecting the ions. An individual cell includes both a cation- and an anion-permeable membrane. Thus, if a potential E is placed across an electrolysis unit with internal electrical resistance R, the current is I = E/R and the total number of equivalent weights of ions collected in the cells is

$$\Delta(QN) = \frac{n}{F_e}\frac{E}{R} \tag{7.105}$$

where Q is the volumetric flowrate and N is the normality of the solution. The efficiency of the separation is then given by

$$\eta_c = \frac{\Delta(QN)}{QN}\frac{nE}{F_e RQN} \tag{7.106}$$

This assumes that the electrical efficiency of the cells is 100%. This is approximately valid in that the efficiency should be 90% or greater in a well-designed electrodialysis unit.

Reverse osmosis is a membrane process that, like filtration, depends on a pressure difference across the membrane to drive transport. This seeming analogy with filtration is misleading. Consider an electrolyte solution and a freshwater solution that are separated by a semi-permeable membrane that rejects the ionic species as shown in Figure 7.25. The freshwater will migrate through the membrane to dilute the electrolyte solution and reduce the concentration gradient across the membrane. This process is called *osmosis*. In a closed system this will continue until the pressure

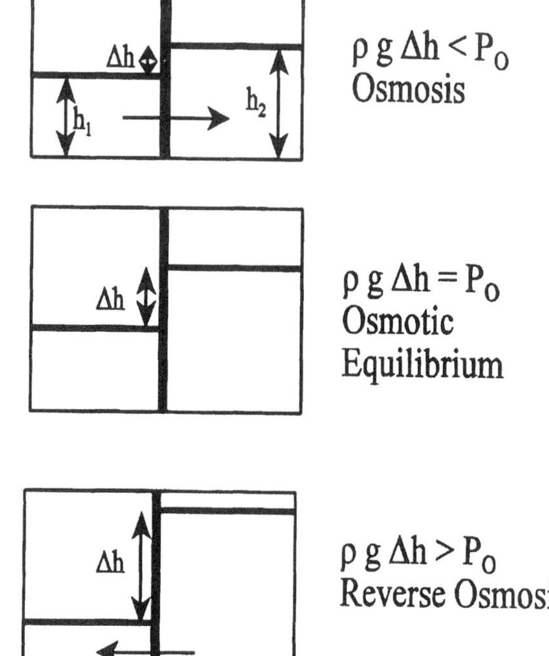

FIGURE 7.25 Illustration of osmosis and reverse osmosis. Equilibrium occurs when the pressure (or head) of the ionic solution is equal to the osmotic pressure, and flow through the membrane toward the ionic solution occurs when its pressure is lower, and flow through the membrane toward freshwater (i.e., more produced fresh water) occurs when its pressure is higher than the osmotic pressure.

in the electrolyte side increases to the *osmotic pressure*, π. At this point the development of a higher pressure in the electrolyte solution exactly offsets the driving force for osmosis. If a higher pressure is applied to the electrolyte solution, osmosis is reversed and water leaves the electrolyte solution. The flux of water from the electrolyte is dependent upon the magnitude of the pressure above the osmotic pressure and is typically modeled by an equation of the form

$$F_w = K_m(P - \pi_o) \tag{7.107}$$

where F_w is the water flux, K_m is the permeability of the membrane (a mass transfer coefficient, e.g., in units of $L \cdot day^{-1} \cdot m^{-2} \cdot kPa^{-1}$), and $P - \pi_0$ is the difference between the applied pressure and the osmotic pressure.

In order to use this model, it is necessary to estimate the osmotic pressure which can be found from

$$\pi_o = \alpha_o NRT \tag{7.108}$$

where α_0 is the osmotic coefficient, N is the normality of the solution (gram equivalent weight per liter), R is the ideal gas constant (in units consistent with the concentration units in normality), and T is absolute temperature. The osmotic coefficient depends upon the electrolyte and its concentration. The osmotic pressure of sea water (salt concentration of 35,000 mg/L) is 2740 kPa at 25°C. If a linear dependence of concentration on the osmotic coefficient is assumed, this means that a change in salt concentration of 1000 mg/L changes the osmotic pressure by 78 kPa.

Although any pressure above the osmotic pressure will cause the movement of water through the membrane from the electrolyte solution, the rate of mass transfer is dependent upon the magnitude of the pressure. Economic operation of the unit generally requires pressures significantly higher than the osmotic pressure, for example, 2 to 5000 kPa above the osmotic pressure.

7.9 SUMMARY

A variety of processes exist capable of removing contaminants from water systems. An almost universal initial step in the treatment of waters is a physical separation step to minimize solids loading on subsequent treatment processes. For organic contaminants, biological processes such as activated sludge systems are widely used and capable of removing contaminants to very low levels. For ionic contaminants, ion exchange or electrodialysis have proven useful. All of the processes summarized herein as well as others have found use for particular contaminant in particular streams. As with the discussions in Chapter 5, our purpose here is to identify some key processes and develop a semi-quantitative understanding of their mode of action and their design. In the final chapter, we will keep to this approach in soil pollution and its control.

PROBLEMS

1. Consider a wide river with a bottom composed of significant vegetation. If the river has an elevation change of about 1 m per 10 km and an average depth of 1 m, estimate the stream velocity and the friction velocity.

2. Repeat Problem 1 for a river with a flat sandy bottom.

3. Estimate the slope required of a drainage ditch with triangular cross section 2 m wide and 1 m deep to drain 1 m^3/s of water.

4. Estimate the volume of runoff after a 10-cm rainfall at a 1 ha construction site composed of low permeability clay soils under average soil moisture conditions. How would this volume change with dry soils? Or with deep sandy soils?

5. Consider a 10 ha area composed of 25% cultivated land, 25% open pasture, and 25% forested land which is exposed to 10 cm of rainfall after a dry period. Tight clay soils underlie the entire area except for the cultivated area which is silty, sandy in makeup. Estimate the volume of runoff water.

6. A 300-m wide area with a slope of 1:100 drains toward a stream running through the center of the property. Estimate the maximum runoff flow and depth of runoff if the land is forested assuming average moisture conditions prior to a rainfall of 5 cm over a 4-h period. The soil is a silty clay.

7. Compare quantitatively the partitioning of dioxin, phenanthrene, and 1,2-dichloroethane to sediment, suspended sediment, water, and biota in a sediment-water system. A lake system is 10 ha in area with an average depth of 5 m. The upper 10 cm of sediment contains 2% organic carbon and is in equilibrium with the lake water as is the fish with an average lipid content of 4%. Suspended sediment is present at 1 g organic matter per liter of water.

8. Compare the water-side mass transfer coefficient at the air-water interface in the open ocean to that of a stream with a depth of 2 m and a velocity of 1 m/s, and a lake 5 m in depth with a 4 m/s wind velocity.

9. A 50,000-L gasoline tank fails and fills a 30 m by 30 m diked area. Consider the gasoline to be 95% octane and 5% benzene. Using mast-transfer coefficients that might apply to the air-water interface estimate the rate of evaporation of the octane (essentially pure) and benzene (dilute component).

10. Characterize the state of mixing and, if possible, estimate the effective diffusivity between the epilimnion and hypolimnion in a lake under the following conditions: epilimnion at 25°C, hypolimnion at 15°C, and a 5-m thick thermocline separating the two layers; epilimnion at 1°C, hypolimnion at 4°C, and a 5-m thick layer separating the two zones; epilimnion of freshwater at 20°C and hypolimnion at 30°C containing 30 part per thousand salt.

11. Consider the Mississippi River under the conditions of Example 7.1. Consider the rate of evaporation and transport to the sediment-water interface of dioxin, phenanthrene, and 1,2-dichloroethane.

12. If 100 kg of a solution containing 100 g of dioxin is introduced to the Mississippi River as in Problem 10 and Example 7.1, predict the water column concentration as a function of distance downstream.

13. If 100 kg of phenanthrene is introduced to the Mississippi River as in Problem 10 and Example 7.1, predict the water column concentration as a function of distance downstream.

14. If 100 kg of 1,2-dichloroethane is introduced to the Mississippi River as in Problem 10 and Example 7.1, predict the water column concentration as a function of distance downstream.

15. If a sudden spill of sodium sulfite caused the oxygen in a stream to drop to 0 mg/L with no excess sodium sulfite, estimate the time and distance downstream required for the stream to recover to 5 mg/L when saturation is 9 mg/L. The stream is 1 m deep and flows at 0.5 m/s.

16. A pollutant is introduced into the stream of Problem 15 that has an equivalent oxygen demand of 100 mg/L and a consumption rate constant of 0.4 h^{-1}. The stream is initially saturated with respect to oxygen. What is the minimum oxygen concentration in the stream and how far downstream does it occur?

17. Repeat Problem 16 for a stream 0.5 m deep and flowing at 1 m/s.

18. Estimate the threshold size for particle resuspension in a stream with a velocity of 0.5 m/s and a depth of 1 m. The bottom of the stream is sandy with a grain density of 2.2 g/cm^3.

19. Estimate the threshold particle size likely to be resuspended at a construction site with an average slope of 1:100 composed of exposed silty clay. During a period of average moisture conditions, a rainfall of 5 cm occurs over 4 h.

20. Estimate the flux out of a sediment bed containing 100 mg/kg of benzene by (a) diffusion, (b) advection (with $q = Q/A = 1$ m/year) and (c) bioturbation (with $D_b = 1$ cm^2/year). Treat each process independently as though it were the only operative process. Compare to the results of Example 7.11.

Sediment conditions:	$\rho_b = 1$ g/cm^3	$\varepsilon = 0.4$	$\omega_{oc} = 0.03$
Contaminant conditions:	$W_s = 100$ mg/kg	$D_w = 8.9 \cdot 10^{-6}$ cm^2/s	$K_{oc} = 10^{2.1}$ L/kg

21. Consider a clarifier 10 m in diameter and 1 m high to which is fed 5 m^3/min of water containing a uniform distribution of biological solids with effective diameters between 100 and 500 μm in size. Estimate the collection efficiency of discrete particles as a function of particle size if the density of the settling solids is 1.1 g/cm^3. Estimate the overall number efficiency and the overall mass efficiency of the clarifier for this waste stream.

22. Estimate the clarifier diameter required to collect all biological solids greater than 100 μm in size in a 10 m^3/min flow. The density of the settlings solids is 1.1 g/cm^3.

23. Estimate the clarifier diameter required to collect particulate matter greater than 10 μm in size in a 10 m^3/min water flow. The density of the settling solids is 2.5 g/cm^3.

24. Given the following settling velocity vs. concentration data for a 10 m^3/min wastewater containing 1000 mg/L solids, estimate the required settler and clarifier area. The design should assume complete collection of the solids by the thickener and an underflow of 0.5 m^3/min.

Solids (g/L)	1	2	4	6	10	15	20	25	30
v_s (cm/min)	13	10	7.5	5.5	2.0	0.50	0.10	0.05	0.02

25. Estimate the biological oxygen demand of the effluent water from an aerated treatment pond fed 1 m^3/min of wastewater containing a BOD of 200 mg/L. The aerated lagoon is designed for a hydraulic retention, or detention, time of 4 days. The kinetic constants are $k_m = 0.5$ day^{-1}, $C_{S1/2} = 75$ mg/L, $Y = 0.5$ kg/kg and $k_d = 0.05$ day^{-1}.

26. In Example 7.14, the substrate concentration corresponding to half of the maximum rate is 140 mg/L. What is the rate constant k_m for the system and the sludge residence time that would result in washout? What is the substrate utilization rate and the food to microorganism ratio?

27. Determine the volume, sludge wastage rate, and recycle rate of a well-mixed activated sludge reactor designed to convert a 5 m^3/min stream containing 200 mg BOD/L to an effluent containing 20 mg/L. Previous experience suggests that the system will operate best with a 10-day biomass residence time and a biomass concentration of 5000 mg/L in the reactor with a substrate utilization coefficient of 0.5. The death rate of organisms is 0.025 day^{-1}. The clarifier/thickener is expected to produce 12500 mg/L sludge in the underflow.

28. The maximum biomass growth rate is 0.5 day^{-1} for the system in Problem 27. Estimate the substrate concentration at half of the maximum growth rate and the sludge residence time that corresponds to washout.

29. Determine the volume, sludge wastage rate, and recycle rate of a well-mixed activated sludge reactor designed to convert a 5 m^3/min stream containing 200 mg BOD/L to an effluent containing 20 mg/L. The maximum growth rate is 5 day^{-1}, death rate is 0.06 day^{-1}, and the concentration at half of the maximum rate is 80 mg/L at the design biomass concentration of 5000 mg/L in the reactor, Y = 0.5. The clarifier/thickener is expected to produce 20,000 mg/L sludge in the underflow.

30. Estimate the sludge age and the sludge age for washout in the reactor of Problem 29.

31. A 0.1 m^3/min wastewater containing 100 μg/L of phananthrene is to be treated by adsorption onto a soil bed. The soil has a bulk density of 1.5 g/cm^3 and an organic carbon content of 10%. Pressure drop considerations suggest that a bed of 10 m^2 and 3 m high should be used. Assuming linear partitioning between the water and the soil, estimate the time required to fully saturate the soil.

32. The batch nature of an adsorption system (i.e., the adsorption for a finite time followed by shutdown and regeneration) can be overcome with a continuous movement of adsorbent through a column. By consideration of the material balances on a continuous *absorber* develop design equations for such a system. Derive a formula for the minimum liquid to solid flowrate.

33. The equilibrium between ammonia and air in a certain range of concentrations can be described by 0.5 (mg NH_3/kg air)/(mg NH_3/kg water). Estimate the minimum required air flow to strip ammonia from a 10 m^3/min wastewater flow. The concentration of ammonia in the wastewater is 100 mg/L.

34. For the sand filter of Example 7.15, make a plot of pressure drop vs. flowrate.

35. For the sand filter of Example 7.15, estimate the pressure drop if the upper 1.5 m were filled with 0.8-mm particles and the lower 1.5 m were filled with the 0.4-mm particles.

36. Estimate the pressure drop of a 1 m^3/min water flow in a sand filter of 0.4 mm particles that is 3 m in diameter and 1 m high. If water were allowed to flow through the filter by gravity only, how deep a layer of water would need to be present at the top of the filter?

37. Estimate the mass of ion exchange resin required to soften a 1 m^3/h flow of water containing 200 mg/L of calcium carbonate hardness for 7 days without regenerating. 250 mEq of hardness/100 g of dry ion exchange resin can be exchanged. If the bulk density of the ion exchange resin is 1.5 g/cm^3, estimate the volume of exchanger required.

38. The osmotic pressure of sea water (containing 35,000 mg/L of dissolved solids/salts) is 2740 kPa at 25°C. Estimate the amount of demineralized water that can be produced per m^2 of membrane if its mass transfer coefficient is 0.2 L/(m^2·day·kPa). The pressure on the salt water side of the membrane is 2000 kPa, while the freshwater side is essentially at atmospheric pressure. Assume that the osmotic pressure is a linear function of the dissolved solids content of the water and the product water contains 1000 mg/L dissolved solids.

REFERENCES

Corbitt, R.A. (1989) *Standard Handbook of Environmental Engineering,* McGraw-Hill, New York.

Haith, D.A. (1980) A mathematical model for estimating pesticide losses in runoff, *J. Environ. Qual., 9,* 3, 428–433.

Knauss, J.A. (1978) *Introduction to Physical Oceanography,* Prentice-Hall,Upper Saddle River, NJ, 338 pp.

Koh, C.Y. and L. N. Fan (1970) Mathematical models for the prediction of temperature distributions resulting from the discharge of heated water into large bodies of water, 16130DW010/70, Water Pollution Control Research Series, Washington, D.C.

Lacey, R.E. (1972) Membrane separation processes, *Chemical Engineering,* McGraw-Hill, New York.

Metcalf and Eddy (1996) *Wastewater Engineering, 3rd ed.,* McGraw-Hill, New York, 1334 pp.

Millington, R.J. and J.M. Quirk (1961) Permeability of porous solids, *Trans. Faraday Soc., 57,* 1200–1207.

Mills, W.B. et al. (1982) Water quality assessment: a screening procedure for toxic and conventional pollutants, U.S. EPA, Environmental Research Laboratory, Athens, GA, EPA 600/6-82/004.

Reynolds, T.D. and P.A. Richards (1996) *Unit Operations and Processes in Environmental Engineering, 2nd ed.,* PWS Publishing, Boston, MA, 798 pp.

Rhoads, D.C. (1974) *Oceanography and Marine Biology Annual Review,* 12, 263 pp.

Soil Conservation Service (1972) *National Engineering Handbooks,* Section 4, Hydrology, U.S. Department of Agriculture.

Thibodeaux, L.J. (1996) *Environmental Chemodynamics,* John Wiley & Sons, New York, 593 pp.

8 Soil Pollution and Its Control

8.1 INTRODUCTION

Contamination of soil was recognized only in the past few decades as a significant pollution problem. The primary paths of exposure to pollutants are via ingestion of food or water or airborne inhalation. Soil does not generally represent a direct path of exposure. Some portion of soil and the contaminants it may contain may be suspended in the air or in water potentially providing some exposure. In addition, contaminants may desorb into an air or water phase from the soil particles to which it is associated and provide a path of exposure.

In this chapter we will examine the basic characteristics of soil, the contaminant fate and transport characteristics in soil, and processes and approaches to remove or otherwise render harmless contaminants in soils. In the previous chapter we touched on the transport of soils by flowing water, so we will focus here on stable soils, that is soils that are largely immobile. In such soils, contaminants can move only by release into the pore fluid (either air or water) and by subsequent pore fluid transport processes. Significant groundwater flows occur in soils that can lead to advection and transport of contaminants. In addition, there is the possibility of unsaturated soils, that is, soils in which the pore space is only partially filled with water. Both of these issues will be addressed in detail in this chapter.

8.2 SOIL CHARACTERISTICS

Soils are of interest as a medium for the transport of fluids because they are *porous* and *permeable.* Porous refers to the pore spaces that constitute the gaps between the soil grains. The permeability of a soil refers to its ability to translate hydraulic forces into fluid flow and it is a function of the total volume and structure of this pore space. The permeability of a soil to movement of a particular fluid also depends on the fraction of the pore space filled by that fluid. If the pore space is *saturated* as in the bottom sediment considered in the previous chapter, the permeability of a medium is at a maximum. If the fluid only fills a portion of the pore space as in the upper layers of soil, the permeability to the fluid decreases and *wettability* becomes important. Wettability is the characteristic of certain fluids to adhere to the surface of the soil grains and it significantly influences the mobility and retention of a fluid by the media.

The basic system of soil and water that is of interest is shown in Figure 8.1. Near the surface, the pore spaces of most soils are partially filled with both air and water. This is the *unsaturated* or *vadose zone* of the soil. Immediately after a rainfall, the water that infiltrates into the soil drains into the deeper soil. Due to the fact that water tends to wet most soils, however, a portion of the water is retained. The quantity of water that remains, typically measured on a volume water per volume total soil basis, is termed the *field capacity* of the soil. Below this zone is often an *aquifer* in which the pore spaces are *saturated,* or filled, with water. This water can either be *confined* by layers of low permeability soil or *unconfined* and free to rise and fall according to the water that infiltrates from above. An unconfined aquifer also is called a *water table* aquifer and the water table refers to the height of the column of water in the aquifer as measured by a well. Due to the wetting of the soil by water, there is additional water above the level that would show in a well, but this water is held by the soil in a *capillary fringe* and not generally available for flow or removal. The water level as measured in a well is a measure of the *hydraulic head* which is the driving force

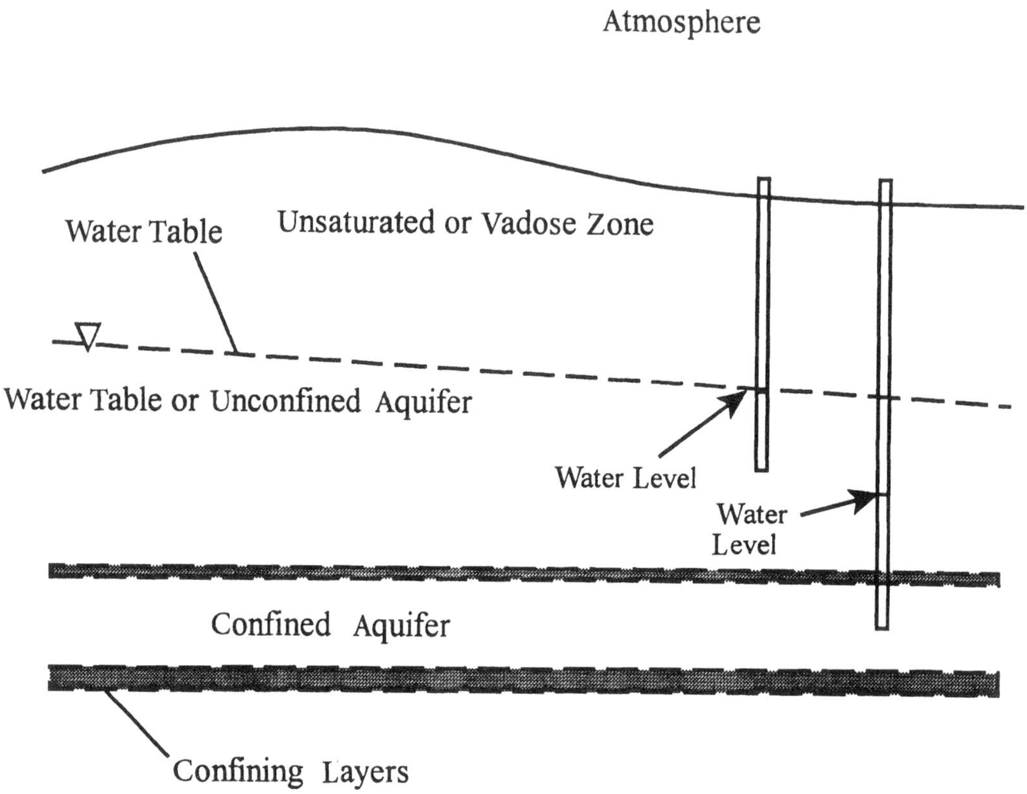

FIGURE 8.1 Depiction of soil-water system of interest showing vadose zone and both an unconfined and confined aquifer. Note that the head or water level in the well in the confined and unconfined aquifer are not equal unless there is a hydraulic connection between the two aquifers.

for water flow. A well used to measure the hydraulic head is termed a *piezometer*. The water in a confined aquifer may be under pressure causing the water level in a well to rise toward the surface. A pressurized confined aquifer is referred to as an *Artesian aquifer*. Both a confined and unconfined aquifer also are depicted in Figure 8.1.

The study of fluid motion in a porous medium will start with a discussion of the soil grains that make up the medium, the pore spaces formed by the space between these grains, and the permeability and wettability of the medium that results. Each of these will be discussed in turn.

8.2.1 Solid Phase Characteristics

The basic composition of soils has been discussed previously in Chapters 4 and 7. Soil is composed of a mixture of sand, silt, and clay in varying proportions. Sand is the description of soil particles of diameter greater than 20 μm (International Soil Science Society) to 50 μm (U.S. Department of Agriculture). It is of high permeability or low resistance to flow and contains minimal organic matter and has a low sorption capacity. Silt (2 to 20–50 μm in size) and clay (<2 μm) exhibit more resistance to flow and account for most of the sorption capacity of a soil. All soils are some mixture of sand, silt, and clay and Figure 8.2 shows a classification of various soil types by their composition. The basic characterization of soils is thus by their particle size distribution. In addition to the textural classes depicted in Figure 8.2, a soil might be characterized by its mean particle size, d_{50}. The notation d_{50} means that 50% of the particles in a soil are equal to or smaller than this diameter. Another common indicator of particle size distribution is the *uniformity coefficient*, which is the

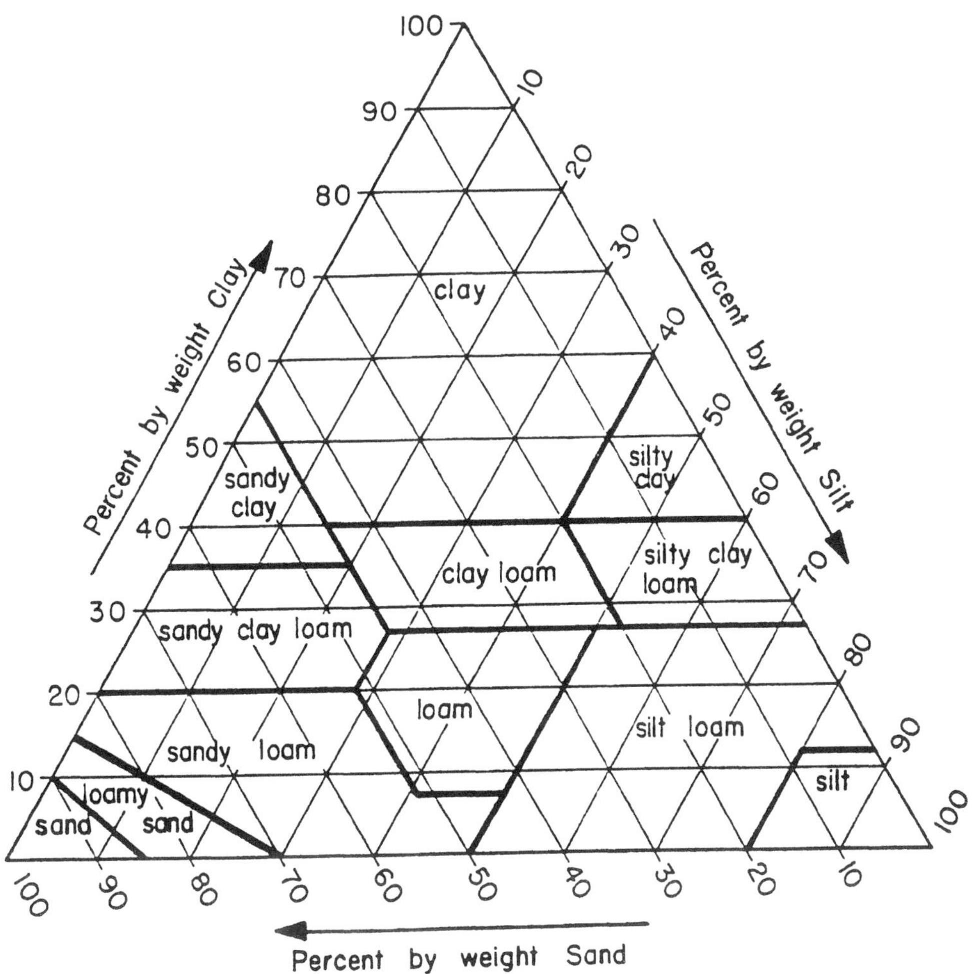

FIGURE 8.2 Soil classification system. (From Hillel, D. (1980) *Fundamentals of Soil Physics*, Academic Press, New York. With permission.)

ratio of d_{60} to d_{10}, or the ratio of the particle size greater than 60% of the particles to that greater than 10% of the particles.

Soils develop via two basic mechanisms, deposition by winds (aeolian soils) and water (alluvial soils) or by weathering of parent bedrock. Soils that form via deposition processes tend to be layered. For example, during high water flow conditions only coarser material may settle while under low flow conditions settling of finer grained material may occur. These layers of differing soil type exhibit different permeabilities and result in the development of preferential flow channels. Soils formed by weathering of bedrock may be layered as well but often not as distinctly. The fine clays in this soil can be washed out of upper layers (the process of eluviation) and washed in to lower layers (the process of illuviation). The upper layers of soil typically contain the bulk of the organic matter in soil due to the presence of decaying plant and animal matter. A surface soil might contain 2 to 4% organic matter although peat and some other high organic matter surface materials may contain considerably more. In deeper soils, the organic matter may be less than a few tenths of a percent.

Of course our primary interest is not in the solid particles but in the void or pore spaces between particles where the mobile fluid, either air or water, resides. The most basic characterization of the void space of a soil is its volume, the voidage or porosity of a soil. The porosity as we have used it previously is defined simply as the ratio of the volume of void space to the total volume of the soil,

$$\varepsilon = \frac{V_v}{V_t} \tag{8.1}$$

The bulk or dry density of a soil, ρ_b, can be related to the porosity through the solid grain density, ρ_s,

$$\rho_b = \rho_s(1-\varepsilon) \tag{8.2}$$

The porosity of most soils does not vary a great deal, remaining in the range of 25 to 50% and typically between 35 and 45%. Only a portion of this porosity may be available for penetration by a mobile fluid but the total porosity typically remains within the range of 25 to 50%. A random packing of identical spheres exhibits a porosity of about 0.40 to 0.44. A regular packing of identical spheres produces a porosity as high as 0.48 (a cubic lattice packing) or as low as 0.26 (a rhombohedral lattice packing). Of course, the presence of different particle sizes can reduce these theoretical porosities as a result of the smaller particles filling the void space between larger particles.

The shape of the void space between particles also is important but it is not easily estimated except in uniform packings of identical spheres. In any real media, the pores are a distribution of sizes just as the solid grains exhibit a distribution of sizes. The pore size distribution is usually measured by the pressure required to force a volume of liquid into the pores and is a function of *wettability*. The pore size distribution is an important consideration in flow through porous media with the finest pores representing the greatest resistance to flow. Thus, as will be indicated later, the permeability of a medium may be dominated by the effect of a relatively small fraction of large pores. As is often the case with irregular flow paths, the hydraulic diameter, 4 times the ratio of the pore volume to the wetted surface area, is often used to characterize the flow through the pore space. Considering a single sphere and the void space surrounding it

$$\begin{aligned}
\text{Volume of soil grain} &= \frac{1}{6}\pi d_p^3 \\
\text{Volume of void space} &= \frac{\varepsilon}{1-\varepsilon}\frac{1}{6}\pi d_p^3 \\
\text{Wetted grain surface area} &= \pi d_p^2 \\
\text{Hydraulic diameter } d_h &= \frac{4\dfrac{\varepsilon}{1-\varepsilon}\dfrac{1}{6}\pi d_p^3}{\pi d_p^2} = \frac{2}{3}\frac{\varepsilon}{1-\varepsilon}d_p
\end{aligned} \tag{8.3}$$

Often the pore space is not filled with the flowing fluid. *Saturation* is the indicator of the relative proportion of the pore space filled with a particular fluid. For example, if ε_w is the fraction of the total volume filled with water (the *volumetric content*) then the water saturation, φ_w, is

$$\varphi_w = \frac{\varepsilon_w}{\varepsilon} \tag{8.4}$$

The hydraulic diameter, as a ratio of (four times) flow volume to wetted perimeter depends on the saturation. This quantity also depends upon the wetting characteristics of the flowing fluid for the soil because some fluids (typically water) may wet the entire perimeter of the pore space at very

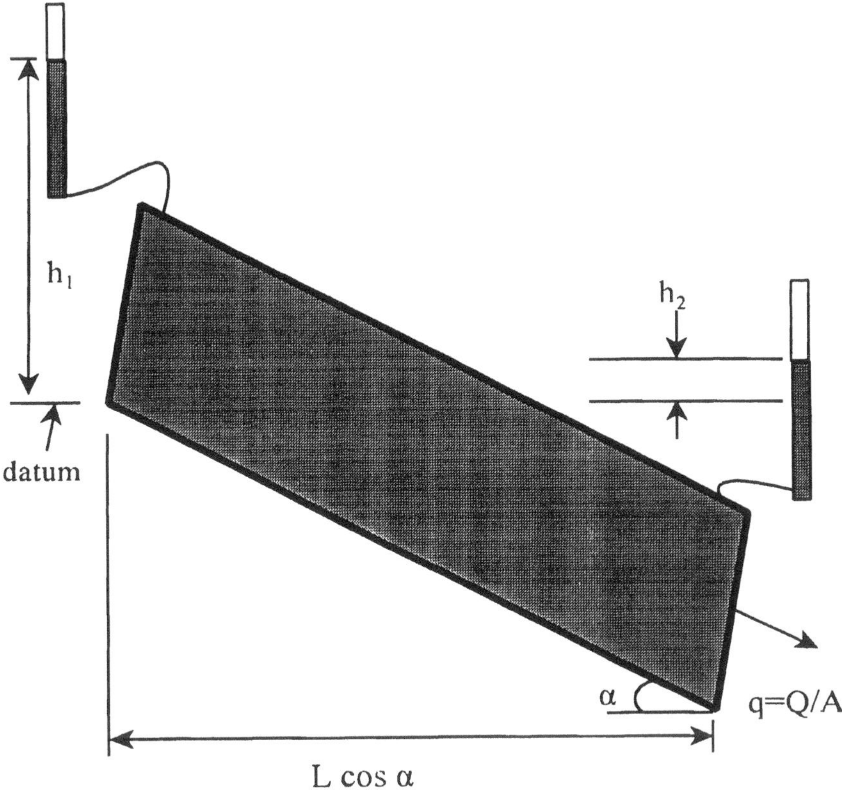

FIGURE 8.3 Soil column for the development of Darcy's Law.

low saturations. Because of their very different behavior, let us discuss flow in water-saturated pore spaces and partially saturated pore spaces separately.

8.2.2 Flow of Fluid in Fully Saturated Media

Fully saturated media are media in which the pore spaces are filled with the flowing fluid of interest, typically water. Permeability is defined by Darcy's Law which states that the volumetric flowrate of the fluid, Q, through an area, A, of a porous medium is proportional to the hydraulic head applied across the medium. Hydraulic head is the height that a column of liquid would reach in a manometer if connected to the porous medium and is measured by the water levels in the wells in Figure 8.1. For analysis of Darcy's Law, let us consider the soil column in Figure 8.3. The different heights of liquid in the manometers at each end of the soil column indicate that a gradient in hydraulic head across the medium equal to $\Delta h = h_2 - h_1$ exists. Darcy's Law then states

$$\frac{Q}{A} = q = -k_p \frac{\Delta h}{L}$$

or more generally (8.5)

$$q = -k_p \frac{dh}{dx}$$

k_p is the hydraulic conductivity or, using a term that is largely out of fashion, the coefficient of permeability. The negative sign recognizes that the change in liquid column height (or head) must be negative (i.e., $h_2 - h_1 < 0$, or $h_2 < h_1$) for a positive flow, that is a flow from left to right in

Figure 8.3. The conductivity has units of velocity and the quantity Q/A or q is often called the Darcy velocity, a superficial velocity that represents the average fluid velocity in the porous medium based on the entire cross-sectional area of flow. In reality, of course, only a portion of the cross-sectional area given by the porosity ε is actually available for flow. Thus, the actual average fluid velocity in the interstitial spaces, u_i, is given by

$$u_i = \frac{q}{\varepsilon} \tag{8.6}$$

Before discussing Darcy's Law further, let us examine the meaning of the driving force, the hydraulic head. The driving force for flow of the fluid in the pores of a soil is its potential energy. Referring back to Chapter 4, the potential of a fluid is the energy it contains relative to some reference state. The contributions to its total energy are enthalpy, potential energy and kinetic energy. If we consider a fluid moving at velocity v, elevation z, and pressure P, its energy content (per unit mass) relative to an unmoving fluid at the reference elevation ($z = z_0$) and reference pressure ($P = P_0$) is

$$\hat{E} = \hat{E}_I + (P - P_0)\hat{V} + g(z - z_0) + \frac{U^2}{2} \tag{8.7}$$

For flow of a fluid in soils under natural gradients, the changes in internal and kinetic energy are generally negligible compared to the effects of pressure differences and gravity. Recognizing that the specific volume, $\hat{V}$ is just the inverse of density, ρ, and using the reference elevation $z_0 = 0$ where the pressure is $P_0 = 0$, the energy per unit mass becomes

$$\hat{E} = \frac{P}{\rho} + gz \tag{8.8}$$

As discussed in Chapter 3, the pressure measured by the manometers in Figure 8.3 is related to the height of liquid in the manometer relative to the connection point, h_p, through $P = \rho g h_p$. Thus, the hydraulic head can be seen as the sum of a *pressure head*, h_p, and an *elevation head*, h_e, and through Equation 8.8 is related to the energy content or potential driving flow through the medium.

$$h = h_p + h_e = \frac{\hat{E}}{g} \tag{8.9}$$

hydraulic head = pressure head + elevation head = flow potential

In a static column of fluid, the hydraulic head is a constant. As one moves down the column of fluid, the pressure head increases to exactly offset the reduction in elevation head, a consequence of the hydrostatic equation, $\Delta P/\rho g = \Delta h_p = \Delta h_e$. Only when these heads do not balance can flow occur. In a water table aquifer where the free surface is at atmospheric pressure, the pressure head can be taken as zero and flow will occur when the elevation head changes with position, that is, when there is a slope to the free surface. In a confined aquifer, the pressure head may be nonzero everywhere in the aquifer and flow may occur due to pressure variations even if the elevation head is constant. In such a case, the variations in pressure head result in the variations in total head required for flow.

The total head in a groundwater aquifer can be measured by the water level in a well. If a well is installed solely for the purpose of measuring elevation and total head, it is termed a *piezometer.*

At the free surface in a well open to the atmosphere, the pressure head relative to atmospheric pressure is zero and the elevation head at the free surface then equals the total head. In a water table aquifer, the water level in the well and the water table or level in the surrounding media will be the same. Measurement of the water level elevation in a well, that is, the total hydraulic head, is then as simple as measuring the depth to water in the well minus the elevation of the measuring point to a common reference datum, such as sea level or any locally convenient reference point. Note that the ground surface does not provide a suitable reference or *datum* since its level will change leading to misinterpretation of the changes in elevation in head of the groundwater aquifer. In a confined aquifer, the water level also can be measured by a piezometer. The water level must continue to rise until the elevation head is equal to the total head since the pressure head at the free surface is zero relative to atmospheric pressure.

This form of Darcy's Law can be understood if one models the porous medium as a bundle of capillary tubes. This is a crude model of the pore structure of a porous medium but it does indicate the basic fluid flow behavior. Considering a single cylindrical pore in the bundle, Newton's second law, or the law of conservation of momentum, says that the time rate of change of momentum in a given direction is balanced by the net force acting in that direction,

$$\frac{d(mv)_i}{dt} = \sum F_i \tag{8.10}$$

Considering steady flow through the pore, then the net force acting on the control volume shown in Figure 8.4 is zero. The forces acting on that control volume in the axial direction include the viscous force associated with the shear stress from Newton's law of viscosity and the pressure force. Flow within a porous medium is one of the few situations in the environment where turbulence is not generally important and viscous effects dominate. Under such conditions, it is possible to derive analytical expressions between velocity and pressure.

The viscous force is the product of the shear stress (τ_{xr}, force per unit area) and the area on which it acts (2 π r dx). At radial position r, this is

$$\text{Viscous force}\big|_r = \tau_{xr} 2\pi r dx\big|_r = \mu \frac{\partial u}{\partial r} 2\pi r dx\big|_r \tag{8.11}$$

The viscous force is thus the "pull" associated with the movement of fluid outside the cylindrical shell transferred to the fluid within the shell via viscosity. As shown in Figure 8.4, the net viscous force is the difference between the force tending to accelerate the fluid in the cylindrical shell from below and the force tending to slow the fluid in the cylindrical shell from above.

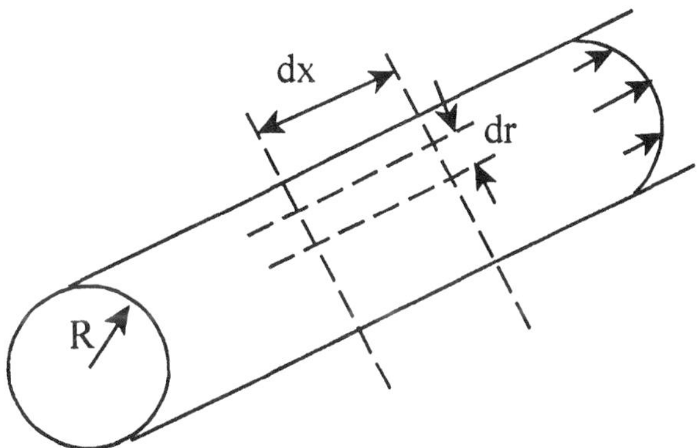

FIGURE 8.4 Control volume for flow in a single capillary in a bundle of capillary tubes model of porous medium.

$$F_v = \tau_{xr}(2\pi r)dx\big|_r - (-\tau_{xr})(2\pi r)dx\big|_{r+dr}$$
$$= -\mu\frac{\partial u}{\partial r}(2\pi r)dx\big|_r - \left(\mu\frac{\partial u}{\partial r}\right)(2\pi r)dx\big|_{r+dr} \qquad (8.12)$$

Similarly the pressure force is the product of the area and the pressure (also a force per unit area and equal to the hydraulic head, h, times ρg). The pressure force acts on the area 2πr dr and thus the net pressure force acting on the cylindrical shell in Figure 8.4 is

$$F_p = \rho g h\, 2\pi r\, dr\big|_x - \rho g h\, 2\pi r\, dr\big|_{x+dx} \qquad (8.13)$$

Summing the pressure and viscous forces, dividing by the product (dr dx) and collecting terms gives

$$\frac{\mu\frac{\partial u}{\partial r}(2\pi r)\big|_{r+dr} - \mu\frac{\partial u}{\partial r}(2\pi r)\big|_r}{dr} = \frac{\rho g h(2\pi r)\big|_{x+dx} - \rho g h(2\pi r)\big|_x}{dx} \qquad (8.14)$$

In the limit as the cylindrical shell becomes a differential element, that is when dr dx → 0.

$$\frac{\partial}{\partial r}(2\pi r)\mu\frac{\partial u}{\partial r} = \frac{\partial(\rho g h)}{\partial x}(2\pi r) \qquad (8.15)$$

Dividing both sides by 2πr, the constant 2 π cancels but r cannot be moved across the differential. Taking also the viscosity as a constant gives

$$\frac{\mu}{r}\frac{\partial}{\partial r}r\frac{\partial u}{\partial r} = \rho g\frac{\partial h}{\partial x} \qquad (8.16)$$

as the equation governing the relationship between the velocity in the capillary tube and the pressure gradient along its length. Note that the length of the capillary is typically longer by a factor L_c/L than the length of the medium, and the pressure gradient is smaller than that based on the length of the medium by the same factor. The factor L_c/L is termed the tortuosity factor, τ, and the pressure gradient is given by

$$\frac{\partial h}{\partial x} = \frac{\partial h}{\partial L}\frac{L}{L_c} = \frac{1}{\tau}\frac{\Delta h}{L} \qquad (8.17)$$

Taking this gradient as constant, the velocity will only depend upon the radius r and the equation becomes the ordinary differential equation

$$\frac{d}{dr}r\frac{du}{dr} = \left(\frac{\rho g\Delta h}{\mu L\tau}\right)r \qquad (8.18)$$

which can be integrated once to give

$$r\frac{du}{dr} = \frac{\rho g \Delta h}{\mu L \tau}\frac{r^2}{2} + C_1 \tag{8.19}$$

where C_1 is a constant of integration. Dividing by r and integrating once again

$$u = \frac{\rho g \Delta h}{\mu L \tau}\frac{r^2}{4} + C_1 \ln r + C_2 \tag{8.20}$$

where C_2 is the second constant of integration. Between r = 0 (the center of the capillary) and r = R (the capillary wall), the velocity must be everywhere finite. That is, C_1 must be zero or else the logarithm term would not be bounded at r = 0. In addition, friction at the tube wall causes the velocity to approach 0 there. Therefore

$$\begin{aligned} 0 &= \frac{\rho g \Delta h}{\mu L \tau}\frac{R^2}{4} + C_2 \\ C_2 &= -\frac{\rho g \Delta h}{\mu L \tau} R_2 \end{aligned} \tag{8.21}$$

This gives the velocity profile in the capillary tube as

$$u = \frac{\rho g \Delta h}{4 \mu L \tau}(r^2 - R^2) \tag{8.22}$$

This implies that the velocity is positive when Δh (= $h_2 - h_1$) is negative (since R > r). This simply means that the pressure at the downstream end of the capillary tube is less than the pressure at the upstream end when the flow is positive (i.e., from upstream to downstream).

Although Equation 8.22 does describe the velocity profile in the capillary tube, rarely are we interested in this level of detail. Instead we are more interested in the relationship between the volumetric flowrate and the pressure drop, as in Darcy's Law. The volumetric flowrate is the product of the velocity and the flow area or

$$\begin{aligned} Q &= \int_0^R u(r)\varepsilon 2\pi r\, dr \\ &= \int_0^R \frac{\varepsilon 2\pi \rho g \Delta h}{4\mu L \tau}(r^3 - rR^2)dr \\ &= \frac{\varepsilon 2\pi \rho g \Delta h}{4\mu L \tau}\left(\frac{r^4}{4} - R^2\frac{r^2}{2}\right)\Bigg|_0^R \\ &= -\frac{\varepsilon \pi \rho g \Delta h}{8\mu L \tau}R^4 \end{aligned} \tag{8.23}$$

This approach recognizes that the flow area is on average equal to ε times the total area. The superficial velocity, or Darcy velocity is then

$$\frac{Q}{A} = \frac{Q}{\pi R^2} = -\frac{\varepsilon R^2 \rho g}{8\mu\tau}\frac{\Delta h}{L} \tag{8.24}$$

or, by comparison to Darcy's Law,

$$k_p = -\frac{\varepsilon R^2 \rho g}{8\tau\mu} \tag{8.25}$$

Thus, the fluid hydraulic conductivity is seen to encompass fluid dependent properties (ρ and μ) as well as media-dependent properties ($\varepsilon R^2/\tau$). The factor of $^1/_8$ depends on the geometry of the pore through which the fluid is moving. For example, in a wide rectangular pore, the factor is $^1/_3$. If we assume that the pore space is formed from identical spheres and replace the capillary radius in Equation 8.25 with the hydraulic radius from Equation 8.3, the hydraulic conductivity is given by

$$k_p = \frac{1}{72\tau}\frac{\varepsilon^3}{(1-\varepsilon)^2}d_p^2\frac{\rho g}{\mu} \tag{8.26}$$

This can be compared to the Kozeny-Carmen equation which predicts the permeability in a bed of uniform spheres

$$k_p = \frac{1}{180}\frac{\varepsilon^3}{(1-\varepsilon)^2}d_p^2\frac{\rho g}{\mu} \tag{8.27}$$

This is equivalent to Equation 8.26 with a tortuosity of 2.5. Saffman (1959) developed a random pore model that arrived at the result that the tortuosity is 3. In practice, neither model can be of much use in predicting permeability except in a clean and uniform sand bed. In a real soil or a fouled sand filter, there is a large variation in pore shapes and sizes which leads to ambiguity, at best, in the analysis of simple permeability relationships.

The separation of fluid and media-dependent properties in the hydraulic conductivity is often explicitly recognized in an alternative formulation of Darcy's Law.

$$\frac{Q}{A} = q = -\kappa_p\frac{\rho g}{\mu}\frac{\Delta h}{L} \tag{8.28}$$

Here, κ_p is the intrinsic permeability of the medium and is, in principle, solely dependent upon the characteristics of the medium. In reality, some fluids may alter the medium such as water flowing through swelling clays and thus there may also be a dependence of the intrinsic permeability on the type of flowing fluid. By comparison to Equation 8.24, the intrinsic permeability is seen to be proportional to the square of the capillary radius ($\kappa_p = 1/180[\varepsilon^3/(1-\varepsilon)^2]d_p^2$ in the Kozeny-Carmen equation) suggesting that finer grained soils exhibit far lower permeabilities. As a result of this phenomenon, groundwater flows and contaminant migration tends to be far greater in coarse fill zones in soil. As examples, old abandoned creek beds, tree root zones, and any sort of man-made construction from mines to foundation piles would tend to contain and be surrounded by more loosely consolidated and coarser grained material than the soil outside these areas. As a result, these zones offer paths for preferential groundwater and contaminant flow. The assessment of contamination at a site can often be reduced to a problem of finding these preferential flow paths.

Intrinsic permeability and hydraulic conductivity are used in a number of different systems of units. Among the most common and their relationships are

TABLE 8.1
Typical Values of Permeability and Hydraulic Conductivity of Soils

Soil Type	Approximate Range of Intrinsic Permeability or Hydraulic Conductivity (Water)			
	κ_p D ~ μm²	κ_p cm²	k_p m/day	k_p gal/day/ft²
Gravel	$100–10^5$	$10^{-6}–10^{-3}$	$10^2–10^5$	$2000–10^7$
Sand	$1–10^3$	$10^{-8}–10^{-5}$	$1–10^3$	20–20000
Silt and sand	0.01–100	$10^{-10}–10^{-6}$	0.01–100	0.2–2000
Silt	$10^{-4}–1$	$10^{-12}–10^{-8}$	$10^{-4}–1$	$2(10)^{-2}–20$
Silty clay	$10^{-6}–0.1$	$10^{-14}–10^{-9}$	$10^{-6}–0.1$	$2(10)^{-5}–2$
Clay	$10^{-7}–10^{-4}$	$10^{-15}–10^{-12}$	$10^{-7}–10^{-4}$	$2(10)^{-6}–0.002$
Sandstone	$10^{-5}–0.1$	$10^{-13}–10^{-9}$	$10^{-5}–0.1$	$2(10)^{-4}–2$
Fractured rocks	$10^{-4}–10$	$10^{-12}–10^{-7}$	$10^{-4}–10$	0.002–200

$$\text{Hydraulic Conductivity} \quad 1\ \text{m/day} = 24.54\ \text{USgal/day/ft}^2$$

$$\text{Intrinsic Permeability} \quad 1\ \text{D} = 9.87 \times 10^{-9}\ \text{cm}^2 \tag{8.29}$$

$$1\ \text{D} \approx 0.835\ \text{m/day (water)} \approx 1\ \mu\text{m}^2$$

The darcy (D) is the permeability that gives a darcy velocity, q, of 1 cm/s for a fluid with a viscosity of 1 cp (approximately that of water) under a pressure gradient of 1 atm/cm. As indicated above, a soil with a permeability of 1 D would provide a water flow velocity of 0.835 m/day if the water level dropped 1 m per m of length and 0.835 cm/day if the water level dropped 1 m per 100 m of length. Table 8.1 summarizes the range of permeabilities of various soil types

As can be seen in the above table, the observed values of hydraulic conductivity or intrinsic permeability vary over a very wide range. It becomes very difficult to assess, even on the basis of extensive measurements, an estimate of the effective permeability of a subsurface formation. Zones of low and high permeability that differ not by factors of 2 or 3 but by as much as 10 orders of magnitude are commonplace. This is the fundamental problem of understanding fluid flow in the subsurface. It influences the rate and direction of contaminant migration in the subsurface in a profound manner.

It remains to emphasize that the conductivity or permeability must still be multiplied by the gradient in the hydraulic head. These are often linked quantities. If a large head gradient existed across a very permeable medium, a large flow would result that would tend to raise the water level and the hydraulic head downstream. The net effect is that the range of groundwater flowrates is typically less than the range of hydraulic permeabilities. Flowrates tend to be higher in more permeable media but the hydraulic head gradients tend to be smaller than in less permeable media. Typical groundwater flowrates vary from a few centimeters per year to a few hundred meters per year.

Darcy's Law also defines the manner in which the head varies in a groundwater system. If we consider steady one-dimensional flow of groundwater in a zone of constant cross-sectional area, a material balance on the differential element of length gives

$$qA|_x - qA|_{x+dx} = 0$$

dividing by dx (8.30)

$$\frac{q|_x - q|_{x+dx}}{dx} = 0$$

as $dx \rightarrow 0$

$$\frac{d}{dx}q = 0$$

Which simply means that the velocity q is a constant. Substituting the general form of Darcy's Law,

$$\begin{aligned} \frac{d}{dx}Ak_p\frac{dh}{dx} &= 0 \\ \frac{d}{dx}k_p\frac{dh}{dx} &= 0 \end{aligned} \tag{8.31}$$

Note that both Equations 8.30 and 8.31 should contain minus signs that can be divided out since the terms are set equal to zero. If the conductivity is uniform in a region, Equation 8.31 can be further written

$$\frac{d^2h}{dx^2} = 0$$

or

$$\frac{dh}{dx} = \text{constant} = \frac{\Delta h}{L} \tag{8.32}$$

$$h = \frac{\Delta h}{L}x + h_0$$

That is, the hydraulic head varies linearly in a steady-state groundwater flow in a homogeneous medium. Example 8.1 illustrates the relationship between head gradient and flow. Note that it is the fact that the head gradient is a constant that allows us to replace the gradient with the difference in hydraulic head over length of travel. It should be emphasized that this is only true in steady (i.e., not time-dependent) flow in a medium with uniform permeability without sources or sinks such as withdrawal or injection wells. If any of these other conditions apply, the material balance must be modified and Equation 8.32 no longer holds.

Example 8.1: Relationship between flow and permeability and variations in hydraulic head

Consider two piezometers that indicate water levels of 2 m and 1 m above sea level, respectively, in a deep water table aquifer. Between the two piezometers is 9 m of sand with a conductivity of 1 m/day and 1 m of silt with a conductivity of 0.01 m/day. What would a piezometer indicate if it were placed at the interface between the sand and the silt? What is the groundwater velocity?

Since the change in height is negligible in the deep water table, the cross-sectional area of groundwater flow is constant and the velocity is also constant. Thus, the difference between the velocity calculated across the sand and silt must be zero.

$$\begin{aligned} q_{sand} - q_{silt} = 0 &= \left(k_p\frac{\Delta h}{L}\right)_{sand} - \left(k_p\frac{\Delta h}{L}\right)_{silt} \\ &= 1\ \text{m/day}\frac{2\ \text{m} - h}{9\ \text{m}} - 0.01\ \text{m/day}\frac{h - 1\ \text{m}}{1\ \text{m}} \end{aligned}$$

Solving this equation for the head at the sand-silt interface, h, h = 1.917. Almost 92% of the total change in head is across the less permeable silt.

The velocity can then be calculated in either the sand or the silt

$$q = \left(k_p \frac{\Delta h}{L}\right)_{sand} = \left(k_p \frac{\Delta h}{L}\right)_{silt}$$

$$= (1 \text{ m/day}) \frac{0.083 \text{ m}}{9 \text{ m}} = (0.01 \text{ m/day}) \frac{0.917 \text{ m}}{1 \text{ m}}$$

$$= 0.00917 \text{ m/day} = 3.35 \text{ m/year}$$

Alternatively, the two resistances in series approach employed to derive the overall mass transfer coefficient in the two-film theory can be employed.

$$\left(\frac{L}{k_p}\right)_{overall} = \left(\frac{L}{k_p}\right)_{sand} + \left(\frac{L}{k_p}\right)_{silt}$$

$$\left(\frac{L}{k_p}\right)_{overall} = 0.00917 \text{ day}^{-1}$$

And the velocity is then again

$$q = \left(\frac{k_p}{L}\right)_{overall} (\Delta h) = (0.00917 \text{ day}^{-1})(1 \text{ m}) = 0.00917 \text{ m/day}$$

Equation 8.31 must be modified when the area through which the groundwater flow changes. This commonly occurs in a shallow water table aquifer as might occur in seepage through a dike or levee. If the depth of the aquifer is not much larger than the change in elevation in the water table, the change in flow area must be considered.

If the flow area is the height of the water table above an impermeable layer, h, times its width, w, Equation 8.31 becomes

$$\frac{d}{dx} qhw = 0$$

$$\frac{d}{dx} k_p wh \frac{dh}{dx} = 0 \tag{8.33}$$

$$\frac{k_p}{2} w \frac{dh^2}{dx} = Q = \text{constant}$$

Integrating between two groundwater levels, h_1 and h_2, and defining $q_1 = Q/(h_1 w)$, this becomes

$$\frac{Q}{h_1 w} = q_1 = \frac{K_p}{2L} \frac{h_1^2 - h_2^2}{h_1} \tag{8.34}$$

Note that although the volumetric flowrate is constant, the groundwater velocity changes. Note also that unlike the linear relationship of Equation 8.32, there is now a parabolic relationship between the volumetric flowrate, Q, and head, h.

Equations 8.31 and 8.34 can be compared by writing $h_1 = h_2 + \Delta h$. Then 8.34 can be written

$$\frac{Q}{h_1 w} = q_1 = \frac{k_p}{2L}\frac{h_1^2 - (h_1 - \Delta h)^2}{h_1}$$

$$= \frac{k_p}{2L}\frac{h_1^2 - (h_1^2 - 2h_1\Delta h + \Delta h^2)}{h_1} \tag{8.35}$$

$$= \frac{k_p}{2L}\frac{(2h_1\Delta h - \Delta h^2)}{h_1}$$

Which for $\Delta h << h_1$, $\Delta h^2 << 2h_1\Delta h$, and

$$\frac{Q}{h_1 w} = q_1 = q_2 = k_p \frac{\Delta h}{L} \tag{8.36}$$

That is, Equations 8.30 through 8.32 are now seen as the limit of Equation 8.34 when the change in elevation of a water table is small compared to the depth of the aquifer. Example 8.2 reworks Example 8.1 assuming a shallow aquifer.

Example 8.2: Flow and head gradients in a shallow unconfined aquifer

Repeat Example 8.1 assuming that the piezometers indicate the water levels relative to an impermeable strata and thus represent the entire depth of the aquifer at that point.

In this case the volumetric flow per unit width of aquifer is constant,

$$\frac{Q_{sand}}{w} = \frac{Q_{silt}}{w}$$

Setting the difference between these two equal to zero and solving for the water level at the interface, h, using 1 to represent the values in sand and 2 to represent the values in silt, the positive root is the only physically meaningful solution and it gives

$$h = \left(\frac{k_{p1}h_1^2 L_2 + k_{p2}h_2^2 L_1}{k_{p1}L_2 + k_{p2}L_2}\right)^{1/2} = 1.937 \text{ m}$$

Thus, the water level is 2 m higher at the interface between the sand and silt. The volumetric flowrate per unit width of aquifer is then given by Equation 8.34,

$$\frac{Q}{w} = 1 \text{ m/day}\frac{(2\text{ m})^2 - (1.937\text{ m})^2}{2(9\text{ m})} = 0.014\frac{\text{m}^3}{\text{m}^2 \cdot \text{day}}$$

Dividing by h_1 to get the velocity at the upgradient edge of the sand gives 0.0068 m/day while dividing by h_2 to get the velocity at the downgradient edge of the silt gives 0.014 m/day. Note that the velocity calculated in Example 8.1 lies between these two values. The velocity calculated by Equations 8.30 through 8.32 provides an "average" for the velocity in a shallow water table aquifer.

8.2.3 Flow of Fluid in a Partially Saturated Media

The hydraulic conductivity and intrinsic permeability discussion above assumes that water or some other liquid fills all of the pore space available to it. Near the surface, water only tends to fill part of the pore space with the remainder filled by air. This has two immediate consequences. First, the effective permeability of the media with respect to water is reduced in that the total flow area is no longer available. As long as the water, or other partially saturating fluid, is continuous through the media, this is often handled by correcting the conductivity with a relative permeability that is a function of the saturation of the flowing fluid. That is, the Darcy velocity of a fluid present at pore volume fraction, or saturation, ϕ_f is given by

$$q = -\kappa_r(\phi_f)\kappa_p \frac{\rho g}{\mu}\frac{\Delta h}{L} \tag{8.37}$$

The relative permeability, κ_r, ranges between 0 and 1. A common and relatively simple model for the relative permeability is that given by Brooks and Corey (1964)

$$\kappa_r = \left[\frac{\varphi_f - \varphi_{ir}}{1-\varphi_{ir}}\right]^{\frac{2+3b}{b}} \tag{8.38}$$

Here, φ_f is the saturation of the flowing liquid, φ_{ir} is the saturation beyond which the saturation cannot be reduced (the irreducible residual), and b is a grain-size distribution parameter that varies from about 2.8 in a uniform sand to more than 10 in clays. Typical relative permeability vs. saturation curves are shown in Figure 8.5.

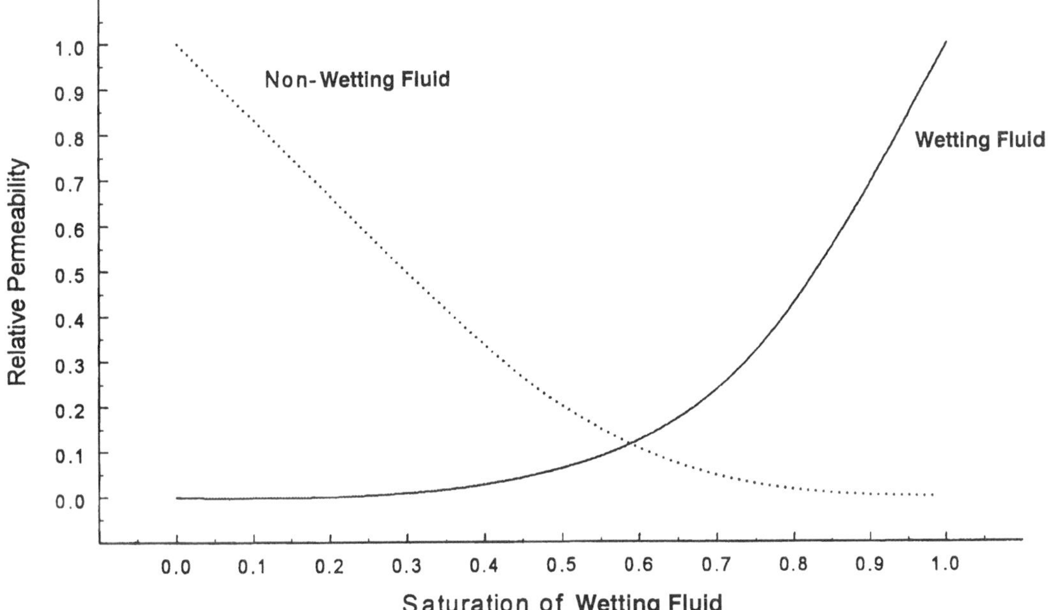

FIGURE 8.5 Relative permeability vs. fluid saturation.

Note that the relative permeability goes to zero well before the saturation goes to zero. This is a reflection of the second consequence of partial saturation. Any liquid will tend to wet the media to some degree. These surface tension forces cause some of the liquid to be retained and a *residual saturation* of that liquid is observed. The suction of the surface tension forces cause the liquid in the unsaturated zone to have a negative pressure head, $h_p<0$. Thus, $h_p>0$ below the water table and in Artesian aquifers, $h_p = 0$ at the water table (relative to atmospheric pressure) and $h_p<0$ above the water table. The negative pressures above the water table give rise to capillary rise and the presence of a *capillary fringe* above the water table.

To understand the causes of the negative pressures in the liquid that wet a porous medium, consider an isolated water droplet in air. Surface tension is an indicator of the amount of work required to change the surface area of the bubble. Consider the force that must be applied across the surface of the bubble to change its radius from r to r + dr and area from $4\pi r^2$ to $4\pi(r + dr)^2$ if the bubble surface tension is σ is given by

$$\text{Work} = \text{Force} \cdot \text{distance} = F dr = \sigma[4\pi(r+dr)^2 - 4\pi r^2]$$

$$F = \frac{\sigma}{dr} 4\pi(r^2 + 2rdr + dr^2 - r^2) \tag{8.39}$$

$$F \approx 8\pi r\sigma$$

At equilibrium this must be balanced by a different in pressures (forces per unit area) across the surface of the bubble.

$$\sum F = 0$$

$$4\pi r^2 P_{in} = 4\pi r^2 P_{out} + 8\pi r\sigma \tag{8.40}$$

$$P_{in} - P_{out} = \frac{2\sigma}{r}$$

This means that the inside pressure of the curved surface is always greater than the pressure on the outside of the curved surface. If we consider the water wetting the sides of an individual capillary, as in Figure 8.6, the air in the region above the interface is *inside* the curved surface and Equation 8.40 suggests that the pressure in the air phase is greater than the pressure in the water. Because the air pressure is atmospheric this means that the pressure in the water is below atmospheric or a negative gage pressure.

Let us evaluate this phenomenon as it applies to capillary rise. To do so, we must consider the point of contact between the liquid and a solid. The effect of interfacial tensions is to minimize surface area. The forces per unit width of surface are as shown in Figure 8.7 where σ_{gs} is the interfacial tension at the gas-solid interface, σ_{sl} is the interfacial tension at the solid-liquid interface, and σ_{lg} is the interfacial tension at the gas-liquid interface. The angle that the water makes with the solid surface is termed the *contact angle*, θ_c. The component of the gas-liquid interfacial tension that is along the solid surface is then $\sigma_{lg} \cos \theta_c$ and a balance of forces gives

$$\sigma_{gs} = \sigma_{sl} + \sigma_{lg} \cos\theta_c$$

or (8.41)

$$\cos\theta_c = \frac{\sigma_{gs} - \sigma_{sl}}{\sigma_{lg}}$$

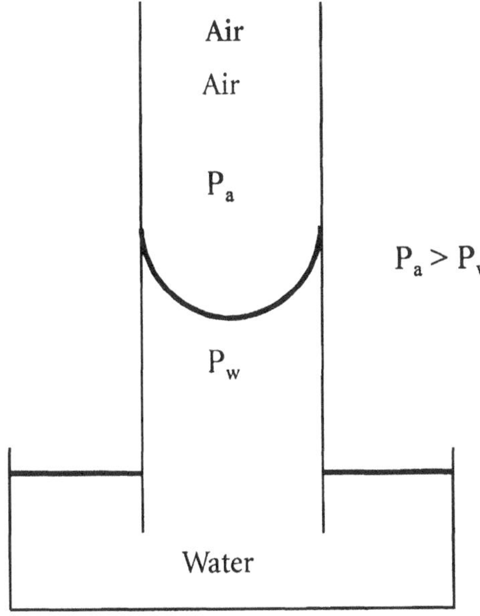

FIGURE 8.6 Capillary rise and pressure in a single capillary. Note that the diameter of the capillary is greatly expanded relative to the water in the bottom tray for purposes of illustration.

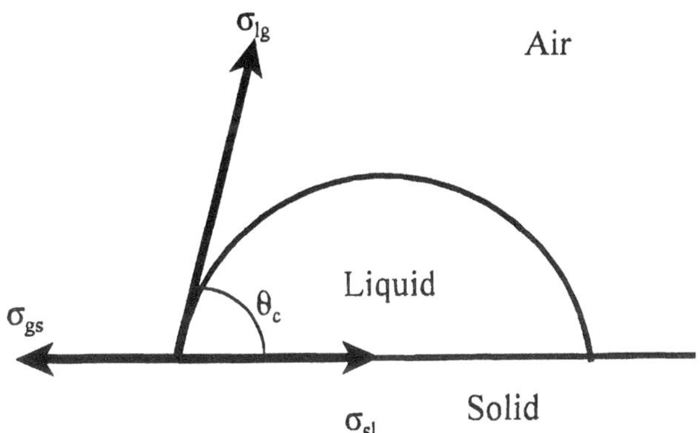

FIGURE 8.7 Contact angle and equilibrium forces acting on a droplet on a solid.

If the water completely wets the surface, the contact angle is 0° and the radius of curvature of the water interface is identical to the capillary radius. In the capillary of Figure 8.6, the water is then at a negative pressure (relative to the atmospheric pressure air) of

$$P_w - P_a = -\frac{2\sigma}{r} \tag{8.42}$$

This suction is available to pull water up the capillary against the force of gravity. At equilibrium, the capillary rise, h_c, is

$$h_c = \frac{2\sigma}{(\rho_w - \rho_a)gr} \tag{8.43}$$

If the contact angle is other than 0°, σ in Equations 8.42 and 8.43 must be replaced by $\sigma \cos\theta_c$. If the contact angle is 90°, there is no capillary rise and if the contact angle were 180°, the liquid would be completely non-wetting and there would be a capillary depression rather than a rise. The liquid phases of interest in soil, however, water and oil or other organic liquids, all tend to wet the surface to some degree. In soils, water tends to be more wetting than an oil or organic phase which, in turn, tends to be more wetting than an air phase.

Equation 8.42 can be generalized to deal with the interface between any wetting (e.g., water) and non-wetting fluid (e.g., oil or air).

$$P_w - P_{nw} = -\frac{2\sigma}{r} \tag{8.44}$$

where σ now represents the interfacial tension between the wetting and non-wetting phase. Note that the pressure difference across the interface, and therefore the capillary suction that gives rise to retention of the wetting liquid and capillary rise, is proportional to the inverse of the capillary radius. That is, the smaller the radius of the capillary, or the smaller the pore space, the greater the curvature and pressure difference across the interface. This causes the more wetting fluid to tend to fill the finer capillaries of a soil and be retained there at greater saturations after drainage. Thus, in the unsaturated zone, water tends to be trapped in the finest capillaries, an organic phase, if present, tends to be in the next largest capillaries, and air is left to the remainder of the pore space. If the soil contains a fine-grained lens, the residual saturations of the liquid phases will tend to be greatest in this zone.

Similarly, if an organic phase is found beneath the water table a pressure difference will occur at the interface between the two liquids depending on the size of the capillaries where they come in contact. Water will again tend to be held in the finest pore spaces since it is the more wetting fluid. An organic phase will tend to be excluded from the fine pore spaces because the pressure difference tends to resist displacement of the water and will, instead be found preferentially in coarser media and pore spaces. After an oil spill to the surface, an organic phase will tend to collect in the finer grained soil regions in the unsaturated zone because it is typically more wetting than air and in the coarser regions in the saturated zone because it is less wetting than water.

As with permeability, the distribution of pore sizes complicates the situation. The capillary rise associated with a uniform pore size is like that shown in Figure 8.8a while in a broad distribution of pore sizes the capillary rise looks more like that in Figure 8.8b. In fine pores, the water rises to a great height while in more coarse pores the capillary rise is less. The capillary rise, of course, is directly related to the magnitude of the negative pressure which is, in turn, inversely related to the size of the pores in the media. The negative pressure where air first begins to enter the media and saturation drops below 100% is called the air entry pressure head, h_a. The height of the bulk of the capillary fringe in a soil is thus characterized by the height associated equivalent to the air entry pressure head.

For a given soil, there is a relationship between the capillary suction pressure head and the water content as shown in Figure 8.9. The relative permeability is a function of fluid saturation and therefore also a function of capillary pressure. As indicated earlier, a residual water exists (or other liquid content) below which it is not possible to reduce the saturation by hydraulic forces. In Figure 8.9 these are indicated by the saturations which remain even at very large negative pressures or when the conductivity is effectively zero. In summary, the effect of pore size distribution and capillary effects is that the water saturation, φ_w, and hydraulic conductivity, K_p, take on the following functional forms.

$$\begin{aligned}
\varphi_w &= \varphi_w(h_p) && h_p < h_a \\
\varphi_w &\approx 1 && h_p \geq h_a \\
k_{eff} &= \kappa_r(h_p)k_p && h_p < h_a \\
k_{eff} &\approx k_p && h_p \geq h_a
\end{aligned} \tag{8.45}$$

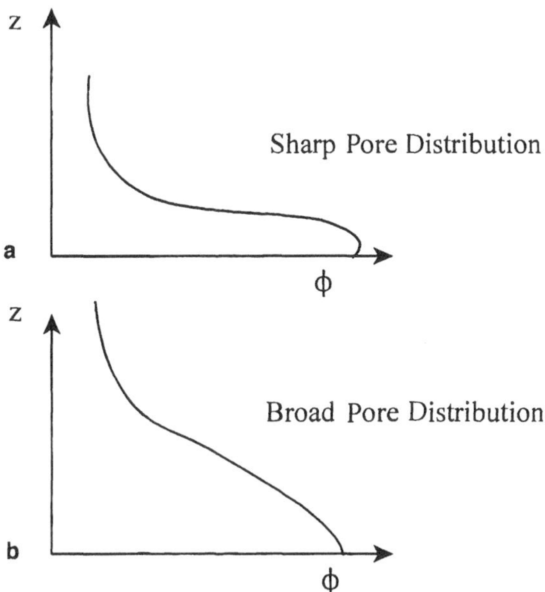

FIGURE 8.8 Capillary head vs. height (coarse and fine soil).

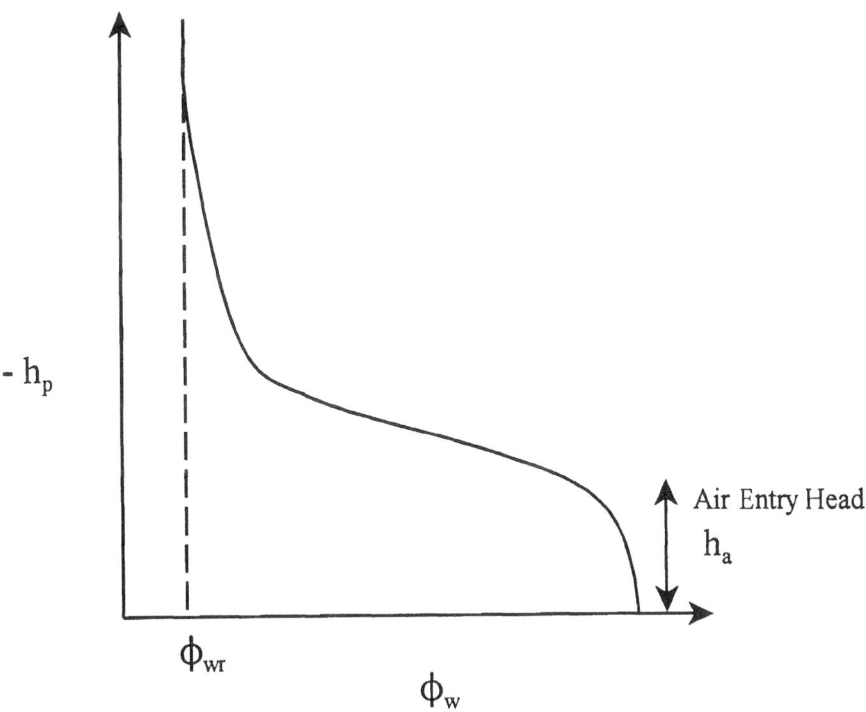

FIGURE 8.9 Capillary pressure as a function of saturation indicating air entry head or displacement pressure.

It is now possible to understand the behavior of water infiltrating after a rainfall. Immediately after a rainfall the upper soil layers are saturated. In a relatively short time the water-filled pores will drain to the soil's field capacity. The combination of low relative permeability and the high capillary suction at low moisture contents limits any further water drainage. Vegetation and evaporation, however, will continue to remove water from the system. Vegetation can continue to remove water until the content approaches the *wilting point,* at which point the water is held so tightly by the soil that it is not possible for vegetation to absorb this water. At the wilting point, the negative pressures exhibited by the water phase cannot be overcome by the suction associated with the uptake by the root systems of vegetation. Because plants can provide suctions of 15 atm, the wilting point represents residual water held very tightly by the soil. Even at these moisture contents water tends to wet the surface of most soil particles and this surface will remain covered with water. Typically, the moisture content must be reduced below 1 to 3% before dry soil surface is exposed. Values of field capacity and wilting point for various types of soils can be found in Table 8.2.

TABLE 8.2
Typical Values of Unsaturated Zone Parameters

Soil	Porosity (ε)	Field Capacity (ε_w)	Wilting Point (ε_w)	h_a (cm of H_2O)	k_p (cm/min)
Sand	0.35–0.45	0.1–0.15	0.05	10	1
Sandy loam	0.38–0.45	0.12–0.15	0.07–0.08	20	0.2
Loam	0.40–0.45	0.15–0.25	0.1	50	0.04
Silty loam	0.42–0.48	0.22–0.30	0.1–0.12	80	0.01
Clay loam	0.45–0.50	0.25–0.35	0.15	60	0.015
Clay	0.48–0.52	0.25–0.35	0.20	40	0.008

Note: Porosity, field capacity, and wilting point shown in volume voids or water per total volume. (Adapted from Clapp, R.B. and G.M. Hornberger (1978) *Water Resour. Res.*, 14 (4): 601–604, and Dunne, T. and L.B. Leopold (1978) *Water in Environmental Planning*, W.H. Freeman and Co., New York.)

8.3 CONTAMINANT TRANSPORT IN SOILS

We are now in a position to evaluate contaminant transport in soils. Let us consider the contamination of a soil by a spill or leak of an organic or oily phase to the surface of the soil. In recognition that this phase is only sparingly soluble in water and remains as a separate phase, it is often referred to as a nonaqueous phase liquid (NAPL). A separate NAPL is sometimes separated into a dense NAPL (DNAPL) if its density is greater than that of water and a lighter than water NAPL (LNAPL). Most chlorinated organics exhibit densities greater than one and are, therefore, DNAPLs in soil while almost all hydrocarbon mixtures such as gasoline, kerosene, most oils, and diesel exhibit a density less than that of water and are LNAPLs.

After a spill or leak of a NAPL, there are four zones of contamination with very different behavior governed by the principles discussed in the preceding section. These include:

1. Bulk NAPL contaminating vadose zone.
2. Vapor phase contaminant from NAPL contaminating vadose zone.
3. Bulk NAPL contaminating initially water-saturated zone.
4. Dissolved phase contaminant from NAPL contaminating aquifer.

The development of these four zones is depicted in Figure 8.10 as a result of a spill or leak of a nonaqueous phase liquid to the surface. Initially the NAPL moves through the unsaturated zone, driven by gravity and capillary forces. Because there is normally always water present in the soil, the finest pore spaces tend to be unavailable to the NAPL. The NAPL is more wetting than the air

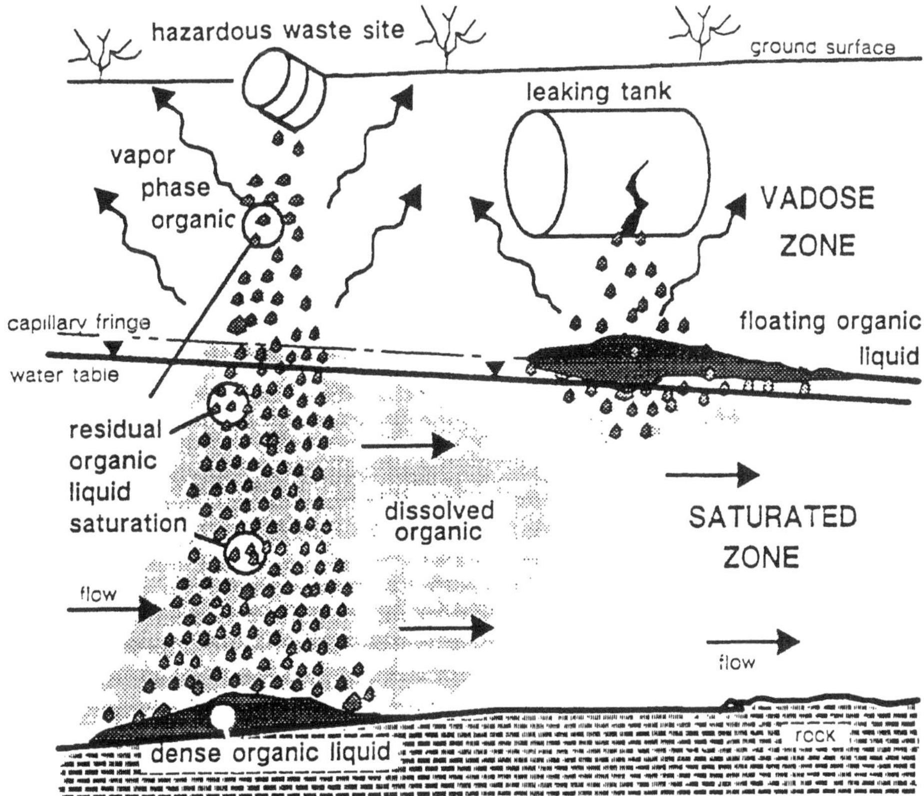

FIGURE 8.10 Migration pattern for an organic liquid more dense than water (left), and less dense than water (right). (From Wilson, J.L., et al. (1990) Laboratory Investigation of Residual Liquid Organics, EPA/600/6-90/004.)

phase, however, and will therefore leave a residual typically representing a saturation of 10 to 20% of the pore space. This largely immobile separate phase residual is the first zone of contamination. Although the bulk liquid phase is largely immobile, the residual can evaporate into the soil vapor space and move as a vapor throughout the vadose zone. The contaminated vapor in the vadose zone is a second zone of contamination.

If there is a sufficient volume of the NAPL, it will ultimately migrate downward to the water table. For a small volume spill or a long distance to the water table, the capillary entrapment of NAPL in the water-unsaturated zone may reduce the flowing volume to the extent that the water table is not reached. This is sometimes referred to as capillary exhaustion of the infiltrating NAPL. If the water table is reached, however, very different behavior is observed with an LNAPL vs. a DNAPL. The LNAPL will tend to spread on the surface of the water table and any further migration might be in the direction of the slope of the water table. The DNAPL, however, will tend to penetrate the water table. As it moves through the originally water-saturated zone, a residual is left behind due to capillary entrapment (although again left to the larger pore spaces due to the typically more wetting behavior of the water). Ultimately, if a sufficient volume of DNAPL is introduced to the subsurface, the phase may progress all the way through the water table aquifer until its motion is arrested by the presence of a low permeability strata. The fine pore spaces in this low permeability strata holds water tightly and typically the nonaqueous phase cannot displace this water and penetrate the strata. As a result, it tends to collect and pool on the surface of the low permeability material. If the strata is tilted, the DNAPL may migrate downslope, regardless of the direction of the groundwater flow. The state of the nonaqueous phase in the originally water-saturated zone, whether pooled at the water table as with the LNAPL or spread as a residual and pooled on low permeability strata at the base of the water table aquifer, as in a DNAPL, is the third zone of contamination.

Ultimately, the motion of the NAPL ceases, either due to pooling on an essentially flat water table or strata or due to "capillary exhaustion" from the residual held by capillary forces. Subsequent contaminant movement is the result of dissolution into the groundwaters and transport by their motion. This is the fourth zone of contamination.

The nature and form of the contamination defines the mobility of the contaminants and appropriate remedial approaches. The quantitative description of contaminant transport in each zone of contamination and the effects of these zones on selecting appropriate remedial options will be discussed in the subsequent sections.

8.3.1 First Zone of Contamination — NAPL Residual in the Vadose Zone

As indicated in the above discussion, a NAPL introduced to the soil via a spill or leak first migrates via gravity and capillary forces through the vadose zone leaving behind a residual. To understand the process that results in the residual, largely immobile contamination of the vadose zone, we must first explore the infiltration of the NAPL. Let us consider the situation in Figure 8.11 in which a body of contaminant liquid is spilled to the surface of the soil where it exhibits a "ponded depth" of height h. This is the depth of the pool of liquid and therefore its elevation head above the surface. Let us assume that the penetration into the soil can be modeled by consideration of the media as a bundle of capillary tubes, a single example of which is shown in Figure 8.11. Note that due to the wetting characteristics of the invading contaminant liquid and the curvature of the liquid-air interface, the pressure in the NAPL is less than the pressure in the air phase being displaced. The pressure head in the liquid is thus negative near the location of the advancing front and increases to zero (gauge) at the free surface of the spilled liquid. This is also depicted in Figure 8.11. Because the air is being displaced from the soil pores and is being pushed downward by the advancing front of the penetrating liquid, it also exhibits a pressure gradient but because of the low viscosity of air and the typically slow rate of infiltration of liquid, this is generally negligible. Thus, the hydraulic head (the sum of the pressure and elevation heads) at the advancing front of the penetrating liquid (position, z_f) is

$$\begin{aligned} h(z_f) &= h_p + h_e \\ &= -h_c - z_f \end{aligned} \tag{8.46}$$

where h_c is the effective head of the capillary suction at the infiltrating front and is approximated by Equation 8.43 for pores of uniform radius r. In a real soil, capillary pressures are strongly saturation dependent and not so easily characterized by a single effective capillary suction. In a relatively uniform sand or in a medium in which the finer pores have been filled with the typically more wetting water, the air-entry pressure provides a reasonable estimate of this parameter. Note that this air-entry pressure would be associated with the liquid phase being displaced (i.e., the NAPL) and not the more commonly measured air-entry pressure into water-filled pores.

The hydraulic head at the soil surface assuming a pooled liquid of height h is

$$h(0) = h \tag{8.47}$$

assuming again that the pressure head in the atmosphere is 0 gauge pressure. Substituting into Darcy's Law for unsaturated flow taking the saturation of the infiltrating NAPL as ϕ_n,

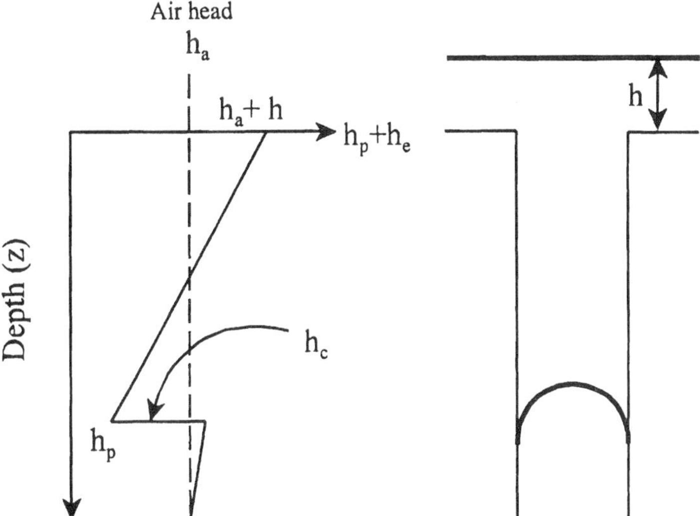

FIGURE 8.11 Model of infiltration illustrated by movement into a single pore assuming a ponded depth h at the surface. The total head, the sum of the pressure head (h_p), and elevation head (h_e) are also shown. The effective capillary pressure (h_c) is the difference in head between air and liquid at the infiltrating liquid front.

$$\frac{Q}{A} = q = \kappa_r(\varphi_n)k_p \frac{z_f + h_c + h(0)}{z_f} \tag{8.48}$$

Remembering that the Darcy velocity is the interstitial velocity, $u_t = dz_f/dt$ times the porosity, or that fraction of the porosity filled by the infiltrating NAPL, $\phi_n\varepsilon$, this can be rewritten

$$q = \varepsilon\varphi_n \frac{dz_f}{dt} = k_r(\varphi_n)k_p \frac{z_f + h_c + h(0)}{z_f} \tag{8.49}$$

This is a differential equation that describes the rate of penetration of the infiltrating liquid. It is only useful if the capillary suction head, h_c, can be approximated as a constant or known function. Variations in pore sizes and liquid saturation may cause this to vary significantly with position. Often the basic features of the infiltration are reasonably well described by a simple constant, however. The infiltrating liquid phase will generally fill all the pore space available to it. Because water is normally present at field capacity or higher levels and filling the finer pore spaces, an infiltrating nonaqueous phase is normally limited to the larger pores of the media. This serves to reduce the effects of capillarity and improve the accuracy of the model. The model has also been used to describe water infiltration after a rainfall (a so-called Green and Ampt model after the first to propose such a model). Reible et al. (1990) describe various solutions to the model. Let us consider here only the condition that the effects of capillarity and the ponded head can be neglected, for example, due to a spill on a coarse sand. Under these conditions, the model reduces to

$$q = \varepsilon\varphi_n \frac{dz_f}{dt} = k_r(\varphi_n)k_p \frac{z_f}{z_f} = \kappa_r(\varphi_n)k_p$$
$$z_f = \frac{k_r(\varphi_n)k_p}{\varepsilon\varphi_n} t \tag{8.50}$$

and the rate of penetration of the infiltrating liquid is a constant as long as the saturation is essentially uniform.

After the liquid infiltrates completely into the soil, the soil will begin to drain leaving a residual saturation behind the main body of liquid. If the effective capillary suction in this partially saturated region is the same as the effective capillary suction at the leading edge of the liquid, Equation 8.50 would also govern the drainage rate. Penetration at this constant velocity would continue until the entire volume of the infiltrating liquid phase was contained within the largely immobile residual. If V_n is the volume and φ_{nr} is the residual saturation with respect to the nonaqueous phase, the maximum penetration is given by

$$z_f\Big|_{\max} = \frac{V_n}{\varphi_{nr}\varepsilon A} \tag{8.51}$$

This is the depth to which the liquid will infiltrate before being exhausted by capillary retention. If this depth is greater than the depth to the water table, then interactions with the water table must be considered. For an LNAPL, the remaining liquid will spread in the form of a thick "pancake" on the water table surface. For a DNAPL, further penetration of the water table is likely as discussed further in the third zone of contamination.

The value of the residual nonaqueous phase saturation, φ_{nr}, should be related to the interfacial tension since the pressure resulting in capillary retention is $2\sigma/r$. Because water is always present, the appropriate interfacial tension is the tension between the nonaqueous phase and water. This can be crudely estimated if we assume that water and the nonaqueous phase are pure and mutually insoluble by replacing the solid phase in Figure 8.7 with a water phase. The nonaqueous phase droplet tends to wet the water suggesting $\cos\theta_c = 1$ and therefore

$$\sigma_{aw} = \sigma_{wn} + \sigma_{an}\cos\theta_c$$

or (8.52)

$$\sigma_{wn} = \sigma_{aw} - \sigma_{an}$$

where the subscripts a, w, and n reflect the air, water and nonaqueous phases, respectively. Thus, the interfacial tension between the water and the nonaqueous phase is simply the difference between liquid-air surface tensions. Because the capillary retention should scale with the interfacial tension, an estimate of the residual NAPL saturation can be deduced from the residual water saturation in an otherwise air-filled soil (field capacity) by

$$\varphi_{nr} \approx \varphi_{wr}\frac{\sigma_{nw}}{\sigma_{aw}} \approx \sigma_{wr}\frac{\sigma_{aw} - \sigma_{an}}{\sigma_{aw}} \tag{8.53}$$

Benzene, for example, exhibits a surface tension with air of 28.8 dyn/cm. Water exhibits a surface tension of 72 dyn/cm. The interfacial tension between water and air can then be estimated as 43.2 dyn/cm and the residual saturation of benzene in a water-wetted soil would be expected to be of the order of 60% of the residual water saturation. This approach neglects the effect of the different pore sizes that would be filled by the benzene vs. the residual water, so it should be considered a crude approximation at best. Organic residual saturations, however, do tend to be less than observed field capacities or residual saturations of water.

The same basis could be used to estimate the effective capillary suction or air-entry pressure into a porous medium filled with an NAPL rather than water. That is,

$$(h_c)_{nw} \approx (h_a)_{nw} \approx (h_a)_w \frac{\sigma_{nw}}{\sigma_{aw}} \approx (h_a)_w \frac{\sigma_{aw} - \sigma_{an}}{\sigma_{aw}} \tag{8.54}$$

where the subscript nw represents the nonaqueous phase values and the subscript w represents the corresponding quantities in water-filled pores.

The form and distribution of the residual NAPL and not just its volume or average saturation also is important to its subsequent mobility. For example, a finely divided residual exhibits more surface area to volume than large "blobs" and is thus more likely to evaporate or dissolve in infiltrating rainwater. This would be especially important for soluble or volatile components of a largely insoluble or nonvolatile mixture. These contaminants must diffuse to the surface of the bulk mixture before they are available for dissolution/vaporization. The larger the blobs of the bulk mixture the longer the diffusion path and the slower the process of release.

The calculation of the infiltration of a nonaqueous liquid phase is illustrated in Example 8.3. Example 8.4 illustrates some approximate characteristics of the residual ganglia or blobs left behind.

Example 8.3: Rate and extent of infiltration of an organic phase in the unsaturated zone

Estimate the rate of infiltration of 1000 L of benzene over 10 m^2 of soil with a water conductivity of 1 cm/min. The residual water saturation is 30%.

The saturated conductivity of benzene is given by

$$(k_p)_{Bz} = (k_p)_w \frac{\rho_n}{\mu_n} \frac{\mu_w}{\rho_w} = 1 \text{ cm/min} \frac{0.88 \text{ g/cm}^3}{0.0064 \text{ g/(cm}\cdot\text{s)}} \frac{0.01 \text{ g/(cm}\cdot\text{s)}}{1 \text{ g/cm}^3} = 1.375 \text{ cm/min}$$

The residual benzene saturation is estimated to be

$$\varphi_{nr} = \varphi_{wr} \frac{\sigma_n}{\sigma_w} = 0.3 \frac{43.2 \text{ dyn/cm}}{72 \text{ dyn/cm}} = 0.18$$

The relative permeability of the benzene is then

$$\kappa_r = \left(\frac{0.7 - 0.18}{1 - 0.18}\right)^{\frac{2+3(2.8)}{2.8}} = 0.184$$

The maximum infiltration rate is $\kappa_r K_p = 15.2$ cm/h. The position of the infiltrating benzene as a function of time is then

$$z_f(t) = \frac{qt}{\varepsilon\varphi_n} = \frac{15.2 \text{ cm/h}}{(0.4)(0.7)} t = (54 \text{ cm/h})t$$

The maximum infiltration depth (assuming a homogeneous soil) is

$$z_{\max} = \frac{V_n}{\varepsilon\varphi_{nr} A} = \frac{1 \text{ m}^3}{0.4 \cdot 0.18 \cdot 10 \text{ m}^2} = 13.9 \text{ m}$$

Example 8.4: Ganglia size in the unsaturated zone

Estimate the maximum size of the residual nonaqueous phase ganglia (or blobs) in the unsaturated zone

The maximum size of a ganglon is dependent on a balance between the capillary forces ($2\sigma/R$ per unit area) stabilizing the ganglia vs. the gravitation or buoyancy forces per unit area (ρ_n g L) destabilizing the ganglia. Setting these two forces per unit area equal, solving for the ganglia size L, and evaluating for benzene in a 0.1-mm pore radius (fine sand),

$$L \approx 2\frac{\sigma_{nw}}{R\rho_n g} = 2\frac{43.2 \text{ dyn/cm}}{(0.01 \text{ cm})(0.88 \text{ g/cm}^3)(9800 \text{ cm/s}^2)} = 1 \text{ cm}$$

For a pore radius more consistent with silt or clay this would increase by an order of magnitude or more. As this suggests the size of these ganglia can be quite large, slowing transport to and from the residual nonaqueous phase. Actual ganglia tend to occur in a range of sizes smaller than this.

8.3.2 Second Zone of Contamination — Contaminant Vapors in Vadose Zone

The residual that remains after penetration of the NAPL through the subsurface can serve as a source of vapors that can then migrate rapidly through the subsurface. At any location where free NAPL exists, the local pore space is likely saturated with vapors from the NAPL. If the NAPL were a pure organic phase, the pore space would exhibit a vapor concentration associated with the pure component vapor pressure of the organic phase. If the NAPL were a mixture, the pore space would exhibit a vapor concentration of a particular compound associated with the vapor pressure of the compound above that mixture. If the NAPL were an ideal mixture of organic compounds, Raoult's Law would hold and the pore space would exhibit a partial pressure of a particular compound equal to the product of the pure component vapor pressure and the mole fraction abundance of that compound in the mixture. Thus the partial pressure in the pore space adjacent to a residual NAPL under the two most common scenarios are given by

$$p_i = \begin{cases} P_v & \text{pure NAPL} \\ x_i P_v & \text{ideal NAPL mixture} \end{cases} \tag{8.55}$$

These contaminant vapors will then migrate away from the zone of liquid residual contamination.

The primary mechanism whereby vapors travel in the subsurface is via diffusion. There are some advective processes but they are generally only important under specialized situations. If the NAPL introduced to the subsurface is extremely volatile (vapor pressures greater than 1 atm), for example, its evaporation could significantly increase the vapor density in the pore space and cause an advective flow away from the point of evaporation. Methyl bromide, for example, is a soil fumigant that is introduced to soil as a liquid under pressure. Rapid vaporization of this compound would likely increase local vapor phase pressures causing a pressure driven flow away from the point of injection. For lower but still high volatility compounds (vapor pressures between 0.1 and 1 atm), the concentration of the vapors in the pore space could be sufficient to cause a buoyancy-driven flow. Air containing 10% or more of a compound with a molecular weight different than that of air, 29 g/mol, would cause such buoyancy-driven flows. For low vapor pressure compounds present in the vapor space at low concentrations, there may still be effects of advective processes if there are significant local variations in pressure. For example, a buried cavity could equilibrate with a high pressure atmosphere and then pressure-driven advective flow could result if the atmospheric pressure was reduced as the result of frontal passage. Finally advective flow could be driven by imposing a pressure gradient across the unsaturated zone, as might be attempted in certain remedial approaches such as soil vacuum extraction.

Under most situations of interest, however, the primary mechanism for vapor transport in soils is diffusion. Repeating Fick's first and second laws of diffusion in one dimension,

$$\begin{aligned} &\text{Fick's First Law} && q_m = -D_{sv}\frac{dC_{sv}}{dz} \\ &\text{Fick's Second Law} && \frac{\partial C_T}{\partial t} = -D_{sv}\frac{\partial^2 C_{sv}}{dz} \end{aligned} \tag{8.56}$$

where C_{sv} is the concentration of a volatile contaminant of interest in the soil vapor phase within the soil pores and D_{sv} is an effective diffusion coefficient for the vapors in the soil media. The term C_T is the total concentration of the contaminant in the soil including contaminant on the soil and water phases. C_T is the total concentration — sum of mass of contaminant in all soil phases divided by the volume of the soil.

In a porous medium, the effective diffusion coefficient is the product of the diffusivity in the fluid filling the pore spaces, in this case, air, and a correction for the fraction of void space available to diffusion, ε or ε_a, and the effective length of the diffusion path, the tortuosity, τ.

$$D_{sv} = D_a\frac{\varepsilon}{\tau} \tag{8.57}$$

The Millington and Quirk (1961) model is a commonly used predictor of the effective diffusion coefficient and it has the form, in a partially saturated medium, of

$$D_{sv} = D_a\frac{\varepsilon_a^{10/3}}{\varepsilon^2} \tag{8.58}$$

where ε_a is the air-filled porosity and ε is the total porosity in the soil. This model suggests that the effective diffusion coefficient in the soil vapors is significantly affected by the presence of residual water as shown in Example 8.5. It should be emphasized that there is no generally reliable model of effective diffusion coefficients in the soil media. The Millington and Quirk model is completely employed, simple, and has been shown to be at least approximately correct in granular media such as sands.

Example 8.5: Soil-vapor diffusivity — the effect of moisture content

Estimate the diffusivity of a vapor in a 40% porosity soil with varying water contents. The diffusivity of the compound in air is 0.1 cm²/s.

Let us consider water saturations of 20, 40, 60, and 80%. The effective diffusivity at 20% water saturation is given by

$$D_{sv} = D_a\frac{\left[\varepsilon(1-\varphi)_w\right]^{10/3}}{\varepsilon^2} = 0.1\ \text{cm}^2/\text{s}\,\frac{[(0.4\cdot(1-0.2)]^{10/3}}{0.4^2} = 0.014\ \text{cm}^2/\text{s}$$

The diffusivity at all water saturations are given below.

φ_w	$D_{sv,}$ cm^2/s
0	0.029
0.2	0.014
0.4	0.0054
0.6	0.0014
0.8	0.00014
1	0 (diffusion in water phase only)

The flux of a contaminant through the pore space is a function of the effective diffusion coefficient. The time required for complete evaporation and recovery of a contaminated zone, however, depends upon the portion of the contaminant that is mobile vs. that which is sorbed to the soil. As indicated previously the ratio of the total concentration to the mobile phase concentration is the retardation factor. Here

$$\frac{C_T}{C_{sv}} = R_f = \varepsilon_a + \frac{\varepsilon_w}{K_{aw}} + \frac{\rho_b K_{sw}}{K_{aw}} \tag{8.59}$$

the air-water partition coefficient, K_{aw}, is assumed to relate vapor and water phase concentrations and if sorption onto only the residual water and onto the soil surface is considered. The first term in Equation 8.59 represents the portion of the contaminant in the vapor space, the second the portion of the contaminant in the residual water, and the third the portion sorbed to the solid in contact with the residual water.

The accumulation in the phases other than the vapor is important when transient, or time-dependent diffusion occurs. Often, however, the time required to set up a steady-state concentration profile and therefore a steady-state diffusive flux is small compared to the time required to deplete the liquid contaminant. Keep in mind that a liquid is typically 1000 times more dense than a vapor phase. As a result, steady-state diffusion in the vapor phase is applicable which implies, in one dimension,

$$\frac{d^2C_a}{dz^2} = 0$$

$$\frac{dC_a}{dz} = \alpha \qquad \text{(Flux is constant)} \tag{8.60}$$

$$C_a = \alpha z + \beta \qquad \text{(linear in } z\text{)}$$

The amount of that flux is given by Fick's first law in Equation 8.56 with an estimate of D_s given by Equation 8.58. Example 8.6 examines steady-state vapor transport in the unsaturated zone of a soil.

Example 8.6: Vapor diffusion through a clean soil cap

Consider a clean soil "cap" 50 cm thick over a contaminated soil containing 100 mg/kg benzene. The average moisture content in the cap is 20%, its total porosity is 40%, bulk density of 1.5 g/cm^3, and organic carbon content of 3%.

The concentration in the soil water and soil vapor at the bottom of the cap are in equilibrium with the contaminated soil with $K_{sw} = K_{oc}\,\omega_{oc} = 2.61$ L/kg, and $K_{aw} = 0.224$ (for benzene).

$$C_{sw} = \frac{W_s}{K_{sw}} = \frac{100 \text{ mg/kg}}{2.61 \text{ L/kg}} = 38.2 \text{ mg/L}$$

$$C_{sv} = C_w K_{aw} = 38.2 \text{ mg/L } 0.224 = 8.6 \text{ mg/L}$$

From Example 8.5, the diffusivity in the 20% saturated soil is about 0.14 times the free air diffusivity (D_B = 0.088 cm²/s) or 0.012 cm²/s. The steady-state flux through the soil cap, assuming the air is effectively free of benzene is

$$q_m = \frac{D_{sv}}{h} C_{sv} = \frac{0.012 \text{ cm}^2/\text{s}}{50 \text{ cm}} (8.6 \text{ mg/L}) = 75.9 \frac{\text{mg}}{\text{m}^2 \cdot \text{h}}$$

The time required to reach this steady-state is dependent upon the retardation factor defined by Equation 8.59.

$$R_f = (0.32) + \frac{0.08}{0.224} + (1.5 \text{ g/cm}^3)\frac{2.61 \text{ L/kg}}{0.224} = 18.2$$

Thus, at any location in the soil cap the mass of benzene in the vapor phase is only 1/18.2 of the total mass. This significantly delays the onset of steady conditions which can be demonstrated by examining the characteristic time to reach steady conditions at the top of the cap.

$$\tau_{diff} = h^2 \frac{R_f}{D_{sv}} = \frac{(50 \text{ cm})(18.2)}{0.012 \text{ cm}^2/\text{s}} = 42.8 \text{ day}$$

Whether this time period is important or not depends upon the amount of benzene in the subsurface (that is, for a large spill, a delay of 42 days to reach steady conditions is insignificant). If the cap material were less sorbing, for example, contained less water or less organic material in the soil, the delay to steady state would be much less. The characteristic time to reach steady state in a dry non-sorbing cap ($R_f = \varepsilon_a = 0.4$) is only 0.9 days.

8.3.3 Third Zone of Contamination — NAPL Residual in Initially Water-Saturated Zone

That portion of the NAPL that ultimately reaches the water table forms the third zone of contamination. If the NAPL is lighter than water, the liquid will tend to pool on the water table surface, although some depression of the water table is likely to occur due to the static head of the NAPL. This pooled NAPL may migrate down the groundwater gradient and impact other locations if the initial spill or leak was of sufficient volume.

If the NAPL is a DNAPL, it will tend to penetrate the water table and progress downward until impeded by low permeability geological strata. The NAPL is generally less wetting of the media than water and thus displacement of water from fine-grained strata is unlikely to occur. The low permeability of such a strata would further reduce the rate of any penetration that is likely to occur.

The residual nonaqueous phase left behind is similar in many respects to that left behind in the vadose zone. The residual nonaqueous phase saturation tends to be somewhat higher than in the vadose zone in large part due to the tendency of the nonaqueous phase to coalesce into larger blobs beneath the water table. As with the vadose zone, the form and distribution of the nonaqueous phase is as important to the subsequent migration of the contaminant as the absolute volume of residual. The particular distribution of the residual is a strong function of the details of the soil heterogeneities and is almost impossible to predict or even measure even with numerous wells and coring of the soils. Unlike the vadose zone, the nonaqueous phase shows a strong preference for coarse-grained media. Due to the difference in wetting characteristics, the nonaqueous phase can only displace water from the coarser media, and then it is effectively trapped there since further migration would require displacement of water from finer-grained media.

If the spill is of sufficient volume, enough NAPL may penetrate to the bottom of the water table aquifer so that a pool of liquid occurs on the underlying strata. This pool is subject to further migration due to the negative buoyancy of the liquid in the direction of the downhill tilt of the water table. This migration downslope may occur regardless of the direction of ground water flow, potentially causing contamination to migrate "up-gradient." This is especially difficult to understand if the presence of the nonaqueous phase pool has not been identified. Developing a conceptual model of the form and distribution of the contamination at a site can be an incredibly daunting task requiring a great deal of sampling and interpretation. At many sites, this site is never completed to the satisfaction of the investigators and surprises in the form of unexpected contamination in a new sample is always possible. Example 8.7 compares the stable size of the ganglia or blobs left behind in the saturated zone to those calculated in Example 8.4 for the unsaturated zone. The stability of the nonaqueous phase residual during water flushing is illustrated in Example 8.8.

Example 8.8: Ganglia size as a function of key parameters in the saturated zone

Estimate the maximum size of the residual nonaqueous phase ganglia (or blobs) in the saturated zone.

The maximum size of a ganglia in the saturated zone is dependent on a balance between the capillary forces ($2\sigma/R$ per unit area) stabilizing the ganglia vs. the pressure gradient forces destabilizing the ganglia. From Darcy's Law, the pressure difference across a ganglia of size L is given by

$$q = \frac{\kappa_p}{\mu_w}\frac{dP}{dz} \Rightarrow \Delta P = \frac{qL\mu_w}{\kappa_p}$$

Setting these two forces per unit area equal and solving for the ganglia size L, a relationship can be found between the ganglia size and the flow and soil parameters. If a fine sand or sandy loam is assumed with a pore radius of the order of 0.01 cm with benzene as the NAPL, the following estimate for ganglia size can be made.

$$L \approx 2\frac{\sigma_{nw}\kappa_p}{Rq\mu_w} = 2\frac{(43.2\ \text{dyn/cm})(3.4\cdot 10^{-8}\ \text{cm}^2)}{(0.01\ \text{cm})(0.2/60\ \text{cm/s})(0.01\ \text{g/(cm}\cdot\text{s)})} = 9\ \text{cm}$$

This estimate assumes a hydraulic conductivity of 0.2 cm/min which corresponds to a medium permeability of $3.4\ (10)^{-8}$ cm^2. The estimate also assumes unit hydraulic gradient such that the groundwater velocity is equal to the hydraulic conductivity. The estimated size is larger than that expected in the unsaturated zone and if a lesser hydraulic gradient were applied such that $q < K_p$, an even larger stable size would result. Again the size of a ganglion can significantly slow mass transport between the NAPL and water

Example 8.8: Volume of water required to dissolve a nonaqueous phase residual

Consider a soil with a residual nonaqueous phase liquid saturation of 20% and a total porosity of 40%. Estimate the pore volumes of water required to completely dissolve the liquid if the liquid is (1) 1,2-

dichloroethane, (2) pure benzene, or (3) a 1% mixture of benzene in an essentially insoluble bulk NAPL with density 0.8 g/cm³ with an average molecular weight of 140.

Let us employ a basis of 1 m³ of soil. The total mass of NAPL to be removed from the soil is then

$$M_n = \rho_n \varphi_n \varepsilon (1\ \text{m}^3) \qquad \rho(\text{g/cm}^3) = \begin{bmatrix} 1.25 \\ 0.88 \\ 0.8 \end{bmatrix} \qquad M_n(\text{kg}) = \begin{bmatrix} 100 \\ 70.4 \\ 64 \end{bmatrix}$$

The volume of water required depends upon the solubility in water of 1,2-dichloroethane and benzene (5500 and 1780 mg/L, respectively) and the solubility times mole fraction benzene for the mixture (assuming Raoult's Law or an ideal nonaqueous phase). The mole fraction and solubility of benzene in the mixture is

$$x_B = \frac{0.01/78}{0.01/78 + 0.99/140} = 0.018 \qquad C_B = x_B S_w = (0.018)(1780\ \text{mg/L}) = 31.7\ \text{mg/L}$$

The minimum volume of water required to remove the contaminants from the soil and the minimum volume divided by the pore volume in the soil (i.e., the number of pore volumes) are

$$V_w = \frac{M_n}{C_w} \qquad V_w(\text{m}^3) = \begin{bmatrix} 18.2 \\ 39.6 \\ 3600 \end{bmatrix}$$

$$V_p = \frac{V_w}{\varepsilon\ 1\ \text{m}^3} \qquad V_p = \begin{bmatrix} 45.5 \\ 98.9 \\ 8990 \end{bmatrix} \begin{matrix} \text{1,2-dichloroethane} \\ \text{Benzene} \\ \text{1\% mixture of benzene} \end{matrix}$$

Thus, very large flushing volumes of water are required to dissolve the nonaqueous phase ganglia, especially for dilute mixtures which reduces the dissolved phase concentration even below the solubility limit.

8.3.4 Fourth Zone of Contamination — Dissolved Contaminant in Subsurface Water

The nonaqueous phase residual or pools present in, on, or beneath the water column serve as a subsequent source of contamination to the aqueous phase. It is this contamination that is generally of most interest in terms of risk to potential users of the ground water. The contamination can migrate to surface waters fed by the groundwater or to drinking water wells leading to exposures by contact or ingestion. If the nonaqueous phase is largely immobile, this migration and exposure results from dissolution and *miscible displacement* of the contaminant.

At any location where NAPL exists, the local pore space is likely saturated with dissolved constituents. If the NAPL were a pure organic phase, the pore space would exhibit a dissolved concentration equal to the solubility of the contaminant. If the NAPL were a mixture, the pore space would exhibit a dissolved concentration according to the partitioning of the compound from that mixture. If the NAPL were an ideal mixture of organic compounds, Raoult's Law would hold. For a hydrophobic organic, the mole fraction of the component times the pure component solubility would be observed in the water in the local pore space. The dissolved concentration in the pore space adjacent to a residual NAPL under the two most common scenarios are given by

$$C_w = \begin{cases} S_w & \text{pure NAPL} \\ x_n S_w & \text{NAPL mixture} \end{cases} \tag{8.61}$$

where S_w is the tabulated solubility of the contaminant in water and x_n is the mole fraction of the component in the nonaqueous phase. Equation 8.61 estimates only the purely dissolved contaminant. Any dissolved or suspended particulate organic carbon can increase the amount of hydrophobic organic contaminant contained in the porewater. Using the models for estimating the proportion on dissolved or suspended particulate organic carbon discussed in Chapter 3, the total porewater concentration is given by

$$\begin{aligned} C_{sw} &= C_w + C_{oc} \\ &= C_w + \rho_{oc} K_{oc} C_w \\ &= C_w (1 + \rho_{oc} K_{oc}) \end{aligned} \tag{8.62}$$

Here we are differentiating between the total concentration, the total soil water (sw) or porewater concentration, and the dissolved concentration. The definition of these are as follows:

C_T— Total concentration is the sum of mass on soil, dissolved and other suspended contaminant per unit volume of soil (as above).

C_{sw} — Soil water concentration is the sum of contaminant dissolved or otherwise suspended in pore water per volume of pore water.

C_w— Dissolved concentration is the sum of dissolved contaminant per volume of porewater.

The dissolved and suspended contaminants will migrate away from the zone of nonaqueous residual. The primary mechanism whereby these contaminants migrate is generally advection. Unlike vapor transport in the vadose zone, diffusive transport in the groundwater is quite slow (vapor diffusivities are of the order of 10,000 times larger than liquid diffusivities). In addition, there is usually a significant groundwater transport rate. The ratio of the advective transport to the diffusive transport is governed by the Peclet number,

$$N_{Pe} = \frac{QL}{AD_{sw}} = \frac{qL}{\left(D_w \dfrac{\varepsilon}{\tau}\right)} \tag{8.63}$$

Again the effective diffusion coefficient is again often estimated using the model of Millington and Quirk which for a water-saturated medium becomes

$$D_{sw} = D_w \varepsilon^{4/3} \tag{8.64}$$

Using the Millington and Quirk estimate of effective diffusion coefficient in a water-saturated medium with a porosity of 40% for a compound with a water diffusivity of 10^{-5} cm²/s, $D_{sw} = (10^{-5}$ cm/s) $0.4^{4/3} = 0.29 \cdot 10^{-5}$ cm/s. For even transport distances of 1 m, the Peclet number exceeds unity and advection is dominant for Darcy velocities as low as $0.29 \cdot 10^{-7}$ cm/s = 0.91 cm/year. Thus, diffusion is negligible for most groundwater flow velocities of interest.

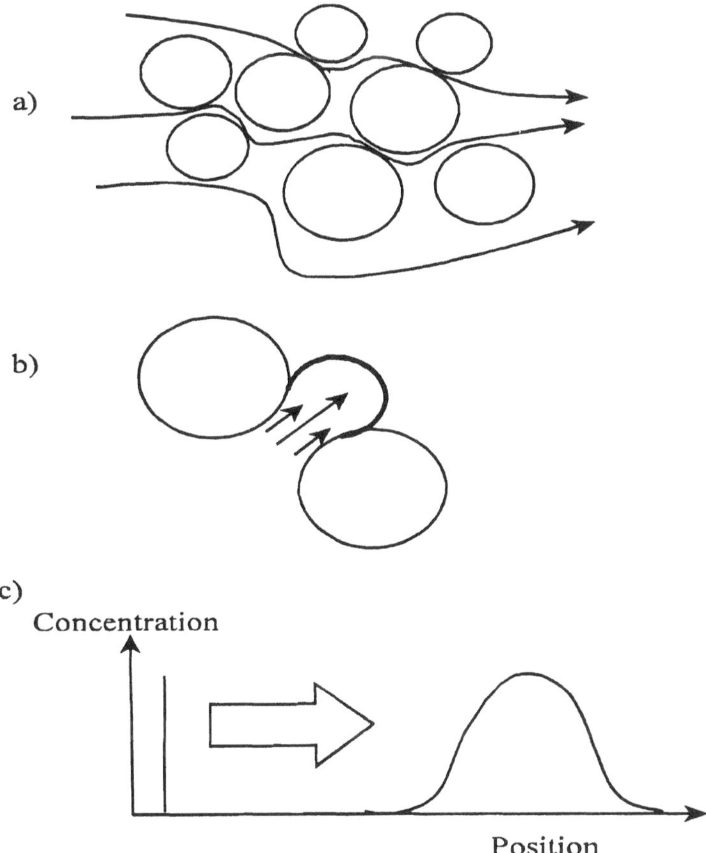

FIGURE 8.12 Mechanism and effect of microscale dispersion: (a) tortuous path fluid takes through media; (b) variations in velocity within a pore; and (c) spreading due to these processes.

This does not mean that diffusive-like processes can be neglected, however. The groundwater flow is limited to moving through the interstitial pore space and as a result follows a tortuous path as shown in Figure 8.12a. In addition, the velocity within the interstitial space is variable with effectively zero velocities at the soil grains and a maximum velocity in the center of the pores, also as shown in Figure 8.12b. As a result of these differences in path and velocity, parcels of fluid are spread in a manner similar to that caused by diffusion during their progress through any porous media. If we examine fluid marked with a tracer entering a sequence of pore spaces, for example, the fluid moves at different velocities and along different paths through the pore spaces. As a result the arrival of the tracer at the exit of the pore sequence is blurred and spread as shown in Figure 8.12c. The spread of the contaminant concentration with distance is referred to as *dispersion*. It is a microscale phenomenon that results from a coupling of advection and diffusion processes causing spreading on the macroscale.

Just as heterogeneities at the pore level cause dispersion, macroscale heterogeneities associated with large scale variations in pore or grain sizes and permeabilities also cause a similar dispersion process in that again, fluid moves at different velocities and along different paths through the media. This is illustrated in Figure 8.13. A finite zone of low permeability material behaves in much the same way as a single soil grain acts at the microscale. Two layers of differing permeability result in spreading the time of arrival of a tracer in a withdrawal well.

Note that in all cases, dispersion is associated directly with the motion of the fluid through the media. Because the effect is associated with spreading in a diffusion-like manner, dispersion is

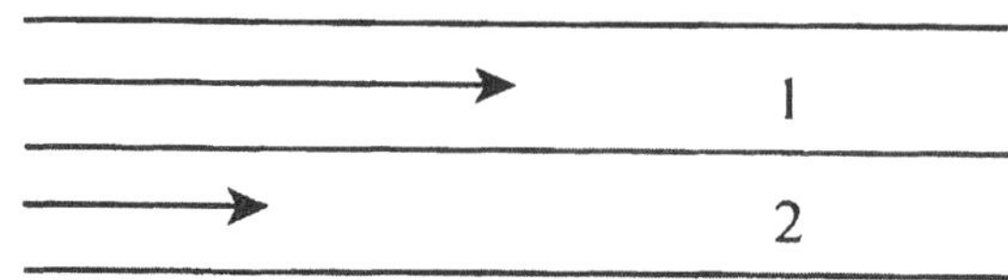

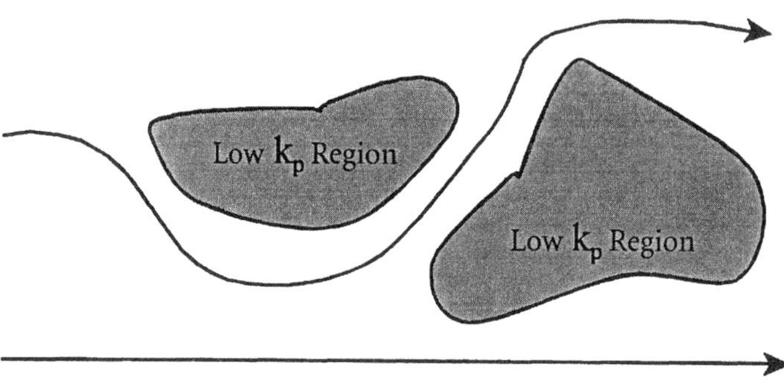

FIGURE 8.13 Illustration of enhanced lateral spreading or dispersion associated with macroscale heterogeneities.

normally modeled with an effective diffusion coefficient, or dispersion coefficient, that depends on the fluid velocity. At the low Darcy velocities observed in most soils, the effective diffusion coefficient resulting from dispersion, or *dispersion coefficient*, K_D, is often observed to exhibit a relationship with velocity of the following form

$$K_D = D_w \frac{\varepsilon}{\tau} + \alpha_D q \tag{8.65}$$

where α_D is termed the dispersivity. Unfortunately, some authors prefer to refer to what we have here identified as K_D as the dispersivity. We will hold to the usage that the quantity referred to as K_D is a dispersion coefficient and that α_D is a dispersivity. The dispersivity is a measure of the scale of the heterogeneities in a medium and has units of length.

For a uniform bed of packed spheres, α_D is approximately equal to the particle diameter in the direction of flow and approximately $^1/_{10}$ of that in the direction normal to the flow. In a real soil, the transverse dispersivity is still of the order of 10% of that in the direction of flow, but the values of both tend to be much greater than in laboratory columns filled with sand. The largest heterogeneities encountered by the flow tend to grow with distance traveled and as a result the observed dispersivity under field conditions also tends to grow with distance. A commonly used approximation is to assume that the dispersivity is of the order of 5 to 10% of the travel distance. Thus, if the distance from a contaminated soil zone to a drinking water well is 1 km, the effective dispersivity over that distance is 50 to 100 m. A similar approach was suggested by Neumann (1990) based on field observations of dispersivity. He proposed the relationship

$$\alpha_D \approx 0.017\,\mathrm{L}^{1.5} \qquad \mathrm{L}, \alpha_D \text{ in m} \tag{8.66}$$

where L is the travel distance of the contaminant in meters. Again the dispersivity in the transverse direction perpendicular to the groundwater flow direction is typically about 10% of this value.

Example 8.9 compares the dispersivity and the dispersion coefficient suggested by this relationship to molecular diffusion. It should be emphasized that the actual dispersivity may vary greatly from this prediction due to variations in local geology.

Example 8.9: Values of the dispersivity and dispersion coefficient

Estimate the dispersivity and dispersion coefficient for groundwater flow between a contaminant source area and a water well potentially at risk 100 m downgradient. Assume a soil porosity of 40%, a molecular diffusivity of 10^{-5} cm²/s, and a Darcy velocity of 10 m/year.

By the Millington and Quirk model the effective molecular diffusion coefficient in the soil is

$$D_{sw} = D_w \varepsilon^{4/3} = 2.95 \cdot 10^{-6}\ \text{cm}^2/\text{s}$$

Estimating the dispersivity by Equation 8.66

$$\alpha_D = 0.017\ \text{L}^{1.5} = 0.017\ (100)^{1.5} = 17\ \text{m}$$

$$K_D = D_{sw} + \alpha_D q = 2.95 \cdot 10^{-10}\ \text{m}^2/\text{s} + (17\ \text{m})(10\ \text{m/year})$$

$$\approx 170\ \text{m}^2/\text{year} = 0.054\ \text{cm}^2/\text{s}$$

The effect of molecular diffusion on horizontal transport is clearly negligible compared to the dispersion. The Peclet number **based upon the dispersion coefficient** is

$$N_{Pe} = \frac{qL}{K_D} = \frac{(10\ \text{m/year})(100\ \text{m})}{170\ \text{m}^2/\text{year}} = 5.9$$

Although advection is the more significant process ($N_{Pe} > 1$), it does not completely dominate and significant mixing via dispersion occurs. If only molecular diffusion were operable, advection without any significant mixing would be observed.

Describing miscible displacement or contaminant migration as a result of groundwater movement requires solution of an advection-diffusion equation. The equations for flux and the concentration equivalent to Fick's first and second laws are thus

$$\begin{aligned} q_m &= qC_{sw} - K_D \frac{dC_{sw}}{dz} \\ \frac{\partial C_T}{\partial t} + q\frac{dC_{sw}}{dx} &= K_D \frac{\partial^2 C_{sw}}{dz^2} \end{aligned} \tag{8.67}$$

Where K_d is the dispersion coefficient as described above.

For movement in groundwater with linear, reversible sorption onto the solid organic carbon fraction as well as sorption onto suspended or particulate organic matter, the total concentration is given by

$$\begin{aligned} C_T &= \left(\varepsilon + \rho_b K_{sw}^{obs}\right) C_{pw} \\ &\approx \left(\varepsilon + \rho_b \omega f_{oc} K_{oc} + \rho_{oc} K_{oc}\right) C_w \end{aligned} \tag{8.68}$$

In the first relationship the observed partition coefficient between the porewater and the solid phase is employed. This is defined by

$$K_{sw}^{obs} = \frac{W_s}{C_{sw}} \tag{8.69}$$

The second relation of Equation 8.68 attempts to estimate this partition coefficient through the assumption that the dissolved fraction of the hydrophobic organic partitions onto the solid organic carbon and fine suspended particulate organic carbon according to the organic carbon based partition coefficient. Often simply measuring the partition coefficient (Equation 8.69) might be more useful in indicating what is sorbed to the immobile solid phase since even filtered water samples normally include the quantity of contaminant sorbed to the suspended fine particulate fraction and not just the truly dissolved. The corresponding definitions of the retardation factor, the ratio of the total concentration to the mobile phase concentration is given by

$$\begin{aligned}\frac{C_T}{C_{sw}} &= R_f = \varepsilon + \rho_b K_{sw}^{obs} \\ &\approx (\varepsilon + \rho_p \omega_{oc} K_{oc} + \rho_{oc} K_{oc}) \frac{C_w}{C_{sw}} \\ &\approx \frac{\varepsilon + \rho_p \omega_{oc} K_{oc} + \rho_{oc} K_{oc}}{1 + \rho_{oc} K_{oc}}\end{aligned} \tag{8.70}$$

The last relationship recognizes that by the linear partitioning to the suspended organic carbon model that $C_{sw} = C_w(1 + \rho_{oc} K_{oc})$. Using this definition of the retardation factor, Equation 8.67 becomes

$$\begin{aligned} q_m &= qC_{sw} - K_D \frac{dC_{sw}}{dz} \\ \frac{\partial C_{sw}}{\partial t} + \frac{q}{R_f}\frac{dC_{sw}}{dx} &= \frac{K_D}{R_f}\frac{\partial^2 C_{sw}}{dz^2}\end{aligned} \tag{8.71}$$

Note that the flux is not different while the transient advection-diffusion equation is changed by the addition of the quantity sorbed onto the immobile soil fraction. Effectively the retardation factor reduces the speed of the contaminant's migration in the porewater. The effective velocity and dispersion coefficient is given by

$$\begin{aligned} u &= \frac{q}{R_f} \\ K_D &= D_w \frac{\varepsilon}{R_f \tau} + \frac{\alpha_D q}{R_f} \\ &= D_w \frac{\varepsilon}{R_f \tau} + \alpha_D u \end{aligned} \tag{8.72}$$

and both are reduced by the retardation factor. Example 8.10 illustrates the effective dispersion coefficient and velocity as reduced by retardation.

Example 8.10: Transit time for advective transport with sorption-related retardation

Estimate the transit time for contaminants between a source area and a water well 100 m away through soil of bulk density 1.5 g/cm³, porosity of 40%, and organic carbon of 1%. Consider the contaminants (1) chloride ion (nonsorbing), (2) benzene (log K_{oc} = 2.1), (3) hexchlorobutadiene (log K_{oc} = 3.67), and (4) pyrene (log K_{oc} = 5.0). The groundwater velocity is 10 m/year.

Based upon the given data the soil-water partition coefficient and retardation factor is given by

$$K_{sw} = K_{oc}\omega_{oc} = \begin{bmatrix} 0 \\ 1.3 \\ 46.8 \\ 1000 \end{bmatrix} \text{L/kg} \qquad R_f = \varepsilon + \rho_b K_{sw} = \begin{bmatrix} 0.4 \\ 2.29 \\ 70.6 \\ 1500 \end{bmatrix}$$

The time required to reach the well by advection only is then given by

$$\tau_{adv} = \frac{L}{q_{eff}} = \frac{LR_f}{q} = R_f \cdot 10 \text{ year} = \begin{bmatrix} 4 \\ 22.9 \\ 706 \\ 1500 \end{bmatrix} \text{year} \quad \begin{matrix} \text{Chloride ion} \\ \text{Benzene} \\ \text{Hexachlorobutadiene} \\ \text{Pyrene} \end{matrix}$$

Clearly the more sorbing compounds, such as pyrene, move very slowly through the subsurface due to the retardation associated with sorption. Note, however, that the chloride ion moves faster than the Darcy velocity, that is at an average velocity of 25 m/year. This is due to the previously recognized fact that the interstitial velocity in the pore spaces is greater than the Darcy or superficial velocity based upon the entire area of the soil.

Note that since the retardation factor arises from the time-dependent term in the advection-diffusion equation, there is no retardation of steady-state transport. The differences between transient and steady-state processes are explored in the examples. Note that the fate and transport of contaminants is an especially important part of evaluating the exposure and risk associated with contaminated soils in that without migration processes contaminated soils generally exhibit no risk. Based upon the release and transport processes, however, it may be important to remediate soils, that is, restore them to an effectively pristine state.

Example 8.11 illustrates the application of the miscible transport models developed above to indicate the rate of subsurface movement of sorbing contaminants.

Example 8.11: Concentrations at a well with advection and dispersion

Employing the dispersion coefficients from Example 8.9 and the retardation factors from Example 8.10, estimate the time required for the well to reach 10% of the concentration at the source area for each of the compounds in Example 8.10.

Assuming a simple one-dimensional advection-diffusion problem, the concentration as a function of time and distance away from the source area can be estimated with Equation 5.136 with K_D replacing D and q replacing U. In addition, $C_0 = 0$.

$$\frac{C_f(z,t)}{C_f(0)} = \frac{1}{2}\left[erfc\left(\frac{R_f z - qt}{\sqrt{4K_D R_f t}} \right) + \exp\left(\frac{qz}{K_D} \right) erfc\left(\frac{R_f z + qt}{\sqrt{4K_D R_f t}} \right) \right]$$

We are using z = 100 m, q = 10 m/year, and K_D = 170 m²/year (assumed constant). R_f is given by the preceding examples. The time required to reach 10% of the source concentration (in the water, i.e., $C_f(z,t)/C_f(0) = 0.1$) is summarized in the table below. The calculation is conducted iteratively by guessing a time and then calculating concentration and continuing this process until the concentration is 10% of that at the source.

Compound	R_f	τ_{adv} Example 8.10	t for $C_f/C_f(0) = 0.1$ Equation 5.136
Chloride ion	0.4	4	1.7 year
Benzene	2.29	22.9	9.8 year
Hexachlorobutadiene	70.6	706	302 year
Pyrene	1500	15000	6430 year

The contaminant also disperses laterally and vertically which significantly reduces the concentration progressing toward the well and increases the time required to achieve any particular concentration at a downwind location. Equation 5.136 is still useful, however, to provide an indication of rate of travel. Equation 5.143 can be used to estimate the steady concentration achieved at the well (which is not subject to retardation). The effective lateral and vertical dispersion coefficients might be 10% of that in the direction of groundwater flow (17 m²/year). For a constant concentration source, the flux at the source is q C(0) and Q_m = q C(0) A. Thus, the concentration at the well (assumed along x or directly downgradient) as a percentage of the concentration at the source and per unit area of source is

$$\frac{C}{C(0)A_s} = \frac{q}{4\pi\sqrt{D_y D_z}\, x}$$

For the conditions of this problem, the steady-state concentration at the well is 0.047% of that at the source per m² of source area.

8.3.5 Natural Attenuation of Contaminants in Soils

It is important to recognize that transport processes do not always control the exposure to contaminants in groundwaters. Many subsurface contaminants undergo natural attenuation processes, generally microbial degradation processes, that render them harmless before they can migrate to a possible exposure point. It has become increasingly clear that we are generally unable to intervene in contaminated soil sites and return them to a pristine, precontamination state. This has caused increased attention to be focused on the natural degradative pathways that might, given sufficient time and appropriate enhancements, allow these sites to ultimately return to a pristine state after the conclusion of whatever human interventions were feasible.

Aerobic degradation of petroleum hydrocarbons and light aromatic compounds has long been recognized and has been discussed previously in aqueous environments. In soil, these processes are generally much slower due to a variety of factors including:

- Presence of soil phase and sorption of degrading compounds onto that phase
- Limited quantities of water, oxygen and nutrients in the subsurface environment

- Trace soil or water components that might cause inhibition or toxicity among the microorganisms
- Presence of separate phase contaminants of limited solubility, slowing degradation which must generally take place in the soil water, and the high contaminant concentrations can lead to inhibition or toxicity effects

More recently it has become clear that chlorinated organics also undergo natural attenuation processes, generally via anaerobic pathways. These processes occur via a more interesting and difficult chemistry, but in many cases they appear to hold the only significant promise for ultimately achieving a return to a near pristine state for soils contaminated with these materials. The mechanisms whereby chlorinated aliphatic compounds including the common solvent contaminants — the chlorinated ethenes and ethanes — are degraded include:

Reductive dehalogenation — In reductive dehalogenation, the chlorinated compound is not used as food by the organism (i.e., as a carbon source) as are the petroleum hydrocarbons in an aerobic degrading environment. Instead the chlorinated compound serves as an electron acceptor, giving up a chlorine atom for a hydrogen atom. Generally this process occurs sequentially, with tetrachlorinated ethane (PCE) dechlorinating to trichloroethylene (TCE), to dichloroethylene (DCE), and finally to vinyl chloride.

$$C_2Cl_4 \Rightarrow C_2Cl_3H \Rightarrow C_2Cl_2H \Rightarrow C_2Cl \qquad (8.73)$$

The rate of dechlorination slows as the number of chlorine atoms remaining decreases. This means that vinyl chloride tends to be a relatively stable product of reductive dechlorination, a result that is unfortunate in that vinyl chloride is the most mobile of these compounds and has been demonstrated to be a potent human carcinogen.

Electron Donor Reactions — It is possible to degrade some chlorinated organic compounds aerobically during which the less oxidized chlorinated compounds, such as vinyl chloride, serve as electron donors and food for the microorganisms. Under the right conditions, it is possible to achieve complete mineralization of the vinyl chloride to carbon dioxide. The difficulty of course is that these are not the same conditions that give rise to the dehalogenation of the more chlorinated compounds.

Cometabolism — The final mechanism for the degradation of chlorinated organic compounds is during the degradation of other compounds. Enzymes or other substances are produced to enable the microbe to degrade one compound, but this same enzyme or factor also enables the degradation of the chlorinated compound. In such a situation, the microorganisms appear to gain no benefit from the coincidental degradation of the chlorinated compound.

Although reactive dechlorination will occur with only natural organic carbon available as a food source for the microbial population, the most rapid degradation occurs when there is more readily degradable anthropogenic carbon available. This is, of course, commonly the case in a contaminated site where residual petroleum hydrocarbons may coexist with the chlorinated contaminants. The ideal situation is rapid dechlorination in an anaerobic zone with significant anthropogenic carbon followed by movement of the contaminant plume into a region with less natural or anthropogenic carbon and oxygen concentrations greater than 1 mg/L. Then aerobic degradation pathways will take over and the partially reduced chlorinated contaminants such as vinyl chloride will be utilized as food via electron donor reactions.

In this manner the chlorinated solvent contaminant may be eliminated prior to arrival at a potential exposure site. It is also possible, at least in principle, to encourage this transition from reductive dechlorination to aerobic mineralization by the addition of dissolved oxygen to the migrating contaminant plume. Active remediation efforts to remove subsurface contamination are the subject of the next section.

8.4 REMEDIATION OF CONTAMINATED SOILS

Until about 20 years ago contaminated soil was largely ignored as an environmental concern. Only when demonstrated migration to points of contact, ingestion, or inhalation led to human and ecological risks was this problem identified and addressed. The first responses to acknowledged soil contamination issues, and still a widely used response, is removal and placement in a more secure landfill environment. Although this simply moves contaminated soil from one place to another, it can be of significant benefit due to improvements in landfill design. Often early landfills were sited in wetlands or adjacent to rivers and encouraged contaminant migration and ultimately exposure to at-risk populations. Wastes could be stabilized after removal and before or during placement to further reduce mobility after placement. Stabilization might include solidification with concrete or a similar material or direct chemical treatment of certain contaminants.

In certain cases, incineration or thermal treatment of the contaminated soil could be used to eliminate organic contaminants susceptible to destruction or removal by these means. Because these processes also necessarily entail vaporization of water and treatment or destruction of certain innocuous organic materials in the soil, the processes are energy-intensive and expensive. As a result, the approach is generally only appropriate and cost effective in soils that cannot be remediated by other means. These approaches are generally not applicable to inorganic contaminated soils.

A variety of other processes have been employed to treat contaminated soils once excavated and removed from a site. Included among these are biological degradation in dedicated bioreactors and sophisticated extraction schemes, for example, supercritical extraction, followed by the application of destruction processes to the effluent. All such options are hindered by the need to remove the soils, with its associated costs, potential disruption of surface activities, and habitat destruction.

An alternative to removal options of remediating soil is the use of *in situ* means that do not require soil removal. These are generally the options of choice if they can be demonstrated effective at reducing the volume, toxicity, or exposure to the wastes. Unfortunately, *in situ* treatments are necessarily less subject to engineering controls that can be implemented quite effectively on above-ground treatment and destruction processes. For deep contamination, contamination beneath sensitive land uses, and some other contamination issues, *in situ* treatment may be the only viable remedial approach.

Let us examine both removal and nonremoval options for soil remediation in more detail.

8.5 REMOVAL OPTIONS FOR SOIL REMEDIATION

8.5.1 INCINERATION

Incineration of solid wastes and contaminated soils are very affective means of destroying organic wastes. Efficiencies of destruction of much greater than 99% of most organic compounds are routinely observed by maintaining temperatures above 2000°F and residence times in excess of 2 s. The effectiveness of incineration is generally measured in terms of the destruction removal efficiency (DRE), measured by

$$DRE = \frac{(Q_m)_{in} - (Q_m)_{out}}{(Q_m)_{in}} X100 \tag{8.74}$$

where Q_m is the mass flowrate of the particular component for which destruction is desired. A reference to "four nine's" refers to a DRE of 99.99%.

A number of problems are often cited for incineration, however, including:

- Products of incomplete combustion (PICs)
- Generation of toxic emissions due to the presence of other contaminants

- Residue of non-combustible contaminants (e.g., metals)
- High cost of soil treatment

Incomplete combustion generally occurs when the residence time and temperature of the combustion chamber are insufficient to completely convert organic compounds to carbon dioxide and water. Some starting and intermediate compounds, however, are inherently difficult to burn completely, especially in the complex soil matrix which places severe demands upon furnace design.

Generally of more concern is the presence of other contaminants, e.g., sulfur or chlorine, that can produce toxic contaminants in the stack gases. For example, 95% or more of the sulfur will oxidize to sulfur dioxide in a combustion chamber, requiring desulfurization of the flue gases or release of additional sulfur dioxide into the environment. Since the amount of sulfur dioxide produced in a small waste or soil incinerator is very little compared to a coal-fired power plant, this may not be of significant concern. The presence of chlorinated wastes in the soil means that combustion will release chlorine in the form of other chlorinated contaminants and hydrochloric acid (HCl). The HCl can cause severe corrosion problems in the incinerator system, but the release of other chlorinated compounds, such as dioxin (2,37,8-tetrachlorinated dibenzo-*p*-dioxin, TCDD) is generally more troubling to the broader community. Some dioxin can always be found in the incinerator flue gases when chlorinated compounds are burned. Since dioxin is considered to be one of the most potent carcinogens known, this problem has significantly hindered the application of incineration to many wastes.

Incinerators also are of little use for dealing with noncombustible contaminants such as metals. The metals typically remain with the ash residue that requires landfilling or some other treatment method. If there is a significant reduction in volume of the material being burned, the metals in the residual could be concentrated significantly, making disposal more difficult.

Finally, the cost of incineration means that it is viable only in a few situations. Clearly incineration of a produced NAPL is appropriate because the residual volume is small, mass transfer limitations associated with a solid phase are not present, and the high heating value of such a waste means that additional fuel usage can be minimized or eliminated. For a soil containing small amounts of combustible material and large amounts of water to vaporize, however, incineration can be extremely expensive due to additional fuel requirements.

8.5.2 Landfilling

Isolating contaminated soil and wastes in a landfill is the oldest means of reducing exposure. Unfortunately, an insecure landfill simply moves the problem and does not reduce the volume or toxicity of the contaminated material. Most of the contaminated soil sites of greatest concern are poorly designed treatment sites. Some of the initial efforts to deal with wastes responsibly have been found in retrospect to be wanting.

The primary goal of an acceptable landfill system is elimination of opportunities for contaminants in the landfilled materials to be released to the air and water and migrate offsite. Figure 8.14 illustrates some of the techniques designed to achieve this. The landfill is lined with liners which exhibit low permeability to the percolation of water from the landfill mass. Liners may be constructed from clay, commercial stabilizers such as concrete, or soil-additive mixtures such as bentonite, or synthetic fabrics or geomembranes. Clay liners are placed by applying thin layers, providing compaction, and then adding additional layers up to a total depth of about 1 m. Soil-additive liners are prepared in a similar manner with a stabilizing additive such as bentonite being added to thin layers and providing mixing and compaction before adding additional layers. The total thickness is typically similar to that for a clay liner. Geomembranes, such as high density polyethylene, are extremely thin (e.g., 30 mils, 0.76 mm) and placed at one time. Care must be taken to avoid puncturing the thin membrane or its water retention properties are lost. All liners are typically placed with a layer of protective soil above.

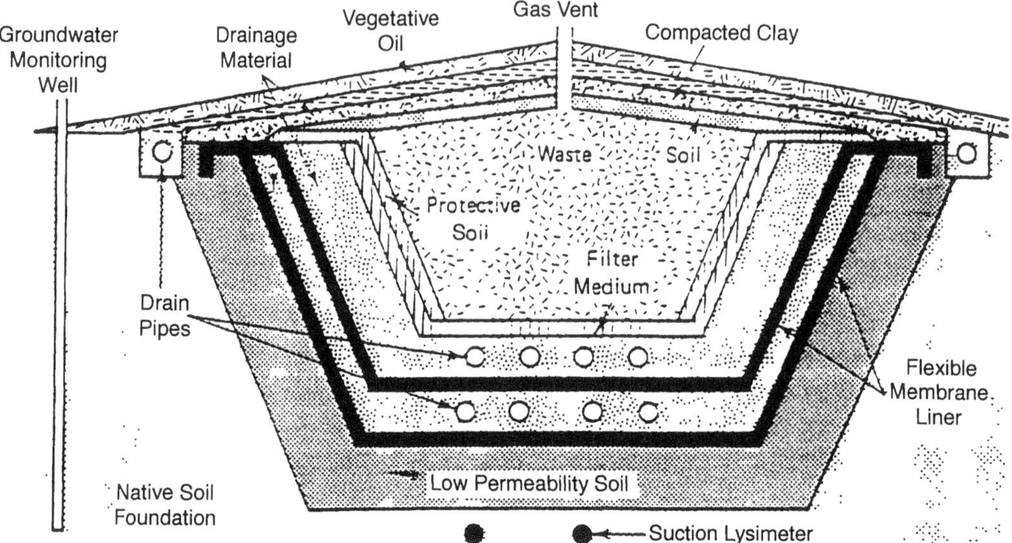

FIGURE 8.14 Schematic of a hazardous waste landfill. (From Masters, G.M. (1994) *Introduction to Environmental Science and Engineering*, Prentice-Hall, Upper Saddle River, NJ. With permission.)

The permeability of the liner is such that water should require significant time to penetrate. A water collection system, referred to as a leachate collection system, is placed above the liner to remove any water before that penetration can occur. The water collected in such a system has percolated through the waste leaching contaminants and therefore treatment of the produced waters is necessary. The type of treatment depends upon the contaminants present in the leachate waters.

In order to minimize the demands placed upon the leachate collection system, wastes are no longer placed in landfills in liquid form and a landfill cover is normally used to reduce infiltration of rainwater into the landfill. The characteristics of the cover are generally very similar to a liner — a layer of low permeability clay or similar material designed to eliminate water penetration. The landfill cover has the additional advantage of reducing vapor migration out of the landfill as illustrated previously in Example 8.6.

8.5.3 Stabilization/Solidification

Prior to landfilling solid wastes or soil in a landfill, solidification/stabilization processes are often used to further guard against contaminant release. This is especially true if the waste is primarily in the form of liquids or if the landfilled solids have a high water content. A wide variety of solidification/stabilization processes exist, but they all attempt to solidify the waste or soil and make associated contaminants less available to a leaching water phase. Portland cement, in combination with soluble silicate or fly ash, lime, and fly ash or cement of lime kiln dust, are all used to solidify and stabilize wastes. All of these reagents contain silica (SiO_2), oxides of calcium and magnesium ($CaO + MgO$), and alumina and iron oxide ($Al_2O_3 + Fe_2O_3$) in varying proportions. Portland cement is the result of the reaction of silica and quicklime (calcium oxide, CaO) to form calcium orthosilicates, Ca_2SiO_4. The Portland cement reduces the mobility of any contaminants contained within its solidified matrix.

Often a bulking agent is added to the mixture to reduce the volume of cementing agent needed to perform a particular solidification. This reduces costs and in some cases a bulking agent exhibits pozzolanic activity. A pozzolan is defined as a material that exhibits cementing ability when mixed with other materials. A pozzolan will encourage solidification as well. The cementing agents that are appropriate for a solidification process and the effectiveness of the solidified product must be evaluated by experimentation with the actual contaminated materials requiring solidification.

The effectiveness of a stabilization system is dependent upon the cementing materials employed and the nature of the contaminants in the soil. The desired reduction in contaminant availability is dependent upon the formation of a tight matrix that keeps water out and provides little exposure of the contaminant. Large amounts of organic materials such as oil and grease may deter the solidification reactions and reduce the effectiveness of the product. Soluble salts of magnesium, tin, zinc, copper or lead may cause swelling or cracking and increase the availability of the contaminants. Soluble sulfates may lead to deterioriation of the cement and release of the contaminants. In addition a variety of constituents may retard settling leading to exposure to leachate water, for example, before the solidification process is complete. Finally, the setting reactions evolve significant amounts of heat and this will tend to drive volatiles out of the system. The solidified system or leachate water may show low levels of contaminants, but this may be due simply to vaporization during the solidification process. Several inventors have promoted solidification treatment processes, which upon closer investigation have proven to simply be vaporization processes.

8.5.4 *Ex Situ* Bioremediation

The final *ex situ* treatment option that we mention here is off-site bioremediation. In principle this is no different from a wastewater treatment system in that the process is generally operated with sufficient water to generate a slurry. Biological activity is entirely within the aqueous phase and the presence of soil simply slows the process by reducing the amount of contaminant available in the aqueous phase. By processing the soil in an above-ground facility careful control of nutrients, oxygen, and mixing requirements can be maintained. In principle, it is possible to provide sufficient mixing to maintain the aqueous slurry at equilibrium with the soil phase. For a hydrophobic organic exhibiting a partition coefficient of K_{sw} with the soil, the concentration in the aqueous phase which is available for biodegradation can be written

$$C_w = \frac{C_{w0}}{(\varepsilon + \rho_b K_{sw})} \tag{8.75}$$

where C_{w0} is the concentration that would be in the aqueous phase if none of the contaminant were sorbed to the solid fraction and ρ_b is the bulk density of the solid phase. As we have employed previously, ε represents the void fraction, here likely to be much greater than the 30 to 45% porosity observed *in situ*. Thus, the presence of the slurry reduces the rate of a first order reaction by the retardation factor, $\varepsilon + \rho_b K_{sw}$. In reality, the reduction may be much greater due to the inability to reach equilibrium saturation with respect to the sorbing chemical.

8.6 *IN SITU* SOIL REMEDIATION PROCESSES

In situ processes avoid the costs and exposure associated with removal of the contaminated soil. Treating the soil *in situ*, however, limits the range of remedial options available and their ultimate effectiveness. Two general options exist for *in situ* remedial processes — extraction of the contaminants or destruction in-place. Extraction below the water table can take place by pumping groundwater and treatment at the surface for the removal of contaminants followed by reinjection of the water. This extraction can be enhanced by thermal or chemical means (with surfactants) to increase the solubility of the contaminants in the extracting water phase. Above the water table, extraction is normally conducted by creating a vacuum to remove vapors from the unsaturated zone. With this vapor comes the volatilized contaminants. This has proven to be an especially useful process for the hazardous components of gasolines and other light hydrocarbon fuels. These components, primarily benzene, toluene, ethyl benzene, and xylene (BTEX) are volatile and easily removed by this process. The remaining constituents pose a much reduced hazard to groundwater and to direct exposure by vapor inhalation than the BTEX compounds. For specific contaminants other extractive technologies also exist. For ionic species, for example, voltage placed across the soil will result in migration of charged species to a collecting cell.

In situ destruction methods typically focus on bioremediation. Microbes exist in large numbers in near-surface soils and it becomes simply a question of encouraging the right microbes and making nutrients and oxygen available to provide the desired contaminant-degrading environment. Some abiotic processes also exist to destroy certain contaminants *in situ* including *in situ* vitrification, which uses electrical power to vitrify, or convert to a glassy state, the soil and its contaminants. This process also tends to drive off volatile species due to the high temperatures that are generated.

Let us examine both extractive and destructive *in situ* remedial processes in more detail.

8.6.1 Pump and Treat Extraction of Contaminated Groundwater

Groundwater pump and treat refers to the effort to remove contaminated groundwater or separate contaminant phases (NAPL) via withdrawal wells for above-ground treatment. In this manner the mobile component of the subsurface contaminants could be treated directly and without removal and treatment of the surrounding soil. Direct removal of a nonaqueous contaminant phase always leaves a residual that can be difficult to locate and may serve as a continuing source of contamination of groundwater. In addition, the vast majority of sparingly soluble contaminant resides in the soil phase and vast volumes of contaminated water must typically be removed to eliminate the subsurface contamination. The cleanup of a site via pump and treat technology was generally expected to require years to decades of pumping of marginally contaminated water. Even these estimates of the required time, however, proved overly optimistic. A variety of problems including site heterogeneity, the presence of long term contaminant sources, and low concentrations of contaminant in the withdrawn water combined to slow the rate of soil and groundwater recovery. It soon became clear that the cleanup of aquifers by simply pumping and treating groundwater to levels approaching that required of drinking water was generally not feasible.

Even though the objective of soil and groundwater remediation was soon recognized as unachievable, groundwater removal often resulted in reversal of groundwater gradients and thus the direction of groundwater flow. Because the primary route of exposure and risk to the ecosystem and humans was generally the result of off-site migration of contaminated liquids, reversal of the groundwater flows could eliminate or minimize these risks. Containment of off-site migration and risk became the primary objective at many pump and treat sites. Unfortunately, this commits a cleanup to operate indefinitely unless some fate processes render the contaminant harmless over time.

The contaminant processes during pump and treat remediation can be conveniently divided into (1) contaminant release from a source area (typically the NAPL residual) and (2) contaminant migration from the source area to the withdrawal point. It is important to recognize that the second of these processes emphasizes that the limitations of pump and treat remediation are not solely associated with release from the NAPL residual. In those occasions where all of the contaminant exists in the aqueous phase and no NAPL residual remains, pump and treat remediation is still limited by the ability of the withdrawn water to carry the contaminant. Contaminants of interest typically sorb to the soil phase, reducing the fraction of contaminant that can be removed by displacing the water from the pore space.

It is often assumed that the soil and the adjacent porewaters are in a state of chemical equilibrium. For the hydrophobic, or sparingly soluble, organic contaminants of interest here, the capacity of the solid phase is governed by the organic carbon fraction of the soil as discussed previously. The equilibrium relations allow one to predict the water or mobile phase concentration that can be swept toward a withdrawal well based on the loading on the soil. Linear partitioning gives a reasonable approximation of the concentration of contaminant in the adjacent water as long as that predicted concentration is well below the compound's solubility. The water concentration cannot exceed the compound's solubility and the soil loading corresponding to this maximum water concentration is sometimes referred to as the critical mass. If the volumetric withdrawal rate from a well is Q, the maximum rate of removal of mass is then given by

$$\begin{aligned} Q_m &= QC_w \\ &= Q\frac{W_s}{\omega_{oc}K_{oc}} \qquad \text{if } W_s < W_{critical} \\ &= QS_w \qquad \text{if } W_s > W_{critical} \end{aligned} \tag{8.76}$$

These relations continue to assume that suspended or colloidal organic carbon does not contribute significantly to the total water concentrations. Colloidal organic carbon can cause the total water concentration to exceed the water solubility which is only a dissolved concentration.

The ratio of the total concentration in a soil system to that in the mobile phase is the retardation factor

$$R_f = \frac{C_{sw}\varepsilon + \rho_b W_s}{C_{sw}} = \varepsilon + \rho_b K_{sw} \tag{8.77}$$

This retardation factor indicates the additional mass that must be transported in the mobile phase (the porewater) to fill the immobile phase. In a transient transport process it indicates the ratio of the actual time required for a contaminant to travel between the source area and the withdrawal well and the time that would be required for a non-sorbing tracer. The minimum volume of water that must be removed to extract a non-sorbing tracer from a volume of contaminated groundwater is the same as the volume of water between the source area and the withdrawal well, or one pore volume (V_p). The minimum volume of water (V_e) or time (T_e) required to extract a sorbing contaminant (V_e) is then

$$\begin{aligned} V_e &= R_f V_p \\ \tau_e &= R_f \tau_p \end{aligned} \tag{8.78}$$

Thus, a sorbing contaminant can require significantly more water to be extracted before the contaminant can be eliminated from the system. This was explored previously in Example 8.8. Consider a soil with a porosity of 40%, a bulk density of 1.5 g/cm^3, and containing 1% organic carbon. Benzene has an organic carbon based partition coefficient of 83 cm^3/g. The retardation factor for benzene in this soil is estimated to be

$$\begin{aligned} R_f &= \varepsilon + \rho_b K_{sw} = \varepsilon + \rho_b \omega_{oc} K_{oc} \\ &= 0.4 + (1.5)(0.01)(83) \\ &= 1.645 \end{aligned} \tag{8.79}$$

A minimum of 1.645 pore volumes of water is required to flush water contaminated with benzene from the subsurface in this example. Note, however, that this is valid only as long as the soil phase concentration of benzene is less than the critical loading or, given a water solubility for benzene of 1780 g/m^3,

$$W_{critical} = S_w f_{oc} K_{oc} = 1480 \text{ mg/kg} \tag{8.80}$$

The actual number of pore volumes or time is considerably greater than given by the above equations in any real soil environment due to two considerations

1. Not all of the contaminant (perhaps only a very small fraction of the total) is present in the porewater at the initiation of flushing. The material that is sorbed or not in contact with the mobile porewater must first be released from the source areas.
2. Uncontaminated water also is withdrawn from the well diluting the effluent and increasing the total volume of water that must be removed before cleanup of the porewater.

In practice, the initial concentrations observed in a groundwater extraction well is due to water that has long been in contact with the contaminated source areas and is equilibrated with those areas. The initial concentrations measured in a withdrawal well are relatively high. Once this water has been removed, however, the subsequent water is generally not in contact with the contaminants sufficiently to reach equilibrium and the effluent concentration decreases. This is shown in Figure 8.15. Ultimately, the concentration drops so low that the pumping is stopped. At one time, it was assumed that the aquifer had been completely remediated when this occurred. Experience has now shown us that if pumping is restarted after a period of no operation once again high concentrations will be observed because the withdrawn water will have had an opportunity to equilibrate once again with the contaminants. Continued pumping will once again result in reduction of the concentration as before. This increase upon the reinitiation of pumping is termed the rebound effect. The inability to achieve lasting low concentration groundwater as a result of pump and treat remediation is the reason that the process is considered appropriate for containment of groundwater, but not for full remediation of the aquifer and adjacent soils.

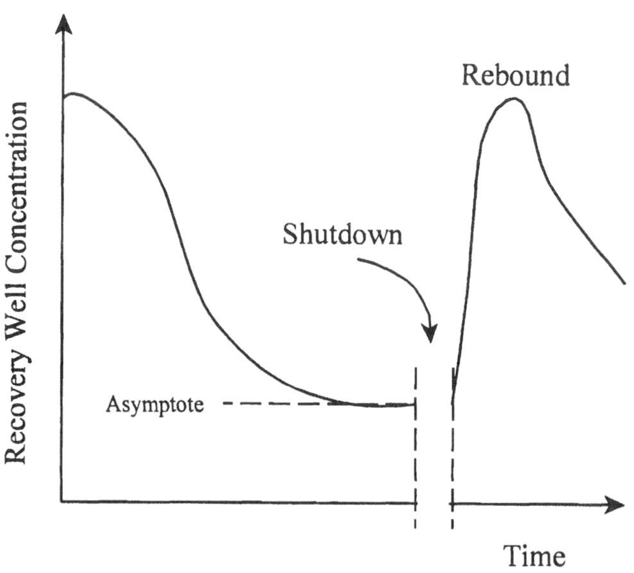

FIGURE 8.15 Concentrations in recovery well as a function of time during a groundwater removal effort. The concentration often approaches an asymptote whereupon the cleanup limit might be assumed to be reached and the recovery operation shutdown. Later restarting of the recovery operation may find a rapid rebound to near the initial well concentrations due to recontamination of the groundwater by previously inaccessible contaminants.

8.6.2 Enhancement of Pump and Treat Processes

Pump and treat methods of remediation of soils are of limited usefulness when significant quantities of NAPLs exist. Due to the low solubility of most soil contaminants, large volumes of water are required to remove contaminants present in a separate phase even if it were possible to maintain the water at saturation. Because the solubility of organic compounds in water increases significantly with temperature, however, the use of steam or other sources of heating has been proposed to encourage soil remediation. As with incineration, it is necessary to raise the temperature of a large volume of inert soil and water in order to heat the contaminants. The process is energy intensive although generally the vaporization of soil water is generally avoided easing the energy demands to some extent.

Surfactants, or surface-active agents, are similarly used to increase the solubility of contaminants in the groundwater. Surfactants present at concentrations of 1 to 5% can increase the solubility of hydrophobic organic compounds by several orders of magnitude. Surfactants also reduce interfacial tension between an oily phase and the adjacent water, however. Because of the Laplace relationship

$$\Delta P = \frac{2\sigma}{r} \tag{8.81}$$

the reduction in interfacial tension translates directly into a reduction in the pressure required to mobilize the NAPL from a pore of a given radius. Thus, surfactants not only contribute to dissolution of the contaminants but also to their mobilization. The mobilization must be controlled through the application of an appropriate hydraulic gradient or by the presence of low permeability supporting strata. Any uncontrolled mobilization which results in further downward contaminant phase migration or migration toward a sensitive receptor is undesirable. As a result of this phenomenon, surfactant-enhanced remediation has received much attention and study but it has only been applied under field conditions at very few sites. Heat addition in the form of steam injection also can increase the mobility of a residual nonaqueous phase to some extent due to the corresponding reduction in viscosity of the contaminant phase. This means that a reduced pressure gradient can be applied to achieve significant mobilization.

8.6.3 Vacuum Extraction in the Unsaturated Zone

A process that is similar conceptually to pump and treat of groundwater is soil vacuum extraction (SVE) in the water-unsaturated zone. In this process, a vacuum is applied to the unsaturated zone by placing a vacuum pump on a well screened in the unsaturated zone. This pulls vapors through the soil, removing any volatile components that have volatilized in the subsurface. Because the viscosity of air is low and the relative capacity of many volatile contaminants in air is quite high (the capacity of atmospheric pressure air for benzene is 0.125 atm or 12.5% at 25°C), the process is potentially quite effective. Just as we did with pump and treat of groundwater, it is possible to make some simple estimates, at least providing a minimum time required to clean a contaminated unsaturated zone via vacuum extraction. Example 8.12 considers the evaporation of a volatile organic into the flushing air phase.

Example 8.12: Volume of air required to evaporate a nonaqueous phase residual

Consider a soil with a residual NAPL saturation of 20% and a total porosity of 40%. Estimate the pore volumes of air required to completely evaporate the liquid if the liquid is (1) 1,2-dichloroethane, (2) pure benzene, or (3) a 1% mixture of benzene in an essentially insoluble bulk NAPL with density 0.8 g/cm^3 with an average molecular weight of 140.

Let us employ a basis of 1 m³ of soil as in Example 8.8. The total mass of NAPL to be removed from the soil is then

$$M_n = \rho_n \varphi_n \varepsilon (1\ \mathrm{m}^3) \qquad \rho(\mathrm{g/cm^3}) = \begin{bmatrix} 1.25 \\ 0.88 \\ 0.8 \end{bmatrix} \qquad M_n(\mathrm{kg}) = \begin{bmatrix} 100 \\ 70.4 \\ 64 \end{bmatrix}$$

$$x_B = \frac{0.01/78}{0.01/78 + 0.99/140} = 0.018 \qquad C_B = x_B S_w = (0.018)(1780\ \mathrm{mg/L}) = 31.7\ \mathrm{mg/L}$$

The volume of air required depends upon the partial pressure in air of pure 1,2-dichloroethane (P_v = 0.24 atm), benzene (P_v = 0.125 atm), and benzene from the mixture (pB = 0.018·0.125 atm = 0.00222 atm). The minimum volume of air required to remove the contaminants from the soil and the minimum volume divided by the pore volume in the soil (i.e., the number of pore volumes) are

$$V_w = \frac{M_n}{C_a} = \frac{M_n}{\dfrac{p_a \cdot MW}{RT}} \qquad V_w(\mathrm{m^3}) = \begin{bmatrix} 103 \\ 177 \\ 9030 \end{bmatrix}$$

$$V_p = \frac{V_w}{\varepsilon \cdot 1\ \mathrm{m}^3} \qquad V_p = \begin{bmatrix} 258 \\ 442 \\ 22600 \end{bmatrix} \begin{matrix} \text{1,2-dichloroethane} \\ \text{Benzene} \\ \text{1\% mixture of benzene} \end{matrix}$$

Note that an even larger volume of air is required to flush the NAPL from the pore space than for water in Example 8.8. The difference is that it is much easier (and cheaper) to pump large volumes of low viscosity and low density air through soil than water, as is shown in Example 8.13.

A model of the dynamics of soil vacuum extraction can also be developed. That the air moves as a plug through the soil as used in groundwater pump and treat is even less likely to be an appropriate assumption. Higher air flowrates and mixing and lack of control over the entry point of the air suggests that it is more appropriate to consider the vacuum extraction of a volume of soil as though it were taking place in a zone of approximately uniform concentration. The initial contaminant mass in the soil is M, which is assumed to be distributed in the local air, water, non-aqueous, and soil phases via equilibrium relationships. If ε_a represents the air-filled porosity, ε_w, and ρ_b the bulk density of the soil, the relationship between the air concentration and the total mass of contaminant in a volume of soil, V, is

$$\begin{aligned} \frac{M}{V} &= \varepsilon_a C_a + \varepsilon_w C_w + \varepsilon_n C_n + \rho_b W_s \\ &= \varepsilon_a C_a + \varepsilon_w \frac{C_a}{K_{aw}} + \varepsilon_n K_{nw} \frac{C_a}{K_{aw}} + \rho_b K_{sw} \frac{C_a}{K_{aw}} \end{aligned} \tag{8.82}$$

where K_{nw} is the partition coefficient between the NAPL and water and all other terms have been defined previously. K_{nw} can be estimated by

$$K_{nw} = \frac{C_n}{C_w} = \frac{\rho_n}{S_w} \tag{8.83}$$

A material balance on the assumed well-mixed soil zone then gives

$$\frac{d(MV)}{dt} = -QC_a$$

$$\frac{d}{dt}\left(\varepsilon_a + \frac{\varepsilon_w}{K_{aw}} + \varepsilon_n \frac{K_{nw}}{K_{aw}} + \rho_b \frac{K_{sw}}{K_{aw}}\right) C_a V = -QC_a \tag{8.84}$$

$$\frac{dC_a}{dt} = -\frac{Q}{VR_f} C_a$$

where

$$R_f = \left(\varepsilon_a + \frac{\varepsilon_w}{K_{aw}} + \varepsilon_n \frac{K_{nw}}{K_{aw}} + \rho_b \frac{K_{sw}}{K_{aw}}\right)$$

This equation has the solution

$$C_a = C_a(0) e^{-\frac{Qt}{VR_f}} = \frac{M_0}{VR_f} e^{-\frac{Qt}{VR_f}} \tag{8.85}$$

Alternatively, solving for the time of remediation

$$t = -V \frac{R_f}{Q} \ln\left(\frac{C_a}{C_a(0)}\right) \tag{8.86}$$

Thus, for 99% remediation of the site gives $\ln(c_a/C_a(0)) = -4.605$ and the time is given by

$$t_{99} = 4.605\ V \frac{R_f}{Q} = 4.605\ V \frac{\left(\varepsilon_a + \frac{\varepsilon_w}{K_{aw}} + \varepsilon_n \frac{K_{nw}}{K_{aw}} + \rho_b \frac{K_{sw}}{K_{aw}}\right)}{Q} \tag{8.87}$$

For the movement of the incompressible groundwater through the soil in an pump and treat system, the volumetric flowrate can be found from Darcy's Law recognizing that the head gradient is linear. For a compressible phase such as air, however, the equivalent of the pressure is governed, in one dimension, by

$$\frac{d^2(P^2)}{dx^2} = 0$$

$$P = \left(P_0^2 - \frac{P_0^2 - P_1^2}{L} x\right)^{1/2} \tag{8.88}$$

Thus, the square of the pressure driving the vapor flow is linear. Taking Darcy's Law in terms of pressure and solving for the volumetric gas flow at the withdrawal well,

$$q = -\kappa_r \frac{\kappa_p}{\mu}\frac{dP}{dx}$$

$$= \frac{\kappa_r \frac{\kappa_p}{\mu}(P_0^2 - P_1^2)}{2LP_1} \tag{8.89}$$

The relative permeability of the air in the soil, κ_r, is in general different from the relative permeability defined by Equation 8.38. Equation 8.38 defines the relative permeability for a wetting phase such as water displacing oil or air or oil displacing air from a medium. Air, however, is always a non-wetting phase and the Brooks and Corey model for such a phase, **in terms of the wetting phase saturation and residual saturation**, is given by

$$(\kappa_r)_a = \left[\frac{1-(\varphi_w - \varphi_{wr})}{1-\varphi_w}\right]^2 \left\{1 - \left[\frac{\varphi_w - \varphi_{wr}}{1-\varphi_{wr}}\right]^{\frac{2+b}{b}}\right\} \tag{8.90}$$

The wetting fluid can be either water or an NAPL.

In this model, note that only the volumetric flowrate at the exit well is required by our assumption of well mixedness in the soil. The loss of contaminant from the soil is based solely on the exit concentrations and flowrate. Example 8.13 explores this model for benzene.

Example 8.13: Dynamics of remediation via soil vacuum extraction

Consider remediation of benzene-contaminated soil in the unsaturated zone via soil vacuum extraction. The saturations of water and benzene are each 25%. Assume that the residual water saturation is 5% and the organic fraction is 3%. $K_{aw} = 0.224$ and $K_{sw} = 2.61$ L/kg as in Example 8.6. Assume a soil with an intrinsic permeability of 10^{-8} cm^2 and a parameter b = 2.8. Assume the contaminated zone is 1 m^2 in area and 10 m in length. Then 1 atm pressure is held at one end of the contaminated zone and a vacuum well at the other end applies a pressure of 0.2 atm (0.8 atm vacuum).

The NAPL-water partition coefficient can be defined by considering the pure benzene phase-water equilibrium

$$K_{nw} = \frac{\rho_n}{S_w} = \frac{0.88 \text{ g/cm}^3}{1780 \ \mu\text{g/cm}^3} = 494$$

From Equation 8.84, the retardation factor is then

$$R_f = 0.4(1-0.25) + \frac{(0.4)(0.25)}{0.224} + (0.4)(0.25)\frac{494}{0.224} + (1.5 \text{ gm/cm}^3)\frac{2.61 \text{ L/kg}}{0.224} = 239$$

The relative permeability from Equation 8.90 is 0.58. The air flow is given by Equation 8.89 to be 67 m/day. The time required to remove 99% of the contamination is then

$$t_{99} = 4.605\frac{R_f V}{qA} = 4.605\frac{(239)(10 \text{ m}^3)}{(67 \text{ m/day})(1 \text{ m}^2)} = 163 \text{ days}$$

This is quite a rapid rate of removal compared to what can be accomplished by water flushing below the water table. For this and economic reasons, soil vacuum extraction has become the standard response to volatile contaminated unsatured soils.

Soil vapor stripping can also be enhanced by thermal means. Because the change in vapor pressure of a compound with temperature is generally much greater than an increase in water solubility with temperature, Henry's Law constant increases with temperature and the effect of retardation by sorption into the water and soil phases decreases. Thus, more of the contaminant becomes available in the vapor phase for removal by the stripping process. As with thermally enhanced contaminant removal by pumping groundwater, the additional cost of the energy supplied may offset any benefit gained. The relative merits of heating depend, as with thermally enhanced groundwater pumping, on the particular compound and soil setting.

8.6.4 *In Situ* Bioremediation of Soils

Perhaps the most desirable of all treatment processes is *in situ* biodegradation to render the soil harmless and to naturally recycle the contaminants. This desirable goal has proved elusive. There are a number of compounds that undergo detoxification by microbial processes at rates that are sufficient to justify natural recovery of contaminated soils. That is, the rate of recovery of the soil by these processes is such that the contaminant will not have traveled sufficiently far off-site to pose a risk to human or ecological health before being rendered harmless by natural degradative processes. In other cases, the addition of nutrients, such as nitrogen or phosphorus or an oxygen source, may provide the enhancement needed to achieve acceptable rates of degradation. This has proven the case for many petroleum hydrocarbons that are degraded by a variety of natural microorganisms under aerobic conditions. In other cases, the presence of easily degradable compounds can lead to *cometabolism* and the resulting destruction of other more refractory compounds. This is the case with high molecular weight polycyclic aromatic hydrocarbons in the presence of lower molecular 2- and 3-ring polycyclic aromatics. Although biodegradation of these petroleum hydrocarbons is routinely accomplished, it is important to recognize that the rate of degradation depends upon a number of factors that are not easily quantified including temperature, desorption rates from the soil for soil-sorbed contaminants, and the concentration and degree of acclimation of microorganisms. Field trials and monitoring must supplement any efforts to predict the rate of the biological degradation processes.

Chlorinated solvents generally pose more difficult problems for *in situ* bioremediation than petroleum hydrocarbons. As discussed previously, dechlorination, the first steps in degradation, typically occurs via anaerobic pathways. Anaerobic microbial processes typically occur at a slower rate than aerobic processes and, in addition, the dechlorination of chlorinated ethanes can lead to the formation of vinyl chloride as discussed previously. This is a serious problem in that vinyl chloride is more mobile than its more chlorinated precursors, and more toxic. Vinyl chloride is a known human carcinogen and its presence in groundwater can lead to significant concerns about the health and safety in exposed populations. This is one of the few examples where biodegradation leads to contaminants that are more toxic than the parent compounds. Generally biodegradation leads to lower molecular weight, less toxic compounds, and under aerobic conditions, more oxidized compounds.

A recent development in the *in situ* treatment of contaminated groundwater overcomes this problem by combining both anaerobic treatment for the dechlorination of multichlorinated organic compounds followed by aerobic degradation of vinyl chloride and other aerobically degraded contaminants. This process is conducted via a process in which contaminated groundwater is forced via a gate into the treatment zone as shown in Figure 8.16. Iron filings are often used to create anaerobic, reducing conditions in the first layers of the gate for the reductive dechlorination of the multichlorinated organic species. This is followed by a zone where oxygen

12. Plot the Brooks and Corey relative permeability function for various values of b between sand (b = 2.8) and a fine clay (b ~10). How does the soil type cause this function to vary?

13. If a fine sand has a capillary fringe of 20 cm above the water table, what is the effective capillary diameter to which this would correspond? Does this seem reasonable given the actual size of pores in such a media?

14. If the fine sand in Problem 13 has an intrinsic permeability of 10^{-8} cm^2, what is the conductivity of an oil with a viscosity of 1000 cp and a specific gravity of 0.8? What is the expected capillary rise and an estimate of the residual oil saturation if the surface tension of the oil is about 20 dyn/cm? The field capacity (or residual water saturation) of the soil is 20% of the total soil volume.

15. If the oil of Problem 14 (σ_{an} = 20 dyn/cm and μ_n = 1000 cp) is spilled onto the surface of the fine sand, estimate the rate of infiltration into the soil as a function of time neglecting the effect of any ponded depth at the surface. If the total spill is 10,000 L over a 10 m^2 area and the soil porosity is 40%, estimate the depth to which the spill will penetrate.

16. If a soil contains a gasoline with a vapor pressure of 0.3 atm at 25°C, estimate the initial rate of evaporation immediately after a spill, assuming the air-side mass transfer coefficient is 3000 cm/h. Estimate the reduction in this rate of evaporation if the gasoline is covered with a 25-cm layer of sand with a total porosity of 40% if the sand is dry, i.e., contains no water, if the sand contains 20% water by volume (water saturation of 50%), and if a sprinkler system at the surface maintains the sand at close to 90% saturation.

17. In the situation of Problem 16, the gasoline also contains 5% benzene. Estimate the benzene evaporation rate immediately after the spill and for a sand cap with each of the moisture conditions evaluated in Problem 16.

18. Estimate a characteristic time ($T_{diff} \approx h^2 R_f / D_{sv}$) for benzene vapors to penetrate a newly placed soil cap 25 cm thick (h) under the following conditions: (1) dry sand cap with 40% porosity; (2) moist sand cap with a volumetric content of 20% (water saturation of 50%), and moist sand cap (as in condition 2) with 5% organic carbon added to increase sorption. The bulk density of the cap soil is 1.5 g/cm^3.

19. The gasoline spill referred to in the preceding problems has migrated to the water table. Estimate the maximum concentration of benzene that would be found in the water in contact with the floating pool of gasoline (benzene is 5% of the gasoline).

20. If 100 L of 1,2-dichloroethane (EDC) is distributed over a cubic region 10 m on a side and the average groundwater flow through this region is 30 m/year, estimate the minimum amount of time required to completely dissolve the trichloroethylene.

21. If the soil through which the groundwater moves in Problem 20 is 40% porosity sand with essentially no sorption potential, estimate the time required for the EDC to be detected at concentrations exceeding 1 μg/L at a drinking water well located 100 m downgradient. Repeat the calculation if the soil contains 5% organic carbon (i.e., if sorption cannot be neglected).

22. Compare the various disposal options for organic-contaminated soil after removal from a contaminated site. How might the feasibility of these options change if the contaminant is a metal?

23. Under the conditions of Problem 19, plot the minimum removal time as a function of groundwater flowrate.

24. Is it feasible to remove the EDC discussed in Problem 19 by placement of withdrawal and injection water wells and increasing the local groundwater flow? Fully support your answer.

25. If 100 L of gasoline (P_v = 0.3 atm at 25°C) is trapped within an unsaturated zone of a cubic soil region 10 m on a side, estimate the minimum time required for removal if a vacuum system applies a pressure differential of 0.25 atm across this region. Make the calculation for each of the following soils. In each soil the total porosity is 40%. Sandy soil with κ_p = 10^{-8} cm^2, b = 2.8, φ_w = 0.2. Silty soil with κ_p = 10^{-10} cm^2, b = 5, φ_w = 0.4. Silty, clayey soil with κ_p = 10^{-12} cm^2, b = 10, φ_w = 0.6.

26. If 100 L of gasoline containing 5% benzene is trapped within an unsaturated zone of a cubic soil region 10 m on a side, estimate the minimum time required for removal of the benzene if a vacuum system applies a pressure differential of 0.25 atm across this region. Make the calculation for each of the following soils. In each soil the total porosity is 40%. Sandy soil with κ_p = 10^{-8} cm^2, b = 2.8, φ_w = 0.2. Silty soil with κ_p = 10^{-10} cm^2, b = 5, φ_w = 0.4. Silty, clayey soil with κ_p = 10^{-12} cm^2, b = 10, φ_w = 0.6.

27. If 100 L of EDC is trapped within an unsaturated zone of a cubic soil region 10 m on a side, estimate the minimum time required for removal if a vacuum system applies a pressure differential of 0.25 atm across this region. Make the calculation for each of the following soils. In each soil the total porosity is 40%. Sandy soil with $\kappa_p = 10^{-8}$ cm^2, b = 2.8, $\varphi_w = 0.2$ Silty soil with $\kappa_p = 10^{-10}$ cm^2, b = 5, $\varphi_w = 0.4$. Silty, clayey soil with $\kappa_p = 10^{-12}$ cm^2, b = 10, $\varphi_w = 0.6$.

28. Estimate a characteristic time for the transport of EDC to an air withdrawal well located 10 m away from the contaminated region. Consider each of the following soils. The total porosity is 40% and the bulk density is 1.5 g/cm^3 of each soil. Sandy soil with $\kappa_p = 10^{-8}$ cm^2, b = 2.8, $\varphi_w = 0.2$, $\omega_{oc} = 0.005$. Silty soil with $\kappa_p = 10^{-10}$ cm^2, b = 5, $\varphi_w = 0.4$, $\omega_{oc} = 0.02$. Silty, clayey soil with $\kappa_p = 10^{-12}$ cm^2, b = 10, $\varphi_w = 0.6$, $\omega_{oc} = 0.05$.

REFERENCES

Brooks, R.H. and A.T. Corey (1964) Hydraulic properties of porous media, Hydrology Paper No. 3, Colorado State University, Fort Collins, CO.

Clapp, R.B. and G. M. Hornberger (1978) Empirical equations for some soil hydraulic properties, *Water Resour. Res.*, 14 (4): 601–604.

Dunne, T. and L.B. Leopold (1978) *Water in Environmental Planning*, W.H. Freeman and Co., San Francisco.

Freeze, R.A. and J.A. Cherry (1979) *Groundwater*, Prentice-Hall, Upper Saddle River, NJ, 604 pp.

Hillel, D. (1970) *Soil Physics,* Academic Press, New York.

Masters, G.M. (1990) *Introduction to Environmental Science and Engineering*, Prentice-Hall, Upper Saddle River, NJ.

Millington, R.J. and J.M. Quirk (1961) Permeability of porous solids, *Trans. Faraday Soc., 57*, 1200–1207.

Neuman, S. P. (1990) Universal scaling of hydraulic conductivities and dispersivities in geologic media, *Water Resour. Res., 26*(8) 1749-1758.

Reible, D.D., T.H. Illangasekare, D.V. Doshi and M.E. Malhiet (1990) "Infiltration of Immiscible Contaminants in the Unsaturated Zone," *Ground Water,* 685–692, August–September.

Saffman, P.G. (1959) *J. Fluid Mech.*, 6, 21.

Wilson, J.L., S.H. Conrad, W.R. Mason, W. Peplinski and E. Hagan (1990) Laboratory Investigation of Residual Liquid Organics, EPA/600/6-90/004.

Appendix

Selected Physical Property Data

Compound	Formula	Mol. Wt.	SG (20/4)	T_m (°C)	$\Delta H_m(T_m)$ (kJ/mo)	log K_{ow}	log K_{oc} (L/kg)	S_w(25°C) (mg/L)	P_v(25°C) (atm)	K_{aw} (conc. ratio)	H (atm* m³/mo)	T_b (C)	$\Delta H_v(T_b)$ (kJ/mol)	T_c (K)	P_c (atm)	ΔG_f° (kJ/mol)	ΔH_f (kJ/mol)	ΔS_f° (kJ/mol)	ΔH_c° (kJ/mol)
Acetic acid	CH_3COOH	60.1	1.049	16.6	12.09	-0.17	—	Soluble	0.016	4.12E-06	1.01E-07	118	24	594.8	57.1	-377	-435	-0.20	-920
Acetone	C_3H_6O	58.1	0.791	-95.0	5.69	-0.24	0.34	Soluble	0.355	8.41E-04	2.06E-05	56.0	30.2	508.0	47	-153	-218	-0.22	-1821
Ammonia	NH_3	17.0	—	-77.8	5.653		0.12	Soluble	—	—	—	-33.43	23.4	405.5	111.3	-16	-46	-0.10	-383
Aniline	C_6H_7N	93.1	1.022	-6.3	—	0.94	1.15	34.000	3.95E-04	7.35E-05	1.80E-06	184.2	41.9	699	52.4	167	87	-0.27	—
Anthrecene	$C_{14}H_{10}$	178.2	1	216		4.45	4.15	0.045	2.57E-07	7.76E-04	1.90E-05	340	57	883		—	225	—	—
Benzene	C_6H_6	78.1	0.879	5.53	9.837	2.12	1.92	1780	0.125	0.23	0.00559	80.1	30.765	562.6	48.6	130	83	-0.16	-3302
Benzoic acid	$C_7H_6O_2$	122.1	1.06615	122.2	—	1.87	0.18	3400	3.28E-06	5.31E-06	1.30E-07	249.8	50.66	752	45	-211	-290	-0.27	-3227
n-Butane	C_4H_{10}	58.1	—	-138.3	4.661	2.89	—	151	24.679	38.48	0.94	-0.6	22.305	425.17	37.47	-17	-126	-0.37	-2879
Carbon (graphite)	C	12.0	2.26	3600	46							4200	—	—	—	0(c)	0(c)	0(c)	-394(c)
Carbon dioxide	CO_2	44.0	—	-56.6	8.33	—	—	1452	3.46E-04	1.19	0.03	—	—	304.2	73	-395	-394	2.95E-03	—
Carbon disulfide	CS_2	76.1	$1.261^{20/20}$	-112.1	4.39	2.00	1.73	2940	0.474	0.50	0.0123	46.25	26.8	552.0	78.0	67	117	0.17	-1103
Carbon monoxide	CO	28.0	—	-205.1	0.837	—	—	26	—	39.38	0.96	-191.5	6.042	133.0	34.5	-137	-111	09	-283
Carbon tetrachloride	CCl_4	153.8	1.595	-22.9	2.51	2.64	2.70	757	0.149	1.23	0.0302	76.7	30.0	556.4	45.0	-58	-100	-0.14	-385
Chlorine	Cl_2	70.9	—	-101.00	6.406	—	—	—	7.500	—	—	-34.06	20.4	417.0	76.1	0(g)	0(g)	0(g)	—
Chlorobenzene	C_6H_5Cl	112.6	1.107	-45	—	2.84	2.52	466	0.015	0.15	0.00372	132.10	36.4	632.4	44.6	99	52	-0.16	—
Chloroform	$CHCl_3$	119.4	1.489	-63.7	—	1.97	1.70	8200	0.199	0.12	0.00287	61.0	30	536.0	54.0	-69	-101	0.11	-373(l)
Copper	Cu	63.5	8.92	1083	13.01	—	—	—	—	—	—	2595	305	—	—	0(c)	0(c)	0(c)	—
Cyclohexane	C_6H_{12}	84.2	0.779	6.7	2.677	3.44	3.23	49	0.133	10.24	0.251	80.7	30.1	553.7	40.4	32	-123	-0.52	-3953
Cyclopentane	C_5H_{10}	70.1	0.745	-93.4	0.609	3.00	—	161	0.424	7.50	0.18	49.3	27.30	511.8	44.5	39	-77	-0.39	-3320
n-Decane	$C_{10}H_{22}$	142.3	0.730	-29.9	—	6.70	—	0.038	0.002	285.25	6.99	173.8	—	619.0	20.8	33	-250	-0.95	-6830
1,2-dichlorobenzene	$C_6H_4Cl_2$	147.0	1.306	-17		3.60	3.23	100	0.001	0.079	0.00193	181	40	697	41	83	30	-0.18	—
1,1-dichloroethane	$C_2H_4Cl_2$	99.0	1.168	-97		1.79	1.48	5500	0.239	0.222	0.00543	57	29	523	50	-73	-130	-0.19	—
1,2-dichloroethane	$C_2H_4Cl_2$	99.0	1.250	-36		1.48	1.15	8520	0.084	0.040	0.000978	83.6	32	561	53	-74	-130	-0.19	—
Dioxin-TCDD	$C_{12}H_4Cl_4O_2$	322.0	1.827	305	—	6.72	6.52	2.00E-05	4.61E-12	6.53E-04	1.60E-05	412	—	—	—	—	—	—	—
Ethane	C_2H_6	30.1	—	-183.3	2.859	1.81	1.60	60	60.526	30.69	0.752	-88.6	14.72	305.4	48.2	-33	-85	-0.17	-1560
Ethyl acetate	$C_4H_8O_2$	88.1	0.901	-83.8	—	0.73	0.52	77.600	0.146	0.01	0.000183	77.0	32	523.1	37.8	-328	-443	-0.39	-2246(l)
Ethyl alcohol	C_2H_5OH	46.1	0.789	-114.6	5.021	-0.31	-0.70	275.000	0.077	4.94E-04	1.21E-05	78.5	38.58	516.3	63.0	-168	-235	-0.22	-1409
Ethyl benzene	C_8H_{10}	106.2	0.867	-94.97	9.163	3.15	3.04	152	0.009	0.26	0.00643	136.2	35.98	619.7	37.0	131	30	-0.34	-4607
Ethyl chloride	C_2H_5Cl	64.5	0.903^{15}	-138.3	4.452	1.43	1.22	5740	1.579	6.04	0.148	13.1	24.7	460.4	52.0	-60	-112	-0.17	—
Ethylene glycol	$C_2H_6O_2$	62.1	1.113^{19}	-13	-11.23	-1.93	-1.57	582.000	8.500E-05	4.07E-07	9.97E-09	197.2	56.9	645	76	-305	-390	-0.28	-1180(l)
Formaldehyde	H_2CO	30.0	0.815^{-20}	-92	—	0.35	0.14	37.200	4.605	0.15	0.00372	-19.3	24.48	408	65	-110	-116	-0.02	-563
n-Heptane	C_7H_{16}	100.2	0.684	-90.59	14.03	4.40	4.19	5.6	0.068	65.71	1.61	98.43	31.69	540.2	27.0	8.00	-187.91	-0.66	-4854
Hexachlorobenzene	C_6Cl_6	284.8	2.05	230.0	—	5.23	3.59	0.006	1.43E-08	0.028	6.81E-04	332.00	—	—	—	—	—	—	—
n-Hexane	C_6H_{14}	86.2	0.659	-95.32	13.03	3.84	3.66	19	0.234	65.31	1.6	68.74	28.85	507.9	29.9	-0.25	-167.31	-0.56	-4195

Compound	Formula	Mol. Wt.	SG (20/4)	T_m (°C)	$\Delta H_m(T_m)$ (kJ/mo)	log K_{ow}	log K_{oc} (L/kg)	S_w(25°C) (mg/L)	P_v(25°C) (atm)	K_{aw} (conc. ratio)	H (atm• m³/mo)	T_b (C)	$\Delta H_v(T_b)$ (kJ/mol)	T_c (K)	P_c (atm)	ΔG_f° (kJ/mol)	ΔH_f (kJ/mol)	ΔS_f° (kJ/mol)	ΔH_c° (kJ/mol)
Hydrogen	H_2	2.0	—	-259.19	0 12	—	—	0 22	—	50.73	1.24	-252.76	0.904	33.3	12.8	0	0	0	-286
Hydrogen chloride	HCl	36.5	—	-114.2	1.99	—	—	—	—	1.64E-05	4 02E-07	-85.0	16.1	324.6	81.5	-95.34	-92.37	0 01	—
Hydrogen fluoride	HF	20 0	0.97	-83	—	1 00	—	—	—	—	—	20	6.70	503.2	64.00	-273.41	-271.32	0 01	—
Hydrogen sulfide	H_2S	34.1	—	-85.5	2.38	0.96		3410		36.75	0.90	-60.3	18.67	363.6	88.9	-33.08	-20.18	0.04	-563
Lead	Pb	207.2	11 337[20/20]	327.4	5.10	—	—	—	—	—	—	1750	179.9	—	—	0(c)	0(c)	0(c)	—
Mercury	Hg	200 6	13.546	-38.87	—	—	—	0 03	2 632E-06	0.05	0 00114	-356.9	—	—	—	0(c)	0(c)	0(c)	—
Methane	CH_4	16 0	—	-182.5	0 94	1.12	—	6560	76.012	28.52	0 70	-161.5	8.179	190.70	45.8	-50.87	-74.91	-0.08	-890
Methyl alcohol	CH_3OH	32.0	0 792	-97.9	3 167	-0.77	-0 36	300.000	0.174	8.16E-04	2E-05	64.7	35.27	513.20	78.50	-162.62	-201.31	-0 13	-764
Methyl ethyl ketone	C_4H_8O	72.1	0.805	-87.1	—	0.27	0 65	268.000	0.102	1.112E-03	2.74E-05	78.2	32 0	535.60	41.00	-146.17	-238.53	-0 31	-2436(l)
Naphthalene	$C_{10}H_8$	128.2	1.145	80.0	—	3.23	2.97	32	1.14E-04	0 029	0.00072	217.8	43.29	748.40	40.00	223.75	151.07	-0.24	-5157(g)
Nitrobenzene	$C_6H_5O_2N$	123.1	1 203	5.5	—	1.85	1 56	1900	1.97E-04	9.80E-04	2.4E-05	210.7	—	—	—	—	—	—	3093(l)
Nitrogen	N_2	28.0	—	-210	0.720	0 92	—	17 4	—	62.41	1.53	-195.8	5.577	126.20	33.5	0(g)	0(g)	0(g)	—
Nitrogen dioxide	NO_2	48.0	—	-9.3	7.335	—	—	—	—	4.03	0 10	21.3	14 73	431.0	100.0	52.00	33 87	-0.06	—
Nitric oxide	NO	30 0	—	-163.6	2.301	—	—	54 000	—	21.64	0.53	-151.8	13.78	179.20	65.0	86.75	90 44	0.01	—
Nitrous oxide	N_2O	44.0	1.226[-89]	-91.1	—	—	—	1056	—	1.60	0.04	-88.8	16.56	309.5	71.70	103.71	81.60	-0.07	—
n-Nonane	C_9H_{20}	128.2	0.718	-53.8	—	5.45	5.25	0 47	0.009	144.49	3.54	150.6	36.94	595	23 0	24.83	-229.20	-0.85	-6171
n-Octane	C_8H_{18}	114.2	0 703	-57.0	—	5.18	4 97	1 1	0.023	142 86	3 5	125.5	34.44	595.0	22.5	16.41	-208.60	-0.76	-5512
Oxygen	O_2	32.0	—	-218.75	0 444	—	—	8.3		32.01	0.78	-182.97	6 82	154 4	49.7	0	0	0	—
Pentachorophenol	C_6HCl_5O	266.3	1.98	190.00	—	5.00	4.72	14.0	1.45E-07	1.12E-04	2 75E-06	310.00	—	—	—	—	—	—	—
n-Pentane	C_5H_{12}	72.2	0 63[18]	-129.6	8 393	3 62	—	40	0 691	50.73	1.24	36.07	25.77	469.80	33.3	-8.37	-146.55	-0.46	-3536
Iso-pentane	C_5H_{12}	72.2	0.62[18]	-160.1	—	3.24	2.90	48	0.908	83.67	2.05	27.7	24.70	461.00	32.9	-14.82	-154.58	-0.47	-3529
Phenanthrene	$C_{14}H_{10}$	178.2	—	100.70		4.46	4 15	1 3	8.95E-07	9.39E-04	2.30E-05	339.60	55.69	878.00	—	—	202.65	—	—
Phenol	C_6H_5OH	94.1	1.071[25]	42.5	11.43	1.46	1.15	93.000	4 49E-04	5.31E-05	1.30E-06	181.4	45.64	692.1	60.5	-32.91	-96.43	-0.21	-3064(s)
Propane	C_3H_8	44.1	—	-187.69	3.52	2.36	2.15	390	10 013	5 06	0.124	-42.07	18 77	369.9	42.0	-23.49	-103.92	-0.27	-2220
Pyrene	$C_{16}H_{10}$	202.3	1.27	156.00		5.32	4 58	0.13	3.29E-09	2.06E-04	5 04E-06	393 00	—	—	—	—	—	—	—
Sodium cyanide	$NaCN$	49.0	—	562	16.7	-0.25	—	Soluble	0.816	0.13	0.0032	1497	155	—	—	—	-89.79(c)	—	—
Sodium hydroxide	$NaOH$	40 0	2 130	319	8.34	—	—	42.000	—	—	—	1390	—	—	—	—	-426 6(c)	—	—
Sulfur dioxide	SO_2	64.1	—	-75.48	7.402	—	—	96.150	—	0.034	821E-04	10.02	24.91	430.7	77.8	-300.38	-297.07	0 01	—
Sulfuric acid	H_2SO_4	98.1	1 834[18]	10.35	9.87	—	—	Soluble	—	—	—		—	—	—	—	-811.32(l)	—	—
Tetrachloroethylene	C_2Cl_4	165 8	1.62	-22.00		3 03	2 82	2900	0.007	0.020	0.000489	121.30	34.75	620.00	44.00	22.61	-12.14	-0.12	—
1,1,1-trichloroethane	C_2HCl_3	131 4	1.46	-86.20		2.50	2 18	1500	0.162	0 201	0.00492	87 40	31.40	571.00	48.50	19.89	-5 86	-0 09	—
Toluene	C_7H_8	92.1	0.866	-94.99	6 619	2 73	2.48	535	0 037	0.260	0.00637	110.62	33.47	593.9	40 3	122.09	50 03	-0.24	-3948
Water	H_2O	18.0	1.00[4]	0.00	6.0095	—	—	—	0 032	—	—	100.00	40.656	647.4	218.3	-228.78	-242 01	-0.04	—
m-Xylene	C_8H_{10}	106.2	0.864	-47.87	11.569	3.26	2.48	130	0.013	0.437	0.0107	139.10	36.40	619	34.6	118.95	17.25	-0.34	-4595

Selected Physical Property Data (continued)

Compound	Formula	Mol. Wt.	SG (20/4)	T_m (°C)	$\Delta H_m(T_m)$ (kJ/mo)	log K_{ow}	log K_{oc} (L/kg)	S_w(25°C) (mg/L)	P_v(25°C) (atm)	K_{aw} (conc. ratio)	H (atm* m³/mo)	T_b (C)	$\Delta H_v(T_b)$ (kJ/mol)	T_c (K)	P_c (atm)	ΔG_f° (kJ/mol)	ΔH_f (kJ/mol)	ΔS_f° (kJ/mol)	ΔH_c° (kJ/mol)
O-Xylene	C_8H_{10}	106.2	0 880	-25.18	13 598	2 95	2 41	175	0 013	0 325	0.00797	144.42	36.82	631.5	35 7	122.18	19 01	-0 35	-4596
p-Xylene	C_8H_{10}	106.2	0.861	13 26	17 11	3 15	2.38	198	0.013	0 302	0.0074	138.35	36 07	618	33 9	121.21	17.96	-0 35	-4595

Note: Quantities are measured at 25°C and represent the pure compound as an ideal gas unles otherwise indicated (l, liquid; c, condensed; or s, solid state).

From Felder, R.M. and R.W. Rousseau (1986) *Elementary Principles of Chemical Processes*, 2nd ed., John Wiley & Sons, New York.

Reid, R.C., J.M., Prausnitz and T. Sherwood (1977) *The Properties of Gasses and Liquids*, 3rd ed., McGraw-Hill, New York.

Montgomery, J.H. and L.M. Welkom (1990) *Groundwater Chemicals Disk Reference*. Lewis Publishers, Ann Arbor, MI.

Multimedia Environmental Pollutant Assessment System, Batelle, 1995.

Properties of Air at Atmospheric Pressure

Temperature (T, K)	Density, ρ_a (kg/m^3)	Viscosity, μ_a (kg/m s $\times 10^5$)
100	3.601	0.6924
150	2.3675	1.0283
200	1.7684	1.3289
250	1.4128	1.488
300	1.1774	1.983
350	0.998	2.075
400	0.8826	2.286
450	0.7833	2.484
500	0.7048	2.671
550	0.6423	2.848
600	0.5879	3.018
650	0.543	3.177
700	0.503	3.332
750	0.4709	3.481
800	0.4405	3.625
850	0.4149	3.765
900	0.3925	3.899
950	0.3716	4.023
1000	0.3524	4.152
1100	0.3204	4.44
1200	0.2947	4.69
1300	0.2707	4.93
1400	0.2515	5.17
1500	0.2355	5.4
1600	0.2211	5.63
1700	0.2082	5.85
1800	0.197	6.07
1900	0.1858	6.29
2000	0.1762	6.5
2100	0.1682	6.72
2200	0.1602	6.93
2300	0.1538	7.14
2400	0.1458	7.35
2500	0.1394	7.57

From Holman, J. (1986) *Heat Transfer*. 2nd ed., McGraw-Hill, New York.

Properties of Water (Saturated Liquid)

Temperature (°C)	Density, ρ_w (kg/m³)	Viscosity, μ_w (kg/ms)
0	999.8	1.79×10^{-3}
4.44	999.8	1.55
10	999.2	1.31
15.56	998.6	1.12
21.11	997.4	9.8×10^{-4}
26.67	995.8	8.6
32.22	994.9	7.65
37.78	993	6.82
43.33	990.6	6.16
48.89	988.8	5.62
54.44	985.7	5.13
60	983.3	4.71
65.55	980.3	4.3
71.11	977.3	4.01
76.67	973.7	3.72
82.22	970.2	3.47
87.78	966.7	3.27
93.33	963.2	3.06
104.4	955.1	2.67
115.6	946.7	2.44
126.7	937.2	2.19
137.8	928.1	1.98
148.9	918	1.86
176.7	890.4	1.57
204.4	859.4	1.36
232.2	825.7	1.2
260	785.2	1.07
287.7	735.5	9.51×10^{-5}
315.6	678.7	8.68

From Holman, J. (1986) *Heat Transfer.* 2nd ed., McGraw-Hill, New York.

Selected Conversion Factors

Length

1 m =	0.001 km
	1000 mm
	100 cm
	3.2808 ft
	39.37 in
	10^6 µm (microns)
	10^{10} Å (angstrom)
1 in =	2.54 cm
1 ft =	0.3048 m
1 mi =	1.6093 km
	5280 ft
1 km =	0.6214 mi

Area

1 m^2 =	10^{-4} ha
	10^{-6} km^2
	10^4 cm^2
	10.764 ft^2
	1550 in^2
	$2.471 \cdot 10^{-4}$ acre
	$3.861 \cdot 10^{-7}$ mi^2
1 ha =	10^4 m^2
	0.01 km^2
	107,600 ft^2
	2.471 acre
	0.00386 mi^2
1 acre =	43,560 ft^2
	4047 m^2
	0.4047 ha
	0.004047 km^2
	0.00156 mi^2

Volume

1 m^3 =	1000 liters (L)
	10^6 cm^3 or mL
	264.2 gal
	35.32 ft^3
1 ft^3 =	0.028 m^3
	28.32L
	28,320 cm^3
	7.481 gal
1 Barrel (bbl) =	42 gal
	5.615 ft^3
	0.1590 m^3

Mass

1kg =	1000 gm
	2.2046 lb_m
1 lb_m =	453.59 gm
	0.4359 kg
1 metric ton =	1000 kg
1 short ton =	2000 lb_m
1 long ton =	2240 lb_m
	1016 kg
1 mole =	$6.023 \cdot 10^{23}$ molecules
	(molecular weight)·grams
	0.0022046 lb-mole

Density and concentrations

1 kg/m^3 =	1 gm/L
	10^3 mg/L
	10^6 µg/L
	0.001 gm/cm^3
	0.06243 lb_m/ft^3
1 lb_m/ft^3 =	16.02 kg/m^3
	0.01602 gm/cm^3

Pressure

1 atm =	760 mmHg (0°C) or torr
	29.921 in Hg (0°C)
	14.696 lb_f/in^2
	$1.0133 \cdot 10^5$ N/m^2 or Pascals (Pa)
1 bar =	10^5 Pa(N/m^2)
	750.1 torr
	14.50 lb_f/in^2
1 $lb_f/in.^2$ =	6895 Pa (N/m^2)
	51.71 torr

Viscosity

1 gm/(cm·s) =	1 poise)
	100 centipoise (cp)
	0.1 $N{\cdot}s/m^2$
	0.1 Pa·s
	0.0672 $lb_m/(ft{\cdot}s)$

Diffusivity, *Kinematic* Viscosity, Thermal Diffusivity

1 cm^2/s =	10^{-4} m^2/s
	8.64 m^2/day
	3156 $m^2/year$
	3.875 ft^2/h

Energy

1 Joule (J) =	1 N-m
	1 kg·m²/s
	9.41·10⁻⁴ British Thermal Units (BTU)
	0.7376 ft·lb$_f$
	0.2390 cal
1 calorie =	4.184 J
1BTU =	1.055 kJ
	252.2 cal
	778.2 ft·lb$_f$

Power

1 Waff(W) =	1 J/s
	0.001 kW
	0.239 cal/s
	0.000948 BTU/s
	0.7376 (ft·lb$_f$)s
1 horsepower =	745.7 W
	178.1 cal/s
	0.7068 BTU/s
	550 (ft·lb$_f$)s

Ideal Gas Constant

R =	82.05 $cm^3 \cdot atm \cdot K^{-1} \cdot mol^{-1}$
	0.08205 $L \cdot atm \cdot K^{-1} \cdot mol^{-1}$
	62.36 $L \cdot torr \cdot K^{-1} \cdot mol^{-1}$
	8.313 $kg \cdot m^{2} \cdot K^{-1} \cdot mol^{-1}$
	8.313 $J \cdot K \cdot mol^{-1}$
	8.313 $L \cdot kPa \cdot K^{-1} \cdot mol^{-1}$
	1.986 $cal \cdot K^{-1} \cdot mol^{-1}$
	1.986 $BTU \cdot R^{-1} \cdot lb{-}mol^{-1}$
	10.731 $ft^3 \cdot (lb_f/in.^2) \cdot R^{-1} \cdot lb{-}mol^{-1}$

Index

B

D

F

G

H

I

K

N

O

P

R

T

U

V

Z

Milton Keynes UK
Ingram Content Group UK Ltd.
UKHW052026071024
449327UK00027B/2439

9 780367 400279